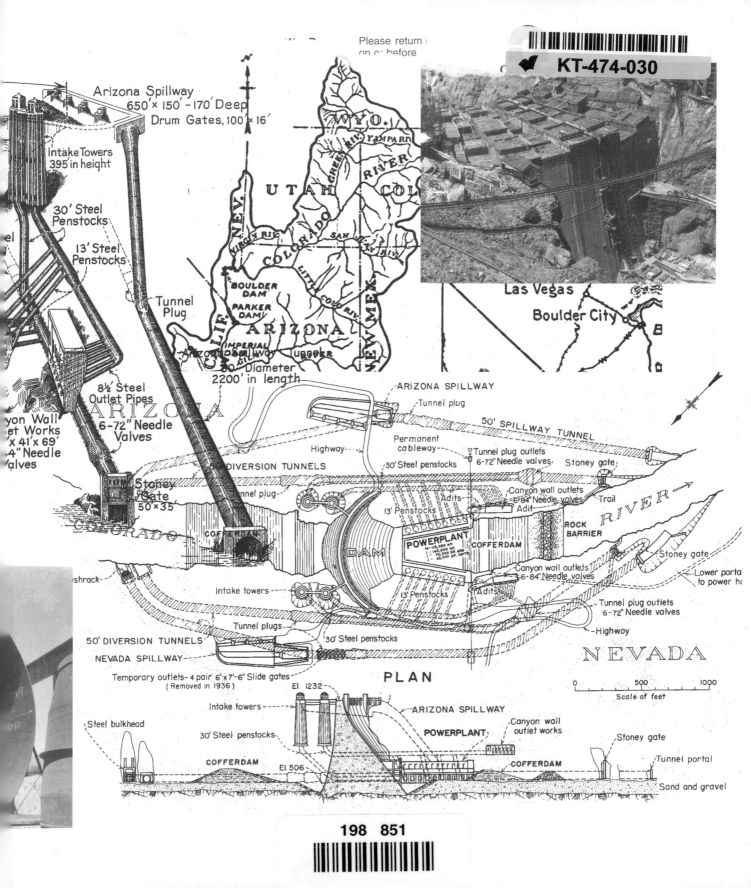

Engineering Design Graphics

Graphics AutoCAD 2004®
Eleventh Edition

James H. Earle

Professor Emeritus
Texas A&M University

PEARSON
Prentice
Hall

Upper Saddle River, New Jersey 07458

Library of Congress Cataloging-in-Publication Data on File

Vice President and Editorial Director, ECS: *Marcia Horton*
Vice President and Director of Production and Manufacturing, ESM: *David W. Riccardi*
Executive Editor: *Eric Svendsen*
Associate Editor: *Dee Bernhard*
Editorial Assistant: *Brian Hoehl*
Executive Managing Editor: *Vince O'Brien*
Managing Editor: *David A. George*
Production Editor: *Rose Kernan*
Director of Creative Services: *Paul Belfanti*
Manager of Electronic Composition and Digital Content: *Jim Sullivan*
Assistant Manager of Electronic Composition and Digital Content: *Allyson Graesser*
Electronic Composition: *William Johnson*
Creative Director: *Carole Anson*
Cover Designer: *James H. Earle*
Art Editor: *Xiaohong Zhu*
Manufacturing Manager: *Trudy Pisciotti*
Manufacturing Buyer: *Lisa McDowell*
Marketing Manager: *Holly Stark*

© 2004 Pearson Education, Inc.
Pearson Prentice-Hall
Pearson Education, Inc.
Upper Saddle River, New Jersey 07458

Pearson Prentice Hall® is a trademark of Pearson Education, Inc.

AutoCAD and the AutoCAD logo are registered trademarks of Autodesk, Inc., 111 McInnis Parkway, San Rafael, CA 94903.

The author and publisher of this book have used their best efforts in preparing this book. These efforts include the development, research, and testing of the theories to determine their effectiveness. The author and publisher shall not be liable in any event for incidental or consequential damages with, or arising out of, the furnishing, performance, or use of these programs.

Printed in the United States of America

10 9 8 7 6 5 4 3 2 1

ISBN: 0-13-142573-0

Pearson Education Ltd., *London*
Pearson Education Australia Pty. Ltd., *Sydney*
Pearson Education Singapore, Pte. Ltd.
Pearson Education North Asia Pte. Ltd., *Hong Kong*
Pearson Education Canada, Inc., *Toronto*
Pearson Educación de Mexico, S.A. de CV.
Pearson Education—Japan, *Tokyo*
Pearson Education Malaysia Pte. Ltd.
Pearson Education, Inc., *Upper Saddle River, New Jersey*

Dedicated to my father,
Hubert Lewis Earle,
October 25, 1900–October 22, 1967

Preface

Engineering Design Graphics has been revised and refined in an eleventh edition that we believe is its best edition in all categories: Content, format, readability, clarity, quality of illustrations, and economy. Every effort has be made to write and illustrate the book to make it easier for the student to learn and the teacher to teach. Also, the content was chosen to serve today's courses and fit tomorrow's needs.

Successfully meeting these goals is a difficult chore for an author in any discipline, but it is especially awesome in the area of engineering design graphics where 2,000 illustrations, 1,000 problems, and a multitude of topics must be merged into a cohesive textbook as compactly as possible. Content must include fundamentals, design, computer graphics, industrial applications, and meaningful problems. We hope you agree that we have met these goals.

Content

Every paragraph and illustration has been revisited for improvement or elimination so every item makes a worthwhile contribution to the learning process. No space has been squandered to make room for exotic illustrations that are far beyond the scope of a beginning freshman course. Instead, that valuable space has been used to better illustrate and present concepts in an understandable format to reduce the amount of classroom tutoring needed by the student.

Major content areas covered in this text are:

design and creativity,
computer graphics,
engineering drawings,
descriptive geometry, and
problem solving.

Design and Creativity

The first eight chapters are devoted to the introduction of design and creativity. Case studies with examples of worksheets and drawings applying the steps of design guide the student through the design process.

Care has been taken to offer realistic design problems that are within the grasp of beginning students rather than overwhelming them with problems beyond their capabilities. Since the primary objective of design instruction is to teach the *process of design*, meaningful design assignments are given to make the process fun and to encourage the application of creativity and intuition. Most problems are adequately challenging to encourage creative and inspirational solutions which may lead to patentable products.

Chapter 9 has a variety of design problems (115 problems) that can be used for quickie problems, short assignments, or as semester-long design projects. Additional design exercises are included at the ends of the chapters throughout the book.

Computer Graphics

AutoCAD® 2004 is presented in a step-by-step format to aid the student in learning how to use this popular software. Steps of each illustrations show the reader what will be seen on the screen as the example is followed. Chapter 37 gives an introduction to two-dimensional computer graphics and Chapter 38 covers three-dimensional computer graphics, solid modeling, and rendering.

The main purpose of *Engineering Design Graphics* is to help students learn the principles of graphics, whether done on the drawing board or on the computer. This book can be used in courses where the entire course is done by computer, none is done by computer, or partly done by computer.

A Format Worth the Effort

Much effort has been devoted to the creation of illustrations separated into multiple steps to present the concepts as clearly and simply as possible. A second color is applied as a functional means of emphasizing sequential steps, key points, and explanations, not just as decoration. Explanatory information and text is closely associated with the steps of each example.

Many illustrations have been drawn, modified, and refined to aid the reader in visualizing and understanding the example at hand. Actual industrial parts and products have been merged with explanatory examples of principles being covered.

The author has personally developed and drawn the illustrations in this book. Years of classroom experience and trial-and-error testing have been applied to create a format that enables the student and teacher to cover more content with fewer learning obstacles.

A Book to Keep

Some material in this book may not be formally covered in the course for which it was adopted due to a variations in courses offered from campus to campus. These lightly covered topics may be the ones that will be needed in later courses or in practice; therefore, this book should be retained as a convenient reference for the engineer, technologist, or technician.

A Teaching System

Engineering Design Graphics used in combination with the supplements listed below comprises a complete teaching system.

Textbook Problems: Approximately 1000 problems are offered to aid the student in mastering the principles of graphics and design.

Solution Manual: A solution manual containing the solutions to most of the problems in this book is available to assist the teacher with grading.

Problem Manuals: Nineteen problem books and teachers' guides (with outlines, problem solutions, tests, and test solutions) that are keyed to this book are available from the author. To obtain these manuals please refer to the listing and information on the inside back cover of this book. New problem manuals are in development as well. Fifteen of the problem books are designed to allow problem solution by computer, by sketching, or on the drawing board.

Acknowledgments

We are grateful for the assistance of many who have influenced the development of this volume. Many industries have furnished photographs, drawings, and applications that have been acknowledged in the corresponding legends. The Engineering Design Graphics staff of Texas A&M University have been helpful in making suggestions for the revision of this book.

Professor Tom Pollock provided valuable information on metallurgy for Chapter 19. Professor Leendert Kersten of the University of Nebraska, Lincoln, kindly provided his descriptive geometry computer programs for inclusion; his cooperation is appreciated.

We are indebted to Denis Cadu for his assistance and cooperation. We are appreciative of the assistance of David Ratner of the Biomechanics Corporation Inc. for providing HUMANCAD® software. Also giving support were Michael Flanagan of Bresslergroup and Henry Keck of Keck-Craig, Inc.

We are appreciate of the fine editorial and production team assembled by Prentice Hall: under the guidance of Executive Editor Eric Svendsen, Production Editor Rose Kernan and Art Editor Xiaohong Zhu worked closely in expertly managing 2,000 illustrations to see that they were properly reproduced, sized, and laid out.

Above all we appreciate the many institutions who have thought enough of our publications to adopt them for classroom use. This is the highest honor that can be paid an author. We are hopeful that this textbook will fill the needs of engineering and technology programs. As always, comments and suggestions for improvement and revision will be appreciated.

JIM EARLE
College Station, Texas

Contents

Engineering Design Graphics

AutoCAD 2004®
Eleventh Edition

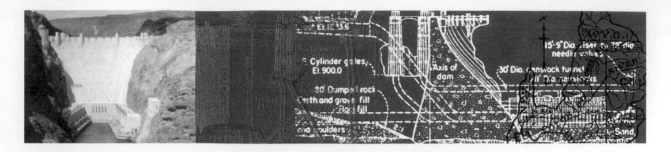

1

Engineering and Technology

1.1 Introduction

This book deals with the field of engineering design graphics and its application to the design process. Engineering graphics is the primary medium for developing and communicating design concepts. The solution of most engineering problems requires a combination of organization, analysis, problem-solving principles, graphics, skill, and communication (**Figure 1.1**).

This book will help you use your creativity and develop your imagination because inno-

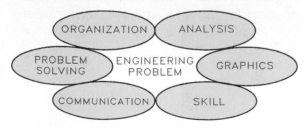

1.1 This text illustrates the total approach to engineering, with the engineering problem as the focal point.

vation is essential to a successful career in engineering and technology. Albert Einstein said, "Imagination is more important than knowledge, for knowledge is limited, whereas imagination embraces the entire world . . . stimulating progress, or, giving birth to evolution."

1.2 Engineering Graphics

Engineering graphics covers the total field of graphical problem solving within two major areas of specialization: **descriptive geometry** and **documentation drawings**. Other areas of application are nomography, graphical mathematics, empirical equations, technical illustration, vector graphics, data analysis, and computer graphics. Graphics is one of the designer's most effective tools for developing design concepts and solving three-dimensional problems. It is also the designer's best means of communicating ideas to others.

Descriptive Geometry

Gaspard Monge (1746-1818), the "father of descriptive geometry," used graphical methods to solve design problems related to fortifications and battlements while a military student in France. His headmaster scolded him for not using the usual long, tedious mathematical process. Only after lengthy explanations and demonstrations of his technique was he able to convince the faculty that graphical methods (now called descriptive geometry) produced solutions in less time.

Descriptive geometry was such an improvement over mathematical methods that it was kept as a military secret for fifteen years before the authorities allowed it to be taught as part of the civilian curriculum. Monge went on to become a scientific and mathematical aide to Napoleon.

Descriptive geometry is the projection of three-dimensional figures on the two-dimensional plane of paper in a manner that allows geometric manipulations to determine lengths, angles, shapes, and other geometric information about the figures.

1.3 Technological Milestones

Many of the technological advancements of the twentieth century are engineering achievements. Since 1900, technology has taken us from the horse-drawn carriage to the moon and back, and more advancements are certain in the future.

Figure 1.2 shows a few of the many technological mileposts since 1900. It identifies products and processes that have provided millions of jobs and a better way of life for all. Other significant achievements include building a railroad from Nebraska to California that met at Promontory Point, Utah, in 1869 in less than four years; constructing the Empire State Building with 102 floors in a mere thirteen and a half months in 1931; and retooling

```
MILESTONES OF THE 20TH CENTURY

1900  Vacuum cleaner     1950  A-bomb tests
      Airplane                 Optical fibers
      Dial telephone           Soviet satelite
      Light bulb               Microchip
      Model T Ford       1960  Commun. satelite
1910  Washing machine          Indus. robot
      Refrigerator             Nuclear reactor
      Wireless phone           Heart transplant
1920  Radio broadcasts         Man on moon
      Telephone service  1970  Silicon chip
      35mm camera              Personal computer
      Cartoons & sound         Videocassette record.
1930  Tape recorder            Supersonic jet
      Atom split               Neutron bomb
      Jet engine         1980  Stealth bomber
      Television               Space shuttle
1940  Elect. computer          Artificial heart
      Missile                  Soviet space station
      Transistor         1990  Computer voice
      Microwave                  recoginition
      Polaroid camera          Artificial intelligence
                               E-Mail and the
                                 internet
```

1.2 This chronology lists some of the significant technological advances of the twentieth century.

industry in 1942 for World War II to produce 4.5 naval vessels, 3.7 cargo ships, 203 airplanes, and 6 tanks each day while supporting 15 million Americans in the armed forces.

One of the "miracle projects" of the 1990s was the construction of the thirty-one mile "Chunnel" that connects England and France under the English Channel for high-speed shuttle trains. It consists of three tunnels drilled 131 feet under the channel floor; two of the tunnels are 24 feet in diameter. The trip from London to Paris can be made in 3-1/2 hours.

1.4 The Technological Team

Technology and design have become so broad and complex that teams of specialists, rather than individuals, undertake most projects (**Figure 1.3**). Such teams usually consist of one or more scientists, engineers, technologists, technicians, and craftspeople, and may include designers and stylists as well (**Figure 1.4**).

Scientists

Scientists are researchers who seek to discover new laws and principles of nature through experimentation and scientific testing (**Figure 1.5**). They are more concerned

1.3 Technological and design team members, with their varying experiences and areas of expertise, must communicate and interact with each other. (Courtesy of Honeywell, Inc.)

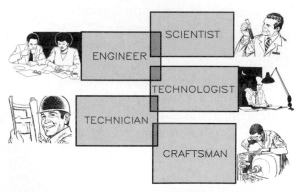

1.4 A ranking of the typical technological team, from the most theoretical level (scientists) to the least technical level (craftspeople).

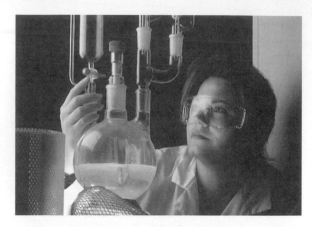

1.5 This scientist is working to develop a new drug to protect the body from contracting AIDS. (Courtesy of FMC Corporation.)

1.6 Engineers and technologists combine their knowledge to work on an avionics modification. (Courtesy of Cessna Aircraft Company.)

with the discovery of scientific principles than with the application of those principles to products and systems. Their discoveries may not find applications until years later.

Engineers

Engineers receive training in science, mathematics, and industrial processes to prepare them to apply the findings of the scientists (**Figure 1.6**). Thus engineers are concerned with converting raw materials and power sources into needed products and services. Creatively applying scientific principles to develop new products and systems is the design process, the engineer's primary function. In general, engineers use known principles and available resources to achieve a practical end at a reasonable cost.

Technologists

Technologists obtain backgrounds in science, mathematics, and industrial processes. Whereas engineers are responsible for analysis, overall design, and research, technologists are concerned with the application of engineering principles to planning, detail design, and production (**Figure 1.6**). Technologists apply their knowledge of engineering principles, manufacturing, and testing to assist in the implementation of projects and production. They also provide support and act as liaisons between engineers and technicians.

Technicians

Technicians assist engineers and technologists at a less theoretical level than technologists and act as liaisons between technologists and craftspeople (**Figure 1.7**). They have backgrounds in mathematics, drafting, computer programming, and materials testing. Their work varies from conducting routine laboratory experiments to supervising craftspeople in manufacturing or construction.

Craftspeople

Craftspeople are responsible for implementing designs by fabricating them according to engineers' specifications. They may be machinists who make product parts or electricians who assemble electrical components. Their ability to produce a part according to design specifications is as necessary to the success of a project as an engineer's ability to design it. Craftspeople include electricians, welders, machinists, fabricators, drafters, and members of many other occupational groups (**Figure 1.8**).

Designers

Designers may be engineers, technologists, inventors, or industrial designers who have special talents for devising creative solutions to design problems. Designers do not necessarily have engineering backgrounds, especially in newer technologies where there is little design precedent. Thomas A. Edison (**Figure 1.9**), for example, had little formal

1.7 An engineering technician works on a phased-array radar antenna. (Courtesy of Northrop Grumman Corporation.)

1.8 A craftsman assembles a portion of an experimental aircraft. (Courtesy of Cessna Aircraft Company.)

1.9 Thomas Edison had no formal education, yet he changed the world with his knowledge and creativity.

education, but he created some of the world's most significant inventions.

Stylists

Stylists are concerned with the appearance and market appeal of a product rather than its fundamental design (**Figure 1.10**). They may design an automobile body or the exterior of an electric iron. Automobile stylists, for example, consider the car's appearance, driver's vision, passengers' enclosure, power unit's space requirement, and so on. However, they

1.10 The stylist develops a product's outward appearance in order to make it as marketable and attractive as possible. (Courtesy of Datum-X.)

are not involved with the design of the car's internal mechanical functions, such as the engine, steering linkage, and brakes. Stylists must have a high degree of aesthetic awareness and an instinct for styling that appeals to the consumer.

1.5 Engineering Fields

Recent changes in engineering include the emergence of technologists and technicians and the growing number of women pursuing engineering careers. More than 15% of today's freshman engineering students are women; 15% of master's degrees and 10% of doctorate degrees in engineering are awarded to women.

1.6 Aerospace Engineering

Aerospace engineering has progressed from the Wright brothers' first flight at Kittyhawk, North Carolina, in 1903 to the penetration of outer space. It deals with all aspects (speeds and altitudes) of flight. Aerospace engineering assignments range from developing complex vehicles capable of traveling millions of miles into space to hover aircraft that can transport and position large construction components. In the space exploration branch of this profession, aerospace engineers work on all types of aircraft and spacecraft—missiles, rockets, propeller-driven planes, and jet-powered planes (**Figure 1.11**).

Second only to the auto industry in sales, the aerospace industry contributes immeasurably to national defense and to the economy. Specialized areas include aerodynamics, structural design, instrumentation, propulsion systems, materials, reliability testing, and production methods.

Aerospace engineers specialize in one of two major areas: research engineering or design engineering. Research engineers investigate known principles in search of new

1.11 The aerospace engineer must deal with all aspects of aircraft such as this polar lander that went to Mars. (Courtesy of National Aeronautics and Space Administration.)

1.12 Agricultural engineers design food processing systems such as this one for processing oranges in Florida. (Courtesy of FMC Corporation.)

ideas and concepts; design engineers translate them into workable applications. The professional society for aerospace engineers is the American Institute of Aeronautics and Astronautics (AIAA).

1.7 Agricultural Engineering

Agricultural engineers are trained to serve the world's largest industry—agriculture—in which they deal with the production, processing, and handling of food and fiber.

Mechanical Power Agricultural engineers who work with manufacturers of farm equipment are concerned with gasoline and diesel engine equipment, including pumps, irrigation machinery, and tractors. Machinery must be designed for the electrical curing of hay, milk and fruit processing, and heating environments for livestock and poultry (**Figure 1.12**). Farm machinery designed by agricultural engineers has been largely responsible for the vastly increased productivity in U.S. agriculture. Today's farmer produces enough food for

120 people, whereas one hundred years ago, a farmer was able to feed only 4 people.

Farm Structures The construction of barns, shelters, silos, granaries, processing centers, and other agricultural buildings requires specialists in agricultural engineering. They must understand heating, ventilation, and chemical changes that might affect the storage of crops.

Electrical Power Agricultural engineers design electrical systems and select equipment that will operate efficiently and meet the requirements of many situations. They may serve as consultants or designers for manufacturers or processors of agricultural products.

Soil and Water-Control Agricultural engineers are responsible for devising systems to improve drainage and irrigation systems, resurface fields, and construct water reservoirs. They may perform activities in conjunction with the U.S. Department of Agriculture, the U.S. Department of the Interior, state agricultural universities, consulting engineering firms, or irrigation companies.

Characteristics Most agricultural engineers are employed in private industry, especially by manufacturers of heavy farm equipment and specialized lines of field, barnyard, and household equipment; by electrical service companies; and by distributors of farm equipment and supplies. Although few agricultural engineers live on farms, they need to understand agricultural problems—farming, crops, animals, and farmers themselves.

The professional society for agricultural engineers is the American Society for Agricultural Engineers (ASAE).

1.8 Chemical Engineering

Chemical engineering involves the design and selection of equipment used to process and manufacture large quantities of chemicals (**Figure 1.13**). Chemical engineers develop and design methods of transporting fluids through ducts and pipelines, transporting solid material through pipes or conveyors, transferring heat from one fluid or substance to another through plate or tube walls, absorbing gases by bubbling them through liquids, evaporating liquids to increase concentration of solutions, distilling mixed liquids to separate them, and handling many other chemical processes. Process control and instrumentation are important specialties in chemical engineering. With the measurement of quality and quantity by instrumentation, process control is fully automatic.

Chemical engineers often utilize chemical reactions of raw products, such as oxidation, hydrogenation, reduction, chlorination, nitration, sulfonation, pyrolysis, and polymerization, in their work. They develop and process chemicals such as acids, alkalies, salts, coal-tar products, dyes, synthetic chemicals, plastics, insecticides, and fungicides for industrial and domestic uses. Chemical engineers help develop drugs and medicines, cosmetics, explosives, ceramics, cements, paints, petroleum products, lubricants, synthetic fibers, rubber, and detergents. They also design equipment for food preparation and canning plants.

Approximately 80% of chemical engineers work in manufacturing industries, primarily the chemical industry. The other 20% work for government agencies, independent research institutes, and as independent consultants. New fields requiring chemical engineers are nuclear sciences, rocket fuel development, and environmental pollution control. The professional society for chemical engineers is the American Institute of Chemical Engineers (AIChE).

1.9 Civil Engineering

Civil engineering, the oldest branch of engineering, is closely related to virtually all our daily activities. The buildings we live and work in, the transportation we use, the water we drink, and the drainage and sewage systems we rely on are all the result of civil engineering.

Construction Civil engineers manage the workers, finances, and materials used on con-

1.13 Chemical engineers developed this closed-system container for insecticides that is the industry standard. (Courtesy of FMC Corporation.)

struction projects. Structural engineers design and supervise the construction of buildings, harbors, airfields, tunnels, bridges, stadiums, and other types of structures.

City Planning Civil engineers working as city planners develop plans for the growth of cities and systems related to their operation. They are involved in street planning, zoning, residential subdivisions, and industrial site development.

Hydraulics Civil engineers work with the behavior of water and other fluids from their conservation to their transportation. Civil engineers design wells, canals, dams, pipelines, flood control and drainage systems, and other methods of controlling and using those resources (**Figure 1.14**).

Transportation Civil engineers design and supervise construction, modification, and maintenance of railroad and mass transit systems. They design and supervise construction of airport runways, control towers, passenger and freight stations, and aircraft hangars. They are involved in all phases of developing the national systems and local networks of

1.14 Civil engineers are involved with the control and conversion of water into electrical energy, as shown in this hydro-power plant. (Courtesy of Kaiser Engineers.)

highways and interchanges for moving automobile traffic, including the design of tunnels, culverts, and traffic control systems.

Sanitary Engineering Civil engineers maintain public health by designing pipelines, treatment plants, and other facilities for water purification and water and air pollution control. They also are involved in solid waste disposal activities.

Other Areas of Specialization Civil engineers are involved in geotechnical engineering, which deals with the study of soils. They also work in environmental engineering, which relates to all aspects of the environment and methods of preserving it.

Characteristics Many civil engineers hold positions in administration and municipal management and are associated with federal, state, and local government agencies and the construction industry. Many work as consulting engineers for architectural firms and independent consultants. The remainder work for public utilities, railroads, educational institutions, and various manufacturing industries.

The professional society for civil engineers is the American Society of Civil Engineers (ASCE). Founded in 1852, it is the oldest engineering society in the United States.

1.10 Electrical Engineering

Electrical engineers are concerned with power, which deals with providing the large amounts of energy required by cities and large industries, and electronics, which deals with the small amounts of power used for communications and automated operations that have become part of everyday life.

Power Power generation, transmission, and distribution pose many electrical engineering problems—from the design of generators for producing electricity to the development of transmission equipment. Power applications in homes are quite numerous—for computers, washers, dryers, vacuum cleaners, and other appliances. Only about one-quarter of electric energy is consumed in the home. About half is used by industry for metal refining, heating, motor drives, welding, machinery controls, chemical processes, plating, and electrolysis. Illumination (lighting) is required in nearly every phase of modern life. Improving its efficiency is a challenging area for electrical engineers.

Electronics Computers have contributed to the development of a gigantic industry that is the domain of electrical engineers. Used with industrial electronics (such as robotics), computers have changed industry's manufacturing and production processes, resulting in greater precision and less manual labor (**Figure 1.15**).

Communications Systems These engineering applications are devoted to the improvement of radio, telephone, telegraph, and television

1.15 The electrical engineer, in conjunction with the computer programer, have revolutionized the workplace. The operations of Seattle's SAFECO Field are powered by computer, from ticket sales to concessions. Play-by-play statistics are updated on the web instantaneously. (Courtesy Lucent Technologies.)

systems, the nerve centers of most industrial operations.

Instrumentation These engineering applications involve systems of electronic instruments used in industrial processes. Electrical engineers make extensive use of the cathode-ray tube and the electronic amplifier in industry and nuclear reactors. Increasingly, engineers are using instrumentation for applications in medical diagnosis and therapy.

Military Electronics This field encompasses most weapons and tactical systems ranging from the walkie-talkie to radar networks for detecting enemy aircraft. Remote-controlled electronic systems are used for navigation and interception of guided missiles.

Characteristics There are currently more electrical engineers than any other type of engineer. The increasing need for electrical equipment, automation, and computerized systems is expected to sustain the rapid growth of this field.

The professional society for electrical engineers is the Institute of Electrical and Electronic Engineers (IEEE). Founded in 1884, the IEEE is the world's largest professional society.

1.11 Industrial Engineering

Industrial engineering, one of the newer engineering fields, differs from other branches of engineering in that it relates primarily to people, their performance, and their working conditions. Industrial engineers often manage people, machines, materials, methods, and money in the production and marketing of goods.

Industrial engineers may be responsible for plant layout, development of plant processes, or determination of operating standards that will improve the efficiency of a plant operation.

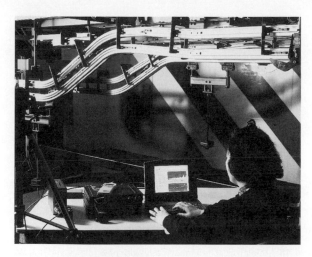

1.16 Industrial engineers are concerned with productivity and safety in the workplace. This engineer is conducting a noise-reduction analysis study of a new conveyor product. (Courtesy of Jervis B. Webb Company.)

They also design and supervise systems for improved safety of personnel and increased production at lower costs (**Figure 1.16**).

Areas of industrial engineering include management, plant design and engineering, electronic data processing, systems analysis and design, control of production and quality, performance standards and measurements, and research. Industrial engineers are increasingly involved in implementing automated production systems.

People-oriented areas include the development of wage-incentive systems, job evaluation, work measurement, and environmental system design. Industrial engineers are often involved in management-labor agreements that affect operations and production.

More than two-thirds of all industrial engineers are employed in manufacturing industries. Others work for insurance companies, construction and mining firms, public utilities, large businesses, and government agencies.

The professional society for industrial engineers is the American Institute of Industrial Engineers (AIIE), which was organized in 1948.

1.12 Mechanical Engineering

Mechanical engineering's major areas of specialization are power generation, transportation, manufacturing, power services, and atomic energy.

Power Generation Mechanical engineers develop and design prime movers (machines that convert natural energy into work) to power electric generators that produce electricity. They are involved in the design and operation of steam engines, turbines, internal combustion engines, and other prime movers.

Transportation Mechanical engineers participate in the design of trucks, buses, automobiles, locomotives, marine vessels, and aircraft. In aeronautics, they develop aircraft engines, controls, and internal environmental systems. Mechanical engineers design marine vessels powered by steam, diesel, or gas-turbine engines, as well as power services throughout the vessels, such as lighting, water, refrigeration, and ventilation (**Figure 1.17**).

Manufacturing Mechanical engineers design new products and the equipment and factories needed to manufacture them economically and at a uniformly high level of quality.

The professional society for manufacturing engineers is the Society of Manufacturing Engineers (SME).

Power Services Mechanical engineers in this area must have a knowledge of pumps, ventilation equipment, fans, and compressors. They apply this knowledge to the development of methods for moving liquids and gases through pipelines and the design and

1.17 Mechanical engineers design unique traffic systems such as this Airtrans system for the Dallas-Fort Worth airport. (Courtesy of Vought Aeronautics.)

1.18 Metallurgical engineers develop new alloys for special applications and new products. (Courtesy of Jones & Laughlin Steel Corporation.)

installation of refrigeration systems, elevators, and escalators.

Nuclear Energy Mechanical engineers develop and handle protective equipment and materials and assist in constructing nuclear reactors.

The professional society for mechanical engineers is the American Society of Mechanical Engineers (ASME).

1.13 Mining and Metallurgical Engineering

Mining engineers are responsible for developing methods of extracting minerals from the earth and preparing them for use by manufacturing industries (**Figure 1.18**). Working with geologists to locate ore deposits, which are exploited through the construction of tunnels and underground operations or surface strip mining, mining engineers must understand safety, ventilation, water supply, and communications. Mining engineers who work at mining sites usually are employed near small, out-of-the-way communities, whereas those in research and consulting most often work in large urban areas.

The two main areas of metallurgical engineering are extractive metallurgy (the extraction of metal from raw ores to form pure metals) and physical metallurgy (the development of new products and alloys). Many metallurgical engineers work on the development of machinery for electrical equipment and in the aircraft industry. The need to develop new lightweight, high-strength materials for spacecraft, jet aircraft, missiles, and satellites will require more metallurgical engineers in the future.

The professional society for mining and metallurgical engineers is the American Institute of Mining, Metallurgical, and Petroleum Engineering (AIME).

1.14 Nuclear Engineering

The earliest work in nuclear engineering involved military applications. Nuclear power for domestic needs was developed for a time, but the new emphasis in nuclear engineering is on medical applications.

Peaceful applications of nuclear engineering fall into two major areas: radiation and nuclear power reactors. Radiation is the prop-

1.19 Nuclear engineers will design the nuclear reactors to economically produce much of the electic power of the future.

1.20 Petroleum engineers will be responsible for projects of this type, which will extract gas from 5,300 feet below sea level in the Gulf of Mexico. (Courtesy FMC Corporation.)

agation of energy through matter or space in the form of waves. In atomic physics, radiation includes fast-moving particles (alpha and beta rays, free neutrons, and so on), gamma rays, and X-rays. Nuclear science is closely allied with botany, chemistry, medicine, and biology.

The use of nuclear energy to produce mechanical or electric power is a major peaceful application of nuclear engineering (**Figure 1.19**). In the production of electric power, nuclear energy is the fuel used to produce steam to drive turbine generators.

Most nuclear engineering training focuses on the design, construction, and operation of nuclear reactors. Other areas include the processing of nuclear fuels, thermonuclear engineering, and the use of various nuclear by-products.

The professional society for nuclear engineers is the American Nuclear Society.

1.15 Petroleum Engineering

The recovery of petroleum and natural gas is the primary concern of petroleum engineers, but they also develop methods for transporting and separating various petroleum products. They also are responsible for improving drilling equipment and ensuring its economical operation. In exploring for petroleum,

petroleum engineers are assisted by geologists and by instruments such as the airborne magnetometer, which indicates uplifts in the earth's subsurfaces that could hold oil or gas.

Petroleum engineers develop equipment to remove oil from the ground most efficiently and supervise oil-well drilling (**Figure 1.20**). They also design systems of pipes and pumps to transport oil to shipping points or to refineries. Development, design, and operation of petroleum processing facilities are done jointly with chemical engineers.

The Society of Petroleum Engineers (SPE) is a branch of AIME, which includes mining and metallurgical engineers and geologists.

1.16 Graphics/Drafting

Drafters help engineers and designers select materials. They also prepare construction documents and specifications as well as translate designs into working drawings and technical illustrations. Working drawings are visual instructions for fabricating products and erecting structures in all fields of engineering.

Technical illustration, the most artistic area of engineering design graphics, visually depicts projects and products for operations manuals and presentations.

In addition, drafters prepare maps, geological sections, and plats from data given to them by engineers, geologists, and surveyors. These drawings show the locations of property lines, physical features, strata, rights-of-way, building sites, bridges, dams, mines, and utility lines.

The three levels of certification for drafters are drafters, design drafters, and engineering designers:

- Drafters are graduates of a two-year, post-high school curriculum in engineering design graphics.

- Design drafters complete two-year programs at an approved junior college or technical institute.

- Engineering designers are graduates of a four-year college course in engineering design graphics who can become certified as technologists.

Industry has embraced computer graphics as a way to improve drafters' productivity. However, the use of computer graphics systems does not lessen the need for a knowledge of graphics. The principles of graphics remain the same; only the medium—the computer—is different (**Figure 1.21**).

1.21 The drafter of today may use the Apollo workstation for documentation and design needs. (Courtesy of Hewlett Packard Company.)

Problems

1. Write a report that outlines the specific duties of and relationships among the scientist, engineer, technologist, technician, craftsperson, designer, and stylist in an engineering field of your choice. For example, explain this relationship for an engineering team involved in an aspect of civil engineering. Your report should be supported by factual information obtained from interviews, brochures, or library references.

2. Investigate and write a report on the employment opportunities, job requirements, professional challenges, and activities of your chosen branch of engineering or technology. Illustrate this report with charts and graphs, where possible, for easy interpretation. Compare your personal abilities and interests with those required by the profession.

3. Arrange a personal interview with a practicing engineer, technologist, or technician in your field of interest. Discuss with that person the general duties and responsibilities of the position in order to gain a better understanding of this field. Summarize your interview in a written report.

4. Write to the professional society in your field of study for information about it. Prepare a notebook of these materials for easy reference. Include in the notebook a list of books that provide career information for that field.

5. Write a one-page report that gives your reasons for selecting engineering or technology as your field of study. List the type of work that you visualize yourself doing upon graduation, where you will work, and for whom. Also, list your personal attributes that you believe will make you successful in this career. Summarize by explaining why your career selection is a good one.

6. Have you considered self-employment? There are many rewards for the person who "invents a new mousetrap" and has the commitment to independently bring it into being as Dr. Land did with the Land Polaroid camera. Write a one-page report listing the advantages and disadvantages of being self-employed with employees on your payroll. Summarize by assessing the "fit" of your personal attributes to being your own boss.

7. If you were to graduate with an engineering or technology degree today, you would probably look for a job where most graduates go—manufacturing companies, government branches, and large, well-established companies in industry. Make a list of the various options that you visualize yourself having because of your technical training, and give a brief explanation of each. Try to think of as many unique and nonconventional opportunities as you can. Limit your best list to no more than one page.

Addresses of Professional Societies

Publications and information from these societies were used in preparing this chapter.

The American Ceramic Society
65 Ceramic Drive, Columbus, OH 43214

The American Institute of Aeronautics and Astronautics
1290 Avenue of the Americas, New York, NY 10019

American Institute of Chemical Engineers
345 East 47th Street, New York, NY 10017

American Institute for Design and Drafting
3119 Price Road, Bartlesville, OK 74003

The American Institute of Industrial Engineers
345 East 47th Street, New York, NY 10017

American Institute of Mining, Metallurgical Engineering
345 East 47th Street, New York, NY 10017

American Nuclear Society
555 N. Kensington Ave, LaGrange Park, IL 60526

American Society of Agricultural Engineers
2950 Niles Road, St. Joseph, MI 49085

American Society of Civil Engineers
1801 Alexander Bell Drive, Reston, VA 20191

American Society for Engineering Education
11 DuPont Circle, Suite 200, Washington, DC 20036

American Society of Mechanical Engineers
345 East 47th Street, New York, NY 10017

The Institute of Electrical and Electronic Engineers
444 Hoes Lane, PO Box 459, Piscataway, NJ 08855

National Society of Professional Engineers
1420 King Street, Alexandria, VA 22314

Society of Petroleum Engineers (SPE)
P.O. Box 833836, Richardson, TX 75083

Society of Women Engineers
120 Wall Street, 11th Floor, New York, NY 10005

2

The Design Process

2.1 Introduction

The design process is the method of devising innovative solutions to problems that will result in new products or systems. Engineering graphics is the primary medium of design that is used for developing designs from initial concepts to final working drawings. Initially, a design consists of sketches that are refined, analyzed, and developed into precise detail drawings and specifications. They, in turn, become part of the contract documents for the parties involved in funding and implementing a project.

At first glance, the solution to a design problem may appear to involve merely the recognition of a need and the application of effort toward its solution, but most engineering designs are more complex than that. The engineering and design efforts may be the easiest parts of a project.

For example, engineers who develop roadway systems must deal with constraints such as

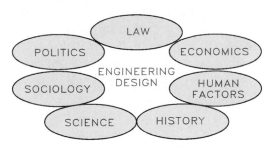

2.1 An engineering project may involve the interaction of people representing many professions and interests, with engineering design as the central function.

ordinances, historical data, human factors, social considerations, scientific principles, budgeting, and politics (**Figure 2.1**). Engineers can readily design driving surfaces, drainage systems, overpasses, and other components of the system. However, adherence to budgetary limitations is essential, and funding for public projects is closely related to politics.

Traffic laws, zoning ordinances, environmental impact statements, right-of-way

16

acquisition, and liability clearances are legal aspects of roadway design that engineers must deal with. Past trends, historical data, human factors (including driver characteristics), and safety features affecting the function of the traffic system must be analyzed. Social problems may arise if proposed roadways will be heavily traveled and will attract commercial development such as shopping centers, fast-food outlets, and service stations. Finally, designers must apply engineering principles developed through research and experience to obtain durable roads, economical bridges, and fully functional systems.

2.2 Types of Design Problems

Most design problems fall into one of two categories: **product design** and **systems design**.

Product Design

Product design is the creation, testing, and manufacture of an item that will usually be mass produced, such as an appliance, a tool, a toy, or larger products such as automobiles (**Figure 2.2**). In general, a product must have a sufficiently broad appeal for meeting a specific need and performing an independent function to warrant its production in quantity. Designers of products, whether automobiles or bicycles, must consider current market needs, production costs, function, sales, and distribution methods, as well as profit predictions (**Figure 2.3**).

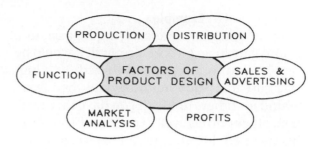

2.3 These are some of the major factors that must be considered when developing a product design.

Products can perform one or many functions. For instance, the primary function of an automobile is to provide transportation, but it also contains products that provide communications, illumination, comfort, entertainment, and safety. Because it is mass produced for a large consumer market and can be purchased as a unit, the automobile is regarded as a product. However, because it consists of many products that perform various functions, the automobile is also a system.

Systems Design

Systems design combines products and their components into a unique arrangement and provides a method for their operation. A residential building is a system of products consisting of heating and cooling, plumbing, natural gas, electrical power, sewage, appliances, entertainment, and others that form the overall system as shown in **Figure 2.4**.

2.2 Product design seeks to develop a product that meets a specific need, that can function independently, and that can be mass produced.

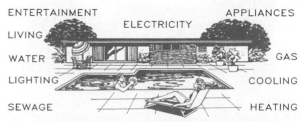

2.4 The typical residence is a system composed of many components and products.

Systems Design Example

Suppose that you were carrying luggage to a distant gate in an airport terminal. It would be easy for you to recognize the need for a luggage cart that you could use and then leave behind for others to use. If you have this need, then others do as well. The identification of this need could prompt you to design a cart like the one shown in **Figure 2.5** to hold luggage, and even a child, and to make it available to travelers. The cart is a product.

How could you profit from providing such a cart? First you would need a method of holding the carts and dispensing them to customers, such as the one shown in **Figure 2.6**. You would also need a method for users to pay

2.7 The coin-operated gate (left) releases a cart from the cart dispenser. A vending machine (right) makes change for $5, $10, and $20 bills and issues baggage cards that can be used at any other airport terminal that utilizes the same cart system. (Courtesy of Smarte Carte.)

2.5 The luggage cart in an airport terminal is a product as well as part of a system. (Courtesy of Smarte Carte.)

2.6 Luggage carts are dispensed from coin-operated centers at major entrances of airport terminals. (Courtesy of Smarte Carte.)

for cart rental, so you could design a coin-operated gate for releasing them (**Figure 2.7**).

For an added customer convenience, you could provide a vending machine that would take bills both to make change and to issue cards entitling customers to multiple use of carts at this and other terminals (Figure 2.7). A mechanism to encourage customers to return carts to another conveniently located dispensing unit would also be helpful and efficient. A coin dispenser (Figure 2.7) that gives a partial refund on the rental fee to customers returning carts to a dispensing unit at their destination might work.

The combination of these products and the method of using them is a system design. Such a system of products is more valuable than the sum of the products alone.

2.3 The Design Process

Design is the process of creating a product or system to satisfy a set of requirements that has multiple solutions by using any available resources. In essentially all cases, the final

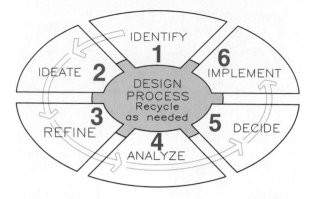

2.8 The design process consists of six steps, each of which can be repeated as necessary.

design must be completed at a profit or within a budget.

The steps of the design process (**Figure 2.8**) are

1. problem identification,

2. preliminary ideas (ideation),

3. refinement,

4. analysis,

5. decision, and

6. implementation.

Designers should work sequentially from step to step but should review previous steps periodically and rework them if a new approach comes to mind during the process.

Problem Identification

Most engineering problems are not clearly defined at the outset and require identification before an attempt is made to solve them (**Figure 2.9**). For example, air pollution is a concern, but we must identify its causes before we can solve the problem. Is it caused by automobiles, factories, atmospheric conditions that harbor impurities, or geographic features that trap impure atmospheres?

Another example is traffic congestion. When you enter a street where traffic is

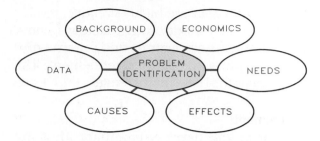

2.9 Problem identification requires that the designer accumulate as much information about a problem as possible before attempting a solution. The designer also should keep product marketing in mind at all times.

unusually congested, can you identify the reasons for the congestion? Are there too many cars? Are the signals poorly synchronized? Are there visual obstructions? Has an accident blocked traffic?

Problem identification involves much more than simply stating, "We need to eliminate air pollution." We need data of several types: opinion surveys, historical records, personal observations, experimental data, physical measurements from the field, and more. It is important that the designer resist the temptation to begin developing a solution before the identification step has been completed.

Preliminary Ideas

The second step of the design process is the development of as many ideas for solving the problem as possible (**Figure 2.10**). A brainstorming session is a good way to collect ideas

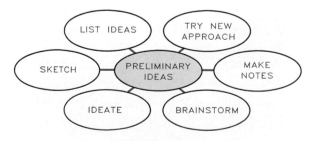

2.10 The designer gathers ideas from a brainstorming session and develops preliminary ideas for a problem's solution. Ideas should be listed, sketched, and noted in order to have a broad range of ideas with which to work.

at the outset that are highly creative, revolutionary, and even wild. Rough sketches, notes, and comments can capture and preserve preliminary ideas for further refinement. The more ideas the better at this stage.

Refinement

Several of the better preliminary ideas are selected for refinement to determine their merits. The rough preliminary sketches are converted into scale drawings for spatial analysis, determination of critical measurements, and the calculation of areas and volumes affecting the design (**Figure 2.11**). Descriptive geometry aids in determining spatial relationships, angles between planes, lengths of structural members, intersections of surfaces and planes, and other geometric relationships.

Analysis

Analysis is the step during which engineering and scientific principles are used most intensively to evaluate the best designs and compare their merits with respect to function, strength, safety, cost, and optimization (**Figure 2.12**). Graphical methods play an important role in analysis as well. Data can be analyzed graphically, forces can be analyzed as graphical vectors, and empirical data can be analyzed, integrated, and differentiated by other graphical methods. The analysis phase is less creative than the previous steps.

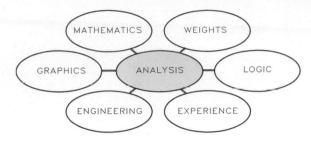

2.12 All available methods, from science to technology to graphics to experience, should be used to analyze a design.

Decision

After analysis, a single design, which may be a compromise among several designs, is decided upon as the solution to the problem (**Figure 2.13**). The designer alone, or a team, may make the decision. The outstanding aspects of each design usually lend themselves to graphical comparisons of manufacturing costs, weights, operational characteristics, and other data essential in decision making.

Implementation

The final design must be described and detailed in working drawings and specifications from which the project can be built, whether it is a computer chip or a suspension bridge (**Figure 2.14**). Workers must have precise instructions for the manufacture of each component, often measured within thousandths of an inch to ensure proper fabrication assembly. Working drawings must be sufficiently explicit in order to serve as part of the legal contract with the successful bidder on the job.

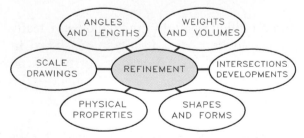

2.11 Refinement begins with the construction of scale drawings of the best preliminary ideas. Descriptive geometry and graphics are used to describe geometric characteristics.

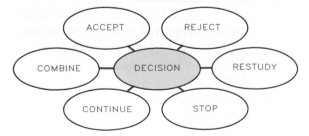

2.13 Decision involves the selection of the best design or design features to implement. This step may require an acceptance, rejection, or a compromise of the proposed solution.

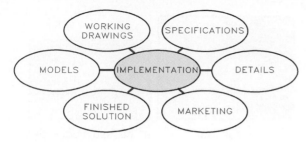

2.14 Implementation consists of the preparation of drawings, specifications, and documentation from which the product can be made. The product is produced and marketing is begun.

2.4 Graphics and Design

Whether by freehand sketching, with instruments, or on a computer, graphics is the medium of design. Engineering, scientific, and analytical principles must be applied throughout the process, and graphics is the medium that is employed in each step, from problem identification to implementation.

An example of how graphics is used by a design firm to develop a product is shown in **Figure 2.15** through **Figure 2.19**. A freehand sketch of the basic concept of an eye-shower device, with notes explaining how it is to be used to rinse out the eye of the user by spraying a solution, is shown in **Figure 2.15**.

The designer makes a series of four sketches in **Figure 2.16** to study how the tank housing is snapped into position. These sketches represent the designer's process as well as a means of recording his ideas and for communicating with others.

A three-dimensional pictorial (**Figure 2.17**) illustrates additional design features as the product evolves. The preliminary designs are studied and evaluated by several designers as a means of "trouble shooting" the design and applying all available experience and expertise to the eye shower.

Additional design details are explained in the sketch in **Figure 2.18**. The movements of the adjacent parts can be shown better in a three-dimensional drawing than in multiview drawings. These concepts must be refined

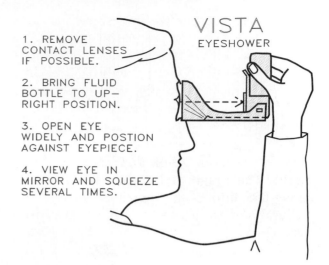

2.15 This freehand sketch with its accompanying notes describes the concept for an eye-shower device. (Courtesy Keck-Craig, Inc.)

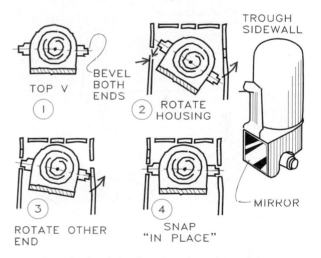

2.16 These freehand sketches show how the tank housing is snapped into position. (Courtesy Keck-Craig, Inc.)

and modified to attain the final design, ready for production as a prototype as shown in **Figure 2.19**. A more detailed coverage of the steps of the design process will be given in the following chapters.

2.5 Application of the Design Process

The following example illustrates the application of the design process to a simple problem.

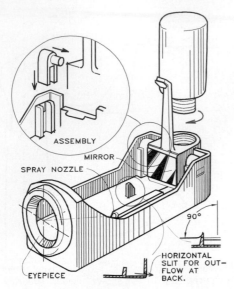

2.17 This three-dimensional pictorial clarifies additional production details that must be provided for the eye-shower device. (Courtesy Keck-Craig, Inc.)

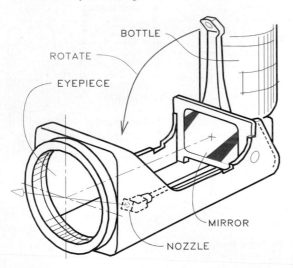

2.18 This pictorial shows the motion of the tank and the relationship of the mirror and nozzle to the eyeshower. (Courtesy Keck-Craig, Inc.)

Hanger Bracket Problem

A bracket is needed to support a 2 inch diameter hot-water pipe from a beam or column 9 inches from the mounting surface. This design project is typical of an in-house assignment by an engineer of a company specializing in pipe hangers and supports.

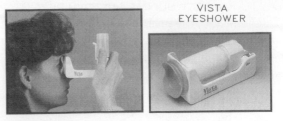

VISTA EYESHOWER

A. OPEN FOR USE   B. CLOSED FOR STORAGE

2.19 Photographs show how the eyeshower is used and closed for carrying in a pocket or purse. (Courtesy Keck-Craig, Inc.)

Problem Identification

First, write a statement of the problem and a statement of need (**Figure 2.20**). List limitations and desirable features and make descriptive sketches to better identify the requirements (**Figure 2.21**). Even if much of this information

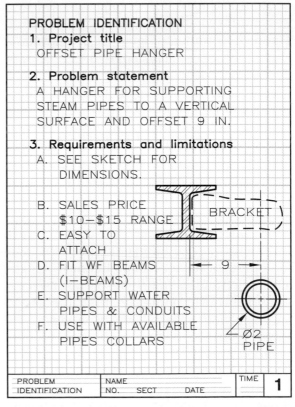

PROBLEM IDENTIFICATION
1. **Project title**
   OFFSET PIPE HANGER

2. **Problem statement**
   A HANGER FOR SUPPORTING STEAM PIPES TO A VERTICAL SURFACE AND OFFSET 9 IN.

3. **Requirements and limitations**
   A. SEE SKETCH FOR DIMENSIONS.
   B. SALES PRICE $10–$15 RANGE
   C. EASY TO ATTACH
   D. FIT WF BEAMS (I–BEAMS)
   E. SUPPORT WATER PIPES & CONDUITS
   F. USE WITH AVAILABLE PIPES COLLARS

   BRACKET

   9

   Ø2 PIPE

   PROBLEM IDENTIFICATION | NAME | NO. SECT DATE | TIME | 1

2.20 This worksheet shows aspects of problem identification, the first step of the design process.

may be obvious, writing statements and making sketches about the problem will help you "warm up" to the problem and begin the creative process. Also, you must begin thinking about sales outlets and marketing methods as well.

Preliminary Ideas

Brainstorm alone or with others for possible solutions to the problem. List the ideas obtained on a worksheet (**Figure 2.22**), and summarize the best ideas and design features on a separate worksheet (**Figure 2.23**). Then

2.21 (Sheet 2) This worksheet shows information needed before the designer can proceed, including sources from which the information can be obtained.

2.22 (Sheet 3) A member of the brainstorming team records the ideas from the session.

2.23 (Sheet 4) The best ideas are selected from the brainstorming session to be developed as preliminary ideas.

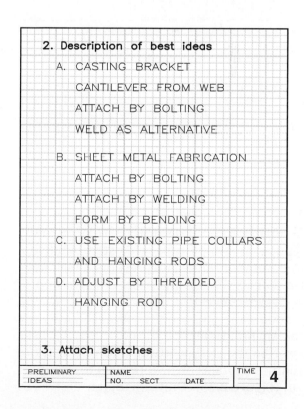

```
PRELIMINARY IDEAS
1. Brainstorming ideas
   A. HANG BY A CABLE
   B. WELD TO BOTTOM OF BEAM
   C. BOLT TO BEAM'S WEB
   D. MAKE FROM SHEET METAL
   E. MAKE AS A CASTING
   F. ATTACH BY CLAMPING VISE
   G. MAKE BRACKET RIGID
   H. MAKE BRACKET FLEXIBLE
   I. ADJUST HEIGHT BY HEX NUT
   J. SUPPORT PIPE IN TROUGH
   K. SUSPEND ON COIL SPRING
   L. CLIP TO UPPER FLANGE
   M. CASTING ATTACHED TO WEB
   N. CANTILEVER EXTENSION
   O. USE FOR LIGHTING CONDUITS
   P. ROLLERS FOR EXPANSION
   Q. SPRINGS FOR EXPANSION
   R. HANGER RODS WITH SPRINGS
   S. USABLE ON COLUMNS TOO
   T. USABLE ON CHANNELS
```
PRELIMINARY IDEAS | NAME NO. SECT DATE | TIME | **3**

```
4. Needed information
   A. NEED DIMENSIONS OF WF
      BEAMS (I-BEAMS)
      CHECK MANUALS
   B. WHAT PIPE COLLARS ARE
      AVAILABLE
      REVIEW-COMMERCIAL SOURCES
   C. DOES A BRACKET EXIST ON THE
      MARKET NOW? CHECK CATALOGS

5. Market considerations
   A. WHAT IS BEING USED NOW?
      CHECK COMPETITORS' CATALOGS
   B. WHAT IS MARKET POTENTIAL?
      CHECK WITH COMPETITORS
      INTERVIEW CUSTOMERS
      CHECK WITH COMPANY REPS
   C. WHAT IS THE PRICE RANGE?
      CHECK WITH OUR FIELD REPS
```
PROBLEM IDENTIFICATION | NAME NO. SECT DATE | TIME | **2**

```
2. Description of best ideas
   A. CASTING BRACKET
      CANTILEVER FROM WEB
      ATTACH BY BOLTING
      WELD AS ALTERNATIVE
   B. SHEET METAL FABRICATION
      ATTACH BY BOLTING
      ATTACH BY WELDING
      FORM BY BENDING
   C. USE EXISTING PIPE COLLARS
      AND HANGING RODS
   D. ADJUST BY THREADED
      HANGING ROD

3. Attach sketches
```
PRELIMINARY IDEAS | NAME NO. SECT DATE | TIME | **4**

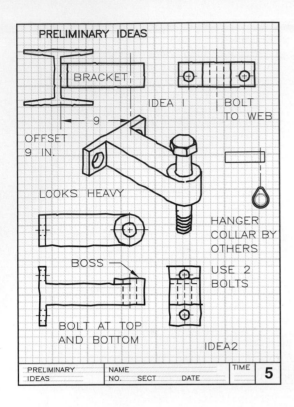

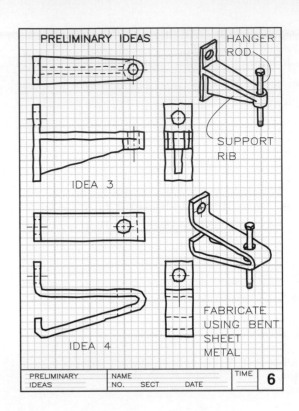

2.24 (Sheet 5) Preliminary ideas are sketched and noted for further development. This is the most creative step of the design process.

2.25 (Sheet 6) Additional preliminary design solutions are sketched here.

2.26 (Sheet 7) Refinement of preliminary ideas begins with written descriptions of the better ideas.

translate and expand these verbal ideas into rapidly-drawn freehand sketches (**Figure 2.24** and **Figure 2.25**). You should develop as many ideas as possible during this step because a large number of ideas represents a high level of creativity. This is the most creative step of the design process.

Problem Refinement

Describe the design features of one or more preliminary ideas on a worksheet for comparison (**Figure 2.26**). Draw the better designs to

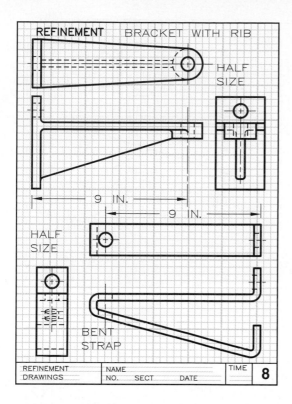

HALF
SIZE

9 IN.

9 IN.

HALF
SIZE

BENT
STRAP

| REFINEMENT DRAWINGS | NAME NO.   SECT   DATE | TIME | 8 |

ANALYSIS
1. Function
   A. EXTENDS HANGER ROD
      OUT 9 IN. FROM VERTICAL
   B. HOLDS 3/8 IN. ROD
   C. MOUNTS ON 1/2 BOLT

2. Human engineering
   A. EASY & QUICK TO INSTALL
   B. REQUIRES NO SPECIAL TOOLS
   C. PROVIDES THE REQUIRED
      SAFETY

3. Market and consumer acceptance
   A. KEEP PRICE UNDER $20
   B. CAN BE SOLD AS AN
      ADDITION TO OUR LINE
   C. SHOULD BE ADDED TO OUR
      PRODUCT CATALOG
   D. NEED TO FIND ESTIMATE OF
      OFFSET BRACKETS SOLD PER YR.

| DESIGN ANALYSIS | NAME NO.   SECT   DATE | TIME | 9 |

2.27 (Sheet 8) Scale drawings of two designs are made to describe the designs. Almost no dimensions are needed.

2.28 (Sheet 9) The beginning of the analysis step.

2.29 (Sheet 10) The next two steps of the analysis.

scale in preparation for analysis; you need to show only a few dimensions at this stage (**Figure 2.27**).

Use instrument-drawn orthographic projections, computer drawings, and descriptive geometry to refine the designs and ensure precision. Let's say that you select ideas 3 and 5 for analysis. Figure 2.27 depicts orthographic and auxiliary views of the two designs.

Analysis
Use an analysis worksheet to analyze the cast-iron bracket design (**Figures 2.28–2.31**). If you

4. Physical description
   A. BRACKET: 10—11 IN. LONG
      WITH RIB FOR STRENGTH
   B. ATTACHED WITH A SINGLE
      BOLT AT UPPER SIDE
   C. WEIGHT: ABOUT 3—4 LBS
   D. BOSS FOR EXTRA STRENGTH
      WHERE HANGER ROD ATTACHES
   E. FORMED AS A CASTING
      CAST IRON

5. Strength
   A. WILL SUPPORT ABOUT 200 LBS
      WITH A SAFETY FACTOR OF 5
   B. 1/2 BOLT SUFFICIENT FOR
      ATTACHMENT TO BEAM OR
      OR COLUMN
   C. 3/8 DIA HANGER ROD MORE
      THAN ADEQUATE TO SUPPORT
      200 LBS

| DESIGN ANALYSIS | NAME NO.   SECT   DATE | TIME | 10 |

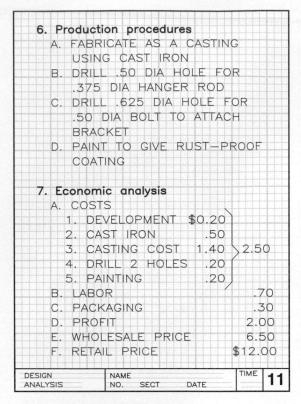

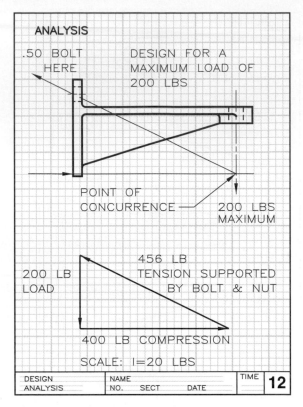

2.30 (Sheet 11) This portion of the analysis focuses on the production and economic considerations.

2.31 (Sheet 12) Graphics and descriptive geometry can be used to determine the load carried by the support bolt.

are considering more than one solution, analyze each design.

By using the maximum load of 200 lbs. and the geometry of the bracket, it is possible to graphically determine the angle of the reaction that the bolt must carry, 456 lbs, when the bracket is fully loaded (**Figure 2.31**). Again, graphics is used as an important design tool.

Decision
The decision table (**Figure 2.32**) compares two designs: the bracket with a rib versus the bent strap. Assign weight factors to be analyzed by assigning points to them that add up to 10.

You can then rank the designs by their overall scores from highest to lowest.

Draw conclusions and summarize the features of the recommended design, along with a projection of its marketability (**Figure 2.33**). In this case, let's suppose we decide to implement the bracket with a rib.

Implementation
Detail the bracket with a rib design in a working drawing that graphically describes and dimensions each individual part (**Figure 2.34**).

Use notes to specify standard parts—the nuts, bolts, and hanger rod—but it will be

DECISION

1. Decision table for evaluation
DESIGN 1: BRACKET WITH RIB
DESIGN 2: BENT STRAP
DESIGN 3:
DESIGN 4:
DESIGN 5:

MAX	FACTORS	1	2	3	4	5
2	FUNCTION	2	1.5			
2	HUMAN FACT.	1.5	1.5			
1	MARKET ANAL	.5	.2			
1	STRENGTH	1	.5			
1	PRODUCTION	.5	.5			
1	COST	.5	.2			
2	PROFITABILITY	1.0	1.0			
0	APPEARANCE	0	0			
10	TOTALS	7	5.4			

DESIGN DECISION	NAME NO. SECT DATE	TIME	13

2.32 (Sheet 13) A decision table is used to evaluate alternative designs in order to arrive at the final selection.

CONCLUSIONS

THE BRACKET WITH A RIB DESIGN APPEARS TO BE THE BEST AND MOST MARKETABLE SOLUTION

IT WOULD BE MOST SUITED TO THE MARKET NEEDS

GOOD POSSIBILITY OF SELLING AS ADDITIONAL PRODUCT ACCESSORY

RETAIL	$12.00
SHIPPING EXPENSES	2.00
NUMBER TO SELL TO BREAK EVEN	1000
MANUFACTURED IN HOUSE	$3.50
PROFIT PER UNIT	$2.00

RECOMMEND IMPLEMENTATION AND PRODUCTION OF THE PRODUCT AND ADD TO LINE.

DESIGN DECISION	NAME NO. SECT DATE	TIME	14

2.33 (Sheet 14) The decision step is summarized to conclude this step of the design process.

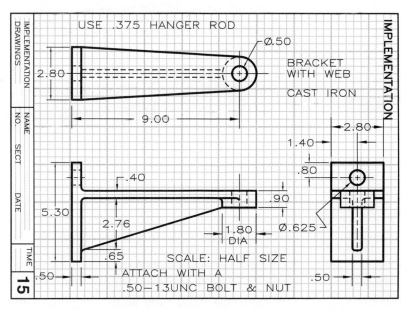

2.34 The completed swing set anchor ready to market.

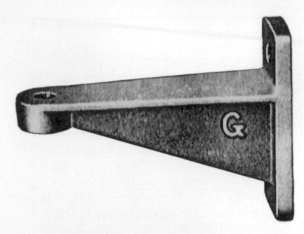

2.35 The completed bracket is shown here ready to be marketed. (Courtesy of Grinnell Company, Inc.)

unnecessary to draw them because they will be bought as standard parts. This working drawing shows how the bracket will be made. Now, you need to build a prototype or model of the product and test it for function.

Figure 2.35 shows the final product, the cast-iron bracket with a rib. After determining how it will be packaged and distributed, your next task will be to add it to your company's product line, list it in your catalog, and introduce it to the marketplace.

Problems

Most of the following problems are to be solved on 8-1/2" x 11" paper, using instruments or by sketching freehand, as assigned. Use a paper with a 0.20-inch printed grid or plain paper and an engineer's scale to lay out the problems.

Endorse each problem sheet with your name and file number, date, and problem number. Letter all points, lines, and planes with 1/8-inch letters, using guidelines. Letter the answers to essay problems with approved single-stroke Gothic lettering. (See Chapter 11.)

1. Outline your plan of activities for the weekend. Indicate aspects of your plans that you feel display creativity or imagination. Explain why.

2. Write a short report (not to exceed two pages) on the engineer or engineering achievement that you believe exhibits a high degree of creativity. Justify your selection by outlining the creative aspects of your choice.

3. Test your ability to recognize the need for new designs. List as many improvements as you can think of for the typical automobile. Make suggestions for implementing these improvements. Follow the same procedure for another product of your choice.

4. List as many systems as you can that affect your daily life. Separate several of these systems into their components (subsystems).

5. Subdivide the following items into their individual components: (a) a classroom, (b) a wristwatch, (c) a movie theater, (d) an electric motor, (e) a coffee percolator, (f) a golf course, (g) a service station, and (h) a bridge.

6. Indicate which of the items in Problem 5 are systems and which are products. Explain your answers.

7. Make a list of products and systems that you believe would be necessary for life on the moon.

8. You have been assigned the responsibility for organizing and designing a skateboard installation where skateboarders can practice that will be self-supporting. Write a paragraph on each of the six steps of the design process to explain how you would apply each step to the problem. For example, explain the action you would take to identify the problem.

9. Suppose that you are responsible for designing a motorized wheelbarrow to be marketed for home use. Write a paragraph on each of the six steps of the design process, explaining how you would apply each step to the problem. For example, what action would you take to identify the problem?

10. List and explain a sequence of steps that you believe would be adequate for, yet different from, the design process given in this chapter. Your version of the design process may contain any of the steps discussed.

11. Design a simple device for holding a fishing pole in a fishing position while the person fishing rows the boat. Make sketches and notes to describe your design.

12. Design a doorstop to keep a door from slamming into the wall behind it. Make rapid freehand sketches and notes using the six design steps. Do not spend more than thirty minutes on this problem. Indicate any information you would need at the decision and implementation steps that you may not have now.

13. List factors to consider during the problem identification step for designing (a) a skillet, (b) a bicycle lock, (c) a handle for a piece of luggage, (d) a method for improving your grades, (e) a child's toy, (f) a stadium seat, (g) a desk lamp, (h) an umbrella, and (i) a hot dog stand.

14. Hanger Bracket identification: Make a worksheet that could follow Sheet 2 (Fig. 2.16) to provide additional identification information. For example, how much per linear foot will a 2" XXX pipe weigh (refer to Chapter 35) when it is full of water (refer to the Appendix 45 for the weight of water). Show your calculations on the worksheet. How far apart should hangers be spaced if each is to carry no more than 200 lbs with a safety factor of 5 (the capacity to carry 5 times the design load)? Make sketches to clarify this information.

15. Floor Bracket: You are responsible for designing a support bracket to serve the same purpose as the one designed in Figures 2.15 through 2.30 with the following exception. The bracket must attach to a concrete floor and support the bottom of the pipe 12 inches above the floor. Make sketches to describe your design.

16. Refer to Section 2.1 and Figure 2.1 and develop a worksheet that will show the various interactions that should be considered for the following projects: (A) adding a room to your home, (B) building a swimming pool, (C) opening a restaurant, (D) planning a garage sale, (E) developing a bike trail.

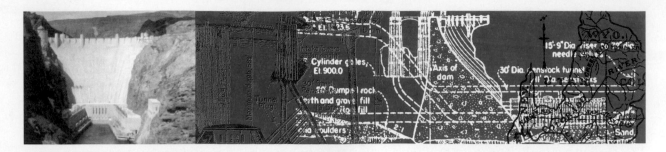

3
Problem Identification

3.1 Introduction

Problem identification is the initial step that a designer takes to solve a problem. It first involves recognizing a need and then proposing design criteria (**Figure 3.1**).

Recognizing a need may begin with the observation of a problem or a defect in an existing product or a system.

Or, you may have one of those rare events where you think to yourself, "Why hasn't someone invented a gadget that . . . ?" The need may be for an improved automobile safety belt, a system allowing full wheelchair access to public buildings, or a new exercise apparatus. The solution may be a product or a system improvement so that it will be more reliable and perhaps more profitable.

Proposing design criteria follows the recognition of a need. Here the designer proposes the specifications a new product or system must meet.

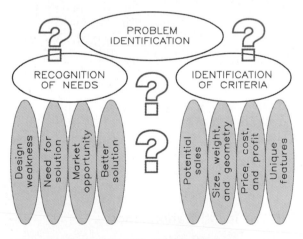

3.1 The two aspects of problem identification.

3.2 Example: Ladder Attachment

When you use a ladder for house repairs, it works well enough when placed against the roof or fascia at the eaves of the house (**Figure 3.2**). However, if a gutter is attached to the fascia, the ladder damages the gutter when

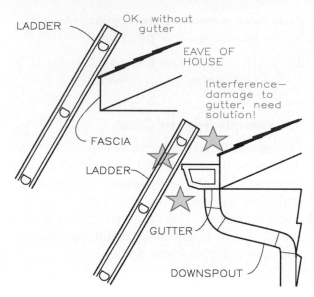

3.2 A ladder placed against the fascia or roof works fine, but when a gutter is attached to the fascia, the ladder damages the gutter. This problem needs a solution.

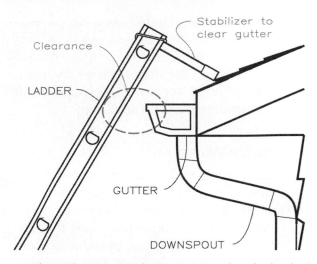

3.3 The problem recognized in Figure 3.2 can be solved with a stabilizer attached to the ladder, which helps the ladder clear the gutter.

placed against it. The need is to prevent gutter damage by the ladder.

Now, you must propose design criteria before proceeding with a solution. You need to know the geometry and dimensions of typical gutters, pitches of roofs, angles of ladder placement, and the potential market for the solution. Later in the design process, you might design a product that attaches to the ladder, allowing it to clear the gutter (**Figure 3.3**). Other solutions are possible, but they all begin with recognizing the need.

3.3 The Identification Process

Problem identification requires the designer to determine requirements, limitations, and other background information before becoming involved in a problem solution. The designer should take the following steps during problem identification (**Figure 3.4**).

1. **Problem statement.** Describe the problem in order to facilitate the thinking process.

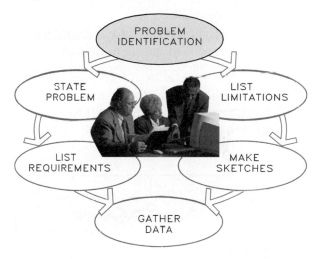

3.4 Problem identification requires a thorough investigation of background factors before attempting a design solution.

2. **Requirements.** List the conditions that the design must meet. Most will be questions to be answered after data are gathered.

3. **Limitations.** List the factors affecting design specifications, such as maximum weight or size.

4. **Sketches.** Make sketches and add notes and dimensions to identify geometric and physical characteristics.

5. Data collection. Gather data on population trends, related designs, physical characteristics, sales records, and market studies. Graph the data for interpretation, as in **Figure 3.5**, for easier interpretation of raw numbers and data tabulation.

3.4 Design Worksheets

Designers must make numerous notes and sketches on worksheets throughout the design process. Worksheets serve to document what has been done and allow for periodic review of earlier ideas, in order to avoid overlooking previously identified concepts. Moreover, a written and visual record of design work helps establish ownership of patentable ideas.

The following materials aid in maintaining permanent records of design activities.

1. *Worksheets* (8-1/2 by 11 inches). Sheets can be either grid-lined or plain and should be three-hole punched for a notebook or binder.

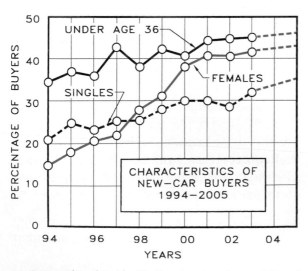

3.5 Data gathered to identify the changing characteristics of new-car buyers can be interpreted more easily when it has been graphed.

2. *Pencils.* A medium-grade pencil (F or HB) is adequate for most purposes.

3. *Binder* or *envelope.* Keep worksheets in a binder or envelope for reference.

3.5 Example: Exercise Bench

Design an exercise bench that can be used by those who lift weights for body fitness. This apparatus should be as versatile as possible and at the low end of the price range for exercise equipment. The contents of the worksheets shown in **Figures 3.6–3.9** identify the problem in a typical manner.

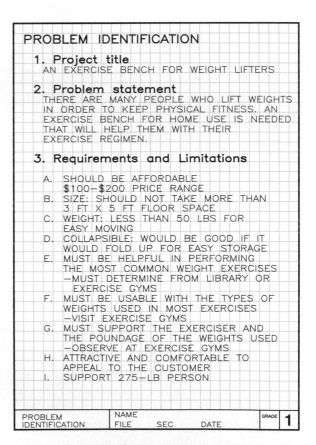

3.6 The problem-identification step gives the project a title, describes what the product will be used for, and lists basic requirements and limitations for the design.

```
4.  Needed information
   A.  NUMBER OF PEOPLE WHO EXERCISE?
       —LIBRARY, INTERVIEW GYM OWNERS
   B.  AGES OF EXERCISERS? OBSERVE GYM;
       HEALTH MAGAZINES
   C.  WHAT IS THE POUNDAGE OF THE
       WEIGHTS USED BY MOST EXERCISERS
   D.  WHERE DO MOST PEOPLE EXERCISE?
       HOME OR GYM? —LIBRARY AND INTER—
       VIEW GYM OPERATORS
   E.  WHAT FEATURES WOULD BE DESIRABLE
       ON AN EXERCISE BENCH? —INTERVIEW
       USERS AT GYM
   F.  IS HOME EXERCISING DONE INSIDE OR
       OUTSIDE? —INTERVIEW USERS AT GYM
   G.  HOW OFTEN DO PEOPLE EXERCISE?
       —INTERVIEW USERS AT GYM

5.  Market considerations
   A.  SUITABLE MARKET PRICE? —CHECK
       RETAIL OUTLETS AND CATALOGS FOR
       COMPETING PRICES
   B.  HOW MUCH IS SPENT ON EXERCISING
       AT GYMS? —CHECK YELLOW PAGES TO
       DETERMINE THE NUMBER OF GYMS
       AND RELATE POPULATION SIZE
   C.  BRISKNESS OF SALES FOR EXERCISING
       BENCHES? —INTERVIEW SALESMEN AT
       SPORTING GOODS STORES
   D.  WHAT FEATURES DO BUYERS OF
       EQUIPMENT LIKE? —QUESTION SALESMEN
       AT SPORTING STORES
   E.  IS THERE A VOID IN THE EXERCISE
       APPARATUS MARKET? —QUESTION SALES—
       PEOPLE AND INTERVIEW USERS AT GYMS
   F.  HOW MANY PEOPLE HAVE MEMBERSHIPS
       AT GYMS? SPOT CHECK SEVERAL GYMS.

PROBLEM                NAME                          GRADE
IDENTIFICATION         FILE    SEC    DATE              2
```

3.7 Continuation of problem identification for the exercise bench includes gathering information on market potential.

```
PROBLEM IDENTIFICATION
A.  CHECK YELLOW PAGES FOR GYMS
    —8 GYMS FOR POPULATION OF 100,000
    —ONE GYM PER 12,500 PEOPLE
B.  GYM MEMBERSHIP FOR STILLMAN'S GYM
    YEAR      MEN        WOMEN
    1994      75         20
    1996      70         36
    1998      110        70
    2000      175        82
    2002      180        120
    2004      210        135

C.  INTERVIEW RETAILERS OF EQUIPMENT
    —3 DEALERS POSITIVE ABOUT BENCH
    —2 DEALERS NEUTRAL
    —1 DEALER NEGATIVE ABOUT PROSPECTS

D.  AGES OF EXERCISERS (BY OBSERVATION)
    —20% UNDER 20
    —40% BETWEEN 20 AND 30
    —30% BETWEEN 30 AND 50
    —10% OVER 50

E.  PRICES AT STORES
    1.  COMPLEX MULTI—USE        $1000
    2.  MEDIUM—RANGE EQUIPMENT    700
    3.  LIGHTWEIGHT BENCHES       130
    4.  WEIGHTS                   100

F.  TYPICAL BRANDS OF EQUIPMENT
    ON THE MARKET?
    1.  WEIDER
    2.  NAUTILUS
    3.  BODY WONDERFUL
    4.  HEALTH PLUS

PROBLEM                NAME                          GRADE
IDENTIFICATION         FILE    SEC    DATE              3
```

3.8 This worksheet shows the data collected for use in designing the exercise bench.

First, give the title of the project and a brief problem statement to describe the problem better. Then list requirements and limitations and add sketches as necessary. You will have to list some requirements as questions for the time being, but in all cases, make estimates and give sources for the answers (**Figure 3.6**). Make estimates as you go along; for example, it must cost between $100 and $200; it must weigh less than 50 pounds; and it must at least support a person weighing 275 pounds. Give estimates as ranges of prices or weights rather than exact numbers. Use catalogs offering similar products as sources for prices, weights, and sizes.

Next, make a list of the questions that need answers. Follow each question with a source for its answer and give a preliminary answer. How many people exercise? What are their ages? Do they buy exercise equipment? What are the most popular weight exercises? You may obtain this type of information from interviews, product catalogs, the library, and sporting-goods stores. Market considerations include the average income of a typical exerciser. How much does he or she spend on physical fitness per year? The opinions of sporting-goods dealers are helpful, and they should be able to direct you to other sources of information (**Figure 3.7**).

3.5 EXAMPLE: EXERCISE BENCH • 33

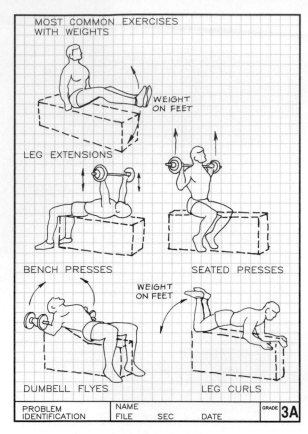

MOST COMMON EXERCISES WITH WEIGHTS

LEG EXTENSIONS — WEIGHT ON FEET

BENCH PRESSES   SEATED PRESSES

DUMBELL FLYES — WEIGHT ON FEET   LEG CURLS

| PROBLEM IDENTIFICATION | NAME FILE   SEC   DATE | GRADE 3A |

3.9 Sketches of the most popular weight exercises.

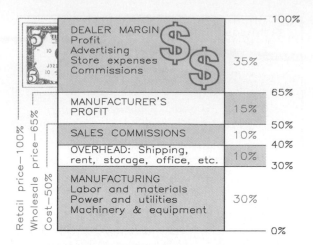

3.10 This chart shows the breakdown of costs involved in the retail price of a product.

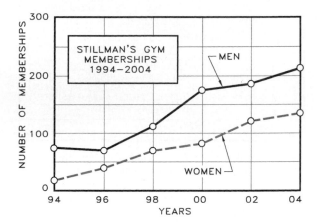

3.11 This graph describes visually the trends in potential customers for the exercise apparatus.

Record the data that you gather by interviewing gym managers, looking at catalogs, and visiting sporting-goods stores to learn about the people who exercise and the equipment they use (**Figure 3.8**). Determine the most popular weight exercises of those for whom this bench is to be designed by surveying users, coaches, and sporting-goods outlets. Sketch those exercises on a worksheet (**Figure 3.9**).

Think about costs and pricing the product even during problem identification. **Figure 3.10** shows how products may be priced from wholesale to retail, but these percentages vary by product. For instance, profit margins are smaller for food sales than for furniture sales. If an item retails for $50, the production and overhead cost cannot be more than about $20 to maintain the necessary margins.

Data are easier to interpret if presented graphically (**Figure 3.11**). Thorough problem identification includes graphs, sketches, and schematics that improve the communication of your findings.

Problem identification is not complete at this point. However, this example should give you a basic understanding of the process.

3.6 Organization of Effort

The designer should prepare a schedule of required design activities after completing problem identification. The **project evalua-**

tion and review technique (**PERT**), developed by project managers for projects requiring coordination of many activities, aids in scheduling tasks. PERT is a means of scheduling activities in sequence and reviewing progress made toward their completion.

The **critical path method** (**CPM**) of scheduling evolved from PERT and is used in conjunction with it. The tasks that must be completed before others can begin are critical and should be given priority. The critical path is the sequence of tasks requiring the longest time from start to finish and has the least flexibility. Activities not in the critical path receive less emphasis.

3.7 Planning Design Activities

The chart in **Figure 3.12** illustrates the steps involved in planning a job (project): (1) list the tasks on a form called a Design Schedule and Progress Record; (2) prepare an Activities Network Diagram, arranging the tasks in sequence; and (3) prepare an Activity Sequence Chart that graphs the tasks in the sequence shown in the network.

Design Schedule and Progress Record

The designer separates the job into tasks, numbers them, and lists them in the **Design Schedule and Progress Record** (**DS&PR**) without concern for their sequence (**Figure 3.13**). Next, the designer estimates the amount of

Design Schedule & Progress Record

TEAM __4__ PROJECT __PRODUCT DESIGN__

WORK PERIODS __II__ MAN HOURS __70__ FINISHING DATE __4-15__

JOB	ASSIGNMENT	Est. Hrs.	Act. Hrs.	PERCENT COMPLETE 0 20 40 60 80
7	WRITE REPORT—ALL	15		
6	BRAINSTORM—ALL	.5	1	
3	WRITE MFGRS—JHE	2		
10	GRAPH DATA—HLE	2		
5	MARKET ANALYSIS—DL	3	3	
12	COLLECT DATA—TT	2		

3.13 The Design Schedule and Progress Record (DS&PR) shows typical entries for a product design project.

time required for each task and enters it in the third column. The designer adjusts the amount of time for each task so that the total matches the time allotted for the job.

Activities Network

The designer prepares an **Activities Network** (**AN**) by arranging the tasks from the DS&PR in their proper sequence in either note or symbol form (**Figure 3.14**). The note form identifies activities by task name; the symbol form identifies activities by task number. Arrows connect activities, showing their sequence. Labels on the arrows indicate the amount of time needed to complete the activities. For example, the first activity in preparing the technical report is to assemble preliminary notes, which requires 0.5 hour (**Figure 3.14A**).

Dummy activities simply indicate connections between activities that involve no work

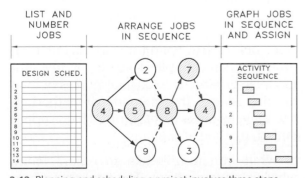

3.12 Planning and scheduling a project involves three steps.

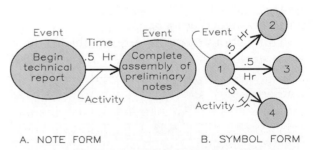

3.14 Two ways of preparing an Activities Network are (A) the note form, and (B) the symbol form, both of which graphically arrange activities in sequence.

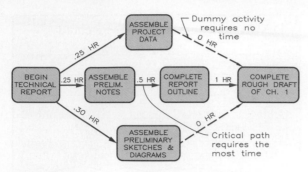

3.15 The critical path is the path from start to end of the project that requires the most time; a portion of the product design project is shown here.

and no time, because activities must be connected in the network. The two dummy activities (dashed lines) in **Figure 3.15** show sequential connections but no expenditure of time. In other words, the project data and the preliminary sketches and diagrams that were assembled were not needed to complete the draft of Chapter 1 but are available for later use in the project.

The critical path identified in the Activities Network (**Figure 3.15**) is the longest time path from the project's beginning to its end. For the partial Activities Network shown, 1.75 hours are required to complete the draft of Chapter 1 of the technical report.

Activity Sequence Chart

The designer next lists the activities on the **Activity Sequence Chart** (**ASC**) (**Figure 3.16**),

Activity Sequence Chart

JOB	MEMBER	PROJECT HOURS							
		0	2	4	6	8	10	12	14	16
7	ALL									
6	SMITH									
3	PRISK									
10	REED									
5	ALL									
12	FLYNN									

3.16 The activity sequence chart (ASC) contains tasks, assignments, and initial time estimates from the DS&PR and activity sequence from the Activity Network.

which shows task numbers and team member assignments from the DS&PR. A bar graph shows the sequence of and the time allocated to each task. For example, 0.5 hour is scheduled for brainstorming, or task 2, from the DS&PR and is to be done first. The ASC also is the overall project schedule.

As work progresses, the designer graphs the status of each activity on the DS&PR (**Figure 3.13**). The actual hours required to complete tasks appear in column 4, for easy comparison with the original time estimates. When the extra time required becomes greater than that scheduled, the designer has to make adjustments in the ASC.

Problems

Problems should be presented on 8½-by-11-inch paper, grid or plain. Notes, sketches, drawings, and graphs should be neatly executed. Written matter should be typed or lettered on paper with ⅛-in. guidelines.

General

1. Identify a need for a design solution that could be completed as a short class assignment requiring fewer than three hours to complete. Submit a proposal outlining this need and your general plan for its design solution. Limit the written proposal to two pages.

2. You are marooned on an uninhabited island with no tools or supplies. Identify the major problems that you would have to solve. List the factors that you would have to consider before attempting a solution for each problem. For example: There is the need for food. Determine (a) available sources of food on the island, (b) methods of gathering/catching, (c) methods of storing a food supply, and (d) method of cooking.

3. While you were walking to class today, what irritants or discomforts did you recognize? Were the sidewalks too narrow? Were the entrances to the buildings inconvenient? Using worksheets, identify the cause of these problems, write a problem statement (including need recognition), and list the requirements for and limitations on solutions.

4. Repeat Problem 3, but use irritants or discomforts found in your (a) living quarters, (b) classroom, (c) recreation facilities, (d) dining facilities, or (e) another environment with which you are familiar.

Product Design Problems

5. Reconstruct the designer's approach to development of the tab-opening can. Even though the problem has been solved and its solution marketed, identify the need for the product and propose specifications for it. Is the existing solution the most appropriate one, or does your problem identification statement suggest others? Explain your answer.

6. Repeat Problem 5 for a travel iron for pressing clothes.

7. List the problems involved in designing a motorized wheelbarrow.

8. Suppose that you recognize the need for a device that could be attached to a bicycle to allow it to be ridden over street curbs to sidewalk level. Identify the problems involved in determining the marketability of such a device.

9. Identify the problems you might encounter in developing a portable engineering travel kit to give engineers the capability of making engineering calculations, notes, sketches, and drawings. The kit might include a carrying case, calculator, computer, drawing instruments, paper, reference material, and other accessories.

10. Use Section 3.7 as a guide and prepare a Design Schedule & Progress Record, an Activities Network, and an Activity Sequence Chart for the portion of your design project as assigned below:

(A) An overview of the entire project as you anticipate it at the present.

(B) The first three steps of the design process.

(C) The problem identification step only.

Use as many sheets as necessary.

11. Develop your ability to make educated guesses and estimates regarding your world. For example, can you guess how many McDonald's hamburger stores there are in a neighboring city? If you looked in the yellow pages of your town, counted the McDonald's there, and divided it into the local population to find the population per store, would this factor help you with your estimate? Check a phone book from the neighboring city to see.

Using this approach, go to your library, look in the yellow pages, and determine the business outlets per 1000 of population for any of the following categories: banks, gas stations, movie theaters, bookstores, fitness centers, restaurants, department stores, or other businesses that are of interest to you.

Learning to make intelligent guesses will help you to be a better engineer and entrepreneur.

12. Pricing products is similar to making estimates, as covered in Problem 11. Use common sense and instincts (that you may not realize you have) to explain why mark-up and profit margins vary among various types of products. Give your explanation in a brief outline form that will make your key points easy to process by the reader.

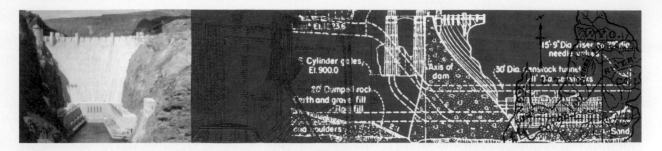

4

Preliminary Ideas

4.1 Introduction

Creativity is highest during the preliminary idea step of the design process because there are no limitations on being innovative, experimental, and daring. During subsequent steps of the design process, freedom of creativity diminishes and the need for information and facts increases. **Figure 4.1** shows this relationship between creativity and information accumulation during the design process.

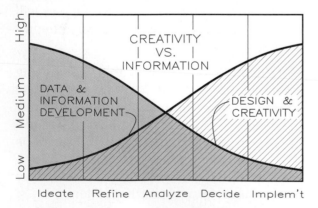

4.1 Creativity is highest during the early stages of the design process (excluding problem identification), and information accumulation increases as the process continues.

A product expected to revolutionize pedestrian traffic is the Segway® Human transporter (**Figure 4.2**), a battery-powered vehicle that will truly give individuals "wheels." It has a top speed of 12.5 miles per hour, a range of 17 miles, and payload of 250 pounds.

Although the Instant Coffee spoon in **Figure 4.3** is a less technical product, it is a unique design that prompts the question, "Why didn't I think of that?" An example of a very simple but highly successful design is the Post-it®, the "sticky" removable note that was developed by the 3M Corporation.

Another advanced design is the prototype voice/data/fax mobile terminal (**Figure 4.4**), which is being developed for trucking and transportation companies. A short time ago these ideas were considered futuristic fantasies. Progress in the future will be limited only by our imagination in applying technology.

4.2 The Segway (the Human Transporter) is expected to revolutionize pedestrian transportation. It is designed to move in response to leaning one's body forward or backward. (Courtesy of Segway LLC.)

4.3 A simple but unique design that solves several problems is the coffee spoon: it serves as a container for the instant coffee and also functions as a spoon. (Courtesy of Aluminum Company of America.)

4.4 This voice/data/fax mobile terminal that operates from satellite signals is being developed for trucking and transportation, utility, and resource companies. (Courtesy of BCE, Inc.)

4.2 Individual Versus Team Methods

Designers work both as individuals and as members of design teams during the process of developing products. Both methods have their advantages and disadvantages.

Individual Approach

Designers working alone must make sketches and notes to communicate first with themselves and then with others. Their primary goal is to generate as many ideas as possible, because better ideas are more likely to come from long lists than from short lists. Rapidly-drawn sketches can capture fleeting thoughts that might otherwise be lost during the ideation phase.

Team Approach

The team approach brings diversity and a broad range of ideas to the design process, but along with it comes problems of management and coordination. Groups perform better with a leader with the authority to guide their activities and make assignments.

Teams should alternate between individual and group work to take advantage of both approaches. For example, each team member could individually develop preliminary ideas, bring them to a team meeting, compare solutions, merge ideas, and return to individual work with a renewed outlook.

4.3 Plan of Action

The following steps are suggested for completing the preliminary ideas step of the design process: (1) hold brainstorming sessions, (2) prepare sketches and notes, (3) research background data, and (4) conduct surveys (**Figure 4.5**). Periodically reviewing notes and worksheets made during problem identification ensures that efforts will stay focused on the design objectives.

4.4 Brainstorming

Brainstorming is a problem-solving technique in which members of a group spontaneously contribute ideas. The best session is one that adheres to the following rules.

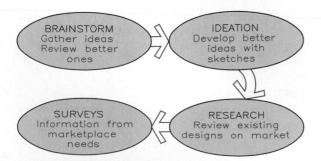

4.5 A plan of action is needed for the preliminary ideas step of the design process, which will probably involve most of the steps shown here.

Rules of Brainstorming

The guidelines for a brainstorming session are:*

1. *Criticism is ruled out.* Judgment of ideas must be withheld until later.

2. *Freewheeling is encouraged.* The wilder the idea, the better; it is easier to tame down than to think up. A good solution may emerge from a suggestion that was made as a joke.

3. *Quantity is wanted.* The larger the number of ideas, the greater will be the likelihood of useful ideas.

4. *Combination and improvement are sought.* Participants should seek ways of improving the ideas of others.

Organization of a Brainstorming Session

The organization of a brainstorming session involves selecting the panel, becoming familiar with the problem, selecting a moderator and recorder, holding the session, and following up.

Panel selection. The optimum number of participants in a brainstorming session is twelve people. For diversity, they should be people both with and without knowledge of the subject. Because supervisors may restrict the flow of ideas, panels should be composed of non-supervisory professionals of similar status.

Preliminary work. An information sheet about the session should be given to panel members a couple of days before the session to allow ideas to incubate.

*From Alex Osborn, *Applied Imagination*, NYC: Scribner, 1963.

The problem. The problem to be brainstormed should be defined concisely. Instead of presenting the problem as "how to improve our campus," it should be presented by category, such as "how to improve student parking," or "how to improve food services."

The moderator and recorder. The panel selects a moderator to be in charge of the session and a recorder to keep track of the ideas presented (**Figure 4.6**).

The session. The moderator introduces the problem and recognizes the first member holding up a hand to respond. The person responding should state an idea as briefly as possible. The moderator then recognizes the next person holding up a hand, and the process continues. A suggestion made by one member often stimulates ideas in others, who snap their fingers to signify that they want to "hitchhike" on the previous idea. This interaction is central to a brainstorming session. The moderator's most important job is to keep the ideas flowing and to prevent participants from giving long, drawn-out responses that dampen the spontaneous and "fun" aspects of the session.

Length. A session should move at a brisk pace and end when ideas slow to an unproductive rate. An effective session can last from a few minutes to an hour, but twenty minutes is considered about the best length.

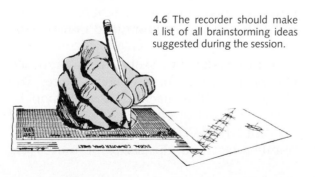

4.6 The recorder should make a list of all brainstorming ideas suggested during the session.

Follow-up. The recorder should reproduce the list of ideas gathered during the session and distribute them to the participants. As many as a hundred ideas may be gathered during a twenty-minute session. The designer should pare them down to those having the most merit.

4.5 Sketching and Notes

Sketching is an effective medium of design, development, and communication whether it is a simple part or a complex product (**Figure 4.7**). There are many instances where modification of designs and their descriptions are almost impossible without the ability to make simple and rapid freehand sketches. Sketching allows the designer's ideas to take form as three-dimensional pictorials or as two-dimensional views that are visual extensions of the thinking process. The sketch shown in **Figure 4.8** was one of many used to develop the styling features of an automobile, an advanced application. **Figure 4.9** shows one of many sketches used in the design of the various components and systems of the space

4.7 Designers use sketches and notes to develop preliminary ideas and communicate with themselves and others. (Courtesy of Ford Motor Company.)

4.8 This simple sketch was made within a couple of minutes by a designer as a preliminary idea for the development of an automobile design.

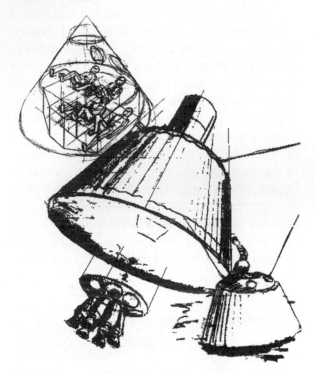

4.9 Sketches are important to the development of solutions to the most complex of engineering and scientific designs. These sketches illustrate preliminary ideas for the manned space capsule. (Courtesy of the National Aeronautics and Space Administration.)

program, one of the most advanced design programs in history.

The preliminary ideas for the Transportable Uni-Lodge (**Figure 4.10**) illustrate the importance of sketches in developing and communicating ideas. These ideas propose that the Uni-Lodge be transported by helicopter to previously unreachable areas and

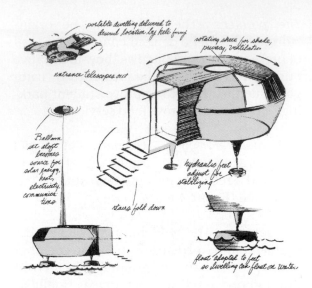

4.10 These preliminary sketches and notes depict a transportable uni-lodge, a mobile dwelling of the future. (Courtesy of Lippincott and Margulies, Inc., and Charles Bruning Company.)

lowered onto the site or the water. Its retractable legs can be equipped with pontoons, allowing it to float. Additional features are noted on the sketches.

Another concept is a self-contained pipelayer (**Figure 4.11**) for laying irrigation pipe in desert areas. The tractor has a cab, sleeping accommodations, radio equipment, power plant, and bulk storage tanks for plastic. The van consists of an extrusion machine, a refrigeration unit, and a control station. The pipelayer transports bulk plastic and machinery that can extrude and lay approximately two miles of pipe from each pair of storage tanks. Empty tanks are discarded and replaced by full tanks that are air-dropped to the crew.

Figure 4.12 shows sketches of a proposed checkout system for a grocery store. The shopper sets the machine in operation by inserting a credit card in a slot. After the card is scanned, the customer receives an order number tag, and the conveyor moves the

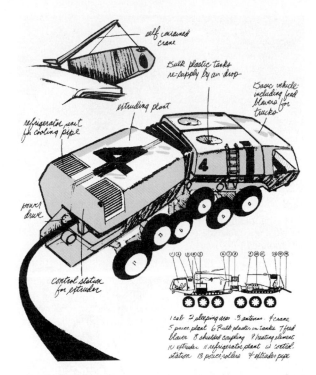

4.11 These sketches show a designer's ideas for a self-contained pipelayer for laying irrigation pipe in large tracts of desert areas. (Courtesy of Donald Desky Associates, Inc., and Charles Bruning Company.)

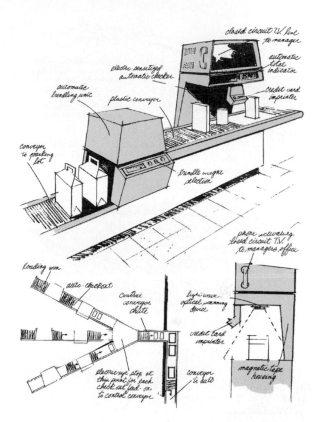

4.12 These sketches describe a concept for an automatic checkout/packaging unit and for a supermarket. (Courtesy of Lester Beall, Inc., and Charles Bruning Company.)

items under a scanner and totals the prices. If a question arises, the customer can stop the machine and talk to a store employee by lifting the phone. The conveyor moves the items to a unit where they are packaged in plastic containers and marked with the customer's number. A central conveyor transports large orders to an exterior pickup point near the customer's parking place.

Sketches in **Figure 4.13** describe a teaching flotilla, mobile floating classrooms that are connected to a master tracking center from which programs are transmitted by lasers and satellite relays. Each helio-hover craft is equipped with advanced technology including two-dimensional and three-dimensional projectors and televisions. The com-

puterized tracking center can transmit presentations to six different floating classrooms.

The concepts in these examples could not have been developed without the use of sketching as a medium of design, and as a means of thinking. Rapid sketches capture a person's thought process, imagination, and creativity in a form that can be shared by others.

4.6 Design Sketching: Application

The previous sketches have ranged from simplistic to complicated, but they are typical of the drawings used to create designs for automobiles, spacecraft, and all other products. Remember that sketching is the medium for developing concepts and details throughout

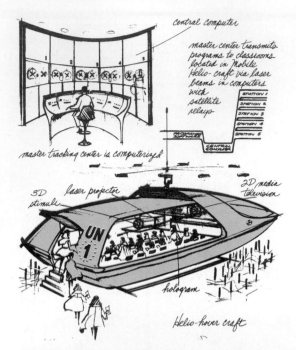

4.13 These sketches illustrate a flotilla of classrooms controlled from a master tracking center by lasers and computer relays. (Courtesy of Raymond Loewy/William Snaith, Inc., and Charles Bruning Company.)

4.14 The designer uses sketching as his primary means of developing preliminary design concepts. (Courtesy I.N. Incorporated.)

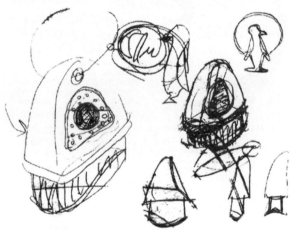

4.15 Design concepts of the shoe brush are brainstormed and sketched as two- and three-dimensional views. (Courtesy Bresslergroup and Shoe Store Supplies.)

the design process from idea to final design. The designer must develop numerous preliminary solutions to design problems, most of which will be rapidly-drawn freehand sketches (**Figure 4.14**).

A series of sketches (**Figure 4.15** through **Figure 4.19**) illustrate how sophisticated sketches are developed as the preliminary idea process progresses. The product illustrated is a shoe brush for Shoe Store Supplies.

Figure 4.15 shows three-dimensional pictorials of one of the initial concepts, in three different positions, much as you would hold it in your hand and look at it from different angles.

Figure 4.16 illustrates another preliminary concept for a shoe brush. Again, three-dimensional pictorials are used by the designer to develop the design and communicate the ideas

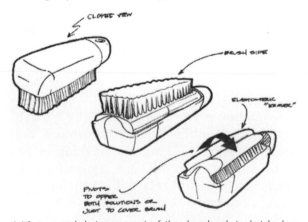

4.16 A second design concept of the shoe brush is sketched as three-dimensional views. (Courtesy Bresslergroup and Shoe Store Supplies.)

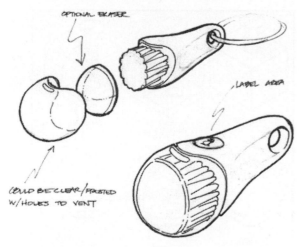

4.17 A third design concept of the shoe brush is sketched as three-dimensional views. (Courtesy Bresslergroup and Shoe Store Supplies.)

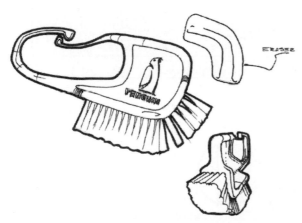

4.18 Another design concept of the shoe brush is sketched as three-dimensional views. (Courtesy Bresslergroup and Shoe Store Supplies.)

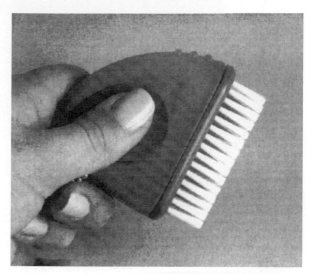

4.19 The final design for the shoe brush is shown here in product form. (Courtesy Bresslergroup and Shoe Store Supplies.)

to others more clearly than with orthographic views. Notes are used to call attention to significant features and color may be added with felt-tip markers to add realism. A third and fourth design concept is sketched in **Figures 4.17** and **4.18**, respectively.

The final design for the shoe brush is shown as a full-size protype in **Figure 4.19**. The concepts shown here are but a few of the many sketches that were made during the preliminary phase of the design process.

The more concepts you have to choose from, the better your chances are of attaining the best design solution.

4.7 Quickie Design

Let's assume that we recognize the need for a product to protect walls and doorknobs where doors swing into walls. First, we make sketches on a worksheet to identify the problem and its geometry (**Figure 4.20**). If we decide to develop a floor-mounted doorstop, we might begin with sketches (made by hand or by computer) of a design that begins as a basic block, which will do the job (**Figure 4.21**). However, we can do better than that; we make additional sketches to refine and develop the design until we find a suitable and marketable solution.

We make more sketches (**Figure 4.22**) to develop the doorstop's details, which we show to others for consultation and evaluation. **Figure 4.23** shows other solutions to this problem that are products already being marketed. These products were developed in a

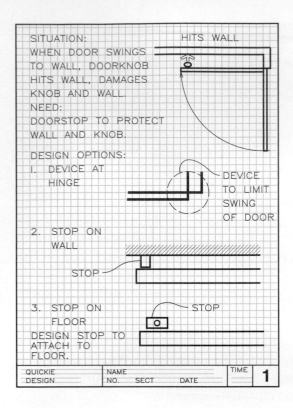

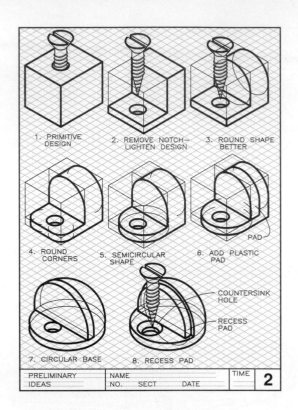

4.20 (Sheet 1) This worksheet of sketches and notes identifies the need for a device to prevent damage caused by a door swinging into a wall.

4.21 (Sheet 2) A series of detailed sketches illlustrates the evolution of the doorstop design.

4.22 (Sheet 3) Detailed sketches are used to refine the design and clarify it for others.

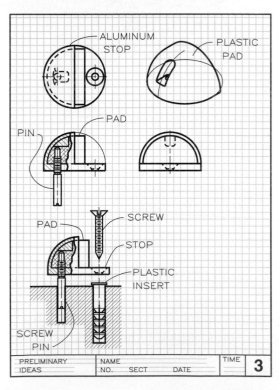

similar manner, by designers using graphics as the medium of design.

4.8 Background Information

One way of gathering preliminary ideas is to look for existing products and designs that are similar to the one being considered. Several sources of background information include magazines, patents, and consultants (**Figure 4.24**).

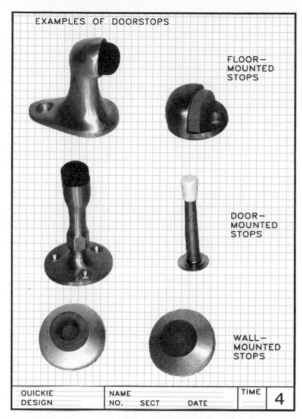

4.23 (Sheet 4) The doorstops shown here are examples of some that are available in today's market.

4.24 The gathering of background information is helpful and necessary to the development of preliminary ideas.

Magazines

Articles in both general and technical magazines often present unique designs, complete with drawings and photographs. Advertisements in these magazines may give helpful information on materials and innovative devices. There are various magazines that specialize in most product areas that your librarian can help you find as a reference.

Patents

Patents from the U.S. Patent Office illustrate the details of all designs that were granted a patent. The designer can use them to understand competing products better and to ensure that there is no infringement on existing patents.

Consultants

Complex projects may require specialists in manufacturing, electronics, materials, and instrumentation. Manufacturers' representatives also provide valuable assistance with projects related to their companies' products. Having no knowledge of a project at the outset can be an advantage, since a fresh perspective is what is needed.

4.9 Opinion Surveys

Designers need to know the attitudes of consumers about a new product at the preliminary design stage. Is there a need for the product? Are consumers excited about the prospects for a particular product being designed? Would they buy the product? What features do they like or dislike? What price range would be acceptable to them? What do retailers think it will sell for? Does size, weight, or color matter?

For a survey to be of value, the population for whom the product is being developed should be identified. Are they homeowners, high-school-age males, or single women? A selected sample of the population can be surveyed by a personal interview, telephone survey, or mail questionnaire.

The Personal Interview

A personal interview survey should be organized to obtain and summarize reliable responses quickly and easily. For this reason, questions should be true-false or multiple choice for ease of tabulation.

Interviewers should introduce themselves, explain the purpose of the interview, and ask for permission to proceed. They should make the interview brief, record responses, and thank the interviewee for participating.

The Telephone Interview

If the opinions of the general public are desired, interviewers can talk to people selected from the telephone book. If opinions from a certain group—say, sporting-goods retailers—are needed, the Yellow Pages provide prospective interviewees.

The Mail Questionnaire

An economical method of contacting a large number of people in many locations is by mail questionnaire. To test the suitability of questions to be asked, the questionnaire can be mailed to a small group of people for their response as a test. The final questionnaire should be mailed to at least three times as many people as the number of responses desired. Inserting a self-addressed, stamped envelope will increase responses.

4.10 Preliminary Ideas: Exercise Bench

We introduced the design of an exercise bench in Chapter 3, where it was taken through the problem identification step. Recall that the bench is to be used by those who lift weights to maintain body fitness. This apparatus should be versatile and at the low end of the price range for exercise equipment.

Now, to apply the preliminary idea step of the design process, begin by holding a brainstorming session with team members to generate and record ideas on a worksheet (**Figure 4.25**). Then select the better ideas, list their features, and summarize them on a worksheet (**Figure 4.26**), even if they have more features than you could possibly use in a single design. The reason is to make sure that you do not forget or lose any concepts.

Using rapid freehand sketching techniques, such as orthographic views and three-dimensional pictorial sketches, sketch ideas on worksheets. Note on the drawings any ideas or questions that come to mind while you sketch. Do not erase and modify your

4.25 (Sheet 4) Ideas gathered during a brainstorming session are listed on a worksheet for future reference.

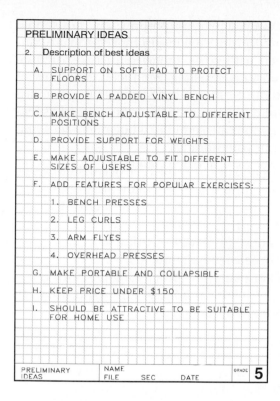

PRELIMINARY IDEAS

2. Description of best ideas

A. SUPPORT ON SOFT PAD TO PROTECT FLOORS

B. PROVIDE A PADDED VINYL BENCH

C. MAKE BENCH ADJUSTABLE TO DIFFERENT POSITIONS

D. PROVIDE SUPPORT FOR WEIGHTS

E. MAKE ADJUSTABLE TO FIT DIFFERENT SIZES OF USERS

F. ADD FEATURES FOR POPULAR EXERCISES:

    1. BENCH PRESSES

    2. LEG CURLS

    3. ARM FLYES

    4. OVERHEAD PRESSES

G. MAKE PORTABLE AND COLLAPSIBLE

H. KEEP PRICE UNDER $150

I. SHOULD BE ATTRACTIVE TO BE SUITABLE FOR HOME USE

PRELIMINARY IDEAS | NAME FILE SEC DATE | GRADE **5**

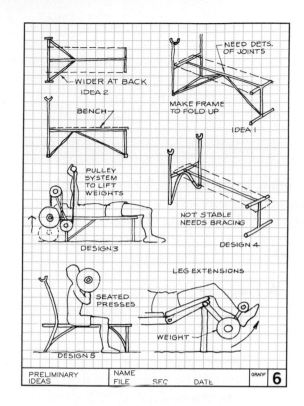

PRELIMINARY IDEAS | NAME FILE SEC DATE | GRADE **6**

4.26 (Sheet 5) The best of the brainstorming ideas are selected and summarized for further development.

4.27 (Sheet 6) Preliminary concepts for the exercise bench are shown as sketches and notes.

4.28 (Sheet 7) Sketches and notes describe additional ideas about the exercise bench.

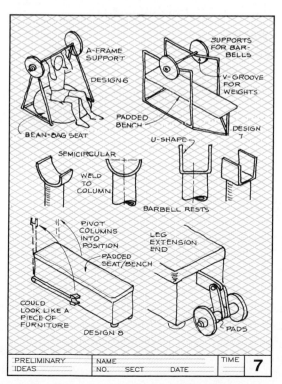

PRELIMINARY IDEAS | NAME NO. SECT DATE | TIME **7**

sketches; instead make new sketches that incorporate revisions and modifications. By so doing, you will be able to review your thinking process from idea to idea. You should occasionally go back to previous steps to be sure that usable concepts have not been overlooked.

In **Figure 4.27**, note the adaptation of ideas from various types of benches and exercise techniques. A number identifies each idea. Another worksheet (**Figure 4.28**) shows other ideas and modifications of previous ideas. Additional sketches can be made of specific details of each design: connections, fabrication details, padding, and so forth.

Problems

Problems should be presented on 8–1/2 × 11-inch paper, grid or plain. All notes, sketches, drawings, and graphs should be neat and accurate. Written matter should be legibly lettered, using 1/8-in. guidelines.

1. Select one or more of the following items and list as many of their uses as possible: (a) empty vegetable cans (3 × 5 inch in diameter), (b) 2,000 sheets of 8–1/2 × 11-inch bond paper, (c) 1 cubic yard of dirt, (d) 3 empty oil drums (24-in. DIA by 36 in.), (e) a load of egg cartons, (f) 25 bamboo poles (10 ft. long), (g) 10 old tires, or (h) old newspapers.

2. If you were going to select an ideal team to develop an engineering solution to a problem, what characteristics would you look for in them? Explain.

3. What are the advantages and disadvantages of working independently on a project? Of working as a member of a design team? Explain and give examples of instances in which each approach would have the advantage.

4. Gather background information on one of the following design problems or on one of your own selection: (a) a one-person canoe, (b) a built-in car jack, (c) an automatic blackboard eraser, (d) a built-in coffee maker for an automobile, (e) a self-opening door to permit a pet to leave or enter the house, (f) an emergency fire escape for a two-story building, (g) a rain protector for people attending outdoor spectator activities, (h) a new household appliance, or (i) a home exerciser. You are concerned with costs, methods of construction, dimensions, existing products, estimates of need, and other information that will assist you in understanding the problem and its feasibility as a project. List your references and present your findings.

5. List and describe the type of consulting services required for the following design projects: (a) a zoning system for a city of 20,000 people, (b) a shopping center, (c) a go-cart operation, (d) a water-purification facility, (e) a hydroelectric system, (f) a nuclear fallout disaster plan, (g) a processing plant for refining petroleum products, and (h) a drainage system for residential and rural areas.

6. Develop a questionnaire to determine the public's attitude toward a particular product of your selection. Explain how you would tabulate the responses to the questionnaire.

7. Organize a group of classmates and hold a brainstorming session to identify problems in need of solutions. Make a list of the ideas.

8. With a team of classmates, make a list of items that are in need of "invention." If you can't think of any, make a list of problems and inconveniences that you are unhappy with. Their solution might lead to a new product or system.

9. Make sketches of a toothbrush holder than can sit on a table surface or be attached to the wall. What options would make it most marketable?

10. Identify the features needed for a dog house that could be sold to the mass market. Make sketches of your better ideas. Are there other products for the pet market that you can think of that would be worthy of development?

4.29 (Problem 11) A chain tensioner with a 6 in. DIA sprocket mounted on a movable shaft. (Courtesy of Brewer Machine and Gear Company.)

11. A tensioner that keeps the chain taut that engages one side of the sprocket (the toothed wheel) is shown in **Figure 4.29**. Tension is applied to the chain by moving the position of the 6 in. DIA sprocket along its shaft by turning the nut on the threaded screw. The entire apparatus must be mounted in a stationary position (either horizontal or vertical) with two bolts, one at each end.

Although some of the parts are not clearly illustrated, make a series of freehand orthographic sketches to describe the parts. Describe them as if you were the designer who is going to explain the design to your drafting department so they can make the finished working drawings of the parts. You will find it helpful to refer to Chapters 13 and 14.

12. Another design for a belt tensioner is shown in **Figure 4.30**, but the pulley (instead of a sprocket) is not shown. Notice that the base has a slot that is an arc of a circle for adjustment. An additional adjustment can be made by moving the arm to different positions. The V-pulley (6 in. DIA) for a flexible belt will be mounted on the shaft passing through the arm and held on by a collar with a setscrew.

Make freehand orthographic sketches of the individual parts to better explain how they are made as well as their relationships to each other. Refer to Chapters 13 and 14 as needed.

13. Problems 11 and 12 give different solutions to the same problems; they are both marketable products. Other products are on the market to serve this same need. Can you think of other solutions? Make freehand sketches of your ideas that could be used to describe your design to a classmate.

4.30 (Problem 12) A V-belt tensioner with a 6 in. DIA V-pulley (not shown) mounted on a shaft in the movable arm. (Courtesy of the Universal Machine Company.)

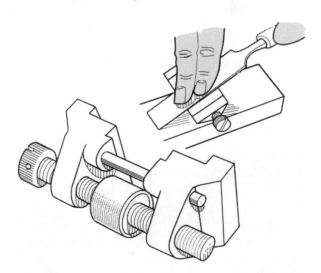

4.31 (Problem 13) Sketch views of parts of this chisel-sharpening guide.

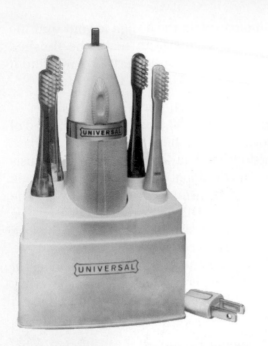

4.32 (Problem 11) Design a toothbrush holder for a cup and six toothbrushes. Sketch your design adequately to communicate your ideas. (Courtesy of General Electric.)

4.33 (Problem 12) Sketch views of the parts of the compressing coupling to better explain it to a classmate.

14. A chisel sharpening guide in Figure 4.31, a product currently on the market, holds a chisel at a constant angle with a whetstone as the chisel is moved back and forth to sharpen it. Make freehand orthographic sketches of the individual parts and determine how they are assembled and how they relate to each other. Make whatever sketches are necessary for explaining this design. Refer to Chapters 13 and 14.

15. Design a different device of your own that could be used to sharpen a chisel as well as the one described in Problem 14. Make freehand sketches as specfied in Problem 14.

16. Design a toothbrush holder (Figure 4.32) that would hold a cup and six toothbrushes (or another combination of your choosing). Sketch your proposed solutions and get a critique from one of your classmates. Give thought as to whether or not a product of this type has an attractive market for an investor. Are there similar products on the market at the present? Look for your competition the next time you visit a department store.

17. The compression coupling (**Figure 4.33**) is shown assembled and exploded. The coupling must fit snuggly with 1.50 diameter shafts. In order to better understand the parts yourself and then communicate this understanding to others, make orthogaphic sketches of the individual parts.

What happens when two shafts meet at the coupling and the hexagon-head bolts are tightened?

18. Design a device that will crush aluminum cans for recycling and saving space as does the one shown in **Figure 4.34**. Consider all options: wall mounted, counter-top usage, attractiveness, durability, and ease of use. Is there a market for such a device; are there competing products on the market. Sketch as many ideas and concepts as possible.

19. The tepee shown in **Figure 4.35** is a home project made of plywood as a child's playhouse. Sketch details of your own design of a similar backyard shelter or apparatus that would be fun for children to play in or on and would be marketable.

20. Make freehand sketches of the handwheel (12 in. outside diameter) shown in **Figure 4.36** to explain its details such as how the handle is attached.

21. Follow the instructions given in Problem 20 to describe the handwheel shown in **Figure 4.37**. Estimate the dimensions, relating them to the hand.

4.35 **(Problem 19)** Design a backyard shelter for children to play in.

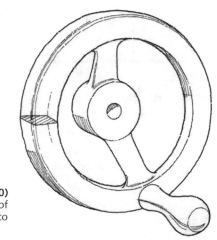

4.36 **(Problem 20)** Make sketches of this handwheel to explain its details.

4.34 **(Problem 18)** Design a can crusher.

4.37 **(Problem 21)** Make sketches of this handwheel to explain its details.

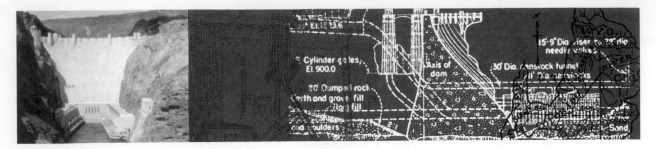

5

Refinement

5.1 Introduction

Refinement of preliminary ideas is the first departure from unrestricted creativity and imagination. The designer must now give primary consideration to function, costs, and practicality.

This step of the design process calls for the designer to make scale drawings with instruments to check dimensions and geometry that cannot be accurately measured in unscaled sketches. However, it is unnecessary to fully dimension these drawings. **Descriptive geometry has its greatest application as a design tool in the idea-refinement step of the design process.**

5.2 Physical Properties

Important in refining an idea is determining the product's physical properties. An example of a refinement drawing is the profile of a Bell D326 Clipper helicopter that shows its overall dimen-

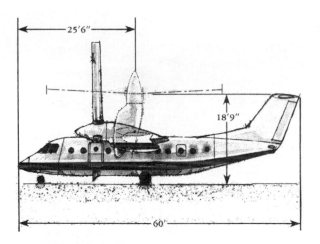

5.1 This scale drawing gives several overall dimensions to describe a helicopter body. It is very important that refinement drawings be drawn to scale, but only major dimensions need be given at this stage. (Courtesy of Bell Helicopter Textron.)

sions (**Figure 5.1**). The positions of the propeller blades are illustrated and dimensioned where they are rotated from vertical to horizontal.

Although most refinement drawings will be rendered in two dimensions that can be accurately scaled and compared, three dimensional drawings can also be used.

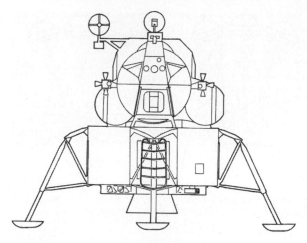

5.2 This single view of the lunar module is drawn to scale so that its geometry can be accurately analyzed and evaluated. Most refinement drawings are two-dimensional orthographic views. (Courtesy of Ryan Aeronautical Company.)

Figure 5.2 is an example of a two-dimensional orthographic view that is drawn to scale so that it can be measured on the drawing with no dimensions given.

The configuration of a spacecraft is refined in the same manner as simpler designs by beginning with orthographic views as shown in **Figure 5.3**. The computer is a significant aid to the designer in taking the refinement a step further. In **Figure 5.4**, the parts of a product are drawn as computer models in order to permit them to be viewed from any direction

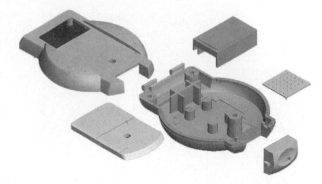

5.4 Computer modeling affords an excellent means of refining a preliminary product design. (Courtesy of Impact Displays, Inc.)

and in any arrangement. Mechanisms with moving parts can be made to operate on the computer's monitor to test their operation. An additional example is a machine part that is rendered as a three-dimensional model with AutoCAD® to permit viewing from any direction (in **Figure 5.5**).

5.3 Application of Descriptive Geometry

Descriptive geometry is the study of points, lines, and surfaces in three-dimensional space, which are the geometric elements that comprise all forms. Before descriptive geome-

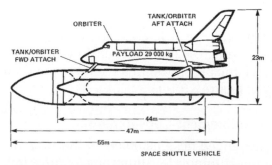

5.3 This two-dimensional, orthographic drawing shows the configuration of an aircraft, the space shuttle vehicle. (Courtesy of National Aeronautics and Space Administration.)

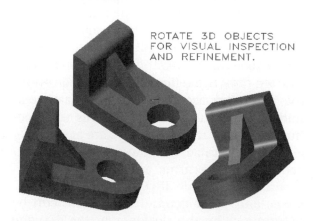

5.5 Designs can be constructed as three-dimensional models and revolved on the screen for visual inspection and refinement, much as if you were holding the part in your hand.

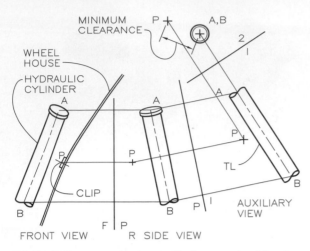

5.6 Descriptive geometry is an effective way to determine clearances between components, such as the clearance between a hydraulic cylinder and a fender.

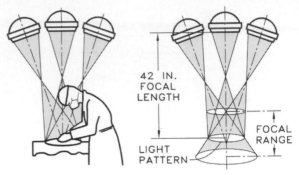

5.7 By using scale drawings developed in the refinement step of the design process, the designer can study the geometry of a surgical lamp. (Courtesy of Sybron Corporation.)

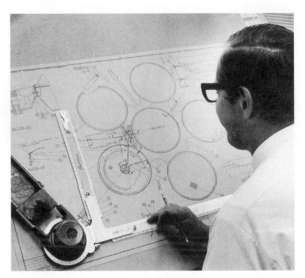

5.8 Refinement drawing shows the overall dimensions of the final design of the surgical lamp. (Courtesy of Sybron Corporation.)

try can be applied, the designer must draw orthographic views to scale, from which auxiliary views can be projected. **Figure 5.6** shows how descriptive geometry is used to determine the clearance between a hydraulic cylinder and the fender of an automobile where the cylinder is attached with a clip. This construction can be performed either by pencil or by computer.

The design of a surgical light requires the application of descriptive geometry as well. The light fixture must provide maximum light on the operating area with the least obstruction, as shown in **Figure 5.7**. This scale drawing depicts the converging beams of light emitted from the reflectors, its position above the operating area, and approximate positions of the surgeons. The beams are positioned so their narrowest rays are at shoulder level to minimize shadows cast by the surgeon's shoulders, arms, and hands.

From these scale drawings, lengths, angles, areas, and other geometric relations can be determined by the designer at the drawing board or at the computer (**Figure 5.8**). When

the three-dimensional geometrical relationships have been determined, the engineering details can be developed for more analysis and testing. The major dimensions of the surgical lamp are shown in the refinement drawing in **Figure 5.9**.

5.4 Refinement Considerations

In advanced designs, such as that of a new model automobile, numerous features must be refined. The exhaust system shown in

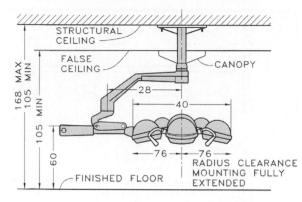

5.9 A well-designed surgical lamp emits light that passes around the surgeon's shoulders with a minimum of shadow. The focal range of this surgical lamp is between 30 and 60 inches. (Courtesy of Sybron Corporation.)

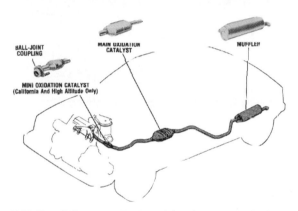

5.10 Descriptive geometry is useful in determining the lengths and angles of an automobile's exhaust system that are necessary to clear the structural members of the chassis. (Courtesy of Chrysler Corporation.)

Figure 5.10 is the result of the many refinement drawings needed to determine its geometry. Descriptive geometry was used to determine the exhaust pipe's bend angles, its length, and clearances required for fitting it to the chassis without interference (**Figure 5.11**).

A designer's refinement drawing of the fuselage of a business jet aircraft is shown as an orthographic section in **Figure 5.12**. This drawing, although simple in concept, is effective in determining clearance, heights, and seat sizes for its passengers.

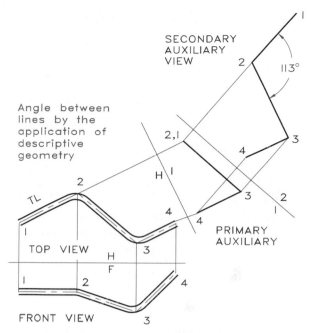

5.11 Descriptive geometry is applied to find the lengths and angles between exhaust pipe segments.

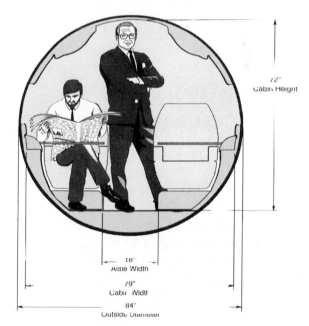

5.12 This refinement drawing is a section drawn through the fuselage of the Hawker Horizon, a business jet aircraft that shows the clearance, heights, and seat sizes of the interior. (Courtesy of Raytheon Aircraft Corporation.)

5.5 Refinement: Exercise Bench

In Chapter 3 the exercise bench design problem was identified and in Chapter 4 preliminary ideas for it were developed. Now, in this chapter, we refine the preliminary ideas for the exercise bench with instrument drawings.

First, list the features to be incorporated into the design on a worksheet (**Figure 5.13**). Then refine a preliminary idea, say, Idea 2 from **Figure 4.17** in an orthographic scale drawing of the seat (**Figure 5.14**). Block in extruded parts, such as the framework members, to expedite the drawing process and omit unneeded hidden lines.

Refinement drawings must be drawn to scale. The use of instruments is important to precisely portray the design from which

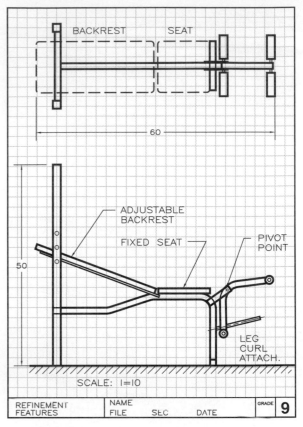

5.14 (Sheet 9) This refinement drawing for the exercise bench is a scale drawing with several of the major dimensions shown.

angles, lengths, shapes, and other geometric elements will be obtained. The drawing shows only overall dimensions and several connecting joints are detailed to explain the design. **Figure 5.15** shows additional design features. These worksheets depict representative types of drawings required to refine a design; additional drawings would be required for a complete refinement of the design.

5.6 Standard Parts

When preparing refinement drawings, specify standard parts whenever possible because they are more economical and they are readily available. Merchandise catalogs, sales brochures, magazine advertisements, news-

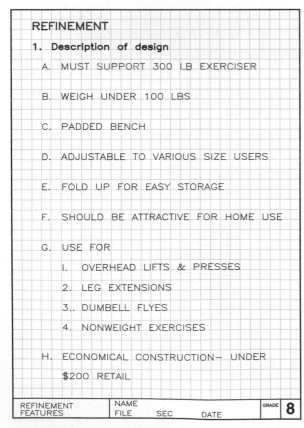

5.13 (Sheet 8) This list of desirable features is a refinement of the exercise bench.

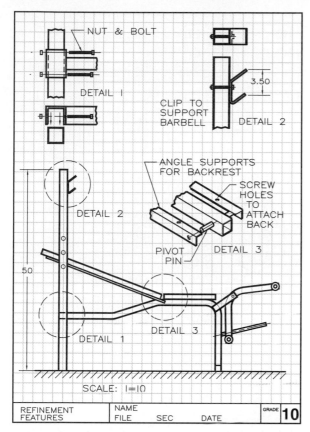

A. LEVELING DEVICES   B. CASTERS FOR MOVABLE EQUIPMENT

5.16 The leveling devices (A) are standard parts for leveling equipment on uneven floors. Casters (B) are attached to equipment that must be moved about a work area. They range from small sizes for TV sets to those that carry over a thousand pounds. (Courtesy of Vlier Industries, and Hamilton Casters.)

5.15 (Sheet 10) This refinement drawing is of another design concept for the exercise bench.

papers, and similar sources contain specifications for standard parts.

Make a practice of keeping files on stock items—such as leveling devices and casters (**Figure 5.16**)—that can be used in developing products and specified in refinement drawings by referring to literature from manufacturers and vendors. You can become a better designer by observing how standard parts and devices are made and how they function. When you see a device that you are unfamiliar with, ask yourself: Why it is made the way it is? How is it used? What application can you think of for it in a design?

Problems

Problems should be presented on 8–1/2 × 11-inch paper, grid or plain. All notes, sketches, drawings, and graphical work should be neatly presented. Written matter should be lettered legibly using 1/8-inch guidelines.

1. When refining a design for a folding lawn chair, what physical properties would a designer need to determine? What physical properties would be needed for a (a) TV set base, (b) golf cart, (c) child's swing set, (d) portable typewriter, (e) shortwave radio, (f) portable camping tent, and (g) warehouse dolly used for moving heavy boxes?

2. Why should scale drawings rather than freehand sketches be used in the refinement of a design? Explain.

3. List five examples of problems involving spatial relationships that could be solved by the application of descriptive geometry. Explain your answers.

4. In the refinement step, how many preliminary designs should be refined? Why?

5. Make a list of refinement drawings that would be needed to develop the installation and design of a 100-foot radio antenna. Make rough sketches of the types of drawings needed, with notes to explain their purposes.

6. If a design is eliminated as a possible solution after refinement drawings have been made, what should be the designer's next step? Explain.

7. For the exercise bench discussed in Section 5.5, what refinement drawings are necessary in addition to those presented? Make freehand sketches of the necessary drawings, with notes to explain what they should show.

Refinement Problems

8. Students often draw wheels as disks with holes at their centers, but even the simplest wheels have more sophistication in their design. Make a refinement drawing of one of the wheels shown in the sketches in **Figure 5.17**. Why do wheels have raised hubs at their centers? Why do they have bushings in the holes through them?

9. Preliminary sketches for a pipe hanger bracket, a device used to support steam pipes from overhead beams, are given in **Figure 5.18**. The sketches are sufficient for you to understand the concepts, but they need further refinement. Make refinement drawings of these sketches and make whatever modifications in the design that you think would improve them.

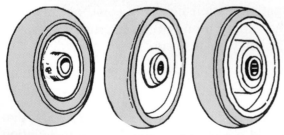

5.17 **(Problem 8)** Why are wheels not disks with holes through their centers? Why do they have raised hubs and bushings?

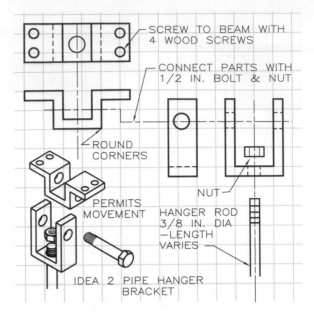

5.18 **(Problem 9)** Preliminary sketches of a pipe hanger for supporting steam pipes from an overhead beam on a hanger rod.

10. The preliminary sketches in **Figure 5.19** illustrate another concept for a pipe hanger bracket. Solve this problem by following the steps in Problem 9.

11. The preliminary sketches in **Figure 5.20** illustrate a pipe clamp assembly that attaches to a hanger rod that attaches to a hanger bracket of the type shown in the two previous figures. Solve this problem by following the steps in Problem 9.

12. Make orthographic refinement drawings of each part of a metal-to-water discharge assembly (**Figure 5.21**) to better understand their relationship.

13. Make orthographic refinement drawings of the parts of a gear puller (**Figure 5.22**). Make separate drawings of each part, and make an assembly drawing showing how the parts fit together.

14. Make orthographic refinement drawings of the casters shown in **Figure 5.16**.

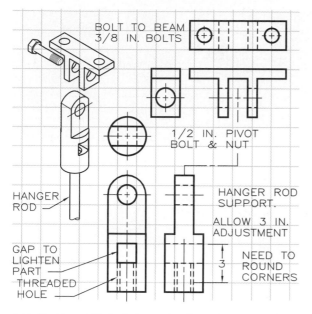

5.19 (Problem 10) Preliminary sketches of concepts for a hanger rod support for suspending steam pipes.

Labels in image 1:
BOLT TO BEAM 3/8 IN. BOLTS
1/2 IN. PIVOT BOLT & NUT
HANGER ROD
GAP TO LIGHTEN PART
THREADED HOLE
HANGER ROD SUPPORT.
ALLOW 3 IN. ADJUSTMENT
NEED TO ROUND CORNERS

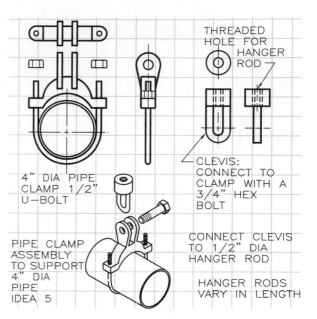

5.20 (Problem 11) Preliminary sketches of a pipe clamp assembly for holding steam pipes and connecting them to an overhead hanger bracket.

Labels in image 2:
THREADED HOLE FOR HANGER ROD
4" DIA PIPE CLAMP 1/2" U–BOLT
CLEVIS: CONNECT TO CLAMP WITH A 3/4" HEX BOLT
PIPE CLAMP ASSEMBLY TO SUPPORT 4" DIA PIPE IDEA 5
CONNECT CLEVIS TO 1/2" DIA HANGER ROD
HANGER RODS VARY IN LENGTH

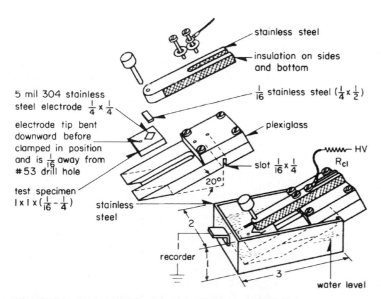

5.21 (Problem 12) A metal-to-water discharge assembly. (Courtesy of the Westinghouse Corporation.)

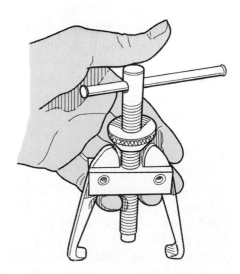

5.22 (Problem 13) A gear puller for removing tightly assembled gears from their shafts.

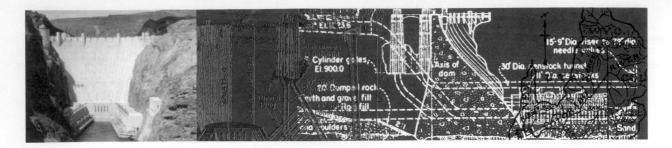

6

Design Analysis

6.1 Introduction

Analysis is the development and evaluation of a proposed design by objective thinking and the application of engineering and technology. For example, bridge designs are analyzed for loads, stresses, travel loads, dimensions, materials, sizes, function, economy, and much more. Less creativity is needed during analysis than during the previous steps of the design process. Analysis is the step in the design process most thoroughly covered in engineering courses.

6.2 Graphics and Analysis

Graphics and geometry are effective tools for analyzing a design in addition to the numerical methods normally used in engineering. Empirical data obtained from laboratory experiments and field observation can be transformed into formats suitable for graphical analysis and evaluation (**Figure 6.1**).

6.1 Designers analyze experimental data and human factors to determine comfortable and safe automobile designs. Graphics is a helpful tool in this step of the design process. (Courtesy of General Motors Corporation.)

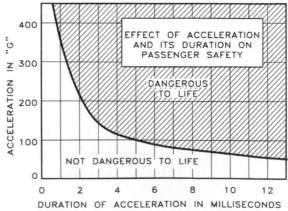

Figure 6.2 shows an example of computer graphics applied to the analysis of a linkage system to determine clearances, limits, and

6.2 Clearance between functional parts and linkage systems can be analyzed efficiently with graphical methods by computer. (Courtesy of Design Technologies International, Inc.)

velocities of its members. Graphics is also an effective way to present and analyze technical information, background data, market surveys, population trends, and sales projections. The graph shown in **Figure 6.3** enables the designer to quickly select the appropriate conveyor for transporting raw material at the desired rate.

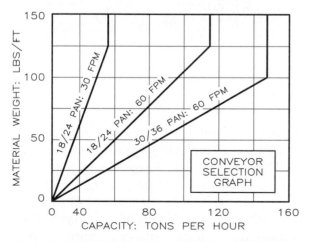

6.3 Graphics can be used to organize and present laboratory and field data to make it much easier to analyze for design applications.

6.3 Types of Analysis

Analysis includes evaluation of the following attributes:

1. Function

2. Human factors

3. Product market

4. Physical specifications

5. Strength

6. Economic factors

7. Models

Function

Function is the most important characteristic of a design because a product that does not function properly is a failure regardless of its other desirable features (**Figure 6.4**). Many products serve a narrow, utilitarian purpose, such as the piston linkage of a gasoline engine (**Figure 6.5**). In those cases, the designer is concerned with function to a much greater extent. Functional analysis usually involves the optimization of several aspects of the design, including safety, economics, durability, appearance, and marketability. For example,

6.4 This vise must function properly to be acceptable. No other features can offset a design that operates poorly. (Courtesy of Wilton Tool Division.)

6.5 Graphical analysis is an efficient way to determine operating limits for a product such as this piston linkage for a gasoline engine. (Courtesy of Knowledge Revolution, Inc.)

A.   B.

6.6 Human dimensions.

A. Leonardo da Vinci analyzed body dimensions and proportions with graphics in the fifteenth century.

B. Engineers used his techniques in the twentieth century to analyze motions by astronauts when restricted by radiation protection vests. (Courtesy of General Dynamics Corporation.)

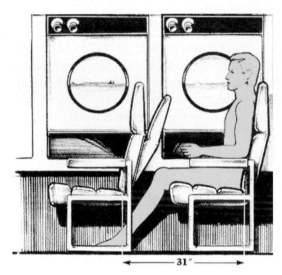

6.7 Analysis of human factors and living environments was part of designing the interior of this helicopter. (Courtesy of Bell Helicopter Corporation.)

despite high fuel costs, most consumers will not accept cars designed to get good mileage at the expense of comfort, safety, and styling. Buyers may be willing to give up some features, but few would give up air conditioning, tape decks, radios, comfortable seats, and safety features for improved gasoline mileage.

Human Factors

Human engineering (ergonomics) is the design of products and workplaces suited to the humans who use and occupy them. Safety and comfort are essential for efficiency, productivity, and profitability. Therefore, the designer must consider the physical, mental, safety, and emotional needs of the user and how to best satisfy them.

Leonardo da Vinci analyzed body dimensions in about 1473 (**Figure 6.6A**). Nearly five hundred years later, NASA performed similar analyses to determine the range of mobility permitted by a radiation protection garment used by astronauts (**Figure 6.6B**). Analysis of human factors is crucial in the space program because even the simplest, most familiar tasks require training and adaptation when astronauts perform them while in a weightless state. It is also important in the configuration of aircraft cabins, which share many similarities with an automobile's interior (**Figure 6.7**). Both must provide comfort and space in

which to function, but the aircraft has the added restriction of less space to work within.

Dimensions and Ranges A design must take into account the sizes, ranges of movement, senses, and comfort zones of the people using the finished product (**Figure 6.8**). Variations in people's physical characteristics conform to the normal distribution curve shown in **Figure 6.9**. Designers must use the

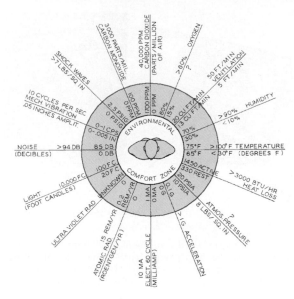

6.8 The inner circle represents the environmental comfort zone and the outer circle represents the bearable limit zone of the human environment. (Courtesy of Henry Dreyfuss, The Measure of Man.)

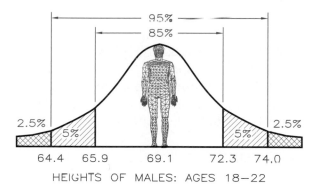

HEIGHTS OF MALES: AGES 18—22

6.9 This chart shows the distribution of average heights in inches of American men from 18 to 22 years of age. Fifty percent of American men in this age range are taller than 69.1 in., and 50 percent are shorter. (Courtesy of HumanCAD.)

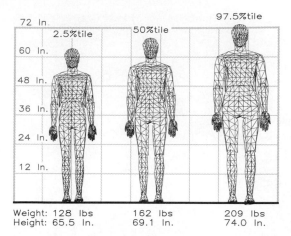

Weight: 128 lbs   162 lbs   209 lbs
Height: 65.5 In.   69.1 In.   74.0 In.

6.10 Men of average build have the body measurements shown here. (Courtesy of HumanCAD.)

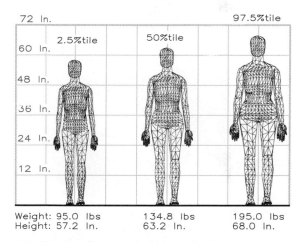

Weight: 95.0 lbs   134.8 lbs   195.0 lbs
Height: 57.2 In.   63.2 In.   68.0 In.

6.11 Women of average build have the body measurements shown here. (Courtesy of HumanCAD.)

body dimensions of the average American man (**Figure 6.10**) and the average American woman (**Figure 6.11**) as the basis for industrial designs. It is a challenging assignment to design accommodations that will be comfortable for the smallest as well as the largest subjects. **Figure 6.12** shows the ranges of worker's

body movements while performing maintenance on a spacecraft.

Motion The study of body motion begins with the understanding of the amount of space required for a person to function comfortably, safely, and efficiently. **Figure 6.13** shows a three-dimensional drawing of a digital air gauge from which a prototype will be made and tested to ensure that it properly fits the hand.

Vision Designs that include gauges and controls must provide the most visually effective means of aiding the operator. The automobile

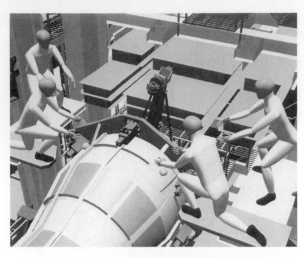

6.12 This computer model permits the analysis of body movements that are required to service a spacecraft. (Courtesy of McDonald Douglas Space & Defense System.)

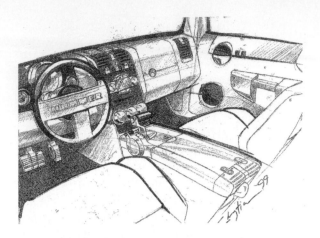

6.14 This concept drawing of the interior of the H2, General Motors' version of the Hummer, is one of many that are necessary to analyze human factors in order to design a safe and efficient instrument panel. (Courtesy of General Motors.)

6.13 A series of three-dimensional drawings are made from which to develop prototypes of a digital tire gauge for testing to ensure that the design is well-adapted to the hand. (Courtesy Bresslergroup and Measurement Specialties, Inc.)

dashboard configuration shown in **Figure 6.14** is a futuristic design, but it is similar enough to traditional dashboards to make the driver feel at home with it. Lights allow the driver to operate the controls easily and to obtain information from the instruments quickly.

Sound Sound must be within specified frequencies so as not to adversely affect a person's stress level and productivity. Many types of sound cause stress and contribute to an unsafe work environment.

Environment Working environments may include an entire industrial plant, a particular workstation, or a specialized location, such as the cockpit of a farm machine. Important environmental factors include temperature, lighting, color, sound, safety, and comfort.

Product Market

Designers study the market for a product during all stages of product development (as discussed in Chapters 3-5) and review it more formally during the analysis step. Areas of product analysis are market prospects, retail outlets, sales features, and advertising.

Market Prospects Market information should be collected to learn about the age groups, income brackets, and geographical locations of prospective purchasers of the product. This information is helpful in planning advertising campaigns to reach potential customers.

Retail Outlets A product may be marketed through existing wholesale and retail channels, from newly established dealerships, or by the manufacturer directly. For example, mainframe computers are not suitable for distribution through retail outlets, so manufacturers' technical representatives work with clients individually. However, an exercise machine can be sold effectively through department stores, television commercials, direct mail, and sporting-goods outlets.

Sales Features Designers should itemize the unique features of a new design that would stimulate interest in the product and attract consumers. They must continually ask questions, such as, "What features make this design better than my competitor's?"

Advertising Manufacturers, wholesalers, and retailers use several media, including personal contact, direct mail, radio, TV, newspapers, and periodicals to advertise their products to potential customers. Advertising costs vary widely and each medium should be analyzed for suitability before one or more is selected.

Physical Specifications

During the refinement step, the designer specifies various measurements, such as lengths, areas, shapes, weight, and angles for the product. During the analysis step, the designer uses the product's geometry and measurements to calculate member sizes and dimensions, weights, volumes, capacities, velocities, operating ranges, packaging and shipping requirements, and similar information (**Figure 6.15**).

Sizes and Dimensions The designer must evaluate product sizes and dimensions to ensure that they meet any standards specified, such as permissible widths, lengths, and weights in automobile design. For products

SEGWAY: THE HUMAN TRANSPORTER

HANDLE BARS

TIRES: Tubless, puncture resistant.

SPEED: 12.5 mph, top speed can be regulated for different users.
RANGE: 17 miles on a single charge.
TURNING RADIUS: Turns in place because of its single axle.
PAYLOAD: 250-pound passenger with 75 pounds of cargo.
WEIGHT: 80 pounds.

6.15 Designs must be analyzed to determine their physical properties, including weights, ranges, geometries, and capacities as shown for the Human Transporter (HT). (Courtesy of Segway LCC.)

that have moving parts, such as a construction crane, the designer must analyze the size of the product when extended, contracted, or positioned differently, as well as weight and balance requirements.

Ranges Many products have ranges of operation, capacities, and speeds that the designer must analyze before finalizing a design. For example, the designer must determine ranges and maximum limits such as seating capacity, miles per gallon, pounds of laundry per cycle, flows in gallons per minute, or power required.

Packaging and Shipping The designer must also be concerned with product packaging and shipping. Packaging relates both to product protection and consumer appeal: how the product is to be shipped—air, rail, mail, or truck—and whether it is to be shipped one at a time or in quantity are important considerations. Shipping and marketing a product assembled, partially assembled, or disassembled requires design attention and analysis, as does the cost of each method.

Strength   Much of engineering is devoted to analyzing a product's strength to support maximum design loads, withstand specified shocks, and endure necessary repetitive motions. **Figure 6.16** illustrates motion analysis for a moving part on a conveyor. The designer plots the data obtained and then uses graphical calculus to find the conveyor's velocity versus time profile as the first step in determining the strength needed by the part.

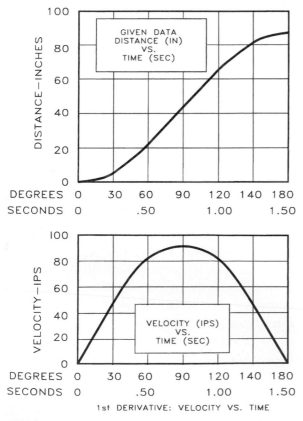

6.16 This graph shows the distance traveled versus the time required for a part on a conveyor. The designer graphically differentiates the plot of the data given to obtain a graph of velocity versus time as the first step in designing parts of sufficient strength for the conveyor.

Economic Factors
Designs must be economically competitive to have a chance of being successful. Therefore, before releasing a product for production, the designer must analyze its cost and expected profit margin. Two methods of pricing a product are **itemizing** and **comparative pricing**.

Itemizing   The process of totaling the costs of each part and its related overhead to determine its final cost is the first step in itemizing a product's price. From the working drawings, the designer (or an estimator) can estimate the costs of materials, manufacturing, labor, overhead, and other items to arrive at the total production expense. The wholesale price is production cost plus profit. Dealer margin plus the wholesale cost gives the retail price. One example of an economic model is shown in **Figure 6.17**; these percentages vary for different areas of manufacturing, marketing, and retailing.

Comparative Pricing   The other method used to estimate the price of a proposed product is to compare it with the prices of similar products. For example, the power tools shown in **Figure 6.18** are priced at approximately $40

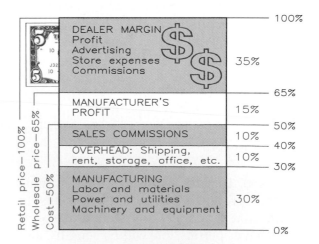

6.17 An economic model that can be used as a guide in the economic analysis and pricing of a product.

6.18 Each of these three products retails for approximately $40. They can be priced comparatively because they are similar in design and have essentially the same market size. (Courtesy of Black & Decker Company.)

6.19 This hunting seat and the exercise bench in Figure 6.20 can be comparatively priced at between $90 and $100 because both have similar manufacturing requirements and market volume potential. (Courtesy of Baker Manufacturing Company, Valdosta, Georgia.)

each. These tools are similar: All use the same type of power source, are made from the same materials, and have the same styling. Most importantly, these products will have about the same market size. Approximately the same number of drills, sanders, and saws are sold; consequently, production costs and retail prices are similar for each.

Another example of comparative pricing are the prices of the hunting seat (**Figure 6.19**) and the exercise bench (**Figure 6.20**), which have somewhat equal market sizes. Since both products' manufacturing requirements and market volumes are similar, both sell for about the same price of $90–$100. However, the baby stroller (**Figure 6.21**) sells for about $60 because of its larger and more competitive market, even though it is very similar in configuration to the hunting seat and the exercise bench.

Manufacturers also use comparative pricing to estimate cost per square foot, per mile, per cubic foot, or per day. These factors yield rough cost estimates as a basis for doing more detailed studies.

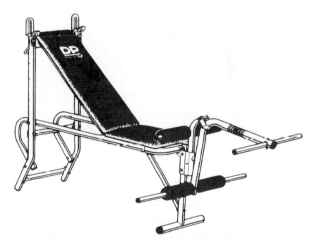

6.20 Priced at $100, this exercise apparatus is similar in manufacture and market appeal to the hunting seat shown in Figure 6.19. (Courtesy of Diversified Products Corporation.)

Miscellaneous Expenses Various expenses incurred in the development of new products can be easily overlooked and thereby affect a product's profitability projections. For example, warehousing and storage costs for finished products must be included in their price.

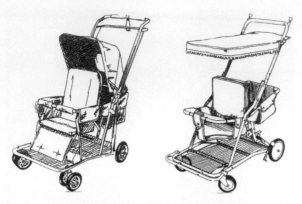

6.21 These baby strollers are priced from $55 to $75, or considerably less than the hunting seat and exercise apparatus, because the stroller market is larger and competition for customers is greater. (Courtesy of Strolee of California.)

6.22 A mock-up of an electro-mechanical product (a computer model in this case) is useful for testing its operation and assembly prior to finalizing its configuration. (Courtesy of Stokes Quality Industrial Designs.)

Associated with warehousing are the costs of insurance, temperature control, shelving, forklifts, and employees.

Models

Models are effective aids for analyzing a design in the final stages of its development. Designers use three-dimensional models to study a product's proportion, operation, size, function, and efficiency. Types of models often used are **conceptual models, mock-ups, prototypes,** and **system layout models**.

Conceptual Models Designers use rough models to analyze a preliminary design or feature concept.

Mock-ups Designers use full-size dummies of the finished design to demonstrate the product's size, appearance, and component relationships (**Figure 6.22**). Mock-ups present a visual impression, not the operation, of a product.

Prototypes Designers use full-size working models to demonstrate the operation of a final product. Because prototypes are made mostly by hand, materials that are easy to fabricate are used instead of those to be used in final production.

System Layout Models Designers use detailed scale models that show the relation-

ships among components of large manufacturing systems, building complexes, and traffic layouts. Designers usually construct system layout models of refineries to supplement working drawings for contractors and construction supervisors during construction (**Figure 6.23**).

Model Materials Designers commonly use balsa wood, cardboard, and clay in model construction because they are easy to shape and require few tools. Standard parts such as

6.23 This system layout model is used to analyze the details of construction of a refinery. (Courtesy of E. I. du Pont de Nemours and Company.)

wheels, tubing, figures, dowels, and other structural shapes can be purchased, rather than made, to save time and effort. Plexiglass can be used to construct models that illustrate both interior and exterior design features. Finished models should give a realistic impression of the design, especially when they are used for sales presentations and displays.

Model Scale A model should be large enough to show the function of the smallest significant moving parts. For example, the student model of a portable home caddy shown in **Figure 6.24** (made of balsa wood) demonstrates a linkage system that permits the wheels to be collapsed for storage. The model's scale is large enough to permit the linkage system to operate as it will in the final product.

Model Testing Using models to test performance is helpful in determining how well a design meets requirements. Aerodynamic characteristics of the rear styling of an automobile can be evaluated by wind tunnel tests. Physical relationships and the functional workings of movable components, as for the hatch and storage area of a car, can be tested in a prototype. Products can be tested on the computer screen to economically obtain

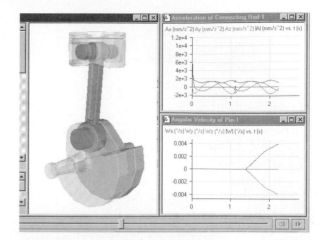

6.25 Computer models can be tested on the computer screen more economically than real models. (Courtesy of Knowledge Revolution, Inc.)

results, as illustrated in **Figure 6.25**. Designers also use models to test consumer reactions to new products before releasing the design for production (**Figure 6.26**).

6.4 Analysis: Exercise Bench

To illustrate a method of analyzing a product design, we return to our problem example, the exercise bench, which was carried through

6.24 A student's model demonstrates how a portable home caddy will fold flat for ease of storage.

6.26 Prototypes are full-size models for demonstrating and testing a design's operation. (Courtesy of General Motors Corporation.)

```
ANALYSIS
1. Function
    A.  PROVIDES SUPPORT FOR BASIC
        EXERCISES
    B.  SUPPORTS BAR BELLS
    C.  ADJUSTABLE TO SUIT INDIVIDUAL

2. Human engineering
    A.  PADDED BENCH
    B.  ADJUSTABLE BACKREST
    C.  CONFORMS TO BODY MOTIONS
    D.  BARBELL BRACKETS FOR SAFETY
    E.  FOAM PADS FOR COMFORT
3. Market and consumer acceptance
    A.  POTENTIAL MARKET
        1.  STATE—56,000
        2.  NATION—7,000,000
    B.  MORE EFFECTIVE EXERCISING
    C.  AFFORDABLE AT $120—$150 RANGE
    D.  USABLE FOR NONWEIGHT EXERCISES
        ALSO
```

DESIGN ANALYSIS	NAME FILE SEC DATE	GRADE 11

```
4. Physical description
    A.  BENCH COMPOSED OF BACKREST & SEAT
    B.  BACKREST ADJUSTABLE FROM 0 DEG.
        TO +30 DEG. WITH HORIZONTAL
    C.  2 BAR HOLDER BRACKETS FOR BARBELLS
    D.  2 BUTTERFLY EXERCISE ATTACHMENTS
    E.  LEG CURL ATTACHMENT
    F.  WEIGHT: 45 LB
    G.  SIZE: 52" LONG X 31" WIDE X 50" TALL

5. Strength
    A.  RECOMMENDED WEIGHT SET: 160 LB
    B.  SUPPORT PERSON WEIGTHING UP TO
        300 LB
    C.  CROSS BRACED TO PROVIDE
        STABILITY
    D.  REPLACEABLE PLASTIC SLEEVES FOR
        BUTTERFLY ATTACHMENTS
    E.  MAX. LEG CURL WEIGHT: 100 LB.
    F.  MAX. RECOMMENDED BUTTERFLY
        WEIGHTS: 50 LB EACH
```

DESIGN ANALYSIS	NAME FILE SEC DATE	GRADE 12

6.27 **(Sheet 11)** A worksheet containing an analysis of function, human engineering, and market considerations for the exercise bench.

6.28 **(Sheet 12)** A worksheet giving the physical description and strength analysis for the exercise bench.

6.29 **(Sheet 13)** A worksheet containing an analysis of the production procedures for and economics of the exercise bench.

the first three steps of the design process in Chapters 3–5. The main areas of analysis listed on the worksheets in **Figures 6.27** through **6.30** will assist you in analyzing the design. Additional worksheets and large sheet sizes for analysis drawings may be used if more space is needed.

Figure 6.30 shows how graphics is used to determine the range of positions of the backrest. Those positions affect the design of the angle-iron supports for the backrest and the

```
6. Production procedures
    A.  STRUCTURAL MEMBERS HOLLOW REC-
        TANGULAR SECTIONS—STEEL, BENT TO
        SHAPE
    B.  PARTS WELDED OR BOLTED TOGETHER
    C.  VINYL SEAT COVERS STAPLED TO
        PLYWOOD SEAT AND BACKREST
    D.  PLASTIC CAPS AT ENDS OF OPEN
        SUPPORT MEMBERS
    E.  METAL PARTS NICKEL PLATED

7. Economic analysis
    MATERIALS            $15
    LABOR                 16
    SHIPPING               7
    WAREHOUSING            1
                TOTAL            $39
    SALES COMMISSION    $ 4
    PROFIT              $20
            WHOLESALE PRICE    $63
            RETAIL PRICE       $90
```

DESIGN ANALYSIS	NAME FILE SEC DATE	GRADE 13

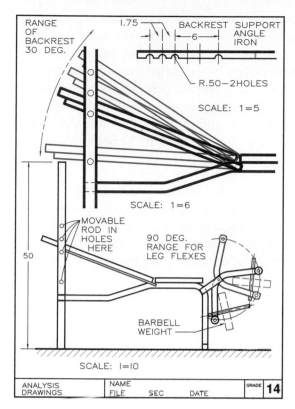

RANGE OF BACKREST 30 DEG.

BACKREST SUPPORT ANGLE IRON

1.75

6

R.50−2HOLES

SCALE: 1=5

SCALE: 1=6

50

MOVABLE ROD IN HOLES HERE

90 DEG. RANGE FOR LEG FLEXES

BARBELL WEIGHT

SCALE: I=IO

ANALYSIS DRAWINGS

NAME
FILE   SEC   DATE

GRADE 14

6.30 (Sheet 14) A worksheet that shows graphical analysis of the range of movements for the adjustable parts of the exercise bench.

6.31 A full-size model of the exercise bench is tested for function and acceptability.

locations of the semicircular holes in the angle irons for a range of settings of 30°.

The leg-exercising attachment at the end of the bench is designed to move through a 90° arc, which is sufficient for leg extensions. By determining the maximum loads on the backrest and the leg exerciser, you can select the member sizes and materials that provide the strength required. For further analysis, construct a model and test the design for suitability. A product that must support body weight plus weights that are being lifted should be rigorously tested to ensure that it is adequately sturdy.

Figure 6.31 illustrates a model of the exercise bench for testing its functional features. The catalog description of the bench shown in

Figure 6.32 lists the physical properties of the Weider® exercise bench to help the consumer understand its features. You should keep a list of descriptive characteristics of your design for the final catalog specifications. These points become very important to consumers as they come closer to buying a product.

Problems

The following problems should be solved on 8–1/2 × 11-inch paper and the solution presented in drawing, note, and text forms. Answers to essay problems can be typed or lettered. All sheets should be placed in a binder or folder.

General
1. Make a list of human factors that must be considered in designing the following items: (a) canoe, (b) hairbrush, (c) water cooler, (d) automobile, (e) wheelbarrow, (f) drawing

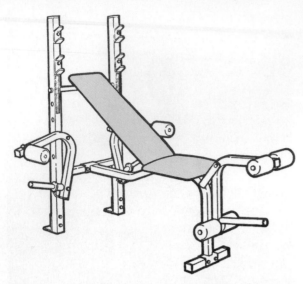

6.32 This catalog description gives the key features of the exercise bench: Weider® bench with butterfly attachment. It features no-pinch supports, multiposition padded back and leg lift, and tubular steel frame. Total weight capacity is 1000 lb; butterfly capacity, 50 lb; leg lift capacity, 65 lb; overall size, 58" × 45" × 41"; weight, 48 lb; and price, $89.99. (Courtesy of Sears, Roebuck and Company.)

table, (g) study desk, (h) pair of binoculars, (i) baby stroller, (j) golf course, (k) seating in a stadium, (l) coffee table, (m) exercise apparatus, and (n) lunch box .

2. What physical quantities have to be determined for the designs in Problem 1?

3. Select one of the items in Problem 1 and outline the steps required to analyze: (a) function, (b) human factors, (c) product market, (d) physical specifications, (e) strength, (f) economic factors, and (g) a prototype model.

Human Engineering
4. Design a computer-graphics table for your body size to meet your own working and comfort needs. Make a drawing indicating the optimum working areas and tilt angle for the computer when you sit at the station. The drawing should also show the most efficient positioning of supplies, materials, and manuals.

5. Using the dimensions for the average man and woman given in **Figures 6.10** and **6.11**, design stadium benches to meet the optimum needs of spectators. A primary consideration is slope of the stadium seating to allow an adequate view of the playing field. Spectator comfort and provision for traffic along aisles in front of the benches also must be considered.

6. Compare the measurements of the male and female students in your class with the averages given in **Figures 6.10** and **6.11**. Tabulate the results and compare them with the percentiles given in the example in **Figure 6.9**.

7. Design a backpack for use on a weekend camping trip. Determine the minimum number of articles a camper should carry; use their weights and volumes in establishing design criteria. Make sketches of the pack and the method of attaching it to the body to provide mobility, comfort, and capacity.

8. State the dimensions, facilities, and provisions needed for a one-person storm shelter to provide protection for 48 hours. Make sketches of the interior in relationship to a person and the supplies.

9. Design a manhole access to an underground facility. Determine the diameter of the manhole required to permit a person to climb a ladder for a distance of ten feet with freedom of movement. Make a sketch of your design and explain your method of solving the problem.

10. Analyze the needs for an observation facility for temporary service in the Arctic. This facility is to be as compact as possible, but it must provide for the needs of one person during a 72-hour duty watch. Make sketches of your design and explain the items considered essential to human survival in that harsh climate.

11. Design an automobile steering wheel that is different from current designs but that is just as functional. Base your design on human factors such as arm position, grip, and vision. Make sketches of your design and list the factors that you considered.

12. Make sketches to indicate safety features that you would build into your automobile to reduce the severity of personal injury in case of a bad accident. Explain your ideas and the advantages of your designs.

13. Assume that you prefer to alternate between sitting and standing when working at a study desk. Determine the ideal height of the table top for working in each position. Indicate how you would devise the table to permit instant conversion from the height for standing to the height for sitting.

14. Identify some human engineering problems that you believe need to be solved. Present several to your instructor for approval. Solve the approved problems. Make a series of sketches and notes to explain your approach.

Market Analysis
15. Conduct a market analysis for the drill shown in **Figure 6.18**, covering the areas mentioned in the text. Assume that this power tool has never been introduced before. Outline the steps you would take in conducting a product market analysis.

16. Make a market analysis of the hunting seat shown in **Figure 6.19**, following the steps suggested in the text. Determine a reasonable price, potential outlets, and other marketing information for the product.

17. Assume that the costs of producing hunting seats are estimated as: 100 seats, $35 each; 200 seats, $20 each; 400 seats, $10 each; 1000 seats, $8.50 each. Using these figures, determine the price at which you could introduce the seats to consumers on a trial basis and still make some money. Explain your plan.

18. List the unique features of the hunting seat that would be important to a sales campaign, including advertising. Make sketches and notes to explain these features.

Strength and Function Analysis
19. The graph in **Figure 6.33** can be used to estimate the capacities of material conveyors at various speeds and mat depths (thickness of the material on the conveyor). By referring to this graph, answer the following questions: (a) For a mat depth of 4 in. and a speed of 120 feet per minute, how many tons per hour are transported? (b) If you wish to have 100 tons per hour transported, what is the slowest speed of the conveyor that would be safe (would avoid the "caution advised" area)? (c) For a mat depth of 5 in. and a speed of 20 ft per minute, what would be the capacity?

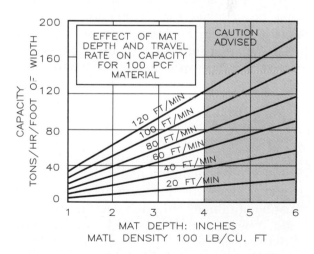

6.33 (Problem 19) These straight-line curves represent conveyor speeds used for transporting materials of 100 lb per cubic foot. The capacity in tons per hour per foot of width of the conveyor is given on the *y* axis and mat depth is shown on the *x* axis. Mat depth is the thickness of the material applied to the conveyor.

7

Decision

7.1 Introduction

After the designer has conceived, developed, refined, and analyzed several designs, one must be selected for implementation. The decision process begins with a presentation by the designer (or design team) of all significant findings, features, estimates, and recommendations. The presentation should be organized in an easy-to-follow form, and it must communicate the designer's conclusions and recommendations because it is the means of gaining support for the project in order for it to become a reality. A committee usually makes the decision when funding must be obtained. Although decision-making is aided by facts, data, and analyses, it is subjective at best.

7.2 Types of Decisions

The purpose of oral presentations and written reports is to present the findings of a project so that a decision can be made whether to implement it. One of three types of decisions may be made:

Acceptance. A design may be accepted in its entirety, which indicates success by the designer.

Rejection. A design may be rejected in its entirety, which does not necessarily mean that the designer has failed. Changes in the economic climate, moves by competitors, or other factors beyond the designer's control may make the design obsolete, premature, or unprofitable.

Compromise. Parts of a design may have weaknesses and compromises may be suggested. For example, the initial production run might be increased or decreased, or various features might be eliminated, modified, merged, or added.

7.3 Decision: Exercise Bench

We have used the exercise bench problem introduced in Chapter 3 to illustrate the first four steps of the design process. We continue to use it here to help explain the decision step.

Decision Table

A decision table like the one shown in **Figure 7.1** can be used to compare designs, where each idea is listed and given a number for identification. Assign maximum values for each factor of analysis, based on your best judgment, so they total to 10 points. Rate each factor for the competing designs by entering points for each.

Sum the columns of numbers to determine the total for each design and compare the scores of each design. Your instincts may disagree with the outcome of this numerical analysis. If so, have enough faith in your judgment to go with your intuition. The scores from the decision table are meant to be a guide for you, not the absolute final word in your decision.

Conclusion

After making a decision, state it and the reasons for it clearly (**Figure 7.2**). Record any additional information, such as number to be produced initially, selling price per unit, profit per unit, estimated sales during the first year, break-even number, and the product's most marketable features, that will help you prepare your presentation.

DECISION

1. Decision for evaluation

DESIGN 1   A-FRAME

DESIGN 2   U-FRAME

DESIGN 3   2-COLUMN FRAME

DESIGN 4

DESIGN 5

MAX	FACTORS	1	2	3	4
3.0	FUNCTION	2.0	2.3	2.5	
2.0	HUMAN FACTORS	1.6	1.4	1.7	
0.5	MARKET ANALYSIS	0.4	0.4	0.4	
1.0	STRENGTH	1.0	1.0	1.0	
0.5	PRODUCTION EASE	0.3	0.2	0.4	
1.0	COST	0.7	0.6	0.8	
1.5	PROFITABILITY	1.1	1.0	1.3	
0.5	APPEARANCE	0.3	0.4	0.4	
10	TOTALS	7.4	7.3	8.5	

DESIGN DECISION   NAME   FILE   SEC   DATE   GRADE 16

7.1 This worksheet shows the decision table used to evaluate the design alternatives for the exercise bench.

CONCLUSIONS

THE 2-COLUMN FRAME IS THE BEST SOLUTION FOR IMPLEMENTATION BECAUSE

1. GOOD MARKET POTENTIAL

2. AIDS WELL IN EXERCISING

3. FULFILLS PRODUCT REQUIREMENTS

4. ATTRACTIVE PRICE

5. MANUFACTURED EASILY

RECOMMEND IMPLEMENTATION AND PRODUCTION OF THE DESIGN

ECONOMIC FORECAST

SALES PRICE $ 90

SHIPPING EXPENSES $ 12

CUSTOMER PRICE $102

MANUFACTURER'S PROFIT $ 20 EACH

BREAK-EVEN AT   500 UNITS

PRODUCTION: RECOMMEND THAT BENCHES BE MADE BY QUALIFIED MANUFACTURER ON A CONTRACT BASIS

PROFITABILITY: BENCH SHOULD YIELD AN ATTRACTIVE RETURN ON INVESTMENT

DESIGN CONCLUSION   NAME   FILE   SEC   DATE   GRADE 17

7.2 This worksheet summarizes the designer's conclusions and recommendations for implementing that design.

If you believe that none of your designs are satisfactory, you should recommend that they not be implemented. A negative recommendation is not a failure of the design process; it means only that your solutions developed so far are not feasible. Going forward with an inadequate solution could waste both money and effort.

Presentation

Until now, your efforts have been self-directed and mostly free from supervision. The work is your own (or that of your team), you have solved the problem to the best of your ability, and you are ready to make recommendations regarding its implementation.

At this point, the project usually involves the input of others besides the designers. These outsiders may be other engineers, managers, administrators, salespeople, company shareholders, investors, or bankers who will loan money for the project. You must prepare a presentation suitable for your audience in order to communicate the important features of your design, the data you gathered and analyzed, and the benefits to be gained by implementing your design.

Present your findings, conclusions, and recommendations as objectively as possible so that the group can make a valid decision. At no time should your enthusiasm for the project outweigh an impartial presentation of the facts.

7.4 Types of Presentations

Presentations may be made to groups ranging from a few knowledgeable design associates to a large number of laypeople unfamiliar with the project and its objectives. Presentations of the first type are usually informal; the second type of presentation is formal.

Informal Presentations

Informal presentations are made to several associates and, perhaps, a supervisor. Although formally prepared visual aids are unnecessary for presentations to a small group, the designer nevertheless needs to graph data, draw pictorials, sketch schematics, and build models to explain design concepts. Ideas and concepts may be sketched on a blackboard or informally discussed in a one-on-one situation (**Figure 7.3**).

Formal Presentations

Formal presentations usually involve large groups that may include various combinations of associates, administrators, and/or laypeople. They may be clients for whom the project is designed, potential investors, or politicians who will vote to approve or disapprove the design. Function and acceptability of a design are the primary concerns of engineering associates, and profitability is most important to investors.

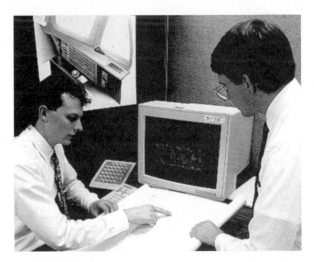

7.3 A decision may be the outcome of an informal presentation to an associate where ideas and designs are discussed one-on-one. (Courtesy of the Cessna Aircraft Company.)

7.5 Organizing a Presentation

An effective method of planning oral and written reports is to use 3-by-5-inch index cards for the ideas to be presented. Placing the cards on a table or tacking them to a bulletin board (**Figure 7.4**) allows an easy choice of sequence and rearrangement as needed. Each card (**Figure 7.5**) should contain the following information:

1. *Number.* The card's position in the presentation sequence.

2. *Illustration.* A sketch of any visual material.

3. *Text.* A brief outline of the points to be covered for that idea.

7.4 Planning cards are useful in preparing the sequence of a presentation by arranging the cards on a planning board (as shown here) or on a table top. (Courtesy of Eastman Kodak.)

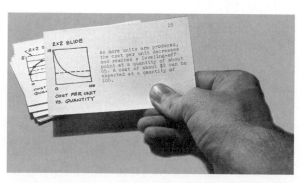

7.5 This layout an a 3-by-5-inch card, showing a sketch of the visual and its accompanying text, illustrates sound planning.

7.6 Visual Aids

Visual aids commonly used in presentations include flip charts, photographic slides, overhead visuals, models, computer images, and videotapes. The following suggestions apply to the preparation of visual aids:

1. Limit each visual to a single concept or point.

2. Reduce statements to key points in order to communicate thoughts clearly and concisely.

3. Make visuals containing text large enough to be readable.

4. Use illustrations, color, and attention-getting devices.

5. Prepare enough visuals so that notes are unnecessary.

Flip Charts

Flip charts consist of bold illustrations drawn on sheets (usually 30 x 36 inches) for presentation to groups no larger than about 30 people (**Figure 7.6**).

Paper Flip charts may be drawn on brown wrapping paper or white newsprint paper attached to a cardboard backing board. A stand or easel is needed to support the cardboard-backed set of sheets.

7.6 The flip chart is an effective method for presentation to small groups.

Lettering Felt-tipped markers, ink, tempera, or sign paints are fine for lettering. When used correctly, felt-tipped markers can yield bold, visible lines in a variety of colors and with sophisticated effects. India ink is an effective medium for lettering and for adding emphasis to a chart.

Color Construction paper cutouts mounted with rubber cement are especially effective for adding color to bar graphs. The use of felt-tipped markers and tempera colors also adds color and interest to a chart.

Assembly Flip chart sheets should be arranged in order of presentation with a title page covered by a blank sheet of paper on top to control audience attention. The sheets are fastened at the top to the backing board.

Presentation Each sheet is flipped in sequence after the presenter has covered the points on it. A pointer should be used to direct the audience's attention to specific items on the sheet. Well-prepared flip charts should require no additional notes.

Photographic Slides

Slides are effective for larger audiences and for showing actual scenes or examples of hardware.

Artwork

The method for proportionally sizing artwork for a 35-mm slide is shown in **Figure 7.7**. An 8 × 12-inch size is suitable for most slides. The artwork should contain color to make the slides more attractive and effective in maintaining audience interest. Colored construction paper, mat board, and other poster materials should be used in preparing slides.

Allow at least an inch margin on all artwork, so the edges will not show when photographed. Uppercase letters are best for slides, with the space between lines of text

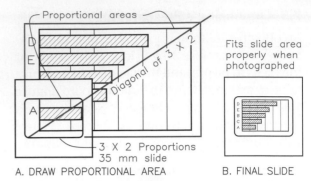

7.7 Use of this method of proportionally sizing artwork ensures that the subject will properly fill a photographic slide.

equal to the height of the letters. Do not use a white background for slide artwork, because it is tiring to the eyes.

Photographing Layouts

To make slides, a camera, copy stand, and lights are required. A 35-mm reflex camera with a through-the-lens view finder is best because the photographer sees exactly what is being photographed. The copy stand holds the camera steady. If all the artwork is uniform in size, the camera can be left in the same position during photography. Small illustrations can be photographed with a close-up lens. The finished slides are reviewed, arranged in sequence, numbered, and loaded in a tray for showing.

Slide Scripts

A slide script is useful when a presentation will be made repeatedly. Photographic copies of the slides attached to the left side of the script serve as prompts for the presenter (**Figure 7.8**).

Overhead Projector Transparencies

Overhead projector transparencies are reproduced on 8-by-10-inch plastic sheets by the heat-transfer or diazo processes, or plotted by

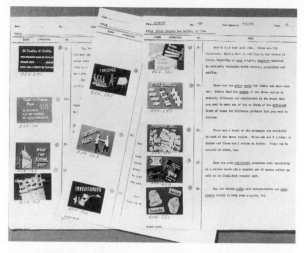

7.8 A slide script is useful for lengthy slide presentations and those that will be given repetitively. (Courtesy of Eastman Kodak.)

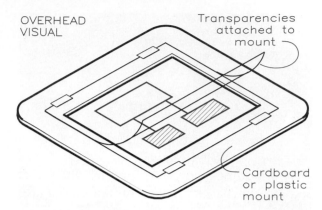

OVERHEAD VISUAL

Transparencies attached to mount

Cardboard or plastic mount

7.9 A transparency used on an overhead projector consists of an 8-1/2" x 11" transparency mounted on a 10" x 12" frame. The projection area within the frame is about 7-1/2" x 9-1/2". To show information sequentially, different-colored overlays can be flipped over, one at a time.

the computer. Tracing paper is the most commonly used drawing surface for preparing art from which transparencies are made. Tracing paper can be used in both the heat-transfer and diazo processes (opaque paper cannot be used in the diazo process). Computer plotting can be done directly onto plastic film with special pens. Diazo transparencies are reproduced on plastic film in the same manner in which blue-line prints are made.

Drawings should be made in black India ink. Stick-on shapes and graphing tapes can give the drawing a professional appearance. Lettering should be at least 0.20 in. high.

Color Overlays

Several color overlays can be hinged to the transparency mount for presentations with multiple steps (**Figure 7.9**). Each color overlay requires a separate piece of artwork, which is drawn on tracing paper placed over the basic layout.

Computer Plotted Transparencies

Computer-generated art and text can be plotted directly on plastic film with fiber-tipped pens,

which come in many colors and match the film. The Romand font of AutoCAD is better suited for large lettering on transparencies than is the Romans font.

Presentation

The presenter stands or sits near the projector in a semilighted room and refers to the transparencies while facing the audience. The presenter can emphasize items on the stage of the projector with a pointer, which is projected onto the screen. In the same manner as multiple overlays are hinged to a mount, paper overlays can be attached to mounts in order to block out parts of the transparency to control the audience's attention.

Models

A model is the most realistic visual aid for showing a final design. Models should be large enough to be seen by all in the audience. A series of photographic close-ups of the model, taken from different angles, can be used to supplement the presentation. Obviously, a full-sized prototype of the completed design (**Figure 7.10**) provides the most

7.10 A model is an effective visual to use in a presentation to a small group (Courtesy of Cessna Aircraft Company.)

7.12 Future presentations may be prepared for the Personal Video Player (PVP) and displayed on its built-in LCD screen or a standard TV. This device is being developed by Sonicblue and the Intel Corporation. (Courtesy of the Intel Corporation.)

accurate description of the product and demonstrates its operation.

Computer Visuals

Software programs such as MicroSoft's *Power-Point* can be used for making slides that can be viewed on the computer monitor for small groups or projected onto a large screen connected to a computer for large groups. **Figure 7.11** shows a dialogue screen used in composing a computer visual with *PowerPoint*. Any drawing or photograph that can be seen on the computer monitor can be used as a visual.

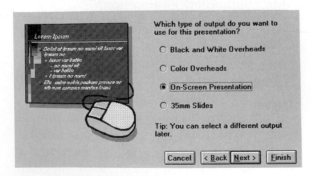

7.11 The dialogue screen illustrates a computer visual and the prompts for composing it. (Courtesy of *PowerPoint*.)

When slides are made by the computer, there is no need for special art supplies, storage problems are eliminated, and the entire slide show can be shown on a lap-top computer. Full color artwork, special effects, and animation can be incorporated into these presentations.

Video

Video presentations or visuals supplemented by a voice-over narration is an effective and sophisticated method. It may include various special effects such as music, sound effects, close-ups, motion, and precise realism.

Formal presentations in the future will be multimedia shows using video and the computer in combination. The PVP (Personal Video Player) a miniature, hand-held player holding up to ten hours of content is an example of a device that will revolutionize presentations of the future (**Figure 7.12**).

7.7 Making a Presentation

You should inspect the room in which the presentation is to be made in advance of the meeting. You also should view projected visuals from various audience locations. Projectors and visual-aid equipment must be positioned and focused before the audience arrives, and remote controls for slide projectors should be readied for use. You should be aware of where to stand so as to not block anyone's view.

Delivery

You should move through the presentation at a moderate pace while emphasizing information on the visual aids with a pointer (**Figure 7.13**). A positive approach in selling ideas should not become deceptive high-pressure salesmanship. The presenter should be frank in pointing out weaknesses in a design and show alternatives that compensate for them.

Conclusions and recommendations should be supported by data and analyses. A recommendation to reject or accept a design should be supported by reasons. A period for questions and answers should follow the presentation for clarification purposes. If available, a technical report should be given to the audience.

Critique

When your team gives a presentation to the class, both your design recommendations and the skill of your presentation will be evaluated and critiqued. The form shown in **Figure 7.14** is typical of the evaluation form that may be used for your critique. It can also be used as a guide in preparing the presentation. The names of the team members are listed at the top of the sheet. As a group, your team must agree on the percent contribution of each member to the project prior to presentation. The sum of the percent contributions of all team members must equal 100 percent. The F-factor for each member is the number of members (N) times the contribution of each (C). The chart in Appendix 44 illustrates how an individual's contribution to the project is translated into his or her individual grade by using the F-factor.

7.13 A presentation should include graphical aids and models to help the speaker communicate with the audience. (Courtesy of Hewlett Packard Corporation.)

The table below should be completed jointly by the team with only the grade column completed by the instructor who will use the chart on page 4 and the factor "F" that was computed for each member.

Oral Report

Team No. **5**   Project: **TOY MANUFACTURING**

	Names	No. (N= **7**)	%Contri-bution (C)	F=NC	GRADE
1.	**Brown, George**		17	119	91
2.	**Prisk,Helen**		14.3	100	87
3.	**Smith, Roger**		20	140	95
4.	**Reed, Ralph**		5.7	40	63
5.	**Potter, Joyce**		14.3	100	87
6.	**Flynn, Errol**		14.3	100	87
7.	**Ross, Lawrence**		14.4	100	87
8.					

	Evaluation by instructor	Max.		Comments:
1.	Introduction of team members	2	2	Good introduction to project
2.	Proper dress of team members	2	2	
3.	Statement of purpose of presentation	5	4	
4.	Use of visuals—point to important points, do not block screen, do not fumble, etc.	10	8	Several visuals to complex to see very well
5.	Adequate number of visual aids	9	9	
6.	Quality of visual aids	15	12	
7.	Clear presentation of recommended design	10	8	Economic analysis could use a little more study
8.	Presentation of alternate solutions considered	2	2	
9.	Consideration of human factors	5	5	
10.	Coverage of economics (manufacturing, shipping, packing, overhead, mark-up, etc.)	10	7	Very good professional manner in giving the presentation
11.	Presentation of an effective conclusion	5	4	
12.	Continuity of presentation	3	3	
13.	Poise and professionalism	2	2	Good conclusion and proposed solution
14.	Team participation (perfect score if all participate)	10	10	
15.	Use of allotted time	10	9	
	TOTAL 100		87	

Instructor comments on back of this sheet.

7.14 An evaluation form for grading oral reports.

7.8 Written Reports

Engineers, technologists, and technicians must know how to prepare a well-written report. The three basic types of written reports are **proposals, progress reports,** and **final reports (Figure 7.15)**.

Proposals

Proposals are written to establish the need for projects and to obtain authorization of funds and support to pursue them. A proposal outlines data, costs, specifications, time schedules, personnel requirements, completion dates, and other information concerning the project. The purpose of the project is stated, with emphasis on its benefit to the client or organization.

Proposals should reflect the interests and language of the reader. For instance, investors are interested in profits and returns, whereas engineers are more concerned with function and feasibility. Typical elements of a proposal include the following:

Statement of the problem. Identify the problem and its purpose.

Method of approach. Outline procedures for attacking the problem.

Personnel needs and facilities. Itemize requirements for equipment, space, and personnel.

Time schedule. Give estimated completion dates for each phase of the project.

7.15 The three basic types of written reports.

Budget. Itemize the funds required for each phase of the project.

Summary. Review the important points.

Progress Reports

Progress reports are periodic reports on the status of a project. They are usually brief and may take the form of a letter or memo. They generally review progress and project the outlook for further progress, including the need for increases or decreases in expenditures or time.

Final Reports

The most comprehensive type of written reports are final reports, which contain five sections:

1. Problem identification.

2. Method.

3. Body.

4. Findings.

5. Conclusions and recommendations.

Some reports present the conclusions at the beginning and others at the end. The order of presentation will vary with the requirements of your instructor or employer.

Report Format Typically, the sequence of a technical report's contents (**Figure 7.16**) is as follows:

1. *Cover.* The report should be inserted into a binder, with its title and author indicated on the cover.

2. *Letter of transmittal* (optional). The first page may be a letter of transmittal, describing briefly the contents of the report and the reasons for the project.

3. *Title page.* The title page contains the title of the report, the name of the person or team that prepared it, and the date.

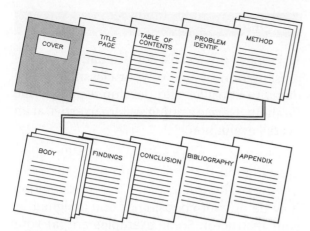

7.16 The elements of a typical technical report.

4. *Table of contents.* The major headings in the report and their page numbers are listed.

5. *Table of illustrations* (optional). A list of the illustrations contained in the report may be included, especially in long, formal reports having many illustrations.

6. *Problem identification.* (Actual heading should be appropriate to the report rather than this general term.) This section explains the importance of and need for a solution to the problem by outlining background information about the problem.

7. *Method.* (Actual heading should be appropriate to the report rather than this general term.) This section should cover the general method used in solving the problem.

8. *Body.* (Actual heading should be appropriate to the report rather than this general term.) This section is the main part of the report. It describes the data collected and analyzed and the steps taken to solve the problem. Subheadings should be used to emphasize the various parts of the report's body.

9. *Findings.* (Actual heading should be appropriate to the report rather than this general term.) Project findings should relate clearly to the data gathered, analyzed, and described in the preceding section.

10. *Conclusions.* Conclusions should be based on the findings and recommendations made.

11. *Bibliography.* The references—books, magazines, brochures, interviews—used in the preparation of the report are listed alphabetically by author.

12. *Appendix.* The appendix includes information (such as drawings, sketches, raw data, brochures, and letters) that supplements (often in more detail) the main sections of the report.

Common Omissions Topics that are often overlooked and omitted from reports by students are sales estimates, advertising methods, shipping costs, product packaging, miscellaneous overhead expenses, and recommendation summaries. Each of these topics must be adequately covered to provide a complete understanding of the total project.

Illustrations Technical reports should be liberally illustrated. Drawings and other illustrations should be numbered and given captions that describe them and relate them to the text. The text should refer to each figure by number and explain the figure's contents. For example, "Figure 6 shows the number of boats sold between 1988 and 1998" identifies the figure being referred to and what it depicts.

Illustrations for the report should be drawn or reproduced on opaque paper rather than tracing paper; ink illustrations are preferred. Drawings should be positioned on the sheet so that they can be read from the bottom or from the right. Large drawings should be folded to 8-1/2 by 11 inches to fit into the binder and they should be easy to unfold for reading.

Evaluating a Written Report   The form shown in **Figure 7.17** is one that may be similar to one used by your instructor to grade your report. The team determines the contribution of each team member and fills in the blanks at the top of the form. The teacher grades the report, and each member's grade is found by using the table in Appendix 44. You will have a good idea of the guidelines for preparing your written report if you review this form before beginning your report.

Written Report

TEAM NO. 5   PROJECT TOY MFGR

NAMES	NO. (N = 7)	% CONTRIBUTION (C)	F = NC	GRADE (G)
1. BROWN, G.		14.3	100	92
2. PRISK, A.		14.3	100	92
3. SMITH, L.		12.0	84	87
4. REED, T.		16.3	114	93
5. POTTER, M.		10.0	70	82
6. FLYNN, O.		8.8	62	79
7. ROSS, N.		14.3	100	92
8.				

EVALUATION BY INSTRUCTOR 100%	Max. Value	Points Earned
1. Use of an appropriate cover	2	2
2. Inclusion of an evaluation sheet	2	2
3. Inclusion of a proper letter of transmittal	2	2
4. Correct title page	2	2
5. Proper table of contents	2	2
6. Sufficient introduction to the the report	5	4
7. Thoroughness in identifying the problem	10	8
8. Continuity and quality of the body the report	10	9
9. Collection and presentation of background data	5	4.5
10. Justification of major decisions	5	5
11. Review of costs, overhead expenses, shipping costs and similar expenses	5	5
12. Arrival at strong conclusion and recommendation	5	4.5
13. Sufficient number of graphs and graphics	10	9
14. Quality of graphics	10	9
15. Bibliography—form and content	5	5
16. Use of footnotes	5	5
17. Appendix—content and form	5	5
18. Form and appearance of report (spelling, punctuation, margins, typing, neatness)	10	9
	100	92

7.17 An evaluation form for grading a written report by a design team.

Problems

1. Prepare a checklist for evaluating an oral presentation by one of your classmates. List items to consider and develop a point scale for them. Devise a rating system to arrive at an overall evaluation.

2. Use 3 × 5-inch cards to plan a flip-chart presentation that will last no more than five minutes. The subject of your presentation may be of your choosing or one assigned by your instructor. Some examples are (a) your career plans for the first two years after graduation; (b) the role of this course in your overall educational program; (c) the importance of effective communication; (d) the need for a design project that you are proposing; and (e) a comparison of engineering with another profession.

3. Prepare graphical aids for an oral presentation using the methods and materials covered in this chapter.

4. Using the planning cards developed in Problem 2, prepare a five minute briefing on a technique that you choose or is assigned by your instructor. Present this briefing to your class.

5. Assume that you are an engineer responsible for developing a proposal for a project that could result in a sizable contract for your firm. Make a list of instructions to give to your assistants for their help in preparing a presentation for a group of 20 people, ranging in background from bankers to engineers.

Your instructions should outline the materials needed, types and number of graphical aids required, method of projection or presentation, assistance needed during the presentation, room seating arrangements, and other factors. Your outline should cover the entire presentation for the length of time you think most desirable. Select a topic or use one assigned by your instructor.

6. When giving a presentation, be sensitive to what the most important and significant points are. Important recommendations and findings must be stressed above all others. Do not get bogged down in ideas and approaches that were attempted but discarded at the expense of the points that are most helpful in attaining acceptance of your recommendations.

Make a list of the major points that should receive emphasis if you were giving one of the following reports: (a) an application for a summer job; (b) a request for a loan from your parents; (c) a proposal to a banker for opening a hotdog stand or a business of your choice, (d) a résumé of introduction to a prospective girl- or boyfriend; (e) a reason for talking your instructor into an excused absence; (f) a list of your qualifications for being elected class president.

7. Many designs and products come onto the market that are mistakes from the beginning—they don't sell and often they may not function well. Poor decisions were made when the "go ahead" was given. Make a list of as many of these "mistakes" as you can think of. For starters, the Susan B. Anthony dollar is one mistake that has been discontinued by the government.

8. One of the best-selling cars of all time was the Ford T-Model touring car (**Figure 7.18**). Make a list of the T-Model's attributes that made it the success that it was at the time it was introduced to the market in 1908. Make another list of the disadvantages that it would have today.

7.18 (Problem 8) The Ford T-Model was one of the best-selling and most popular cars of all time.

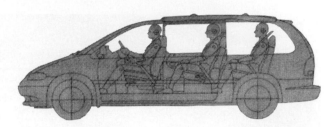

7.19 (Problem 9) One of the best-selling cars of today is the seven-passenger minivan.

9. A best-selling car of today (**Figure 7.19**) is the seven-passenger minivan, a new concept vehicle that has outsold the traditional sedan. Make two lists; one that lists the advantages of the minivan over the sedan, and one that lists the disadvantages of the minivan over the sedan.

10. The sales generated by advertisement spots on TV must be sufficiently effective in order to pay for all other programs, shows, and sports. Therefore, TV commercials are good examples of extremely short reports that must get the viewer's attention, tell a story in a matter of seconds, and be persuasive in selling a concept or product.

Pay close attention to several TV commercials and tabulate your analysis of the following points:

(A) What introduction technique got your attention?

(B) What "plot" was used to direct your thinking?

(C) What method was used to make you receptive to the advertised product?

(D) Was the commercial effective?

(E) List negative factors.

Compare your findings with those of a classmate to determine the level of agreement. Make mental notes of these findings and use them in preparing future presentations.

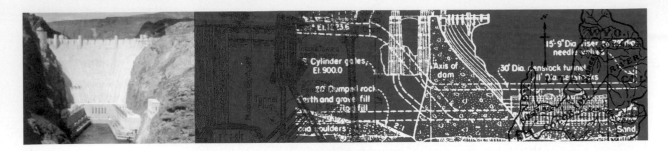

8

Implementation

8.1 Introduction

Implementation is the final step of the design process, in which the design becomes a reality. The designer details the product in working drawings with specifications and notes for its fabrication. **Graphical methods are particularly important during implementation, because all products are manufactured from working drawings and specifications.** Implementation also involves the packaging, warehousing, distribution, and sales of the manufactured product.

8.2 Working Drawings

Working drawings, with orthographic views, dimensions, and notes, describe how to make the individual parts of a product. The wiper hanger in **Figure 8.1** is drawn by computer as a working drawing for implementation in

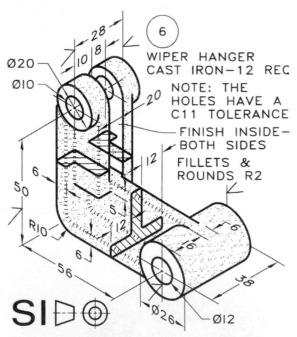

8.1 This wiper hanger is shown in a dimensioned working drawing in Figure 8.2 with three orthographic views.

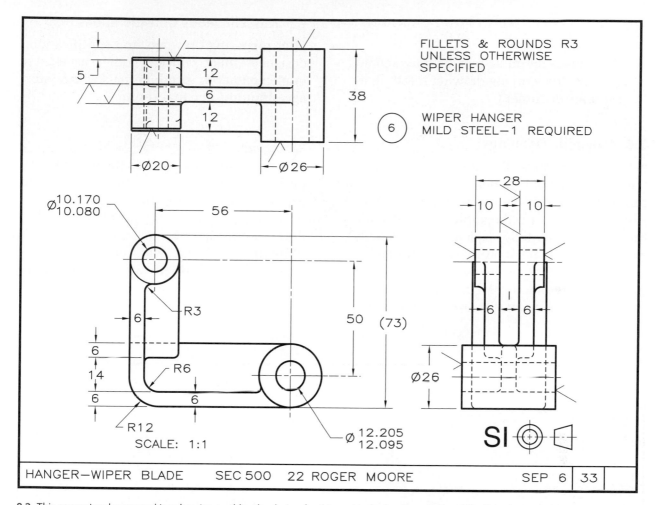

FILLETS & ROUNDS R3
UNLESS OTHERWISE
SPECIFIED

WIPER HANGER
MILD STEEL—1 REQUIRED

HANGER—WIPER BLADE   SEC 500   22 ROGER MOORE   SEP 6 33

8.2 This computer-drawn working drawing enables the design for this part to be implemented as a final product.

Figure 8.2. Properly executed working drawings ensure that the resulting products will be identical when the instructions on the drawings are followed, regardless of the shop in which they are made.

When making working drawings, designers draw several parts on the same sheet without attempting to arrange them in relationship to each other or in the order of their assembly. The names of the parts, their identifying numbers, the quantity required, and the materials to be used in making them are noted near the views.

8.3 Specifications

Specifications are written notes and instructions that supplement the information shown in the drawings. Specifications may be prepared as separate typed documents that accompany drawings or that stand alone when graphical representation is unnecessary. Instructional notes, such as the following, are adequate as written specifications without drawings:

METALLURGICAL INSPECTION IS REQUIRED
BEFORE MACHINING.

or

PAINT WITH TWO COATS OF FLAT BLACK PAINT
(NO. 780) AFTER FINISHING.

When space permits, specifications should be given on the working drawing rather than in a separate document.

8.4 Assembly Drawings

Assembly drawings illustrate how individual parts are to be put together to become the final product. They can be drawn as three-dimensional pictorials or orthographic views that are fully assembled, fully exploded, or partially exploded. **Figure 8.3** shows a partially exploded orthographic view of an assembly with part numbers in balloons. An assembly drawing usually contains a parts list for easy reference.

8.5 Miscellaneous Considerations

After preparing drawings and specifications, designers must consider other aspects of implementation: **product packaging, storage, shipping,** and **marketing.**

Packaging

In some industries, such as the toy industry, packaging is elaborate and may be as expensive as the product. Designers must be aware of packaging problems as they develop a design because a product that is difficult to package will cost more. Many products are shipped partially disassembled to make packaging easier and less expensive.

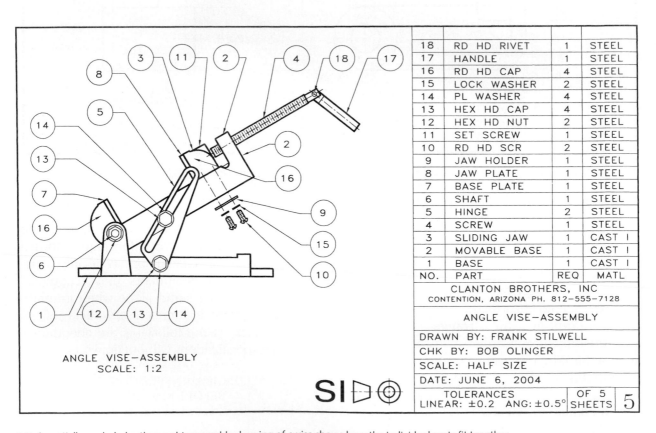

18	RD HD RIVET	1	STEEL
17	HANDLE	1	STEEL
16	RD HD CAP	4	STEEL
15	LOCK WASHER	2	STEEL
14	PL WASHER	4	STEEL
13	HEX HD CAP	4	STEEL
12	HEX HD NUT	2	STEEL
11	SET SCREW	1	STEEL
10	RD HD SCR	2	STEEL
9	JAW HOLDER	1	STEEL
8	JAW PLATE	1	STEEL
7	BASE PLATE	1	STEEL
6	SHAFT	1	STEEL
5	HINGE	2	STEEL
4	SCREW	1	STEEL
3	SLIDING JAW	1	CAST I
2	MOVABLE BASE	1	CAST I
1	BASE	1	CAST I
NO.	PART	REQ	MATL

CLANTON BROTHERS, INC
CONTENTION, ARIZONA PH. 812-555-7128

ANGLE VISE-ASSEMBLY

DRAWN BY: FRANK STILWELL

CHK BY: BOB OLINGER

SCALE: HALF SIZE

DATE: JUNE 6, 2004

TOLERANCES	OF 5	
LINEAR: ±0.2 ANG: ±0.5°	SHEETS	5

ANGLE VISE-ASSEMBLY
SCALE: 1:2

SI▷⊕

8.3 A partially exploded orthographic assembly drawing of a vise shows how the individual parts fit together.

Storage

Most manufacturers maintain an inventory of products for shipment. Therefore, warehousing costs must be figured into the product's final selling price.

Shipping

Industries that locate warehouse facilities in the middle of their market areas have lower shipping costs than those with warehouses at the edges of their market area.

Marketing

Designers must be concerned with all aspects of a product after it enters the marketplace, including its marketability and consumer acceptance. Complaints about a product's reliability and function are important to designers, alerting them to design or manufacturing defects that must be overcome in future versions of the product.

8.6 Implementation: Exercise Bench

To illustrate the implementation of a product design, we return to the exercise bench, which was introduced in Chapter 3 and has been used to demonstrate the application of each step in the design process.

Working Drawings

The two working drawings shown in **Figures 8.4** and **8.5** depict some details of the exercise bench design. Additional working drawings are required to show the other parts of the bench, which are dimensioned in decimal

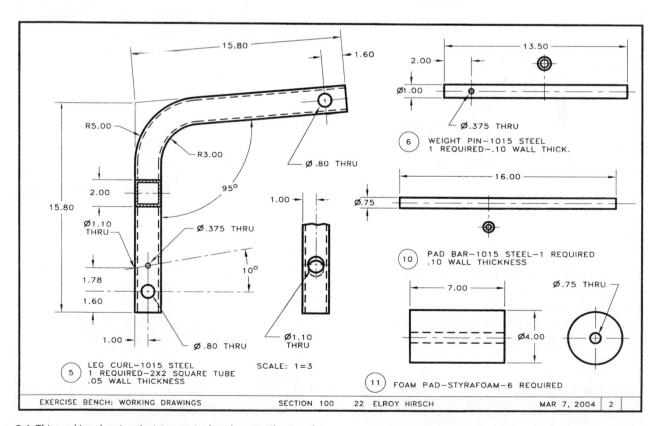

8.4 This working drawing depicts exercise bench parts (Sheet 1 of 5).

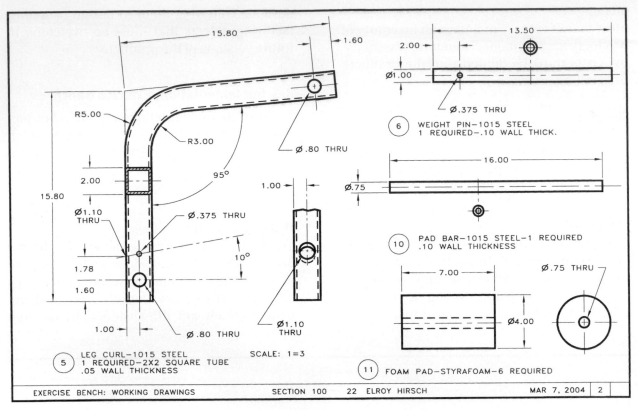

8.5 This working drawing depicts exercise bench parts (Sheet 2 of 5).

inches. Standard parts to be purchased from suppliers are not drawn but are itemized on the drawing, given part numbers, and listed in the parts list on the assembly drawing.[*]

Assembly Drawing

Figure 8.6 shows an assembly drawing that illustrates how the parts are to be assembled after they have been made. The assembly is shown pictorially, with the different parts identified by numbered balloons attached to leaders. The parts list identifies each part by number and describes it generally.

[*]This particular design was developed and patented and is marketed by Weider Health and Fitness, 2100 Erwin Street, Woodland Hills, CA 91367.

Packaging

The Weider exercise bench is packaged in a corrugated cardboard box and weighs approximately 40 pounds. It is shipped unassembled so that it will fit into a smaller carton for ease of handling during shipment (**Figure 8.7**).

Storage

An inventory of benches must be maintained in order to meet retailer demand. The need to hold inventory increases overhead costs for interest payments, warehouse rent, warehouse personnel, and loading equipment.

Shipping

Shipping costs for all types of carriers (rail, motor freight, air delivery, and mail services) must be evaluated. The shipping cost for a

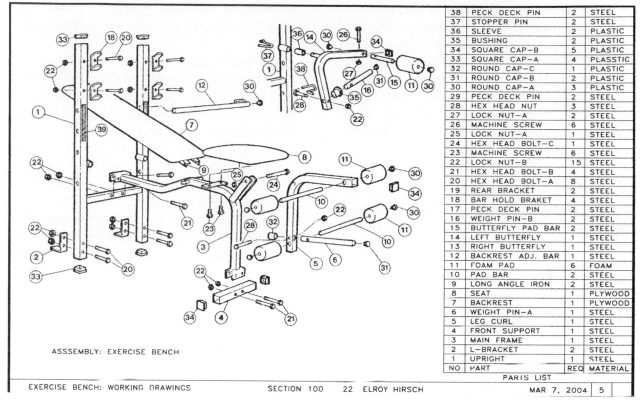

NO	PART	REQ	MATERIAL
38	PECK DECK PIN	2	STEEL
37	STOPPER PIN	2	STEEL
36	SLEEVE	2	PLASTIC
35	BUSHING	2	PLASTIC
34	SQUARE CAP-B	5	PLASTIC
33	SQUARE CAP-A	4	PLASSTIC
32	ROUND CAP-C	1	PLASTIC
31	ROUND CAP-B	2	PLASTIC
30	ROUND CAP-A	3	PLASTIC
29	PECK DECK PIN	2	STEEL
28	HEX HEAD NUT	3	STEEL
27	LOCK NUT-A	2	STEEL
26	MACHINE SCREW	6	STEEL
25	LOCK NUT-A	1	STEEL
24	HEX HEAD BOLT-C	1	STEEL
23	MACHINE SCREW	6	STEEL
22	LOCK NUT-B	15	STEEL
21	HEX HEAD BOLT-B	4	STEEL
20	HEX HEAD BOLT-A	8	STEEL
19	REAR BRACKET	2	STEEL
18	BAR HOLD BRAKET	4	STEEL
17	PECK DECK PIN	2	STEEL
16	WEIGHT PIN-B	2	STEEL
15	BUTTERFLY PAD BAR	2	STEEL
14	LEFT BUTTERFLY	1	STEEL
13	RIGHT BUTTERFLY	1	STEEL
12	BACKREST ADJ. BAR	1	STEEL
11	FOAM PAD	6	FOAM
10	PAD BAR	2	STEEL
9	LONG ANGLE IRON	2	STEEL
8	SEAT	1	PLYWOOD
7	BACKREST	1	PLYWOOD
6	WEIGHT PIN-A	1	STEEL
5	LEG CURL	1	STEEL
4	FRONT SUPPORT	1	STEEL
3	MAIN FRAME	1	STEEL
2	L-BRACKET	2	STEEL
1	UPRIGHT	1	STEEL

PARTS LIST

ASSSEMBLY: EXERCISE BENCH

EXERCISE BENCH: WORKING DRAWINGS SECTION 100 22 ELROY HIRSCH MAR 7, 2004 5

8.6 An assembly drawing for the Weider exercise bench (Sheet 5 of 5). (Courtesy of Weider Health and Fitness.)

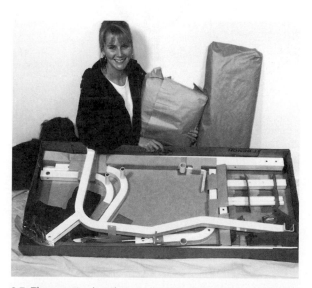

8.7 The exercise bench is packaged unassembled and flat for ease of packaging and handling during shipment.

Weider bench with its accessories is $10–$15, depending on distance, when shipped one at a time by United Parcel Service. The cost per unit is about 50 percent less when units are shipped in bundles of ten to the same destination.

Accessories

Examples of accessories, or add-ons, are the butterfly attachments for arm exercises. Accessories enable buyers to upgrade the basic product in stages, which can increase product marketability and sales.

Prices

The retail price of the Weider bench is about $100. This type of product generally retails for about five or six times the cost of manufacturing them (materials and labor). Retailers receive approximately a 40-percent margin, distributors earn about 10 percent, and the remainder of the price represents advertising costs and the other miscellaneous costs mentioned previously. The consumer pays all of these costs (pro-rated to each exercise bench) as part of the purchase price.

8.7 Patents

Inventors of processes or products should investigate the possibility of obtaining patents on them from the U.S. Patent and Trademark Office (USPTO) before disclosing their inventions. The USPTO issued its first patent in 1836, a patent for traction wheels (**Figure 8.8**). The patent procedure is outlined in *General Information Concerning Patents,* a publication available from the PTO from which the following material was extracted.

What May Be Patented?

Any person who "invents or discovers any new and useful process, machine, manufacture, or composition of matter, may obtain a patent," subject to the conditions and requirements of law. These categories include everything made by humans and the processes for making them. **Figure 8.9** shows the cover page of Thomas Edison's patent for the electric lamp.

Inventions used for the development of nuclear and atomic weapons for warfare are not patentable because they are not considered "useful." Also, a design for a mechanism that will not operate as described is not

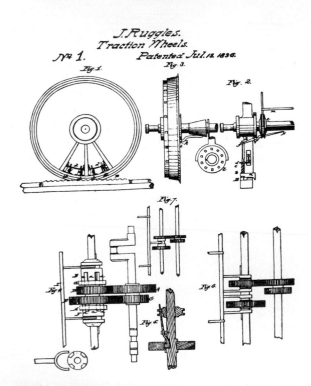

8.8 The U.S. Patent and Trademark Office issued this first patent in 1836.

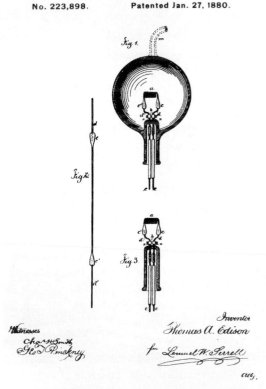

8.9 Thomas Edison received this patent for the electric lamp in 1880, thus beginning the electrical/electronics revolution.

patentable. An idea or concept for a new invention is not patentable; it must be designed and described in detail before it can be considered for patent registration.

Who May Apply for a Patent?

Only the inventor may apply for a patent. A patent given to a person who was not the inventor would be void and the recipient subject to prosecution for perjury. However, the executor of a deceased inventor's estate may apply for a patent, and two or more people may apply for a patent as joint inventors.

Patent Rights

An inventor, who has been granted a patent, has the right to exclude others from making, using, or selling the invention throughout the United States for 20 years from the time of application. At the end of that time, anyone may make, use, or sell the invention without authorization from the patent holder.

Application for a Patent

An inventor applying for a patent must provide the following:

1. A completed form that includes a petition, specification (description and claims), and oath or declaration;

2. A drawing, if a drawing is possible; and

3. The filing fee.

Petition and Oath   In the petition and oath (usually on one form) the inventor asks to be given a patent on the invention and declares that he or she is the original inventor of the device described in the application.

Specification   The inventor must submit a written specification, describing the invention in detail, so that a person skilled in the field to which the invention pertains can pro-

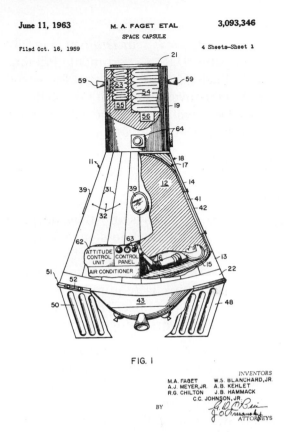

June 11, 1963   M. A. FAGET ETAL   3,093,346
                        SPACE CAPSULE
Filed Oct. 16, 1959   4 Sheets—Sheet 1

FIG. I

                              INVENTORS
M.A. FAGET   W.S. BLANCHARD,JR.
A.J. MEYER,JR.   A.B. KEHLET
R.G. CHILTON   J.B. HAMMACK
        C.C. JOHNSON, JR.
BY
                              ATTORNEYS

8.10 This patent drawing of a space capsule was developed by the National Aeronautics and Space Administration (NASA) in 1963.

duce the item. Drawings should carry figure numbers and contain part numbers for text references (**Figure 8.10**).

Claims   The inventor's claims are a brief description of the invention's features that distinguish it from already patented items. The USPTO studies claims to judge the novelty and patentability of an invention.

Fee   As part of the application for a patent, the inventor must submit a $790 filing fee. Additional fees can be charged based on additional specifications and claims. An issue fee of $1320 is payable when the USPTO grants the patent. As a general rule, patents cost the inventor about $4000–$6000, excluding attorney's fees.

8.8 Patent Drawings

A booklet, *Guide for Patent Draftsmen* (available from the U.S. Government Printing Office), outlines the required format for patent drawings. If the inventor cannot furnish drawings, the USPTO will recommend a drafter who can prepare them at the inventor's expense.

Patent Drawing Standards

Patent drawings must meet the following standards.

Paper and Ink Drawings must be on pure white paper of the thickness of a two- or three-ply Bristol board with a surface that is calendared and smooth to permit erasure and correction. India ink is required for permanence and solid black lines. The use of white pigment to cover errors is not allowed.

Sheet Size and Margins Sheet size must be 8–1/2 × 11 inches or 21.0 × 29.7 cm (DIN size A4). One of the shorter sides is regarded as the top of the sheet. The sheets should not have frames drawn around the usable surface but should have cross hairs (target points) printed on two catecorner margins.

The following specifiations are required:

For 8–1/2 × 11 inch sheets, there must be a top margin of at least 1 inch, a left margin of at least 1 inch, a right margin of at least 5/8 inch, and a bottom margin of at least 3/8 inch, there by enclosing an area no greater than 6–15/16 × 9–5/8 inches.

For 21.0 × 29.7 cm sheets, there must be a top margin of at least 2.5 cm, a left margin of at least 2.5 cm, a right margin of at least 1.5 cm, and a bottom margin of at least 1.0 cm, thereby enclosing an area no greater than 17.0 cm × 26.2 cm.

All sheets in must be the same size and no holes should be made in the sheets (binder holes or staple holes).

Hatching and Shading Hatching lines used to shade the surface of an object should be par-

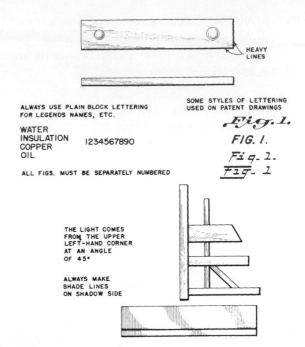

8.11 These are typical examples of lines and lettering recommended for patent drawings.

allel and at least 1/20 inch apart (**Figure 8.11**). Heavy lines are used on the shaded side of the views if they do not confuse the drawing. The light is assumed to come from the upper left-hand corner at an angle of 45°. **Figure 8.12** depicts several types of surface delineation.

Scale The scale must be large enough to show the mechanism without crowding when the drawing is reduced for reproduction. Portions of the mechanism may be drawn at a larger scale to show details.

Reference Characters The drafter should identify different views of a mechanism by consecutive plain, legible numerals at least 1/8-inch high, not encircled, and placed close to their parts. A blank space should be provided on hatched surfaces if numbers are to be placed on them. The same part appearing in more than one view on the drawing should be labeled with the same numeral.

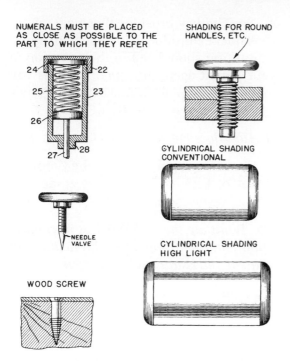

NUMERALS MUST BE PLACED AS CLOSE AS POSSIBLE TO THE PART TO WHICH THEY REFER

24 22
25 23
26
27 28

SHADING FOR ROUND HANDLES, ETC.

CYLINDRICAL SHADING CONVENTIONAL

NEEDLE VALVE

CYLINDRICAL SHADING HIGH LIGHT

WOOD SCREW

8.12 Several techniques of representing surfaces and beveled planes may be used on patent drawings.

Symbols　Symbols used to represent various materials in sections, electrical components, and mechanical devices are recommended by the USPTO and conform to engineering drawing standards.

Signature and Names　The signature or name of the applicant and the signature of the attorney or agent are placed in the lower right-hand corner of each sheet within the marginal lines or below the lower marginal line.

Views　Figures should be numbered consecutively in order of their appearance. Figures may be plan, elevation, section, perspective, or detail views. Exploded views may be used to describe an assembly of multiple parts. Large parts may be broken into sections and drawn on several sheets if this approach is not confusing. Removed sections may be used if the cutting plane is labeled to indicate the

section by number. All sheet headings and signatures are to be placed in the same position on the sheet whether the drawing is read from the bottom or the right of the sheet. Completed drawings should be sent flat, protected by heavy board, or rolled in a suitable mailing tube.

8.9 Patent Searches

A patent can be granted only after USPTO examiners have searched existing patents to verify that the invention has not been patented previously. With more than 6,000,000 patents on record, a search is the most time-consuming part of obtaining a patent. Most inventors employ patent attorneys or agents to do preliminary searches for possible infringement on other patents.

8.10 Questions and Answers

We used the PTO's pamphlet, *Questions and Answers About Patents,* to answer the following questions. Additional information can be found on the web at **www.uspto.gov.**

Nature and Duration of Patents
1. **Q.** *What is a patent?*

   A. A patent is a grant issued by the U.S. Government, giving an inventor the right to exclude all others from making, using, or selling his or her invention within the United States, its territories, and possessions.

2. **Q.** *For how long is a patent granted?*

   A. Twenty years from the date on which an application is filed; except for patents on ornamental designs, which are granted for terms of 3-1/2, 7, or 14 years.

3. **Q.** *May the term of a patent be extended?*

   A. Only by a special act of Congress, which occurs rarely and only under exceptional circumstances.

4. Q. *Does the person granted the patent have any control over the patent after it expires?*

A. No. Anyone has the right to use an invention covered in an expired patent so long as they do not use features covered in other unexpired patents.

5. Q. *On what subject matter may a patent be granted?*

A. A patent may be granted to the inventor or discoverer of any new and useful process, machine, manufacture, or composition of matter, or any new and useful improvement thereof, or on any distinct and new variety of plant, or on any new, original, and ornamental design for an article of manufacture.

6. Q. *What may not be patented?*

A. A patent may not be granted on a useless device, on printed matter, on a method of doing business, on an improvement in a device that would be obvious to a person skilled in the art, or on a machine that will not operate, particularly on alleged perpetual motion machines.

7. Q. *What do "patent pending" and "patent applied for" mean?*

A. They are used by a manufacturer or seller of an article to indicate that a patent application for that article is on file with the U.S. Patent and Trademark Office. Those using these terms falsely to deceive the public can be fined.

8. Q. *I have made some changes and improvements in my invention after my patent application was filed with the USPTO. May I amend my patent application by adding a description or illustration of these features?*

A. No. The law provides that new matter shall not be introduced into a patent application. You should call to the attention of your patent agent any such changes you may make, or plan to make, so steps may be taken for your protection.

9. Q. *How does someone apply for a patent?*

A. By making application to the Commissioner of Patents, Patent and Trademark Office, Washington, DC, 20231.

10. Q. *What are the USPTO's fees in connection with the filing of an application for patent and issuance of the patent?*

A. A filing fee of $790 plus certain additional charges for claims, depending on their number and the manner of their presentation, are required when the application is filed. An issue fee of $1320 plus certain printing charges are required if the patent is to be granted.

11. Q. *Are models required as a part of the application?*

A. Only in the exceptional cases. The USPTO has the authority to require that a model be submitted, but rarely exercises it.

12. Q. *Is it necessary for me to go to the USPTO in Washington to transact business concerning patent matters?*

A. No. Most business is conducted by correspondence. Interviews regarding pending applications can be arranged with examiners if necessary and often are helpful.

13. Q. *Can the USPTO give me advice about whether to apply for a patent?*

A. No. It can only consider the patentability of an invention when an application comes before it.

14. Q. *Is there any danger that the USPTO will give others information contained in my application while it is pending?*

A. No. All patent applications are kept secret until the patent is issued. After the patent is issued, the USPTO file containing the application and all correspondence leading to its issuance is made available in the Patent Office Search Room to anyone, and copies may be purchased from the USPTO.

15. Q. *May I write to the USPTO about my application after it is filed?*

A. The USPTO will answer your inquiries about the status of the application and indicate whether the application has been rejected, allowed, or is awaiting action. However, you should forward correspondence through your patent attorney or agent.

16. Q. *What happens when two inventors apply separately for a patent on the same invention?*

A. The PTO declares an "interference" and requires that testimony be submitted to determine which inventor is entitled to the patent.

8.13 Thomas Alva Edison, the most important inventor in history, was granted over 1,000 patents.

17. Q. *May applications be examined out of their regular order?*

A. No. All applications are examined in the order in which they are filed, except under special conditions.

When to Apply for a Patent

18. Q. *I have been making and selling my invention for the past thirteen months and have not filed any patent application. Is it too late for me to apply?*

A. Yes. A patent may not be obtained if the invention has been in public use or for sale in this country for more than a year prior to application. Your own use and sale of it for more than a year before filing will bar your right to a patent as though someone else had done so.

19. Q. *I published an article describing my invention in a magazine thirteen months ago. Is it too late to apply for a patent?*

A. Yes. The inventor is not entitled to a patent if the invention has been described in a printed publication anywhere in the world more than a year before filing an application.

20. Q. *If two or more people work together on an invention, to whom will the patent be granted?*

A. If each had a share in the ideas forming the invention, they are joint inventors and a patent will be issued to them jointly if an application is filed by them jointly. If one person provided all the ideas and the other has only followed instructions in making the device, the person contributing the ideas is the sole inventor and the patent application and patent should be in his or her name only.

21. Q. *If one person furnishes all the ideas for an invention and someone else employs* *that person or furnishes the money for building and testing the invention, should the patent application be filed by them jointly?*

A. No. The application must be signed, executed, sworn to, and filed in the name of the inventor, who is the person furnishing the ideas, not the employer or the person furnishing the money.

22. **Q.** *May a patent be granted if an inventor dies before filing an application?*

    A. Yes. The application may be filed by the executor or administrator of the inventor's estate.

23. **Q.** *While in England this summer, I found an ingenious article that has not been introduced into the United States or patented. May I obtain a U.S. patent on it?*

    A. No. A U.S. patent may be obtained only by the inventor, not by someone learning of someone else's invention.

24. **Q.** *May the inventor sell or otherwise transfer the right to the patent or patent application to someone else?*

    A. Yes. The inventor may sell all or part of the interest in the patent application or patent to anyone by a properly worded legal assignment. However, the application for a patent must be filed in the name of the inventor, not in the name of the purchaser.

25. **Q.** *Is it advisable to conduct a search of patents and other records before applying for a patent?*

    A. Yes. If the device has been patented previously, making an application is useless. A patent search avoids the expense of filing a needless application.

Technical Knowledge Available from Patents

26. **Q.** *May I obtain information through patents of what has been done by others to solve a particular problem?*

    A. The patents in the Patent Office Search Room in Washington contain a wealth of technical information. It is organized so that you can easily find and review previous work related to your problem or general field of interest. You may review these patents personally, or hire a patent practitioner to do so and send you copies of patents related to your problem.

27. **Q.** *Can I obtain information about patents and the patent process on the World Wide Web?*

    A. Yes. You may contact the Patent Office at **www.uspto.gov** to obtain information about patents and trademarks plus most answers to questions that you will have about patents. Although the patents are accessible by computer, the patent drawings are not.

28. **Q.** *If I obtain a patent on my invention, will that protect me against the claims of others who assert that I am infringing on their patents when I make, use, or sell my own invention?*

    A. No. There may be a patent of a more basic nature on which your invention is an improvement. If your invention is a detailed refinement or feature of a basically protected invention, you may not use it without consent of the patentee, just as no one will have the right to use your patented improvement without your consent.

29. **Q.** *Will the USPTO help me prosecute others if they infringe on the rights granted me by my patent?*

A. No. The USPTO has no jurisdiction over questions relating to the infringement of patent rights. If your patent is infringed upon, you may sue the infringer at your own expense.

Patent Protection in Foreign Countries

30. Q. *Does a U.S. patent give protection in foreign countries?*

A. No. The U.S. patent protects your invention only in this country. If you want to protect your invention in foreign countries, you must file an application in the patent office of each such country within the time required by law.

Problems

1. Prepare working and assembly drawings as the implementation step of the design process for one of the refinement problems at the end of Chapter 5 on 11 × 17-in. sheets of tracing vellum or film.

2. Prepare working and assembly drawings for one of the problems assigned or selected from those at the end of Chapter 23. Produce the drawings on 11 × 17-in. sheets of tracing vellum or film.

Patents

3. Write for a copy of a patent that is of interest to you. List the features used as a basis for the patent.

4. Suggest modifications to the patent obtained in Problem 3. Sketch innovations that would improve the patented mechanism.

5. Write to the U.S. Patent and Trademark Office for patent application forms. Prepare a patent application for a simple invention that has been previously patented, such as a fountain pen, drafting instrument, or similar item. Determine the drawings and materials needed to complete your application.

6. Make a list of ideas for products that you believe to be patentable.

7. Write a technical report on the history and significance of the patent system and its role in our industrial society. Consult your library and available government publications on patents.

8. The U.S. Patent and Trademark Office is a resource available to you for learning more about patents and obtaining free informational booklets that will guide you through the patent process. The regulations and laws governing patents is constantly undergoing change which requires that you stay in touch with the USPTO to remain current.

Access the USPTO on your computer, search their on-line offerings to become acquainted with the services available to you. They can be contacted by mail, fax, e-mail, or by telephone.

U. S. Patent and Trademark Office
General Information Services Division
Crystal Plaza 3, Room 2C02
P.O. Box 1450
Washington, D.C. 20231
Phone 800-786-9199 or 703-308-4357
Fax 703-305-7786
E-mail www.uspto.gov

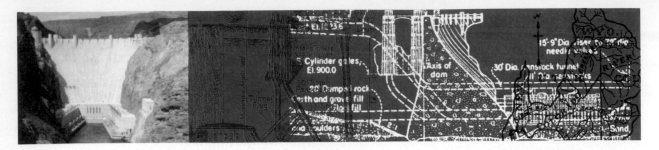

9

Design Problems

9.1 Introduction

This chapter offers problems that are suitable for both individual assignments and team projects to provide experience in applying the methods of creative problem solving presented in this textbook. Graphics have many applications during the design process: All new products begin with sketches at the preliminary idea step and end with documentation drawings at the implementation step.

9.2 The Individual Approach

The solution of short problems (one to two hours) is best suited to students working alone. Although simple design problems may involve fewer details and less depth than comprehensive problems, the same design steps are involved.

9.3 The Team Approach

An effectively organized team working on a problem represents more talent than the typical individual possesses. However, management of talent becomes as much of the process as solving the problem. It is essential that the team process be learned since most engineering activities in industry are done by teams.

Team Size

Student design teams should have from three to eight members. Three is the minimum number needed for a valid team experience, and four is the number needed to minimize the possibility of domination by one or two members.

Team Composition

In practice, an engineering team often consists of representatives of different departments or even different firms, who may be unacquainted. This situation can be advantageous because it reduces the impact of preconceived notions about individuals.

Team Leader

A leader is necessary for teams to function effectively. The leader is responsible for making assignments, ensuring that deadlines are met, and mediating disagreements.

9.4 Selection of a Problem

The best problem for a student design project is one that involves familiar and accessible conditions that can be observed, measured, and inspected. A design for a water-ski rack for an automobile is more feasible than is a design for a support bracket for an airplane. When a student design team selects a design problem, the team should prepare a written proposal identifying the problem and outlining its limits. Assignment of problems by the instructor in the classroom is analogous to assignments by a supervisor in the workplace.

9.5 Problem Specifications

An individual or team may be expected to complete any or all of the following tasks.

Short Problems (One or Two Hours)

1. Worksheets that record development of a design procedure (Chapter 2).

2. Freehand sketches of the design for implementation (Chapters 4 and 13).

3. Instrument drawings of the solution (Chapters 8 and 23).

4. Pictorial sketches (or drawings made with instruments or computer) illustrating the design (Chapters 4, 13, and 25).

5. Visual aids, flip charts, or other media for presentation to a group (Chapter 7).

Comprehensive Problems (40 to 100 Hours)

1. A proposal identifying the problem and outlining an approach for solving it (Chapters 2, 3, and 7).

2. Worksheets documenting the preliminary ideas for a solution (Chapter 4).

3. Schematic diagrams, flowcharts, or other graphics to illustrate refinements of the design (Chapter 5).

4. A market survey evaluating the product's possible acceptance and estimated profit (Chapters 3 and 6).

5. A model or prototype for analysis and/or presentation (Chapter 6).

6. Pictorials to illustrate features of the final design solution (Chapters 8 and 25).

7. Dimensioned working drawings and assembly drawings to give details and specifications (Chapters 8 and 25).

8. A written or oral report, illustrated with graphs and diagrams, to explain the method of solution and present conclusions and recommendations (Chapter 7).

9.6 Scheduling Team Activities

The semester schedule shown in **Figure 9.1** is suggested for a comprehensive design project. Spreading design projects over the semester allows time for thinking about the problem, gathering information, and working on the solution. Refer to the exercise bench example in Chapters 3 through 8 as a guide for carrying out your project.

9.7 Short Design Problems

The following short design problems can be completed in less than two hours.

1. **Lamp bracket.** Design a bracket to attach a desk lamp to a vertical wall for reading in bed. It should be removable for use as a conventional desk lamp.

2. **Towel bar.** Design a towel bar for a kitchen or bathroom. Determine optimum

	MONDAY	WEDNESDAY	FRIDAY
1			
2			
3		Assign teams	
4		Identify problem	
5		Identify problem	
6		Brainstorm	
7		Preliminary ideas	
8		Refinement	
9		Refinement	
10		Analysis	
11		Decision	
12		Implementation	
13		Prepare	
14		Present	
15			
16			

9.1 The semester schedule for the integration of a design project.

9.2 (Problem 6) Base redesign.

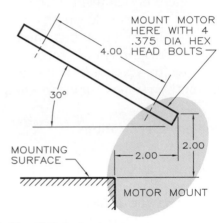

9.3 (Problem 8) Motor bracket.

size and consider styling, ease of use, and method of attachment.

3. **Bicycle rack.** Design a rack (or basket) for a bicycle for carrying books and class materials.

4. **Boot puller.** Design a device for helping you remove cowboy boots from your feet.

5. **Side-mounted mirror.** Design an improved side-mounted rearview mirror for an automobile. Consider aerodynamics, protection from inclement weather, visibility, and other factors.

6. **Part modification.** Modify the base that supports the 2" diameter shaft by changing the square base to a circular base with six holes instead of 4 (**Figure 9.2**). Modify the ribs accordingly.

7. **Pipe column support.** Design a base that can be attached to a concrete slab with bolts that would provide a base for 3-in. diameter pipe columns.

8. **Motor bracket (Figure 9.3).** Design a bracket to support a motor. The plate should be the upper part of the finished bracket.

9. **Pulley bracket clamp (Figure 9.4).** Design a clamp that can be attached to an overhead I-beam to support a pulley without welding or drilling holes in the beam.

10. **Pipe roll stand (Figure 9.5).** Design an alternative design to this roll stand that supports pipe up to 10 in. in diameter with a roller that allows expansion and contraction.

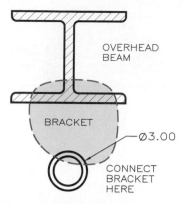

9.4 **(Problem 9)** Pulley bracket clamp.

9.5 **(Problem 10)** Pipe roll stand. (Courtesy of Grinnell Company, Inc.)

11. **Paint can holder.** Paint cans held by their wire bails are difficult to hold as well as to get paint brushes into. Design a holding device that can be attached and removed easily from a gallon-size paint can. Consider weight, grip, balance, and function.

12. **Foot scraper.** Design a device that can be attached to the sidewalk for scraping mud from your shoes.

13. **Audio cassette storage unit.** Design a storage unit for an automobile that will hold several audio cassettes, making them accessible to the driver but not to a thief.

14. **Slide projector elevator.** Design a device for raising a slide projector to the proper angle for projection on a screen. It may be part of the original projector or an accessory to be attached to existing projectors.

15. **Book holder 1.** Design a holder to support a book while reading in bed.

16. **Book holder 2.** Design a holder to support a textbook or reference book at a workstation for ease of reading and accessibility.

17. **Table leg design.** Do-it-yourselfers build a variety of tables using hollow doors or plywood for the tops and commercially available legs. Determine standard heights for various types of tables and design a family of legs that can be attached to table tops with screws.

18. **Leaf cart.** Design a cart or a container that leaves can be raked into while working in the yard and then transported to their final destination. Make it portable, light-weight, and economical.

19. **Pipe clamp (Figure 9.6).** A pipe with a 4-in. diameter must be supported by angles that are spaced 8 ft. apart. Design a clamp that will support the pipe without drilling holes in the angles.

20. **Toothbrush holder.** Design a toothbrush holder for a cup and two toothbrushes that can be attached to a bathroom wall.

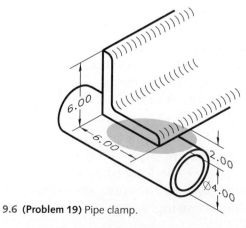

9.6 **(Problem 19)** Pipe clamp.

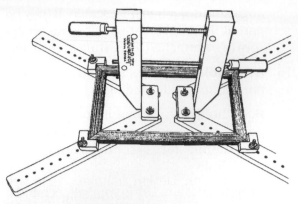

9.7 (**Problem 21**) Framing fixture. (Courtesy of Jorgensen Co.)

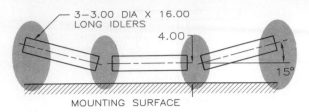

9.8 (**Problem 27**) Roller brackets.

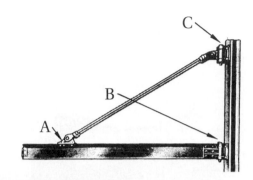

9.9 (**Problem 28**) Jib-crane bracket.

21. **Framing fixture. (Figure 9.7).** Design a device that can be used for the assembly of picture frames.

22. **Clothes hook.** Design a clothes hook that can be attached to a closet door for hanging clothes.

23. **Step stool.** Design a step stool that can be used for standing on to reach various heights as needed for home chores such as changing light bulbs.

24. **Hammock support.** Design a hammock support that will fold up and that will require minimal storage space.

25. **Door stop.** Design a door stop that can be attached to a wall or floor to prevent a door knob from hitting the wall.

26. **Basketball goal.** Design a basketball goal that is easy to install for the 8–10 age range.

27. **Roller brackets (Figure 9.8).** The rollers are to be positioned as shown to support a conveyor belt that carries bulk material. Design brackets to support the rollers.

28. **Jib-crane brackets.** Design the brackets at the joints indicated (**Figure 9.9 A**, **B**, and **C**) to form a jib-crane made of an 8 in. x 4 in. x 8 ft. long I-beam and a steel connecting rod. The crane should have at least a 180° swing.

29. **Drawer handle.** Design a handle for a standard file cabinet drawer.

30. **Paper dispenser.** Design a dispenser that will hold a 6 x 24-in. roll of wrapping paper.

31. **Handrail bracket.** Design a bracket that will support a tubular handrail to be used on a staircase.

32. **TV yoke.** Design a yoke to hang from a classroom ceiling that will support a TV set and that will permit it to be adjusted for viewing from various parts of the room.

33. **Flagpole socket.** Design a flagpole socket that is to be attached to a vertical wall.

34. **Sit-up bench (Figure 9.10).** Design a sit-up bench for exercising. Can you make it serve multiple purposes?

35. **Trashcan cover.** Design a functional lid with an appropriate opening through which to put garbage (**Figure 9.11**).

36. **Conduit connector hanger (Figure 9.12).** Design an attachment for a 3/4-in. con-

9.10 **(Problem 34)** Sit-up bench.

9.11 **(Problem 35)** Trashcan cover.

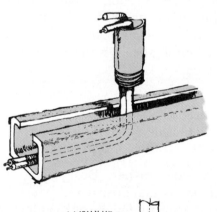

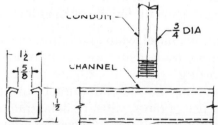

9.12 **(Problem 36)** Conduit connector hanger.

duit to support a channel used as a raceway for electrical wiring.

37. **Hose spool.** Design a spool/rack on which a garden hose can be wound and left neatly near the outside faucet.

38. **Gate hinge.** Design a hinge that can be attached to a 3 in. diameter tubular post to support a 3-ft wide wooden gate.

39. **Cup holder.** Design a holder that will support a soft-drink can or bottle in an automobile.

40. **Channel bracket (Figure 9.13).** The bracket shown is designed to fit on the flat side of the channel. Design a method of attaching the bracket so the inside nut will not drop down inside the channel.

9.8 Systems Design Problems

Systems problems require analysis of the interrelationship of various components as well as the application of design principles.

41. **Multipurpose utility meter.** Residences have meters for electricity, water, and gas, which are checked monthly by separate utility companies. Consider the feasibility of combining all the meters into a single

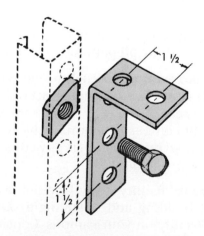

9.13 **(Problem 40)** Channel bracket.

unit that one meter-reading service could read. Describe briefly how to organize and implement such a system.

42. **Portable bleachers.** Design portable bleachers that can be easily assembled, disassembled, and stored. Identify the uses that would justify their production for the general market.

43. **Archery range.** Determine the feasibility of providing an archery range that can be operated profitably. Investigate the potential market for such a facility and factors such as location, equipment needed, method of operation, utilities, concessions, parking, costs, and fees.

44. **Bicycle-rental system.** Investigate the feasibility of a student-operated bicycle rental system. Determine student interest, cost factors, number of bikes needed, prices, personnel needs, storage, maintenance, and so on. Summarize your location, operating cost, and profitability conclusions.

45. **Model-airplane field.** Investigate the need for a model-airplane field, including space requirements, types of surfaces needed, sound control, safety factors, and method of operation. Select a site on or near your campus that is adequate for this facility, and evaluate the equipment, utilities, and site preparation required.

46. **Overnight campsite.** Analyze the feasibility of converting a vacant tract of land near a major highway into sites for overnight campers. Determine the facilities required by the campers as well as the venture's profitability.

47. **Skateboard facility study.** Determine the cost of building and operating a skateboard facility on your campus. Consider where it could be located, how many stu-

dents would use it, and the amounts of equipment and labor necessary to operate it. Would it be financially feasible?

48. **Hot water supply.** Your weekend cottage does not have a hot-water supply, but cold water is available from a private well. Design a system that uses the sun's energy in the summer to heat water for bathing and kitchen use. Devise a system for heating the water in the winter by some other source. Determine whether one or both of these systems could be made portable for showers on camping trips and, if so, how.

49. **Information center.** Design a drive-by information center to help campus visitors find their way around. Determine the best location for it and the informational material needed, such as slides, photographs, maps, sound, and other audio-visual aids.

50. **Golf driving range ball-return system.** Balls at golf driving ranges are usually retrieved by hand or with a specially designed vehicle. Design a system capable of automatically returning balls to the tee area.

51. **Car wash.** Design a car wash facility for your campus that would be self-supporting and would provide the basic needs for washing cars. Think simple and economical.

52. **Underside washer.** Design an attachment for a water hose that can be used for cleaning the underside of an automobile.

53. **Instant motel.** Many communities need temporary housing for celebrations and sporting events. Investigate methods of providing an "instant motel" involving the use of tents, vans, trailers, train cars, or other temporary accommodations. Estimate profitability.

54. **Mountain lodge.** Determine the food supply and other provisions needed for a mountain lodge to accommodate six people who might be snowed-in for two weeks. Identify and explain the features that should be included in the design of the lodge.

55. **Drive-in garbage system.** Design a system for disposing wrappers, boxes, and napkins left over after eating in your car at a hamburger drive-in. It would be preferred if this could be done without getting out of your car.

56. **Computer-wiring system.** Computer installations become cluttered with wiring as accessories are attached and access to connections becomes difficult. Design a system whereby the electrical wiring can be more organized, convenient, and accessible.

57. **Bonfire.** Develop a plan for building a bonfire to be burned the night before a major football game at your school. A few of the questions that must be answered are: Where will logs be obtained? How many people will be required to build it? Where can it be located? How can it be funded? What hazards must be controlled?

58. **Stadium expansion.** Study the attendance figures for your school's stadium (football, basketball, or baseball) and predict future attendance. Recommend whether to enlarge the stadium. If you recommend that it be enlarged, determine the quantity of seating needed and prepare a schedule for adding it.

59. **Shopping checkout system.** A problem for grocery stores and shopping malls is check out, bagging, payment, and delivery of purchases to customers' automobiles. Develop a system that will improve these conditions.

60. **Injury-proof playground.** Design a playground that permits the greatest degree of participation by children with the least risk of injury.

61. **Modification of an existing facility.** Select a facility on your campus or in your community that is inadequate, such as a street intersection, parking lot, recreational area, or classroom. Identify its deficiencies and propose improvements to it.

62. **Recreational facility.** Analyze the various recreational activities on your campus that could be improved with the construction of a multipurpose facility to accommodate activities such as movies, plays, sports, meetings, and dances. The facility should be designed as an outdoor installation with the minimum of structures.

63. **Educational toy production (Figure 9.14).** You are responsible for establishing the production system for manufacturing the educational toy shown. Perform an analysis on what would be needed to produce 500 units per month—space, raw materials, office facilities, manufacturing facilities, people, and warehousing; the workstations required; and so forth.

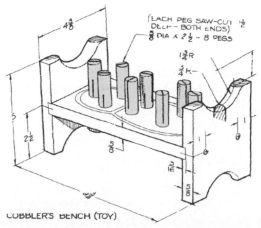

9.14 (Problem 63) Educational toy production.

Calculate the cost of producing the item and the selling price necessary to provide a profit.

64. **Patio table production.** Proceed as in Problem 63 for production of a plywood patio table. Determine the number to be manufactured per month to break even and the selling price. Establish quantity breaks for selling prices for quantities that exceed the break-even level.

9.9 Product Design Problems

Product design involves developing a device that will perform a specific function, be mass produced, and be sold to a large number of consumers.

65. **Hunting blind.** Design a portable hunting blind adequate for hunting geese or ducks that can be easily carried to its site. Consider making the blind of biodegradable materials so that it can be left at the site.

66. **Self-adjusting assembly jig.** A concept for an assembly jig that holds together parts for bonding thin pieces to thick pieces by furnace brazing is shown in **Figure 9.15**.

(A) Make the necesary drawings to explain the concept as shown.

(B) Design an alternative jig of your own.

67. **Mailbox.** Design a residential mailbox that either attaches to the house or that is supported on a pole at the street.

68. **Writing table arm.** Design a writing table arm that can be attached to a folding chair.

69. **Rescue litter (Figure 9.16).** Design a rescue litter than can be folded in a number of positions in order to care for the injured. It should be lightweight and collapsible for convenient storage.

70. **Yard helper (Figure 9.17).** Design a movable container that can be used for gardening and yard work.

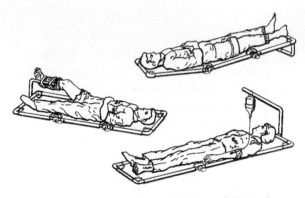

9.16 (Problem 69) Rescue litter. (Courtesy of National Aeronautics and Space Administration.)

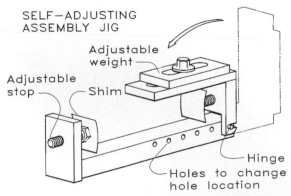

9.15 (Problem 66) Self-adjusting assembly jig. (Courtesy of National Aeronautics and Space Adminstration.)

9.17 (Problem 70) Yard helper.

71. **Computer mount.** Design a device that can be clamped to a desktop for holding a computer, permitting it to be adjusted to various positions while leaving the desktop free to work on.

72. **Workers' stilts.** Design stilts to give workers access to an 8-ft high ceiling, permitting them to nail 4 x 8-ft ceiling panels into position.

73. **Drum truck (Figure 9.18).** Design a truck that can be used for handling 55-gal drums of turpentine (7.28 lb per gallon) one at a time. Drums are stored in a vertical position and used in a horizontal position. The truck should be useful in tipping a drum into a horizontal position (as shown), as well as for moving the drum.

74. **Pole-vault uprights.** Pole-vault uprights must be adjusted for each vaulter by moving them forward or backward 18 in. The crossbar must be replaced at heights of over 18 ft by using poles and ladders. Develop a more efficient set of uprights that can be readily adjusted and allow the crossbar to be replaced easily.

75. **Sportsman's chair.** Design a sportsman's chair that can be used for camping, for fishing from the shore or from a boat, at sporting events, and for other purposes.

76. **Bicycle child carrier.** Design a seat that can be used to carry a small child as a passenger on a bicycle.

77. **Car washer.** Design a garden-hose attachment that can apply water and agitation to wash a car. Suggest other applications for this device.

78. **Universal drill jig.** A concept for a drill jig is shown in **Figure 9.19** that can be used to guide a drill bit at a variety of angles.

   (A) Make the necessary drawings to depict the design as it is given.

   (B) Design your own version of a drill jig of this type. Consider how drills of various diameters could be accommodated.

79. **Power lawn-fertilizer attachment.** The rotary power lawn mower emits a force caused by the rotating blades that might be used to distribute fertilizer during

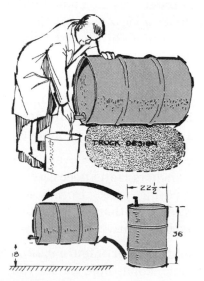

9.18 **(Problem 73)** Drum truck.

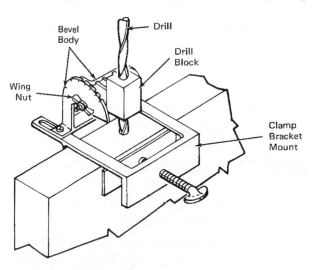

9.19 **(Problem 78)** Universal drill jig. (Courtesy of National Aeronautics and Space Administration.)

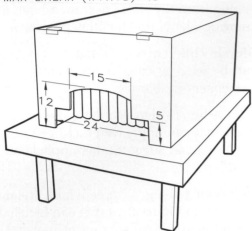

15x9 FOR BAGS
24x5 FOR HANGING BAGS
MAX LINEAR (W+H+D)=45

9.20 (Problem 80) Carry-on luggage template.

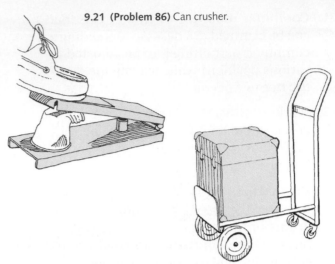

9.22 (Problem 90) Portable hauler.

mowing. Design an attachment for a lawn mower that can be used in this manner.

80. **Carry-on luggage template.** In an attempt to limit the size of carry-on luggage, airlines are testing designs for an effective template that can be used to screen luggage at check-in time. Design a template that would solve this problem using the specifications shown in **Figure 9.20**.

81. **Sawhorse.** Design a portable sawhorse for use in carpentry projects that folds up for easy storage. It should be about 36 in. long and 30 in. high.

82. **Monitor support.** Design a computer monitor arm to support and position the screen for ease of use.

83. **Projector cabinet.** Design a cabinet to serve as an end table or some other function while housing a slide projector and slide trays ready for use.

84. **Heavy appliance mover.** Design a device for moving large appliances—stoves, refrigerators, and washers—about the house for the purposes of rearranging,

cleaning, and servicing them.

85. **Car jack.** The average car jack does not attach itself adequately to the automobile's frame or bumper, which causes a safety problem. Design one that would employ a different method of lifting a car on various types of terrain.

86. **Can crusher (Figure 9.21).** Design a device for flattening aluminum cans.

87. **Map holder.** Design a map holder to give the driver a view of the map in a convenient location in the car while driving. Provide a method of lighting the map that will not distract the driver.

88. **Stump remover.** Design an apparatus that can be attached to a car bumper to remove dead stumps by pushing or pulling.

89. **Gate opener.** An annoyance to farmers and ranchers is the necessity of opening and closing gates. Design a manually operated gate that could be opened and closed by the driver from his vehicle.

90. **Portable hauler (Figure 9.22).** Design a portable, collapsible hauler that can be used for various home applications.

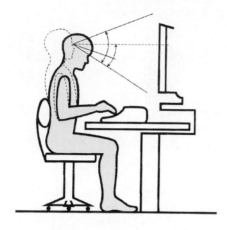

9.23 **(Problem 94)** Computer workstation.

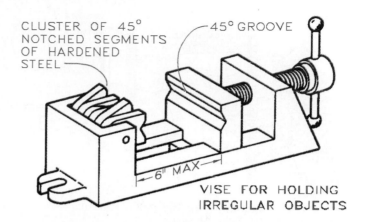

CLUSTER OF 45°
NOTCHED SEGMENTS
OF HARDENED
STEEL

45° GROOVE

6" MAX

VISE FOR HOLDING
IRREGULAR OBJECTS

9.24 **(Problem 97)** Specialty vise. (Courtesy of National Aeronautics and Space Administration.)

91. **Paint mixer.** Design a product for use at paint stores or by paint contractors to mix paint quickly in the store or on the job.

92. **Automobile coffee maker.** Design a device that will provide hot coffee from the dashboard of an automobile. Consider the method of changing and adding water, the spigot system, and similar details.

93. **Baby seat (cantilever).** Design a chair to support a child. It should attach to and be cantilevered from a standard table top. The chair also should be collapsible for ease of storage.

94. **Computer workstation (Figure 9.23).** Design a computer workstation that reduces user discomfort and fatigue.

95. **Miniature-TV support.** Design a device that would support TV sets ranging in size from 6 × 6 in. to 7 × 7 in. for viewing from a bed. Provide adjustments on the device to allow positioning of the set.

96. **Backpack.** Design a backpack that can be used for carrying camping supplies. Adapt the backpack to the human body for maximum comfort over extended periods of time. Suggest other uses for your design.

97. **Specialty vise.** A concept for a vise for holding irregular objects while they are being machined is shown in **Figure 9.24**.

 (A) Make the necessary drawings for explaining the design concept as given.

 (B) Design a solution of your own for this problem.

98. **Panel applicator.** A worker applying 4 × 8-ft plasterboard to a ceiling needs a helper to hold the panel in place while it is nailed. Design a device to hold the panel and eliminate the need for an assistant.

99. **Automobile controls.** Design driving controls that can be easily attached to the standard automobile to permit a car to be driven without using the legs.

100. **Bathing apparatus.** Design an apparatus that would help a wheelchair-bound person to get in and out of a bathtub without assistance from others.

101. **Adjustable TV base.** Design a base to support full-sized TV sets and allow maximum adjustment up and down and rotation about vertical and horizontal axes.

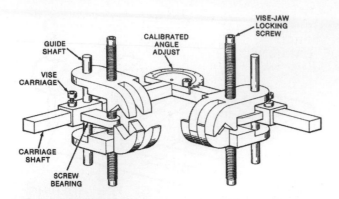

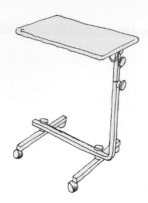

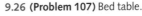

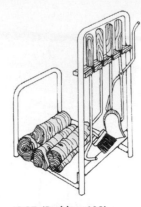

9.25 (Problem 102) Centering vise. (Courtesy of National Aeronautics and Space Administration.)

9.26 (Problem 107) Bed table.

9.27 (Problem 108) Firewood rack.

102. **Centering vise.** A concept for a vise that secures and aligns stainless steel tubing (up to 1" diameter) prior to joining by induction brazing is shown in **Figure 9.25**.

   (A) Make the necessary drawings to detail the concept shown.

   (B) Design a solution of your own.

103. **Log splitter.** Design a device to aid in splitting logs for firewood.

104. **Projector cabinet.** Design a portable cabinet that can remain permanently in a classroom to house a slide projector and/or a movie projector in a ready-to-use position. It should provide both convenience and security from theft.

105. **Projector eraser.** Design a device to erase grease-pencil markings from the acetate roll of a specially equipped overhead projector as the acetate is cranked past the stage of the projector.

106. **Cement mixer.** Design a portable cement mixer that a home owner can operate manually. Such mixers are used only occasionally, so it also should be affordable in order to make it marketable.

107. **Bed table (Figure 9.26).** Modify the design of a hospital table to permit it to be used for studying, computing, writing, and eating in bed.

108. **Firewood rack (Figure 9.27).** Design a rack for holding firewood inside near the fireplace. Can you give it multiple uses?

109. **Deer-hunting seat.** Design a deer-hunting seat that can be carried to the field and attached to a tree trunk.

110. **Punching bag platform.** Design a platform for a speed punching bag that is adjustable to various heights, is portable, and ships in a flat box.

111. **Boat trailer.** Design a trailer from which a boat hangs (rather than riding on top of the trailer) so that the boat can be launched in very shallow water.

112. **Washing machine.** Design a manually operated washing machine. It may be considered an "undesign" of an electrically powered washing machine.

113. **Pickup truck hoist.** Design a manual lift that can be attached to the tailgate of a pickup truck for raising and lowering loads from and to the bed of the truck.

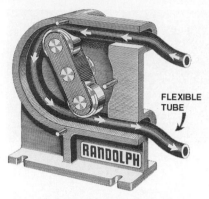

9.28 **(Problem 114)** Rotating pump.

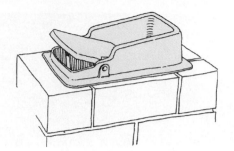

9.29 **(Problem 118)** Chimney cover.

114. **Rotating pump.** A portion of the Randolph pump is shown in **Figure 9.28**. Make the necessary drawings to complete the total design. The flexible tube is 0.50 in. in diameter.

115. **Display booth.** Design a portable display booth for use behind or on an 8-ft long table for displaying your company's name and product information. It must be collapsible so that it can be carried by one person as airplane luggage.

116. **Patio grill.** Design a portable charcoal grill for cooking on the patio. Consider how it would be cleaned, stored, and used. Study competing products already on the market.

117. **Shop bench.** Design an adjustable shop bench that is collapsible for easy storage. Design it to accommodate accessories such as vises, anvils, and electrical tools.

118. **Chimney cover (Figure 9.29).** Design a chimney cover that can be closed from inside the house for repelling rain and reducing temperature loss.

Competition Projects

Whereas design problems have a multitude of solutions, some design solutions can be measured in terms of the strongest, lightest, least expensive, and so forth. The following problems lend themselves to measurement of success and competition between classmates. In each case, the competitors should be limited to an equal supply of materials—paper, soda straws, toothpicks, glue, etc. so that the method of design is tested rather than the materials used. Sketches of the designs should be made before the competion so each can be compared with the results of the tests.

119. **The egg drop.** An often-used problem is the design of a protective encasement for an egg to prevent it from breaking when dropped from an upper-story window or staircase.

120. **Truss span.** Design a truss to span 12 inches between two tables. Gradually load each truss until it breaks. The load can be applied by adding water to a plastic jug attached at its midpoint until it collapses. The winner supports the most weight.

121. **Arched truss.** Same as in Problem 120 except the truss arch must span 12 inches horizontally between two tables and arch over a height of 6 inches above the table elevations at its midpoint.

122. **Powered sled.** Build a sled (no wheels) propelled by a specified type of rubberband that will move a prescribed object (baseball, tennis ball, etc.) across the floor. Determine the winner by the distance traveled.

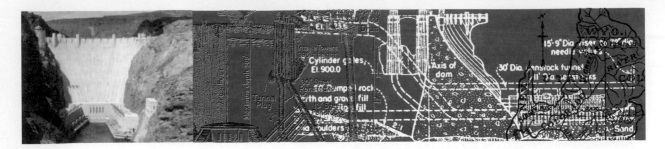

10

Drawing Instruments

10.1 Introduction

The preparation of technical drawings requires the ability to use drawing instruments. Even people with little artistic ability can produce professional technical drawings when they learn to use drawing instruments properly.

The drawing instruments covered in this chapter are traditional ones that are used by hand as opposed to computer instruments. Traditional instruments will always have an application in the development of drawings, but to a lesser degree than in the past.

Computer Instruments

Electronic drawing instruments—computers, plotters, scanners, and similar equipment—are covered in Chapter 37. The ability to use computer graphics and its associated hardware is a necessary skill for engineers because of its extensive use in producing graphics for engineering and technology.

10.2 Drawing Media

Pencils

A good drawing begins with the correct pencil grade and its proper use. Pencil grades range from the hardest, 9H, to the softest, 7B (**Figure 10.1**). The pencils in the medium-grade range, 4H-B, are used most often for drafting work of the type covered in this textbook.

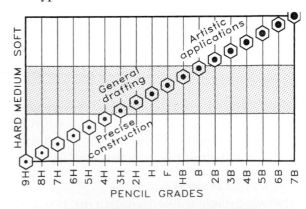

10.1 The hardest pencil lead is 9H, and the softest is 7B. The diameters of the hard leads are smaller than those of the soft leads.

A. LEAD HOLDER

Holds any size lead; point must be sharpened.

B. FINE—LINE HOLDER

Must use different size holder for different lead sizes; does not need to be sharpened.

C. WOOD PENCIL

Wood must be trimmed and lead must be pointed.

D. THE PENCIL POINT

Sharpen point to a conical point with a lead pointer or a sandpaper pad.

10.2 Sharpen the drafting pencil to a tapered conical point (not a needle point) with a sandpaper pad or other type of sharpener.

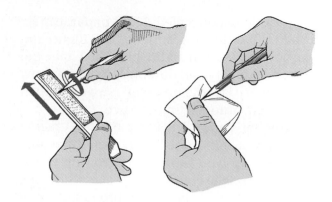

10.3 Revolve the drafting pencil about its axis while stroking the sandpaper pad to form a conical point. Wipe away the graphite from the point with a tissue.

Figure 10.2 shows three standard pencils used for drawing. The leads used in the lead holder shown in **Figure 10.2A** (the best all-around pencil of the three) are marked in white at their ends to indicate their grade.

The fine-line leads used in the lead holder in **Figure 10.2B** are more difficult to identify because their sizes are smaller and are not marked. A different fine-line holder must be used for each size of lead. The common sizes are 0.3 mm, 0.5 mm, and .007 mm, which are the diameters of the leads. A disadvantage of the fine-line pencil is the tendency of the lead to snap off when you apply pressure to it.

Although you have to sharpen it and point its lead, the wood pencil shown in **Figure 10.2C** is a very satisfactory pencil. The grade of the lead is marked on one end of the pencil, therefore, the opposite end should be sharpened so the identity of the grade of lead will be retained. The wood can be sharpened with a knife or a drafter's pencil sharpener to leave about 3/8 inch of lead exposed.

You must sharpen a pencil's lead properly to obtain a point, as shown in **Figure 10.2D**. To obtain a conical point, stroke the pencil lead against a sandpaper board and revolve the pencil about its axis in the process (**Figure 10.3**). Wipe excess graphite from the point with a cloth or tissue.

Although sharpening a pencil point with a sandpaper board may seem outdated, it is a very suitable and practical way to sharpen pencil points and about the only way to sharpen compass points. However, there are a multitude of pencil pointers available, some of which work better than others.

Papers and Films

Sizes Sheet sizes are specified by the letters A-F. These sizes are multiples of either the standard 8-1/2 x 11-inch sheet (used by engineers) or the 9 x 12-inch sheet (used by architects) (**Figure 10.4**). The metric sizes (A4-A0) are equivalent to the 8-1/2 x 11-inch modular sizes.

Detail Paper When drawings are not to be reproduced by the diazo or blue-line process,

	ENGINEERS'	ARCHITECTS'		METRIC		
A	11" X 8.5"	12" X 9"	A4	297	X	21(
B	17" X 11"	18" X 12"	A3	420	X	29;
C	22" X 17"	24" X 18"	A2	594	X	42(
D	34" X 22"	36" X 24"	A1	841	X	59<
E	44" X 34"	48" X 36"	A0	1189	X	84!

10.4 Standard sheet sizes vary by purpose.

an opaque paper, called **detail paper**, can be used as the drawing surface. The higher the rag content (cotton additive) of the paper, the better its quality and durability. You may draw preliminary layouts on detail paper and then trace them onto the final surface.

Tracing Paper Tracing paper, or tracing vellum, is a thin, translucent paper that permits light to pass through it, allowing reproduction by the blue-line process. Tracing papers that yield the best reproductions are the most translucent ones. Vellum is a chemically treated tracing paper to improve its translucency, but vellum does not retain its original quality as long as high-quality, untreated tracing papers do.

Tracing Cloth Tracing cloth is a permanent drafting medium used for both ink and pencil drawings. It is made of cotton fabric and is coated with a starch compound to provide a tough, erasable drafting surface that yields excellent blue-line reproductions. Tracing cloth does not change shape as much as tracing paper with variations in temperature and humidity. Repeated erasures do not damage the surface of tracing cloth.

Polyester Film An excellent drafting surface is polyester film, which is available under several trade names such as *Mylar*. It is more transparent, stable, and tougher than paper or cloth and is waterproof. Mylar film is used for both pencil and ink drawings. A plastic-lead pencil must be used with some films, whereas standard lead pencils may be used with others.

10.3 Drawing Equipment

Drafting Machine

Most professional drafters prefer the mechanical drafting machine (**Figure 10.5**), which is attached to the drawing table top and has fingertip controls for drawing lines at any angle. A modern, fully equipped drafting station is shown in **Figure 10.6**. Today, most offices are

10.5 The drafting machine is used for drawings made by hand. (Courtesy of Keuffel & Esser Company.)

10.6 The professional drafter or engineer may work in this type of environment. (Courtesy of Martin Instrument Company.)

equipped with computer graphics stations, which have replaced much of the manual equipment (**Figure 10.7**).

Triangles

The two types of triangles used most often are the 45° triangle and the 30°–60° triangle. The size of a 30°–60° triangle is specified by the longer of the two sides adjacent to the 90° angle (**Figure 10.8**). Standard sizes of 30°–60°

10.7 A typical engineering workstation used in industry. (Courtesy Jervis B. Webb Co.)

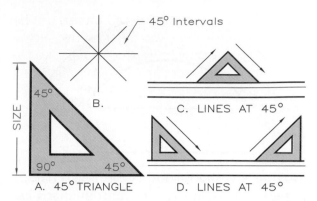

10.9 Use the 45° triangle to draw lines at 45° angles throughout 360°.

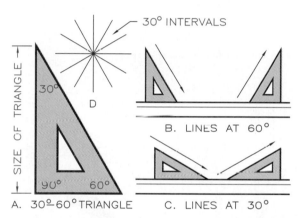

10.8 The 30°–60° triangle is used to draw lines at 30° intervals throughout 360°.

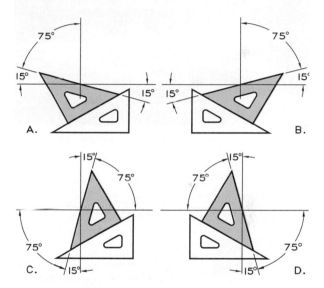

10.10 By using a 30°–60° triangle in combination with a 45° triangle, angles can be drawn at 15° intervals.

triangles range in 2-inch intervals from 4 to 24 inches.

The size of 45° triangle is specified by the length of the sides adjacent to the 90° angle. These range in 2-inch intervals from 4 to 24 inches, but the 6-inch and 10-inch sizes are adequate for most classroom applications. **Figure 10.9** shows the various angles that you may draw with this triangle. By using the 45° and 30°–60° triangles in combination, you may draw angles at 15° intervals throughout 360° (**Figure 10.10**).

Protractor

When drawing or measuring lines at angles other than multiples of 15°, a protractor is used (**Figure 10.11**). Protractors are available as semicircles (180°) or circles (360°). Adjustable triangles with movable edges that can be set at different angles with thumbscrews also are available.

Instrument Set

Figure 10.12 shows a cased set of some of the basic drawing instruments that are available individually as well. The three most important

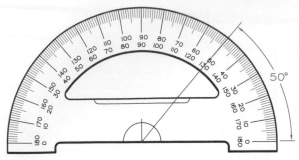

10.11 This semicircular protractor is used to measure angles.

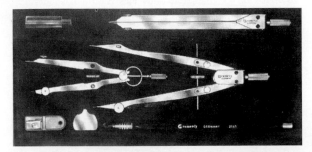

10.12 This is a typical cased set of drawing instruments. (Courtesy of Gramercy Guild.)

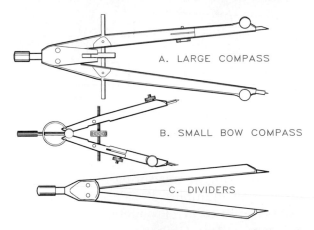

A. LARGE COMPASS

B. SMALL BOW COMPASS

C. DIVIDERS

10.13 Drawing instruments can be purchased individually or as a set.

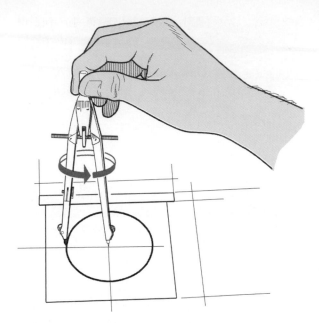

10.14 Use a compass for drawing circles.

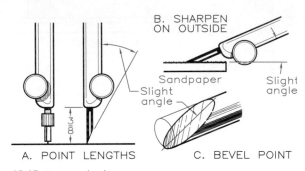

A. POINT LENGTHS

B. SHARPEN ON OUTSIDE

C. BEVEL POINT

10.15 Compass lead.

A Adjust the pencil point to be the same length as the compass point.

B and C Sharpen the lead from the outside with a sandpaper pad.

instruments for hand drawing are the large compass, small compass, and divider shown in **Figure 10.13**.

Compass Use the compass to draw circles and arcs in pencil or ink (**Figure 10.14**). To draw circles well with a pencil compass, sharpen the lead on its outside with a sandpaper board (**Figure 10.15**). A bevel cut of this type gives the best point for drawing circles. You cannot draw a thick arc in pencil with a single sweep; draw a series of thin concentric circles by adjusting the radius of the compass slightly.

When setting the compass pivot point in the drawing surface, insert it just enough for a firm set, not to the shoulder of the point. When the table top has a hard covering, place several sheets of paper under the drawing to provide a seat for the compass point.

Use a small bow compass (**Figure 10.16**) to draw small circles of up to 2 inches in radius. For larger circles, use an extension bar included in most sets to extend the range of the large bow compass. You may draw small circles conveniently with a circle template aligned with the centerlines of the circles (**Figure 10.17**).

Divider The divider looks like a compass without a drawing point. It is used for laying off and transferring dimensions onto a drawing. For example, you can step off equal divisions rapidly and accurately along a line (**Figure 10.18**). As you make each measure-

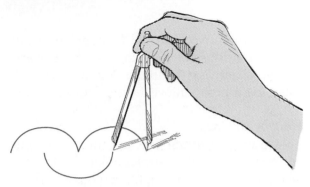

10.18 Use dividers to step off measurements.

ment, the divider's points make a slight impression mark in the drawing surface.

Also, use dividers to transfer dimensions from a scale to a drawing (**Figure 10.19**) or to divide a line into a number of equal parts. Bow dividers (**Figure 10.20**) are useful for transferring smaller dimensions, such as the spacing between lettering guidelines.

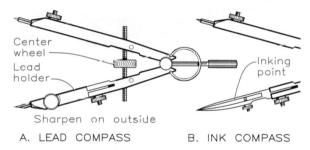

10.16 Use a small bow compass for drawing circles of up to 2-in. in radius in pencil or in ink.

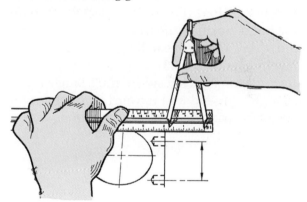

10.19 Use dividers to transfer dimensions from a scale to a drawing.

10.17 Circle templates are convenient for drawing small circles without a compass.

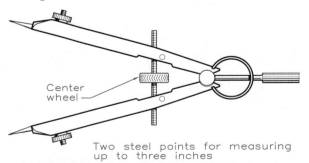

10.20 Bow dividers are used to transfer small dimensions such as spacing for guidelines for lettering.

10.4 Lines

The type of line produced by a pencil depends on the hardness of its lead, drawing surface, and your drawing technique. You must experiment in order to achieve the ideal combination.

Horizontal Lines

To draw a horizontal line, use the upper edge of your horizontal straightedge and make strokes from left to right, if you are right-handed (**Figure 10.21**), and from right to left if you are left-handed. Rotate the pencil about its axis so that its point will wear evenly. Darken pencil lines by drawing over them with multiple strokes. For drawing the best line, leave a small space between the straightedge and the pencil or pen point (**Figure 10.22**).

Vertical Lines

Use a triangle and a straightedge to draw vertical lines. Hold the straightedge firmly with one hand, position the triangle where needed, and draw the vertical lines with the other hand (**Figure 10.23**). Draw vertical lines upward along the left side of the triangle if you are right-handed and upward along the right side of the triangle if you are left-handed.

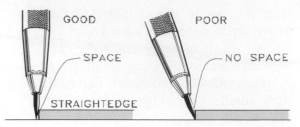

10.22 While drawing, hold the pencil or pen point in a plane perpendicular to the paper, leaving a space between the point and the straightedge.

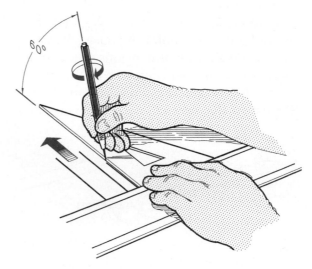

10.23 Draw vertical lines along the left side of a triangle (if you are right-handed) in an upward direction, holding the pencil or pen in a plane perpendicular to the paper and at 60° to the surface.

Irregular Curves

Curves that are not arcs must be drawn with an irregular curve (sometimes called French curves). These plastic curves come in a variety of sizes and shapes, but the one shown in **Figure 10.24** is typical. Here, we used the irregular curve to connect a series of points to form a smooth curve.

Erasing Lines

Always use the softest eraser that will do a particular job. For example, do not use ink erasers to erase pencil lines because ink erasers are

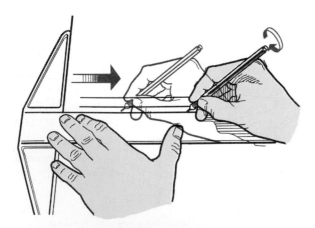

10.21 Draw horizontal lines along the upper edge of a straightedge while holding the pencil in a plane perpendicular to the paper and at 60° to the surface and rotate it about its axis.

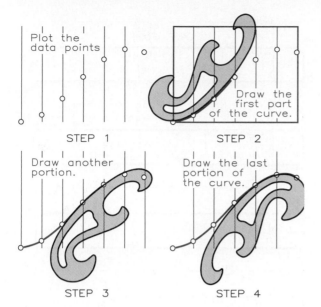

STEP 1

STEP 2

Plot the data points

Draw the first part of the curve.

STEP 3

STEP 4

Draw another portion.

Draw the last portion of the curve.

10.24 Using the irregular curve:

Step 1 Plot data points.

Step 2 Position the curve to pass through as many points as possible and draw that portion of the curve.

Step 3 Reposition the irregular curve and draw another portion of the curve.

Step 4 Draw the last portion to complete the curve.

coarse and may damage the surface of the paper. When working in small areas, you should use an erasing shield to avoid accidentally erasing adjacent lines (**Figure 10.25**). Follow erasing by brushing away the "crumbs" with a dusting brush. Wiping the crumbs away with your hands will smudge the drawing. A typical cordless electric eraser, shown in **Figure 10.26**, can be used with several grades of erasers to meet your needs.

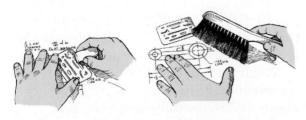

10.25 Use an erasing shield for erasing in tight spots. Use a brush, not your hand, to brush away the erasure crumbs.

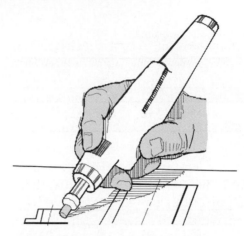

10.26 This cordless electric eraser is typical of those used by professional drafters.

10.5 Measurements

Scales

All engineering drawings require the use of scales for measuring lengths and sizes. Scales may be flat or triangular and are made of wood, plastic, or metal. **Figure 10.27** shows triangular architects', engineers', and metric scales. Most scales are either 6 or 12 inches long.

Architects' Scale

Drafters use architects' scales to dimension and scale features such as room-size, cabinets, plumbing, and electrical layouts. Most indoor measurements are made in feet and inches with the architects' scale. **Figure 10.28** shows how to indicate the scale you are using on a drawing. Place this scale designation in the title block or in a prominent location on the drawing. Because dimensions measured with the architects' scale are in feet and inches, you must convert all dimensions to decimal equivalents (all feet or all inches) before making calculations.

Use the 16 scale for measuring full-size lines (**Figure 10.29A**). An inch on the 16 scale is divided into sixteenths to match the ruler used by carpenters. The measurement shown

A. ARCHITECTS' SCALE

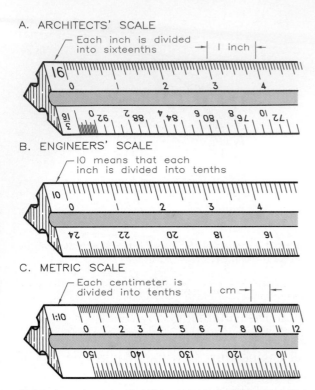

B. ENGINEERS' SCALE

C. METRIC SCALE

10.27 The architects' scale (A) measures in feet and inches. The engineers' scale (B) and the metric scale (C) are calibrated in decimal units.

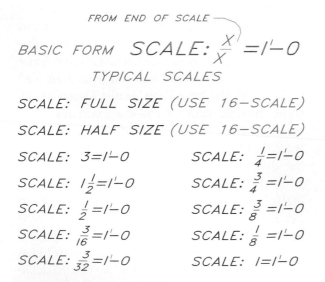

FROM END OF SCALE

BASIC FORM SCALE: $\frac{X}{X}=1'-0$

TYPICAL SCALES

SCALE: FULL SIZE (USE 16-SCALE)

SCALE: HALF SIZE (USE 16-SCALE)

SCALE: $3=1'-0$	SCALE: $\frac{1}{4}=1'-0$
SCALE: $1\frac{1}{2}=1'-0$	SCALE: $\frac{3}{4}=1'-0$
SCALE: $\frac{1}{2}=1'-0$	SCALE: $\frac{3}{8}=1'-0$
SCALE: $\frac{3}{16}=1'-0$	SCALE: $\frac{1}{8}=1'-0$
SCALE: $\frac{3}{32}=1'-0$	SCALE: $1=1'-0$

10.28 Use this basic form to indicate the scale on a drawing made with an architects' scale.

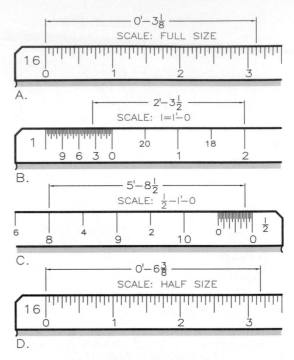

10.29 Lines measured with an architects' scale.

is 3-1/8". When the measurement is less than 1 ft, a zero may precede the inch measurements, with inch marks omitted, or 0'-3-1/2. (Inch marks are omitted since inches are understood to the units of measurement in the English system.)

Figure 10.29B shows the use of the 1=1'-0 scale to measure a line. Read the nearest whole foot (2 ft in this case) and the remainder in inches from the end of the scale (3-1/2 in.) for a total of 2'-3-1/2. Note that, at the end of each scale, a foot is divided into inches for use in measuring fractional parts of feet in inches. The scale 1"=1'-0 is the same as saying that 1 in. is equal to 12 in. or that the drawing is 1/12 the actual size of the object.

When you use the 1/2 = 1'-0 scale, 1/2 in. represents 12 in. on a drawing. Thus the line in **Figure 10.29C** measures 5'-8-1/2.

To obtain a half-size measurement, divide the full-size measurement by 2 and measure it with the 16 scale. Half-size is sometimes specified as SCALE: 6 = 12 (inch marks omitted). The line in **Figure 10.29D** measures to be 0'-6–3/8.

Letter dimensions in feet and inches are shown in **Figure 10.30**, with fractions twice as tall as whole numerals.

Engineers' Scale

On the engineers' scale, each inch is divided into multiples of 10. Because it is used for making drawings of outdoor projects— streets, structures, tracts of land, and other topographical features—it is sometimes called the civil engineers' scale.

Figure 10.31 shows the form for specifying scales when using the engineers' scale. For example, scale: 1 = 10'. With measurements already in decimal form, performing calculations is easy; there is no need to convert from one unit to another as when you use the architects' scale.

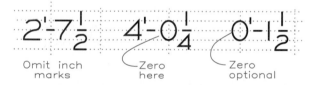

10.30 Omit inch marks but show foot marks (according to current standards). When the inch measurement is less than a whole inch, use a leading zero.

FROM END OF SCALE

BASIC FORM   SCALE: 1= XX
               EXAMPLE SCALES

10	SCALE: 1=1'	SCALE: 1=1,000
20	SCALE: 1=200'	SCALE: 1=20 LB
30	SCALE: 1=3'	SCALE: 1=3,000'
40	SCALE: 1=4'	SCALE: 1=40'
50	SCALE: 1=50'	SCALE: 1=500'
60	SCALE: 1=6	SCALE: 1=0.6'

10.31 Use this basic form to indicate the scale on a drawing made with an engineers' scale.

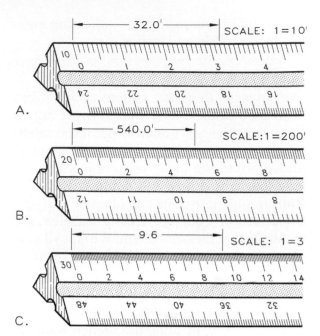

10.32 These lines are measured with engineers' scales.

Each end of the scale is labeled 10, 20, 30, and so on, which indicates the number of units per inch on the scale (**Figure 10.32**). You may obtain many combinations simply by mentally moving the decimal places of a scale.

Figure 10.32A shows the use of the 10 scale to measure a line 32.0 ft long drawn at the scale of 1 = 10'. **Figure 10.32B** shows the use of the 20 scale to measure a line 540.0 ft long drawn at a scale of 1 = 200'. **Figure 10.32C** shows the use of the 30 scale to measure a line of 9.6 in. long at a scale of 1 = 3. **Figure 10.33**

10.33 For decimal fractions in inches, omit leading zeros and inch marks. For feet, leave adequate space for decimal points between numbers and show foot marks.

shows the proper format for indicating measurements in feet and inches.

English System of Units

The English (Imperial) system of units has been used in the United States, Great Britain (until recently), and Canada since it was established. This system is based on arbitrary units (of length) of the inch, foot, cubit, yard, and mile. Because there is no common relationship among these units, calculations are cumbersome. For example, finding the area of a rectangle that measures 25 in. × 6-3/4 yd. first requires conversion of one unit to the other.

Metric System (SI) of Units

France proposed the metric system in the fifteenth century. In 1793, the French National Assembly agreed that the meter (m) would be one ten-millionth of the meridian quadrant of the earth and fractions of the meter would be expressed as decimal fractions. An international commission officially adopted the metric system in 1875.

The worldwide organization responsible for promoting the metric system is the **International Standards Organization (ISO)**. It has endorsed the Systéme International d'Unites (International System of Units), abbreviated **SI**. Prefixes to SI units indicate placement of the decimal, as **Figure 10.34** shows.

Value		Prefix	Symbol	Pronunciation
1 000 000 = 10^6	=	Mega	M	"Megah"
1 000 = 10^3	=	Kilo	k	"Keylow"
100 = 10^2	=	Hecto	h	"Heck tow"
10 = 10^1	=	Deka	da	"Dekah"
1 =				
0.1 = 10^{-1}	=	Deci	d	"Des sigh"
0.01 = 10^{-2}	=	Centi	c	"Cen'–ti"
0.001 = 10^{-3}	=	Milli	m	"Mill lee"
0.000 001 = 10^{-6}	=	Micro	μ	"Microw"

10.34 These prefixes and abbreviations indicate decimal placement for SI measurements.

Metric Scales

The meter is 39.37 inches. The basic metric unit of measurement for an engineering drawing is the millimeter (mm), which is one-thousandth of a meter, or one-tenth of a centimeter. Dimensions on a metric drawing are understood to be in millimeters unless otherwise specified.

The width of the fingernail of your index finger is a convenient way to approximate the dimension of one centimeter, or ten millimeters (**Figure 10.35**). Depicted in **Figure 10.36** is the format for specifying metric scales on a drawing.

Decimal fractions are unnecessary on most drawings dimensioned in millimeters. Thus dimensions are rounded off to whole numbers except for those measurements dimensioned with specified tolerances. For

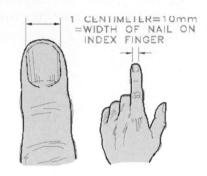

10.35 The nail width of your index finger is approximately equal to 1 centimeter, or 10 millimeters.

```
                            FROM END OF SCALE

BASIC FORM    SCALE:  I= XX
              EXAMPLE SCALES

SCALE: I:I    (Imm=Imm; Icm=Icm)

SCALE: I:2    (Imm=2mm;  Imm=20mm)

SCALE: I:3    (Imm=30mm; Imm=0.3mm)

SCALE: I:4    (Imm=4mm;  Imm=40mm)

SCALE: I:5    (Imm=5mm;  Imm=500mm)

SCALE: I:6    (Imm=6mm;  Imm=60mm)
```

10.36 Use this basic form to indicate the scale of a drawing made with a metric scale.

metric measurements of less than 1, a zero goes in front of the decimal. In the English system, the zero is omitted from measurements of less than an inch (**Figure 10.37**).

Metric scales are expressed as ratios: 1:20, 1:40, 1:100, 1:500, and so on. These ratios mean that one unit represents the number of units to the right of the colon. For example, 1:10 means that 1 mm equals 10 mm or 1 cm equals 10 cm or 1 m equals 10 m. The full-size metric scale (**Figure 10.38**) shows the relationship between the metric units of the decimeter, centimeter, millimeter, and micrometer. The line shown in **Figure 10.39A** measures 59 mm. Use the 1:2 scale when 1 mm represents 2 mm, 20 mm, 200 mm, and so on. The line shown in **Figure 10.39B** measures 106 mm. **Figure 10.39C** shows a line measuring 165 mm, where 1 mm represents 3 mm.

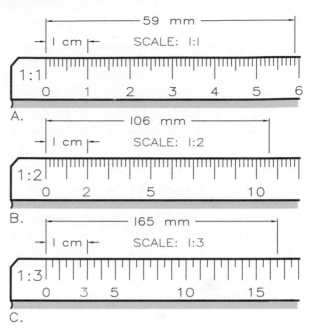

10.39 Measurements with metric scales.

10.37 In the metric system, a zero precedes the decimal. Allow adequate space for it.

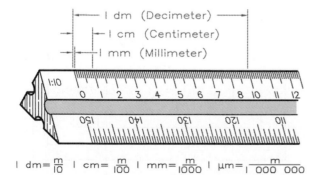

$1 \ dm = \frac{m}{10} \quad 1 \ cm = \frac{m}{100} \quad 1 \ mm = \frac{m}{1000} \quad 1 \ \mu m = \frac{m}{1\,000\,000}$

10.38 The decimeter is one tenth of the meter, the centimeter is one hundreth of a meter, a millimeter is one thousandth of a meter, and a micrometer is one millionth of a meter.

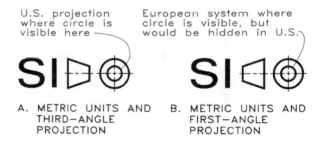

10.40 The large SI indicates that measurements are in metric units. The partial cones indicate whether the views are drawn in (A) the third-angle (U.S. system) or (B) the first-angle of projection (European system).

Metric Symbols To indicate that drawings are in metric units, insert SI in or near the title block (**Figure 10.40**). The two views of the partial cone denote whether the orthographic views were drawn in accordance with the U.S. system (third-angle projection) or the European system (first-angle projection). **Figure 10.41** shows several comparisons of English and SI units.

Expression of Metric Units The general rules for expressing SI units are given in

10.41 These scales show a comparison of the English system units with the metric system units.

Omit commas and group into threes	1 000 000 GOOD	1,000,000 POOR
Use a raised dot for multiplication	N • M GOOD	NM POOR
Precede decimals with zeros	0.72 mm GOOD	.72 mm POOR
Methods of division	kg/m or GOOD	kg • m^{-1} GOOD

10.42 Follow these general rules for showing SI units.

Figure 10.42. Do not use commas to separate digits in large numbers; instead, leave a space between them, as shown.

Scale Conversion

Appendix 1 gives factors for converting English to metric lengths and vice versa. For example, multiply decimal inches by 25.4 to obtain millimeters, and divide millimeters by 0.394 to obtain inches.

Multiply an architects' scale by 12 to convert it to an approximate metric scale. For example, scale: 1/8 = 1'-0 is the same as 1/8 in. = 12 in. or 1 in. = 96 in., which closely approximates the metric scale of 1:100. You cannot convert most met-

ric scales exactly to English scales, but the metric scale of 1:60 does convert exactly to 1 = 5', which is the same as 1 in. = 60 in.

10.6 Presentation of Drawings

The following formats are suggested for the presentation of drawings. Most problems can be drawn and solved on size A (8-1/2 x 11-inch) sheets in a vertical format with a title strip as **Figure 10.43** shows. Size A sheets can also be layed out in a horizontal format as **Figure 10.44** shows.

Figure 10.44 shows the standard sizes of sheets, from size A through size E and an alternative title strip for sizes A through E. Always use guidelines for lettering title strips.

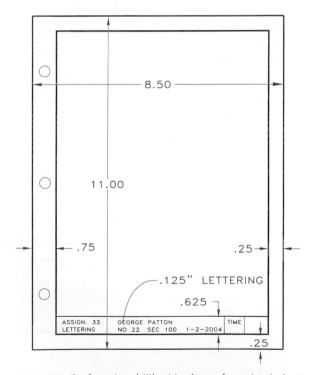

10.43 Use the format and title strip shown for a size A sheet (vertical format) to present the solutions to problems at the end of each chapter.

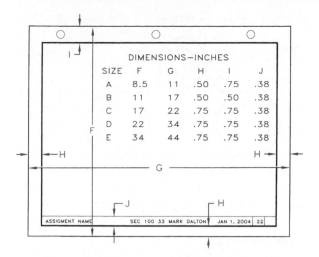

10.44 This is a horizontal format for size A through size E sheets. The numbers in columns F–J are the various layout dimensions for size A through size E sheets.

DIMENSIONS—INCHES					
SIZE	F	G	H	I	J
A	8.5	11	.50	.75	.38
B	11	17	.50	.50	.38
C	17	22	.75	.75	.38
D	22	34	.75	.75	.38
E	34	44	.75	.75	.38

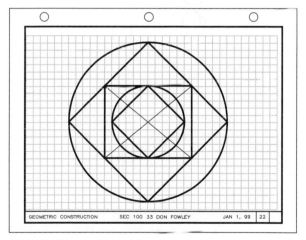

10.45 Problem 1 and sheet format.

Problems

Problems 1–9 (Figures 10.45–10.47): Draw these problems on size A plain sheets or paper with a printed grid, using the format shown in **Figure 10.45**. You may be assigned to solve two half-size drawings per sheet on size A sheets. Each grid represents 0.25 in. or 6 mm.

Problems 10–19 (Figures 10.48–10.57): Draw full-size views on size A sheets (horizontal format) and omit the dimensions.

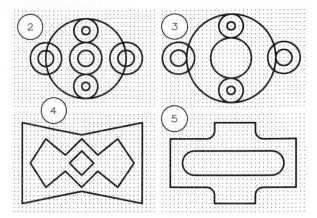

10.46 Problems 2-5.

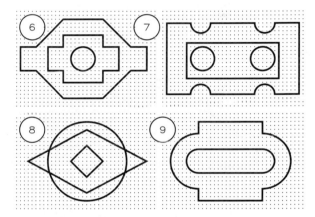

10.47 Problems 6-9.

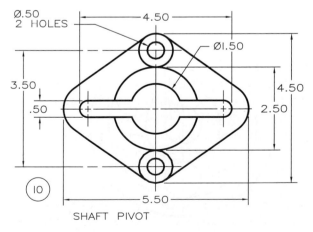

SHAFT PIVOT

10.48 Problem 10.

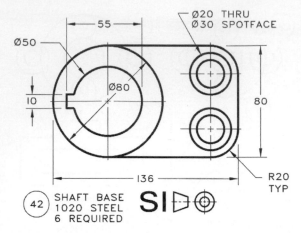

Ø20 THRU
Ø30 SPOTFACE

Ø50

55

Ø80

10

80

136

R20
TYP

(42) SHAFT BASE
1020 STEEL
6 REQUIRED

SI

10.49 Problem 11.

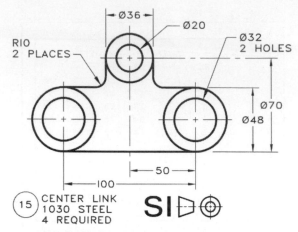

Ø36

Ø20

Ø32
2 HOLES

RIO
2 PLACES

Ø70

Ø48

50

100

(15) CENTER LINK
1030 STEEL
4 REQUIRED

SI

10.52 Problem 14.

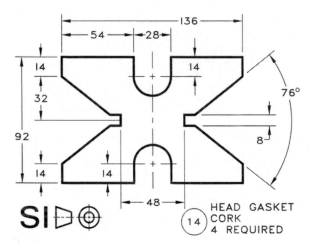

136

54

28

14

14

14

32

76°

92

8

14

14

48

HEAD GASKET
CORK
(14) 4 REQUIRED

SI

10.50 Problem 12.

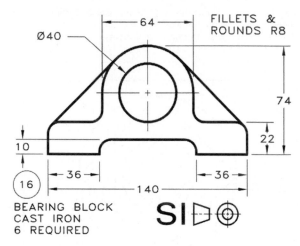

64

FILLETS &
ROUNDS R8

Ø40

74

22

10

36

36

(16)

140

BEARING BLOCK
CAST IRON
6 REQUIRED

SI

10.53 Problem 15.

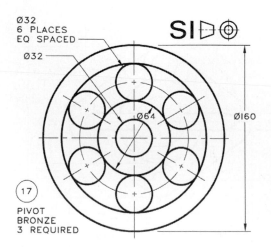

Ø32
6 PLACES
EQ SPACED

SI

Ø32

Ø64

Ø160

(17)

PIVOT
BRONZE
3 REQUIRED

10.51 Problem 13.

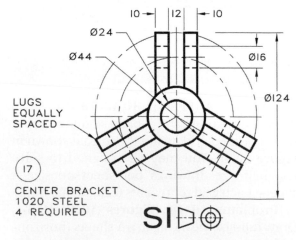

10

12

10

Ø24

Ø44

Ø16

Ø124

LUGS
EQUALLY
SPACED

(17)

CENTER BRACKET
1020 STEEL
4 REQUIRED

SI

10.54 Problem 16.

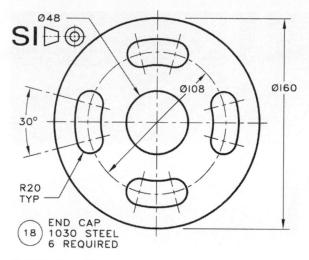

SI ⊟⊚ Ø48

Ø108

Ø160

30°

R20
TYP

(18) END CAP
1030 STEEL
6 REQUIRED

10.55 Problem 17.

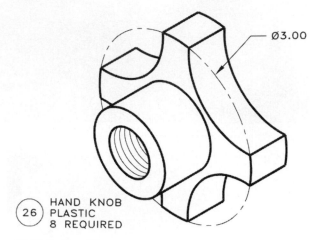

Ø3.00

(26) HAND KNOB
PLASTIC
8 REQUIRED

10.57 Problem 19.

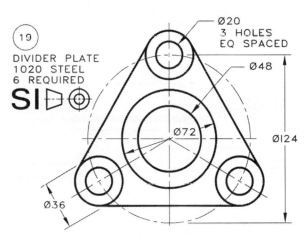

(19)

DIVIDER PLATE
1020 STEEL
6 REQUIRED

SI ⊟⊚

Ø20
3 HOLES
EQ SPACED

Ø48

Ø72

Ø124

Ø36

10.56 Problem 18.

Design 2

Make an instrument drawing of the film reel in **Figure 10.58** as it is shown, or design your own geometry while adhering to the overall dimensions.

Design Application
Design 1

The knob in **Figure 10.57** has a diameter of 3 in. (flat-to-flat). Make an instrument drawing of the descriptive and of this part; estimate the dimensions as if you were its designer .

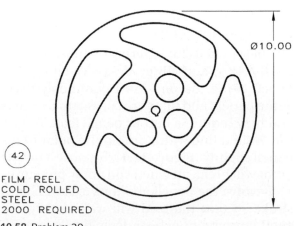

Ø10.00

(42)

FILM REEL
COLD ROLLED
STEEL
2000 REQUIRED

10.58 Problem 20.

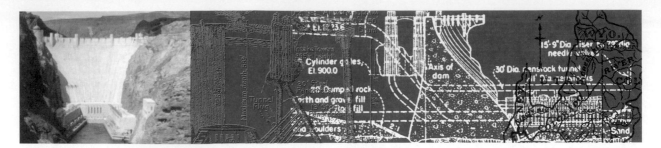

11

Lettering

11.1 Introduction

Notes, dimensions, and specifications, which must be lettered, supplement all drawings. The ability to letter freehand is an important skill to develop because it affects the use and interpretation of drawings. It also displays an engineer's skill with graphics, and may be taken as an indication of professional competence.

11.2 Lettering Tools

The best pencils for lettering on most surfaces are the H, F, and HB grades, with an F grade pencil being the one most commonly used. Some papers and films are coarser than others and may require a harder pencil lead. To give the desired line width, round the point of the pencil slightly (**Figure 11.1**), because a needle point will break off when you apply pressure.

Revolve the pencil about its axis with each stroke so that the lead will wear evenly. For good reproduction, bear down firmly to make

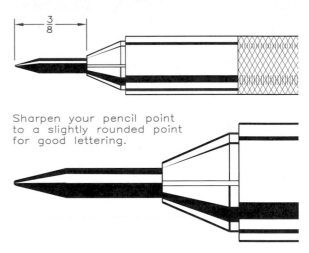

11.1 Good lettering begins with a properly sharpened pencil point. The F grade pencil is good for lettering.

letters black and bright with a single stroke. Prevent smudging while lettering by placing a sheet of paper under your hand to protect the drawing (**Figure 11.2**).

132

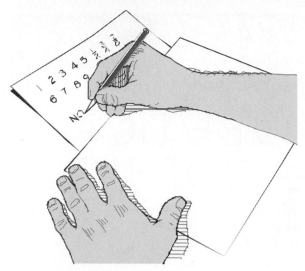

11.2 Place a protective sheet under your hand to prevent smudges and work from a comfortable position for natural strokes.

11.3 Guidelines

The most important rule of lettering is to use guidelines at all times, whether you are lettering a paragraph or a single letter. **Figure 11.3** shows how to draw and use guidelines. Use a sharp pencil in the 2H-4H grade range and draw light guidelines, just dark enough to be seen.

Most lettering on an engineering drawing is done with capital letters that are 18 inches (3 mm) high. The spacing between lines of lettering should be no closer than half the height of the capital letters, or 1/16 inch in this case.

Lettering Guides

Two instruments for drawing guidelines are the **Braddock-Rowe** lettering triangle and the **Ames** lettering instrument.

The Braddock-Rowe triangle contains sets of holes for spacing guidelines (**Figure 11.4**). The numbers under each set of holes represent thirty-seconds of an inch. For example, the numeral 4 represents 4/32 inch or 1/8 inch for making uppercase (capital) letters. Some triangles have millimeter markings. Intermediate holes provide guidelines for

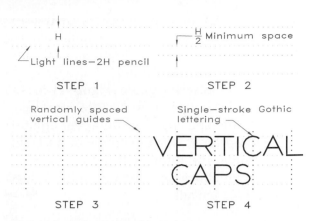

11.3 Method of using lettering guidelines:

Step 1 Lay off letter heights, H, and draw light guidelines with a 2H pencil.

Step 2 Space lines no closer than H/2 apart.

Step 3 Draw vertical guidelines as light, thin, randomly spaced lines.

Step 4 Draw letters with single strokes using a medium-grade pencil: H, F, or HB. Leave the guidelines.

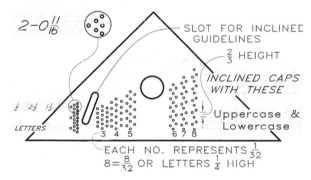

11.4 The Braddock-Rowe triangle is also used for drawing guidelines for lettering. The numbers near the guideline holes represent thirty-seconds of an inch.

lowercase letters, which are not as tall as capital letters.

While holding a horizontal straightedge firmly in position, place the Braddock-Rowe triangle against its upper edge. Insert a sharp 2H pencil point in the desired guideline hole to contact the drawing surface and guide the pencil point across the paper, drawing the guideline while the triangle slides along the straightedge. Repeat this procedure by moving the pencil point to each successive hole to draw other guidelines. Use the slanted slot in the triangle to draw guidelines, spaced randomly, for inclined lettering.

The **Ames lettering guide** (**Figure 11.5**) is a similar device but has a circular dial for selecting guideline spacing. The numbers around the dial represent thirty-seconds of an inch. For example, the number 8 represents 8/32 inch, or guidelines for drawing capital letters that are 1/4 inch tall.

11.4 Gothic Lettering

The lettering recommended for engineering drawings is single-stroke Gothic lettering, so called because the letters are a variation of the Gothic style made with a series of single strokes. Gothic lettering may be vertical or inclined (**Figure 11.6**), but only one style or the other should be used on a single drawing.

Vertical Letters

Uppercase Figure 11.7 shows the alphabet in the single-stroke Gothic uppercase (capital) letters. Each letter is drawn inside a square

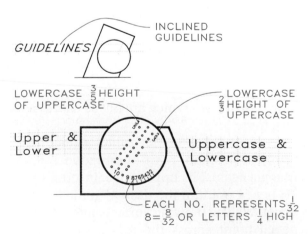

11.5 The Ames lettering guide is used for drawing guidelines for lettering. Set the dial to the desired number of thirty-seconds of an inch for the height of uppercase letters.

11.6 Use either vertical or inclined single-stroke Gothic lettering on engineering drawings.

11.7 This alphabet shows the form of single-stroke Gothic vertical uppercase letters. Letters are drawn inside squares to show their proportions.

box of guidelines to show their correct proportions. Draw straight lines with a single stroke; for example, draw the letter A with three single strokes. Letters composed of curves can best be drawn in segments; for example, draw the letter O by joining two semicircles.

The shape (form) of each letter is important. Small wiggles in strokes will not detract from your lettering if the letter forms are correct. **Figure 11.8** shows common errors in lettering that you should avoid.

LETTERS POOR

A. LETTERS POORLY DONE

LETTERS THIN

B. STROKES TOO THIN

TOO HEAVY

C. STROKES TOO THICK

TOO LIGHT

D. BLACKER—BEAR DOWN WITH F PENCIL

11.8 Avoid these common errors when lettering.

Lowercase   The alphabet of lowercase letters, shown in **Figure 11.9**, should be either two-thirds or three-fifths as tall as uppercase letters. Both lowercase ratios are labeled on the Ames guide, but only the two-thirds ratio is available on the Braddock-Rowe triangle.

Some lowercase letters, such as the letter b, have ascenders that extend above the body of the letter; some, such as the letter p, have descenders that extend below the body. Ascenders and descenders should be equal in length.

The guidelines in **Figure 11.9** that form squares about the body of each letter are used to illustrate their proportions. Letters that have circular bodies may extend slightly beyond the sides of the guideline squares. **Figure 11.10** shows examples of capital and lowercase letters used together.

Numerals   Vertical numerals for use with single-stroke Gothic lettering are shown in **Figure 11.11**; each numeral is enclosed in a square of guidelines. Numbers should be the same height as the capital letters being used, usually 1/8 inch. The numeral 0 (zero) is an oval, whereas the letter O is a circle in vertical lettering.

Inclined Letters

Uppercase   Inclined uppercase (capital) letters have the same heights and proportions as vertical letters; the only difference is their 68°

A. Lowercase $\frac{2}{3}$ height of caps

B. Lowercase $\frac{3}{5}$ height of caps   From Ames guide

11.10 The ratio of lowercase letters to uppercase letters should be either two-thirds (A) or three-fifths (B). The Ames guide has both ratios, but the Braddock-Rowe triangle has only the two-thirds ratio.

11.9 The alphabet is drawn here in single-stroke Gothic vertical lowercase letters.

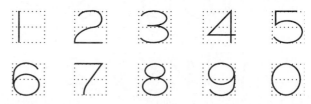

11.11 These numerals are used with single-stroke Gothic vertical letters.

11.12 This is an alphabet of single-stroke Gothic inclined uppercase letters.

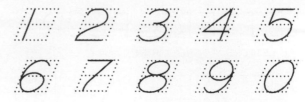

11.14 Single-stroke Gothic inclined numerals.

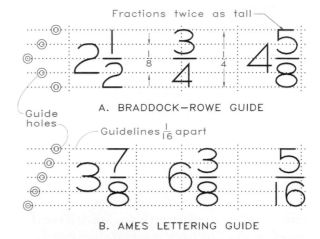

A. BRADDOCK–ROWE GUIDE

B. AMES LETTERING GUIDE

11.15 Inclined common fractions are twice as tall as single numerals. Omit inch marks in dimensions.

inclination (**Figure 11.12**). Guidelines for inclined lettering can be drawn with both the Braddock-Rowe triangle and the Ames guide.

Lowercase Inclined lowercase letters are drawn in the same manner as vertical lowercase letters (**Figure 11.13**), but circular features are drawn as ovals (ellipses). The angle of inclination is 68°, the same as for uppercase letters.

Numerals Examples of inclined numerals and letters used in combination are shown in **Figures 11.14** and **11.15**. The ratio of lower-

11.13 This is an alphabet of single-stroke Gothic inclined lowercase letters.

case to uppercase letters should be either two-thirds (A) or three-fifths (B). The Ames guide has both ratios, but the Braddock-Rowe triangle has only the two-thirds ratio. Inclined letters and numbers are used in combination in **Figure 11.15**. Guidelines are drawn using the Braddock-Rowe triangle or the Ames lettering guide shown in **Figure 11.16**.

Spacing Numerals and Letters

Allow adequate space between numerals for the decimal point and fractions (**Figure 11.17**). Common fractions are twice as tall as single numerals (**Figure 11.17**). Both the Braddock-Rowe triangle and the Ames guide have separate sets of holes spaced 1/16 inch apart for common fractions. The center guideline locates the fraction's crossbar.

When grouping letters to spell words, make the areas between the letters approxi-

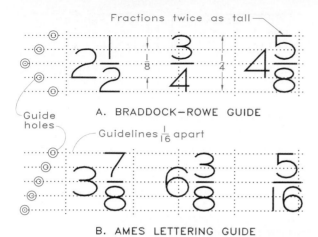

11.16 Guidelines are drawn for fractions so they can be twice as tall as single numerals.

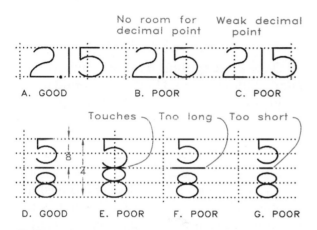

11.17 Avoid making these errors in lettering fractions.

mately equal for the most visually pleasing result (**Figure 11.18**). **Figure 11.19** shows the incorrect use of guidelines and other violations of good lettering practice that you should avoid.

11.5 Computer Lettering

Although AutoCAD offers a wide range of lettering (text) fonts, the ROMANS font shown in **Figure 11.20** is the Gothic single-stroke text recommended for engineering drawings. **DTEXT** (dynamic text found under **DRAW> TEXT> Single Line Text,** or by typing **DTEXT**)

A. GOOD—EQUAL AREAS BETWEEN LETTERS

B. POOR—EQUAL LINEAR SPACING

11.18 Letters should be spaced so that the areas between them are about equal.

11.19 Always use guidelines (vertical and horizontal) whether you are lettering a paragraph or a single letter.

ABCDEFGHIJKLMNO
PQRSTUVWXYZ&%
0123456789abcd
efghijklmnopqrstu
vwxyz

11.20 ROMANS is the text (font) that is most like Gothic lettering and is recommended for engineering drawings. ROMAND is the same letter form drawn with a double stroke which is often better when plotted with a laser printer.

is the command that applies text to a drawing on the screen as it is typed at the keyboard:

Command: (Enter) DTEXT

Specify start point of text or [Justify/Style]: Justify (Enter)

[Align/Fit/Center/Middle/Right/TL/TC/TR/ML/ MC/MR/BL/BC/BR]:

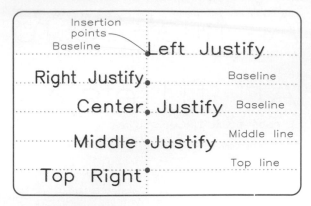

11.21 The options of *Dtext/justify* are available for variations in text (lettering) placement.

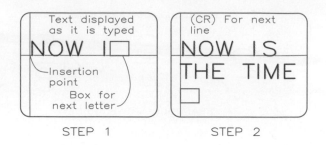

11.23 The DTEXT command:
Step 1 *Command:* <u>DTEXT</u> **(Enter)**
Specify start point of text or [Justify/Style]: (Cursor-select)
Specify height <.18>: <u>.125</u> **(Enter)**
Specify rotation angle of text <0>: <u>0</u> (Enter)
Enter text: <u>NOW IS</u> (Enter)

Step 2 (Box moves to start of next line.)
Enter text: <u>THE TIME</u> (Enter) (Box moves to start of next line. Press (Enter) to save text.)

The abbreviations in the last line above are insertion points for lines of text defined in **Figure 11.21**.

By typing *Style* at the command line (or selecting *Text style* under *Format*), a dialogue box appears on the screen that enables you to name a *New* text file, RS for example (**Figure 11.22**). You can select and assign a font to RD, specify the text height (zero is preferred because it allows you to specify any desired height on the screen), assign a width factor, and specify an obliquing angle.

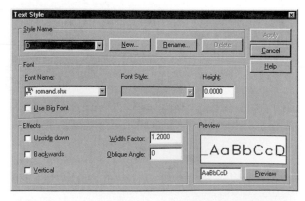

11.22 The *Text style* menu (*Format> Text Style*) can be used to assign various characterics to the letter form (text) being used. (See Chapter 37 for more details.)

Once these settings have been made, type *Dtext* at the command line and respond to the options shown in **Figure 11.23**. At the prompt, **Enter text**, type the text and it will appear on the screen as it is typed.

RD is now the current style with the font of Romand; it has a height of zero, and each letter has a width that is 120% of (1.2 times) the default width. The width of 1.2 matches the recommended letter proportions of Gothic engineering lettering. Assigning zero permits you to change heights by responding when prompted by the *Dtext* command. Had a height been specified, it would remain constant until you modified it using the *Style* command. Additonal information on Auto-CAD lettering is covered in Section 37.39 of Chapter 37.

Problems

Complete exercises on size A (11 x 8-1/2-in., horizontal format) plain paper or paper with a grid using the format shown in **Figure 11.24**.

11.24 The sheet layout for lettering assignments.

11.25 A typical sheet layout for lettering assignments.

1. Draw the alphabet in vertical, uppercase letters as shown in **Figure 11.24**. Draw each letter four times: three As, three Bs, and so on. Use a medium-weight pencil—H, F, or HB.

2. Draw vertical numerals and the alphabet in lowercase letters, as shown in **Figure 11.25**. Construct each letter and numeral three times: two 1s, two 2s, two a's, two b's, and so on. Use a medium-weight pencil—H, F, or HB.

3. Draw the alphabet in inclined uppercase letters as shown in **Figure 11.12**. Construct each letter four times with a medium-weight pencil—H, F, or HB—as shown in **Figure 11.24**.

4. Draw the vertical numerals and the alphabet in lowercase letters as shown in **Figures 11.9** and **11.11**. Draw each letter three times (**Figure 11.25**). Use a medium-weight pencil—H, F, or HB.

5. Construct guidelines for 1/8-in. capital letters starting 1/4 in. from the top border of a sheet similar to the one shown in **Figure 11.25**. Each guideline should end 1/2 in. from the left and right borders. Letter the first paragraph of the text of this chapter using all vertical capitals. Spacing between the lines should be 1/8 in.

6. Repeat Problem 5, but use inclined capital (uppercase) letters. Use inclined guidelines to help you slant letters uniformly.

7. Repeat Problem 5 but use vertical capital and lowercase letters in combination. Capitalize only those words capitalized in the text.

8. Repeat Problem 5, but use inclined capital and lowercase letters in combination. Capitalize only those letters that are capitalized in the text. Use single-stroke Gothic vertical lowercase letters.

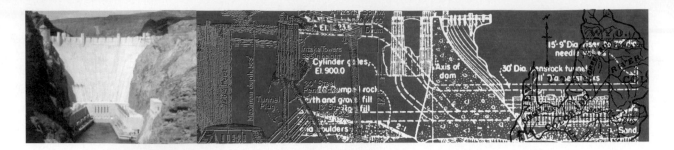

12

Geometric Construction

12.1 Introduction

The solution of many graphical problems requires the use of geometry and geometric construction. Because mathematics was an outgrowth of graphical construction, the two areas are closely related. The proofs of many principles of plane geometry and trigonometry can be developed by using graphics. Graphical methods can be applied to solve some types of problems in algebra and arithmetic and virtually all types of problems in analytical geometry.

12.2 Polygons

A **polygon** is a multisided plane figure of any number of sides. If the sides of a polygon are equal in length, the polygon is a **regular polygon**. A regular polygon can be inscribed in a circle and all its corner points will lie on the circle (**Figure 12.1**). Other regular polygons not pictured are the **heptagon** (7 sides), the

SQUARE 4 sides   PENTAGON 5 sides   HEXAGON 6 sides   OCTAGON 8 sides

12.1 Regular polygons can be inscribed in circles.

nonagon (9 sides), the **decagon** (10 sides), and the **dodecagon** (12 sides).

The sum of the angles inside any polygon (interior angles) is S = (n - 2) x 180°, where n is the number of sides of the polygon.

Triangles

A **triangle** is a three-sided polygon. The four types of triangles are **scalene, isosceles**, **equilateral**, and **right** triangles (**Figure 12.2**). The sum of the interior angles of a triangle is 180° [(3-2)x180°].

140

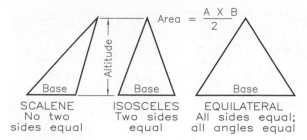

12.2 Types of triangles and their definitions.

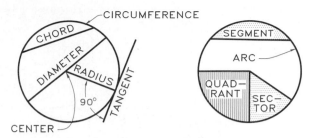

12.4 Elements of a circle and their definitions.

Quadrilaterals

A **quadrilateral** is a four-sided polygon of any shape. The sum of the interior angles of a quadrilateral is 360° [(4-2)x180°]. **Figure 12.3** shows the various types of quadrilaterals and the equations for their areas.

12.3 Circles

Figure 12.4 gives the names of the elements of a circle that are used throughout this textbook. A circle is constructed by swinging a radius from a fixed point through 360°. The area of a circle equals πR^2.

12.4 Geometric Solids

Figure 12.5 illustrates the various types of solid geometric shapes and their characteristics.

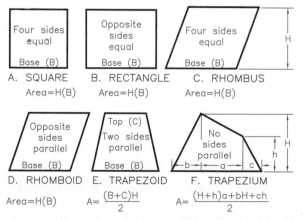

12.3 Types of quadrilaterals, their definitions, and how to calculate their areas.

Polyhedra

A polyhedron is a multisided solid formed by intersecting planes. If its faces are regular polygons (having sides of equal length), it is a regular polyhedron. Five regular polyhedra are the tetrahedron (4 sides), the hexahedron (6 sides), the octahedron (8 sides), the dodecahedron (12 sides), and the icosahedron (20 sides).

Prisms

A **prism** has two parallel bases of equal shape connected by sides that are parallelograms. The line from the center of one base to the center of the other is the axis. If its axis is perpendicular to the bases, the prism is a **right prism**. If its axis is not perpendicular to the bases, the prism is an **oblique prism**. A prism that has been cut off to form a base not parallel to the other is a **truncated prism**. A **parallelepiped** is a prism with bases that are either rectangles or parallelograms.

Pyramids

A **pyramid** is a solid with a polygon as a base and triangular faces that converge at a vertex. The line from the vertex to the center of the base is the **axis**. If its axis is perpendicular to the base, the pyramid is a **right pyramid**. If its axis is not perpendicular to the base, the pyramid is an **oblique pyramid**. A truncated pyramid is called a **frustum** of a pyramid.

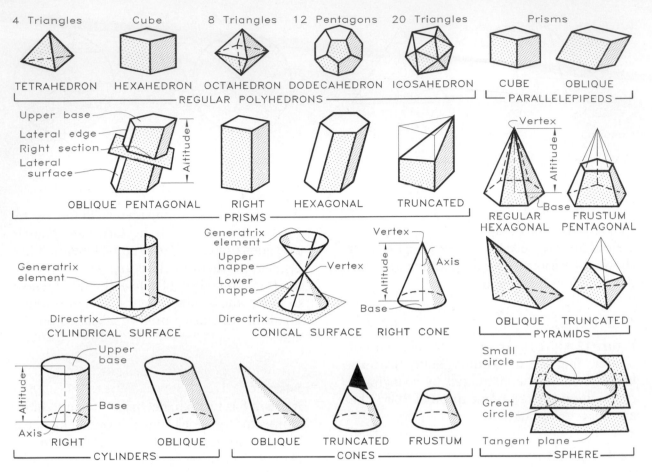

12.5 Various types of geometric solids and their elements.

Cylinders

A **cylinder** is formed by a line or an element (called a *generatrix*) that moves about the circle while remaining parallel to its axis. The axis of a cylinder connects the centers of each end of a cylinder. If the axis is perpendicular to the bases, it is the altitude of a **right cylinder**. If the axis does not make a 90° angle with the base, the cylinder is an **oblique cylinder**.

Cones

A **cone** is formed by a generatrix, one end of which moves about the circular base while the other end remains at a fixed vertex. The line from the center of the base to the vertex is the axis. If its axis is perpendicular to the base, the cone is a **right cone**. A truncated cone is called a **frustum** of a cone.

Spheres

A **sphere** is generated by revolving a circle about one of its diameters to form a solid. The ends of the axis of revolution of the sphere are **poles**.

12.5 Constructing Polygons

A regular polygon (having equal sides) can be inscribed in or circumscribed about a circle. When it is inscribed, all corner points will lie on the circle (**Figure 12.6**). For example, con-

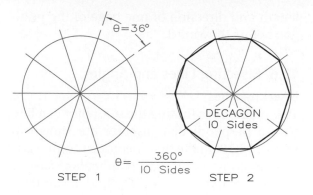

STEP 1 $\theta = \dfrac{360°}{10 \ \text{Sides}}$ STEP 2

θ=36°

DECAGON
10 Sides

12.6 A regular polygon (sides of equal length):

Step 1 Divide circle into the correct number of sectors.

Step 2 Connect the division points with straight lines where they intersect the circle.

structing a 10-sided polygon involves dividing the circle into 10 sectors and connecting the points to form the polygon.

Triangles

When you know the lengths of all three sides of a triangle, you may construct it with a compass by triangulation, as **Figure 12.7** shows.

Hexagons

The **hexagon**, a six-sided regular polygon, can be inscribed in or circumscribed about a circle (**Figure 12.8**). Use a 30°-60° triangle to

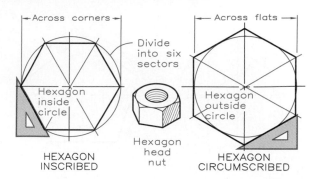

HEXAGON
INSCRIBED

Hexagon
head
nut

HEXAGON
CIRCUMSCRIBED

12.8 A hexagon can be inscribed in or circumscribed about a circle with a 30°–60° triangle.

draw the hexagon. The circle represents the distance from corner to corner for an inscribed hexagon and from flat to flat when the hexagon is circumscribed about a circle.

Octagons

The **octagon**, an eight-sided regular polygon, can be inscribed in or circumscribed about a circle (**Figure 12.9**). Use a 45° triangle in both circumscribing and inscribing an octagon about a circle.

Pentagons

The **pentagon**, a five-sided regular polygon, can be inscribed in or circumscribed about a circle. **Figure 12.10** shows the steps of con-

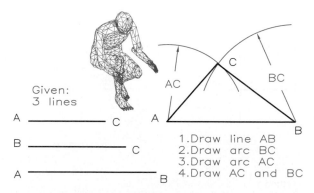

Given:
3 lines

1. Draw line AB
2. Draw arc BC
3. Draw arc AC
4. Draw AC and BC

12.7 A triangle can be constructed by triangulation with a compass when three sides are given.

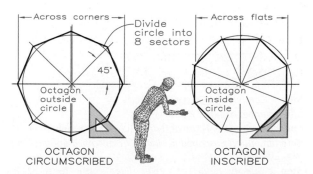

OCTAGON
CIRCUMSCRIBED

OCTAGON
INSCRIBED

12.9 An octagon can be inscribed in or circumscribed about a circle with a 45° triangle.

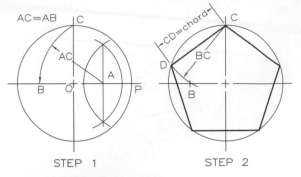

12.10 Constructing an inscribed pentagon:

Step 1 Bisect radius OP to locate point A. With A as the center and radius AC, locate point B on the diameter.

Step 2 With point C as the center and BC as the radius, locate point D. Use line CD as the chord to locate the other corners of the pentagon.

structing a pentagon with a compass and straightedge where the vertices lie on the circle.

Computer Method You may use one of two *Polygon* options from under the *Draw* command to draw polygons (**Figure 12.11**). The *Center* option asks you to give the number of sides, select the center, specify the radius, and indicate whether the polygon is to be inscribed in or circumscribed about an imaginary circle. The *Edge* option allows you to specify the number of sides and specify the

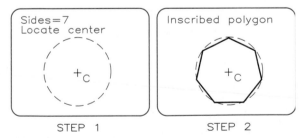

12.11 A polygon by computer: (Draw Menu)

Step 1 *Command:* <u>POLYGON</u> (Enter)

Enter number of sides: <u>7</u> (Enter)

Specify center of polygon or [Edge]: (Locate with cursor.)

Enter an option [Inscribed in circle/Circumscribed about circle] <I>: <u>I</u> (Enter)

Step 2 *Specify radius of circle:* (Drag cursor to select R. Polygon is inscribed inside the imaginary circle.)

length and direction of one edge of the polygon before drawing it.

12.6 Bisecting Lines and Angles

Bisecting Lines

Two methods of finding the midpoint of a line with a perpendicular bisector are shown in **Figure 12.12**. In the first method, a compass is used to construct the perpendicular bisector to the line. In the second method, a standard triangle and a straightedge are used.

Computer Method The midpoint of a line may be found (**Figure 12.13**) by using the *Midpoint* mode of *Osnap* while drawing a line from any point, P, to the line. The line will snap to the line's midpoint.

Bisecting Angles

You may bisect angles by using a compass and drawing three arcs, as shown in **Figure 12.14**.

Computer Method Use the *Arc* command (with *Osnap* set to Midpoint) and draw an arc

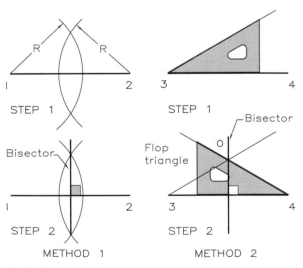

12.12 Bisecting a line:

Method 1 Use a compass and any radius.

Method 2 Use a standard triangle and a straightedge.

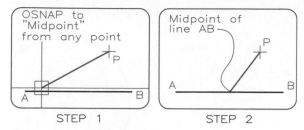

12.13 Midpoint of a line by computer: (DRAW menu)

Step 1 *Command:* LINE (Enter)
Specify first point: P (Locate P anywhere with cursor.)
Specify next point or [Undo]: MID (Enter) (Midpoint mode.)

Step 2 *of* (Select any point on line AB.) The line from point P is drawn to the midpoint of the line AB.

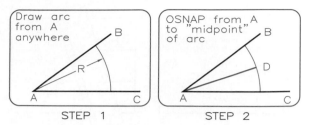

12.15 Bisecting an angle by computer: (DRAW menu)

Step 1 Use the ARC command, any radius, and center A to draw an arc that OSNAPs nearest to AC and AB.

Step 2 *Command:* LINE (Enter)
Specify first point or [Undo]: MID (Enter) (Midpoint option.)
(Select arc. Line AD is the bisector of the angle.)

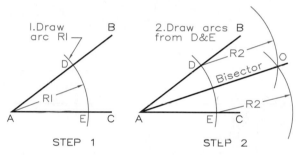

12.14 Bisecting an angle:

Step 1 Swing an arc R1 to locate points D and E.

Step 2 Draw equal arcs from D and E to locate point O. Line AO is the bisector of the angle.

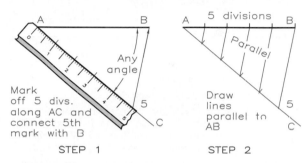

12.16 Dividing a line:

Step 1 To divide line AB into five equal lengths, lay off five equal divisions along line AC, and connect point 5 to end B with a construction line.

Step 2 Draw a series of five construction lines parallel to 5B to divide line AB.

of a convenient radius between the two lines, with its center at vertex A (**Figure 12.15**). Use the *Line* command (with *Osnap* set to Midpoint) to draw a line from the vertex, A, to the arc's midpoint, D. Line AD is the bisector.

12.7 Division of Lines

Dividing a line into several equal parts often is necessary. **Figure 12.16** shows the method used to solve this type of problem where the line AB is divided into five equal lengths by using a scale to lay off the five divisions.

The same principle applies to locating equally spaced lines on a graph (**Figure 12.17**). Lay scales with the desired number of

units (0 to 3 and 0 to 5, respectively) across the graph up and down and then left to right. Make marks at each whole unit and draw vertical and horizontal index lines through these points. These index lines are used to show data in a graph.

12.8 An Arc Through Three Points

An arc can be drawn through three points by connecting the points with two lines and drawing perpendicular bisectors through each line to locate the center of the circle at C

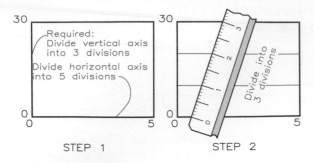

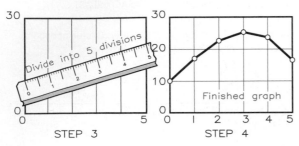

12.17 Dividing axes on a graph:

Step 1 Draw the outline of the graph.

Step 2 To divide the y axis into three equal segments, lay a scale so that three units of measurement span the graph with the 0 and 3 located on the top and bottom lines. Make marks at points 1 and 2 and draw horizontal lines through them.

Step 3 To divide the x axis into five equal segments, lay a scale so that five units of measurement span the graph with the 0 and 5 on the left- and right-hand vertical lines. Make marks at points 1, 2, 3, and 4 and draw vertical lines through them.

Step 4 Plot the data points and draw the curve to complete the graph.

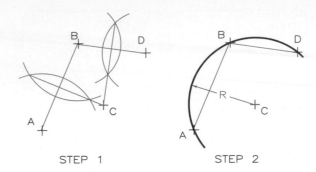

12.18 An arc through three points:

Step 1 Connect points A, B, and D with two lines and construct their perpendicular bisectors that intersect at the center, C.

Step 2 Using the center, C, and the distance to the points as the radius, R, draw the arc through the points.

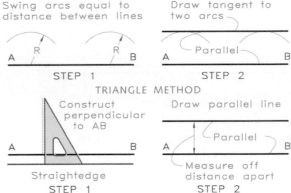

12.19 Drawing parallel lines:

Compass Method

Step 1 Swing two arcs from line AB.

Step 2 Draw the parallel line tangent to the arcs.

Triangle Method

Step 1 Draw a line perpendicular to AB.

Step 2 Measure the desired distance, R, along the perpendicular and draw the parallel line through it.

(**Figure 12.18**). Draw the arc, and lines AB and BD become chords of the arc.

To find the center of a circle or an arc, reverse this process by drawing two chords that intersect at a point on the circumference and bisecting them. The perpendicular bisectors intersect at the center of the circle.

12.9 Parallel Lines

You may draw one line parallel to another by using either method shown in **Figure 12.19**. In the first method, use a compass and draw two

arcs having radius R to locate a parallel line at the desired distance (R) from the first line. In the second method, measure the desired perpendicular distance R from the first line, mark it, and draw the parallel line through it with your drafting machine.

12.10 Tangents

Marking Points of Tangency

A point of tangency is the theoretical point at which a line joins an arc or two arcs join without crossing. **Figure 12.20** shows how to find the point of tangency with triangles by constructing a perpendicular line to the tangent line from the arc's center. **Figure 12.21** shows the conventional methods of marking points of tangency.

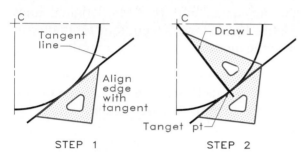

STEP 1   STEP 2

12.20 Locating a tangent point:

Step 1 Align a triangle with the tangent line while holding it against a firmly held straightedge.

Step 2 Hold the triangle in position, locate a second triangle perpendicular to it, and draw a line from the center to locate the tangent point.

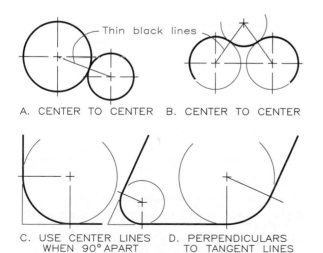

A. CENTER TO CENTER   B. CENTER TO CENTER

C. USE CENTER LINES   D. PERPENDICULARS
   WHEN 90° APART   TO TANGENT LINES

12.21 Use thin, black lines that extend from the centers slightly beyond the arcs to mark tangency points.

Line Tangent to an Arc

The point of tangency may be found by using a triangle and a straightedge as shown in **Figure 12.22**. The classical compass method of finding the point of tangency between a line and a point is shown in **Figure 12.23**. Connect point A to the arc's center and bisect

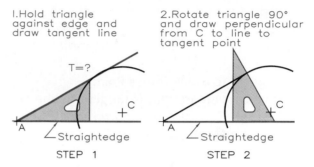

STEP 1   STEP 2

12.22 A tangent to an arc from a point:

Step 1 Hold a triangle against a straightedge and draw a line from point A that is tangent to the arc.

Step 2 Rotate your triangle 90° and locate the point of tangency by drawing a line through center C.

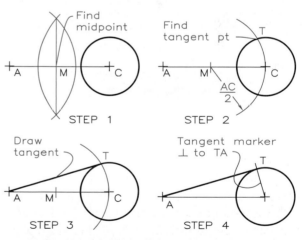

STEP 1   STEP 2

STEP 3   STEP 4

12.23 A line tangent to an arc from a point:

Step 1 Connect point A with center C and locate point M by bisecting AC.

Step 2 Using point M as the center and MC as the radius, locate point T on the arc.

Step 3 Draw the line from A to T tangent to the arc at T.

Step 4 Draw the tangent marker perpendicular to TA from the center past the arc as a thin, dark line.

line AC (Step 1); swing an arc from point M through point C locating tangent point T (Step 2); draw the tangent to T (Step 3); and mark the tangent point (Step 4).

Computer Method   A line can be drawn from a point tangent to an arc by using the *Osnap Tangent* option (**Figure 12.24**). When prompted for the second point, select a point on the arc near the tangent point and the line will be drawn to the true tangent point.

Arc Tangent to a Line from a Point

To construct an arc that is tangent to line DE at T and that passes through point P (**Figure 12.25**), draw the perpendicular bisector of line TP. Draw a line perpendicular to line DE at point T to locate the center at point C and swing an arc with radius CT.

Arc Tangent to Two Lines

Figure 12.26 shows how to construct an arc of a given radius tangent to two nonparallel lines that form an acute angle. The same steps apply to constructing an arc tangent to two lines that form an obtuse angle (**Figure 12.27**). In both cases, the points of tangency are located with thin, dark lines drawn from their centers perpendicular to and past the original lines. **Figure 12.28** shows a technique for finding an arc tangent to lines that are perpendicular.

Computer Method   Draw an arc tangent to two nonparallel lines with the *Fillet* command (*Modify* menu) (**Figure 12.29**). When the

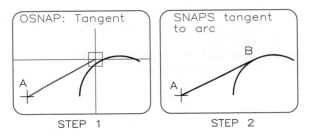

12.24 A line tangent to an arc by computer:
Step 1 *Command:* <u>LINE</u> (Enter)
Specify first point or [Undo]: <u>A</u> (Select point A)
Specify next point or [Undo]: <u>TAN</u> (Enter)
Step 2 *to:* (Select point on arc near tangent point. Line AB is drawn tangent to the arc.)

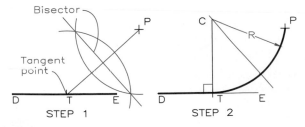

12.25 An arc through two points:
Step 1 To draw an arc through point P tangent to line DE at point T, draw the perpendicular bisector of TP.
Step 2 Construct a perpendicular to line DE at point T to intersect the bisector at point C and draw the arc from C with radius CT.

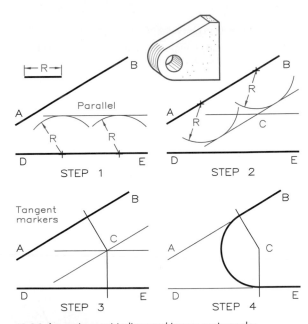

12.26 An arc tangent to lines making an acute angle:
Step 1 Construct a light construction line parallel to line DE with radius R.
Step 2 Draw a second light line parallel to and R distance from line AB to locate center C.
Step 3 Draw thin, dark lines from center C perpendicular to lines AB and DE to locate the tangency points.
Step 4 Draw the tangent arc and darken lines.

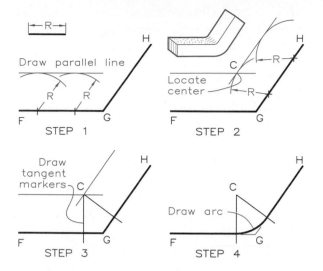

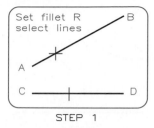

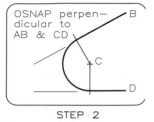

12.29 An arc tangent to two lines by computer:

Step 1 *Command:* <u>FILLET</u> (Enter)
Select first object or [Polyline/Radius/Trim]: <u>R</u> (Enter)
Specify fillet radius <0.0000>: <u>.75</u> (Enter)
Command: (Enter)
Select first object or [Polyline/Radius/Trim]: (Select points on AB and CD.)

Step 2 *Command:* <u>LINE</u> (Enter)
Line from first point: <u>CENTER</u> (Enter) of (Select point on the arc.)
Specify next point or [Undo]: <u>PERPEND</u> (Enter) to (Select line AB; a perpendicular is drawn to the point of tangency. Locate the tangent point on CD in the same manner.)

12.27 An arc tangent to lines making an obtuse angle:

Step 1 Using radius R, draw a light line parallel to FG.

Step 2 Construct a light line parallel to line GH that is R distance from it to locate center C.

Step 3 Draw thin lines from center C perpendicular to lines FG and GH to locate the tangency points.

Step 4 Draw the tangent arc and darken your lines.

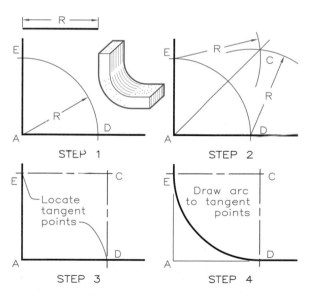

12.28 An arc tangent to perpendicular lines:

Step 1 Using radius R and center A, locate D and E.

Step 2 Find C by swinging two arcs with radius R.

Step 3 Perpendiculars CE and CD locate tangent points.

Step 4 Draw the tangent arc and darken your lines.

radius length has been set and a point selected on each line, the arc is drawn and the lines trimmed. To mark tangent points, snap to the center of the arc with the *Center* option of *Osnap* and draw two lines perpendicular (use *Osnap*'s *Perpend* option) to lines AB and CD from the center.

Arc Tangent to an Arc and a Line

Figure 12.30 shows the steps for constructing an arc tangent to an arc and a line. **Figure 12.31** shows a variation of this technique for an arc drawn tangent to a given arc and line with the arc reversed.

Arc Tangent to Two Arcs

Figure 12.32 shows how to draw an arc tangent to two arcs. Lines drawn between the centers locate the points of tangency. The resulting tangent arc is concave from the top. Drawing a convex arc tangent to the given arcs requires that its radius be greater than the

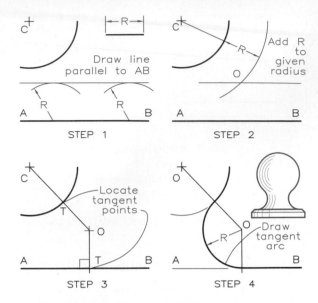

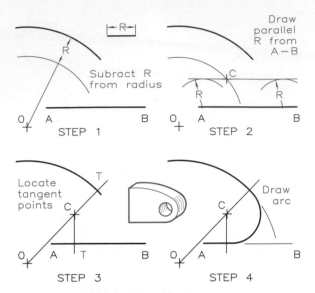

12.30 An arc tangent to an arc and a line:

Step 1 Draw a line parallel to AB R distance from it.

Step 2 Add radius R to the radius from center C. Swing the extended radius to find the center O.

Step 3 Lines OC and OT locate the tangency points.

Step 4 Draw the tangent arc between the points of tangency with radius R and center O.

12.31 An arc tangent to an arc and a line:

Step 1 Subtract radius R from the radius through center O. Draw a concentric arc.

Step 2 Draw a line parallel to line 1-2 and R distance from it to locate the center at C.

Step 3 Locate the tangency points with lines O C and from C perpendicular to 1-2.

Step 4 Draw the tangent arc between the tangent points with radius R and center C.

radius of either of the given arcs, as shown in **Figure 12.33**.

One variation of this problem (**Figure 12.34**) is to draw an arc of a given radius tangent to the top of one arc and the bottom of the other. Another (**Figure 12.35**) is to draw an arc tangent to a circle and a larger arc.

12.32 A concave arc tangent to two arcs:

Step 1 Extend the radius by adding the radius R to it. Draw a concentric arc with the extended radius.

Step 2 Extend the radius of the other circle by adding radius R to it. Use this extended radius to construct a concentric arc to locate center C.

Step 3 Connect center C with centers C1 and C2 with thin, dark lines to locate the tangency points.

Step 4 Draw the tangent arc between the points of tangency using radius R and center C.

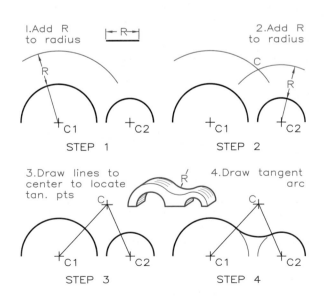

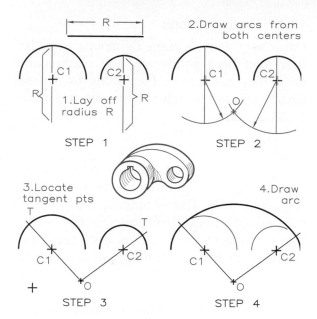

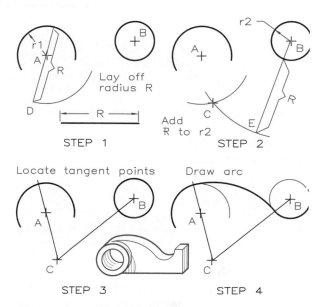

12.33 A convex arc tangent to two arcs:

Step 1 Extend each radius from the arc past its center by a distance of R, the radius, along these lines.

Step 2 Use the distance from each center to the ends of the extended radii to swing arcs to locate center O.

Step 3 Draw thin, dark lines from center O through centers C1 and C2 to locate the points of tangency.

Step 4 Draw the tangent arc between the tangent points using radius R and center O.

12.34 An arc tangent to two circles:

Step 1 Lay off radius R from the arc along an extended radius to locate point D, swing arc AD.

Step 2 Extend the radius from center B and add radius R to it to point E. Use radius BE to locate center C.

Step 3 Draw thin lines from center C through centers A and B to locate the points of tangency.

Step 4 Draw the tangent arc between the tangent points using radius R and center C.

Computer Method Use the *Fillet* command (**Figure 12.36**) to draw an arc tangent to two arcs. After entering the command, specify the radius when prompted. Press the carriage return (Enter) twice to return to the *Command* mode, and the *Fillet* command is ready for use.

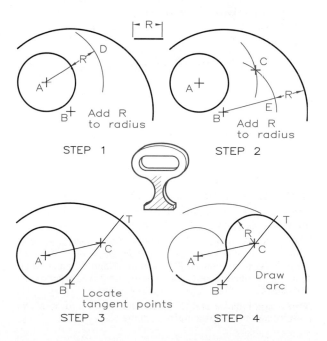

12.35 An arc tangent to two arcs:

Step 1 Add radius R to the radius from A. Use radius AD to draw a concentric arc from center A.

Step 2 Subtract radius R from the radius through B. Use radius BE to draw an arc to locate center C.

Step 3 Draw thin lines to connect the centers and mark the points of tangency.

Step 4 Draw the tangent arc between the tangency points using radius R and center C.

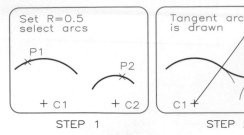

STEP 1   STEP 2

12.36 An arc tangent to two arcs by computer:

Step 1 *Command:* <u>FILLET</u> (Enter)
Select first object or [Polyline/Radius/Trim]: R (Enter)
Specify fillet radius <0.0000>: .5 (Enter)

Step 2 *Command:* (Enter)
Select first object or [Polyline/Radius/Trim]: P1
Select second object or [Polyline/Radius/Trim]: P2 (Tangent arc is drawn. Locate tangent points with lines between the centers.)

Select the two arcs with your cursor; the tangent arc is drawn and the arcs are trimmed at the points of tangency. Mark the tangency points by drawing lines from centers C1 and C2.

Ogee Curves

The **ogee curve** is an S curve formed by tangent arcs. The ogee curve shown in **Figure 12.37** is the result of constructing two arcs tangent to three intersecting lines.

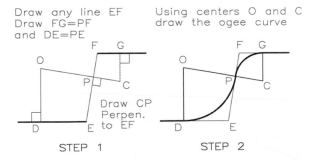

STEP 1   STEP 2

12.37 An ogee curve:

Step 1 To draw an ogee curve between two parallel lines, draw a light line EF at any angle. Locate P anywhere on EF. Find the tangent points by making FG equal to FP and DE equal to EP. Draw perpendiculars at G and D to intersect the perpendicular at O and C.

Step 2 Use radii CP and OP at centers O and C to draw two tangent arcs to complete the ogee curve.

12.11 Conic Sections

Conic sections are plane figures that can be described both graphically and mathematically; they are formed by passing imaginary cutting planes through a right cone, as **Figure 12.38** shows.

Ellipses

The ellipse is a conic section formed by passing a plane through a right cone at an angle (**Figure 12.38B**). Mathematically, the ellipse is the path of a point that moves in such a way that the sum of the distances from two focal points is a constant. The largest diameter of an ellipse—the major diameter—is always the true length. The shortest diameter—the minor diameter—is perpendicular to the major diameter.

Revolving the edge view of a circle yields an ellipse (**Figure 12.39**). The ellipse template shown in **Figure 12.40** is used to draw the same ellipse. The angle between the line of sight and the edge of the circle is the angle of the ellipse template (or the one closest to this

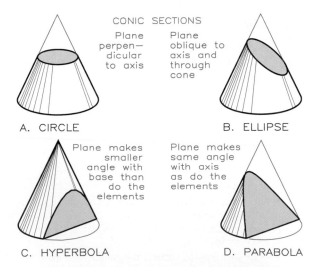

12.38 The conic sections are the (A) circle, (B) ellipse, (C) hyperbola, and (D) parabola. They are formed by passing cutting planes through a right cone.

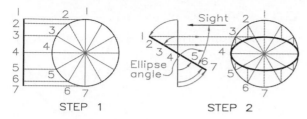

12.39 An ellipse by revolution:

Step 1 When the edge of a circle is perpendicular to the projectors between its adjacent view, it appears as a circle. Mark equally spaced points around the circle's circumference and project them to the edge.

Step 2 Revolve the edge of the circle and project the points to the circular view. Project the points vertically downward to obtain the elliptical view.

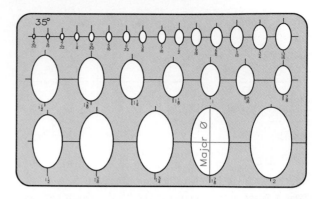

12.41 Ellipse templates are calibrated at 5° intervals from 15° to 60° with size increments of 1/8 inch.

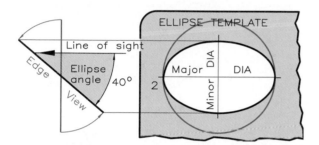

12.40 When the line of sight is not perpendicular to a circle's edge, it appears as an ellipse. The angle between the line of sight and the edge of the circle is the ellipse template angle.

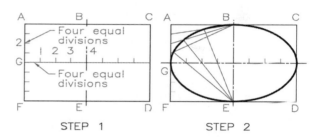

12.42 An ellipse by the parallelogram method:

Step 1 Draw an ellipse inside a rectangle or parallelogram by dividing the horizontal line GH and vertical line AG into the same number of equal divisions.

Step 2 The curve construction is shown for one quadrant. Sets of rays from E and B cross at points on the curve.

size) that should be used. Ellipse templates are available in 5° intervals and in major diameter sizes that vary in increments of about 1/8 inches (**Figure 12.41**).

You may construct an ellipse inside a rectangle or parallelogram by plotting a series of points to form the ellipse (**Figure 12.42**).

An ellipse can be drawn on x and y axes by plotting x and y coordinates from the equation of an ellipse. The mathematical equation of an ellipse is

$$\frac{x^2}{a^2} + \frac{y^2}{b^2} = 1, \text{ where a, b are not 0}$$

Parabolas

The **parabola** is defined as a plane curve, each point of which is equidistant from a straight line (called a directrix) and a focal point. The parabola is the conic section formed when the cutting plane and an element on the cone's surface make the same angle with the cone's base as shown in **Figure 12.38D**.

Figure 12.43 shows construction of a parabola by using its mathematical definition as is done in analytical geometry. A second method of drawing a parabola, which involves the use of

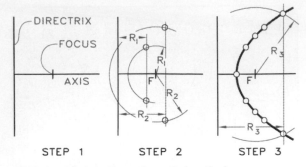

STEP 1 STEP 2 STEP 3

12.43 A parabola by the mathematical method:

Step 1 Draw an axis perpendicular to the directrix (a line). Choose a point for the focus, F.

Step 2 Use a series of selected radii to find points on the curve. For example, draw a line parallel to the directrix and R2 from it. Swing R2 from F to intersect the line and plot the point.

Step 3 Continue the process with a series of arcs of varying radii until you find an adequate number of points to complete the curve.

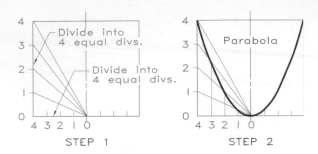

STEP 1 STEP 2

12.44 A parabola by the parallelogram method:

Step 1 Draw a parallelogram or rectangle to contain the parabola; draw its axis parallel to the sides through 0. Divide the sides into equal segments; draw rays from 0.

Step 2 Draw lines parallel to the sides (vertical in this case) to locate points along the rays from 0 and draw a smooth curve through them.

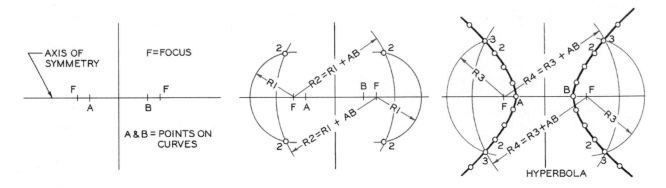

HYPERBOLA

12.45 A hyperbola:

Step 1 Draw a perpendicular through the axis of symmetry. Locate focal points F equidistant from it on both sides. Locate points A and B equidistant from the perpendicular at a distance of your choice but between the focal points.

Step 2 Use radius R1 to draw arcs using focal points F as the centers. Add R1 to AB (the nearest distance between the hyperbolas) to find R2. Draw arcs using radius R2 and the focal points as centers. R1 and R2 locate point 2.

Step 3 Select other radii and add them to AB to locate additional points as shown in Step 2. Draw a smooth curve through the points with an irregular curve to draw the two hyperbolic curves.

a rectangle or parallelogram, is shown in **Figure 12.44**. The mathematical equation of the parabola is:

$$y = ax^2 + bx + c, \text{ where a is not } 0$$

Hyperbolas

The hyperbola is a two-part conic section defined as the path of a point that moves in such a way that the difference of its distances from two focal points is a constant (**Figure 12.38C**). **Figure 12.45** shows construction of a

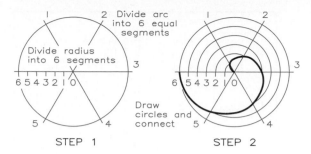

12.46 Constructing a spiral:

Step 1 Draw a circle and divide it into equal parts. Divide the radius into the same number of equal parts (six in this case).

Step 2 Begin inside and draw arc 0-1 to intersect radius 0-1. Then swing arc 0-2 to radius 0-2, and continue to point 6 on the original circle, and connect the points.

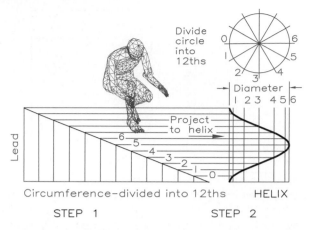

12.47 A cylindrical helix:

Step 1 Divide the top view of the cylinder into equal parts and project them to the front view. Layout the circumference and the lead (pronounced *leed*) as the cylinder. Divide the circumference into the same number of equal parts tranferred from the top view.

Step 2 Project the points along the inclined rise to their respective elements on the diameter and connect them with a smooth curve.

hyperbola according to this definition. By selecting a series of radii until enough points have been located, the hyperbolic curve can be accurately drawn.

12.12 Spirals

The **spiral** is a coil lying in a single plane that begins at a point and becomes larger as it travels around the origin. **Figure 12.46** shows the steps for constructing a spiral. The number of divisions selected in this construction depends upon the degree of accuracy desired.

12.13 Helixes

The **helix** is a three-dimensional curve that coils around a cylinder or cone at a constant angle of inclination. Applications of helixes are corkscrews and the threads on a screw. **Figure 12.47** shows the construction of a cylindrical helix.

Problems

Present your solutions to these problems on size A (8-1/2 x11 inch) vertical format sheets. The printed grid represents 0.20-in. intervals, so you can use your engineers' 10 scale to lay out the problems. By equating each grid interval to 5 mm, you also can use your full-sized metric scale to lay out and solve the problems. Show your construction and mark all points of tangency, as recommended in the chapter.

1. Basic constructions (**Sheet 1**):

   (**A**) Draw triangle ABC using the given sides.

   (**B-C**) Inscribe an angle in the semicircles with the vertexes at point P.

   (**D**) Inscribe a three-sided regular polygon inside the circle.

   (**E**) Circumscribe a four-sided regular polygon about the circle.

2. Construction of regular polygons (**Sheet 2**):

   (**A**) Circumscribe a hexagon about the circle.

   (**B**) Inscribe a hexagon in the circle.

   (**C**) Circumscribe an octagon about the circle.

   (**D**) Construct a pentagon inside the circle using the compass method.

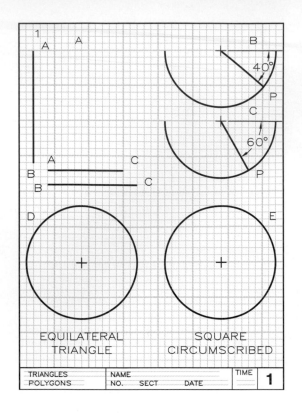

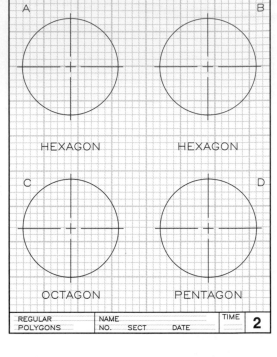

3. Basic constructions (**Sheet 3**):

   (A) Bisect the lines.

   (B) Bisect the angles.

   (C) Bisect the sides of the triangle.

4. Line division and tangencies (**Sheet 4**):

   (A) Divide AB into seven equal parts using a construction line through point A.

   (B) Divide the space between the two vertical lines into four equal segments. Draw three vertical lines at the division points that are equal in length to the given lines.

   (C) Construct an arc with radius R that is tangent to the line at J and that passes through point P.

   (D) Construct an arc with radius R that is tangent to the line and passes through P.

5. Tangency construction (**Sheet 5**):

   (A) Using the compass method, draw a line from P tangent to the arc. Mark the points of tangency.

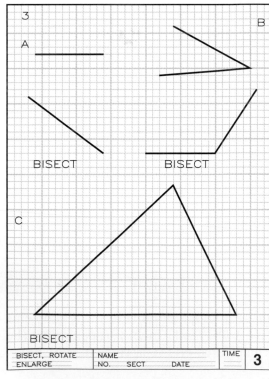

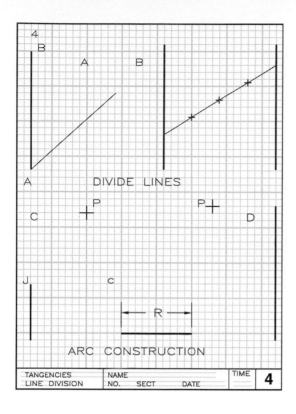

DIVIDE LINES

ARC CONSTRUCTION

| TANGENCIES LINE DIVISION | NAME NO. SECT DATE | TIME | **4** |

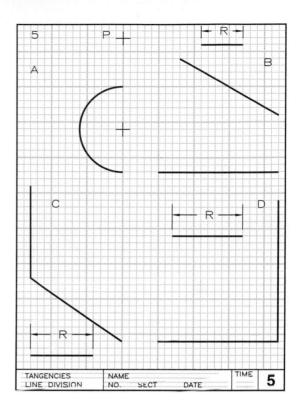

| TANGENCIES LINE DIVISION | NAME NO. SECT DATE | TIME | **5** |

(B-D) Construct arcs with the given radii tangent to the lines.

6. Tangency construction (**Sheet 6**):

(A-D) Construct arcs that are tangent to the arcs or lines shown. The radii are given for each problem.

7. Ogee curve construction (**Sheet 7**):

(A-D) Construct ogee curves that connect the ends of the given lines and pass through points P.

8. Tangency construction (**Sheet 8**):

(A-B) Using the given radii, connect the circles with the tangent arcs as indicated in the sketches.

9. Ellipse construction (**Sheet 9**):

(A-C) Construct ellipses inside the rectangles given. Use enough divisions to make it possible to draw accurate ellipses with your irregular curve.

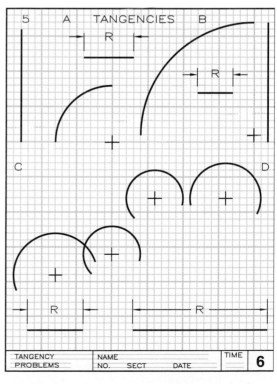

TANGENCIES

| TANGENCY PROBLEMS | NAME NO. SECT DATE | TIME | **6** |

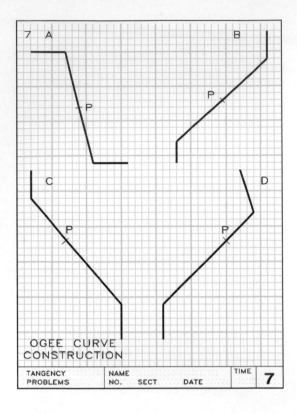

OGEE CURVE
CONSTRUCTION

TANGENCY PROBLEMS	NAME			TIME	**7**
	NO.	SECT	DATE		

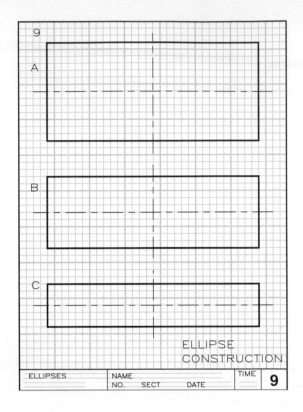

ELLIPSE
CONSTRUCTION

ELLIPSES	NAME			TIME	**9**
	NO.	SECT	DATE		

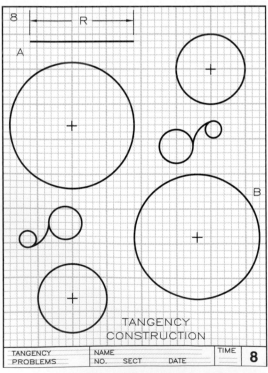

TANGENCY
CONSTRUCTION

TANGENCY PROBLEMS	NAME			TIME	**8**
	NO.	SECT	DATE		

10. Ellipse and parabola (**Sheet 10**):

   (**A**) Construct an ellipse inside the circle when the edge view has been rotated as shown.

   (**B**) Using the focal point F and the directrix, plot and draw the parabola formed by these elements.

11. Hyperbola and parabola (**Sheet 11**):

   (**A**) Using the focal point F, points A and B on the curve, and the axis of symmetry, construct the hyperbola.

   (**B**) Construct a parabola by using the parallelogram method and five divisions along each axis for each half of the parabola.

12. Helix and spiral (**Sheet 12**):

   (**A**) Construct a helix that has a rise equal to the height of the cylinder.

   (**B**) Construct a spiral by using the six divisions marked along the radius.

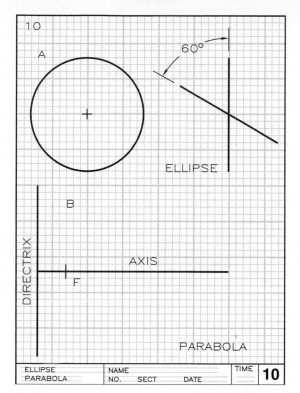

10

A

60°

ELLIPSE

B

DIRECTRIX

AXIS

F

PARABOLA

| ELLIPSE | NAME | | TIME | 10 |
| PARABOLA | NO. SECT DATE | | | |

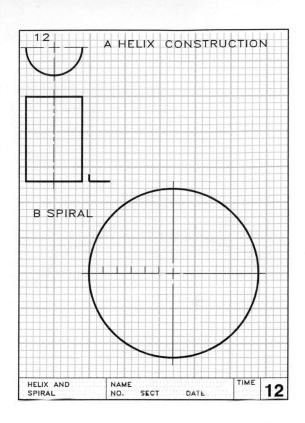

12

A HELIX CONSTRUCTION

B SPIRAL

| HELIX AND | NAME | | TIME | 12 |
| SPIRAL | NO. SECT DATE | | | |

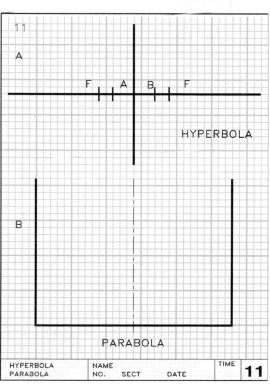

11

A

F | A | B | F

HYPERBOLA

B

PARABOLA

| HYPERBOLA | NAME | | TIME | 11 |
| PARABOLA | NO. SECT DATE | | | |

Practical applications (**Figures 12.48–12.58**). Construct the given shapes on size A sheets, one problem per sheet. Select the scale that will best fit the problem to the sheet. Mark all points of tangency and strive for good line quality.

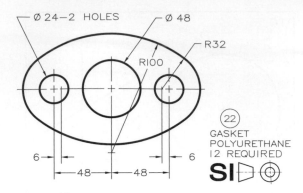

12.50 Problem 15.

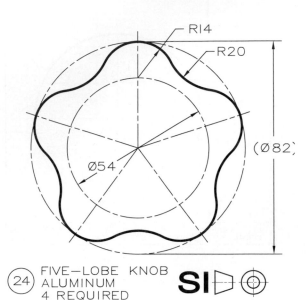

12.48 Problem 13.

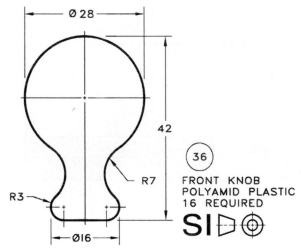

12.51 Problem 16.

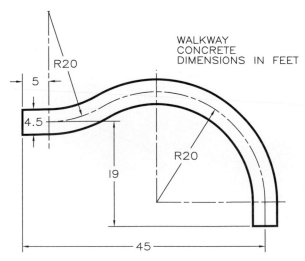

12.49 Problem 14.

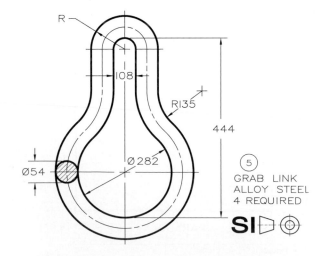

12.52 Problem 17.

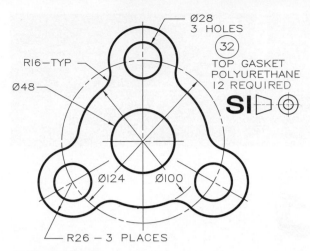

12.53 Problem 18.

Ø28
3 HOLES
32
TOP GASKET
POLYURETHANE
12 REQUIRED
SI
R16—TYP
Ø48
Ø124
Ø100
R26 – 3 PLACES

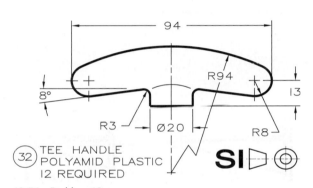

12.54 Problem 19.

94
R94
13
8°
R3
Ø20
R8
32
TEE HANDLE
POLYAMID PLASTIC
12 REQUIRED
SI

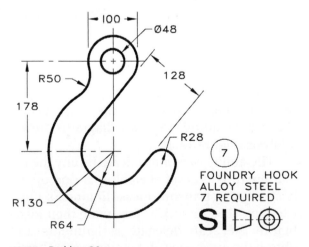

12.55 Problem 20.

100
Ø48
R50
128
178
R28
R130
R64
7
FOUNDRY HOOK
ALLOY STEEL
7 REQUIRED
SI

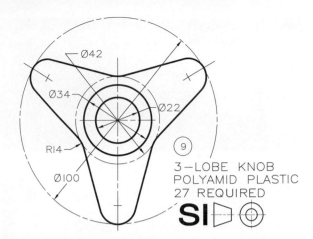

12.56 Problem 21.

Ø42
Ø34
Ø22
R14
Ø100
9
3–LOBE KNOB
POLYAMID PLASTIC
27 REQUIRED
SI

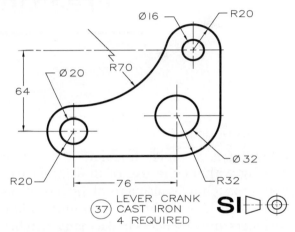

12.57 Problem 22.

Ø16
R20
Ø20
R70
64
Ø32
R20
76
R32
37
LEVER CRANK
CAST IRON
4 REQUIRED
SI

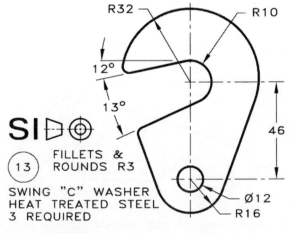

12.58 Problem 23.

R32
R10
12°
13°
46
SI
FILLETS &
ROUNDS R3
13
Ø12
R16
SWING "C" WASHER
HEAT TREATED STEEL
3 REQUIRED

13

Freehand Sketching

13.1 Introduction

Sketching is a rapid, freehand method of drawing without the use of drawing instruments. Sketching is also a thinking process as much as it is a method of communication. Designers and engineers make many sketches as a method of developing ideas before arriving at the final solution. Many new products and projects have begun as sketches made on the back of an envelope or on a napkin at a restaurant (**Figure 13.1**). Sketching is used by

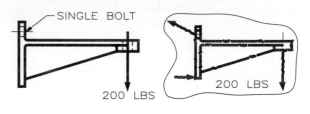

13.2 Sketching is a necessary skill used in all aspects of design, from free-body diagrams through design documentation.

the engineer throughout the engineering process including free-body diagrams during analysis (**Figure 13.2**).

The ability to communicate by any means is a great asset, and sketching is one of the best ways to transmit ideas. Engineers must use their sketching skills to explain their ideas before they can delegate assignments and obtain the assistance of their team members.

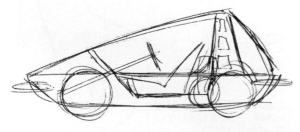

13.1 Designs begin with rough, freehand sketches as a means of developing concepts. (Courtesy of Chrysler Corporation.)

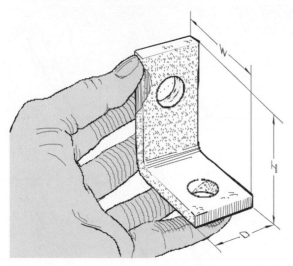

13.3 How can you sketch this angle bracket to convey its shape effectively?

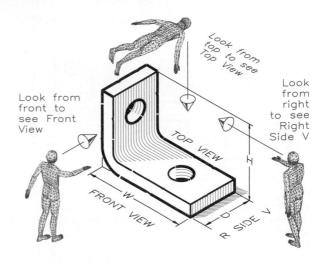

13.4 These positions give the viewpoints for three orthographic views of the angle bracket: top, front, and right side.

13.2 Shape Description

Although the angle bracket in **Figure 13.3** is a simple three-dimensional object, describing it with words is difficult. Most untrained people would think that drawing it as a three-dimensional pictorial would be a challenge. To make drawing such objects easier, engineers devised a standard system, called **orthographic projection**, for showing objects in different views.

In orthographic projection, separate views represent the object at 90° intervals as the viewer moves about it (**Figure 13.4**). **Figure 13.5** shows two-dimensional views of the bracket from the front, top, and right side. The top view is drawn above the front view, and both share the dimension of width. The right-side view is drawn to the right of the front view, and both share the dimension of height.

The views of the bracket are drawn with three types of lines: **visible lines**, **hidden lines**, and **centerlines**. Visible lines are the thickest. Thinner hidden lines (dashed lines) represent features that are invisible, or hidden, in a view. The thinnest lines are centerlines, which are imaginary lines composed of

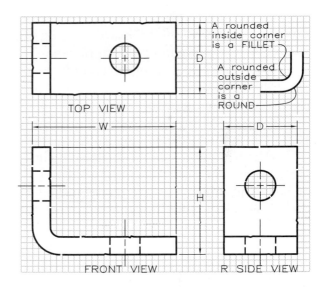

13.5 This sketch shows three orthographic views of the angle bracket.

long and short dashes to show the centers of arcs and the axes of cylinders.

The space between views may vary, but the views must be positioned as shown here. This arrangement is logical, the views are easiest to interpret in this order, and the drawing process is most efficient because the views project from each other. **Figure 13.6** illustrates the lack of

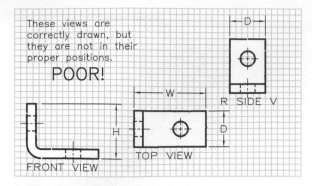

13.6 Views must be sketched in their standard orthographic positions. If they are incorrectly positioned, the object cannot be readily understood.

clarity when views are incorrectly positioned, even though each view is properly drawn.

13.3 Sketching Techniques

You need to understand the application of line types used in sketching (freehand) orthographic views before continuing with the principles of projection. The "alphabet of lines" for sketching is presented in **Figure 13.7**. All lines, except construction lines, should be black and dense. Construction lines are drawn lightly so that they need not be erased. The other lines are distinguished by their line widths (line thicknesses), but they are equal in darkness.

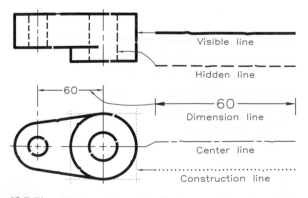

13.7 The alphabet of lines for sketching is shown here. The lines at the right are full size.

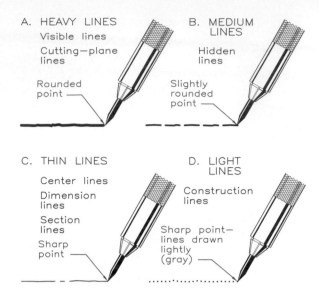

13.8 An F pencil is a good choice for sketching all lines if you sharpen it for varying line widths.

Medium-weight pencils, such as H, F, or HB grades, are best for sketching the lines shown in **Figure 13.8**. By sharpening the pencil point to match the desired line width, you may use the same grade of pencil for all these lines. Lines sketched freehand should have a freehand appearance; do not attempt to make them appear mechanical. Using a printed grid or laying translucent paper over a printed grid can aid your sketching technique (**Figure 13.9**).

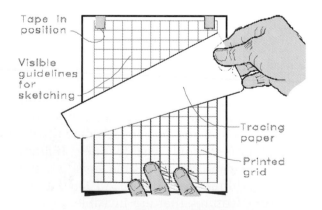

13.9 A grid placed under a sheet of tracing paper will provide guidelines as an aid in freehand sketching.

When you make a freehand sketch, lines will be vertical, horizontal, angular, and/or circular. By not taping your drawing to the table top, you can position the sheet for the most comfortable strokes, usually from left to right (**Figure 13.10**). Examples of correctly sketched lines are contrasted with incorrectly sketched ones in **Figure 13.11**.

13.4 Six-View Sketching

The maximum number of principal views that can be drawn in orthographic projection is six, as the viewer changes position at 90° inter-

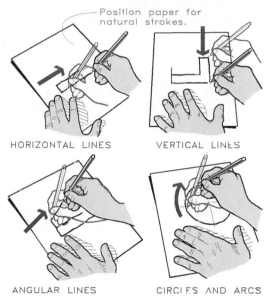

13.10 Sketch lines as shown here for the best results; rotate your drawing sheet for comfortable sketching positions.

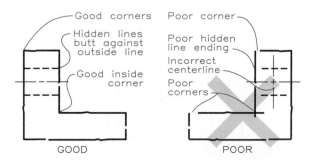

13.11 For good sketches, follow the examples of good technique and avoid the common errors of poor technique shown.

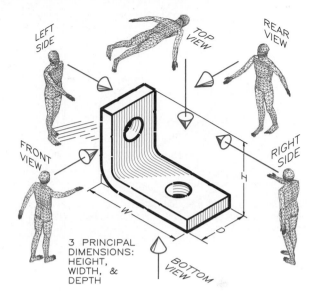

13.12 Six principal views of the angle bracket can be sketched from the viewpoints shown.

vals (**Figure 13.12**). In each view, two of the three dimensions of **height**, **width**, and **depth** are seen.

These views must be sketched in their standard positions (**Figure 13.13**). The width dimension is shared by the top, front, and bottom views. The height dimension is shared with the right-side, front, left-side, and rear views. Note the simple and effective dimensioning of each view with two dimensions. Seldom is an object so complex that it requires six orthographic views.

13.5 Three-View Sketching

You can adequately describe most objects with three orthographic views; usually the top, front, and right-side views. **Figure 13.14** shows a typical three-view sketch of a T-block with height, width, and depth dimensions and the front, top, and right-side views labeled.

The object shown in **Figure 13.15** is represented by three orthographic views on a grid in **Figure 13.16**. To obtain those views, first sketch the overall dimensions of the object, then sketch the slanted surface in the top view and

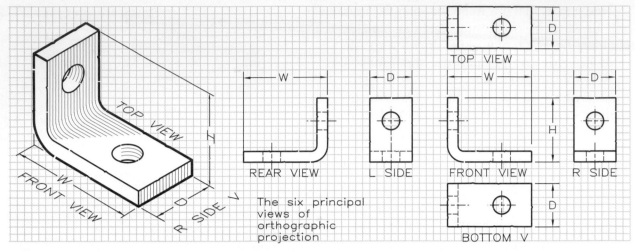

13.13 This six-view sketch of the angle bracket shows the six principal views of orthographic projection. Note the placement of dimensions on the views.

The six principal views of orthographic projection

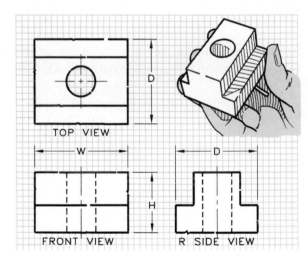

13.14 This sketch shows the standard orthographic arrangement for three views of a T-nut, with dimensions and labels.

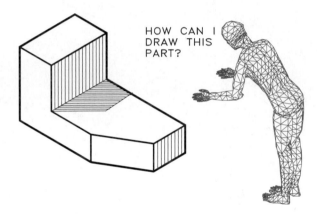

HOW CAN I DRAW THIS PART?

13.15 Sketches of three orthographic views describe this fixture block in Figure 13.16.

project it to the other views. Finally, darken the lines, label the views, and letter the overall dimensions of height, width, and depth.

Slanted surfaces will appear as **edges** or **foreshortened** (not-true-size) planes in the principal views of orthographic projection (**Figure 13.17**). In **Figure 13.17C**, two intersecting planes of the object slope in two directions; thus both appear foreshortened in the front, top, and right-side views.

A good way to learn orthographic projection is to construct a missing third view (the front view in **Figure 13.18**) when two views are given. In **Figure 13.19**, we construct the missing right-side view from the given top and front views. To obtain the depth dimension for the right-side view, transfer it from the top view with dividers; to obtain the height dimension, project it from the front view.

Figure 13.20 shows a fixture pad sketched in three views. The pad has a **finished surface**, indicated by √ marks in the two views where

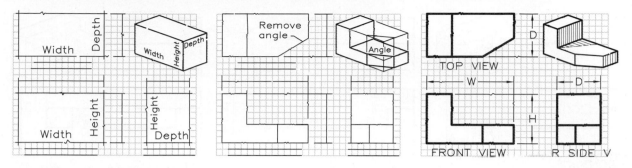

13.16 Three-view sketching:

Step 1 Block in the views with light construction lines. Allow proper spacing for labeling and dimensioning the views.

Step 2 Remove the notches and project from view to view.

Step 3 Check for correctness, darken the lines, and letter the labels and dimensions.

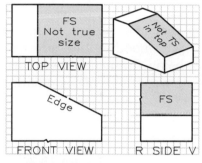

FORESHORTENED IN TOP

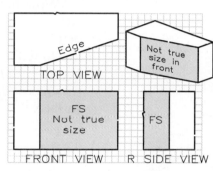

FORESHORTENED IN FRONT

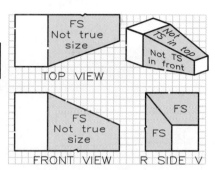

FORESHORTENED IN ALL

13.17 Views of planes.

A The plane appearing as an angular edge in the front view is foreshortened in the top and side views.

B The plane appearing as an angular edge in the top view is foreshortened in the front and side views.

C Two sloping planes appear foreshortened in all views and neither appears as an edge in either the top, side, or front views.

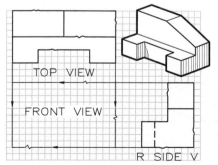

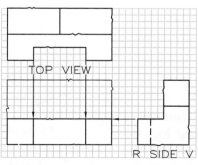

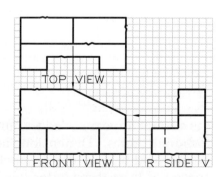

13.18 Sketching a missing front view:

Step 1 To sketch the front view, begin by blocking it in with light construction lines that will not need to be erased.

Step 2 Project the notch from the top view to the front and side views and sketch the lines as final lines.

Step 3 Project the ends of the angular notch from the top and right-side views, check the views, and darken the lines.

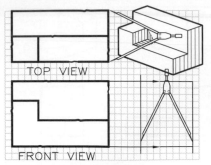
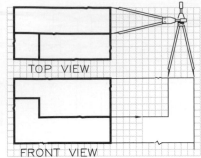
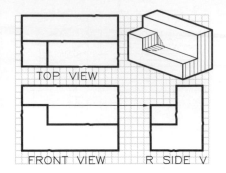

13.19 Sketching a missing side view:

Step 1 Transfer the depth with dividers and project the height from the front. Block in the view with construction lines.

Step 2 Locate the notch in the side view with your dividers and project its base from the front view. Use light construction lines.

Step 3 Project the top of the notch from the front view, check for correctness, darken the lines, and label the views.

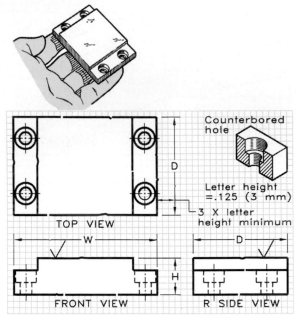

13.20 Three orthographic views adequately describe the rest pad. Space dimension lines at least three letter heights from the views. Finish marks (√ marks) indicate that the top surface has been machined to a smooth finish. Counterbored holes allow bolt heads to be recessed.

the surface appears as edges, and four **counterbored** holes. Dimension lines for the height, width, and depth labels should be spaced at least three letter heights from the views. For example, when you use 1/8-inch letters, position them at least 3/8-inch from the views.

Apply the finish mark symbol to the edge views of any finished surfaces, visible or hidden, to specify that the surface is to be machined to make it smoother. The surface in **Figure 13.21** is being finished by grinding, which is one of many methods of smoothing a surface.

13.6 Circular Features

The pulley shaft depicted in **Figure 13.22** in two views is composed of circular features. **Centerlines** are added to better identify these cylindrical features. **Figure 13.23** shows how

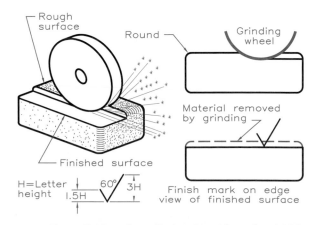

13.21 Place a finish mark on all edge views of a surface (visible or hidden) that have to be smoothed by machining. Grinding is one of the methods used to finish a surface.

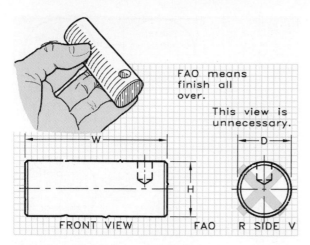

13.22 This pulley shaft is a typical cylindrical part that can be represented adequately by one view.

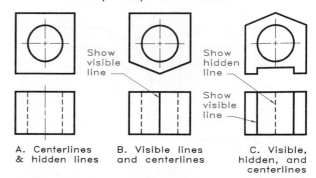

First priority: Visible lines
Second priority: Hidden lines
Third priority: Centerlines

A. Centerlines & hidden lines

B. Visible lines and centerlines

C. Visible, hidden, and centerlines

13.24 When visible lines coincide with hidden lines, show the visible lines. When hidden lines coincide with centerlines, show the hidden lines.

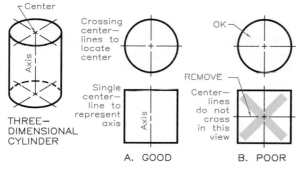

13.23 Centerlines identify the centers of circles and axes of cylinders. Centerlines cross only in the circular view and extend about 1/8 inch beyond the outside lines.

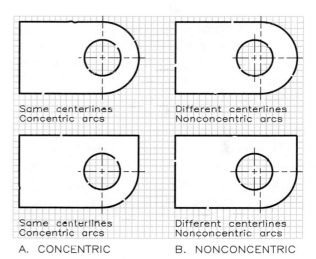

A. CONCENTRIC

B. NONCONCENTRIC

13.25 Centerlines.

A Extend centerlines beyond the last arc that has the same center.

B Sketch separate centerlines when the arcs are not concentric.

to apply centerlines to indicate the center of the circular ends of a cylinder and its vertical axis. Perpendicular centerlines cross in circular views to locate the center of the circle and extend beyond the arc by about 1/8 inch. Centerlines consist of alternating long and short dashes, about 1 inch and 1/8 inch in length, respectively.

When centerlines coincide with visible or hidden lines, the centerline should be omitted because object lines are more important and centerlines are imaginary lines. **Figure 13.24** shows the precedence of lines.

The centerlines shown in **Figure 13.25** clarify whether the circles and arcs are concentric (share the same centers). **Figure 13.26** shows the correct manner of applying centerlines to orthographic views of an object composed of concentric cylinders.

13.6 CIRCULAR FEATURES • 169

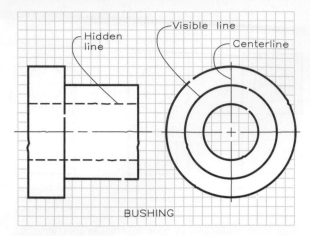

13.26 This orthographic sketch depicts the application of centerlines to concentric cylinders and the relative weights of various lines.

Sketching Circles

Circles can be sketched by either of the methods shown in **Figure 13.27** by using light guidelines and dark centerlines to block in the circle. Drawing a freehand circle in one continuous arc is difficult, so draw arcs in segments with the help of the guidelines.

A typical part having circular features is represented by two sketched views in **Figure 13.28**. Note the definitions of a **round** and a **chamfer**. **Figure 13.29** shows the steps involved in constructing three orthographic views of a part having circular features.

13.7 Pictorial Sketching: Obliques

An **oblique pictorial** is a three-dimensional representation of an object's height, width, and depth. It approximates a photograph of an object, making the sketch easier to understand at a glance than do orthographic views. Sketch the front of the object as a true-shape orthographic view (**Figure 13.30**). Sketch the receding axes at an angle of between 20° and 60° oblique with the horizontal in the front view. Lay off the depth dimension as its true

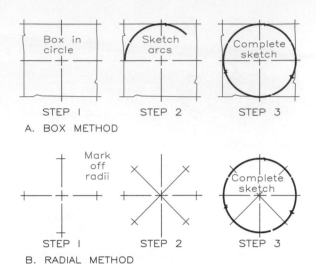

13.27 Sketching circles:

Box Method

Step 1 Block in the diameter of the circle about the centerlines.
Step 2 Sketch an arc tangent through two tangent points.
Step 3 Complete the circle with other arcs.

Radial Method

Step 1 Mark off radii on the centerlines.
Step 2 Mark off radii on two construction lines drawn at 45°.
Step 3 Sketch the circles with arcs passing through the marks.

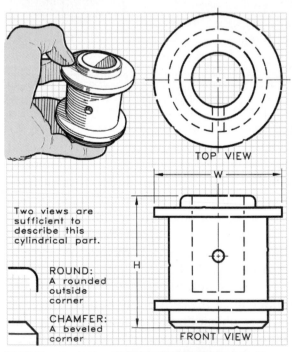

13.28 Two views adequately describe this cylindrical pivot base.

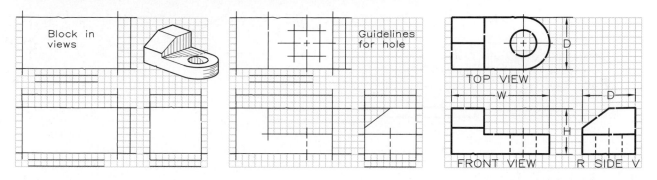

13.29 Sketching circular features:

Step 1 Begin by blocking in the overall dimensions with construction lines. Leave room for labels and dimensions.

Step 2 Draw the centerlines and squares that block in the diameter of the circle. Find the slanted surface in the side view.

Step 3 Sketch the arcs, darken the lines, label the views, and show the dimensions W, D, and H between the views.

length along the receding axes. When the depth is laid off true length, the oblique is a **cavalier** oblique.

The major advantage of an oblique pictorial is the ease of sketching circular features as circular arcs on the true-size front plane. **Figure 13.31** shows an oblique sketch of a shaft block. Circular features on the receding planes appear as ellipses, requiring slanted guidelines, as shown.

13.8 Pictorial Sketching: Isometrics

Another type of three-dimensional representation is the **isometric pictorial**, in which the axes make 120° angles with each other (**Figure 13.32**). Specially printed isometric grids with lines intersecting at 60° angles make isometric sketching easier (**Figure 13.33**). Simply transfer the dimensions from the squares in the orthographic views to the isometric grid.

You cannot measure angles in isometric pictorials with a protractor; you must find them by connecting coordinates of the angle laid off along the isometric axes. In **Figure 13.34**, locate the ends of the angular plane by using the coordinates for width and height. When a part has two sloping planes that intersect (**Figure 13.35**), you must sketch them one

at a time to find point B. Line AB is found as the line of intersection between the planes. A more thorough coverage of isometric drawing is given in Chapter 25.

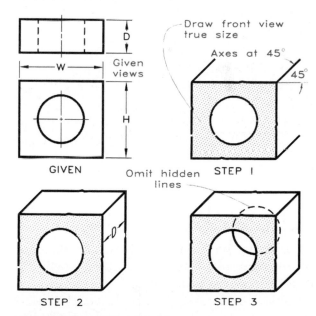

13.30 Sketching oblique pictorials:

Step 1 Sketch the front of the part as an orthographic front view and the receding lines at 45° to show the depth dimension.

Step 2 Measure the depth along the receding axes and sketch the back of the part.

Step 3 Locate the circle on the rear plane, show the visible portion of it, and omit the hidden lines.

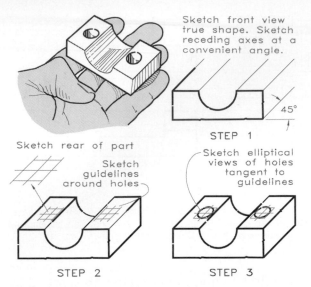

Sketch front view true shape. Sketch receding axes at a convenient angle.

45°

STEP 1

Sketch rear of part

Sketch guidelines around holes

Sketch elliptical views of holes tangent to guidelines

STEP 2

STEP 3

13.31 Sketching arcs in oblique pictorials:

Step 1 Sketch the front view of the mounting bracket saddle as a true front view. Sketch the receding axes from each corner.

Step 2 Sketch the rear of the part by measuring its depth along the receding axes. Sketch guidelines about the holes.

Step 3 Sketch the circular features as ellipses on the upper planes tangent to the guidelines.

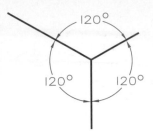

120°

120°   120°

A. THE ISOMETRIC AXES

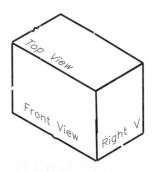

Top View

Front View

Right V

B. ISOMETRIC DRAWING

13.32 An isometric sketch:

A Begin an isometric pictorial by sketching three axes spaced 120° apart. One axis usually is vertical.

B Sketch the isometric shape parallel to the three axes and use its true measurements as the dimensions.

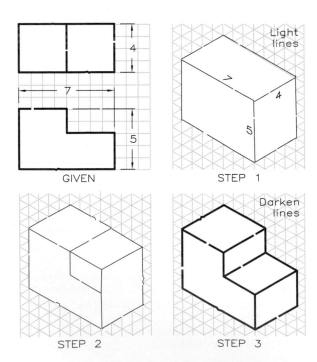

GIVEN

4

7

5

STEP 1

Light lines

7

4

5

STEP 2

STEP 3

Darken lines

Circles in Isometric

Circles appear as ellipses in isometric pictorials. When you sketch them, begin with their centerlines and construction lines enclosing their diameters, as shown in **Figure 13.36**. The end of the block is semicircular in the front view, so its center must be equidistant from the top, bottom, and end of the front view.

13.33 Sketching isometric pictorials:

Step 1 Use an isometric grid, transfer dimensions from the given views, and sketch a box having those dimensions.

Step 2 Locate the notch by measuring over four squares and down two squares, as shown in the orthographic views.

Step 3 Finish the notch and darken the lines.

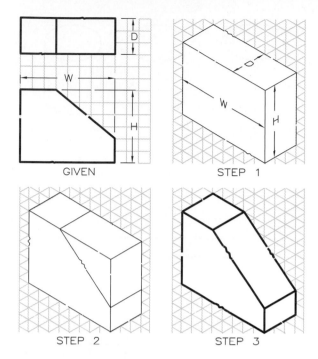

13.34 Sketching angles in isometric pictorials:

Step 1 Sketch a box from the overall dimensions given in the orthographic views.

Step 2 Angles cannot be measured with a protractor. Find each end of the angle with coordinates measured along the axes.

Step 3 Connect the ends of the angle and darken the lines.

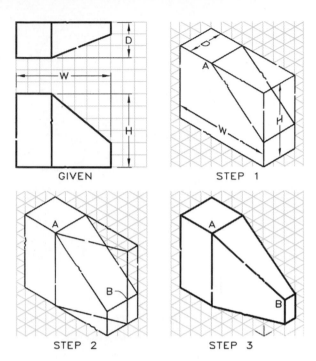

13.35 Sketching double angles in isometric pictorials:

Step 1 This object has two sloping angles that intersect. Begin by sketching the overall box and draw one of the angles.

Step 2 Find the second angle, which locates point B, the intersection line between the planes.

Step 3 Connect points A and B and darken the lines. Line AB is the line of intersection between the two sloping planes.

Circles and ellipses are easier to sketch if you use construction lines.

Figure 13.37 shows how to use centerlines and construction lines to draw ellipses in the three isometric planes: frontal, horizontal (top), and profile (side) views. This technique is used to sketch a cylinder in **Figure 13.38** and an object having semicircular ends in **Figure 13.39**. Hidden lines are usually omitted in isometric drawings.

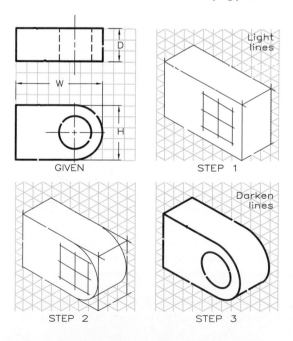

13.36 Sketching circles in isometric pictorials:

Step 1 Sketch a box using the overall dimensions given. Sketch the centerlines and a rhombus blocking in the circular hole.

Step 2 Sketch the isometric arcs tangent to the box. These arcs are elliptical rather than circular.

Step 3 Sketch the hole and darken the lines. Hidden lines usually are omitted in isometric pictorial sketches.

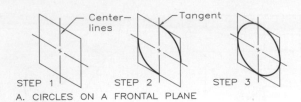

Center—
lines
Tangent

STEP 1   STEP 2   STEP 3
A. CIRCLES ON A FRONTAL PLANE

STEP 1   STEP 2   STEP 3
B. CIRCLES ON A HORIZONTAL PLANE

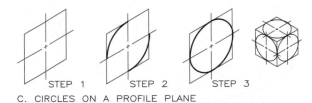

STEP 1   STEP 2   STEP 3
C. CIRCLES ON A PROFILE PLANE

13.37 Sketching circular features in isometric pictorials:

Step 1 Lay out centerlines and guidelines.

Step 2 Sketch two opposite arcs of the ellipse.

Step 3 Connect the ends of the arcs to complete the ellipses.

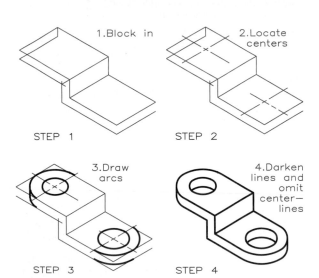

1.Block in

2.Locate centers

STEP 1   STEP 2

3.Draw arcs

4.Darken lines and omit center— lines

STEP 3   STEP 4

13.39 Sketching circular features in isometric pictorials:

Step 1 Block in the isometric shape of the object with light lines.

Step 2 Locate the centerlines of the holes and the rounded ends.

Step 3 Sketch the semicircular ends of the part and the holes.

Step 4 Draw the bottoms of the holes and darken the lines.

174 • CHAPTER 13 FREEHAND SKETCHING

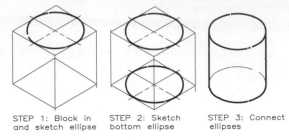

STEP 1: Block in   STEP 2: Sketch   STEP 3: Connect
and sketch ellipse   bottom ellipse   ellipses

13.38 Sketching a cylinder as an isometric pictorial:

Step 1 Block in the cylinder and sketch the upper ellipse.

Step 2 Sketch the lower ellipse.

Step 3 Connect the ellipses with lines tangent to the elliptical ends and darken the lines.

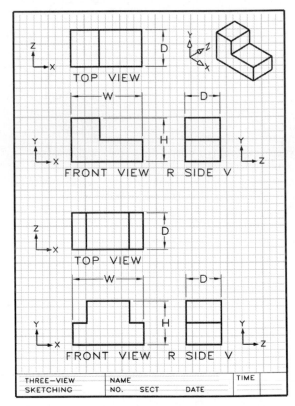

13.40 Use this layout of a size A sheet for sketching problems. You may sketch two problems on each sheet.

Problems

Sketch these problems (**Figures 13.41–13.43**) on size A (8-1/2 x 11-inch) paper, with or without a printed grid as shown in **Figure 13.40**. Each grid is equal to 0.20 in. or approximately 5 mm.

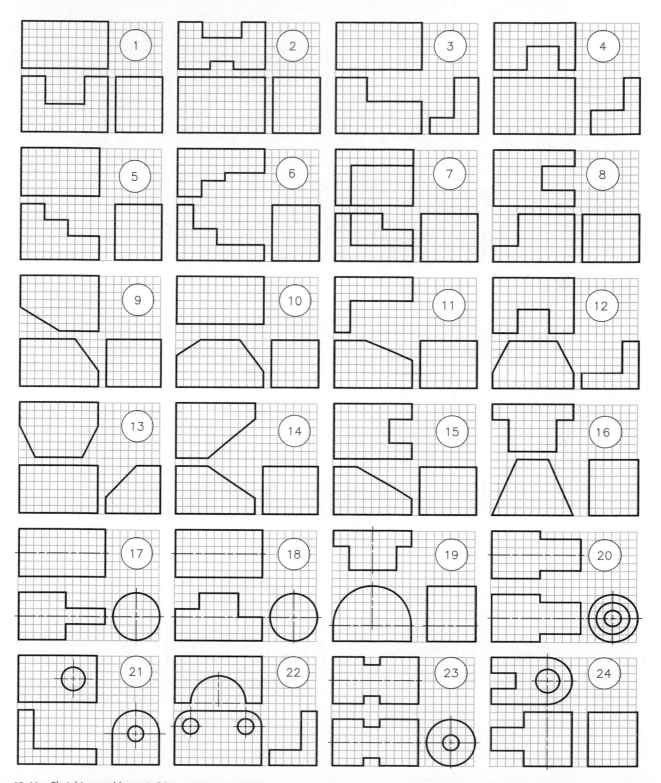

13.41 Sketching problems: 1–24

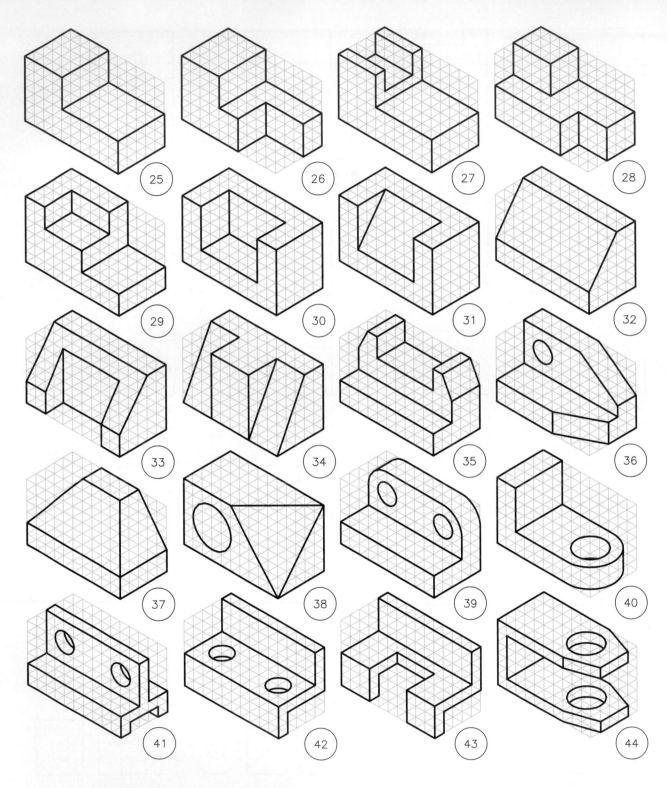

13.42 Sketching problems: 25–44

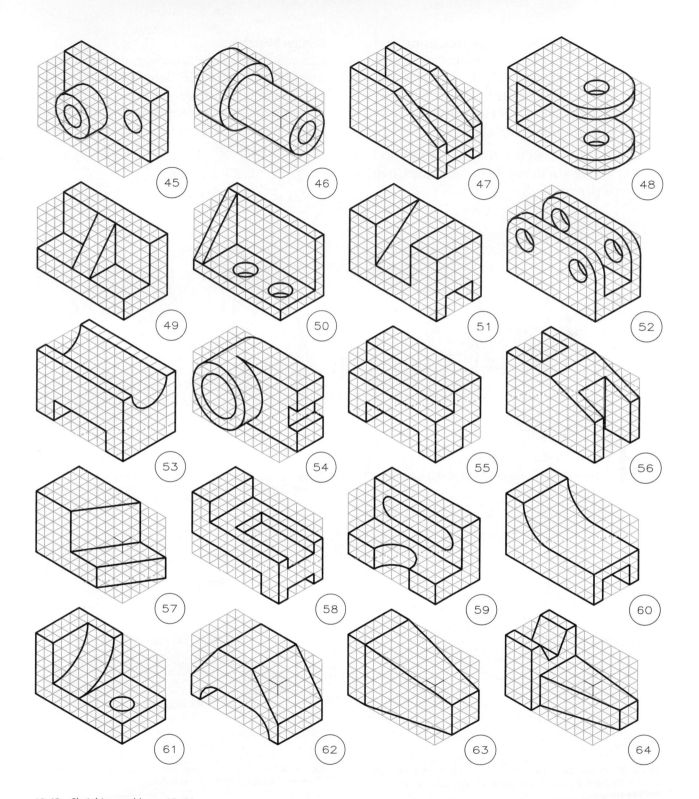

13.43 Sketching problems: 45–64

Problems 1-24: (A) Sketch the top, front, and right-side views of the problems assigned; supply lines that may be missing from all views. (B) Sketch oblique pictorials of the problems assigned. (C) Sketch isometric pictorials of the problems assigned.

Problems 25-64: Sketch the top, front, and right-side views of the problems assigned, two problems per sheet.

Design Sketching

General dimensions are given on the following problems. You must determine all missing information as if you were the original designer of the parts. Sketch your solutions on size A sheets (8-1/2 x 11), either a vertical or horizontal format.

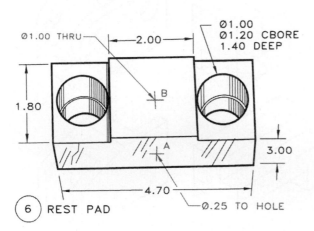

Design 1: Sketch the necessary views of the rest pad and add the two new holes as specified.

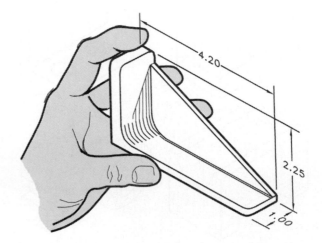

Design 3: Modify the aircraft bracket to have the triangular rib moved to the center of the part. Show fillets and rounds. Sketch the necessary orthographic views.

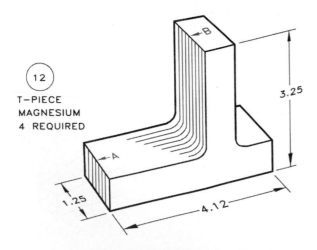

Design 2: Modify the T-piece to have a 0.25 in.-thick rib from point A to point B. Sketch the necessary orthographic views to describe the part.

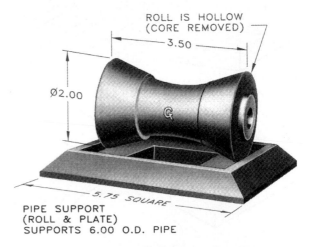

Design 4: Sketch orthographic views of both parts of the pipe support. The inside of the roll is hollow.

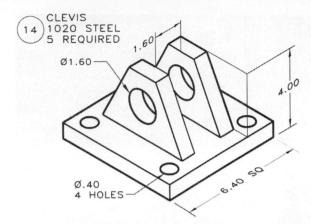

CLEVIS
(14) 1020 STEEL
5 REQUIRED

Ø1.60

1.60

4.00

Ø.40
4 HOLES

6.40 SQ

Design 5: Modify the clevis to have semicircular ends about the 1.60 DIA holes and make concentric rounded corners at each 0.40 DIA hole. Sketch the necessary views.

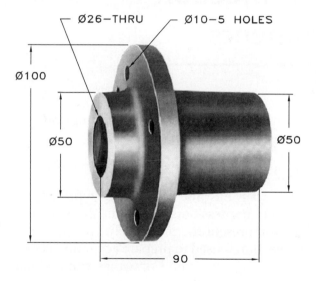

Ø26–THRU   Ø10–5 HOLES

Ø100

Ø50

Ø50

90

Design 7: Modify the socket by moving the flange to the center of the part's length. Sketch the necessary views to describe the part on a size A sheet.

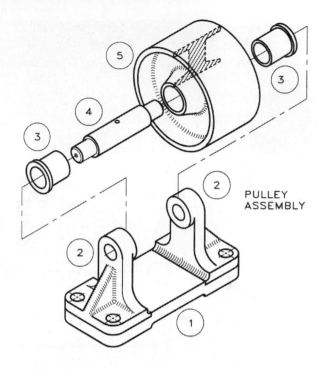

PULLEY
ASSEMBLY

Design 6: There are five parts in the pulley assembly in Design 6, which has a 1 in. DIA shaft (Part 4). Sketch one part per A size sheet following the instructions below.

Option 1: Make a two-view sketch of the shaft (Part 4). Do you know why there are holes in the shaft?

Option 2: Make a two-view drawing of the bushing (Part 3). What are bushings and why are they used? What type of material is best for a bushing?

Option 3: Make a two-view drawing of the base (Part 1). How are the brackets (Part 2) connected to the base?

Option 4: Make a three-view drawing of the shaft bracket (Part 2). How do the brackets support the shaft?

Option 5: Make a two-view drawing of the pulley (Part 5). Why does it have a boss (hub) protruding from it through which the shaft (Part 4), passes?

Option 6: Redesign the bracket (Part 2) and show your proposed modification in a three-view sketch.

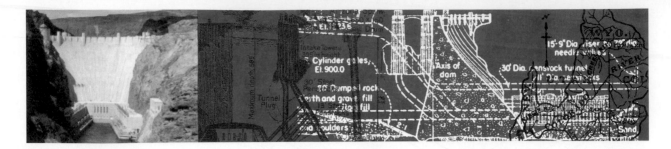

14

Orthographic Projection
with Instruments

14.1 Introduction

In Chapter 13, you were introduced to ortho-graphic projection by freehand sketching, which is an excellent way to develop a design concept. Now, you must convert these sketches into orthographic views drawn to scale with instruments (or by computer) to more precisely define your design. Afterwards, you will add dimensions, notes, and specifications to convert these drawings into working drawings from which the design will become a reality.

Orthographic drawings are three-dimen-sional objects represented by separate views arranged in a standard manner that are read-ily understood by the technological team. Because multiview drawings usually are exe-cuted with instruments and drafting aids, they are often called **mechanical drawings**. They are called **working drawings**, or **detail drawings**, when sufficient dimensions, notes, and specifications are added to enable the

product to be manufactured or built from the drawings.

14.2 Orthographic Projection

An artist has the option of representing objects impressionistically, but the engineer must represent them precisely. Orthographic projection is used to prepare accurate, scaled, and clearly presented drawings from which the project depicted can be built.

Orthographic projection is the system of drawing views of an object by projecting them perpendicularly onto projection planes with parallel projectors. **Figure 14.1** illustrates this concept of projection by imagining that the object is inside a glass box and three of its views are projected to planes of the box.

Figure 14.2 illustrates the principle of orthographic projection where the front view is projected perpendicularly onto a vertical projection plane, called the frontal plane, with parallel projectors. The projected front view is

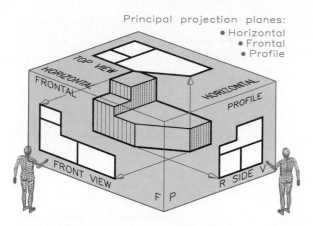

Principal projection planes:
• Horizontal
• Frontal
• Profile

14.1 Orthographic projection is the system of projecting views onto an imaginary glass box with parallel projectors to the three mutually perpendicular projection planes.

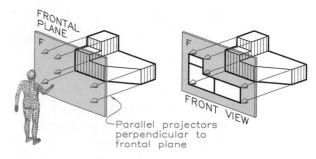

14.2 An orthographic view is found by projecting from the object to a projection plane with parallel projectors that are perpendicular to the projection plane.

two dimensional because it has only width and height and lies in a single plane. Similarly, the top view is projected onto a horizontal projection plane, and the side view is projected onto a second vertical projection plane.

Imagine that the box is opened into the plane of the drawing surface. **Figure 14.3A** illustrates how three planes of a glass box are

opened into a single plane (**Figure 14.3B**) to yield the standard positions for the three orthographic views. These views are the front, top, and right-side views.

The principal projection planes of orthographic projection are the **horizontal (H)**, **frontal (F)**, and **profile (P)** planes. Views projected onto these principal planes are principal views. The dimensions used to give the sizes of principal views are **height (H)**, **width (W)**, and **depth (D)**.

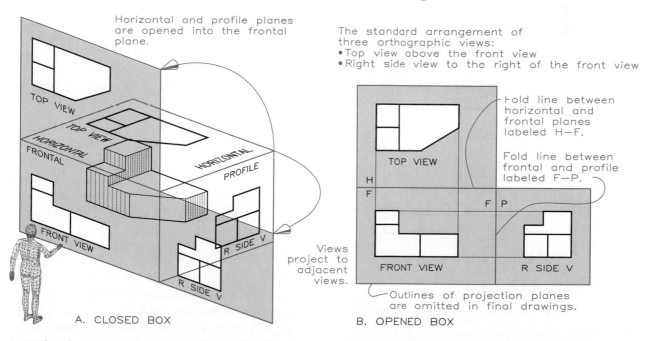

Horizontal and profile planes are opened into the frontal plane.

A. CLOSED BOX

Views project to adjacent views.

The standard arrangement of three orthographic views:
• Top view above the front view
• Right side view to the right of the front view

Fold line between horizontal and frontal planes labeled H—F.

Fold line between frontal and profile labeled F—P.

Outlines of projection planes are omitted in final drawings.

B. OPENED BOX

14.3 When the imaginary glass box is opened, the orthographic views and labeling are drawn in this format shown at B.

14.3 Alphabet of Lines

Draw all orthographic views with dark and dense lines as if drawn with ink. Only the line widths should vary, except for guidelines and construction lines, which are drawn very lightly for layout and lettering. **Figure 14.4** gives examples of lines used in orthographic projection and the recommended pencil grades for them. The lengths of dashes in hidden lines and centerlines are drawn longer as a drawing's size increases. **Figure 14.5** further describes these lines.

Computer Lines Computer graphics plotters use pens with points of varying widths, usually 0.7 mm and 0.3 mm wide, to draw lines of different thicknesses. To vary the lengths of dashes and the spaces between them in dashed lines, use *Ltscale* (**Figure 14.6**). When *Ltscale* is used, **all** noncontinuous lines (centerlines and hidden lines) on the drawing are changed at the same time.

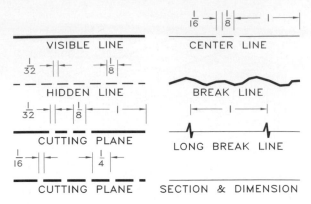

14.5 Full-size line weights recommended for drawing orthographic views are shown above.

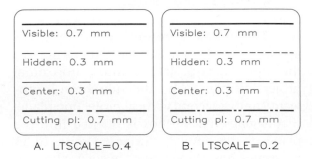

14.6 These lines were drawn using AutoCAD's standard LINETYPEs and two pens (P.3 and P.7 points). (**A**) The LTSCALE factor of 0.4 gives relatively long dashes and spaces between dashes. (**B**) By contrast, the LTSCALE factor of 0.2 produces dashes and spaces that are half as long as the factor of 0.4.

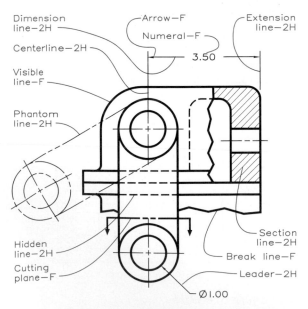

14.4 The alphabet of lines and recommended pencil grades for drawing orthographic views are shown here.

The laser and dot-matrix printers have come into widespread usage and have replaced many pen plotters. The ability to print in color with a high degree of sharpness is the major advantage of these printers over the pen plotter. However, drawings in the 36 in. range are plotted mostly on the pen plotter.

14.4 Six-View Drawings

When you imagine that an object is inside a glass box you will see two horizontal planes, two frontal planes, and two profile planes (**Figure 14.7**). Therefore, the maximum number of principal views that can be used to rep-

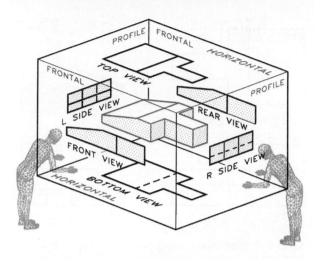

14.7 Six principal views of an object can be drawn in orthographic projection. Imagine that the object is in a glass box with the views projected onto its six planes.

resent an object is six. The top and bottom views are projected onto horizontal planes, the front and rear views onto frontal planes, and the right- and left-side views onto profile planes.

To draw the six views on a sheet of paper, imagine the glass box is opened up into the plane of the drawing paper as shown in **Figure 14.8**. Place the top view over and the bottom view under the front view; place the right-side view to the right and the left-side view to the left of the front view; and place the rear view to the left of the left-side view.

Projectors align the views both horizontally and vertically about the front view. Each side of the fold lines of the glass box is labeled **H**, **F**, or **P** (horizontal, frontal, or profile) to identify the projection planes on each side of the imaginary fold lines (**Figure 14.8**).

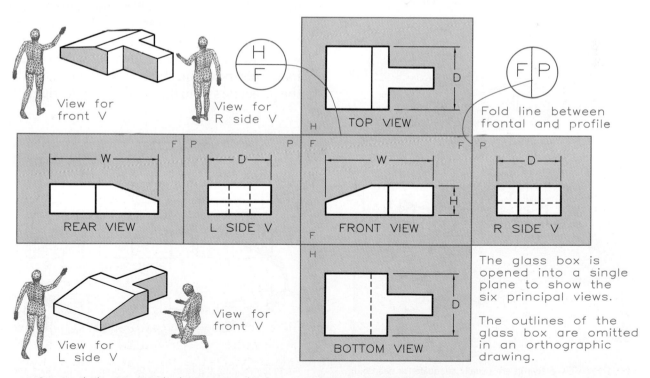

14.8 Opening the box into a single plane positions the six views as shown to describe the object.

Height (H), width (W), and depth (D), the three dimensions necessary to dimension an object, are shown in their recommended positions in **Figure 14.8**. The standard arrangement of the six views allows the views to share dimensions by projection. For example, the height dimension, which is shown only once between the front and right-side views, applies to the four horizontally aligned views. The width dimension is placed between the top and front views, but applies to the bottom view also.

14.5 Three-View Drawings

The most commonly used orthographic arrangement of views is the three-view drawing, consisting of front, top, and right-side views. Imagine that the views of the object are projected onto the planes of the glass box (**Figure 14.9**) and the three planes are opened

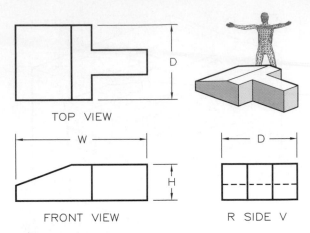

14.10 This three-view drawing depicts the object shown in Figure 14.9.

into a single plane, the frontal plane. **Figure 14.10** shows the resulting three-view drawing where the views are labeled and dimensioned with H, W, and D.

In addition to using the glass-box approach for visualizing orthographic projection, imagine that the object is revolved until the desired view is obtained. For example, if the front view of the part in **Figure 14.11** is revolved 90°, you will see its side view true shape (TS). Placed in its proper position, the side view aligns with the projectors from the front view at its right.

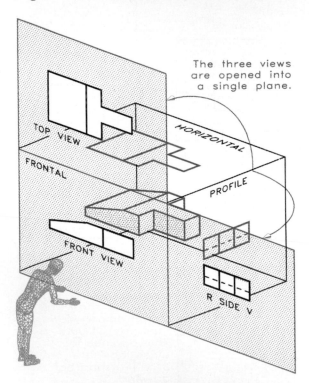

14.9 Three-view drawings are usually adequate for describing most small objects, such as machine parts.

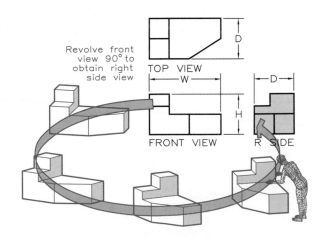

14.11 By rotating the front view 90° about a vertical axis, the side view is found.

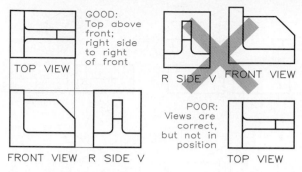

14.12 Orthographic views must be arranged in their proper positions in order for them to be interpreted correctly.

14.6 Arrangement of Views

Figure 14.12 shows the standard positions for a three-view drawing: The top and side views are projected from and aligned with the front view. Improperly arranged views that do not project from view to view are also shown. **Figure 14.13** illustrates the rules of projection and shows the proper alignment of dimensions. Orthographic projection shortens layout time, improves readability, and reduces the number of dimensions required because they are placed between and are shared by the views to which they apply.

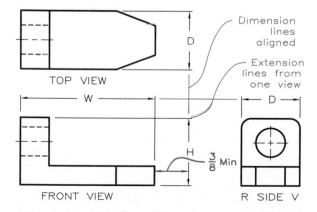

14.13 Dimension and extension lines used in three-view orthographic projection should be aligned. Draw extension lines from only one view when dimensions are placed between views.

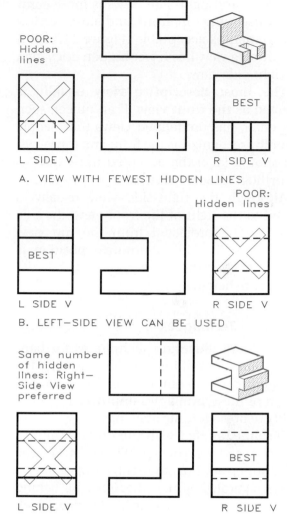

14.14 Selection of views.

A Select the sequence of views with the fewest hidden lines.

B Select the left-side view because it has fewer hidden lines than the right-side view.

C When both views have an equal number of hidden lines, select the right-side view.

14.7 Selection of Views

Select the sequence of orthographic views with the fewest hidden lines. **Figure 14.14A** shows that the right-side view is preferable to the left-side view because it has fewer hidden lines. Although the three-view arrangement of top, front, and right-side views is more commonly used, the top, front, and left-side view arrangement is acceptable (**Figure 14.14B**) if the left-side view has fewer hidden lines than the right-side view.

The most descriptive view usually is selected as the front view. If an object, such as a chair, has predefined views that people generally recognize as the front and top views, you should label the accepted front view as the orthographic front view.

Although the right-side view usually is placed to the right of the front view, the side view can be projected from the top view (**Figure 14.15**). This alternative position is advisable when the object has a much larger depth than height.

14.8 Line Techniques

Figure 14.16 illustrates techniques for handling most types of intersecting lines, hidden lines, and arcs in combination. Proper application of these principles improves the readability of orthographic drawings.

Become familiar with the order of importance (precedence) of lines (**Figure 14.17**). The most important line, the visible object line, is shown regardless of any other line lying behind it. Of next importance is the hidden line, which is more important than the centerline.

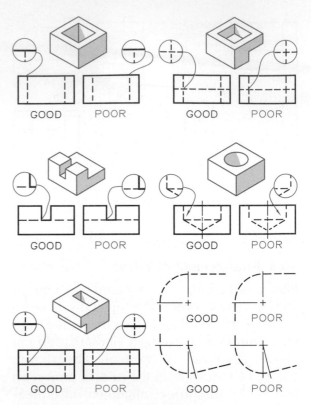

14.16 These drawings show proper intersections and other line techniques in orthographic views.

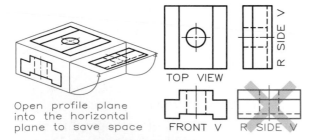

14.15 The side view can be projected from the top view instead of the front view. This alternate position saves space when the depth of an object is considerably greater than its height.

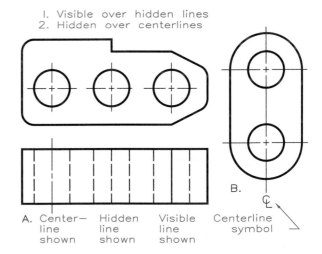

14.17 When lines conincide with each other, the more important lines take precedence (cover up) the other lines. The order of importance is: visible lines, hidden lines, and center lines.

14.9 Point Numbering

Some orthographic views are difficult to draw due to their complexity. By numbering the endpoints of the lines of the parts in each view as you construct it (**Figure 14.18**), the location of the object's features will be easier. For example, using numbers on the top and side views of this object aids in the construction of the missing front view. Projecting points from the top and side views to the intersections of the projectors locates the object's front view.

14.10 Lines and Planes

A line can appear **true length (TL)**, **foreshortened (FS)**, or as a **point (PT)** in an orthographic view (**Figure 14.19**). A line that appears true length is parallel to the reference line in the adjacent view. The reference line in

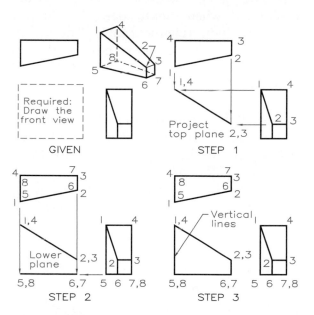

14.18 Point numbering:

Required: Find the front view.

Step 1 Number the corners of plane 1-2-3-4 in the top and side views and project these points to the front view.

Step 2 Number the corners of plane 5-6-7-8 in the top and side views and project these points to the front view.

Step 3 Connect the numbered lines to complete the front view.

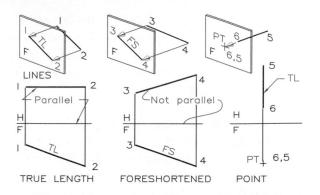

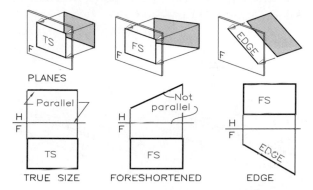

14.19 A line will appear in orthographic projection as true length, foreshortened, or as a point. A plane in orthographic projection will appear as true size, foreshortened, or as an edge.

these examples represents the fold line between the horizontal and frontal planes and therefore is labeled H-F. A plane can appear true size (TS), foreshortened (FS), or as an edge in orthographic projection.

Lines and planes that are true length or true size can be measured with a scale. Foreshortened lines and planes are not true length and are less than full size.

14.11 Drawing with Triangles

Two triangles can be effectively used to make instrument drawings on 8-1/2 x 11 sheets without taping the sheet to the drawing surface. It is better that it can be moved about to comfortably position the triangles.

Parallel lines The 45° triangle or the 30°-60° triangle can be used to draw parallel lines as shown in **Figure 14.20**. One triangle or a straightedge is held in position while the other triangle is moved to where the parallel line is drawn.

Perpendiculars To draw a line perpendicular to AB in **Figure 14.21**, align the 30°-60° triangle's hypotenuse side with AB and against the lower triangle or straightedge. Hold the lower triangle in postion and rotate the triangle so that the hypotenuse side is perpendicular to AB and draw CD.

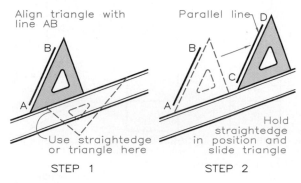

STEP 1   STEP 2

14.20 Drawing parallel lines:

Step 1 Align the upper triangle with AB and in contact with the lower triangle (or straightedge).

Step 2 Hold the lower straightedge in position and slide the upper triangle to where CD is drawn parallel to AB.

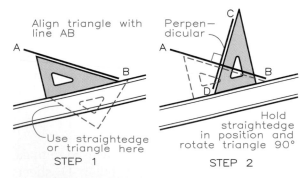

STEP 1   STEP 2

14.21 Drawing perpendiculars:

Step 1 Align your triangle with line AB and in contact with the lower triangle (or straightedge).

Step 2 Hold the lower straightedge in position, rotate your triangle, and draw CD perpendicular to AB.

Angles To draw a line making a 30° angle with AB, use the steps shown in **Figure 14.22**. Two triangles can be used in other combinations as a means of making instrument drawings in this informal manner, and yet with a sufficient degree of accuracy.

14.12 Views by Subtraction

Figure 14.23 illustrates how three views of a part are drawn by beginning with a block having the overall height, width, and depth of the finished part and removing volumes from it. This drawing procedure is similar to the steps of making the part in the shop.

14.13 Three-View Drawings

The depth dimension applies to both the top and side views, but these views usually are positioned where depth does not project between them (**Figure 14.24**). The depth dimension can be transferred between the top and side views with dividers or by using a 45° miter line.

Layout Rules The basic rules of making orthographic drawings are summarized below. Refer to the examples in **Figure 14.25**

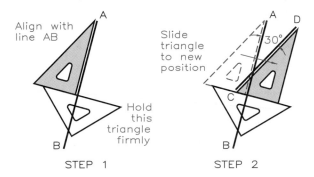

STEP 1   STEP 2

14.22 Drawing a 30° angle:

Step 1 Hold the 30°-60° triangle aligned with AB and in contact with the lower triangle (or straightedge).

Step 2 Hold the triangle in position and slide the triangle and draw CD at 30° to AB.

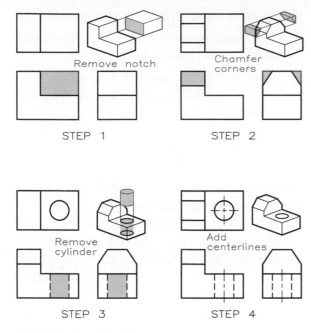

STEP 1 — Remove notch

STEP 2 — Chamfer corners

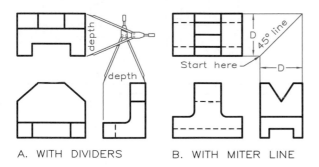

STEP 3 — Remove cylinder

STEP 4 — Add centerlines

14.23 Views by subtraction:

Step 1 Block in the views of the object using overall dimensions of H, W, and D. Create the notch by removing the block.

Step 2 Remove the triangular volumes at the corners.

Step 3 Form the hole by removing a cylindrical volume.

Step 4 Add centerlines to complete the views.

A. WITH DIVIDERS

B. WITH MITER LINE

14.24 Transferring depth:

A Transfer the depth dimension to the side view from the top view with your dividers.

B Use a 45° miter line to transfer the depth dimension between the top and side views by projection.

thru **Figure 14.29** and observe how these rules have been used. Notice how the dimensions have been applied and how the views have been labeled.

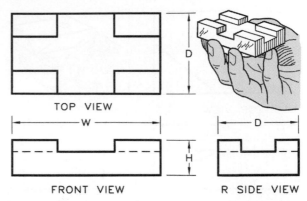

TOP VIEW

FRONT VIEW   R SIDE VIEW

14.25 This three-view drawing depicts an object that has only horizontal and vertical planes.

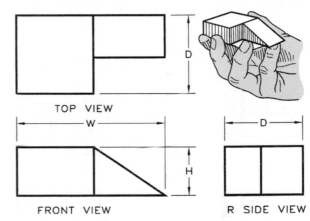

TOP VIEW

FRONT VIEW   R SIDE VIEW

14.26 This three-view drawing shows an object that has a sloping plane.

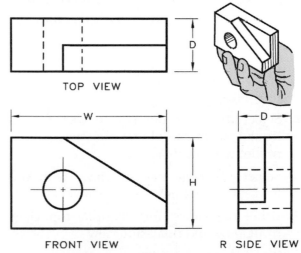

TOP VIEW

FRONT VIEW   R SIDE VIEW

14.27 This three-view drawing shows an object that has a sloping plane with a cylindrical hole through it.

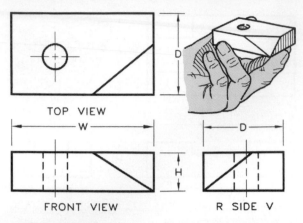

14.28 This three-view drawing depicts an object that has a plane with a compound slope.

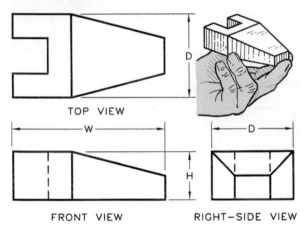

14.29 This three-view drawing shows an object that has planes with compound slopes.

1. Draw orthographic views in their proper positions.

2. Select the most descriptive view as the front view, if the object does not have a predefined front view.

3. Select the sequence of views with the fewest hidden lines.

4. Label the views; for example, top view, front view, and right-side view.

5. Place dimensions between the views to which they apply.

6. Use the proper alphabet of lines.

7. Leave adequate room between the views for labels and dimensions.

8. Draw the views necessary to describe a part. Sometimes fewer and more views are required.

14.14 Views by Computer

The steps of drawing three orthographic views of an object are shown in **Figure 14.30** where the *Line* command is used to draw the overall outlines of the three views. Other visible and hidden lines are added in by projecting from view to view. The *Dtext* command is used to label the views if this is desired. Additional details of using the *Line* command can be found in Chapter 37.

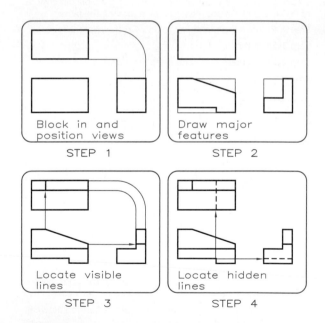

14.30 Three views by computer:

Step 1 Use the *Line* command and draw the outlines of the views.

Step 2 Use orthographic projection and draw the major features.

Step 3 Draw the visible lines.

Step 4 Draw the hidden lines.

Three-dimensional (3D) Solids The object in **Figure 14.31A** was used as the example in **Figure 14.32** to show how a 3D solid can be constructed and *rendered* as in **Figure 14.31B**. Once the part is drawn as a 3D solid, it can be viewed from different directions to obtain the standard orthographic views.

In order to make a three-view orthographic layout, set *Tilemode* to 0, which changes the

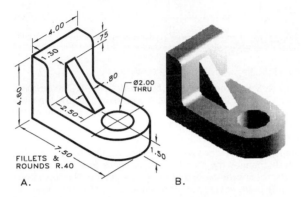

A.   B.

14.31 The given problem (left) was constructed as a 3D solid and rendered as shown at the right by using AutoCAD.

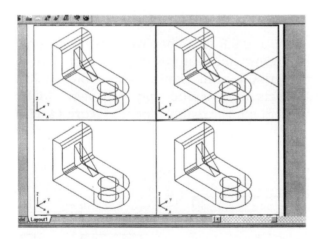

14.32 Create floating viewports:

Step 1 *Command:* Tilemode> 0. (The screen is set to *Paper Space (PS)* with a triangle icon in the lower left corner.)

Step 2 *Command:* Mview> (Select 4 to specify 4 viewports)> Fit (The screen is filled with four viewports with the part shown in each one.)An *X-Y-Z Model-Space* icon appears in each viewport. Type MS (model space) and select a viewport with your cursor to make it the active one.

screen to *Paper Space (PS)* and blanks the screen. Before the 3D solid can be seen, you must create *Floating Viewports* with the *Mview* command (make view) and select *4* and the *Fit* option to create 4 viewports on the screen (**Figure 14.32**). Type Ms to enter *Model Space* and select the upper right port to make it active. Type UCS (*User Coordinate System*) and select the option of *Save*, and after *Desired UCS name*, type ISO to name and keep this isometric view (**Figure 14.33**).

Select the upper left port with the cursor, type *plan* to make X-Y icon parallel to the top of the object. Type *UCS*, select *Save*, and name the view TOP. The X-Y icon must be rotated 90° to be parallel with the front of the part; pick *UCS> X> 90*. Then type *Plan* to display the front view. Type *UCS> Save> Name> FRONT.*

Select the lower right port to make it active and make the X-Y icon parallel to the side view (**Figure 14.34**): Type UCS> Y> 90. Type PLAN to get the side view; Type UCS> Save> Name> SIDE.

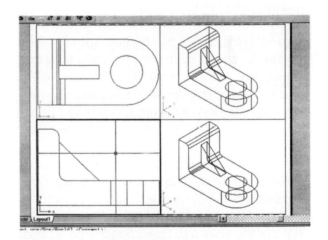

14.33 Creating top and front views:

Step 1 Select the upper right viewport, pick *UCS> Save> Desired UCS name:* ISO (Enter).

Step 2 Select the upper left viewport and type Plan to obtain a view parallel to the X-Y axes, the top view. Pick *UCS> Save>Desired UCS name:* TOP (Enter). This saves the top view UCS.

Step 3 Select the lower left viewport; pick *UCS> X> Rotation> 90* to make the X-Y axes parallel to the front of the object. Type Plan to get a front view; pick *UCS> Save> Name>* Front (Enter) to save the UCS.

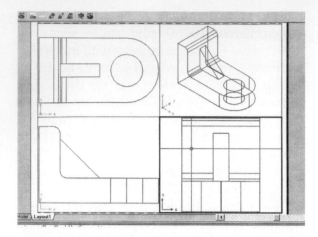

14.34 Create a side view:

Step 1 Select the lower right viewport and pick *UCS> X> 90* to make the X–Y icon line in vertical plane. Type *UCS> Y> 90* to make the X–Y axes parallel to the side of the part. Type <u>Plan</u> to obtain the side view.

Step 2 Pick *UCS> Save> Name > SIDE*, to save the *UCS* named <u>SIDE</u> for the side view. Now the three orthographic views are obtained.

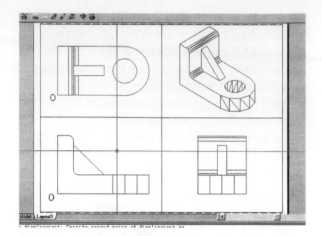

14.35 Scaling and aligning the views:

Step 1 Select the top view by cursor; pick *UCS> Restore> Name?> TOP*, to restore the X–Y icon parallel to the top view. Type *Zoom> 1.2* and the top view is sized to properly fill the viewport. Repeat this step for the front and side views to size the views equally.

Step 2 Align the views by typing *Mvsetup* (Multi View Set Up), which changes the screen to *Paper space (PS)*. Select *Align> Vertical> Specify basepoint>* (select a point in the top view)> *Specify point in viewport to be panned:>* (select the same corner in the front view to align the top and front views). Repeat but use the horizontal option to align the front and side views.

The three views that have been found are probably shown at different sizes and are not aligned correctly as orthographic views should be. Select each view one at a time and restore the saved views. For example, select the top viewport, type UCS> Restore> Name> Top, and the X-Y icon returns parallel to the top view. Type Zoom> 1.2 to size the top view to appropriately fill the viewport. Select the front- and side-view ports, restore the views by name, and use Zoom to 1.2 to make them equal in size to each other.

Type <u>MVsetup</u> (Model View Setup), which changes the screen to Paper Space (PS), to align the views properly as shown in **Figure 14.35**. Type MVsetup> Align> Select base point> (Select a corner point of the top view)> Other point> (Pick the same point in the adjacent viewport) and the views are automatically aligned. Now you have three orthographic views drawn from a single 3D solid. Refer to Chapter 38 for more coverage of solid modeling.

14.15 Two-View Drawings

Time and effort can be saved by drawing only the views and features that are necessary to describe a part. **Figure 14.36** shows typical objects that require only two views to be described. The fixture block in **Figure 14.37** is

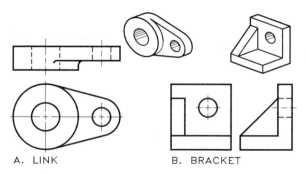

A. LINK   B. BRACKET

14.36 These objects can be adequately described with two orthographic views.

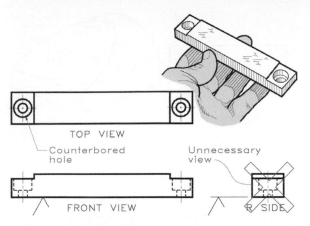

14.37 Two views adequately describe this part.

another example of a part needing only two views to be adequately described. You can see that the top and front views are the best for this part. The front and side views would not be as good.

14.16 One-View Drawings

Simple cylindrical parts and parts of a uniform thickness can be described by only one view as shown in **Figure 14.38**. Supplementary notes clarify features that would have been shown in the omitted views.

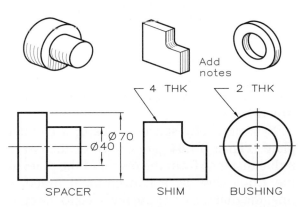

14.38 Objects that are cylindrical or of a uniform thickness can be described with only one orthographic view and supplementary notes.

Diameters are labeled with diameter signs and thicknesses are noted.

14.17 Simplified and Removed Views

The right- and left-side views of the part in **Figure 14.39** would be harder to interpret if all hidden lines were drawn by rigorously following the rules of orthographic projection. Simplified views in which confusing and unnecessary lines have been omitted are better and more readable.

When it is difficult to show a feature with a standard orthographic view because of its location, a **removed view** can be drawn (**Figure 14.40**). The removed view, indicated

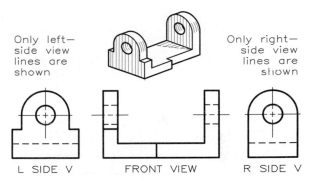

14.39 Use simplified views with unnecessary and confusing hidden lines omitted to improve clarity.

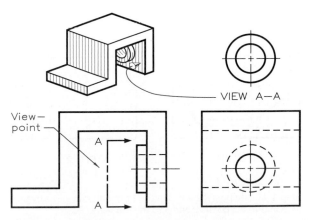

14.40 Use a removed view, indicated by the directional arrows, to show hard-to-see views in removed locations.

by the directional arrows, is clearer when moved to an isolated position.

14.18 Partial Views

Partial views of symmetrical or cylindrical parts may be used to save time and space. Omitting the rear of the circular top view in **Figure 14.41** saves space without sacrificing clarity. To clarify that a part of the view has been omitted, a conventional break is used in the top view.

14.19 Curve Plotting

An irregular curve can be plotted by following the rules of orthographic projection as shown in **Figure 14.42**. Begin by numbering the points in the given front and side views along the curve.

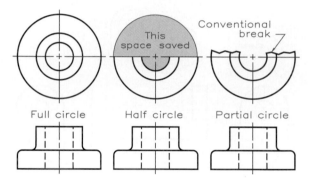

14. 41 Save space and time by drawing the circular view of a cylindrical part as a partial view.

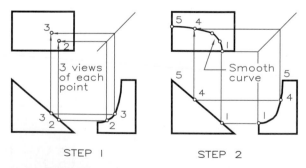

STEP 1   STEP 2

14.42 Curve plotting:

Step 1 Locate points 2 and 3 in the front and side views by projection. Project points 2 and 3 to the top view.

Step 2 Locate the remaining points in the three views and connect the points in the top view with a smooth curve.

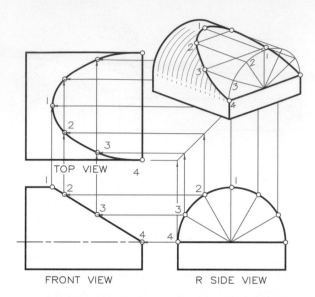

FRONT VIEW   R SIDE VIEW

14.43 The ellipse in the top view was found by numbering points in the front and side views and projecting them to the top view.

Next, project from the points having the same numbers in the front and side views to the top where the projectors intersect. Continue projecting in this manner and connect the points in the top view with a smooth curve drawn with an irregular curve. **Figure 14.43** shows an ellipse plotted in the top view by projecting points from front and side-views. It is best to number them one at a time as they are transferred to avoid getting lost in your construction.

14.20 Conventional Practices

The readability of an orthographic view may be improved if the rules of projection are violated. Violations of rules customarily made for the sake of clarity are called **conventional practices**.

Symmetrically spaced holes in a circular plate (**Figure 14.44**) are drawn at their true radial distance from the center of the plate in the front view as a conventional practice. Imagine that the holes are revolved to the centerline in the top view before projecting them to the front view.

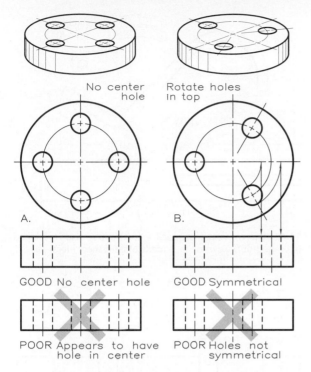

A.

GOOD No center hole

POOR Appears to have hole in center

B.

GOOD Symmetrical

POOR Holes not symmetrical

14.44 Placement of holes.

A Omit the center hole found by true projection that gives an impression that a hole passes through the center of the plate.

B Use a conventional view to show the holes located at their true radial distances from the center. They are imagined to be rotated to the centerline in the top view.

This principle of revolution also applies to symmetrically positioned features such as ribs, webs, and the three lugs on the outside of the part shown in **Figure 14.45**. **Figure 14.46** shows the applications of conventional practices to holes and ribs in combination.

Another conventional revolution is illustrated in **Figure 14.47** where the front view of an inclined arm is revolved to a horizontal position so that it can be drawn true size in the top view. The revolved arm in the front view is not drawn because the revolution is imaginary.

Figure 14.48 shows how to improve views of parts by conventional revolution. By revolving the top views of these parts 45°, slots and holes no longer coincide with the centerlines

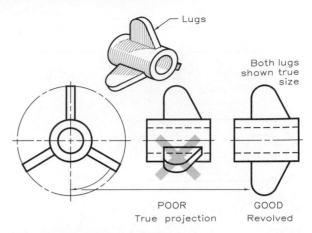

14.45 Symmetrically positioned external features, such as webs, ribs, and these lugs, are imagined to be revolved to their true-size positions for the best views.

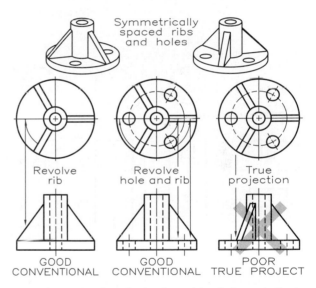

14.46 Conventional methods of revolving holes and ribs in combination improve clarity.

and can be seen more clearly. Draw the front views of the slots and holes true size by imagining that they have been revolved 45°.

Another type of conventional view is the true-size development of a curved sheet-metal part drawn as a flattened-out view (**Figure 14.49**). The top view shows the part's curvature.

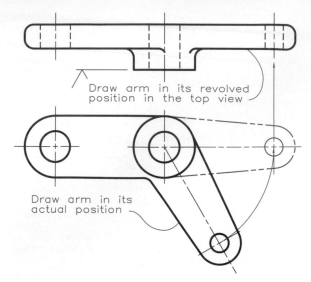

14.47 Imagine that the front view of the arm is revolved so its true length can be drawn in the top view as a conventional practice.

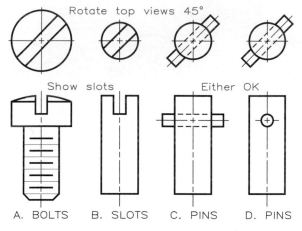

14.48 It is conventional practice to draw the slots and holes at 45° in the top view and true size in the front view.

14.21 Conventional Intersections

In orthographic projection, lines are drawn to represent the intersections (fold lines) between planes of object. Wherever planes intersect, forming an edge, this line of intersection is projected to its adjacent view. Examples showing where lines are required are given in **Figure 14.50**.

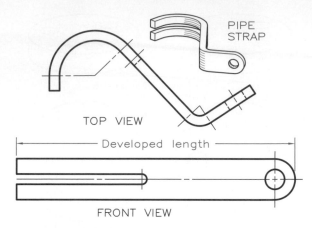

14.49 It is conventional practice to use true-size developed (flattened-out) views of parts made of bent sheet metal.

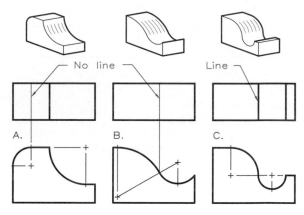

14.50 Object lines are drawn only where there are sharp intersections or where arcs are tangent at their centerlines as at C.

Figure 14.51 shows how to draw intersections between cylinders rather than plotting more complex, orthographically correct lines of intersection. **Figures 14.51A** and **C** show conventional intersections, which means they are approximations drawn for ease of construction while being sufficiently representative of the object. **Figure 14.51B** shows an easy-to-draw intersection between cylinders of equal diameters, and this is a true intersection as well. **Figures 14.52** and **14.53** show other cylindrical intersections, and **Figure 14.54** shows conventional practices for depicting intersections formed by holes in cylinders.

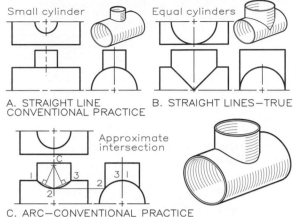

A. STRAIGHT LINE
CONVENTIONAL PRACTICE

B. STRAIGHT LINES—TRUE

Approximate
intersection

C. ARC—CONVENTIONAL PRACTICE

14.51 Intersections between cylinders.

A and **C** Use these methods of construction.

B Equal size cylinders have straight-line intersections.

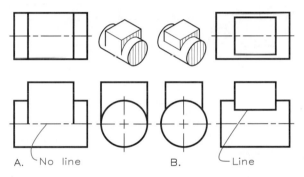

A. No line   B. Line

14.52 These are true intersections between cylinders and prisms.

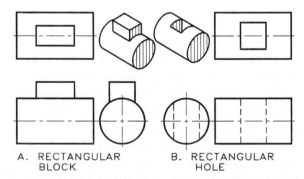

A. RECTANGULAR
BLOCK

B. RECTANGULAR
HOLE

14.53 These are conventional intersections between cylinders and prisms.

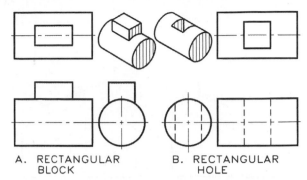

A. RECTANGULAR
BLOCK

B. RECTANGULAR
HOLE

14.54 Conventional methods of describing holes in cylinders are easy to draw and to understand.

14.22 Fillets and Rounds

Fillets and rounds are rounded intersections between the planes of a part that are used on castings, such as the body of the pillow block in **Figure 14.55**. A fillet is an inside rounding and a round is an external rounding on a part. The radii of fillets and rounds usually are small, about 1/4 inch. Fillets give added strength at inside corners, rounds improve appearance, and both remove sharp edges (**Figure 14.56**).

A casting will have square corners when its surface has been finished, which is the process of machining away part of the surface to a smooth finish (**Figure 14.57**). Finished surfaces are indicated by placing a finish mark √ on all edge views of finished surfaces whether the edges are visible or hidden. **Figure 14.58** shows four types of finish marks. A more detailed surface texture symbol is presented in Chapter 21.

14.55 The edges of this pillow block are rounded with fillets and rounds. The surface of the casting is rough except where it has been machined. (Courtesy of Dodge Mfgr. Company.)

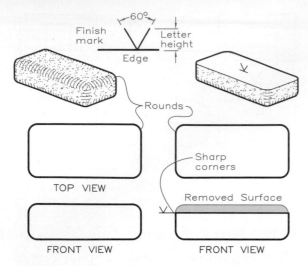

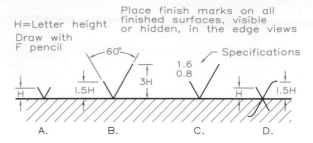

14.58 Any of these finish marks are placed on all views of finished surfaces, visible or hidden.

14.56 When a surface is finished (machined), the cut removes the rounded corners and leaves sharp corners. The finish mark placed on the edge of the surface indicates that it is to be finished.

Figure 14.59 illustrates several techniques for showing fillets and rounds on orthographic views with a circle template.

Computer Method Fillets and rounds are drawn by computer as shown in **Figure 14.60**.

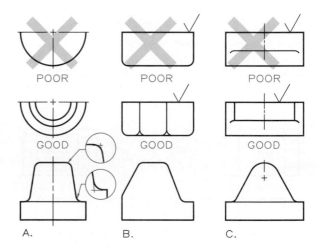

14.59 These examples show both poorly drawn and conventionally drawn fillets and rounds.

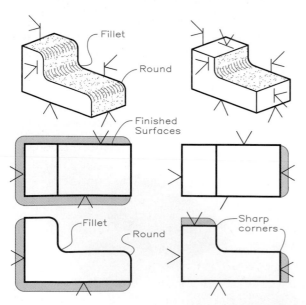

14.57 Fillets and rounds are rounded inside and outside corners respectively that are standard features on castings. When surfaces are finished, fillets and rounds are removed as shown here.

The object is first drawn with angular intersections, and the *Fillet* command is used to round the corners for both fillets and rounds.

Figure 14.61 gives a comparison of intersections and runouts of parts with and without **fillets** and **rounds**. Large runouts are constructed as an eighth of a circle with a compass as shown in **Figure 14.62**. Small runouts are drawn with a circle template. Runouts on orthographic views reveal much about the details of an object. For example, the runout in the top view of **Figure 14.63A** tells us that the rib has rounded corners, whereas the top view of **Figure 14.63B** tells us the rib is completely round. **Figures 14.64** and

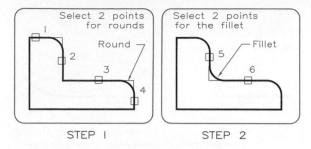

STEP 1　　　　　STEP 2

14.60 Fillets and rounds by computer:

Step 1 *Command: FILLET* (Enter)
Select first object or [Polyline/Radius/Trim]: R (Enter)
Specify fillet radius <0.000>: .50 (Enter)
Command: (Enter)
Select first object or [Polyline/Radius/Trim]:(Select 1 and 2 ; round is drawn.)(Enter) (Select 3 and 4 to draw round.)

Step 2 *Command:* (Enter)(Select 5 and 6 when prompted and the fillet is drawn. The lines are trimmed at the fillets and rounds.)

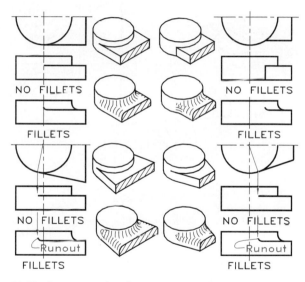

14.61 These examples show conventional intersections and runouts on cylindrical features of parts. Runouts result from fillets and rounds intersecting cylinders.

14.65 illustrate other types of filleted and rounded intersections.

Computer Method A drawing of a part with runouts at its tangent points is shown in **Figure 14.66**. The runouts are plotted by using

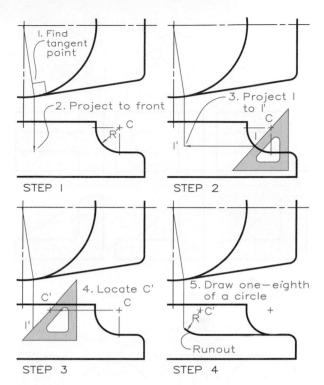

STEP 1　　　　　STEP 2

STEP 3　　　　　STEP 4

14.62 Plotting runouts:

Step 1 Find the tangency point in the top view and project it to the front view.

Step 2 Find point 1 with a 45° triangle and project it to 1'.

Step 3 Move the 45° triangle to locate point C' on the horizontal projector from center C.

Step 4 Use the radius of the fillet to draw the runout with C' as its center. The runout arc is equal to one-eighth of a circle.

the *Line* command, then using the *Arc* command, and finally pressing (Enter) to obtain the first point of an arc tangent to and connected to the end of the line. Locate the other end of the arc to complete the runout.

14.23 First-Angle Projection

The examples in this chapter are third-angle projections in which the top view is placed over the front view and the right-side view is placed to the right of the front view, as shown in **Figure 14.67**. This method is used in the United States, Great Britain, and Canada.

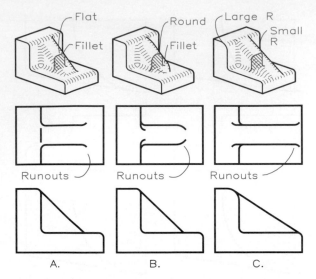

14.63 Examples of typical runouts of edges with fillets and rounds.

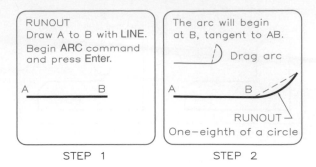

STEP 1   STEP 2

14.66 Runouts by computer:

Step 1 *Command:* LINE (Enter)
Specify first point: A (Enter)
Specify next point or [Undo]?: B (Enter)
Step 2 *Command:* ARC (Enter)
Specify start point or arc or [CEnter]: (Enter)
Specify end point of arc: C (Drag to point C.)

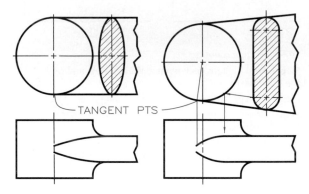

14.64 Conventional representation of runouts.

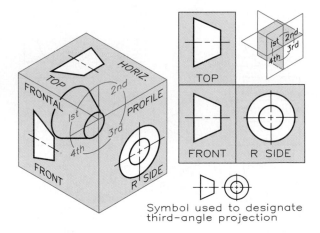

Symbol used to designate third-angle projection

14.67 Third-angle projection is used for drawing orthographic views in the U.S., Great Britain, and Canada. The top view is placed over the front view and the right-side view is placed to the right of the front view. The truncated cone is the symbol used to designate third-angle projection.

However, most of the world uses first-angle projection.

The first-angle system is illustrated in **Figure 14.68**, in which an object is placed above the horizontal plane and in front of the frontal plane. When these projection planes are opened onto the surface of the drawing paper, the front view projects over the top view, and the right-side view to the left of the front view.

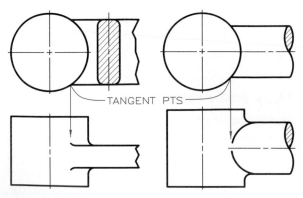

14.65 These are conventional runouts for different cross sections.

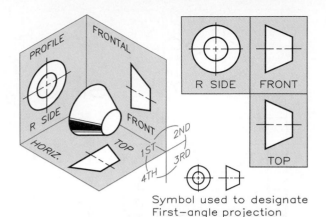

Symbol used to designate
First—angle projection

14.68 First-angle projection is used in most of the world. It shows the right-side view to the left of the front view and the top view under the front view. The truncated cone designates first-angle projection.

The angle of projection used in making a drawing is indicated by placing the truncated cone in or near the title block (**Figure 14.69**). When metric units of measurement are used, the SI symbol is given in combination with the cone on the drawing.

Problems

1–7. (**Figures 14.70–14.76**) Draw the given views on size A sheets, two per sheet, using the dimensions given and draw the missing top, front, or right-side views. Lines may be missing in the given views.

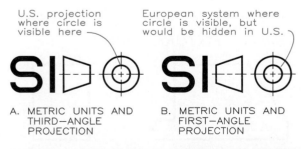

U.S. projection where circle is visible here

European system where circle is visible, but would be hidden in U.S.

A. METRIC UNITS AND THIRD—ANGLE PROJECTION

B. METRIC UNITS AND FIRST—ANGLE PROJECTION

14.69 These symbols are placed on drawings to specify first-angle or third-angle projection and metric units of measurement.

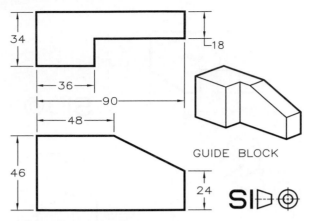

14.70 Problem 1.

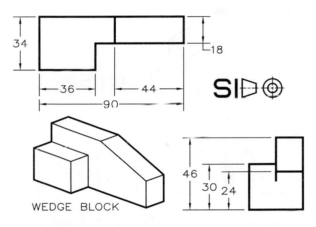

14.71 Problem 2.

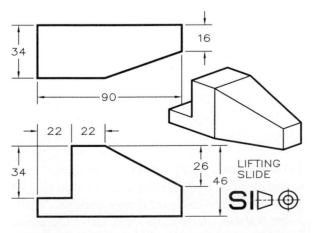

14.72 Problem 3.

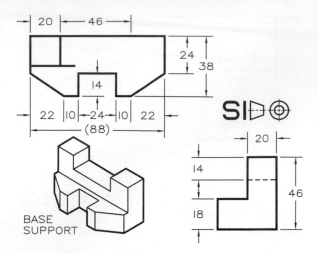

14.73 Problem 4.

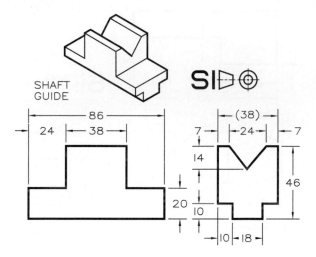

14.74 Problem 5.

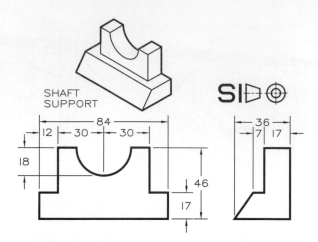

14.75 Problem 6.

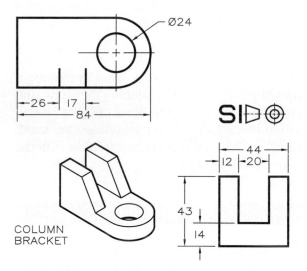

14.76 Problem 7.

8–14. (Figures 14.77–14.83) Draw three views of the objects. Each square grid is equal to 0.20 inches or 5 mm. Two problems can be placed on a size A sheet (Vertical format). Label the views and show the overall dimensions as W, D, and H.

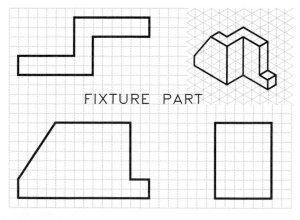

14.77 Problem 8.

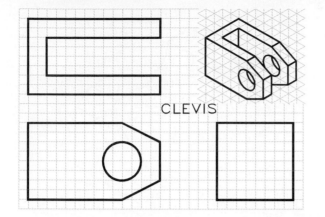

14.78 Problem 9.

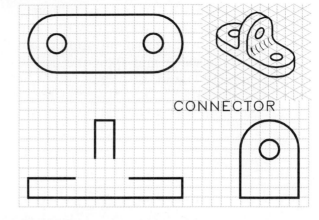

14.81 Problem 12.

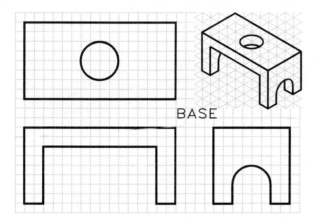

14.79 Problem 10.

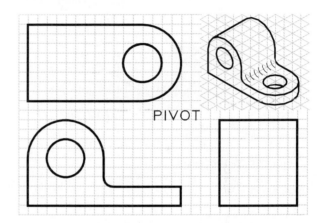

14.82 Problem 13.

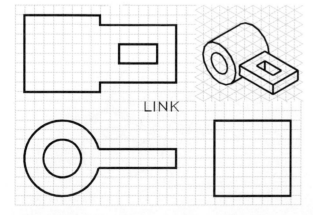

14.80 Problem 11.

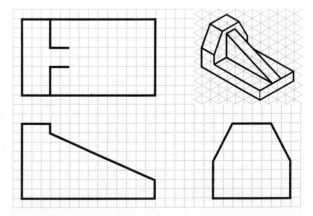

14.83 Problem 14.

15–36. (Fig. 14.84–14.105) Construct the necessary orthographic views to describe the objects on B-size sheets at an appropriate scale. Label the views and show the overall dimensions of W, D, and H.

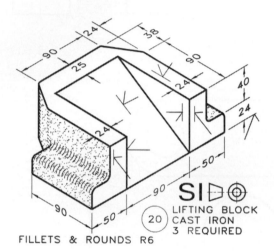

14.84 Problem 15.

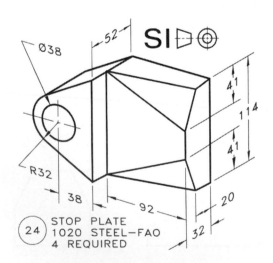

14.85 Problem 16.

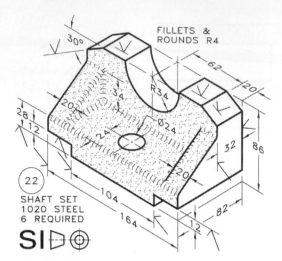

14.86 Problem 17.

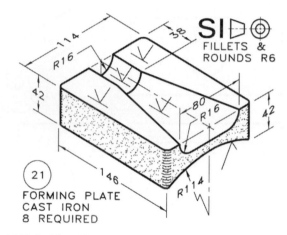

14.87 Problem 18.

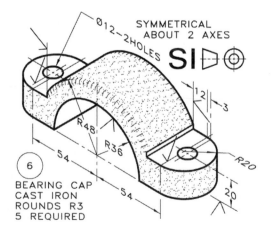

14.88 Problem 19.

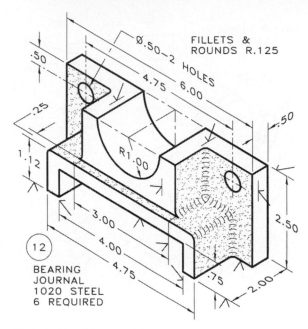

FILLETS &
ROUNDS R.125

Ø.50–2 HOLES

4.75
6.00

.50

.25

.50

.50

1.12

R1.00

3.00

4.00

4.75

.75

2.00

2.50

(12)

BEARING
JOURNAL
1020 STEEL
6 REQUIRED

14.89 Problem 20.

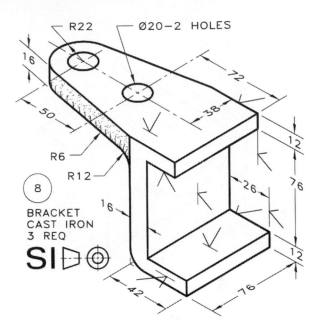

R22

Ø20–2 HOLES

16

50

72

38

R6

R12

16

26

12

76

(8)

BRACKET
CAST IRON
3 REQ

SI ▷⊕

42

76

12

14.91 Problem 22.

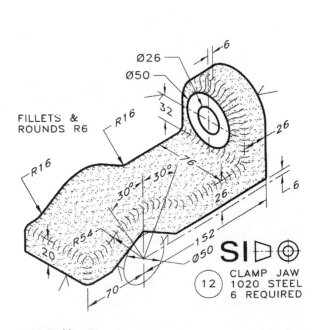

Ø26

Ø50

6

FILLETS &
ROUNDS R6

R16

32

26

R16

6

30°

30°

6

26

R54

152

6

20

Ø50

70

SI ▷⊕

(12)

CLAMP JAW
1020 STEEL
6 REQUIRED

14.90 Problem 21.

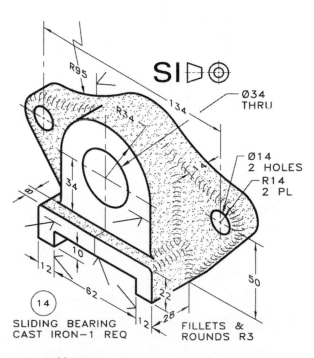

R95

SI ▷⊕

134

Ø34
THRU

R34

Ø14
2 HOLES

R14
2 PL

34

8

10

12

62

22

28

12

50

(14)

SLIDING BEARING
CAST IRON–1 REQ

FILLETS &
ROUNDS R3

14.92 Problem 23.

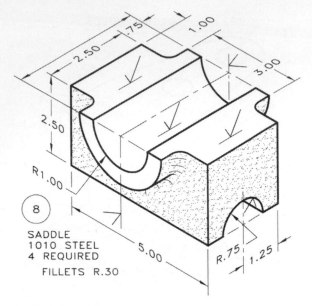

14.93 Problem 24.

(8)
SADDLE
1010 STEEL
4 REQUIRED
FILLETS R.30

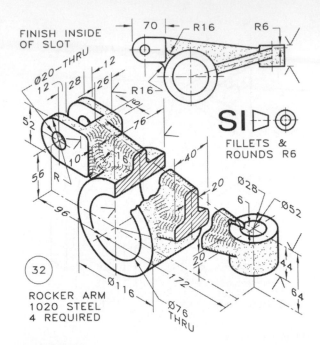

14.95 Problem 26.

FINISH INSIDE
OF SLOT

FILLETS &
ROUNDS R6

(32)
ROCKER ARM
1020 STEEL
4 REQUIRED

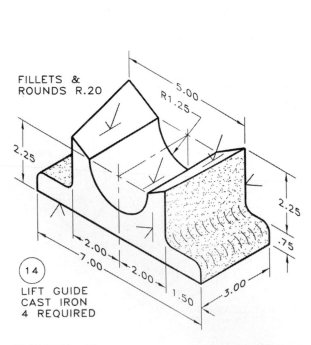

14.94 Problem 25.

FILLETS &
ROUNDS R.20

(14)
LIFT GUIDE
CAST IRON
4 REQUIRED

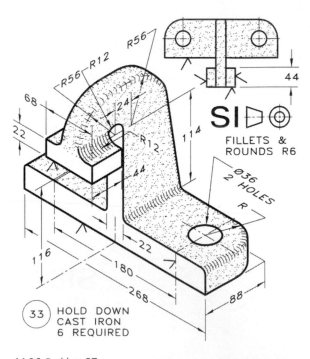

14.96 Problem 27.

FILLETS &
ROUNDS R6

(33)
HOLD DOWN
CAST IRON
6 REQUIRED

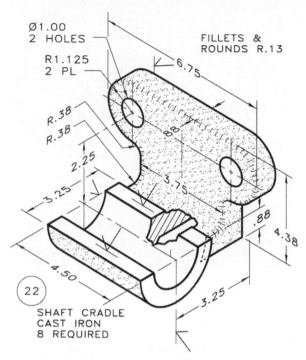

Ø1.00
2 HOLES

R1.125
2 PL

R.38
R.38

FILLETS &
ROUNDS R.13

6.75

88

2.25

3.25

3.75

.88

4.38

4.50

3.25

22

SHAFT CRADLE
CAST IRON
8 REQUIRED

14.97 Problem 28.

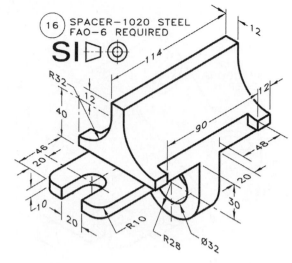

16 SPACER–1020 STEEL
FAO–6 REQUIRED

SI

12

R32

12

40

114

12

46

20

90

48

10

20

R10

R28

Ø32

20

30

14.99 Problem 30.

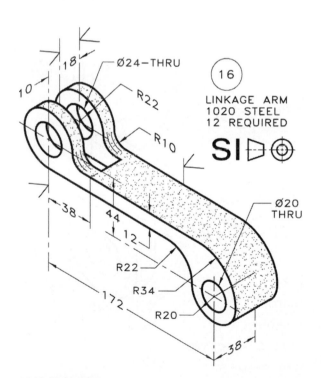

18

10

Ø24–THRU

R22

16

LINKAGE ARM
1020 STEEL
12 REQUIRED

SI

R10

Ø20
THRU

38

44

12

R22

172

R34

R20

38

14.98 Problem 29.

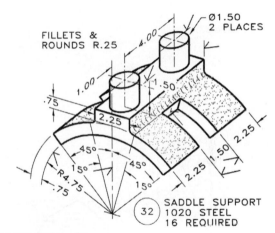

FILLETS &
ROUNDS R.25

Ø1.50
2 PLACES

4.00

1.00

1.50

.75

2.25

45°

45°

15°

15°

R4.75

.75

2.25

2.25

1.50

1.50

2.25

32 SADDLE SUPPORT
1020 STEEL
16 REQUIRED

14.100 Problem 31.

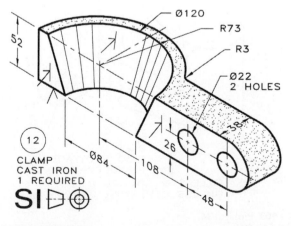

Ø120

R73

R3

Ø22
2 HOLES

52

12

CLAMP
CAST IRON
1 REQUIRED

SI

Ø84

26

108

38

48

14.101 Problem 32.

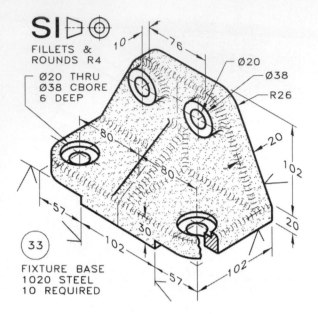

SI ⬠⊕
FILLETS &
ROUNDS R4

Ø20 THRU
Ø38 CBORE
6 DEEP

Ø20
Ø38
R26

10
76
80
80
57
30
102
57
102
102
20
102
20

(33)

FIXTURE BASE
1020 STEEL
10 REQUIRED

14.102 Problem 33.

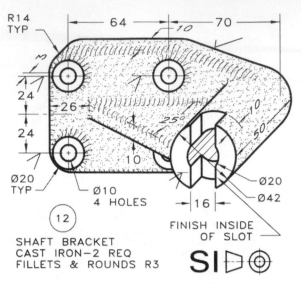

R14
TYP

64
10
70

3
24
26
24

25°
10
50
10

Ø20
TYP

Ø10
4 HOLES

Ø20
Ø42

16

(12)

SHAFT BRACKET
CAST IRON−2 REQ
FILLETS & ROUNDS R3

FINISH INSIDE
OF SLOT

SI ⬠⊕

14.104 Problem 35.

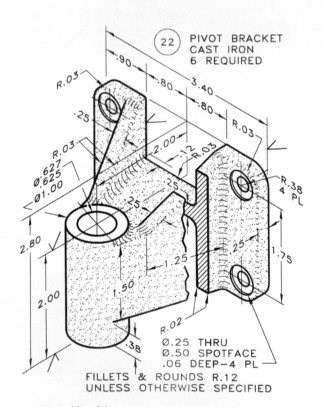

(22) PIVOT BRACKET
CAST IRON
6 REQUIRED

.90
.80
3.40
.80
R.03
R.03
R.03
R.03
.25
2.00
.12
R.38
4 PL
Ø.627
Ø.625
Ø1.00
.25
.25
.25
2.80
1.25
1.75
2.00
1.50
.25
R.02
.38

Ø.25 THRU
Ø.50 SPOTFACE
.06 DEEP−4 PL

FILLETS & ROUNDS R.12
UNLESS OTHERWISE SPECIFIED

14.103 Problem 34.

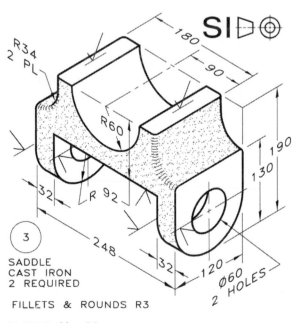

SI ⬠⊕

R34
2 PL

180
90

R60

32
R 92

248
32
120
Ø60
2 HOLES

190
130

(3)

SADDLE
CAST IRON
2 REQUIRED

FILLETS & ROUNDS R3

14.105 Problem 36.

Design Problems:

Follow the instructions for each of the partially-dimensioned problems given as if you were the original designer of the products.

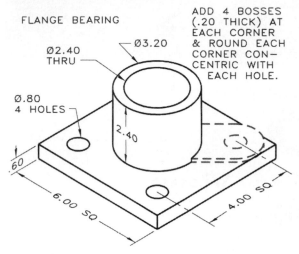

FLANGE BEARING

Ø2.40 THRU
Ø3.20
ADD 4 BOSSES (.20 THICK) AT EACH CORNER & ROUND EACH CORNER CONCENTRIC WITH EACH HOLE.
Ø.80 4 HOLES
2.40
.60
6.00 SQ
4.00 SQ

Design 1: Redesign the flange bearing to have 4 bosses and round each corner as noted. Draw the necessary views to describe the part.

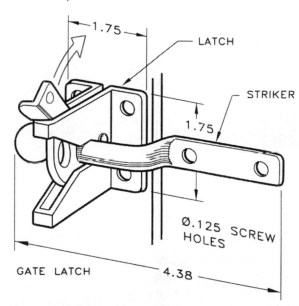

1.75
LATCH
STRIKER
1.75
Ø.125 SCREW HOLES
GATE LATCH
4.38

Design 2: Gate latch.

Option 1: Make instrument drawings of the parts of the partially dimensioned gate latch on size A sheet. (More than one sheet may be required.)

Option 2: Design a different gate latch and make the necessary instrument drawings to describe your design.

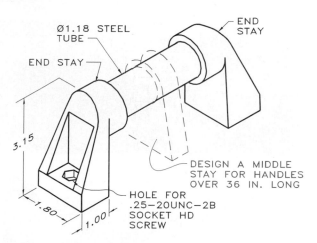

Ø1.18 STEEL TUBE
END STAY
END STAY
3.15
1.80
1.00
DESIGN A MIDDLE STAY FOR HANDLES OVER 36 IN. LONG
HOLE FOR .25–20UNC–2B SOCKET HD SCREW

Design 3: Handle assembly.

Option 1: Make the necessary orthographic views of the end stays of the door handle to adequately describe them.

Option 2: Design a middle stay that can be used for handles over 36 in. and make the necessary instrument drawings (orthographic views) to describe your design.

Option 3: Design a completely different door handle and make the necessary instrument drawings (orthographic views) to describe your design.

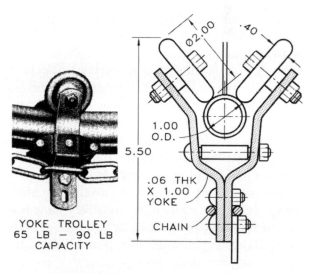

Ø2.00
.40
1.00 O.D.
5.50
.06 THK X 1.00 YOKE
YOKE TROLLEY 65 LB – 90 LB CAPACITY
CHAIN

Design 4: Yoke trolley.

Option 1: Make the necessary orthographic freehand sketches of the views of the parts of the trolley.

Option 2: Make the necessary orthographic view of the parts of the trolley using instruments.

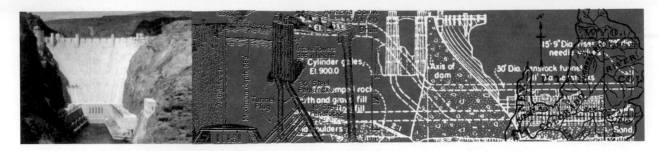

15

Primary Auxiliary Views

15.1 Introduction

Objects often are designed to have sloping or inclined surfaces that do not appear true size in principal orthographic views. A plane of this type is not parallel to a principal projection plane (horizontal, frontal, or profile) and is, therefore, a **nonprincipal plane**. Its true shape must be projected onto a plane that is parallel to it. This view is called an **auxiliary view**.

An auxiliary view projected from a primary view (principal view) is called a primary auxiliary view. An auxiliary view projected from a primary auxiliary view is a secondary auxiliary view. By the way, get out your dividers; you must use them all the time in drawing auxiliary views.

The inclined surface of the part shown in **Figure 15.1A** does not appear true size in the top view because it is not parallel to the horizontal projection plane. However, the inclined surface will appear true size in an auxiliary

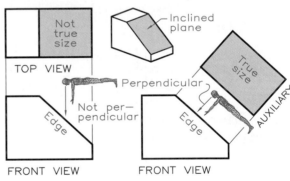

15.1 A surface that appears as an inclined edge in a principal view can be found true size by an auxiliary view. (A)The top view is foreshortened, but (B) the inclined plane is true size in the auxiliary view.

view projected perpendicularly from its edge view in the front view (**Figure 15.1B**).

The relationship between an auxiliary view and the view it was projected from is the same as that between any two adjacent orthographic views. **Figure 15.2A** shows an auxil-

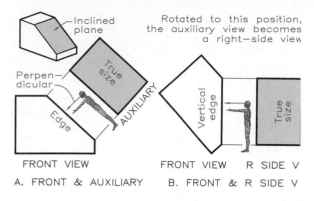

A. FRONT & AUXILIARY B. FRONT & R SIDE V

15.2 An auxiliary view has the same relationship with the view it is projected from as that of any two adjacent principal views.

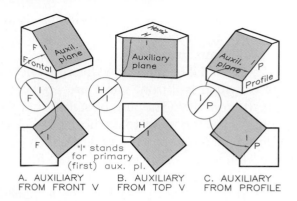

A. AUXILIARY FROM FRONT V B. AUXILIARY FROM TOP V C. AUXILIARY FROM PROFILE

15.3 A primary auxiliary plane can be folded from the frontal, horizontal, or profile planes. The fold lines are labeled F-1, H-1, and P-1, with 1 on the auxiliary plane side and P on the principal-plane side.

iary view projected perpendicularly from the edge view of the sloping surface. By rotating these views (the front and auxiliary views) so that the projectors are horizontal, the views have the same relationship as regular front and right-side views (**Figure 15.2B**).

15.2 Folding Line Principles

The three principal orthographic planes are the **frontal** (F), **horizontal** (H), and **profile** (P) planes. An auxiliary view is projected from a principal orthographic view (a top, front, or side view), and a primary auxiliary plane is perpendicular to one of the principal planes and oblique to the other two.

Think of auxiliary planes as planes that fold into principal planes along a folding line (**Figure 15.3**). The plane in **Figure 15.3A** folds at a 90° angle with the frontal plane and is labeled F-1 where F is an abbreviation for frontal, and 1 represents first, or primary, auxiliary plane. **Figures 15.3B** and **15.3C** illustrate the positions for auxiliary planes that fold from the horizontal and profile planes, labeled H-1 and P-1, respectively.

It is important that reference lines be labeled as shown in **Figure 15.3**, with the

numeral 1 placed on the auxiliary side and the letter H, F, or P on the principal-plane side.

15.3 Auxiliaries from the Top View

By moving your position about the top view of a part as shown in **Figure 15.4**, each line of sight is perpendicular to the height dimension. One of

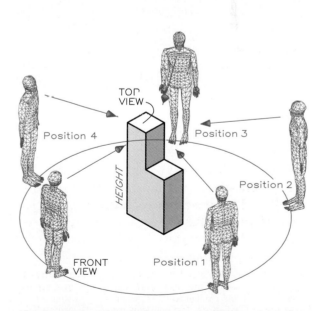

15.4 By moving your viewpoint around the top view of an object, you will see a series of auxiliary views in which the height dimension (H) is true length in all of them.

the views, the front view, is a principal view while the other positions see nonprincipal views called **auxiliary views**.

Figure 15.5 illustrates how these five views (one of which is a front view) are projected from the top view. The line of sight for each auxiliary view is parallel to the horizontal projection plane; therefore the height dimension is true length in each view projected from a top view. The height (H) dimensions are transferred from the front to each of the auxiliary views by using your dividers.

Folding-Line Method

The inclined plane shown in **Figure 15.6** is an edge in the top view and is perpendicular to the horizontal plane. If an auxiliary plane is

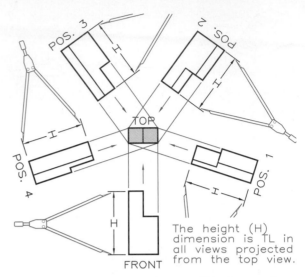

15.5 The views shown in Figure 15.4 would be drawn as shown here with the same height dimensions common to each view.

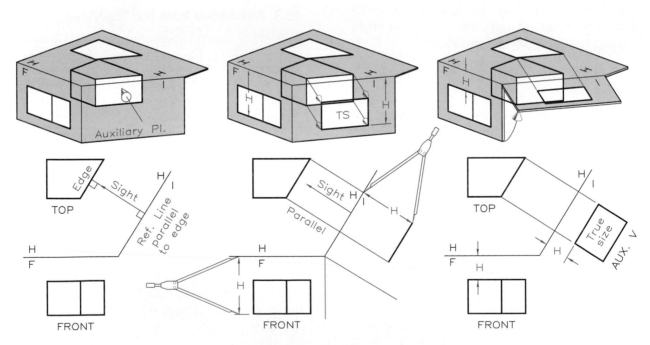

15.6 Auxiliary view from the top:

Step 1 Draw the line of sight perpendicular to the edge view of the inclined surface. Draw the H-1 line parallel to its edge and draw the H-F reference line between the top and front views.

Step 2 Project from the edge view of the inclined surface parallel to the line of sight. Transfer the H dimensions from the front to locate a line in the auxiliary view.

Step 3 Locate the other corners of the inclined surface by projecting to the auxiliary view and locating the points by transferring the height (H) from the front view.

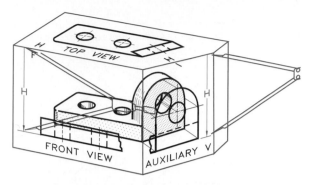

15.7 If you imagine that the object is inside a glass box, you can see the relationship of the auxiliary plane, on which the true-size view is projected, and the horizontal projection plane.

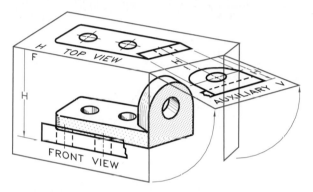

15.8 Fold the auxiliary plane into the horizontal projection plane by revolving it about the H-1 fold line.

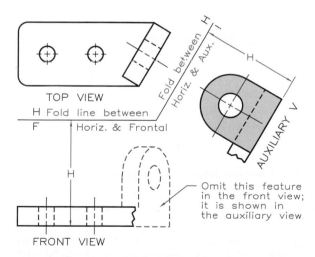

15.9 The front and auxiliary views are drawn as partial views to avoid drawing its elliptical features, which are shown in the auxiliary view as true arcs..

drawn parallel to the inclined surface, the view projected onto it will be a true-size view of the inclined surface. **A surface must appear as an edge in a principal view in order for it to be found true size in a primary auxiliary view.** When the auxiliary view is projected from the top view, the height dimensions in the front view must be transferred to the auxiliary view with your dividers.

15.4 Auxiliaries from the Top: Application

Figure 15.7 illustrates how the folding-line method is used to find an auxiliary view of a part that is imagined to be in a glass box. The semicircular end of the part does not appear true size in front or side views, making these views difficult to draw and interpret if they were drawn. However, because the inclined surface appears as an edge in the top view, it can be found true size in a primary auxiliary view projected from the top view. The height dimension (H) in the frontal view will be the same as in the auxiliary plane, because both planes are perpendicular to the horizontal projection plane. Height is transferred from the front to the auxiliary view with your dividers. In **Figure 15.8**, the auxiliary plane is rotated about the H-1 fold line into the plane of the top view, the horizontal projection plane. This rotation illustrates how the placement of the views is arrived at when drawing the views.

When drawn on a sheet of paper, the views of this object appear as shown in **Figure 15.9**. The top view is a complete view, but the front view is drawn as a partial view because the omitted portion would have been hard to draw and would not have been true size. The auxiliary view also is drawn as a partial view because the front view shows the omitted features better, which saves drawing time and space on a drawing.

Reference-Plane Method

A second method of locating an auxiliary view uses reference planes instead of the folding-line method. **Figure 15.10A** shows a horizontal reference plane (HRP) drawn through the center of the front view. Because this view is symmetrical, equal height dimensions on both sides of the HRP can be conveniently transferred from the front view to the auxiliary view and laid off on both sides of the HRP.

The reference plane can be placed at the base of the front view as shown in **Figure 15.10B**. In this case, the height dimensions are measured upward from the HRP in both the front and auxiliary views. You may draw a reference plane (the HRP in this example) in any convenient position in the front view: through the part, above it, or below it.

A similar example of an auxiliary view drawn with a horizontal reference plane is shown in **Figure 15.11**. In this example, the hole appears as a true circle in the auxiliary view instead of an ellipse.

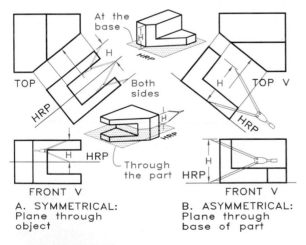

15.10 A horizontal reference plane (HRP) can be positioned through the part or in contact with it. The dimension of height (H) is measured from the HRP and transferred to the auxiliary view with your dividers.

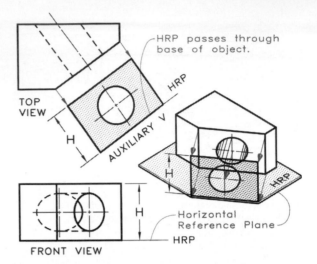

15.11 An auxiliary view projected from the top view is used to draw a true-size view of the inclined surface using a horizontal reference plane. The HRP is drawn through the bottom of the front view.

15.5 Rules of Auxiliary Construction

Now that several examples of auxiliary views have been discussed, it would be helpful to summarize the general rules of construction, which are outlined in **Figure 15.12**.

1. An auxiliary view that shows a surface true size must be projected perpendicularly from the edge view of the surface. Usually, the inclined surface, or a partial view, is all that is needed in the auxiliary view, but the entire object can be drawn in the auxiliary view if desired, as shown here.

2. Draw the sight line perpendicular to the inclined edge of the plane you wish to find true size (TS).

3. Draw the reference line, H-1 for example, parallel to the edge view of the inclined plane that will be perpendicular to the line of sight.

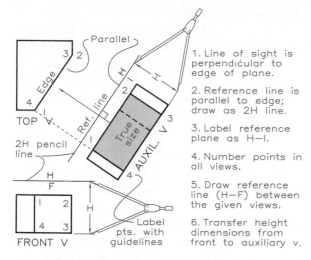

1. Line of sight is perpendicular to edge of plane.

2. Reference line is parallel to edge; draw as 2H line.

3. Label reference plane as H–I.

4. Number points in all views.

5. Draw reference line (H–F) between the given views.

6. Transfer height dimensions from front to auxiliary v.

15.12 Rules of auxiliary view construction:

Step 1 Draw a line of sight perpendicular to the edge of the inclined surface. Draw the H-1 fold line parallel to the edge of the inclined surface and draw an H-F fold line between the given views.

Step 2 Find points 1 and 2 by transferring the height (H) dimensions with your dividers from the front view to the auxiliary view.

Step 3 Find points 3 and 4 in the same manner by transferring the H dimensions.

4. Draw a reference line between the given views (front and top in this example). Reference lines (fold lines) should be drawn as thin black lines with a 2H or 3H pencil.

5. If an auxiliary is projected from the front view, it will have an F-1 reference line; if it is projected from the horizontal view (top view), it will have an H-1 reference line; and if projected from the side view (profile view), it will have a P-1 reference plane.

6. Transfer measurements from the other given view with your dividers (not the view you are projecting from), height in the front view in this example.

7. It would be very helpful to number the points one at a time in the primary views and the auxiliary views as they are plotted.

8. Do your lettering in a professional manner with guidelines.

9. Connect the points with light construction lines and use light gray projectors that do not have to be erased with a pencil in the 2H-4H range.

10. Draw the outlines of the auxiliary view as thick visible lines the same as visible lines in principal views, with an F or HB pencil.

15.6 Auxiliaries from the Front View

By moving about the front view of the part as shown in **Figure 15.13**, you will be looking parallel to the edge view of the frontal plane. Therefore the depth dimension (D) will appear true size in each auxiliary view projected from the front view. One of the positions gives a principal view, the right-side view, and position 1 gives a true-size view of the inclined plane. **Figure 15.14** illustrates the relationship between the auxiliary views projected from the front view.

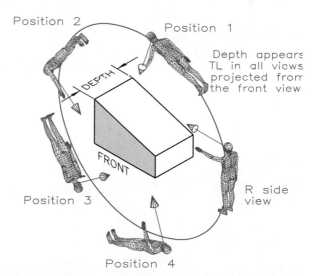

15.13 By moving your viewpoint around the frontal view of an object, you will see a series of auxiliary views in which the depth dimension (D) is true length in all views.

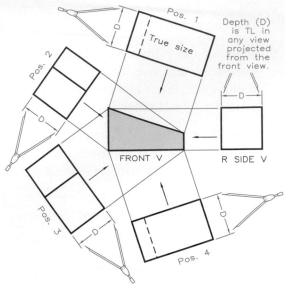

15.14 The auxiliary views shown in Figure 15.13 would be seen in this arrangement when viewing the front view.

Folding-Line Method

A plane of an object that appears as an edge in the front view (**Figure 15.15**) is true size in an auxiliary view projected perpendicularly from it. Draw fold line F-1 parallel to the edge view of the inclined plane in the front view at a convenient location.

Draw the line of sight perpendicular to the edge view of the inclined plane in the front view. Observed from this direction, the frontal plane appears as an edge; therefore measurements perpendicular to the frontal plane depth dimensions (D) will be seen true length. Transfer depth dimensions from the top view to the auxiliary view with your dividers.

The object in **Figure 15.16** is imagined to be enclosed in a glass box and an auxiliary plane is folded from the frontal plane to be

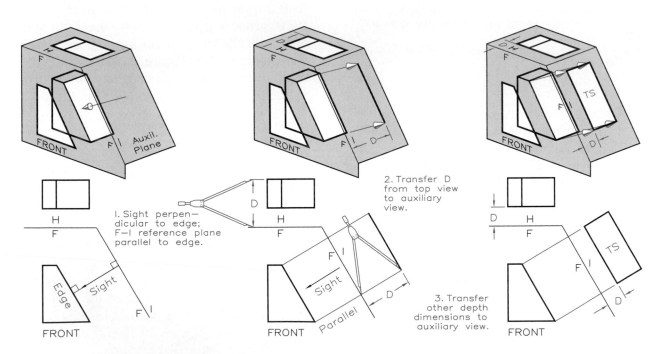

15.15 Auxiliary from the front: folding-line method:

Step 1 Draw the line of sight perpendicular to the edge of the plane and draw the F-1 line parallel to it. Draw the H-F fold line between the top and front views.

Step 2 Project perpendicularly from the edge view of the inclined surface and parallel to the line of sight. Transfer the depth dimensions (D) from the top to the auxiliary view with your dividers.

Step 3 Locate the other corners of the inclined surface by projecting to the auxiliary view. Locate the points by transferring the depth dimensions (D) from the top to the auxiliary view.

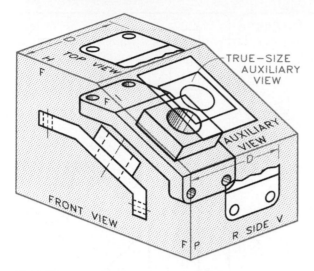

15.16 This part is shown in an imaginary glass box to illustrate the relationship of the auxiliary plane, on which the true-size view of the inclined surface is projected, with the principal planes.

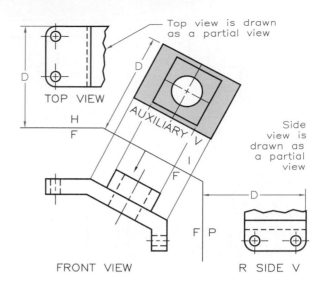

15.17 The layout and construction of an auxiliary view of the object shown in Figure 15.16 is shown here.

parallel to the inclined surface. When drawn on a sheet of paper, the views appear as shown in **Figure 15.17**. The top and side views are drawn as partial views because the auxiliary view eliminates the need for drawing complete views. The auxiliary view, located by transferring the depth dimension measured perpendicularly from the edge view of the frontal plane in the top view and transferred to the auxiliary view, shows the surface's true size.

Computer Method Find the true-size view of the inclined surface that appears as an edge in the front view (**Figure 15.18**) by using the *Lisp* commands, *Parallel* and *Transfer* (see Section 27.2) (This program is not a regular part of AutoCAD, but it is an excellent addition to have available for solving auxiliary problems. It was developed by Professor Leendert Kersten of the University of Nebraska. Once copied as a LISP file, it is accessed by typing (Load "ACAD") at the command line; be sure to use the parentheses.)

While in AutoCAD's drafting mode, type *Parallel* to draw the reference line parallel to edge AB. Type *Transfer* to obtain prompts for transferring measurements from the top view as if you were using your dividers. Connect the circular points to complete the auxiliary view.

Reference-Plane Method
The object shown in **Figure 15.19** has an inclined surface that appears as an edge in the front view; therefore this plane can be found true size in a primary auxiliary view. It is helpful to draw a reference plane through the center of the symmetrical top view because all depth dimensions can be located on each side of the frontal reference plane. Because the reference plane is a frontal plane, it is labeled FRP in the top and auxiliary views. In the auxiliary view, the FRP is drawn parallel to the edge view of the inclined plane at a convenient distance from it. By transferring depth dimensions from the FRP in the top view to the FRP in the auxiliary view, the symmetrical view of the part is drawn.

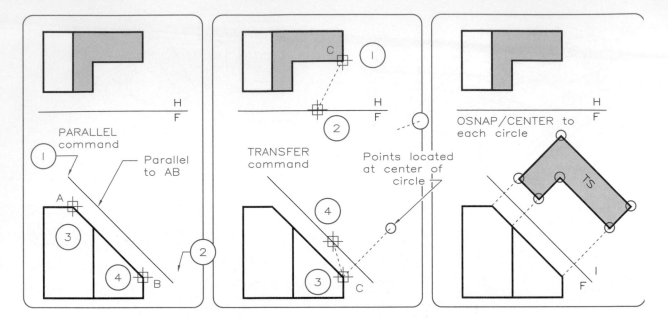

15.18 Auxiliary view by AutoCAD:

Step 1 Type *Parallel* to receive the prompts for the first end of the reference line (1) and its approximate second end-point (2). You are then prompted for the ends of line AB, to which the reference line is parallel.

Step 2 Type *Transfer*; you are prompted for a point in the top view (1) and its distance from the H-F line (2) to transfer. You are prompted for the front view of the point to project (3) and the reference line (4). Point C is projected to the auxiliary view.

Step 3 Continue using *Transfer* to locate the other corner points of the inclined plane. Connect the points using the CENTER option of OSNAP to snap to the centers of the circles. ERASE the circles after connecting their centers.

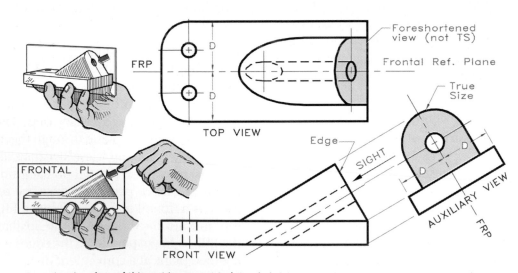

15.19 Because the inclined surface of this part is symmetrical, it is helpful to use a frontal reference plane (FRP) that passes through the object. Project the auxiliary view perpendicularly from the edge view of the plane in the front view. The FRP appears as an edge in the auxiliary view, and depth dimensions (D) are transferred from each side of it in the top view to locate points on the true-size view of the inclined surface.

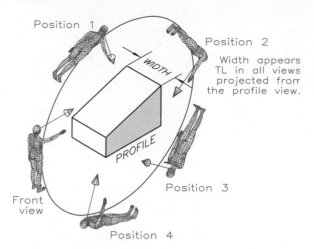

15.20 By moving your viewpoint around a profile (side) view of an object you will obtain a series of auxiliary views in which the width dimension (W) is true length.

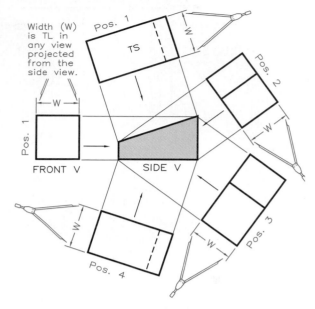

15.21 The auxiliary views shown in Figure 15.20 would be seen in this arrangement when projected from the side view.

15.7 Auxiliaries from the Profile View

By moving your position about the profile view (side view) of the part as shown in **Figure 15.20**, you will be looking parallel to the edge view of the profile plane. Therefore, the width dimension will appear true size in each auxiliary view projected from the side view. One of the positions gives a principal view, the front view, and position 1 gives a true-size view of the inclined plane. **Figure 15.21** illustrates the arrangement of the auxiliary views projected from the side view if they were drawn on a sheet of paper.

Folding-Line Method

Because the inclined surface in **Figure 15.22** appears as an edge in the profile plane, it can be found true size in a primary auxiliary view projected from the side view. The auxiliary fold line, P-1, is drawn parallel to the edge view of the inclined surface. A line of sight perpendicular to the auxiliary plane shows the profile plane as an edge. Therefore width dimensions (W) transferred from the front view to the auxiliary view appear true length in the auxiliary view.

Reference-Plane Method

The object shown in **Figure 15.23** has an inclined surface that appears as an edge in the right-side view, the profile view. This inclined surface may be drawn true size in an auxiliary view by using a profile reference plane (PRP) that is a vertical edge in the front view. Draw the PRP through the center of the front view because the view is symmetrical. Then find the true-size view of the inclined plane by transferring equal width dimensions (W) with your dividers from the edge view of the PRP in the front view to both sides of the PRP in the auxiliary view.

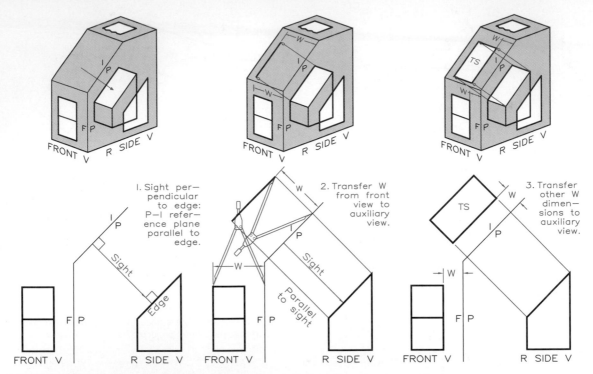

15.22 From the side: Folding-line method:

Step 1 Draw a line of sight perpendicular to the edge of the inclined surface. Draw the P-1 fold line parallel to the edge view, and draw the F-P fold line between the given views.

Step 2 Project the corners of the edge view parallel to the line of sight. Transfer the width dimensions (W) from the front view to locate a line in the auxiliary view.

Step 3 Find the other corners of the inclined surface by projecting to the auxiliary view. Locate the points by transferring the width dimensions (W) from the front view to the auxiliary view.

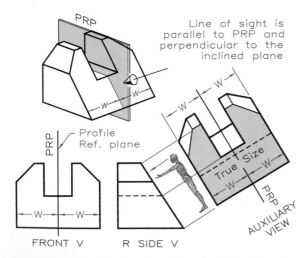

15.23 An auxiliary view is projected from the right-side view by using a profile reference plane (PRP) to show the true-size view of the inclined surface.

15.8 Curved Shapes

The cylinder shown in **Figure 15.24** has an inclined surface that appears as an edge in the front view. The true-size view of this plane can be seen in an auxiliary view projected from the front view.

Because the cylinder is symmetrical, a frontal reference plane (FRP) is drawn through the center of the side view so that equal dimensions can be laid off on both sides of it. Points located about the circular side view are projected to its edge view in the front view.

In the auxiliary view, the FRP is drawn parallel to the edge view of the plane in the front view, and the points are projected perpendicularly from the edge view of the plane.

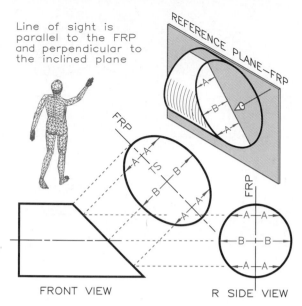

Line of sight is parallel to the FRP and perpendicular to the inclined plane

REFERENCE PLANE—FRP

FRP

FRONT VIEW

R SIDE VIEW

15.24 The auxiliary view of this elliptical surface was found by locating a series of points about its perimeter. The frontal reference plane (FRP) is drawn through its center in the side view since the object is symmetrical.

Dimensions A and B are shown as examples of depth dimensions used for locating points in the auxiliary view. To construct a smooth elliptical curve, more points than shown are needed.

A true-size auxiliary view of a surface bounded by an irregular curve is shown in **Figure 15.25**. Project points from the curve in the top view to the front view. Locate these points in the auxiliary view by transferring depth dimensions (D) from the FRP in the top view to the auxiliary view.

15.9 Partial Views

Auxiliary views are used as supplementary views to clarify features that are difficult to depict with principal views alone. Consequently, portions of principal views and auxiliary views may be omitted, provided that the partial views adequately describe the part. The object shown in **Figure 15.26** is composed of a complete front view, a partial auxiliary view, and a partial top view. These partial

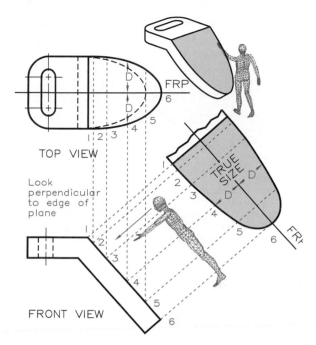

TOP VIEW

Look perpendicular to edge of plane

FRONT VIEW

15.25 The auxiliary view of this curved surface required that a series of points be located in the top view, be projected to the front view, and then projected to the auxiliary view. The FRP was passed through the top view.

views are easier to draw and are more descriptive without sacrificing clarity.

15.10 Auxiliary Sections

In **Figure 15.27**, a cutting plane labeled A-A is passed through the part to obtain the auxiliary section labeled section A-A. The auxiliary section provides a good and efficient way to describe features of the part that could not be as easily described by additional principal views.

15.11 Secondary Auxiliary Views

Figure 15.28 shows how to project a secondary auxiliary view from a primary auxiliary view. An edge view of the oblique plane is found in the primary auxiliary view by finding the point view of a true-length line (2-3) that lies on the oblique surface. A line of sight perpendicular to the edge view of the plane gives a secondary auxiliary view that shows the oblique plane as true size.

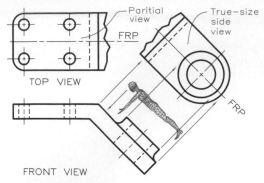

15.26 Partial views with foreshortened portions omitted can be used to represent objects. The FRP reference line is drawn through the center of the object in the top view because the object is symmetrical to make point location easier.

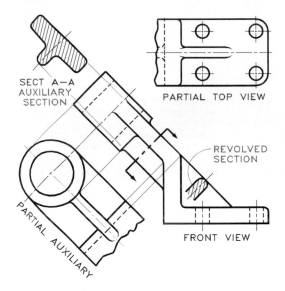

15.27 A cutting plane labeled A-A is passed through the object and the auxiliary section, section A-A, is drawn as a supplementary view to describe the part. The top and front views are drawn as partial views.

Note that the reference line between the primary auxiliary view and the secondary auxiliary view is labeled 1-2 to represent the fold line between the primary plane (1) and the secondary plane (2). The 1 label is placed on the primary side and the 2 label is placed on the secondary side.

Figure 15.29 illustrates the construction of a secondary auxiliary view that gives the true-size view of a surface on a part using these

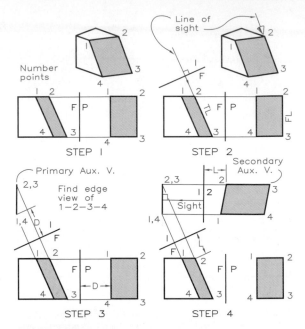

15.28 Secondary auxiliary views:

Step 1 Draw a fold line F-P between the front and side views. Label the corner points in both views.

Step 2 Line 2-3 is a true-length frontal line in the front view. Draw reference line F-1 perpendicular to line 2-3 at a convenient location with a line of sight parallel to line 2-3.

Step 3 Find the edge view of plane 1-2-3-4 by transferring depth dimensions (D) from the side view.

Step 4 Draw a line of sight perpendicular to the edge view of 1-2-3-4 and draw the 1-2 fold line parallel to the edge view. Find the true-size auxiliary view by transferring the dimensions (L) from the front view to the auxiliary view.

same principles and a combination of partial views. A secondary auxiliary view must be used in this case because the oblique plane does not appear as an edge in a principal view.

Find the point view of a line on the oblique plane to find the edge view of the plane in the primary auxiliary view. The secondary auxiliary view, projected perpendicularly from the edge view of the plane in the primary auxiliary view, gives a true-size view of the plane. In this example, all of the views are drawn as partial views.

15.12 Elliptical Features

Occasionally, circular shapes will project as ellipses, which must be drawn with an irregu-

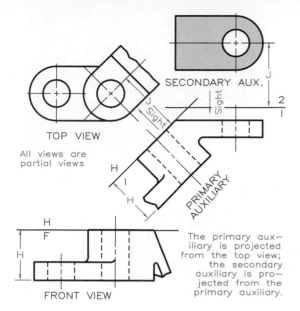

15.29 A secondary auxiliary view projected from a primary auxiliary view that was projected from the top view is shown here. All views are drawn as partial views.

SECONDARY AUX.

TOP VIEW

All views are partial views

PRIMARY AUXILIARY

FRONT VIEW

The primary auxiliary is projected from the top view; the secondary auxiliary is projected from the primary auxiliary.

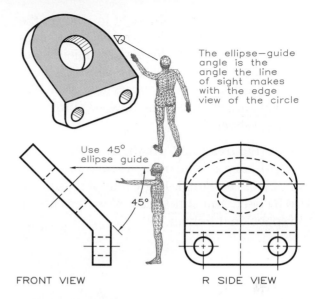

The ellipse–guide angle is the angle the line of sight makes with the edge view of the circle

Use 45° ellipse guide

FRONT VIEW

R SIDE VIEW

15.30 The ellipse guide angle is the angle that the line of sight makes with the edge view of the circular feature. The ellipse angle for the right-side view is 45°.

lar curve or an ellipse template. The ellipse template (guide) is by far the most convenient method of drawing ellipses. The angle of the ellipse template is the angle the line of sight makes with the edge view of the circular feature. In **Figure 15.30**, the angle is found to be 45° where the curve is an edge in the front

view, so the right-side view of the curve is drawn as a 45° ellipse.

Problems

1–13. (**Figure 15.31**) Using the example layout, change the top and front views by substituting

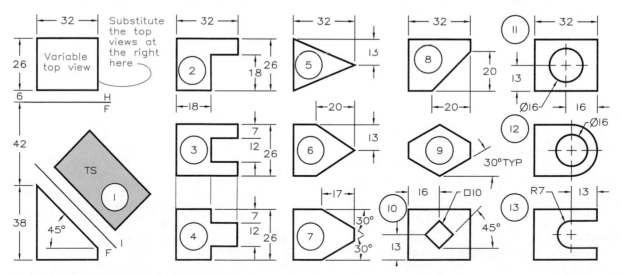

15.31 (Problems 1-13) Primary auxiliary views.

the top views given at the right in place of the one given in the example. The angle of inclination in the front view is 45° for all problems, and the height is 38 mm (1.5 inches) in the front view. Construct auxiliary views that show the inclined surface true size. Draw two problems per size A sheet.

14–31. (**Figures 15.32–15.49**) Draw the necessary primary and auxiliary views to describe the objects assigned. Draw one per size A or size B sheet. Adjust the scale of each to utilize the space on the sheet.

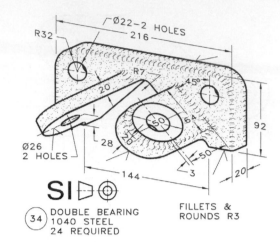

15.34 Problem 16.

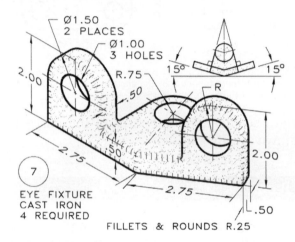

15.32 Problem 14.

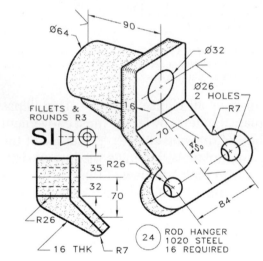

15.35 Problem 17.

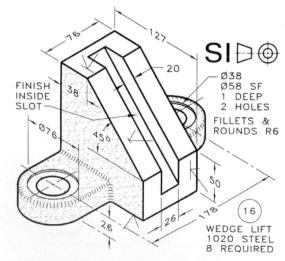

15.33 Problem 15.

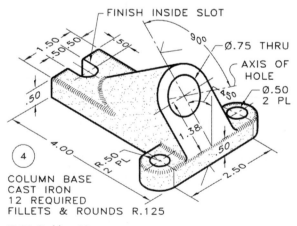

15.36 Problem 18.

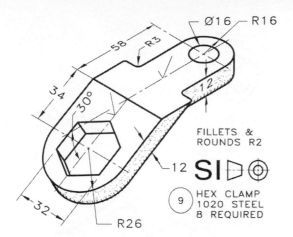

15.37 Problem 19.

Ø16 — R16
58 — R3
34
30°
12
FILLETS &
ROUNDS R2
12
SI ⊳⊙
HEX CLAMP
1020 STEEL
8 REQUIRED
R26
32

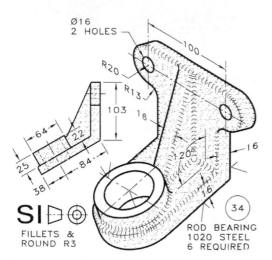

15.38 Problem 20.

Ø16
2 HOLES
100
R20
R13
103
16
64
22
25
84
38
120°
16
16
SI ⊳⊙
FILLETS &
ROUND R3
34
ROD BEARING
1020 STEEL
6 REQUIRED

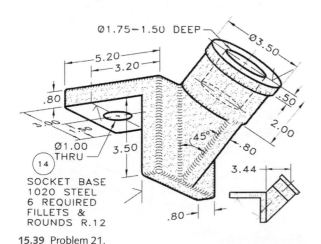

15.39 Problem 21.

Ø1.75—1.50 DEEP
Ø3.50
5.20
3.20
.80
.50
2.00
45°
.80
Ø1.00
THRU
3.50
14
SOCKET BASE
1020 STEEL
6 REQUIRED
FILLETS &
ROUNDS R.12
3.44
.80

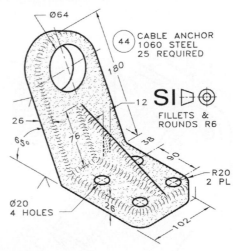

15.40 Problem 22.

Ø64
44
CABLE ANCHOR
1060 STEEL
25 REQUIRED
180
12
SI ⊳⊙
FILLETS &
ROUNDS R6
26
65°
76
38
90
R20
2 PL
Ø20
4 HOLES
26
102

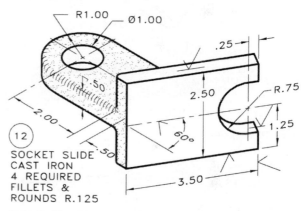

15.41 Problem 23.

R1.00 — Ø1.00
.25
.50
2.50
R.75
2.00
60°
1.25
.50
12
SOCKET SLIDE
CAST IRON
4 REQUIRED
FILLETS &
ROUNDS R.125
3.50

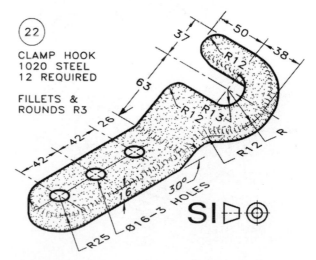

15.42 Problem 24.

22
CLAMP HOOK
1020 STEEL
12 REQUIRED

FILLETS &
ROUNDS R3
37
50
38
R12
63
R12 R13
R12 R
26
42
16
R12
30°
Ø16–3 HOLES
R25
SI ⊳⊙

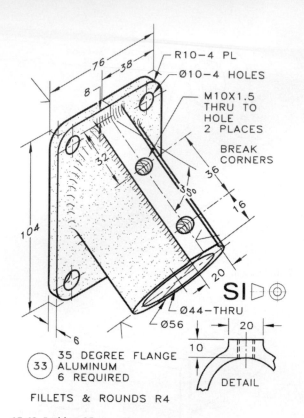

R10—4 PL
Ø10—4 HOLES
M10X1.5
THRU TO
HOLE
2 PLACES

BREAK
CORNERS

76
38
8
32
35°
36
16
104
20
6
SI
Ø44—THRU
Ø56

20
10
DETAIL

33 35 DEGREE FLANGE
ALUMINUM
6 REQUIRED

FILLETS & ROUNDS R4

15.43 Problem 25.

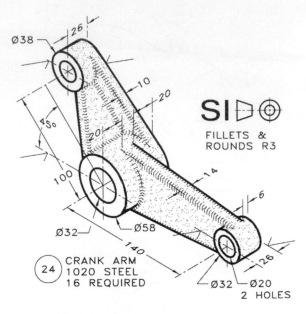

Ø38
26
10
20
45°
20
100
14
6
SI
FILLETS &
ROUNDS R3
Ø32
Ø58
140
26
Ø32 Ø20
2 HOLES

24 CRANK ARM
1020 STEEL
16 REQUIRED

15.45 Problem 27.

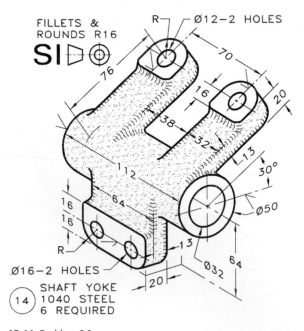

FILLETS &
ROUNDS R16
SI
R
Ø12—2 HOLES
76
70
16
20
38
32
13
30°
112
Ø50
16
64
16
R
13
Ø32
64
20
Ø16—2 HOLES

14 SHAFT YOKE
1040 STEEL
6 REQUIRED

15.44 Problem 26.

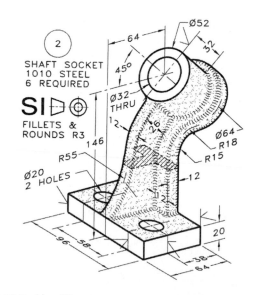

2 SHAFT SOCKET
1010 STEEL
6 REQUIRED

SI
FILLETS &
ROUNDS R3

64
Ø52
32
45°
Ø32
THRU
12
26
146
Ø64
R18
R15
R55
Ø20
2 HOLES
12
96
58
38
64
20

15.46 Problem 28.

226 • **CHAPTER 15   PRIMARY AUXILIARY VIEWS**

15.47 (Problem 29) Lay out the necessary orthographic views of the oblique bracket on a B size sheet. Construct the true-size auxiliary view that shows the inclined surface true size. Select the appropriate scale.

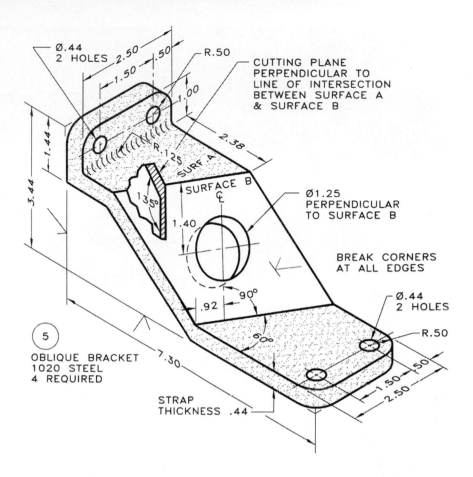

Ø.44
2 HOLES

2.50

.50

R.50

CUTTING PLANE
PERPENDICULAR TO
LINE OF INTERSECTION
BETWEEN SURFACE A
& SURFACE B

1.50

1.00

1.44

R.1.25

2.38

SURF. A

SURFACE B
C

3.44

135°

1.40

Ø1.25
PERPENDICULAR
TO SURFACE B

BREAK CORNERS
AT ALL EDGES

Ø.44
2 HOLES

R.50

.92

90°

60°

⑤

OBLIQUE BRACKET
1020 STEEL
4 REQUIRED

7.30

1.50

.50

2.50

STRAP
THICKNESS .44

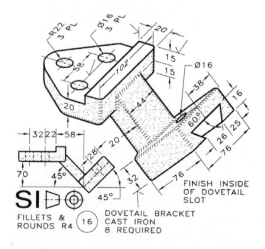

R22
3 PL

Ø16
3 PL

20

15

15

Ø16

38

16

52

102

44

60°

26

25

20

76

20

32|22|58

128

76

FINISH INSIDE
OF DOVETAIL
SLOT

70

45°

32

45°

SI ▷⊕
FILLETS &
ROUNDS R4

⑯

DOVETAIL BRACKET
CAST IRON
8 REQUIRED

SI ▷⊕
FILLETS &
ROUNDS R10

Ø36
2 HOLES

26

245

25°

60

60 80

100

26

28

48

180

㉒

ANGLE BRACKET
CAST IRON
2 REQUIRED

Ø32-2 HOLES
2 HOLES

15.48 Problem 30.

15.49 Problem 31.

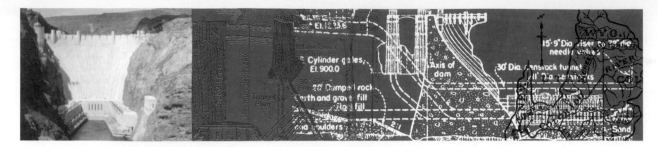

16

Sections

16.1 Introduction

Correctly drawn orthographic views that show all hidden lines may not clearly describe an object's internal details. The gear housing shown in **Figure 16.1** is an example that is better understood when a section has been cut from it. The technique of constructing imaginary cross-sectional cuts through a drawing of a part results in an orthographic view called a **section**.

16.2 Basics of Sectioning

In **Figure 16.2A**, standard views of a cylinder are shown as a top and front view where its interior features are drawn as hidden lines. If you imagined a knife edge cutting through the top view, the front view would become a **section**. This section is a **full section** since the cutting plane passes fully through the part (**Figure 16.2B**). The portion of the part that was cut by the imaginary plane is crosshatched and hidden lines usually are omitted in sectional views because they are not needed.

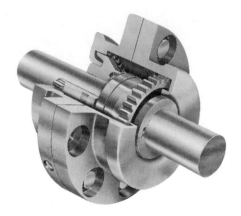

16.1 This gear housing has many internal features that cannot be described clearly in a standard orthographic view. Sections are used to clarify interior parts.

Figure 16.3 shows two types of cutting planes. Either is acceptable although the one with pairs of short dashes is most often used. The spacing and proportions of the dashes depend on the size of the drawing. The line thickness of the cutting plane is the same as the visible object line. Letters placed at each

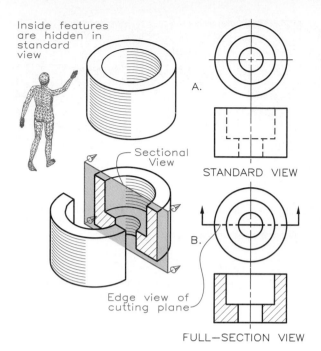

16.2 A part with internal features can be better shown with sectional views than with the standard views with hidden lines.

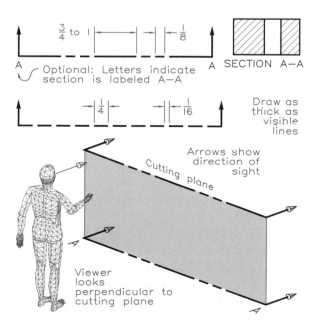

16.3 Cutting planes can be thought of as knife edges that pass through views to reveal interior features in sections. The cutting plane marked A-A results in a section labeled SECTION A-A.

end of the cutting plane are used to label the sectional view, such as SECTION A-A.

The sight arrows at the ends of the cutting plane are always perpendicular to the cutting plane. In the sectional view, the observer is looking in the direction of the sight arrows, perpendicular to the surface of the cutting plane.

Figure 16.4 shows the three basic positions of sections and their respective cutting planes. In each case, perpendicular arrows point in the direction of the line of sight. For example, the cutting plane in **Figure 16.4A** passes through and removes the front of the top view and the line of sight is perpendicular to the remainder of the top view.

The top view appears as a section when the cutting plane passes through the front view and the line of sight is downward (**Figure 16.4B**). When the cutting plane passes vertically through the side view (**Figure 16.4C**), the front view becomes a section.

16.3 Sectioning Symbols

The hatching symbols used to distinguish between different materials in sections are shown in **Figure 16.5**. Although these symbols may be used to indicate the materials in

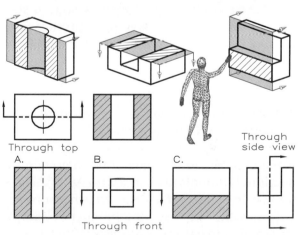

16.4 The three examples of cutting planes that pass through the principal views, (A) top, (B) front, and (C) side views, are shown here. The arrows point in the direction of your line of sight for each section.

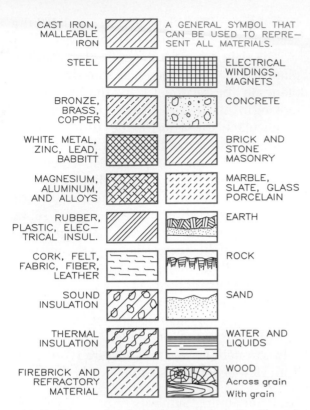

16.5 Use these symbols for hatching parts in section. The cast-iron symbol may be used for any material.

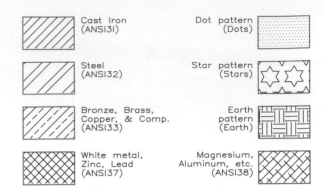

16.6 These are a few of the hatching symbols available in AutoCAD. The spacing and size of the symbols can be varied by setting the pattern scale factor.

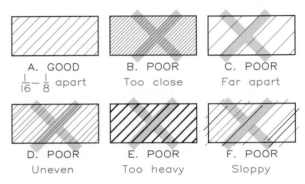

16.7 Hatching techniques.

A Section lines are thin black lines drawn 1/16 to 1/8 in. apart.

B-F Avoid these common errors of section lining.

a section, you should provide supplementary notes specifying the materials to ensure clarity.

The cast-iron symbol (evenly spaced section lines) may be used to represent any material and is the symbol used most often. Draw cast-iron symbols with a 2H pencil, slant the lines upward and to the right at 30°, 45°, or 60° angles, and space the lines about 1/16 inch apart (close together in small areas and farther apart in larger areas).

Computer Method A few of the many cross-sectional symbols available with AutoCAD are shown in **Figure 16.6**. The spacing between the lines and the dash lengths can be changed by setting the pattern scale factor.

Properly drawn section lines are thin and evenly spaced, as shown in **Figure 16.7**. **Figure 16.7B–F** shows common errors of section lining.

Thin parts such as sheet metal, washers, and gaskets are sectioned by completely blacking in their areas (**Figure 16.8**) because space does not permit the drawing of section lines. On the other hand, large parts are sectioned with an **outline section** to save time and effort.

Sectioned areas should be hatched with symbols that are neither parallel nor perpendicular to the outlines of the parts lest they be confused with serrations or other machining treatments of the surface (**Figure 16.9**).

Computer Method The most basic principal of applying section symbols to an area with

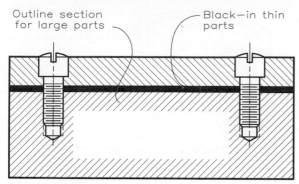

16.8 Large sectional areas are hatched with outline sectioning (around their edges) and thin parts are blacked in solid.

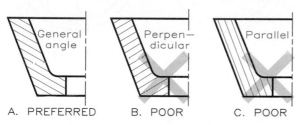

A. PREFERRED   B. POOR   C. POOR

16.9 Draw section lines that are neither parallel nor perpendicular to the outlines of the part, so that they are not misunderstood as machining features.

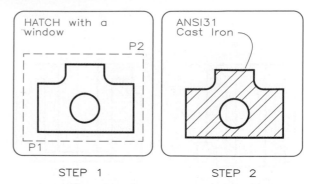

STEP 1   STEP 2

16.10 Hatching by computer:

Step 1 *Command:* HATCH (Enter)
Enter pattern name or (? /Solid/User defined) <ANSI31>: ANSI32 (Enter)
Specify a scale for the pattern <1.00>: .50 (Enter)
Specify an angle for pattern <0>: 0 (Enter)
Step 2 *Select objects:* W (window option)
Specify first corner: P1
Specify opposite corner: P2 (Enter) (The hatching is applied.)

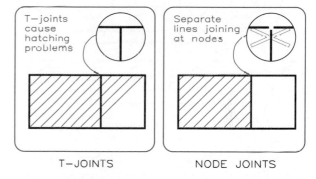

T-JOINTS   NODE JOINTS

16.11 For the HATCH command to work properly, the outlines of areas to be section lines must be drawn with perfect closing outlines. T-joints, overlaps, or gaps at intersections can result in irregular results.

AutoCAD is shown in **Figure 16.10**. After assigning the proper hatch symbol with the *Hatch* command, select the area to be sectioned with a window and the hatch lines are drawn. To vary the spacing of the section lines, change the patterns scale of the *Hatch* command. *Bhatch* and other hatching commands are covered in Chapter 37 and Chapter 38.

The lines that outline areas to be hatched must intersect perfectly at each corner point; no T-joints are permitted (**Figure 16.11**). Poor intersections may cause hatching symbols to fill the desired area improperly.

16.4 Sectioning Assemblies of Parts

When sectioning an assembly of several parts, draw section lines at varying angles to distinguish the parts from each other (**Figure 16.12A**). Using different material symbols in an assembly also helps distinguish between the parts and their materials. Crosshatch the same part at the same angle and with the same symbol even though portions of the part may be separated (**Figure 16.12B**).

16.5 Full Sections

A cutting plane passed fully through an object and removing half of it forms a **full section** view. **Figure 16.13** shows two orthographic

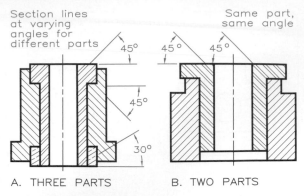

A. THREE PARTS   B. TWO PARTS

16.12 Hatching parts in assembly.

A Draw section lines of different parts in an assembly at varying angles to distinguish the parts.

B Draw section lines on separated portions of the same part (both sides of a hole here) in the same direction.

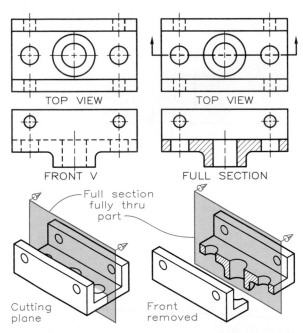

16.13 A full section is found by passing a cutting plane fully through the top view of this part, removing half of it. The arrows on the cutting plane give the direction of your sight. The front sectional view shows the internal features clearly.

views of an object with all its hidden lines. We can describe the part better by passing a cutting plane through the top view to remove half of it. The arrows on the cutting plane indicate the direction of sight. The front view becomes a

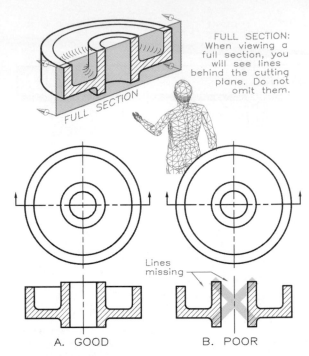

16.14 Full section: Cylindrical part.

A When a front view of a cylinder is shown as a full section as shown here, visible lines will be seen behind the cut surface.

B By showing only the lines of the cut surface (the section), the view will be incomplete.

full section, showing the surfaces cut by the cutting plane. **Figure 16.14** shows a full section through a cylindrical part, with half the object removed. **Figure 16.14A** shows the correctly drawn sectional view. A common mistake in constructing sections is omitting the visible lines behind the cutting plane (**Figure 16.14B**).

Omit hidden lines in sectional views unless you consider them necessary for a clear understanding of the view. Also, omit cutting planes if you consider them unnecessary. **Figure 16.15** shows a full section of a part from which the cutting plane was omitted because its path is obvious.

Parts Not Requiring Section Lining Many standard parts, such as nuts and bolts, rivets, shafts, and set screws, do not require section lining even though the cutting plane passes

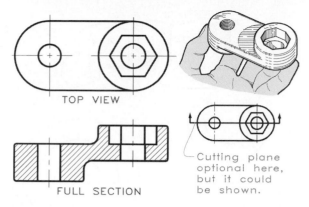

TOP VIEW

FULL SECTION

Cutting plane optional here, but it could be shown.

16.15 The cutting plane of a section can be omitted if its location is obvious.

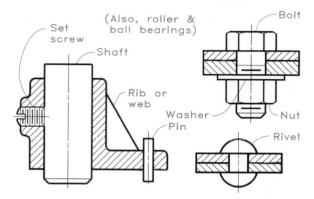

Set screw

(Also, roller & ball bearings)

Shaft

Rib or web

Pin

Bolt

Washer

Nut

Rivet

16.16 By conventional practice, these parts are not section lined even though cutting planes pass through them.

through them (**Figure 16.16**). These parts have no internal features, so sections through them would be of no value. Other parts not requiring section lining are roller bearings, ball bearings, gear teeth, dowels, pins, and washers.

Ribs Ribs are not section lined when the cutting plane passes flatwise through them (**Figure 16.17A**), because to do so would give a misleading impression of the rib. But ribs do require section lining when the cutting plane passes perpendicularly through them and shows their true thickness (**Figure 16.17B**).

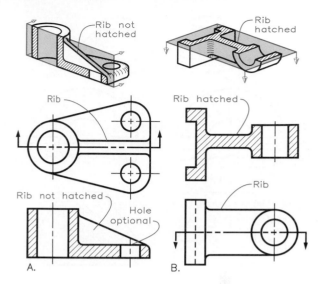

Rib not hatched

Rib hatched

Rib

Rib hatched

Rib not hatched

Hole optional

Rib

A.

B.

16.17 Ribs in section.

A Do not hatch a rib cut in a flatwise direction.

B Ribs are hatched when cutting planes pass through them, showing their true thicknesses.

By not section lining the ribs in **Figure 16.18A**, we provided a descriptive section view of the part. Had we hatched the ribs, the section would give the impression that the part was solid and conical (**Figure 16.18B**).

Figure 16.19 shows an alternative method of section lining webs and ribs. The outside ribs in **Figure 16.19A** do not require section lining because the cutting plane passes flatwise through them and they are well identified. As a rule, webs do not require crosshatching, but the webs shown in **Figure 16.19B** are not well identified in the front section and could go unnoticed. Therefore using alternate section lines as shown in **Figure 16.19C** is better. Here, extending every-other section line through the webs ensures that they can be identified easily.

16.6 Partial Views

A conventional method of representing symmetrical views is the half view that requires less space and less time to draw than full views (**Figure 16.20**). A half top view is sufficient

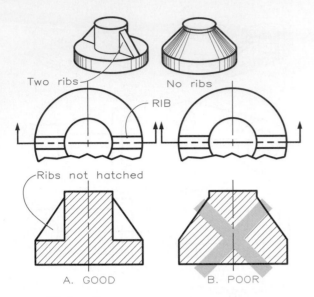

16.18 Ribs in section.

A Ribs are not hatched in section to better describe the part.

B If ribs were hatched in section, a misleading impression of the part would be given.

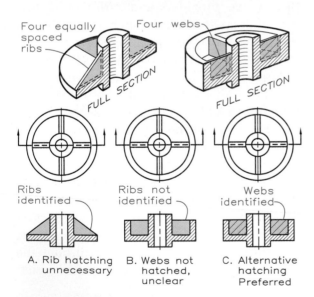

16.19 Ribs and webs in section.

A These ribs are well defined in this section and are not hatched.

B These webs are poorly defined when not hatched in the sectional view.

C Alternate hatch lines call attention to poorly-defined webs.

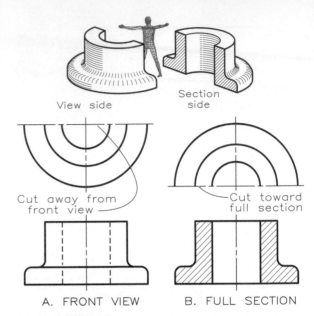

16.20 Half views of symmetrical parts can be used to conserve space and drawing time as an approved conventional practice. (A) The external portion of the half view is toward the front view. (B) The internal portion of the half view is toward the front view when it is a section. In half-sections, the omitted half view can be either toward or away from the section.

when drawn adjacent to the section view or front view. For half views (not sections), the removed half is away from the adjacent view (**Figure 16.20A**). For full sections, the removed half is the half nearest the section (**Figure 16.20B**). When drawing partial views with half sections, you may omit either the near or the far halves of the partial views.

16.7 Half-Sections

A **half-section** is a view obtained by passing a cutting plane halfway through an object and removing a quarter of it to show both external and internal features. Half-sections are used with symmetrical parts and with cylinders, in particular, as shown in **Figure 16.21**. By comparing the half-section with the standard front view, you can see that both internal and external features show more clearly in a half-section than in a view. Hidden lines are unnecessary,

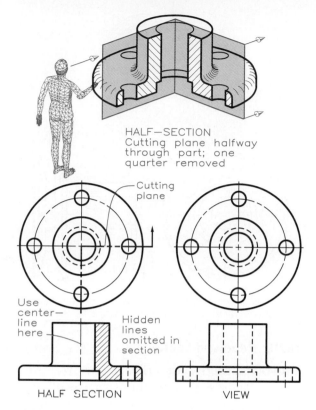

HALF—SECTION
Cutting plane halfway through part; one quarter removed

Cutting plane

Use center—line here

Hidden lines omitted in section

HALF SECTION   VIEW

16.21 In a half-section, the cutting plane passes halfway through the object, removing a quarter of it, to show half the outside and half the inside of it. Omit hidden lines in sectional views unless they are needed for clarity.

and we've omitted them to simplify the section. **Figure 16.22** shows a half section of a pulley.

Note the omission of the cutting plane from the half section shown in **Figure 16.23** because the cutting plane's location is obvious. Because the parting line of the half-section is not at a centerline, you may use a solid line or a centerline to separate the sectional half from the half that appears as an external view.

16.8 Offset Sections

An **offset section** is a full section in which the cutting plane is offset to pass through important features that do not lie in a single plane. **Figure 16.24** shows an offset section in which

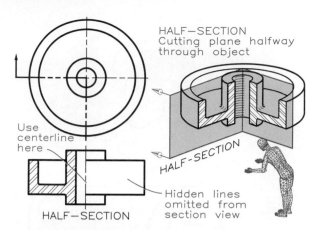

HALF—SECTION
Cutting plane halfway through object

Use centerline here

Hidden lines omitted from section view

HALF—SECTION

16.22 This half-section describes the part that is shown orthographically and pictorially.

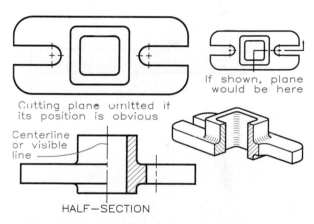

If shown, plane would be here

Cutting plane omitted if its position is obvious

Centerline or visible line

HALF—SECTION

16.23 The cutting plane can be omitted when its location is obvious. The parting line between the section and the view may be a visible line or a centerline if the part is not cylindrical.

the plane is offset to pass through the large hole and one of the small holes. The cut formed by the offset is not shown in the section because it is imaginary.

16.9 Broken-Out Sections

A **broken-out section** shows a partial view of a part's interior features. The broken-out section of the part shown in **Figure 16.25** reveals details of the wall thickness to describe the part better. The irregular lines representing the

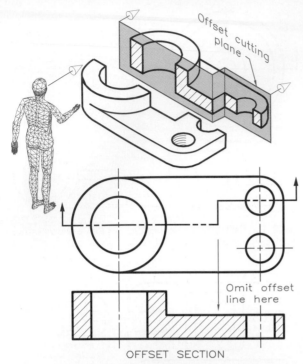

16.24 An offset section is formed by a cutting plane that must be offset to pass through features not in a single plane.

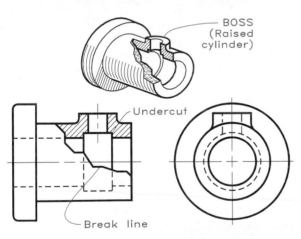

16.25 A broken-out section is one in which a part of the object has been broken away to show internal features.

break are conventional breaks (discussed later in this chapter).

The broken-out section of the pulley in **Figure 16.26** clearly depicts the keyway and threaded hole for a setscrew. This method shows the part efficiently, with the minimum of views.

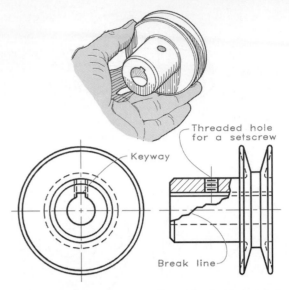

16.26 This broken-out section shows the keyway and the threaded hole for the set screw in the pulley.

16.10 Revolved Sections

A **revolved section** describes a part when you revolve its cross section about an axis of revolution and place it on the view where the revolution occurred. Note the use of revolved sections to explain two cross sections of the shaft shown in **Figure 16.27** (with and without conventional breaks). Conventional breaks

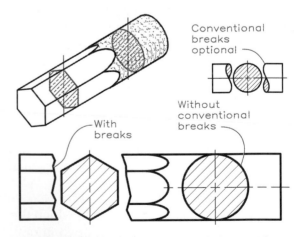

16.27 Revolved sections show cross-sectional features of a part and eliminate the need for additional orthographic views. Revolved sections may be superimposed on the given views or conventional breaks can be used to separate them from the given view.

are optional; you may draw a revolved section on the view without them.

A revolved section helps to describe the part shown in **Figure 16.28**. Imagine passing a cutting plane through the top view of the part (Step 1). Then imagine revolving the cutting plane in the top view and projecting it to the front (Step 2). The true-size revolved section is completed in Step 3. Conventional breaks could be used on each side of the revolved section.

Figure 16.29 demonstrates how to use typical revolved sections to show cross sections

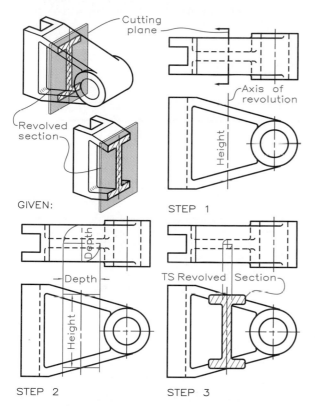

16.28 Revolved section construction:

Given: The part is shown pictorially with a cutting plane that shows the cross section that will be revolved.

Step 1 An imaginary cutting plane is located in the top view and the axis of rotation is located in the front view.

Step 2 The depth in the top view is rotated in the top view and is projected to the front view. The height and depth in the front view give the overall dimensions of the revolved view.

Step 3 The revolved section is drawn, fillets and rounds are added, and section lines are applied to finish the view.

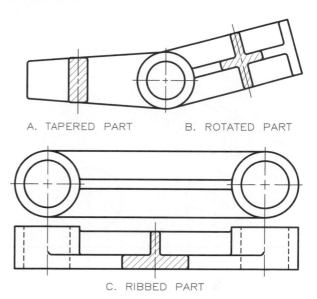

16.29 These revolved sections describe the cross sections of the two parts that would be difficult to depict in supplementary orthographic views, such as a side view.

through parts without having to draw additional orthographic views.

16.11 Removed Sections

A **removed section** is a revolved section that is shown outside the view in which it was revolved (**Figure 16.30**). Centerlines are used as axes of rotation to show the locations from which the sections are taken. Where space does not permit revolution on the given view (**Figure 16.31A**), removed sections must be used instead of revolved sections (**Figure 16.31B**).

Removed sections do not have to be positioned directly along an axis of revolution adjacent to the view from which they were revolved. Instead, removed sections can be located elsewhere on a drawing if they are properly labeled (**Figure 16.32**). For example, the plane labeled with an A at each end identifies the location of section A-A; the same applies to section B-B.

When a set of drawings consists of multiple sheets, removed sections and the views from which they are taken may appear on different sheets. When this method of layout is necessary,

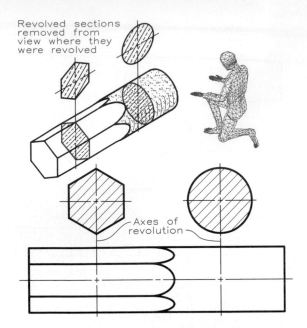

16.30 Removed sections are revolved sections that are drawn ouside the object and along their axes of revolution.

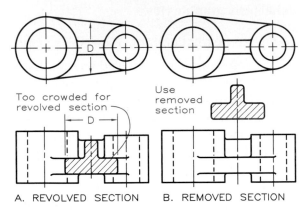

16.31 Removed sections are necessary when space does not permit the revolved section to be superimposed on the part.

label the cutting plane in the view from which the section was taken and the sheet on which the section appears (**Figure 16.33**).

16.12 Conventional Revolutions

In **Figure 16.34A**, the middle hole is omitted because it does not pass through the center of the circular plate. However, in **Figure 16.34B**, the hole does pass through the plate's center

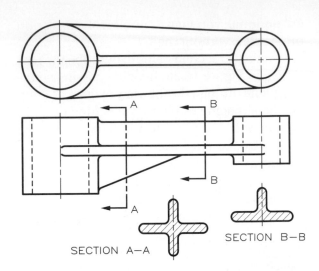

16.32 Lettering each end of a cutting plane (such as A-A) identifies the removed section labeled SECTION A-A drawn elsewhere on the drawing.

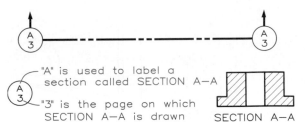

16.33 When a removed section is placed on another page of a set of drawings, label each end of the cutting plane with a letter and a number. The letters identify the section and the numerals indicate the page on which it is drawn.

and is shown in the section. Although the cutting plane does not pass through one of the symmetrically spaced holes in the top view (**Figure 16.34C**), the hole is revolved to the cutting plane to show the full section.

When ribs are symmetrically spaced about a hub (**Figure 16.35**), it is conventional practice to revolve them so that they appear true size in both views and sections. **Figure 16.36** illustrates the conventional practice of revolving both holes and ribs (or webs) of symmetrical parts. Revolution gives a better description of the parts in a manner that is easier to draw.

A cutting plane may be positioned in either of two ways shown in **Figure 16.37**. Even

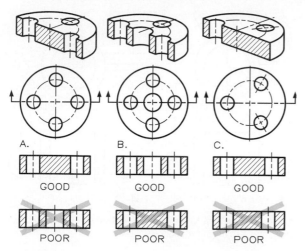

16.34 Symmetrically spaced holes are revolved to show their true radial distance from the center of a circular part in sectional views. (A) Omit the middle holes; it is not at the center of the plate. (B) Show the middle hole; it is at the center of the plate. (C) Rotate the holes to the centerline and project to make a symmetrical section.

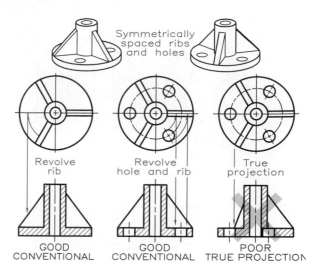

16.36 Symmetrically spaced ribs and holes should be shown in sections with the ribs rotated to show them true size and the holes rotated to show them at their true radial distance from the center.

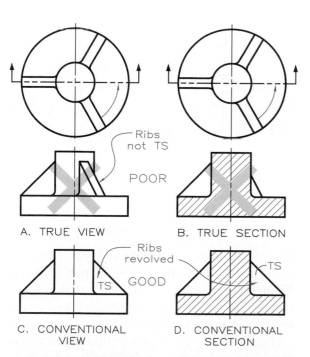

16.35 Symmetrically spaced ribs are revolved and drawn true size in their orthographic and sectional views to show them true size as a conventional practice.

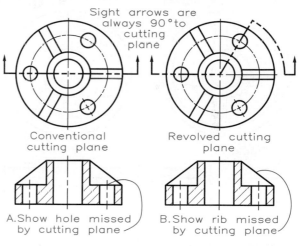

16.37 Show symmetrically spaced ribs true size whether or not the cutting plane passes through them. As an alternative, the cutting plane can be drawn to pass through the ribs for clarity.

though the cutting plane does not pass through the ribs and holes in **Figure 16.37A**, they may appear in the section as if the cutting plane passed through them. The path of the cutting plane also may be revolved, as shown in **Figure 16.37B**. In this case, the ribs are revolved to their true-size position in the section view, although the plane does not cut through them.

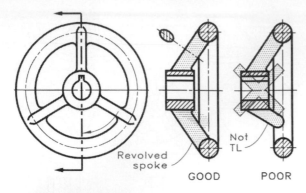

16.38 Revolve spokes to show them true size in section. Do not section line (hatch) spokes.

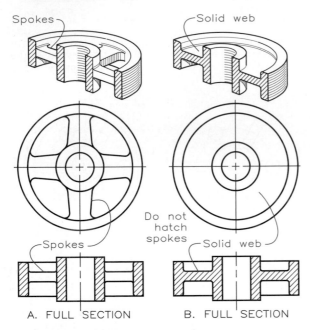

16.39 Spokes and webs in section.

A Spokes are not hatched even though they have been cut.

B Webs are hatched when cut by the cutting plane.

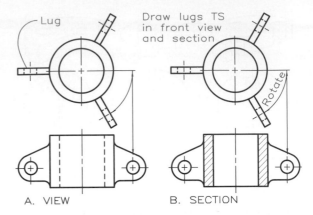

16.40 Lugs are revolved to show their true size in (A) the front view, and also in (B) the sectional view.

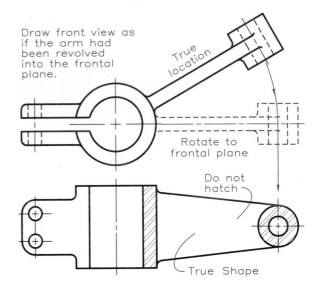

16.41 It is conventional practice to revolve parts with inclined features in order to show them true size in both sections and regular orthographic views.

The same principles apply to symmetrically spaced spokes (**Figure 16.38**). Draw only the revolved, true-size spokes and do not section line them. If the spokes shown in **Figure 16.39A** were hatched, they could be misunderstood as a solid web, as shown in **Figure 16.39B**.

Revolving the symmetrically positioned lugs shown in **Figure 16.40** gives their true size in both the front view and sectional view. The same principles of rotation apply to the part shown in **Figure 16.41**, where the inclined arm appears in the section as if it had been revolved to the centerline in the top view and then projected to the sectional view. These conventional practices save time and space on a drawing and also make the views more understandable by the reader.

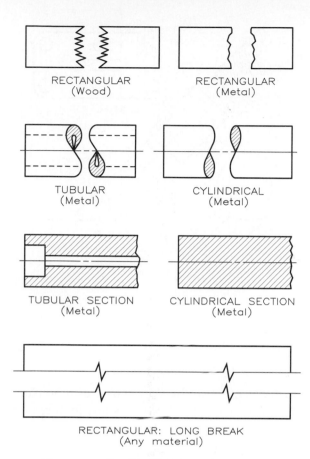

RECTANGULAR
(Wood)

RECTANGULAR
(Metal)

TUBULAR
(Metal)

CYLINDRICAL
(Metal)

TUBULAR SECTION
(Metal)

CYLINDRICAL SECTION
(Metal)

RECTANGULAR: LONG BREAK
(Any material)

16.42 These conventional breaks indicate that a portion of an object has been omitted.

16.13 Conventional Breaks

Figure 16.42 shows types of conventional breaks to use when you remove portions of an object. You may draw the "figure-eight" breaks used for cylindrical and tubular parts freehand (**Figure 16.43**) or with a compass when they are larger, as shown in **Figure 16.44**.

Conventional breaks can be used to shorten a long piece by removing the portion between the breaks so that it may be drawn at a larger scale (**Figure 16.45**). The dimension specifies the true length of the part, and the breaks indicate that a portion of the length has been removed.

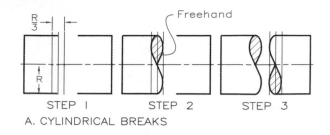

STEP 1   STEP 2   STEP 3

A. CYLINDRICAL BREAKS

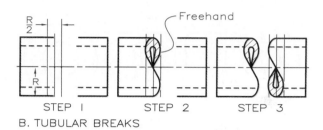

STEP 1   STEP 2   STEP 3

B. TUBULAR BREAKS

16.43 It is essential that you use guidelines in drawing conventional breaks for both (A) solid cylinders and (B) tubular cylinders freehand. The radius R is used to determine the widths.

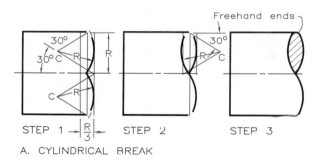

STEP 1   STEP 2   STEP 3

A. CYLINDRICAL BREAK

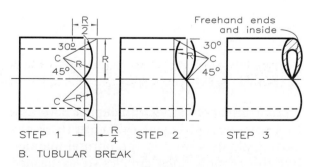

STEP 1   STEP 2   STEP 3

B. TUBULAR BREAK

16.44 These steps can be followed to draw conventional breaks with a compass.

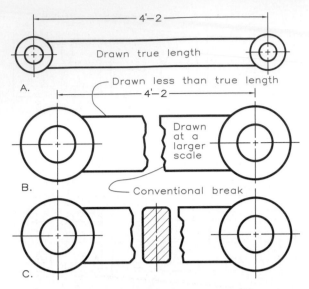

16.45 By using conventional breaks, part of the object can be removed so the part to be drawn at a larger scale or space can be saved. A revolved section can also be inserted between the breaks to further describe the part.

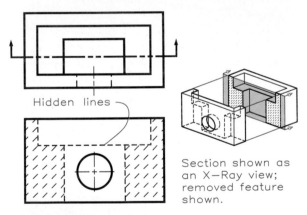

16.46 Phantom sections give an "X-ray" view of a part to show features on both sides of the cutting plane. Section lines are drawn as dashed lines.

16.14 Phantom (Ghost) Sections

A **phantom** or **ghost section** depicts parts as if they were being X-rayed. In **Figure 16.46**, the cutting plane is used in the normal manner, but the section lines are drawn as dashed lines. If the object were shown as a regular full section, the circular hole through the front

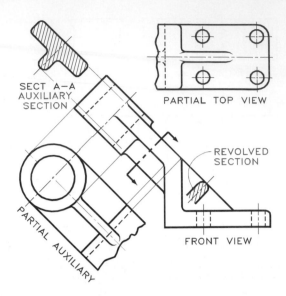

16.47 Auxiliary sections are helpful in clarifying the details of inclined features of a part.

surface could not be shown in the same view. A phantom section lets you show features on both sides of the cutting plane.

16.15 Auxiliary Sections

You may use **auxiliary sections** to supplement the principal views of orthographic projections (**Figure 16.47**). Pass auxiliary cutting plane A-A through the front view and project the auxiliary view from the cutting plane as indicated by the sight arrows. Section A-A gives a cross-sectional description of the part that would be difficult to depict by other principal orthographic views.

Problems

Solve the problems shown in **Figure 16.48** on size A sheets by drawing two solutions per sheet. Each grid space equals 0.20 in., or 5 mm.

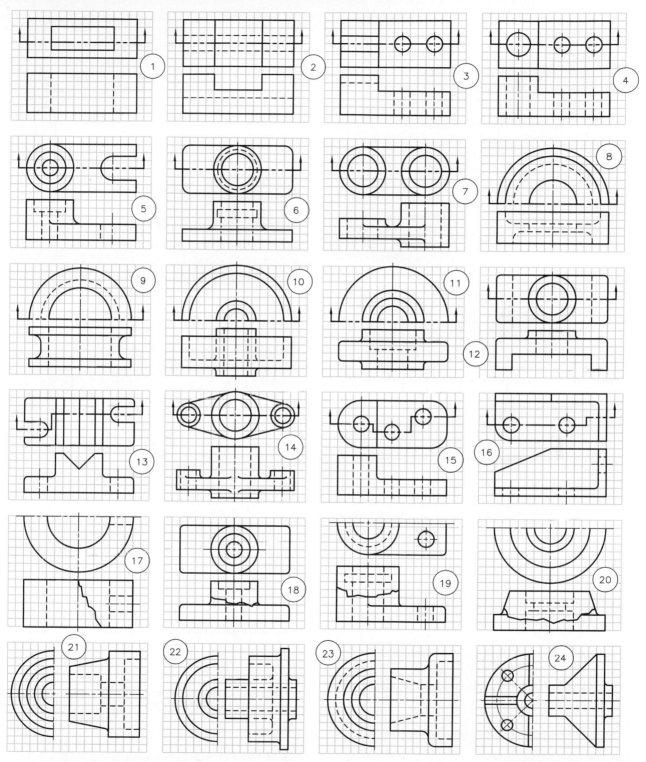

16.48 (Problems 1–24) Introductory sections.

25–35. (Figures 16.49–16.59) Complete these drawings as full sections. Draw one problem per size A sheet (horizontal format). Each grid space equals 0.20 in. or 5 mm. Show the cutting planes in each problem.

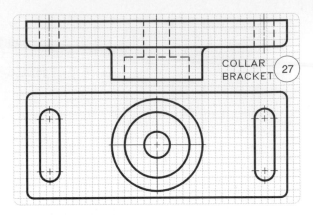

16.51 **(Problem 27)** Full section.

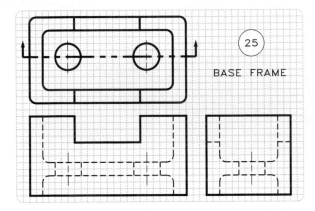

16.49 **(Problem 25)** Full section.

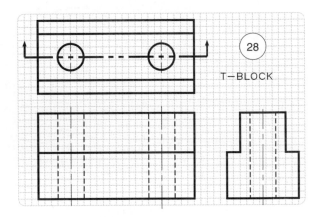

16.52 **(Problem 28)** Full section.

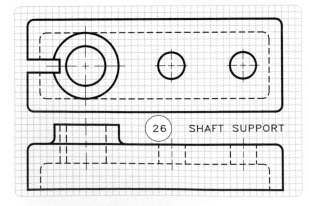

16.50 **(Problem 26)** Full section.

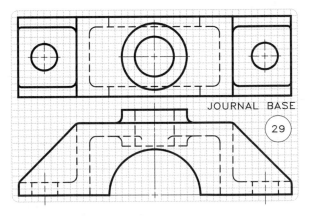

16.53 **(Problem 29)** Full section.

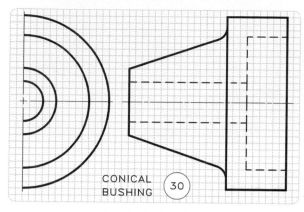

16.54 (Problem 30) Half-section.

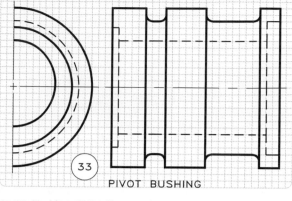

16.57 (Problem 33) Half-section.

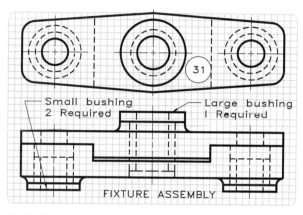

16.55 (Problem 31) Half-section.

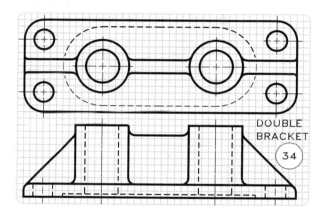

16.58 (Problem 34) Offset section.

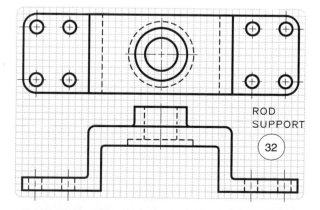

16.56 (Problem 32) Half-section.

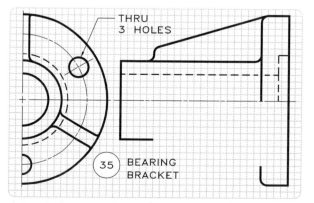

16.59 (Problem 35) Offset section.

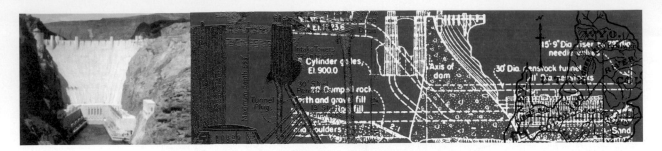

17

Screws, Fasteners, and Springs

17.1 Introduction

Screws provide a fast and easy method of fastening parts together, adjusting the position of parts, and transmitting power. **Screws, sometimes called threaded fasteners, should be purchased rather than made as newly designed parts for each product**. Screws are available through commercial catalogs in countless forms and shapes for various specialized and general applications. Such screws are cheap, interchangeable, and easy to replace.

The types of threaded parts most often used in industry are covered by current ANSI Standards and include both Unified National (UN) and International Organization for Standardization (ISO) threads. Adoption of the UN thread in 1948 by the United States, Great Britain, and Canada (sometimes called the ABC Standards), a modification of the American Standard and the Whitworth thread, was a major step in standardizing threads. The

ISO developed metric thread standards for even broader worldwide applications.

Other types of fasteners include **keys**, **pins**, and **rivets**. **Springs** resist and react to forces and have applications varying from pogo sticks to automobiles. Springs are also available in many forms and styles from specialty manufacturers, who supply most of them, to industry.

17.2 Thread Terminology

Understanding threaded parts begins with learning their terminology, which is used throughout this chapter.

External thread: a thread on the outside of a cylinder, such as a bolt (**Figure 17.1**).

Internal thread: a thread cut on the inside of a part, such as a nut (**Figure 17.1**).

Major diameter: the largest diameter on an internal or external thread (**Figure 17.2**).

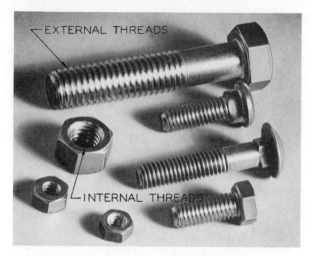

17.1 This photograph shows examples of internal and external threads and three head types. (Courtesy of Russell, Burdsall & Ward Bolt and Nut Company.)

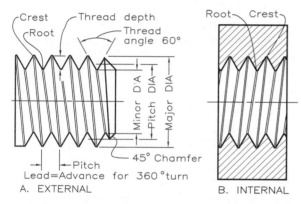

17.2 Most of the definitions of thread terminology are labeled for (A) external and (B) internal threads.

Minor diameter: the smallest diameter on an internal or external thread (**Figure 17.2**).

Crest: the peak edge of a screw thread (**Figure 17.2**).

Root: the bottom of the thread cut into a cylinder to form the minor diameter (**Figure 17.2**).

Depth: the depth of the thread from the major diameter to the minor diameter; also measured as the root diameter (**Figure 17.2**).

Thread angle: the angle between threads cut by the cutting tool, usually 60° (**Figure 17.2**).

Pitch (thread width): the distance between crests of threads, found by dividing 1 inch by the number of threads per inch of a particular thread (**Figure 17.2**).

Pitch diameter: the diameter of an imaginary cylinder passing through the threads at the points where the thread width is equal to the space between the threads (**Figure 17.2** and **Figure 17.4**).

Lead (pronounced leed): the distance a screw will advance when turned 360°.

Form: the shape of the thread cut into a threaded part (**Figure 17.3**).

Series: the number of threads per inch for a particular diameter, grouped into coarse, fine, extra fine, and eight constant-pitch thread series.

Class: the closeness of fit between two mating parts. Class 1 represents a loose fit and Class 3 a tight fit.

Right-hand thread: one that will assemble when turned clockwise. A right-hand external thread slopes downward to the right when its

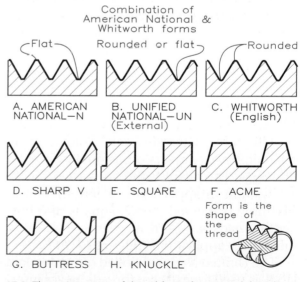

17.3 The various types of thread forms for external threads are shown here.

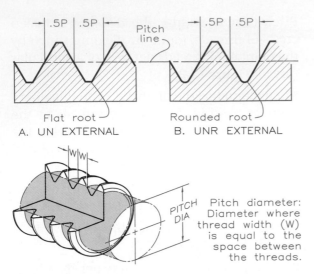

17.4 UN and UNR thread forms.

A The UN external thread has a flat root (rounder root is optional) and a flat crest.

B The UNR thread has a rounded root formed by rolling rather than by cutting. The UNR form does not apply to internal threads.

axis is horizontal and in the opposite direction on internal threads.

Left-hand thread: one that will assemble when turned counterclockwise. A left-hand external thread slopes downward to the left when its axis is horizontal and in the opposite direction on internal threads.

17.3 English System Specifications

Form Thread form is the shape of the thread cut into a part (**Figure 17.3**). The Unified National form, denoted by UN in thread notes, is the most widely used form in the United States. The American National form, denoted by N, appears occasionally on older drawings. The sharp V thread is used for set screws and in applications where friction in assembly is desired. Acme, square, and buttress threads are used in gearing and other machinery applications (**Figure 17.3**).

The Unified National Rolled form, denoted UNR, is used for external threads, never internal threads, because internal threads cannot be formed by rolling. The standard UN form has a flat root (a rounded root is optional) (**Figure 17.4A**), and the UNR form (**Figure 17.4B**) has a rounded root formed by rolling a cylinder across a die. The UNR form can be used instead of the UN form where precision of assembly is less critical.

Series The thread series designates the spacing of threads that vary with diameter. The American National (N) and the Unified National (UN/UNR) forms include three graded series: **coarse** (**C**), **fine** (**F**), and **extra fine** (**EF**). Eight **constant-pitch** series (4, 6, 8, 12, 16, 20, 28, and 32 threads per inch) are also available.

Coarse Unified National forms are denoted UNC or UNRC, which is a combination of form and series designation. The coarse thread (UNC/UNRC or NC) has the largest pitch of any series and is suitable for bolts, screws, nuts, and general use with cast iron, soft metals, and plastics when rapid assembly is desired. An American National (N) form for a coarse thread is written NC.

Fine threads (NF or UNF/UNRF) are used for bolts, nuts, and screws when a high degree of tightening is required. Fine threads are closer together than coarse threads, and their pitch is graduated to be smaller on smaller diameters.

Extra-fine threads (UNEF/UNREF or NEF) are suitable for sheet metal screws and bolts, thin nuts, ferrules, and couplings when the length of engagement is limited and high stresses must be withstood.

Constant-pitch threads (4 UN, 6 UN, 8 UN, 12 UN, 16 UN, 20 UN, 28 UN, and 32 UN) are used on larger diameter threads (beginning near the 1/2-inch size) and have the same pitch size regardless of the diameter size. The most commonly used constant-pitch threads are 8 UN, 12 UN, and 16 UN members of the series, which are used on threads of about 1 inch in diameter and larger.

Constant-pitch threads may be specified as UNR or N thread forms. The ANSI table in Appendix 6 shows constant-pitch threads for larger thread diameters instead of graded pitches of coarse, fine, and extra fine.

Class of Fit The class of fit is the tightness between two mating threads as between a nut and bolt, and is indicated in the thread note by the numbers 1, 2, or 3 followed by the letters A or B. For UN forms, the letter A represents an external thread and the letter B represents an internal thread. The letters A and B do not appear in notes for the American National form (N).

Class 1A and **1B** threads are used on parts that assemble with a minimum of binding and precision. **Class 2A** and **2B** threads are general-purpose threads for bolts, nuts, and screws used in general and mass-production applications. **Class 3A** and **3B** threads are used in precision assemblies where a close fit is required to withstand stresses and vibration.

Single and Multiple Threads A single thread (**Figure 17.5A**) is a thread that advances the distance of its pitch in a revolution of 360°; that is, its pitch is equal to its lead. The crest lines have a slope of 1/2 P since only 180° of the revolution is visible in the view.

Multiple threads are used where quick assembly is required. A double thread is composed of two threads that advance a distance of 2P when turned 360° (**Figure 17.5B**); that is, its lead is equal to 2P. The crest lines have a slope of P because only 180° of the revolution is visible in the view. A triple thread advances 3P in 360° with a crest line slope of 1–1/2 P in the view where 180° of the revolution is visible (**Figure 17.5C**).

17.4 English Thread Notes

Drawings of threads are only symbolic representations and are inadequate to give the details of a thread unless accompanied by

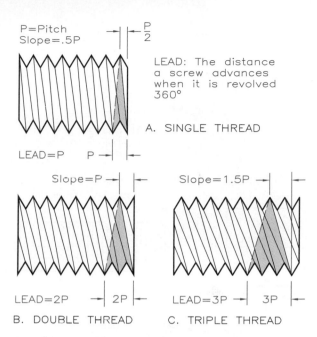

17.5 Threads can be (A) single, (B) double, or (C) triple, which represents the ratio that each advances when turned 360°.

notes (**Figure 17.6**). In a thread note, the major diameter is given first, then the number of threads per inch, the form, series, the class of fit, and the letter A or B to denote external or internal threads, respectively. For a double or triple thread, include the word double or triple in the note, and for left-hand threads add the letters LH.

Figure 17.7 shows a UNR thread note for the external thread. (UNR does not apply to internal threads.) When inches are the unit of measurement, fractions can be written as decimal or as common fractions. The information for thread notes comes from ANSI tables in Appendix 5.

Using Thread Tables

A portion of Appendix 5 is shown in **Figure 17.8**, which gives the UN/UNR thread table from which specifications for standardized interchangeable threads can be selected. Note that a 1–1/2-inch-diameter bolt with fine

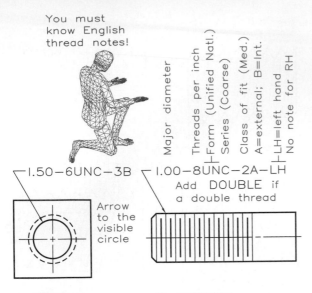

You must know English thread notes!

Major diameter
Threads per inch
Form (Unified Natl.)
Series (Coarse)
Class of fit (Med.)
A=external; B=Int.
LH=left hand
No note for RH

1.50−6UNC−3B

Arrow to the visible circle

1.00−8UNC−2A−LH
Add DOUBLE if a double thread

A. INTERNAL

B. EXTERNAL

17.6 Thread notes of the English form that are applied to (A) internal and (B) external threads are defined here.

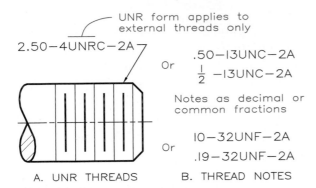

UNR form applies to external threads only

2.50−4UNRC−2A

Or .50−13UNC−2A
$\frac{1}{2}$ −13UNC−2A

Notes as decimal or common fractions

Or 10−32UNF−2A
.19−32UNF−2A

A. UNR THREADS

B. THREAD NOTES

17.7 Thread notes.
A The UNR thread note is applied to only external threads.
B Diameters in thread notes can be given as decimal fractions or common fractions.

thread (UNF) has 12 threads per inch, and its thread note is written as

1.500–12 UNF–2A or $1\frac{1}{2}$–12 UNF–2A.

If the thread were internal (nut), the thread note would be the same but the letter B would be used instead of the letter A. For constant-

Nominal Diameter	Basic Diameter	Coarse NC & UNC		Fine NF & UNF		Extra Fine NEF/UNEF	
		Thds per In.	Tap Drill DIA	Thds per In.	Tap Drill DIA	Thds per In.	Tap Drill DIA
1	1.000	8	.875	12	.922	20	.953
1−1/16	1.063	...	...	...	...	18	1.000
1−1/8	1.125	7	.904	12	1.046	18	1.070
1−3/16	1.188	...	...	...	...	18	1.141
1−1/4	1.250	7	1.109	12	1.172	18	1.188
1−5/16	1.313	...	...	...	...	18	1.266
1−3/8	1.375	6	1.219	12	1.297	18	1.313
1−7/16	1.438	...	...	...	...	18	1.375
1−1/2	1.500	6	1.344	12	1.422	18	1.438

17.8 This is a portion of the ANSI tables for UN and UNR threads in Appendix 5.

pitch thread series, selected for larger diameters, write the thread note as

1.750–12 UN–2A or $1\frac{3}{4}$–12 UN–2A.

For the UNR thread form (for external threads only) substitute UNR for UN in the last three columns; for example, UNREF for UNEF (extra fine). **Figure 17.9** shows the preferred placement of thread notes (with leaders) for external and internal threads.

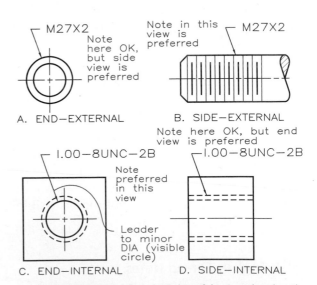

M27X2

Note here OK, but side view is preferred

Note in this view is preferred

M27X2

A. END−EXTERNAL

B. SIDE−EXTERNAL

Note here OK, but end view is preferred

1.00−8UNC−2B

Note preferred in this view

1.00−8UNC−2B

Leader to minor DIA (visible circle)

C. END−INTERNAL

D. SIDE−INTERNAL

17.9 Notes for external threads are best if they are placed on the rectangular view of the threads. Notes for internal threads are best applied in the circular view if space permits. Metric notes are given at (A) and (B), and English notes are given at (C) and (D).

17.5 Metric Thread Notes

Metric thread notes are usually given as a basic designation suitable for general applications. However, for applications where the assembly of threaded parts is crucial, a complete thread designation should be noted. The ISO thread table in Appendix 7, a portion of which is shown in **Figure 17.10**, contains specifications for metric thread notes.

Basic Designation   **Figure 17.11** shows examples of metric screw thread notes. Each note begins with the letter M, designating the note as metric, followed by the major diameter size in millimeters and the pitch in millimeters separated by the multiplication sign and the number 3.

Complete Designation   The first part of a complete designation note (**Figure 17.12**) is identical to the basic designation, to which is added the tolerance class designation separated by a dash. (A tolerance is a specified maximum variation in size that ensures assembly of the threaded parts.) The 5g represents the pitch diameter tolerance, and the 6g represents the crest diameter tolerance.

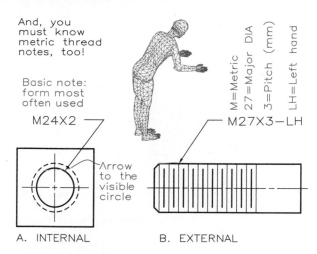

17.11 This is the basic thread note that will be used for (A) internal and (B) external threads.

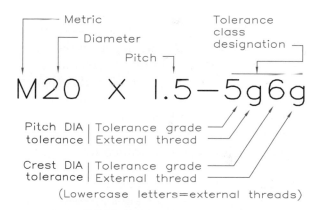

17.12 A complete designation note for metric threads adds tolerance specifications to the basic note.

The numbers 5 and 6 are thread tolerance grades (variations in size from the basic diameter) from **Figure 17.13**. These grades are for the pitch diameter and the major and minor diameters of medium general-purpose thread similar to class 2A and 2B threads in the UN system. Grades of less than 6 are best suited to fine-series fits and short lengths of engagement. Grades greater than 6 are best suited to coarse-series fits and long lengths of engagement.

The letters following the grade numbers denote tolerance positions, which means either external or internal threads. Lowercase

COARSE		FINE	
MAJ. DIA & THD PITCH	TAP DRILL	MAJ. DIA & THD PITCH	TAP DRILL
M20 X 2.5	17.5	M20 X 1.5	18.5
M22 X 2.5	19.5	M22 X 1.5	20.5
M24 X 3	21.0	M24 X 2	22.0
M27 X 3	24.0	M27 X 2	25.0
M30 X 3.5	26.5	M30 X 2	28.0
M33 X 3.5	29.5	M33 X 2	31.0
M36 X 4	32.0	M36 X 2	33.0
M39 X 4	35.0	M39 X 2	36.0
M42 X 4.5	37.5	M42 X 2	39.0

17.10 This portion of the metric (ISO) thread tables in Appendix 7 shows specifications of metric thread notes.

External Thread		Internal Thread	
Major Diameter	Pitch Diameter	Minor Diameter	Pitch Diameter
–	3	–	–
4	4	4	4
–	5	5	5
6	6	6	6
–	7	7	7
8	8	8	8
–	9	–	–

(FINE — MEDIUM — COARSE)

17.13 Tolerance grades for major and minor diameters and pitch diameters of each range from fine to coarse. Tolerance grades combined with position symbols (such as 6g) become tolerance classes.

EXTERNAL THREADS (Lowercase letters)

e = Large allowance
g = Small allowance
h = No allowance

INTERNAL THREADS (Uppercase letters)

G = Small allowance
H = No allowance

LENGTH OF ENGAGEMENT

S = Short   N = Normal   L = Long

EXAMPLE: Refer to Figure 17.13
Position   Lower–case=external threads
               Upper–case=internal threads
Allowance   e, g, h, G, H=amount of allowance
Engagement   S, N, & L Columns of Fig. 17.13

17.14 These symbols represent position, allowance, and engagement length. Position means either external or interior threads. A lowercase letter indicates an external thread and an uppercase letter signifies an internal thread.

letters designate external threads (bolts) as shown in **Figure 17.14**. The letters e, g, and h represent large allowance, small allowance, and no allowance, respectively. (Allowance is the permitted variation in size from the basic diameter.) Uppercase letters designate internal threads (nuts). The letters G and H, placed after the tolerance grade number, denote small allowance and no allowance, respectively. For example, 5g designates a medium tolerance with small allowance for the pitch diameter of an external thread, and 6H designates a medium tolerance with no allowance for the minor diameter of an internal thread.

Tolerance classes are **fine, medium,** and **coarse** (**Figure 17.15**). They represent combinations of tolerance grades, tolerance positions, and lengths of engagement—**short** (**S**), **normal** (**N**), and **long** (**L**). A table of lengths of engagement (the actual length of the assembled thread in mating parts) is given in the

Quality	External Threads (bolts)									Internal Threads (nuts)					
	Tolerance position e (large allowance)			Tolerance position g (small allowance)			Tolerance position h (no allowance)			Tolerance position G (small allowance)			Tolerance position H (no allowance)		
	Length of engagement			Length of engagement			Length of engagement			Length of engagement			Length of engagement		
	Group S	Group N	Group L	Group S	Group N	Group L	Group S	Group N	Group L	Group S	Group N	Group L	Group S	Group N	Group L
Fine							3h4h	4h	5h4h				4H	5H	6H
Medium		6e	7e6e	5g6g	6g	7g6g	5h6h	6h	7h6h	5G	6G	7G	5H	6H	7H
Coarse					8g	9g8g					7G	8G		7H	8H

17.15 Tolerance classes for large and medium allowances and for no allowance for internal and external threads are shown here. Select the most commonly used tolerance class, commercial threads (in boxes) first, then the classes in bold print, medium-size print, and small-size print, respectively.

ANSI Standards, but not in this text. After deciding whether to use a fine, medium, or coarse class of fit, select the thread designation first from those in boxes (commercial threads), second from those in bold print, third from those in medium-size print, and fourth from those in small print. The 6H class is comparable to the 2B class of fit in the UN system for an interior thread, and the 6g class is similar to the 2A class of fit.

Figure 17.16 shows variations for complete designation thread notes. When the minor and pitch diameters have identical grades, the tolerance class symbol consists of one number and letter, such as 6H (**Figure 17.16A**). The uppercase H indicates that the position is internal and no allowance is added to the basic thread note. If necessary, add the length of engagement symbol (S, N, L) to the tolerance class designation (**Figure 17.16B**). For unknown lengths of thread engagement, use group N (normal).

Specify the fit between mating threads with notes as shown in **Figure 17.17**, with a slash separating the tolerance class designations of internal and external threads. Additional information about ISO threads may be obtained from *ISO Metric Screw Threads,* a booklet of standards published by ANSI that was used as the basis for most of this section.

g=External thds.
and small allow.
6=Toler. grade

H=Internal thds.
and no allowance
6=Toler. grade

M6XI−6H/6g

A.

g=External thds. and
small allow. 6=Tol.
grade for Crest DIA

g=External thds. and
small allow. 5=Tol.
grade for Pitch DIA

M20X2−6H/5g6g

B.

17.17 A slash mark is used to separate the tolerance class designations of mating internal and external threads.

17.6 Drawing Threads

Threads may be represented by **detailed**, **schematic**, and **simplified** symbols (**Figure 17.18**). Detailed symbols represent a thread most realistically, simplified symbols represent a thread least realistically, and schematic symbols are a compromise between the two. Detailed and schematic symbols can be used for drawing larger threads on a drawing (1/2" diameter and larger) and simplified symbols are best for smaller threads.

17.7 Detailed Symbols

UN/UNR Threads Detailed thread symbols for external threads in view and in section are shown in **Figure 17.19**. Instead of drawing helical curves, straight lines are used to depict crest and root lines. Variations of detailed thread symbols for internal threads drawn in views and sections are shown in **Figure 17.20**.

If pitch and crest
DIA tolerances are
equal, use only
one tolerance symbol

Letters S, N, or L
are used to in−
dicate length of
thread engagement

M22X1.5−6H   M24X3−7g6gL

A. NOTE WITH EQUAL
TOLERANCE GRADES

B. LENGTH OF THREAD
ENGAGEMENT INDICATED

17.16 Complete metric notes.

A When both pitch and crest diameter tolerance grades are the same, show the tolerance class symbol only once.

B The letters S, N, and L indicate the length of the thread engagement.

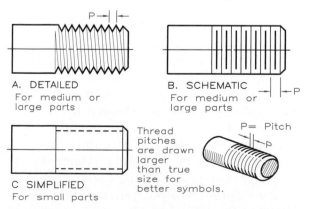

A. DETAILED
For medium or
large parts

B. SCHEMATIC
For medium or
large parts

C SIMPLIFIED
For small parts

Thread
pitches
are drawn
larger
than true
size for
better symbols.

P= Pitch

17.18 The three types of thread symbols used for drawing threads are (A) detailed, (B) schematic, and (C) simplified.

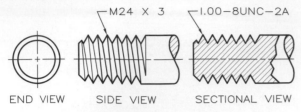

END VIEW   SIDE VIEW   SECTIONAL VIEW

17.19 These detailed symbols represent external threads in view and section.

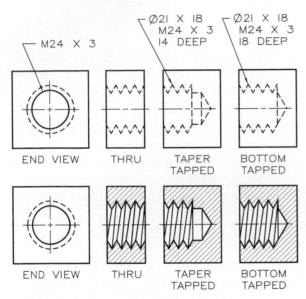

END VIEW   THRU   TAPER TAPPED   BOTTOM TAPPED

END VIEW   THRU   TAPER TAPPED   BOTTOM TAPPED

17.20 These detailed symbols represent internal threads. Approximate the minor diameter as 75% of the major diameter. Tap drill diameters (21) are found in Appendix 21.

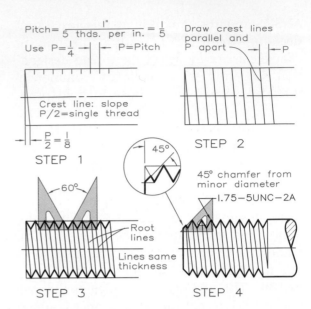

17.21 Detailed representation of threads:

Step 1 To draw a detailed drawing of a 1.75–5 UNC–2A thread, find the pitch by dividing 1 inch by the number of threads per inch, or 5 in this case. Use a pitch of 1/4 instead of 1/5 to space the threads apart. Lay off the pitch along the length of the thread and draw a crest line at a slope of P/2, or 1/8 inch in this case.

Step 2 Draw the other crest lines as dark, visible lines parallel to the first crest line.

Step 3 Find the root lines by constructing 60° vees between the crest lines. Draw the root lines from the bottom of the vees. Root lines are parallel to each other but not to crest lines.

Step 4 Construct a 45° chamfer at the end of the thread from the minor diameter. Darken all lines and add a thread note.

The steps of drawing a detailed thread representation, whether English or metric threads, are shown in **Figure 17.21**. When using the English tables, calculate the pitch by dividing 1 inch by the number of threads per inch. In the metric system, pitch is given in the tables. **Draw the spacing between crest lines larger than the actual pitch size to avoid "clogged-up" lines.**

Computer Method You can draw detailed thread symbols by computer (**Figure 17.22**) and duplicate them with the *Copy* command's *Multiple* option. The program produces a typical set of threads in Step 1 and then copies it repetitively in Step 2.

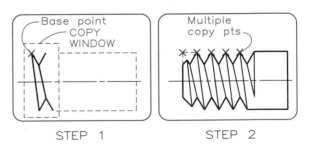

STEP 1   STEP 2

17.22 Detailed thread symbols by computer:

Step 1 Draw a typical detailed thread symbol at the end of the screw using the *Line* command.

Step 2 Duplicate a typical set of threads with the *Copy* command and the *Multiple* option along the predetermined snap points of the screw.

Square Threads **Figure 17.23** shows how to draw and note a detailed drawing of a square thread. Follow the same basic steps to draw views and sections of square internal threads (**Figure 17.24**). In section, draw both the internal crest and root lines, but in a view, draw only the outline of the threads. Place thread notes for internal threads in the circular view, whenever possible, with the leader pointing toward the center and stopping at the visible circle.

When a square thread is long, it can be represented by using phantom lines, without drawing all the threads (**Figure 17.25**). This conventional practice saves time and effort without reducing the drawing's effectiveness.

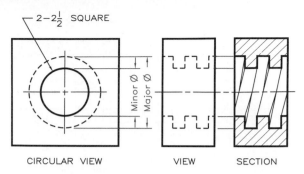

17.24 This drawing shows internal square threads in view and section.

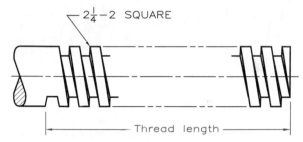

17.25 The conventional method of showing square threads is to draw sample threads at each end and to connect them with phantom lines.

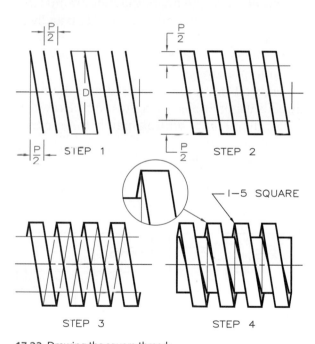

17.23 Drawing the square thread:

Step 1 Lay out the major diameter. Space the crest lines 1/2P apart and slope them downward to the right for right-hand threads.

Step 2 Connect every other pair of crest lines. Find the minor diameter by measuring 1/2P inward from the major diameter.

Step 3 Connect the opposite crest lines with light construction lines to establish the profile of the thread form.

Step 4 Connect the inside crest lines with light construction lines to locate the points on the minor diameter where the thread wraps around the minor diameter. Darken the final lines.

Acme Threads A modified version of the square thread is the Acme thread, which has tapered (15°) sides for easier engagement than square threads. The steps involved in drawing detailed Acme threads are shown in **Figure 17.26**. Acme threads are heavy threads that are used to transmit force and power in mechanisms such as screw jacks, leveling devices, and lathes. Appendix 8 contains the table for Acme thread specifications and dimensions.

Internal Acme threads are shown in view and in section in **Figure 17.27**. Left-hand internal threads in section appear the same as right-hand external threads.

17.8 Schematic Symbols

Figure 17.28 shows schematic representations of external threads with metric notes. Because schematic symbols are easy to draw and adequately represent threads, it is the

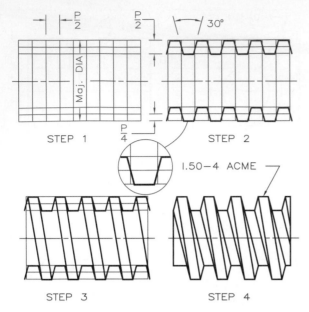

17.26 Drawing the Acme thread:

Step 1 Lay out the major diameter and thread length and divide the shaft into equal divisions 1/2P apart. Locate the minor and pitch diameters by using distances 1/2P and 1/4 P.

Step 2 Draw construction lines at 15° angles with the vertical along the pitch diameter to make a total angle of 30°.

Step 3 Draw the crest lines across the screw.

Step 4 Darken the lines, draw the root lines, and add the thread note to complete the drawing.

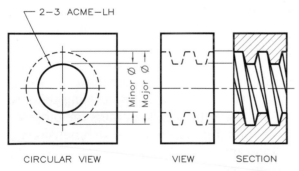

17.27 This drawing shows internal Acme threads in view and section.

thread symbol used most often for medium-size threads. Draw schematic thread symbols by using thin parallel crest lines and thick root lines. Schematic drawings of left-hand and right-hand threads are identical, only the LH in the thread note indicates that a thread is

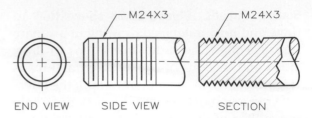

17.28 These schematic symbols represent external threads in view and section.

left-handed. Right-hand threads are not marked RH, but are understood to be right-hand threads.

Figure 17.29 shows threaded holes in view and in section drawn with schematic symbols. The size of the tap drill diameter is approximately equal to the major diameter minus the pitch. However, the minor diameter usually is drawn a bit smaller to provide better separation between the lines representing the major and minor diameters.

Figure 17.30 shows how to draw schematic threads using English specifications. Draw the minor diameter at approximately three-quar-

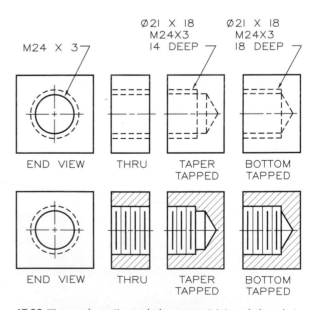

17.29 These schematic symbols represent internal threads in view and section. Tap drill diameters are found in Appendix 21.

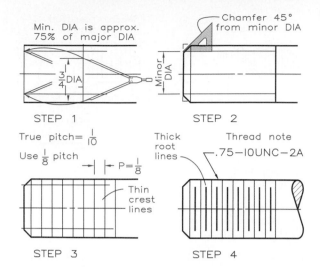

STEP 1

STEP 2

Min. DIA is approx. 75% of major DIA

$\frac{3}{4}$ DIA

Chamfer 45° from minor DIA

Minor DIA

True pitch = $\frac{1}{10}$

Use $\frac{1}{8}$ pitch

P = $\frac{1}{8}$

Thick root lines

Thin crest lines

Thread note
.75–10UNC–2A

STEP 3

STEP 4

17.30 Schematic representation of threads:

Step 1 Lay out the major diameter and locate the minor diameter (about three-quarters of the major diameter). Draw the minor diameter with light construction lines.

Step 2 Chamfer the end of the threads with a 45° angle from the minor diameter.

Step 3 Find the pitch of a .75–10UNC–2A thread (0.1) by dividing 1 inch by the number of threads per inch (10). Use a larger pitch, 1/8 inch in this case, for spacing the thin crest lines.

Step 4 Draw root lines as thick as the visible lines between the crest lines to the construction lines representing the minor diameter. Add a thread note.

ters of the major diameter and the chamfer (bevel) 45° from the minor diameter. Draw crest lines as thin lines and root lines as thick visible lines.

Computer Method You may draw schematic thread symbols by computer (**Figure 17.31**) and duplicate them with the *Copy* command's *Multiple* option. The program produces a typical set of threads in Step 1 and copies it repetitively in Step 2.

17.9 Simplified Symbols

Examples of external threads drawn with simplified symbols and noted with both metric and English formats are shown in **Figure 17.32**. Simplified symbols are the easiest to

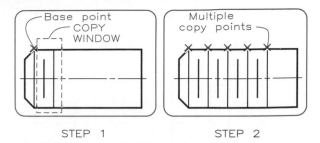

Base point
COPY
WINDOW

Multiple copy points

STEP 1

STEP 2

17.31 Schematic threads by computer:

Step 1 Draw the outline of the threaded shaft having a chamfered end. Produce typical minor and major diameters and window them for a *Multiple Copy* command.

Step 2 Repetitively *Copy* the threads along the screw at the predetermined snap points.

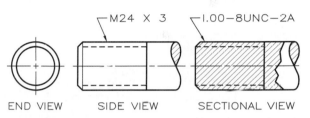

M24 X 3

1.00–8UNC–2A

END VIEW   SIDE VIEW   SECTIONAL VIEW

17.32 These simplified thread symbols represent external threads in view and section.

draw and are the best suited for drawing small threads where drawing using schematic and detailed symbols would be too crowded. Various techniques of applying simplified symbols to threads are shown in **Figure 17.33**. The minor diameter is drawn as hidden lines spaced at about three-quarters the major diameter. **Figure 17.34** shows the steps of drawing simplified threads. With experience, you will be able to approximate the location of the minor diameter of simplified threads by eye.

Drawing Small Threads

Remember, a drawing of a thread is a pictorial symbol, therefore do not try to draw the true spacing of the threads. True spacing will be too close and too hard to read. Instead, select

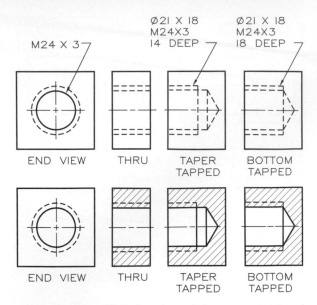

17.33 These simplified thread symbols represent internal threads in view and section. Draw minor diameters at about 3/4 the major diameter.

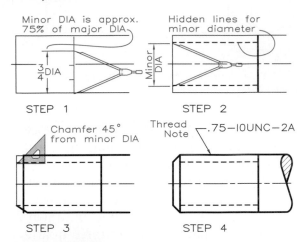

17.34 Simplified representation of threads:

Step 1 Lay out the major diameter. Locate the minor diameter (about three-quarters of the major diameter).

Step 2 Draw hidden lines to represent the minor diameter.

Step 3 Draw a 45° chamfer from the minor diameter to the major diameter.

Step 4 Darken the lines and add a thread note.

a wider spacing between root and crest lines to be easier to read and to draw (**Figure 17.35**). This conventional practice of enlarging the thread's pitch to separate thread symbols is applied in drawing all three types of

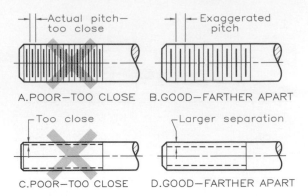

17.35 Most threads must be drawn using exaggerated dimensions instead of actual measurements to prevent the drawing from having lines drawn too closely together.

symbols (simplified, schematic, and detailed). Add a thread note to the drawing to give the necessary detailed specifications.

17.10 Nuts and Bolts

Nuts and bolts (**Figure 17.36**) come in many forms and sizes for many different applications. Some common types of threaded fasteners are shown in **Figure 17.37**. A **bolt** is a threaded cylinder with a head and is used with a nut to hold parts together. A **stud** is a headless bolt, threaded at both ends, that is screwed into one part with a nut attached to the other end.

A **cap screw** usually does not have a nut, but passes through a hole in one part and screws into another threaded part. A hexagon-head **machine screw** is similar to but smaller than a cap screw. Machine screws also come with other types of heads. A **set screw** is used to hold

17.36 This photo shows a nut, bolt, and washers in combination. (Courtesy of Lamson & Sessions.)

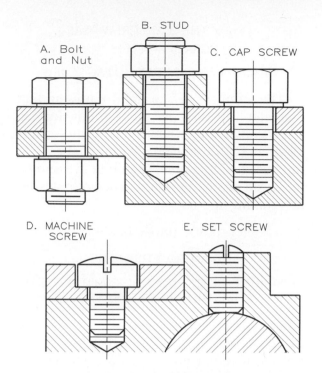

A. Bolt and Nut

B. STUD

C. CAP SCREW

D. MACHINE SCREW

E. SET SCREW

17.37 Types of threaded bolts are illustrated here.

17.38 Examples of types of nuts. (Courtesy of Russell, Burdsall & Ward Bolt and Nut Company.)

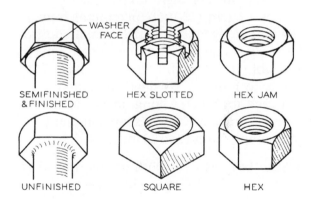

WASHER FACE

SEMIFINISHED & FINISHED

HEX SLOTTED

HEX JAM

UNFINISHED

SQUARE

HEX

17.39 Finished and semifinished bolts and nuts have raised washer faces. Several types of nuts are also shown here.

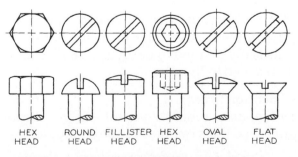

HEX HEAD

ROUND HEAD

FILLISTER HEAD

HEX HEAD

OVAL HEAD

FLAT HEAD

17.40 These are standard types of bolt and screw heads.

one part fixed in place with another, usually to prevent rotation, as with a pulley on a shaft.

Types of heads used on **regular** and **heavy** bolts and nuts are shown in **Figure 17.38** and **Figure 17.39**. Heavy bolts have thicker heads than regular bolts for heavier usage. A **finished head** (or nut) has a 1/64-inch-thick washer face (a circular boss) to provide a bearing surface for smooth contact. **Semifinished bolt heads** and **nuts** are the same as finished bolt heads and nuts. Unfinished bolt heads and nuts have no bosses and no machined surfaces.

A **hexagon jam nut** does not have a washer face, but it is chamfered (beveled at its corners) on both sides. **Figure 17.40** shows other standard bolt and screw heads for cap screws and machine screws.

Dimensions

Figure 17.41 shows a properly dimensioned bolt. The ANSI tables in Appendices 10–18

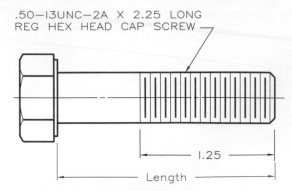

.50–13UNC–2A × 2.25 LONG
REG HEX HEAD CAP SCREW

1.25

Length

17.41 A properly dimensioned and noted hexagon-head bolt.

give nut and bolt dimensions, but you may use the following guides for hexagon-head and square-head bolts.

Overall Lengths Hexagon-head bolts are available in 1/4-inch increments up to 8 inches long, in 1/2-inch increments from 8 to 20 inches long, and in 1-inch increments from 20 to 30 inches long. Square-head bolts are available in 1/8-inch increments from 1/2 to 3/4 inch long, in 1/4-inch increments from 3/4 inch to 5 inches long, in 1/2-inch increments from 5 to 12 inches long, and in 1-inch increments from 12 to 30 inches long.

Thread Lengths For both hexagon-head and square-head bolts up to 6 inches long,

$$\text{Thread length} = 2D + 1/4 \text{ in.}$$

where D is the diameter of the bolt. For bolts more than 6 inches long

$$\text{Thread length} = 2D + 1/2 \text{ inch.}$$

Threads for bolts can be coarse, fine, or 8-pitch threads. The class of fit for bolts and nuts is understood to be 2A and 2B if no class is specified in the note.

Dimension Notes
Designate standard square-head and hexagon-head bolts by notes in one of three forms:

$3/8$–16×3 1–$\frac{1}{2}$ LONG SQUARE BOLT- STEEL;

$\frac{1}{2}$–13×3 HEX CAP SCREW—SAE
GRADE 8-STEEL;

$.75 \times 3$ 5.00 UNC–2A HEX HD LAG SCREW.

The numbers (left to right) represent bolt diameter, threads per inch (omit for lag screws), bolt length, screw name, and material (material designation is optional). When not specified in a note, each bolt is assumed to have a class 2 fit. Three types of notes for designating nuts are

$\frac{1}{2}$–13 SQUARE NUT-STEEL;

$\frac{3}{4}$–16 HEAVY HEX NUT;

1.00–8UNC–2B HEX HD THICK SLOTTED NUT
-CORROSION-RESISTANT STEEL.

When nuts are not specified as heavy, they are assumed to be regular. When the class of fit is not specified in a note, it is assumed to be 2B for nuts.

17.11 Drawing Square Heads

Appendices 10 and 11 give dimensions for square bolt heads and nuts. However, conventional practice is to draw nuts and bolts by using the general proportions shown in **Figure 17.42**. Your first step in drawing a bolt head or nut is to determine whether the view is to be across corners or across flats—that is, whether the lines at either side of the view represent the square's corners or flats. Drawing across corners shows nuts and bolts best, but occasionally you must draw one across flats when the head or nut is truly in this orientation.

17.12 Drawing Hexagon Heads

Figure 17.43 shows the steps of drawing the head of a hexagon bolt across corners by using the bolt's major diameter, D, as the basis for all other proportions. Begin by drawing the top view of the head as a circle of a diameter of 1–1/2 D. For a regular head, the thickness is

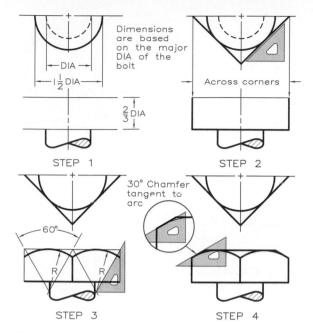

17.42 Drawing the square head:

Step 1 Draw the major diameter, DIA, of the bolt. Use 1.5 DIA to draw the hexagon-head's diameter and 2/3 DIA to establish its thickness.

Step 2 Draw the top view of the square head at a 45° angle to give an across-corners view.

Step 3 Show the chamfer in the front view by using a 30°–60° triangle to find the centers for the radii.

Step 4 Show a 30° chamfer tangent to the arcs in the front view. Darken the lines.

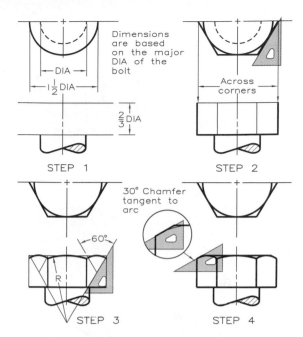

17.43 Drawing the hexagon head:

Step 1 Draw the major diameter, DIA, of the bolt and use it to establish the head diameter (as 1.5 DIA) and thickness (as 2/3 DIA).

Step 2 Construct a hexagon head with a 30°–60° triangle to give an across-corners view.

Step 3 Find arcs in the front view to draw the chamfer of the head.

Step 4 Draw a 30° chamfer tangent to the arcs in the front view. Darken the lines.

2/3 D and for a heavy head it is 7/8 D. Circumscribe a hexagon about the circle. Then draw outside arcs in the rectangular view and tangent chamfers (bevels) to complete the drawing.

Computer Method You may draw a hexagon head for a bolt by computer, as shown in Step 1 of **Figure 17.44**. Use a bolt diameter of 1 inch for easy scaling. You may then scale and rotate the block as desired when you use the *Insert* command. Step 2 shows the drawing *Blocked* and *inserted* at scales of 50% and 75%. To insert a thread with a diameter of 0.50 inch, assign a size factor of 0.50 when prompted by the *Insert* and *Block* commands.

Drawing Nuts

Use the same techniques to draw a square and a hexagon nut (shown across corners in **Figure 17.45**) that you did to draw bolt heads. The difference is that nuts are thicker than bolt heads: The thickness of a regular nut is 7/8 D, and the thickness of a heavy nut is 1 D, where D is the bolt diameter. Hidden lines may be inserted in the front view to indicate threads, or omitted. Exaggerate the thickness of the 1/64-inch washer face on the finished and semifinished hexagon nuts to about 1/32 inch to make it more noticeable. Place thread notes on circular views with leaders when space permits. Square nuts that are not labeled heavy are assumed to be regular nuts.

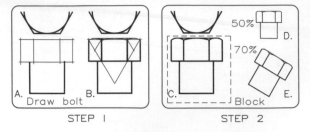

17.44 A bolt head representation drawn by computer:

Step 1 Draw a hexagon head using the steps of geometric construction in Figure 17.43. Base the size of the head on a bolt diameter of 1 inch to form a *Unit Block* for easy scaling.

Step 2 Convert the drawing into a *Block* by using a window. Insert the block at any size and position and scale with the desired factor as shown at D and E.

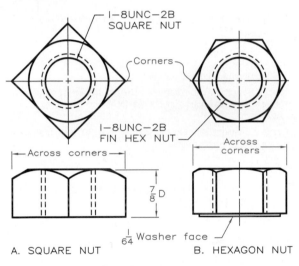

17.45 Drawing square and hexagon nuts across corners involves the same steps used for drawing bolt heads. Add notes to give nut specifications.

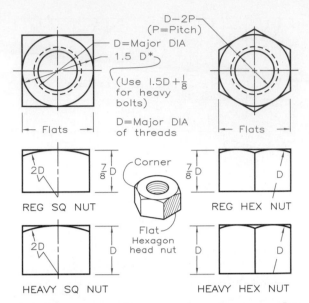

17.46 These square and hexagon nuts are drawn across flats with notes added to give their specifications. Square nuts are unfinished.

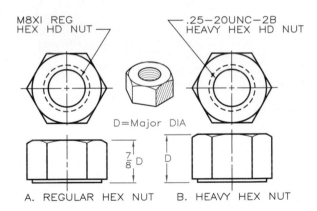

17.47 These regular and heavy hexagon nuts are drawn across corners, with notes added to give their specifications.

Figure 17.46 shows how to construct square and hexagon nuts across flats. For regular nuts, the distance across flats is 1–1/2 × D (D = major diameter of the thread), and 1–5/8 D for heavy nuts. Draw the top views in the same way you did across-corner top views, but rotate them to give across-flat front views. **Figure 17.47** depicts dimensioned and noted hexagon regular and heavy nuts drawn across corners.

Drawing Nut and Bolt Combinations

Nuts and bolts in assembly are drawn in the same manner they are drawn individually (**Figure 17.48**). Use the major diameter, D, of the bolt as the basis for other dimensions. Here, the views of the bolt heads are across corners, and the views of the nuts are across flats, although both views could have been drawn

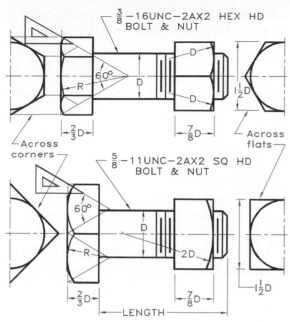

17.48 The proportions and the geometry for drawing square nuts and bolts and hexagon nuts and bolts are shown here.

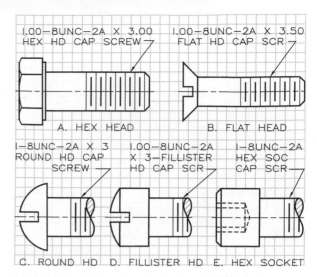

17.49 These cap screws are drawn on a grid to give the proportions for drawing them at different sizes. Notes give thread specifications, length, head type, and bolt name (cap screw).

across corners. The half end views are used to find the front views by projection. Add a note to give the specifications of the nut and bolt.

17.13 Types of Screws

Cap Screws

The cap screw passes through a hole in one part and screws into a threaded hole in the other part so the two parts can be held together without a nut. Cap screws are usually larger than machine screws and they may also be used with nuts. **Figure 17.49** shows the standard types of cap screw heads drawn on a grid that can be used as a guide for drawing cap screws of other sizes. Appendices 14–18 give cap screw dimensions, which can aid in drawing them.

Machine Screws

Smaller than most cap screws, **machine screws** usually are less than 1 inch in diameter. They screw into a threaded hole in a part or into a nut. Machine screws are fully threaded

when their length is 2 inches or less. Longer screws have thread lengths of 2D + 1/4 inch (D = major diameter of the thread). **Figure 17.50** shows four types of machine screws, along with notes, drawn on a grid that may be used as an aid in drawing them without dimensions from a table. Machine screws range in diameter from No. 0 (0.060 inch) to 3/4 inch, as shown in Appendix 19, which gives the dimensions of round-head machine screws.

Set Screws

Set screws are used to hold parts together, such as pulleys and handles on a shaft, and prevent rotation. **Figure 17.51** shows various types of set screws, with dimensions denoted by letters that correspond to the tables of dimensions in Appendix 20.

Set screws are available in combinations of points and heads. The shaft against which the set screw is tightened may have a machined flat surface to provide a good bearing surface for a **dog** or **flat-point** set screw end to press against. The **cup point** gives good gripping when pressed against round shafts. The **cone point**

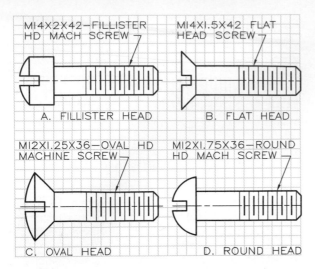

A. FILLISTER HEAD — MI4X2X42—FILLISTER HD MACH SCREW

B. FLAT HEAD — MI4XI.5X42 FLAT HEAD SCREW

C. OVAL HEAD — MI2XI.25X36—OVAL HD MACHINE SCREW

D. ROUND HEAD — MI2XI.75X36—ROUND HD MACH SCREW

17.50 These are standard types of machine screws. The same proportions may be used to draw machine screws of all sizes.

works best when inserted into holes drilled in the part being held. The **headless set screw** has no head to protrude above a rotating part. An exterior square head is good for applications in

which greater force must be applied with a wrench to hold larger set screws in position.

Wood Screws

A wood screw is a pointed screw having sharp coarse threads that will screw into wood making its own internal threads in the process. **Figure 17.52** shows the three most common types of wood screws drawn on a grid to show their relative proportions.

Sizes of wood screws are specified by single numbers, such as 0, 6, or 16. From 0 to 10, each digit represents a different size. Beginning at 10, only even-numbered sizes are standard, that is, 10, 12, 14, 16, 18, 20, 22, and 24. Use the following formula to translate these numbers into the actual diameter sizes:

$$\text{Actual DIA} = 0.06 + (\text{screw number} \times 0.013).$$

For example, the diameter for the No. 7 wood screw shown in **Figure 17.52** is calculated as follows:

$$\text{DIA} = 0.06 + 7(0.013) = 0.151.$$

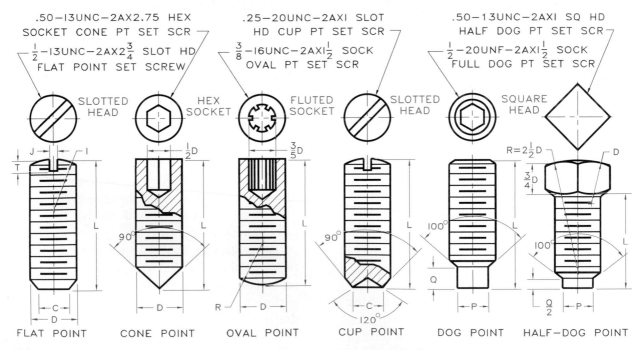

FLAT POINT — .50—13UNC—2AX2.75 HEX SOCKET CONE PT SET SCR — $\frac{1}{2}$—13UNC—2AX2$\frac{3}{4}$ SLOT HD FLAT POINT SET SCREW — SLOTTED HEAD

CONE POINT — HEX SOCKET

OVAL POINT — .25—20UNC—2AXI SLOT HD CUP PT SET SCR — $\frac{3}{8}$—16UNC—2AXI$\frac{1}{2}$ SOCK OVAL PT SET SCR — FLUTED SOCKET

CUP POINT — SLOTTED HEAD

DOG POINT — .50—13UNC—2AXI SQ HD HALF DOG PT SET SCR — $\frac{1}{2}$—20UNF—2AXI$\frac{1}{2}$ SOCK FULL DOG PT SET SCR

HALF—DOG POINT — SQUARE HEAD

17.51 Set screws are available with various combinations of heads and points. Notes give their measurements. (See Appendix 20.)

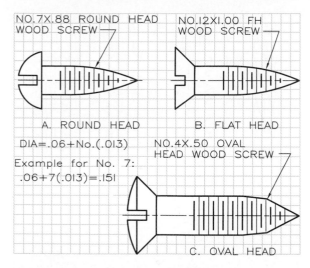

17.52 Standard types of wood screws are drawn on a grid that gives the proportions for drawing them at other sizes.

17.14 Other Threaded Fasteners

Only the more standard types of nuts and bolts are covered in this chapter. **Figure 17.53** illustrates a few of the many other types of threaded fasteners that have their own special applications. Three types of wing screws that are turned by hand are available in incremental lengths of 1/8 inch (**Figure 17.54**). **Figure 17.55** shows two types of thumb screws, which serve the same purpose as wing screws, and **Figure 17.56** shows wing nuts that can be screwed together by fingertip without wrenches or screwdrivers.

17.15 Tapping a Hole

An internal thread is made by drilling a hole with a tap drill with a 120° point (**Figure 17.57**). The depth of the drilled hole is measured to the shoulder of the conical point, not to the point. The diameter of the drilled hole is approximately equal to the root diameter, calculated as the major diameter of the screw thread minus its pitch (Appendix 21). The hole is **tapped**, or threaded, with a tool called a **tap** of one of the types shown.

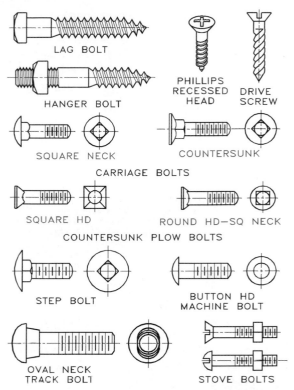

17.53 A few of the many different types of bolts and screws are illustrated here.

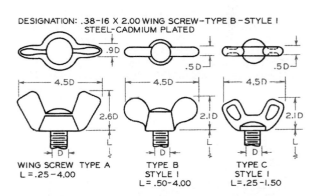

17.54 These wing screw proportions are for screw diameters of about 5/16 inch. The same proportions may be used to draw wing screws of any diameter. Type A screws are available in diameters of 4, 6, 8, 10, 12, 0.25", 0.313", 0.375", 0.438", 0.50", and 0.625". Type B screws are available in diameters of 10 to 0.625". Type C screws are available in diameters of 6 to 0.375".

The **taper, plug,** and **bottoming** hand taps have identical measurements, except for the chamfered portion of their ends. The taper tap

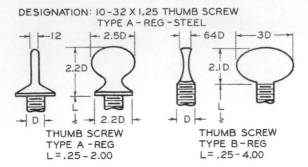

DESIGNATION: 10-32 X 1.25 THUMB SCREW
TYPE A - REG -STEEL

THUMB SCREW
TYPE A - REG
L= .25 - 2.00

THUMB SCREW
TYPE B - REG
L= .25 - 4.00

17.55 These thumb screw proportions are for screw diameters of about 1/4 inch. The same proportions may be used to draw thumb screws of any diameter. Type A screws are available in diameters of 6, 8, 10, 12, 0.25", 0.313", and 0.375". Type B thumb screws are available in diameters of 6 to 0.50".

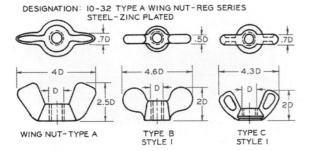

DESIGNATION: 10-32 TYPE A WING NUT-REG SERIES
STEEL-ZINC PLATED

WING NUT-TYPE A

TYPE B
STYLE I

TYPE C
STYLE I

17.56 These wing nut proportions are for screw diameters of 3/8 inch. The same proportions may be used to draw thumb screws of any size. Type A wing nuts are available in screw diameters of 3, 4, 5, 6, 8, 10, 12, 0.25", 0.313", 0.375", 0.438", 0.50", 0.583", 0.625", and 0.75". Type B nuts are available in sizes from 5 to 0.75". Type C nuts are available in sizes from 4 to 0.50".

has a long chamfer (8 to 10 threads), the plug tap has a shorter chamfer (3 to 5 threads), and the bottoming tap has the shortest chamfer (1 to 1–1/2 threads).

When tapping is to be done by hand in open or "through" holes, the taper tap should be used for coarse threads and in harder metals because it ensures straighter alignment and starting. The plug tap may be used in soft metals and for fine-pitch threads. When a hole is tapped to its bottom, all three taps—taper, plug, and bottoming—are used in that sequence on the same internal threads.

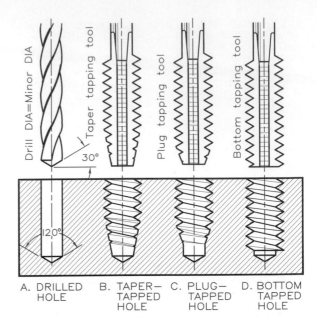

A. DRILLED HOLE B. TAPER—TAPPED HOLE C. PLUG—TAPPED HOLE D. BOTTOM TAPPED HOLE

17.57 Three types of tapping tools are used to thread internal drilled holes: taper tap, plug tap, and bottom tap.

Notes are added to specify the depth of a drilled hole and the depth of the threads within it. For example, a note reading 7/8 DIA–3 DEEP × 1/8 UNC–2A × 2 DEEP means that the hole is to be drilled deeper than it is threaded and that the last usable thread will be 2 inches deep in the hole.

17.16 Washers, Lock Washers, and Pins

Various types of washers are used with nuts and bolts to improve their assembly and increase their fastening strength.

Plain washers are noted on a drawing as

.938 × 1.750 × 0.134 TYPE A PLAIN WASHER,

where the numbers (left to right) represent the washer's inside diameter, outside diameter, and thickness. (See Appendices 26 and 27.)

Lock washers reduce the likelihood that threaded parts will loosen because of vibration and movement. **Figure 17.58** shows several common types of lock washers. Appendix 28 contains a table of dimensions for regular

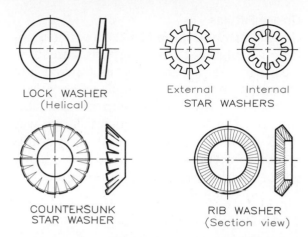

LOCK WASHER
(Helical)

External   Internal
STAR WASHERS

COUNTERSUNK
STAR WASHER

RIB WASHER
(Section view)

17.58 Lock washers are used to keep threaded parts from vibrating apart.

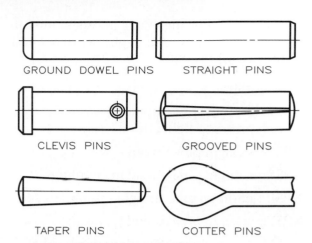

GROUND DOWEL PINS   STRAIGHT PINS

CLEVIS PINS   GROOVED PINS

TAPER PINS   COTTER PINS

17.59 Pins are used to hold parts together in assembly.

helical spring lock washers. Designate them with a note in the form:

HELICAL-SPRING LOCK WASHER
1/4 REGULAR—PHOSPHOR BRONZE,

where the 1/4 is the washer's inside diameter. Designate tooth lock washers with a note in one of two forms:

INTERNAL-TOOTH LOCK WASHER
1/4-TYPE A—STEEL;

EXTERNAL-TOOTH LOCK WASHER
.562-TYPE B—STEEL.

Pins (**Figure 17.59**) are used to hold parts together in a fixed position. Appendix 23 gives dimensions for straight pins. The cotter pin is another locking device that everyone who has had a toy wagon is familiar with. Appendix 22 contains a table of dimensions for cotter pins.

17.17 Pipe Threads and Fittings

Pipe threads are used for connecting pipes, tubing, and various fittings including lubrication fittings. The most commonly used pipe thread is tapered at a ratio of 1 to 16 on its diameter, but straight pipe threads also are

available (**Figure 17.60**). Tapered pipe threads will engage only for an effective length of

$$L = (0.80D + 6.8)P,$$

where D is the outside diameter of the threaded pipe and P is the pitch of the thread.

The pipe threads shown in **Figure 17.60** have a taper exaggerated to 1:16 on radius (instead of on diameter) to emphasize it. Drawing them with no taper obviously is easier. You may use either schematic or simplified symbols to show the threaded features.

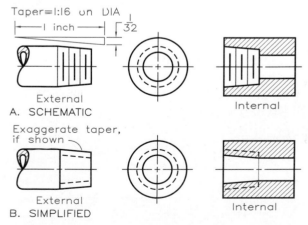

Taper=1:16 on DIA

1 inch

1/32

External
A. SCHEMATIC

Internal

Exaggerate taper, if shown

External
B. SIMPLIFIED

Internal

17.60 Pipe threads are shown with schematic and simplified symbols here.

Use the following ANSI abbreviations in pipe thread notes. All begin with NP (for National Pipe thread).

NPT: national pipe taper

NPTF: national pipe thread (dryseal, for pressure-tight joints)

NPS: straight pipe thread

NPSC: straight pipe thread in couplings

NPSI: national pipe straight internal thread

NPSF: straight pipe thread (dryseal)

NPSM: straight pipe thread for mechanical joints

NPSL: straight pipe thread for locknuts and lock nut pipe threads

NPSH: straight pipe thread for hose couplings and nipples

NPTR: taper pipe thread for railing fittings

To specify a pipe thread in note form, give the nominal pipe diameter (the common-fraction size of its internal diameter), the number of threads per inch, and the thread-type symbol:

$$1\text{-}1/4\text{-}11\text{-}1/2\ NPT$$

Appendix 9 gives a table of dimensions for pipe threads. **Figure 17.61** shows how to present specifications for external and internal threads in note form. Dryseal threads, either straight or tapered, provide a pressure-tight joint without the use of a lubricant or sealer.

17.61 These are typical pipe thread notes.

Grease Fittings

Grease fittings (**Figure 17.62**) allow the application of lubricant to moving parts. Threads of grease fittings are available as tapered and straight pipe threads. The ends where grease is inserted with a grease gun are available straight or at 90° and 45° angles. A one-way valve, formed by a ball and spring, permits grease to enter the fitting (forced through by a grease gun) but prevents it from escaping.

17.18 Keys

Keys are used to attach pulleys, gears, or crank handles to shafts, allowing them to remain assembled while moving and transmitting power. The four types of keys shown in **Figure 17.63** are the most commonly used. Appendices 24 and 25 contain tables of dimensions for keyways, keys, and keyseats.

17.19 Rivets

Rivets are fasteners that permanently join thin overlapping materials (**Figure 17.64**). The rivet is inserted in a hole slightly larger than the diameter of the rivet, and the application of pressure to the projecting end forms the headless end into

Thread size	$\frac{1}{8}$	3mm	$\frac{1}{4}$	6mm	$\frac{3}{8}$	10mm
Overall length	L=in.	mm	L=in.	mm	L=in.	mm
Straight	.625	16	1.000	25	1.200	30
90° Elbow	.800	20	1.250	32	1.400	36
45° Angle	1.000	25	1.500	38	1.600	41

17.62 Three standard types of grease fittings used to lubricate moving parts with a grease gun are shown here.

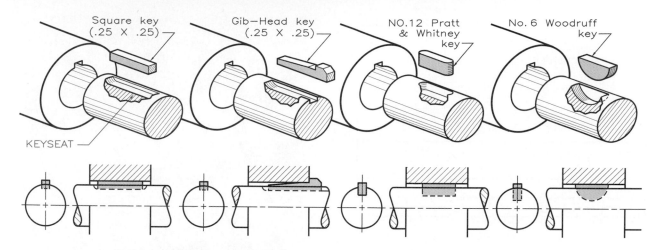

17.63 Standard types of keys used to hold parts on a shaft.

17.64 Rivets are used to permanently fasten structural elements together. (Courtesy of Russell, Burdsall & Ward Bolt and Nut Company.)

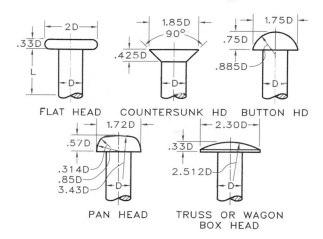

17.65 The proportions of small rivets with shanks up to 1/2 inch are shown here.

shape. Forming may be done with either hot or cold rivets, depending on the application.

Figure 17.65 shows typical shapes and proportions of small rivets that vary in diameter from 1/16 to 1–3/4 inches. Rivets are used extensively in pressure-vessel fabrication, heavy construction (such as bridges and buildings), and sheet-metal construction.

Figure 17.66 shows some of the standard ANSI symbols for representing rivets. Rivets that are driven in the shop are called shop riv-

ets, and those assembled at the job site are called field rivets.

17.20 Springs

Springs are devices that absorb energy and react with an equal force (**Figure 17.67**). Most springs are **helical**, as are bed springs, but they can also be **flat** (leaf), as in an automobile chassis. Some of the more common types of springs are **compression**, **torsion**, **extension**, **flat**, and

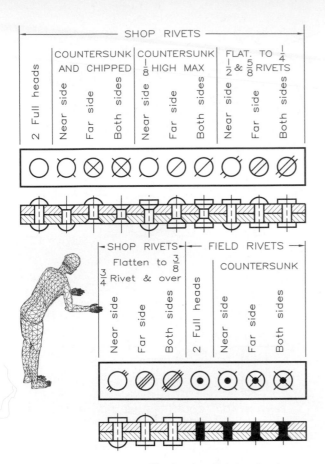

SHOP RIVETS									
2 Full heads	COUNTERSUNK AND CHIPPED			COUNTERSUNK $\frac{1}{8}$ HIGH MAX			FLAT. TO $\frac{1}{4}$ $\frac{1}{2}$ & $\frac{5}{8}$ RIVETS		
	Near side	Far side	Both sides	Near side	Far side	Both sides	Near side	Far side	Both sides

SHOP RIVETS			FIELD RIVETS			
Flatten to $\frac{3}{8}$ $\frac{3}{4}$ Rivet & over			COUNTERSUNK			
Near side	Far side	Both sides	2 Full heads	Near side	Far side	Both sides

17.66 Rivets are represented by these symbols in a drawing.

constant force springs. **Figure 17.68A–C** shows single-line conventional representations of the first three types. **Figure 17.68D–F** represents the types of ends used on compression springs.

Plain ends of springs simply end with no special modification of the coil. Ground plain ends are coils that have been machined by grinding to flatten the ends perpendicular to their axes. Squared ends are inactive coils that have been closed to form a circular flat coil at the end of a spring, which may also be ground.

In **Figure 17.68G**, a conical helical spring is shown in its simplified form. **Figure 17.68H** shows schematic single-line representations of the same types of springs depicted in **Figure 17.68D–G**, with phantom outlines instead of

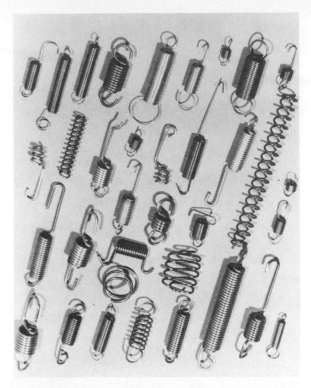

17.67 Springs are available for numerous special applications.

A. COMPRESSION   B. TORSION   C. EXTENSION

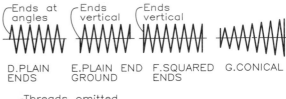

D. PLAIN ENDS   E. PLAIN END GROUND   F. SQUARED ENDS   G. CONICAL

H. SINGLE—LINE REPRESENTATIONS: SIMPLIFIED

17.68 Single-line spring drawings.

A–C These are single-line representations of various types of springs.

D–G These single-line representations of springs show various types of ends.

H These are simplified single-line representations of the springs depicted in D-G.

all the coils. This is a conventional method of drawing springs that saves time and effort.

Working drawing specifications of a compression spring drawn as a double-line representation are shown in **Figure 17.69**. Two coils are drawn at each end of the spring and phantom lines are drawn between them in order to save drawing time. A dimension is given with the diameter and free length of the spring on the drawing. The remaining specifications are given in a table placed near the drawing.

A working drawing of an extension spring (**Figure 17.70**) is similar to that of a compression spring. An extending spring is designed to resist stretching, whereas a **compression spring** is designed to resist squeezing. In a drawing of a helical torsion spring, which resists and reacts to a twisting motion (**Figure 17.71**), angular dimensions specify the initial and final positions of the spring as torsion is applied. Again, dimension the drawing and add specifications to describe the details.

17.21 Drawing Springs

Springs may be represented with single-line drawings (**Figure 17.68**) or as more realistic double-line drawings (**Figure 17.72**). Draw

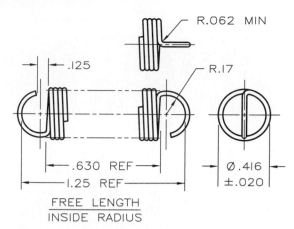

FREE LENGTH
INSIDE RADIUS

WIRE DIA	0.42
DIRECTION OF HELIX OPTIONAL	
TOTAL COILS	14 REF
RELATIVE POSITION OF ENDS	180° ±20°
EXTENDED LENGTH INSIDE ENDS WITHOUT PERMANENT SET	2.45 IN (MAX)
INITIAL TENSION	1.00 LB ±.10 LB
LOAD	4.0 LB ±.4 LB AT 1.56 IN
EXTENDED LG INSIDE ENDS	
LOAD	6.30 LB ±.63 LB AT 1.95

17.70 This conventional double-line drawing shows an extension spring and its specifications.

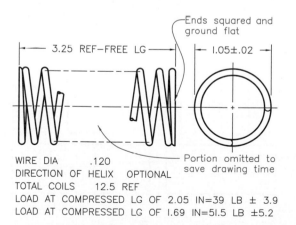

WIRE DIA	.120
DIRECTION OF HELIX	OPTIONAL
TOTAL COILS	12.5 REF
LOAD AT COMPRESSED LG OF 2.05 IN=39 LB ± 3.9	
LOAD AT COMPRESSED LG OF 1.69 IN=51.5 LB ±5.2	

17.69 This conventional double-line drawing is of a compression spring and includes its specifications.

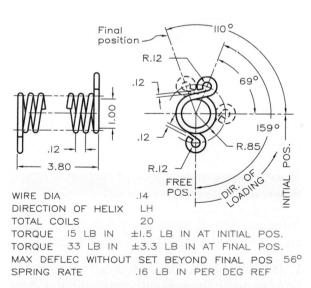

WIRE DIA	.14
DIRECTION OF HELIX	LH
TOTAL COILS	20
TORQUE 15 LB IN	±1.5 LB IN AT INITIAL POS.
TORQUE 33 LB IN	±3.3 LB IN AT FINAL POS.
MAX DEFLEC WITHOUT SET BEYOND FINAL POS	56°
SPRING RATE	.16 LB IN PER DEG REF

17.71 This conventional double-line drawing is of a helical torsion spring and includes its specifications.

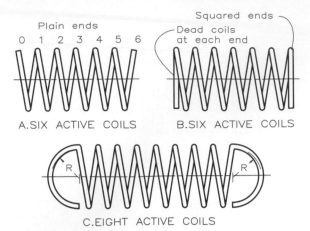

A. SIX ACTIVE COILS   B. SIX ACTIVE COILS

C. EIGHT ACTIVE COILS

17.72 Double-line drawings of springs.

A This double-line drawing shows a spring with six active coils.

B This double-line drawing shows a spring with six coils and a "dead" coil (inactive coil) at each end.

C This double-line drawing shows an extension spring with eight active coils.

each type shown by first laying out the diameters of the coils and lengths of the springs and then dividing the lengths into the number of active coils (**Figure 17.72A**). In **Figure 17.72B**, both end coils are "dead" (inactive) coils, and only six coils are active. **Figure 17.72C** depicts an extension spring with eight active coils.

The steps of drawing a double-line detailed representation of a compression spring are shown in **Figure 17.73**. Springs can be drawn right-hand or left-hand, but like threads, most are drawn as right-hand coils. The ends of the spring in this case are to be squared by grinding the ends to make them flat and perpendicular to the axis of the spring.

Problems

Solve and draw these problems on size A sheets. Each grid space equals 0.20 inch, or 5 mm.

1. (**Figure 17.74**) Draw detailed representations of Acme threads with major diameters of 2 in. Show both external and internal threads

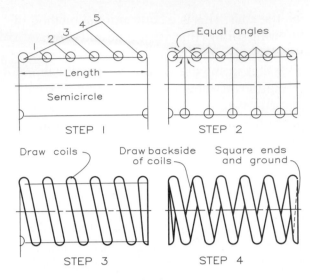

STEP 1   STEP 2

STEP 3   STEP 4

17.73 Drawing a spring in detail:

Step 1 Lay out the diameter and length of the spring and locate the five coils by the diagonal-line technique.

Step 2 Locate the coils on the lower side along the bisectors of the spaces between the coils on the upper side.

Step 3 Connect the coils on each side. This is a right-hand coil; a left-hand spring would slope in the opposite direction.

Step 4 Construct the back side of the spring and the end coils to complete the drawing. The spring has a square end that is to be ground.

as views and sections. Give a thread note by referring to Appendix 8.

2. Repeat Problem 1, but draw internal and external detailed representations of square threads.

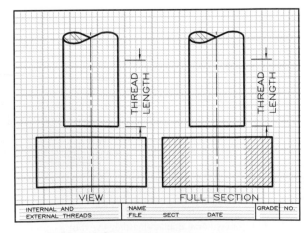

17.74 Problems 1–3.

3. Repeat Problem 1, but draw internal and external detailed representations of UN threads. Give a thread note for a coarse thread with a class 2 fit.

4. Using the notes in **Figure 17.75**, draw detailed representations of the internal threads and holes in section. Provide thread notes on each as specified.

5. Repeat Problem 4, but use schematic symbols.

6. Repeat Problem 4, but use simplified symbols.

7. Using the partial views in **Figure 17.76** and detailed thread symbols, draw external, internal, and end views of the full-size threaded parts. Provide thread notes for UNC threads with a class 2 fit.

8. Repeat Problem 7, but use schematic symbols.

9. Repeat Problem 7, but use simplified symbols.

10. (**Figure 17.77**) Complete the drawing of the finished hexagon-head bolt and a heavy hexagon nut. Draw the bolt head and nut across corners using detailed thread symbols.

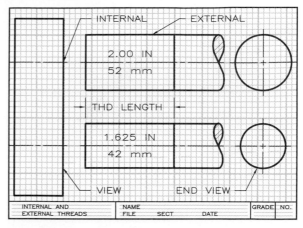

17.76 Problems 7–9.

Provide thread notes in either English or metric forms, as assigned.

11. Repeat Problem 10, but draw the nut and bolt as having unfinished square heads. Use schematic thread symbols.

12. Repeat Problem 10, but draw the bolt with a regular finished hexagon head across flats, using simplified thread symbols. Draw the nut across flats also and provide thread notes for both.

13. Use the notes in **Figure 17.78** to draw the screws in section and complete the sectional view showing all cross-hatching. Use detailed

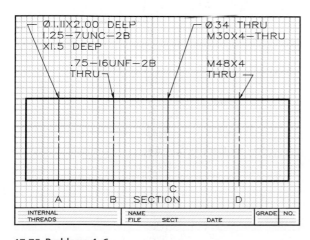

17.75 Problems 4–6.

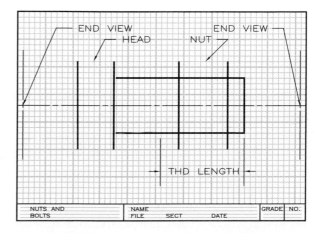

17.77 Problems 10–12.

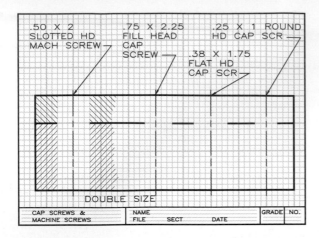

17.78 **Problems 13–15.**

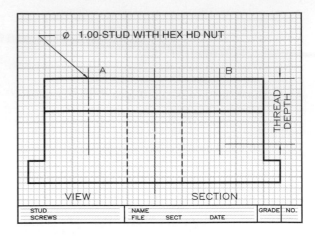

17.79 **Problem 16.**

thread symbols and provide thread notes to the parts.

14. Repeat Problem 13, using schematic thread symbols.

15. Repeat Problem 13, using simplified thread symbols.

16. (**Figure 17.79**) On axes A and B, draw studs having a hexagon-head nut (across flats) that hold the two parts together. The studs are to be fine series with a class 2 fit, and they should not reach the bottom of the threaded hole. Provide a thread note. Show the view as a half section.

17. (**Figure 17.80**) On axes A and B, construct hexagon-head cap screws (across flats), with UNC threads and a class 2 fit. The cap screws should not reach the bottoms of the threaded holes. Convert the view to a half section.

18. (**Figure 17.81**) Draw a 2.00-in. (50-mm) diameter hexagon-head bolt, with its head across flats, using schematic symbols. Draw a plain washer and regular nut (across corners) at the right end. Design the size of the opening in the part at the left end to hold the bolt head so that it will not turn. Use a UNC thread with a series 2 fit and provide a thread note.

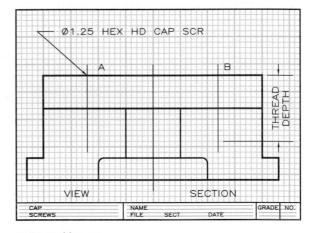

17.80 **Problem 17.**

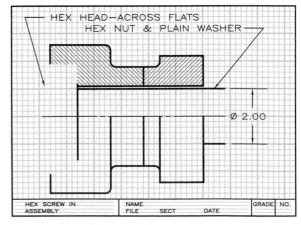

17.81 **Problem 18.**

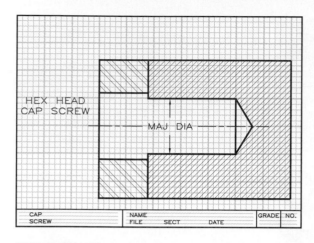

17.82 Problem 19.

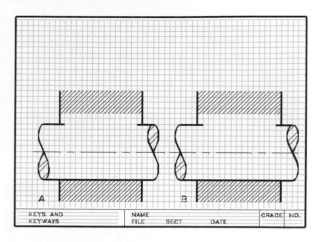

17.83 Problems 20–21.

19. (**Figure 17.82**) Draw a 2.00-in. (50-mm) diameter hexagon-head cap screw that holds the two parts together. Determine the length of the bolt, show the threads with schematic thread symbols, and provide a thread note.

20. (**Figure 17.83**) The part at A is held on the shaft by a square key, and the part at B is held on the shaft by a gib-head key. Using Appendix 25, complete the drawings and provide the necessary notes.

21. Repeat Problem 20, but use Woodruff keys, one with a flat bottom and the other with a round bottom. Using Appendix 24, complete the drawings and provide the necessary notes.

22. (**Figure 17.84**) Make a double-line drawing of the spring with the following specifications: No. of turns 4, Pitch 1, Wire size No. 4 = 0.2253, Inside diameter 3, Right hand.

23. (**Figure 17.85**) The 1/4-in shaft of the pencil pointer fits into a bracket designed to clamp to a desk top. A set screw holds the shaft in position. Make a drawing of the bracket, estimating its dimensions. Show the details and the thread notes involved in the design.

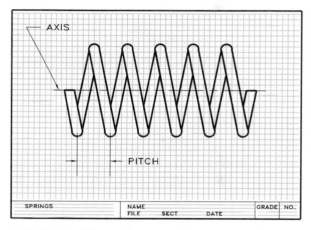

17.84 Problem 22.

17.85 Problem 23.

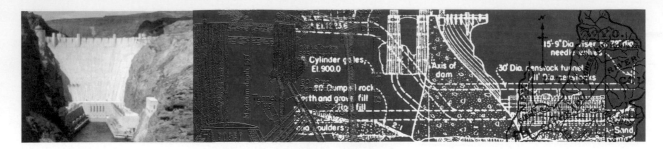

18

Gears and Cams

18.1 Introduction

Gears are toothed wheels whose circumferences mesh to transmit force and motion from one gear to the next. Multiple gears and cams in combination that mesh precisely can be seen in the multiple-spindle machine in **Figure 18.1**. The three most common types are spur gears, bevel gears, and worm gears (**Figure 18.2**).

Cams are irregularly shaped plates and cylinders that control the motion of a follower as they revolve to produce a type of reciprocating action. For example, cams make the needle of a sewing machine move up and down.

18.1 Numerous gears and cams are used in this detail of an Acme-Gridley multiple-spindle bar machine. (Courtesy of the National Acme Company.)

18.2 Spur Gears

Terminology

The **spur gear** is a circular gear with teeth cut around its circumference. Two meshing spur gears transmit power from one shaft to a parallel shaft. When the two meshing gears are unequal in diameter, the smaller gear is called the **pinion** and the larger one the **spur**.

The following terms and corresponding formulas describe the parts of a spur gear, several of which are shown in **Figure 18.3** and **Figure 18.4**.

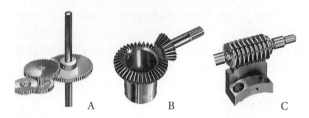

18.2 The three basic types of gears are (A) spur gears, (B) bevel gears, and (C) worm gears. (Courtesy of the Process Gear Company.)

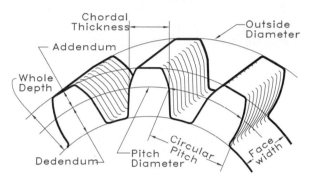

18.3 These terms apply to spur and pinion gears.

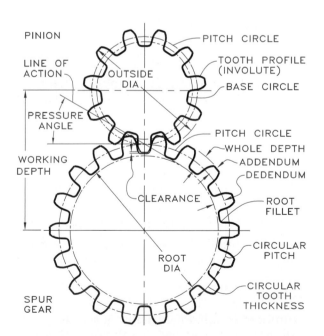

18.4 These terms apply to bevel gears. (Courtesy of Philadelphia Gear Corporation.)

Pitch circle (PC): the imaginary circle of a gear, as if it were a friction wheel without teeth that contacted another circular friction wheel.

Pitch diameter (PD): the diameter of the pitch circle; PD = N/DP, where N is the number of teeth and DP is the diametral pitch.

Diametral pitch (DP): the ratio between the number of teeth on a gear and its pitch diameter; DP = N/PD, where N is the number of teeth, and is expressed as teeth per inch of diameter.

Circular pitch (CP): the circular measurement from one point on a tooth to the corresponding point on the next tooth measured along the pitch circle; CP = 3.14/DP.

Center distance (CD): the distance from the center of a gear to its mating gear's center; CD = $(N_P + N_S)/(2DP)$, where N_P and N_S are the number of teeth in the pinion and spur, respectively.

Addendum (A): the height of a gear above its pitch circle; A = 1/DP.

Dedendum (D): the depth of a gear below the pitch circle; D = 1.157/DP.

Whole depth (WD): the total depth of a gear tooth; WD = A + D.

Working depth (WKD): the depth to which a tooth fits into a meshing gear; WKD = 2/DP, or WKD = 2A or where A = addendum.

Circular thickness (CRT): the circular distance across a tooth measured along the pitch circle; CRT = 1.57/DP.

Chordal thickness (CT): the straight-line distance across a tooth at the pitch circle; CT = $\frac{1.5708}{DP}$, where DP is the dimetral pitch.

Face width (FW): the width across a gear tooth parallel to its axis; a variable dimension, but usually three to four times the circular pitch; FW = 3CP to 4CP.

Outside diameter (OD): the maximum diameter of a gear across its teeth; OD = PD + 2A.

Root diameter (RD): the diameter of a gear measured from the bottom of its gear teeth; RD = PD − 2D.

Pressure angle (PA): the angle between the line of action and a line perpendicular to the centerline of two meshing gears; angles of 14.5° and 20° are standard for involute gears.

Base circle (BC): the circle from which an involute tooth curve is generated or developed; BC = PD cos PA.

Tooth Forms

The most common gear tooth is an involute tooth with a 14.5° pressure angle. The 14.5° angle is the angle of contact between two gears when the tangents of both gears are in contact. Gears with pressure angles of 20° and 25° also are used. Gear teeth with larger pressure angles are wider at the base and thus stronger than the standard 14.5° teeth.

18.3 Gear Ratios

The diameters of two meshing spur gears establish ratios that are important to their function (**Figure 18.5**). If the diameter of a gear is twice that of its pinion (the small gear), the gear has twice as many teeth as the pinion. The pinion then must make twice as many turns as the spur; therefore the revolutions per minute (RPM) of the pinion is twice that of the spur.

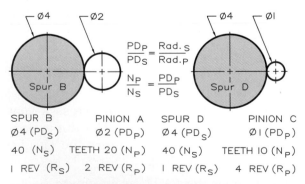

SPUR B | PINION A | SPUR D | PINION C
Ø4 (PD$_S$) | Ø2 (PD$_P$) | Ø4 (PD$_S$) | Ø1 (PD$_P$)
40 (N$_S$) | TEETH 20 (N$_P$) | 40 (N$_S$) | TEETH 10 (N$_P$)
1 REV (R$_S$) | 2 REV (R$_P$) | 1 REV (R$_S$) | 4 REV (R$_P$)

18.5 These are examples of ratios between meshing spur gears and pinion gears.

The relationship between two meshing gears may be determined by finding the velocity of a point on the pinion that is equal to $_\pi$PD × RPM. The velocity of a point on the large gear equals $_\pi$PD × RPM. The velocity of points on each gear must be equal, so

$$_\pi PD_P(RPM_P) = {_\pi}PD_S(RPM_S);$$

therefore

$$\frac{PD_P}{PD_S} = \frac{RPM_S}{RPM_P}.$$

If the radius of the pinion is 1 inch, the diameter of the spur is 4 inches, and the RPM of the pinion is 20, the RPM of the spur is

$$\frac{2(1)}{2(4)} = \frac{RPM_S}{20\ RPM_P}.$$

or

$$RPM_S = \frac{2(20)}{2(4)} = 5\ RPM$$

Thus the RPM of the spur (5) is one-fourth that of the pinion (20).

The number of teeth on each gear is proportional to the diameters of a pair of meshing gears, or

$$\frac{N_P}{N_S} = \frac{PD_P}{PD_S}$$

where NP and NS are the number of teeth on the pinion and spur, respectively, and PD$_P$ and PD$_S$ are their pitch diameters.

Calculations

Before starting a working drawing of a gear, you have to calculate the gear's dimensions.

Problem 1 Calculate the dimensions for a spur that has a pitch diameter of 5 in., a diametral pitch of 4, and a pressure angle of 14.5°.

Solution

Number of teeth: PD × (DP) = 5(4) = 20.
Addendum: 1/4 = 0.25".
Dedendum: 1.157/4 = 0.2893".

Circular thickness: CT = 1.5708/DP = 0.3927"

Outside diameter: OD = PD + 2A = 5 + 0.50
   = 5.50"

Root diameter: RD= PD−2D = 5"− 2(0.2893)
   = 4.421"

Chordal thickness: CT = PD × sin (90°/N)
   = 5 × .0786) = 0.3923

Chordal addendum: CA = A + CT² /(4×PD)
   = 0.25 + 0.3927²/(4 × 5") = 0.2577"

Face width: FW = 3.5 × CP = 3.50" × 0.79 = 2.75"

Circular pitch: CP = 3.14/DP = 3.14/4 = 0.785"

Working depth: WKD = 2A = 2 × 0.25" = 0.50"

Whole depth: WD = A + D
   = 0.250" + 0.289" = 0.539"

Use these dimensions to draw the spur and to provide specifications necessary for its manufacture.

Problem 2 shows the method of determining design information for two meshing gears when you know their working ratios.

Problem 2 Find the number of teeth and other specifications for a pair of meshing gears with a driving gear that turns at 100 RPM and a driven gear that turns at 60 RPM. The diametral pitch for each is 10, and the center-to-center distance between the gears is 6 in.

Solution

Step 1 Find the sum of the teeth on both gears:

Total teeth = 2(center-to-center distance)(DP)
   = 2(6)(10) = 120 teeth.

Step 2 Find the number of teeth for the driving gear:

$$\frac{\text{Driver RPM}}{\text{Driven RPM}} + 1 = \frac{100}{60} + 1 = 2.667,$$

so

$$\frac{\text{Total Teeth}}{\frac{100}{60}+1} = \frac{120}{2.667} = 45 \text{ teeth.}$$

(The number of teeth must be a whole number since there cannot be fractional teeth on a gear.)

Step 3 Find the number of teeth for the driven gear:

Total teeth − teeth on driver =
   teeth on driven gear

120 − 45 = 75 teeth.

Step 4 Calculate the other dimensions for the gears as in Problem 1. Adjusting the center distance to yield a whole number of teeth may be necessary.

18.4 Drawing Spur Gears

Figure 18.6 shows a conventional drawing of a spur gear. Not having to draw the gear teeth in the circular view saves a lot of time. Showing only simplified circular and sectional views of the gear and providing a table of dimensions called **cutting data** is acceptable. Circular phantom lines represent the root circle, pitch circle, and outside circle of the gear in the circular view.

A table of dimensions is a necessary part of a gear drawing (**Figure 18.7**). You may calculate these data or get them from tables of standards in gear handbooks such as *Machinery's Handbook*.

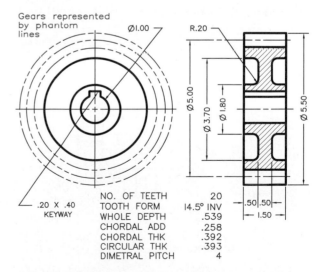

NO. OF TEETH	20
TOOTH FORM	14.5° INV
WHOLE DEPTH	.539
CHORDAL ADD	.258
CHORDAL THK	.392
CIRCULAR THK	.393
DIMETRAL PITCH	4

18.6 This detail drawing of a spur gear contains a table of values that supplements the dimensions shown on the view and section.

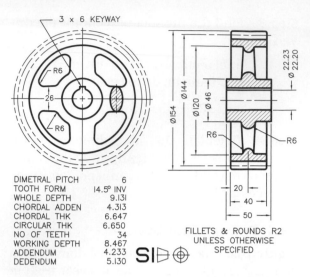

DIMETRAL PITCH	6
TOOTH FORM	14.5° INV
WHOLE DEPTH	9.131
CHORDAL ADDEN	4.313
CHORDAL THK	6.647
CIRCULAR THK	6.650
NO OF TEETH	34
WORKING DEPTH	8.467
ADDENDUM	4.233
DEDENDUM	5.130

FILLETS & ROUNDS R2
UNLESS OTHERWISE
SPECIFIED

18.7 This is a detail drawing of a spur gear that was produced on a computer.

18.5 Bevel Gears

Terminology

Bevel gears have axes that intersect at angles. The angle of intersection usually is 90°, but other angles are also used. The smaller of the two bevel gears is the **pinion**, as with spur gears; the larger is the **gear**.

Figures 18.8 and **18.9** illustrate the terminology of bevel gearing. A further explanation

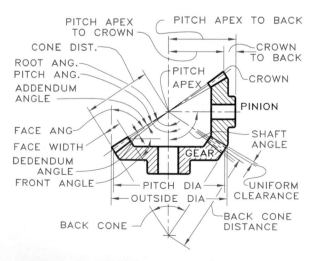

18.8 These terms apply to bevel gears. (Courtesy of Philadelphia Gear Corporation.)

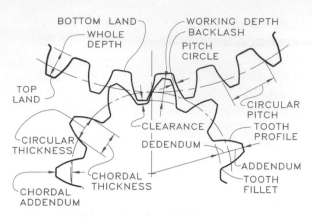

18.9 These additional terms apply to bevel gears. (Courtesy of Philadelphia Gear Corporation.)

of and the corresponding formula for each feature follows. You may also use gear handbooks to find these dimensions.

Pitch angle of pinion (PAp): tan PAp = Np/Ng, where Ng and Np are the number of teeth on the gear and pinion, respectively.

Pitch angle of gear (PAg): tan PAg = Ng/Np.

Pitch diameter (PD): the number of teeth, N, divided by the diametral pitch, DP; PD = N/P.

Addendum (A): measured at the large end of the tooth; A = 1/DP.

Dedendum (D): measured at the large end of the tooth; D = 1.157/DP.

Whole tooth depth (WD): WD = 2.157/DP.

Thickness of tooth (TT): measured at the pitch circle; TT = 1.571/DP.

Diametral pitch: DP = N/PD, where N is the number of teeth.

Addendum angle (AA): the angle formed by the addendum and pitch cone distance; tan AA = A/PCD. PCD = pitch cone distance.

Angular addendum: AK = cos PA × A.

Pitch cone distance (PCD): PCD = PD/(2 sin PA).

Dedendum angle (DA): the angle formed by the dedendum and the pitch cone distance; tan DA = D/PCD.

Face angle (FA): the angle between the gear's centerline and the top of its teeth; FA = 90° − (PCD + AA).

Cutting angle (or root angle) (CA): the angle between the gear's axis and the roots of the teeth; CA = PCD − D.

Outside diameter (OD): the greatest diameter of a gear across its teeth; OD = PD + 2A.

Apex to crown distance (AC): the distance from the crown of the gear to the apex of the cone measured parallel to the axis of the gear; AC = OD/(2 tan FA).

Chordal addendum (CA): CA = A + [(TT^2 cos PA)/4PD].

Chordal thickness (CT): measured at the large end of the tooth; CT = PD (sin 90°/N).

Face width (FW): can vary, but approximately equal to the pitch cone distance divided by 3.5; FW = PCD/3.5.

Calculations

The following example demonstrates use of the preceding formulas. Some of the formulas result in specifications that apply to both gear and pinion.

Problem 5 Two bevel gears intersect at right angles and have a diametral pitch of 3. The gear has 60 teeth and the pinion has 45 teeth and a face width of 4 in. Find the dimensions of the gear and pinon.

Solution

Pitch cone angle of gear:
  tan PCA = 60/45 = 1.33; PCA = 53°7'.

Pitch cone angle of pinion:
  tan PCA = 45/60; PCA = 36°52'.

Pitch diameter of gear: 60/3 = 20.00".

Pitch diameter of pinion: 45/3 = 15.00".

The following calculations yield the same dimensions for both gear and pinion:

Addendum: 1/3 = 0.333".

Dedendum: 1.157/3 = 0.3857".

Whole depth: 2.157/3 = 0.719".

Tooth thickness on pitch circle:
  1.571/3 = 0.5237".

Pitch cone distance:
  20/(2 sin 53°7') = 12.5015".

Addendum angle:
  tan AA = 0.333/12.5015 = 1°32'

Dedendum angle:
  DA = 0.3857/12.5015 = 0.0308 = 1°46'.

Face width: PCD/3 = 4.00".

The following dimensions must be calculated separately for gear and pinion:

Chordal addendum of gear: 0.333" + [(0.5237^2 cos 53°7')/(4 × 20)] = 0.336".

Chordal addendum of pinion:
  0.333" + [(0.5237^2 cos 36°52')/(4 × 15)] = 0.338".

Chordal thickness of gear:
  sin (90°/60) × 20" = 0.5240".

Chordal thickness of pinion:
  sin (90°/45) × 15" = 0.5235".

Face angle of gear:
  90° − (53°7' + 1°32') = 35°21'.

Face angle of pinion:
  90° − (36°52' + 1°32') = 51°36'.

Cutting angle of gear: 53°7' − 1°46' = 51°21'.

Cutting angle of pinion: 36°52' − 1°46' = 35°6'.

Angular addendum of gear:
  0.333" cos 53°7' = 0.1999".

Angular addendum of pinion:
  0.333" cos 36°52' = 0.2667".

Outside diameter of gear:
  20" + 2(0.1999") = 20.4000".

Outside diameter of pinion:
  15" + 2(0.2667") = 15.533".

Apex-to-crown distance of gear:
  (20.400"/2)(tan 35°7') = 7.173".

Apex-to-crown distance of pinion:
  (15.533"/2)(tan 51°36') = 9.800".

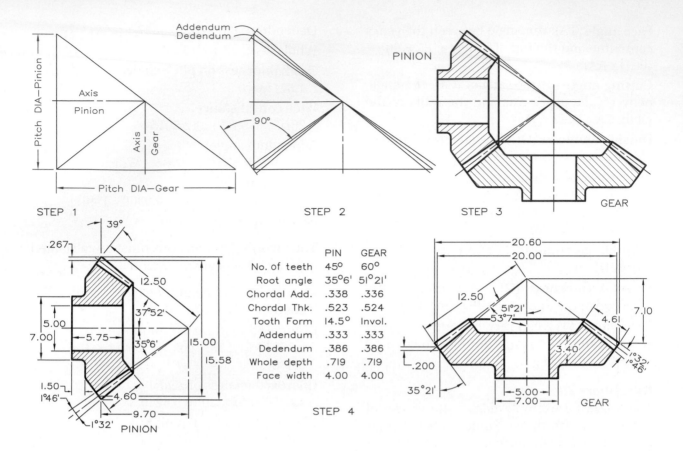

18.10 Drawing bevel gears:

Step 1 Lay out the pitch diameters and axes of the two bevel gears with construction lines.

Step 2 Draw construction lines to establish the limits of the teeth by using the addendum and dedendum dimensions.

Step 3 Draw the pinion and gear using the specified or calculated dimensions.

Step 4 Complete the detail drawings of both gears and provide a table of cutting data.

18.6 Drawing Bevel Gears

Use the calculated dimensions to lay out bevel gears in a detail drawing. Many of these dimensions are difficult to measure with a high degree of accuracy on a drawing. Therefore, providing a table of cutting data for each gear is essential.

The steps involved in drawing bevel gears are shown in **Figure 18.10**. On the final drawing, notice that dimensions on the views are supplemented by a table of additional dimensions and data.

18.7 Worm Gears

A worm gear consists of a threaded shaft called a **worm** and a circular gear called a spider (**Figure 18.11**). When the worm is revolved, it causes the spider to revolve about its axis. **Figures 18.11** and **18.12** illustrate the terminology of worm gearing. The following lists further explain these terms and provide the formulas for calculating their dimensions for worm and spider.

Worm Terminology
Linear pitch (P): the distance from one thread to the next, measured parallel to the worm's axis; P = L/N, where N is the number of threads (1 if a single thread, 2 if a double thread, and so on).

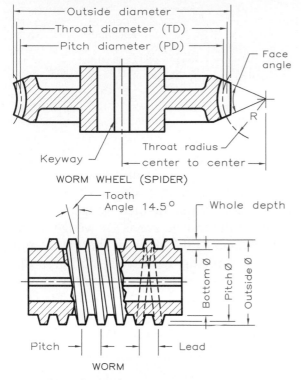

18.11 These terms apply to worm gears.

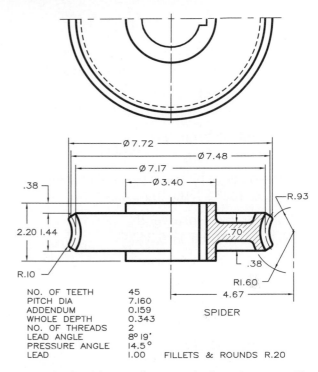

NO. OF TEETH	45
PITCH DIA	7.160
ADDENDUM	0.159
WHOLE DEPTH	0.343
NO. OF THREADS	2
LEAD ANGLE	8° 19'
PRESSURE ANGLE	14.5°
LEAD	1.00

FILLETS & ROUNDS R.20

18.12 This detail drawing shows a spider for a worm gear with a table of cutting data.

Lead (L): the distance a thread advances in a turn of 360°.

Addendum of tooth (AW): AW = 0.3183P.

Pitch diameter (PDW): PDW = OD − 2AW, where OD is the outside diameter.

Whole depth of tooth (WDT): WDT = 0.6866P.

Bottom diameter of worm (BD): BD = OD − 2WDT.

Width of thread at root (WT): WT = 0.31P.

Minimum length of worm (MLW):

$$MLW = \sqrt{8PDS\ (AW)},$$ where PDS is the pitch diameter of the spider.

Helix angle (HA): $\cot \beta = 3.14(PDW)/L$.

Outside diameter (OD): OD = PD + 2A.

Spider Terminology

Pitch diameter of spider (PDS):

PDS = N(P)/3.14, where N is the number of teeth on the spider.

Throat diameter of spider (TD): TD = PDS + 2A.

Radius of spider throat (RST):

RST = (OD of worm/2) − 2A.

Face angle (FA): may be selected between 60° and 80° for the average application.

Center-to-center distance (CD): measured between the worm and spider; CD = (PDW + PDS)/2.

Outside diameter of spider (ODS):

ODS = TD + 0.4775P.

Face width of gear (FW): FW = 2.38P + 0.25.

Calculations

The following example demonstrates use of the preceding formulas to find the dimensions for a worm gear.

Problem 6 Calculate the dimensions for a worm gear (worm and spider). The spider has 45 teeth, and the worm has an outside diameter of 2.50 in., a double thread, and a pitch of 0.5 in.

Solution

Lead: $L = 0.5"(2) = 1"$.

Worm addendum: $AW = 0.3183P = 0.1592"$.

Pitch diameter of worm:
$PDW = 2.50" - 2(0.1592") = 2.1818"$.

Pitch diameter of spider:
$PDS = (45" \times 0.5)/3.14 = 7.166"$.

Center distance between worm and spider:
$CD = (2.182" + 7.166")/2 = 4.674"$.

Whole depth of worm tooth:
$WDT = 0.687(0.5") = 0.3433"$.

Bottom diameter of worm:
$BD = 2.50" - 2(0.3433") = 1.813"$.

Helix angle of worm:
$\cot \beta = 3.14(2.1816)/1 = 8°19'$.

Width of thread at root: $WT = 0.31(1) = 0.155"$.

Minimum length of worm:
$MLW = \sqrt{8(0.1592)(7.1656)} = 3.02"$

Throat diameter of spider:
$TD = 7.1656" + 2(0.1592") = 7.484"$.

Radius of spider throat:
$RST = (2.5/2) - (2 \times 0.1592) = 0.9318"$.

Face width: $FW = 2.38(0.5) + 0.25 = 1.44"$.

Outside diameter of spider:
$ODS = 7.484 + 0.4775(0.5) = 7.723"$.

18.8 Drawing Worm Gears

Draw and dimension the worm and spider as shown in **Figures 18.12** and **18.13**. The preceding calculations yield the dimensions needed for scaling and laying out the drawings and providing cutting data.

18.9 Cams

Plate cams are irregularly shaped machine elements that produce motion in a single plane, usually up and down (**Figure 18.14**). As

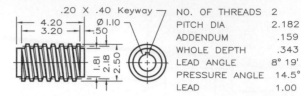

NO. OF THREADS	2
PITCH DIA	2.182
ADDENDUM	.159
WHOLE DEPTH	.343
LEAD ANGLE	8° 19'
PRESSURE ANGLE	14.5°
LEAD	1.00

18.13 This detail drawing of a worm is based on calculated dimensions.

the cam revolves about its center, the cam's shape alternately raises and lowers the follower that is in contact with it. Cams utilize the principle of the inclined wedge, with the surface of the cam acting as the wedge, causing a change in the slope of the plane, and thereby producing the desired motion of the follower. Cams are designed primarily to produce (1) uniform or linear motion (2) harmonic motion, (3) gravity motion (uniform acceleration), or (4) combinations of these motions.

Uniform Motion

The uniform motion depicted in **Figure 18.15A** represents the motion of the cam follower as the cam rotates through 360°. This curve has sharp corners, indicating abrupt changes of velocity that cause the follower to bounce. Therefore uniform motion usually is modified to smooth the changes of velocity. The radius of the modifying arc varies up to a

18.14 This photo shows three types of machined cams. (Courtesy of Ferguson Machine Company.)

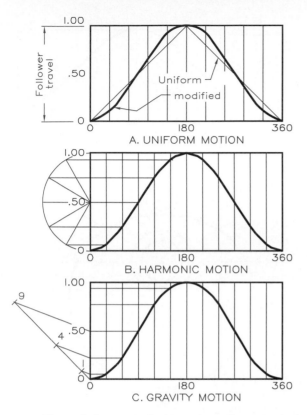

A. UNIFORM MOTION

B. HARMONIC MOTION

C. GRAVITY MOTION

18.15 These displacement diagrams show three standard motions: uniform, harmonic, and gravity.

radius of one-half the total displacement, depending on the speed of operation.

Harmonic Motion

The harmonic motion plotted in **Figure 18.15B** is a smooth, continuous motion based on the change of position of points on a circle. At moderate speeds this displacement gives a smooth operation.

Gravity Motion

The gravity motion (uniform acceleration) illustrated in **Figure 18.15C** is used for high-speed operation. The variation of displacement is analogous to the force of gravity, with the difference in displacement being 1, 3, 5, 5, 3, 1, based on the square of the number. For instance, $1^2 = 1$; $2^2 = 4$; $3^2 = 9$ give a uniform

acceleration. This motion is repeated in reverse order for the remaining half of the follower's motion. Intermediate points are obtained by squaring fractional increments, such as $(2.5)^2$.

Cam Followers

Three basic types of cam followers are the flat surface, roller, and knife edge (**Figure 18.16**). Use of flat-surface and knife-edge followers is limited to slow-moving cams, where minor force will be exerted during rotation. The roller follower is able to withstand higher speeds.

18.10 Designing Plate Cams

Harmonic Motion The steps involved in designing a plate cam for harmonic motion are shown in **Figure 18.17**. Before laying out the drawing of a cam, you must know the motion of the follower, rise of the follower, diameter of the base circle, and direction of rotation. The displacement diagram shown in Step 1 of Figure 18.17 gives the specifications graphically for the cam.

Gravity Motion The steps involved in designing a cam for gravity motion are shown in **Figure 18.18**. The same steps used in designing a cam for harmonic motion apply, but the displacement diagram and knife-edge follower are different.

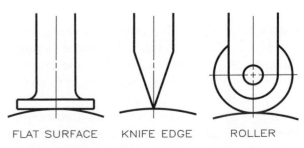

FLAT SURFACE   KNIFE EDGE   ROLLER

18.16 Three basic types of cam followers are the flat surface, roller, and knife edge.

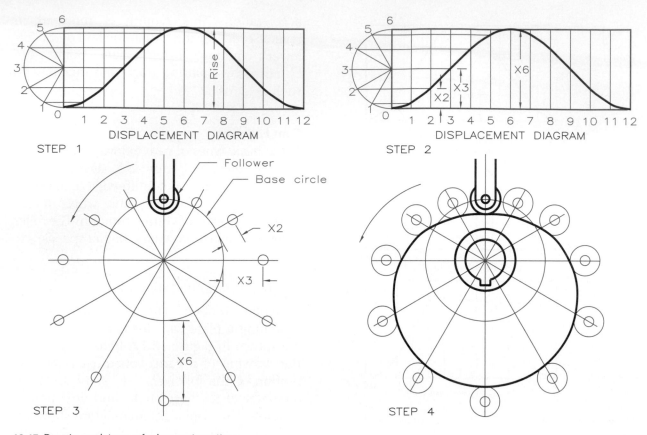

18.17 Drawing a plate cam for harmonic motion:

Step 1 Construct a semicircle on the vertical side of the displacement diagram whose diameter equals the rise of the follower. Divide the semicircle into the same number of segments as there are between 0° and 180° on the horizontal axis of the displacement diagram. Plot the displacement curve.

Step 2 Measure distances of rise and fall (X1, X2, X3, ..., X6) at each interval from the base circle.

Step 3 Construct the base circle, draw the follower, and divide the circle into the same number of sectors as there are divisions on the displacement diagram. Transfer distances X1, X2, ... from the displacement diagram to their respective radial lines of the circle, measuring outward from it.

Step 4 Draw circles to represent the positions of the roller as the cam revolves counterclockwise. Draw the cam profile tangent to all the rollers to complete the drawing.

Cam-with-an-Offset-Follower The cam shown in **Figure 18.19** produces harmonic motion through 360°. In this case, plot the motion directly from the follower rather than from the usual displacement diagram.

Draw a semicircle with its diameter equal to the total motion of the follower beginning at the centerline of the follower roller. Draw

the base circle to pass through the center of the roller of the follower. Extend the centerline of the follower downward and draw a circle tangent to the extension with its center at the center of the base circle. Divide the small circle into 30° intervals to establish points through which to draw construction lines tangent to the circle.

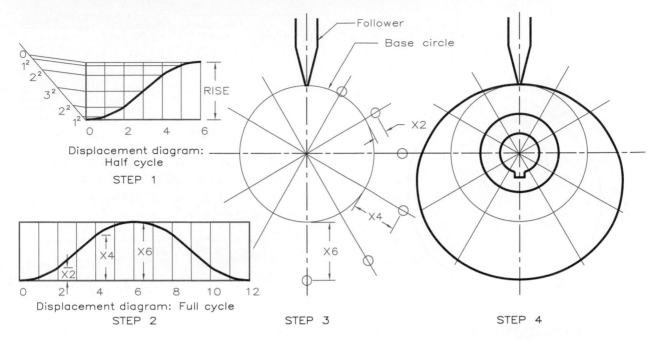

18.18 Drawing a plate cam for uniform acceleration:

Step 1 Construct a displacement diagram to represent the rise of the follower. Divide the horizontal axis into angular increments of 30°. Draw a construction line through point 0; locate the 12, 22, and 32 divisions and project them to the vertical axis to represent half the rise.

Step 2 Use the same construction to find the right half of the symmetrical curve.

Step 3 Construct the base circle and draw the knife-edge follower. Divide the circle into the same number of sectors as there are divisions in the displacement diagram. Transfer distances from the displacement diagram to their respective radial lines of the base circle and measure outward from the base circle.

Step 4 Connect the points found in Step 3 with a smooth curve to complete the cam profile. Show the cam hub and keyway.

Lay out the distances from tangent points to the position points along the path of the follower along the tangent lines drawn at 30° intervals. Locate these points by measuring from the base circle, as shown. For example, point 3 is located distance X from the base circle. Draw the circular roller in all views, and then draw the profile of the cam tangent to the rollers at all positions.

Problems

Gears
Use size A sheets for the following gear problems. Select appropriate scales so that the drawings will effectively use the available space.

1–5. Calculate the dimensions for the following spur gears, and make a detail drawing of each. Give the dimensions and cutting data for each gear. Provide any other dimensions needed.

Problem	Gear Teeth	Diametral Pitch	14.5° Involute
1	20	5	0
2	30	3	0
3	40	4	0
4	60	6	0
5	80	4	0

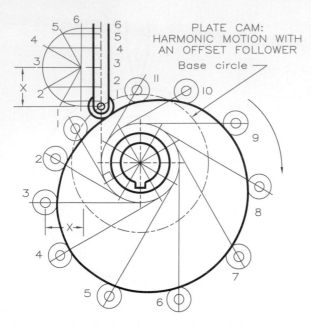

PLATE CAM:
HARMONIC MOTION WITH
AN OFFSET FOLLOWER

Base circle

18.19 Drawing of a plate cam having an offset roller follower.

6–10. Calculate the gear sizes and number of teeth using the following ratios and data.

Problem	RPM Pinion	RPM Gear	Center to Center	Diametral Pitch
6	100 (driver)	60	6.0″	10
7	100 (driver)	50	8.0″	9
8	100 (driver)	40	10.0″	8
9	100 (driver)	35	12.0″	7
10	100 (driver)	25	14.0″	6

11–20. Make a detail drawing of each gear for which you made calculations in Problems 6–10. Provide a table of cutting data and other dimensions needed to complete the specifications.

21–25. Calculate the specifications for the bevel gears that intersect at 90°, and make detail drawings of each, including the necessary dimensions and cutting data.

Problem	Diametral Pitch	No. of Teeth on Pinion	No. of Teeth on Gear
21	3	60	15
22	4	100	40
23	5	100	60
24	6	100	50
25	7	100	30

26–30. Calculate the specifications for the worm gears and make a detail drawing of each, providing the necessary dimensions and cutting data.

Problem	No. of Teeth in Spider Gear	Outside DIA of Worm	Pitch of Worm	Thread of Worm
26	45	2.50	0.50	double
27	30	2.00	0.80	single
28	60	3.00	0.80	double
29	30	2.00	0.25	double
30	80	4.00	1.00	single

Cams

Use size B sheets for the following cam problems. The standard dimensions are base circle, 3.50 in.; roller follower, 0.60-in. diameter; shaft, 0.75-in. diameter; and hub, 1.25-in. diameter. The direction of rotation is clockwise. The follower is positioned vertically over the center of the base circle. Lay out the problems and displacement diagrams as shown in **Figure 18.20** (shown on the next page).

31. Draw a plate cam with a knife-edge follower for uniform motion and a rise of 1.00 in.

32. Draw a displacement diagram and a cam that will give a modified uniform motion to a knife-edge follower with a rise of 1.7 in. Modify the uniform motion with an arc of one-quarter the rise in the displacement diagram.

33. Draw a displacement diagram and a cam that will give a harmonic motion to a roller follower with a rise of 1.60 in.

34. Draw a displacement diagram and a cam that will give a harmonic motion to a knife-edge follower with a rise of 1.00 in.

35. Draw a displacement diagram and a cam that will give uniform acceleration to a knife-edge follower with a rise of 1.70 in.

36. Draw a displacement diagram and a cam that will give a uniform acceleration to a roller follower with a rise of 1.40 in.

37. (**Figure 18.21**) Draw a displacement diagram and a cam with an offset follower that will give a harmonic motion with a rise of 1.60 in.

38. (**Figure 18.21**) Draw a displacement diagram and a cam with an offset follower that will give a uniform motion with a rise of 1.50 in.

39. (**Figure 18.21**) Draw a displacement diagram and a cam with an offset follower that will give a gravity motion with a rise of 1.70 in.

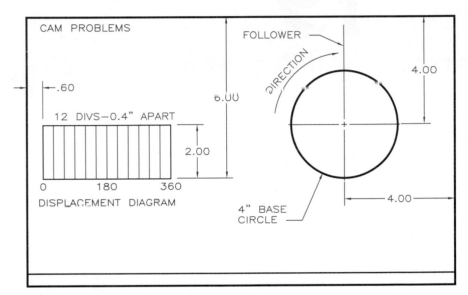

18.20 Layout for Problems 31–36 on size B sheets.

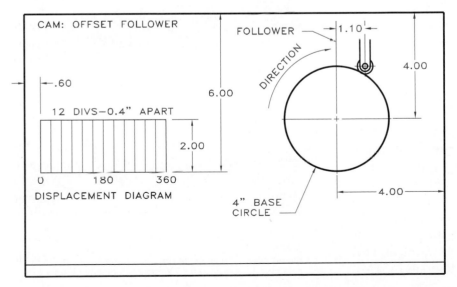

18.21 Layout for Problems 37–39 on size B sheets.

19

Materials and Processes

19.1 Introduction

Various materials and manufacturing processes are commonly used to make parts similar to those discussed in this textbook. A large proportion of parts designed by engineers are made of metal, but other materials (such as plastics, fibers, and ceramics) are available to the designer in increasingly useful applications.

Metallurgy, the study of metals, is a field that is constantly changing as new processes and alloys are developed (**Figure 19.1**). These developments affect the designer's specification of metals and their proper application for various purposes. Three associations have standardized and continually update guidelines for designating various types of metals: the American Iron and Steel Institute (AISI), the Society of Automotive Engineers (SAE), and the American Society for Testing Materials (ASTM).

19.1 These workmen are assembling a sand casting mold to produce a transmission housing weighing 186 pounds of aluminum alloy. The shape in the foreground is part of the mold assembly. (Courtesy of the Aluminum Company of America.)

19.2 Commonly Used Metals

Iron*

Metals that contain iron, even in small quantities, are called **ferrous** metals. Three common types of iron are **gray iron**, **white iron**, and **ductile iron**.

Gray iron contains flakes of graphite, which result in low strength and low ductility which makes it easy to machine. Gray iron resists vibration better than other types of iron. **Figure 19.2** shows designations of and typical applications for gray iron.

White iron contains carbide particles that are extremely hard and brittle, enabling it to withstand wear and abrasion. Because the composition of white iron differs from one supplier to another, there are no designated grades of white iron. It is used for parts on grinding and crushing machines, digging teeth on earthmovers and mining equipment, and wear plates on reciprocating machinery used in textile mills.

Ductile iron (also called nodular or spheroidized iron) contains tiny spheres of

---

*This section on iron was developed by Dr. Tom Pollock, a metallurgist at Texas A&M University.

DESIGNATION OF GRAY IRON (450 LBS/CF)

ATSM Grade (1000 psi)	SAE Grade	Typical Uses
ASTM 25 CI	G 2500 CI	Small engine blocks, pump bodies, clutch plates, transmission cases
ASTM 30 CI	G 3000 CI	Auto engine blocks, heavy castings, flywheels
ASTM 35 CI	G 3500 CI	Diesel engine blocks, tractor transmission cases, heavy and high—strength parts
ASTM 40 CI	G 4000 CI	Diesel cylinders, pistons, camshafts

19.2 These are the numbering designations of gray iron and its typical uses.

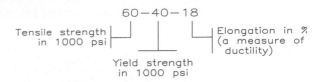

19.3 Ductile iron is specified with numerical notes in this format.

graphite, making it stronger and tougher than most types of gray iron and more expensive to produce. Three sets of numbers (**Figure 19.3**) describe the most important features of ductile iron. **Figure 19.4** shows the designations of and typical applications for the commonly used alloys of ductile iron.

Malleable iron is made from white iron by a heat-treatment process that converts carbides into carbon nodules (similar to ductile iron). The numbering system for designating grades of malleable iron is shown in **Figure 19.5**. Some of the commonly used grades of

DESIGNATIONS OF DUCTILE IRON (490 LBS/CF)

Grade	Typical Uses
60—40—18 CI	Valves, steam fittings, chemical plant equipment, pump bodies
65—45—12 CI	Machine components that are shock loaded, disc brake calipers
80—55—6 CI	Auto crankshafts, gears, rollers
100—70—3 CI	High—strength gears and machine parts
120—90—2 CI	Very high—strength gears, rollers, and slides

19.4 These are the numbering designations of ductile iron and its typical uses.

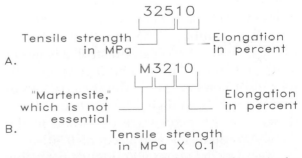

19.5 These notes illustrate the numbering designations for malleable iron.

DESIGNATIONS OF MALLEABLE IRON (490 LBS/CF)

ASTM Grade	Typical Uses
35018 CI	Marine and railroad valves and fittings, "black-iron" pipe fittings (similar to 60-40-18 ductile CI)
45006 CI	Machine parts (similar to 80-55-6 ductile CI)
M3210 CI	Low-stress components, brackets
M4504 CI	Crankshafts, hubs
M7002 CI	High-strength parts, connecting rods, universal joints
M8501 CI	Wear-resistant gears and sliding parts

19.6 These are the numbering designations of malleable iron and its typical uses.

DESIGNATIONS OF STEEL (490 LBS/CF)

Type of steel	Number	Applications
Carbon steels		
Plain carbon	10XX	Tubing, wire, nails
Resulphurized	11XX	Nuts, bolts, screws
Manganese steel	13XX	Gears, shafts
Nickel steel	23XX	Keys, levers, bolts
	25XX	Carburized parts
	31XX	Axles, gears, pins
	32XX	Forgings
	33XX	Axles, gears
Molybdenum	40XX	Gears, springs
Chromium-moly.	41XX	Shafts, tubing
Nickel-chromium	43XX	Gears, pinions
Nickel-moly.	46XX	Cams, shafts
	48XX	Roller bearings, pins
Chromium steel	51XX	Springs, gears
	52XX	Ball bearings
Chrom. vanadium	61XX	Springs, forgings
Silicon manganese	92XX	Leaf springs

19.7 These are the numbering designations of steel and its applications.

1010, 1015, 1020, 1030, 1040, 1070, 1080, 1111, 1118, 1145, 1320, 2330, 2345, 2515, 3130, 3135, 3240, 3310, 4023, 4042, 4063, 4140, and 4320.

malleable iron and their typical applications are shown in **Figure 19.6**.

Cast iron is iron that is melted and poured into a mold to form it by casting, a commonly used process for producing machine parts. Although cheaper and easier to machine than steel, iron does not have steel's ability to withstand shock and force.

Steel

Steel is an alloy of iron and carbon, which often contains other constituents such as manganese, chromium, or nickel. Carbon (usually between 0.20% and 1.50%) is the ingredient having the greatest effect on the grade of steel. The three major types of steel are **plain carbon steels**, **free-cutting carbon steels**, and **alloy steels**. **Figure 19.7** gives the types of steels and their SAE designations by four-digit numbers. The first digit indicates the type of steel: 1 is carbon steel, 2 is nickel steel, and so on. The second digit gives content (as a percentage) of the material represented by the first digit. The last two or three digits give the percentage of carbon in the alloy: 100 equals 1%, and 50 equals 0.50%.

Steel weighs about 490 pounds per cubic foot. Some frequently used SAE steels are

Copper

One of the first metals discovered, copper is easily formed and bent without breaking. Because it is highly resistant to corrosion and is highly conductive, it is used for pipes, tubing, and electrical wiring. It is an excellent roofing and screening material because it withstands the weather well. Copper weighs about 555 pounds per cubic foot.

Copper has several alloys, including brasses, tin bronzes, nickel silvers, and copper nickels. Brass (about 530 pounds per cubic foot) is an alloy of copper and zinc, and bronze (about 548 pounds per cubic foot) is an alloy of copper and tin. Copper and copper alloys are easily finished by buffing or plating; joined by soldering, brazing, or welding; and machined.

Wrought copper has properties that permit it to be formed by hammering. A few of the numbered designations of wrought copper are C11000, C11100, C11300, C11400, C11500, C11600, C10200, C12000, and C12200.

Aluminum

Aluminum is a corrosion-resistant, lightweight metal (approximately 169 pounds per cubic foot) that has numerous applications. Most materials called aluminum actually are aluminum alloys, which are stronger than pure aluminum.

The types of wrought aluminum alloys are designated by four digits (**Figure 19.8**). The first digit (2 through 8) indicates the alloying element that is combined with aluminum. The second digit indicates modifications of the original alloy or impurity limits. The last two digits identify other alloying materials or indicate the aluminum's purity.

Figure 19.9 shows a four-digit numbering system used to designate types of cast aluminum and alloys. The first digit indicates the alloy group, and the next two digits identify

ALUMINUM DESIGNATIONS (169 LBS/CF)

Composition	Alloy Number	Application
Aluminum (99% pure)	1XXX	Tubing, tank cars
Aluminum alloys		
Copper	2XXX	Aircraft parts, screws, rivets
Manganese	3XXX	Tanks, siding, gutters
Silicon	4XXX	Forging, wire
Magnesium	5XXX	Tubes, welded vessels
Magnesium and silicon	6XXX	Auto body, pipes
Zinc	7XXX	Aircraft structures
Other elements	8XXX	

19.8 These are the numbering designations of aluminum and aluminum alloys and their applications.

ALUMINUM CASTINGS AND INGOT DESIGNATIONS

Composition	Alloy Number
Aluminum (99% pure)	1XX.X
Aluminum alloys	
Copper	2XX.X
Silicon with copper and/or magnesium	3XX.X
Silicon	4XX.X
Magnesium	5XX.X
Magnesium and silicon	6XX.X
Zinc	7XX.X
Tin	8XX.X
Other elements	9XX.X

19.9 These are the numbering designations of cast aluminum, ingots, and aluminum alloys.

the aluminum alloy or aluminum purity. The number to the right of the decimal point represents the aluminum form: XX.0 indicates castings, XX.1 indicates ingots with a specified chemical composition, and XX.2 indicates ingots with a specified chemical composition other than the XX.1 ingot. Ingots are blocks of cast metal to be remelted, and billets are castings of aluminum to be formed by forging.

Magnesium

Magnesium is a light metal (109 pounds per cubic foot) available in an inexhaustible supply because it is extracted from seawater and natural brines. Magnesium is an excellent material for aircraft parts, clutch housings, crankcases for air-cooled engines, and applications where lightness is desirable.

Magnesium is used for die and sand castings, extruded tubing, sheet metal, and forging. Magnesium and its alloys may be joined by bolting, riveting, or welding. Some numbered designations of magnesium alloys are M10100, M11630, M11810, M11910, M11912, M12390, M13320, M16410, and M16620.

19.3 Properties of Metals

All materials have properties that designers must utilize to the best advantage. The following terms describe these properties.

Ductility: a softness in some materials, such as copper and aluminum, which permits them to be formed by stretching (drawing) or hammering without breaking.

Brittleness: a characteristic that will not allow metals such as cast irons and hardened steels to stretch without breaking.

Malleability: the ability of a metal to be rolled or hammered without breaking.

Hardness: the ability of a metal to resist being dented when it receives a blow.

Toughness: the property of being resistant to cracking and breaking while remaining malleable.

Elasticity: the ability of a metal to return to its original shape after being bent or stretched.

Modifying Properties by Heat Treatment

The properties of metals can be changed by various types of heat treating. Although heat affects all metals, steels are affected to a greater extent than others.

Hardening: heating steel to a prescribed temperature and quenching it in oil or water.

Quenching: rapidly cooling heated metal by immersing it in liquids, gases, or solids (such as sand, limestone, or asbestos).

Tempering: reheating previously hardened steel and then cooling it, usually by air, to increase its toughness.

Annealing: heating and cooling metals to soften them, release their internal stresses, and make them easier to machine.

Normalizing: heating metals and letting them cool in air to relieve their internal stresses.

Case hardening: hardening a thin outside layer of a metal by placing the metal in contact with carbon or nitrogen compounds that it absorbs as it is heated; afterward, the metal is quenched.

Flame hardening: hardening by heating a metal to within a prescribed temperature range with a flame and then quenching the metal.

19.4 Forming Metal Shapes

Casting

One of the two major methods of forming shapes is casting, which involves preparing a mold in the shape of the part desired, pouring molten metal into it, and cooling the metal to form the part. The types of casting, which differ in the way the molds are made, are **sand casting**, **permanent-mold casting**, **die casting**, and **investment casting**.

Sand Casting In the first step of sand casting, a wood or metal form or pattern is made in the shape of the part to be cast. The pattern is placed in a metal box called a flask and molding sand is packed around the pattern. When the pattern is withdrawn from the sand, it leaves a void forming the mold. Molten metal is poured into the mold through sprues or gates. After cooling, the casting is removed and cleaned (**Figure 19.10**).

Cores formed from sand may be placed in a mold to create holes or hollows within a

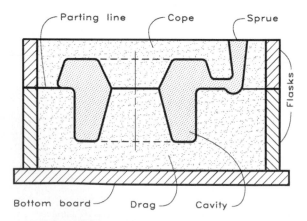

19.10 A two-section sand mold is used for casting a metal part.

casting. After the casting has been formed, the cores are broken apart and removed, leaving behind the desired void within the casting.

Because the patterns are placed in and removed from the sand before the metal is poured, the sides of the patterns must be tapered, called **draft**, for ease of withdrawal from the sand. The angle of draft depends on the depth of the pattern in the sand and varies from 2° to 8° in most applications. **Figure 19.11** shows a pattern held in the sand by a lower flask. Patterns are made oversize to compensate for shrinkage that occurs when the casting cools.

Because sand castings have rough surfaces, features that come into contact with other parts must be machined by drilling, grinding, finishing, or shaping. The tailstock base of a lathe shown in **Figure 19.12** illustrates raised bosses that have been finished. The casting must be made larger than the finished size where metal is to be removed by machining.

Fillets and rounds are used at the inside and outside corners of castings to increase their strength by relieving the stresses in the cast metal (**Figure 19.13**). Fillets and rounds also are used because forming square corners by the sand-casting process is difficult and

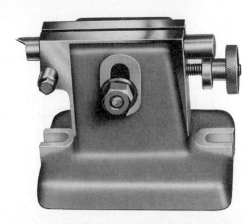

19.12 The tailstock casting for a lathe has raised bosses and contact surfaces that were finished to improve the effectiveness of nuts and bolts. Fillets and rounds were added to the inside and outside corners. (Courtesy L. W. Chuck Company.)

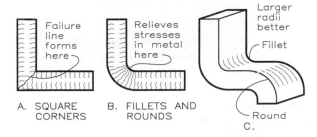

19.13

A Square corners cause a failure line to form, causing a weakness at this point.

B Fillets and rounds make the corners of a casting stronger and more attractive.

C The larger the radii of fillets and rounds, the stronger the casting will be.

because rounded edges make the finished product more attractive (**Figure 19.12**).

Permanent-Mold Casting Permanent molds are made for the mass production of parts. They are generally made of cast iron and coated to prevent fusing with the molten metal poured into them (**Figure 19.14**).

Die Casting Die castings are used for the mass production of parts made of aluminum, magnesium, zinc alloys, copper, and other materials. Die castings are made by forcing molten metal

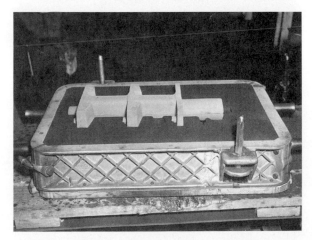

19.11 This pattern is held in the bottom half (the drag) of a sand mold to form a mold for a casting.

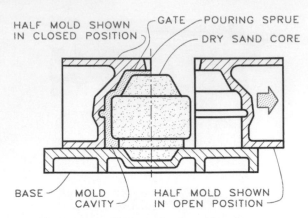

19.14 Permanent molds are made of metal for repetitive usage. Here, a sand core made from another mold is placed in the permanent mold to create a void within the casting.

19.16 An investment casting (lost-wax process) is used to produce complex metal objects and art pieces.

into dies (or molds) under pressure. They are inexpensive, meet close tolerances, and have good surface qualities. The same general principles of sand castings—using fillets and rounds, allowing for shrinkage, and specifying draft angles—apply to die castings (**Figure 19.15**).

Investment Casting Investment casting is used to produce complicated parts or artistic sculptures that would be difficult to form by other methods (**Figure 19.16**). A new pattern must be used for each investment casting, so a mold or die is made for casting a wax master pattern. The wax pattern, identical to the casting, is placed inside a container and plaster or sand is poured (invested) around it. Once the

investment has cured, the wax pattern is melted, leaving a hollow cavity to serve as the mold for the molten metal. After the casting has set, the plaster or sand is broken away from it.

Forgings

The second major method of forming shapes is forging, which is the process of shaping or forming heated metal by hammering or forcing it into a die. Drop forges and press forges are used to hammer metal billets into forging dies. Forgings have the high strength and resistance to loads and impacts required for applications such as aircraft landing gears (**Figure 19.17**).

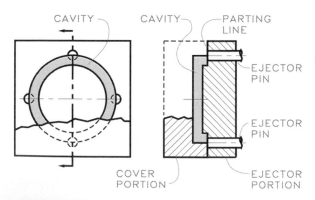

19.15 This die is used for casting a simple part. The metal is forced into the die to form the casting.

19.17 This aircraft landing-gear component was formed by forging. (Courtesy of Cameron Iron Works.)

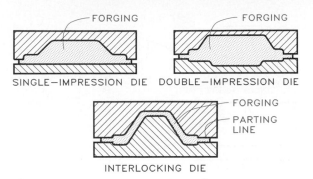

SINGLE—IMPRESSION DIE   DOUBLE—IMPRESSION DIE

INTERLOCKING DIE

19.18 This drawing shows three types of forging dies.

Figure 19.18 shows three types of dies. A single-impression die gives an impression on one side of the parting line between the mating dies, a double-impression die gives an impression on both sides of the parting line, and the interlocking dies give an impression that may cross the parting line on either side. **Figure 19.19** shows how an object is forged with horizontal dies and a vertical ram to hollow the object.

Figure 19.20 illustrates the sequence of forging a part from a billet by hammering it

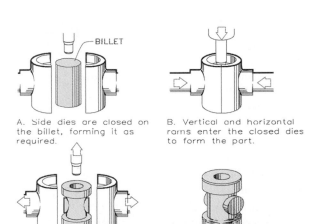

A. Side dies are closed on the billet, forming it as required.

B. Vertical and horizontal rams enter the closed dies to form the part.

C. Rams are withdrawn, the dies open, and the forging extracted.

D. Result: A forging having multiple planes, no flash, and no draft.

19.19 These are the steps involved in forging a part with external dies and an internal ram. (Courtesy of Cameron Iron Works.)

19.20 Steps A through G are required to forge a billet into a finished connecting rod. (Courtesy of the Drop Forging Association.)

into different dies. It is then machined to its proper size within specified tolerances.

Figure 19.21 shows a working drawing for making a forged part. When preparing forging drawings, you must consider (1) draft angles and parting lines, (2) fillets and rounds, (3) forging tolerances, (4) extra material for machining, and (5) heat treatment of the finished forging.

Draft, the angle of taper, is crucial to the forging process. The minimum radii for inside corners (fillets) are determined by the height of the feature (**Figure 19.22**). Similarly, the minimum radii for the outside corners (rounds) are related to a feature's height (**Figure 19.23**). The larger the radius of a fillet or round, the better it is for the forging process.

Some of the standard steels used for forging are designated by the SAE numbers 1015, 1020, 1025, 1045, 1137, 1151, 1335, 1340, 4620,

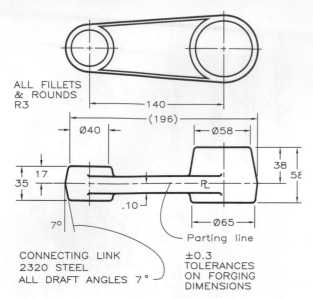

ALL FILLETS & ROUNDS R3

140 (196)

Ø40 Ø58

17 35 38 58

7°

.10 PL

Ø65

Parting line

CONNECTING LINK
2320 STEEL
ALL DRAFT ANGLES 7°

±0.3
TOLERANCES
ON FORGING
DIMENSIONS

19.21 This working drawing for a forging shows draft angles and the parting line (PL) where the dies come together.

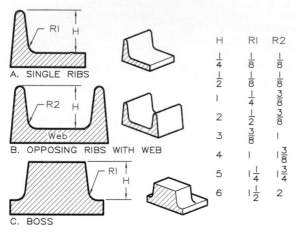

A. SINGLE RIBS RI H

B. OPPOSING RIBS WITH WEB R2 H Web

C. BOSS RI H

H	RI	R2
$\frac{1}{4}$	$\frac{1}{8}$	$\frac{1}{8}$
$\frac{1}{2}$	$\frac{1}{8}$	$\frac{1}{8}$
1	$\frac{1}{4}$	$\frac{3}{8}$
2	$\frac{1}{2}$	$\frac{3}{8}$
3	$\frac{3}{8}$	1
4	1	$1\frac{3}{8}$
5	$1\frac{1}{4}$	$1\frac{3}{4}$
6	$1\frac{1}{2}$	2

19.22 These guidelines are for determining the minimum radii for fillets (inside corners) on forged parts.

5120, and 5140. Iron, copper, and aluminum also can be forged.

Rolling Rolling is a type of forging in which the stock is rolled between two or more rollers to shape it. Rolling can be done at right angles or parallel to the axis of the part (**Figure 19.24**). If a high degree of shaping is required,

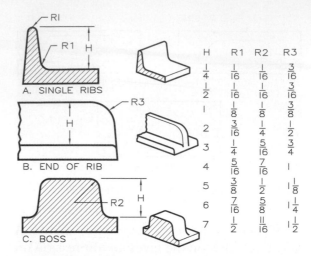

A. SINGLE RIBS R1 H

B. END OF RIB R3 H

C. BOSS R2 H

H	R1	R2	R3
$\frac{1}{4}$	$\frac{1}{16}$	$\frac{1}{16}$	$\frac{3}{16}$
$\frac{1}{2}$	$\frac{1}{16}$	$\frac{1}{16}$	$\frac{3}{16}$
1	$\frac{1}{8}$	$\frac{1}{8}$	$\frac{3}{8}$
2	$\frac{3}{16}$	$\frac{1}{4}$	$\frac{2}{3}$
3	$\frac{1}{4}$	$\frac{5}{16}$	$\frac{3}{4}$
4	$\frac{5}{16}$	$\frac{7}{16}$	1
5	$\frac{3}{8}$	$\frac{1}{2}$	$1\frac{1}{8}$
6	$\frac{7}{16}$	$\frac{5}{8}$	$1\frac{1}{4}$
7	$\frac{1}{2}$	$\frac{11}{16}$	$1\frac{1}{2}$

19.23 These guidelines are for determining the minimum radii of rounds (outside corners) on forged parts.

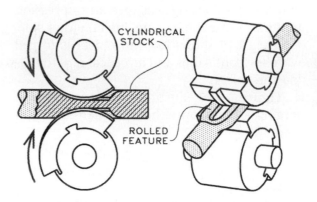

CYLINDRICAL STOCK

ROLLED FEATURE

19.24 Features on parts may be formed by rolling. Here a part is being rolled parallel to its axes.

the stock usually is heated before rolling. If the forming requires only a slight change in shape, rolling can be done without heating the metal, which is called **cold rolling** (CR); CRS means cold-rolled steel. **Figure 19.25** shows a cylindrical rod being rolled.

Stamping

Stamping is a method of forming flat metal stock into three-dimensional shapes. The first step of stamping is to cut out the shapes, called blanks, which are formed by bending

19.25 This cylindrical rod is being rolled to shape.

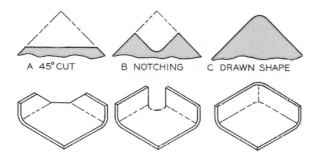

A 45° CUT   B NOTCHING   C DRAWN SHAPE

19.26 Box-shaped parts formed by stamping.

A A corner cut of 45° permits flanges to be folded with no further trimming.

B Notching has the same effect as the 45° cut and is often more attractive.

C A continuous corner flange requires that the blank be developed so that it can be drawn into shape.

and pressing them against forms. **Figure 19.26** shows three types of box-shaped parts formed by stamping, and **Figure 19.27** shows a design for a flange to be formed by stamping. Holes in stampings are made by punching, extruding, or piercing (**Figure 19.28**).

19.5 Machining Operations

After metal parts have been formed, machining operations must be performed to complete them. The machines used most often are

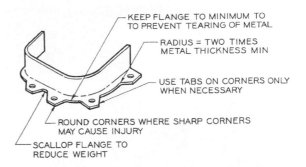

KEEP FLANGE TO MINIMUM TO TO PREVENT TEARING OF METAL

RADIUS = TWO TIMES METAL THICKNESS MIN

USE TABS ON CORNERS ONLY WHEN NECESSARY

ROUND CORNERS WHERE SHARP CORNERS MAY CAUSE INJURY

SCALLOP FLANGE TO REDUCE WEIGHT

19.27 This drawing shows a sheet metal flange design with notes that explain design details.

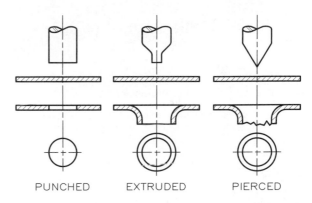

PUNCHED   EXTRUDED   PIERCED

19.28 These three methods are used to form holes in sheet metal by punching.

the **lathe**, **drill press**, **milling machine**, **shaper**, and **planer**. Some of these machines require manual operation; others are computer programmed to run at high speeds automatically, and require minimal or no operator attention.

Lathe

The lathe shapes cylindrical parts while rotating the work piece between its centers (**Figure 19.29**). The fundamental operations performed on the lathe are **turning**, **facing**, **drilling**, **boring**, **reaming**, **threading**, and **undercutting** (**Figure 19.30**).

Turning forms a cylinder with a tool that advances against and moves parallel to the cylinder being turned between the centers of the lathe (**Figure 19.31**). **Facing** forms flat

19.29 This typical metal lathe holds and rotates the work piece between its centers for machining. (Courtesy of the Clausing Corporation.)

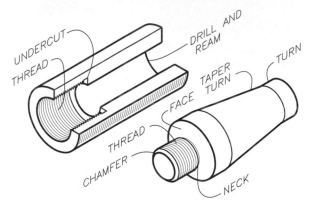

19.30 The basic operations that are performed on a lathe are shown here.

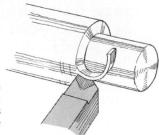

19.31 The most basic operation performed on the lathe is turning, whereby a continuous chip is removed by a cutting tool as the part rotates.

surfaces perpendicular to the axis of rotation of the part being rotated.

Drilling is performed by mounting a drill in the tail stock of the lathe and rotating the work while the bit is advanced into the part (**Figure 19.32**, **A** thru **C**). **Boring** makes large holes by enlarging smaller drilled holes with a tool mounted on a boring bar (**Figure 19.33A**). **Undercutting** is a groove cut inside a cylindrical hole with a tool mounted on a boring bar. The groove is cut as the tool advances from the center of the axis of revolution into the part (**Figure 19.33B**). **Reaming** removes only

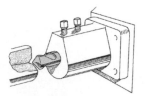

A. Start drilling   B. Twist drilling

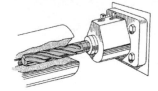

C. Core drilling

19.32 The three steps in drilling a hole in the end of a cylinder are (A) start drilling, (B) twist drilling, and (C) core drilling.

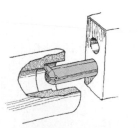

A. Boring   B. Undercutting

19.33 (A) This hole is being bored with a cutting tool attached to a boring bar of a lathe. (B) An undercut is being cut by the tool and boring bar.

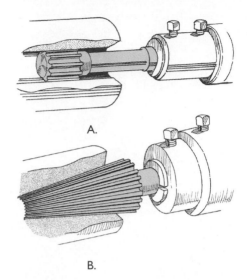

A.

B.

19.34 Fluted reamers can be used to finish inside (A) cylindrical and (B) conical holes within a few thousandths of an inch.

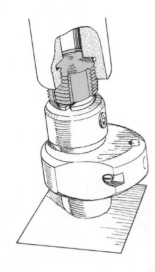

19.36 Internal threads can be cut on a lathe with a die called a tap. A recess, called a thread relief, was formed at the end of the threaded hole.

thousandths of an inch of material inside cylindrical and conical holes to enlarge them to their required tolerances (**Figure 19.34**). **Figure 19.35** shows a close-up view of reaming by honing.

Threading of external shafts and internal holes can be done on the lathe. The die used for cutting internal holes is called a tap (**Figure 19.36**). A tapping die being used to

thread a hole held in the chuck of a lathe is illustrated in **Figure 19.37**.

The **turret lathe** is a programmable lathe that can perform sequential operations, such as drilling a series of holes, boring them, and then reaming them. The turret is a multi-sided tool holder that sequentially rotates each tool into position for its particular operation (**Figure 19.38**).

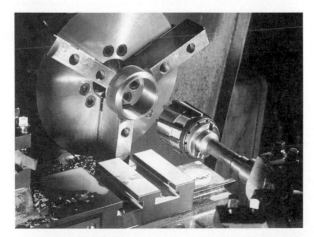

19.35 This hole is being reamed by honing. (Courtesy Barber-Coleman Company.)

19.37 A die is being used on a lathe to cut internal threads in this part. (Courtesy the Landis Machine Company.)

19.38 A turret lathe performs a sequence of operations by revolving the turret on which are mounted various tools.

19.39 This small drill press is used to make holes in parts. (Courtesy of Clausing Corporation.)

Drill Press

The **drill press** is used to drill small- and medium-sized holes (**Figure 19.39**). The stock being drilled is held securely by fixtures or clamps. The drill press can be used for counter-drilling, countersinking, counterboring, spot-facing, and threading (**Figure 19.40**). Multiple-head drill presses can be programmed to perform a series of drilling operations for mass production applications (**Figure 19.41**).

Measuring Cylinders The diameters, not the radii, of cylindrical features of parts made on a drill press or a lathe are measured to determine their sizes. Internal and external micrometer calipers are used for this purpose to measure to within one ten-thousandth of an inch.

Broaching Machine

Cylindrical holes can be converted into square, rectangular, or hexagonal holes with a **broach**

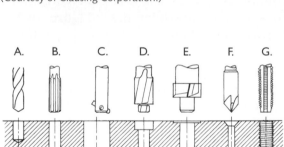

19.40 The basic operations performed on the drill press are (A) drilling, (B) reaming, (C) boring, (D) counterboring, (E) spotfacing, (F) countersinking, and (G) tapping (threading).

mounted on a special machine (**Figure 19.42**). A broach has a series of teeth graduated in size along its axis, beginning with teeth that are nearly the size of the hole to be broached and tapering to the final size of the hole. The broach is forced through the hole by pushing or pulling in a single pass, with each tooth cutting more from the hole as it passes through. Broaches can be used to cut external grooves, such as keyways or slots, in a part.

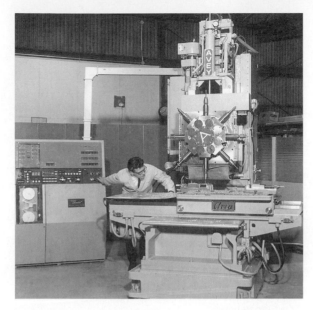

19.41 This multiple-head drill press can be programmed to perform a series of operations sequentially.

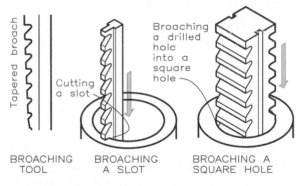

19.42 A broaching tool can be used to cut slots and holes with square corners on the interior and exterior of parts. Other shapes may be broached also.

Milling Machine

The **milling machine** uses a variety of cutting tools, rotated about a shaft (**Figure 19.43**), to form different grooved slots, threads, and gear teeth. The milling machine can cut irregular grooves in cams and finish surfaces on a part within a high degree of tolerance. The cutters revolve about a stationary axis while the work is passed beneath them.

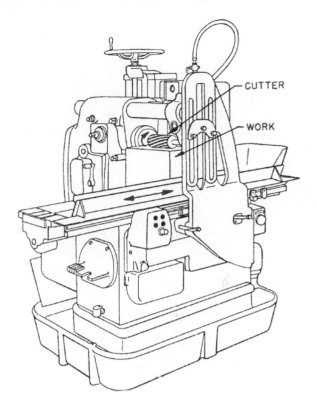

19.43 The milling machine operates by mounting the work on a bed that moves beneath revolving cutters. (Courtesy of the General Motors Corporation.)

Shaper

The shaper is a machine that holds a work piece stationary while the cutter passes back and forth across it to shape the surface or to cut a groove one stroke at a time (**Figure 19.44**). With each stroke of the cutting tool, the material is shifted slightly to align the part for the next overlapping stroke (**Figure 19.45**).

Planer

Unlike the shaper, which holds the work piece stationary, the **planer** passes the piece under the cutters to machine large flat surfaces (**Figure 19.46**). Like the shaper, the planer can cut grooves or slots and finish surfaces that must meet close tolerances.

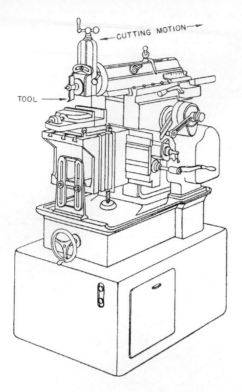

19.44 The shaper holds the work stationary while the machine's cutting tool makes strokes back-and-forth across the part to cut the desired shape in the work piece. (Courtesy of the General Motors Corporation.)

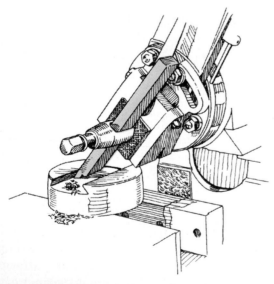

19.45 The shaper moves back and forth across the part, removing metal as it advances, to shape surfaces, cut slots, and perform other operations.

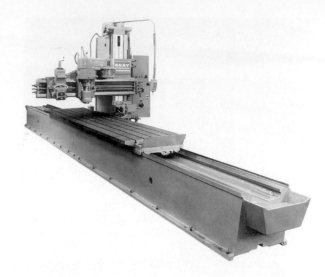

19.46 This planer has stationary cutters and a 30-foot bed. Work is fed past the cutters to finish large surfaces. (Courtesy of Gray Corporation.)

19.6 Surface Finishing

Surface finishing produces a smooth, uniform surface. It may be accomplished by **grinding**, **polishing**, **lapping**, **buffing**, and **honing**.

Grinding involves holding a flat surface against a rotating abrasive wheel (**Figure 19.47**). Grinding is used to smooth surfaces, both cylindrical and flat, and to sharpen edges used for cutting, such as drill bits (**Figure 19.48**). **Polishing** is done in the same way as grinding, except that the polishing wheel is flexible because it is made of felt, leather, canvas, or fabric.

Lapping produces very smooth surfaces. The surface to be finished is held against a lap, which is a large, flat surface coated with a fine abrasive powder that finishes a surface as the lap rotates. Lapping is done only after the surface has been previously finished by a less accurate technique, such as grinding or polishing. Cylindrical parts can be lapped by using a lathe with the lap.

Buffing removes scratches from a surface with a belt or rotating buffer wheel made of

19.47 The operator is grinding the upper surface of this part to a smooth finish with a grinding wheel. (Courtesy of the Clausing Corporation.)

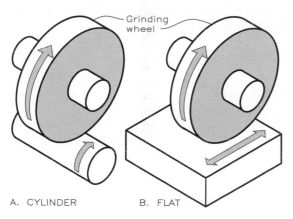

A. CYLINDER   B. FLAT

19.48 Grinding may be used to finish (A) cylindrical and (B) flat surfaces.

wool, cotton, felt, or other fabric. To enhance the buffing, an abrasive mixture is applied to the buffed surface during the process.

Honing finishes the outside or inside of holes within a high degree of tolerance (see

Figure 19.35). The honing tool is rotated as it is passed through the holes to produce the types of finishes found in gun barrels, engine cylinders, and other products requiring a high degree of smoothness.

19.7 Plastics and Other Materials*

Plastics (polymers) are widely used in numerous applications ranging from clothing, containers, and electronics to automobile bodies and components. Plastics are easily formed into irregular shapes, have a high resistance to weather and chemicals, and are available in limitless colors. The three basic types of plastics are **thermoplastics**, **thermosetting plastics**, and **elastomers**.

Thermoplastics may be softened by heating and formed to the desired shape. If a polymer returns to its original hardness and strength after being heated, it is classified as a thermoplastic. In contrast, **thermosetting plastics** cannot be changed in shape by reheating after they have permanently set. **Elastomers** are rubberlike polymers that are soft, expandable, and elastic, which permits them to be deformed greatly and then return to their original size.

Figure 19.49 shows commonly used plastics and other materials, including **glass** and **fiberglass**. The weights and yields of the materials are given, along with examples of their applications.

The motorized wheelchair shown in **Figure 19.50** is made of plastic. It has fewer parts and weighs less than motorized wheelchairs made of metal. Its rounded corners make it safe, eliminate joints, and make it easy to fabricate and clean.

---

*Parts of this section are based on *Manufacturing Engineering and Technology*, 2nd Ed., Serope Kalpakjian. Addison-Wesley, 1992.

E=Excellent
G=Good
F=Fair
P=Poor
A=Adhesives

	MACHINABILITY	FORMABILITY	CASTABILITY	WELDABILITY	CORROSION RES.	ABRASION RES.	LB/CU FT	YIELD: 1000 PSI	Typical Applications
THERMOPLASTICS									
ACRYLIC	G	G	E	A	E	F	74	9	Aircraft windows, TV parts, Lenses, skylights
ABS	G	G	G	A	E	G	66	66	Luggage, boat hulls, tool handles, pipe fittings
POLYMIDES (NYLON)	E	G	G	—	G	E	73	15	Helmets, gears, drawer slides, hinges, bearings
POLYETHYLENE	G	F	G	A	F	F	58	2	Chemical tubing, containers, ice trays, bottles
POLYPROPYLENE	G	G	G	A	E	G	56	5.3	Card files, cosmetic cases, auto pedals, luggage
POLYSTYRENE	G	E	G	A	P	G	67	7	Jugs, containers, furniture, lighted signs
POLYVINYL CHLORIDE	E	E	G	A	G	G	78	4.8	Rigid pipe and tubing, house siding, packaging
THERMOSETS									
EPOXY	F	G	G	—	E	G	69	17	Circuit boards, boat bodies, coatings for tanks
SILICONE	F	G	G	—	G	G	109	28	Flexible hoses, heart valves, gaskets
ELASTOMERS									
POLYURETHANE	G	G	G	A	G	E	74	6	Rigid: Solid tires, bumpers; Flexible: Foam, sponges
SBR RUBBER	—	—	E	—	F	E	39	3	Belts, handles, hoses, cable coverings
GLASSES									
GLASS	F	G	—	—	F	F	160	10+	Bottles, windows, tumblers, containers
FIBERGLASS	G	—	E	A	G	G	109	20+	Boats, shower stalls, auto bodies, chairs, signs

19.49 These are the characteristics of and typical applications for commonly used plastics and other materials.

19.50 The use of Dow plastic in this motorized wheelchair (made by Amigo, Inc.) reduced the number of parts by 97% and weight by 10%. It is also safe and easy to clean. (Courtesy of Dow Chemical Corporation.)

Review Questions

If you generally know the answers to these questions, you have read the chapter pretty well. If not, scan it again.

1. What is a ferrous metal? List the types of ferrous metals. Is steel ferrous or nonferrous?

2. What type of iron contains spheres of graphite that make it stronger than most other types of iron?

3. What type of iron is used for engine blocks because it resists vibration better than other types of iron?

4. What type of iron is known for withstanding wear and abrasion? What does it contain that gives it this characteristic?

5. For a ductile iron that is noted as 80-55-6, what is its tensile strength? What is its yield strength, and what is its percent of elongation?

6. For a malleable iron that is noted as 45006, what is its tensile strength and what is its percent of elongation?

7. What is the main ingredient that makes steel stronger than iron? Explain what the number designation, 1015 steel, means.

8. In general, what is the comparison of the weight of steel and iron?

9. When the number designation of aluminum is 5232, what is the main alloying element used with the aluminum?

10. How much would a 9 in. $\times$ 6 in. $\times$ 8 in. block of aluminum weigh?

11. How much would the block in Question 10 weigh if it were magnesium?

12. If a part were to be made by forging (hammering into shape), what metal characteristic would be most desirable?

13. What are the main differences in brittleness and hardness of metals?

14. In a sand casting mold, what are the top and bottom parts of the two-piece mold called?

15. Fillets and rounds are used to remove sharp corners from a metal part and to make it look more attractive. Are there other reasons for fillets and rounds? Explain.

16. What is the name of the casting process for molding an irregular-shaped piece of an art sculpture?

17. How much would the connecting link in **Figure 19.21** weigh if it were made of steel, brass, or magnesium?

18. What are the major methods of forming sheet metal?

19. What is the difference between boring and drilling?

20. What is a thread relief and why is it important?

21. What is the process for making a square hole?

22. What machining process is used to make interior holes smoother and make more accurate to a higher level of tolerance?

23. What is the difference between a shaper and a planer? Explain.

24. What are the methods of surface finishing? What are the characteristics of each?

25. What is a polymer? What are the three types of plastics and what are their respective applications?

20

Dimensioning

20.1 Introduction

Working drawings show dimensions and notes that convey sizes, specifications, and other information necessary to build a project. Drawings, with their full dimensions and specifications, serve as construction documents which become legal contracts.

The techniques of dimensioning presented here are based primarily on the standards of the American National Standards Institute (ANSI), especially Y14.5M, *Dimensioning and Tolerancing for Engineering Drawings.* Standards of companies such as the General Motors Corporation also are used.

20.2 Terminology

The strap shown in **Figure 20.1** is described in **Figure 20.2** with orthographic views to which dimensions were added. Refer to this drawing as various dimensioning terms are introduced.

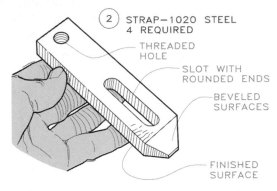

20.1 This tapered strap is a part of a clamping device that is dimensioned in Figure 20.2.

Dimension lines: thin lines (2H–4H pencil) with arrows at each end and numbers placed near their midpoints to specify size.

Extension lines: thin lines (2H–4H pencil) extending from the part and between which dimension lines are placed.

Centerlines: thin lines (2H–4H pencil) used to locate the centers of cylindrical parts such as holes.

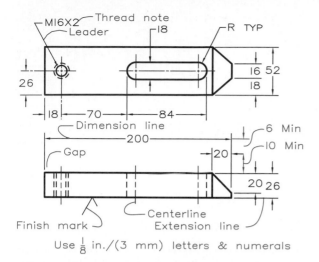

20.2 This typical dimensioned drawing of the tapered strap shown in Figure 20.1 introduces the terminology of dimensioning.

Leaders: thin lines (2H–4H pencil) drawn from a note to the feature to which it applies.

Arrowheads: drawn at the ends of dimension lines and leaders and the same length as the height of the letters or numerals, usually 1/8 inch, as shown in **Figure 20.3**.

Dimension numbers: placed near the middle of the dimension line and usually 1/8-inch high, with no units of measurement (″, in., or mm) shown.

20.3 Units of Measurement

The two commonly used units of measurement are the decimal inch in the English (imperial) system, and the millimeter in the

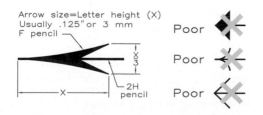

20.3 Draw arrowheads as long as the height of the letters used on the drawing and one third as wide as they are long.

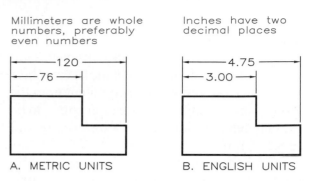

20.4 For the metric system, round millimeters to the nearest whole number. For the English system, show inches with two decimal places, even for whole numbers such as 3.00.

metric system (SI) (**Figure 20.4**). Giving fractional inches as decimals rather than common fractions makes arithmetic easier.

Figure 20.5 demonstrates the proper and improper dimensioning techniques with millimeters, decimal inches, and fractional inches. In general, round off dimensions in millimeters to whole numbers without fractions. However, when you must show a metric dimension of less than a millimeter, use a zero before the

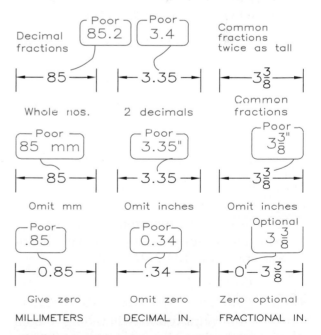

20.5 Basic principles of specifying measurements in SI and English units on a drawing.

decimal point. Do not use a zero before the decimal point when inches are used.

Show decimal inch dimensions with two-place decimal fractions, even if the last numbers are zeros. Omit units of measurement from the dimensions because they are understood to be in millimeters or inches. For example, use 112 (not 112 mm) and 67 (not 67″ or 5′–7″).

Architects use combinations of feet and inches in dimensioning and show foot marks, but usually omit inch marks, for example 7′–2. Engineers use feet and decimal fractions of feet to dimension large-scale projects such as road designs, for example 252.7′.

20.4 English/Metric Conversions

To convert dimensions in inches to millimeters, multiply by 25.4. Similarly, to convert dimensions in millimeters to inches, divide by 25.4.

When millimeter fractions are required as a result of conversion from inches, one-place fractions usually are sufficient, but two-place fractions are used in some cases. Find the decimal digit by applying the following rules.

- Retain the last digit unchanged if it is followed by a number less than 5; for example, round 34.43 to 34.4.

- Increase the last digit retained by 1 if it is followed by a number greater than 5; for example, round 34.46 to 34.5.

- Retain the last digit unchanged if it is even and is followed by the digit 5; for example, round 34.45 to 34.4.

- Increase the last digit retained by 1 if it is odd and is followed by the digit 5; for example, round 34.75 to 34.8.

20.5 Dual Dimensioning

On some drawings you may have to give both metric and English units, called dual dimensioning (**Figure 20.6**). Place the millimeter equiva-

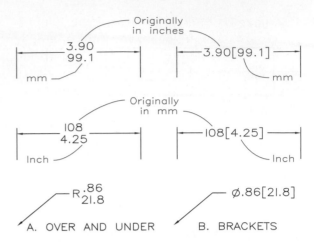

20.6 In dual dimensioning, place size equivalents in millimeters under or to the right of the inches (in brackets). Place the equivalent measurement in inches under or to the right of millimeters (in brackets). Show millimeters converted from inches as decimal fractions.

lent either under or over the inch units, or place the converted dimension in brackets to the right of the original dimension. Be consistent in the arrangement you use on any set of drawings.

(Note: Brief examples of dimensioning by computer are shown throughout this chapter. Refer to Chapter 37 for a more thorough coverage.)

Computer Method As shown in **Figure 20.7**, the dimensioning property *Dimalt* must be

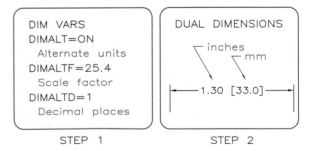

20.7 Alternate (dual) dimensions by computer:

Step 1 Type *Dimalt* and select *On* to set dual dimensioning on, type *Dimaltf* to assign the scale factor, and type *Dimaltd* and specify the number of decimal places.

Step 2 Find the linear dimensions by using the same steps as shown in Figure 20.33. The dimension in brackets is the metric equivalent of the inch dimensions.

set to *On* to obtain alternative (dual) dimensions in brackets following the units originally used. Set the property *Dimaltf* (scale factor) to the value of the multiplier to be used to change the first dimension. Use *Dimaltd* to assign the desired number of decimal places for the second dimension. Then select dimensions by using the *Dimlinear* command the same way you do to find single-value dimensions (**Figure 20.34**). Dimensioning variables can be set also from *Dimension Style* dialogue boxes.

20.6 Metric Units

Recall that in the metric system (SI), the first-angle of projection positions the front view over the top view and the right-side view to the left of the front view (**Figure 20.8**). You should label metric drawings with one of the symbols shown in **Figure 20.9** to designate the angle of projection. Display either the letters SI or the word METRIC prominently in or near the title block to indicate that the measurements are metric.

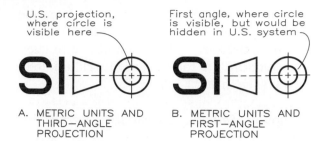

A. METRIC UNITS AND THIRD—ANGLE PROJECTION

B. METRIC UNITS AND FIRST—ANGLE PROJECTION

20.9 The SI symbol:

A The SI symbol indicates that the millimeter is the unit of measurement, and the truncated cone specifies that third-angle projection was used to position the orthographic views.

B Again, the SI symbol denotes use of the millimeter, but the truncated cone designates that the first-angle projection was used.

20.7 Numerals and Symbols

Vertical Dimensions

Vertical numeric dimensions on a drawing may be **aligned** or **unidirectional**. In the unidirectional method, all dimensions appear in the standard horizontal position (**Figure 20.10A**). In the aligned method, numerals are parallel with vertical and angular dimension lines and read from the right-hand side of the drawing, never from the left-hand side

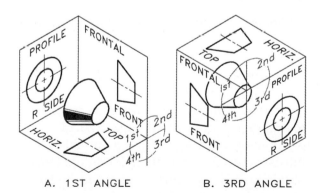

A. 1ST ANGLE   B. 3RD ANGLE

20.8 Projection systems:

A The SI system uses the first angle of orthographic projection, which places the top view under the front view.

B The American system uses the third angle of projection, which places the top view over the front view.

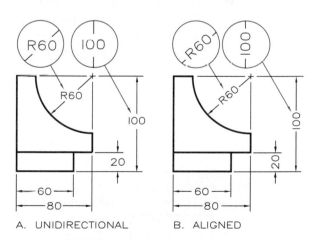

A. UNIDIRECTIONAL   B. ALIGNED

20.10 Unidirectional and aligned dimensions:

A Dimensions are unidirectional when they are horizontal in both vertical and horizontal dimension lines.

B Dimensions are aligned when they are lettered parallel to angular and vertical dimension lines to read from the right-hand side of the drawing (not from the left).

(**Figure 20.10B**). Aligned dimensions are used almost entirely in architectural drawings where dimensions composed of feet, inches, and fractions are too long to fit well unidirectionally (such as 22′–10-1/2).

Computer Method   The variable *Dimtih* (Text inside dimension lines is horizontal), a *Dim Vars* of the *Dim* command of AutoCAD, must be set to *Off* for aligned dimensions and to *On* for unidirectional dimensions. The *Dimtoh* mode controls the position of text lying outside dimension lines where the numerals do not fit within a short dimension line. When *Dimtoh* is *On*, numerals will be horizontal; when *Off*, numerals will align with the dimension line.

Placement

Dimensions should be placed on the most descriptive views of the part being dimensioned. The first row of dimensions should be at least three times the letter height (3H) from the object (**Figure 20.11**). Successive rows of dimensions should be spaced equally at least two times the letter height apart (0.25 inch, or 6 mm, when 1/8-inch letters are used). Use the Braddock-Rowe lettering guide triangle to space the dimension lines (**Figure 20.12**).

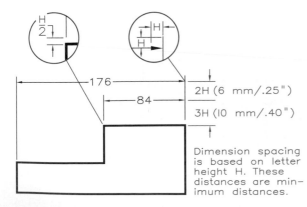

20.11 Place dimensions on a view as shown here, where all dimensioning geometry is based on the letter height (H) used.

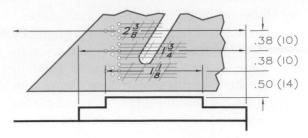

20.12 Draw guidelines for common-fractions in dimensions by aligning the center holes in the Braddock-Rowe triangle with the dimension line.

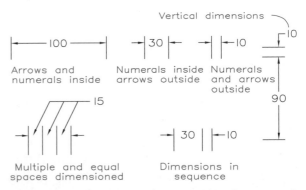

20.13 When space permits, place numerals and arrows inside extension lines. For smaller spaces, use other placements, as shown.

Figure 20.13 illustrates how to place dimensions in limited spaces. Regardless of space limitations, do not make numerals smaller than they appear elsewhere on the drawing.

Computer Method   Dimensioning variables and their minimum settings are shown in **Figure 20.14**. You may change variables set at these proportions at the same time by using *Dimscale*. For example, *Dimscale* = 25.4 would convert all dimensioning variables from inch to millimeter proportions. **Figure 20.15** shows other AutoCAD dimensioning variables. The full list of variables and their definitions can be obtained on the screen by typing *Dim* and then *Status* (see Chapter 37, also).

Dimensioning Symbols

Figure 20.16 shows standard dimensioning symbols and their sizes based on the letter

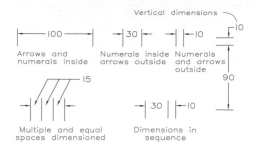

Vertical dimensions

Arrows and numerals inside

Numerals inside arrows outside

Numerals and arrows outside

Multiple and equal spaces dimensioned

Dimensions in sequence

20.14 The dimensioning properties (*Dim>Status*) to be applied by AutoCAD are shown here. They remain active with the file in use. *Dimscale* can be used to enlarge or reduce all of these variables.

Dim Vars	Default	Description
DIMADEC	−1	Decimal places for ang. dims.
DIMALT	OFF	Alternate units selected
DIMALTD	2	Alternate unit decimal places
DIMALTF	25.4	Alternate unit scale factor
DIMALTTD	2	Alternate tolerance dec. places
DIMALTTZ	0	Alternate tolerance zero suppress.
DIMALTU	2	Alternate units
DIMALTZ	0	Alternate unit zero suppression
DIMAPOST	−	Default suffix for alternate text
DIMASO	ON	Create associative dimensions
DIMASZ	.125	Arrow length
DIMAUNIT	0	Angular unit format
DIMBLK	−	Arrow block name
DIMBLK1	−	First arrow block name
DIMBLK2	−	Second arrow block name
DIMCEN	.09	Center mark size
DIMCLRD	BYLAYER	Dimension line color
DIMCLRE	BYLAYER	Extension line & leader color
DIMCLRT	BYLAYER	Dimension & extension color
DIMDEC	4	Decimal places for dimensions
DIMDLE	0	Dimension line extension
DIMDLI	.38	Dim. increment for continuation
DIMEXE	.125	Extension beyond dimension line
DIMEXO	.06	Extension line offset
DIMFIT	3	Fit text
DIMGAP	.06	Justification of text on dim. line
DIMJUST	0	Gap from dimension line to text
DIMLFAC	1	Length factor
DIMLIM	OFF	Gives tolerances in limit form
DIMPOST	−	Character suffix after dimensions
DIMRND	0	Rounding value for distances
DIMSAH	OFF	Separate arrowheads at each end
DIMSCALE	1	Scale factor for all dim. vars.
DIMSD1	OFF	Suppress first dimension line
DIMSD2	OFF	Suppress second dimension line
DIMSE1	OFF	Suppress first extension line
DIMSE2	OFF	Suppress second extension line
DIMSHO	ON	Changes dimens. while dragging
DIMSOXD	OFF	Suppress outside dimension lines
DIMSTYLE	STANDARD	Current dimensioning style
DIMTAD	0	Text placed above dimension line
DIMTDEC	4	Tolerance decimal places
DIMTFAC	1	Tolerance text scale factor
DIMTIH	ON	Text inside extension lines horiz.
DIMTIX	OFF	Text forced inside extension lines
DIMTM	0	Minus tolerance value
DIMTOFL	OFF	Forces dim. line inside, text out
DIMTOH	ON	Text outside ext. lines is horiz
DIMTOL	OFF	Applies tolerances to dimensions
DIMTOLJ	1	Tolerance vertical justification
DIMTP	0	Plus tolerance value
DIMTSZ	0	Tick size
DIMTVP	0	Text over or under dimen. line
DIMTXSTY	STANDARD	Text style
DIMTXT	.125	Text height
DIMTZIN	0	Tolerance zero suppression
DIMUNIT	2	Unit format
DIMUPT	OFF	User positioned text
DIMZIN	0	Zero suppression

20.15 Most of AutoCAD's dimensioning variables (*Dim>Status*) are shown here. Chapter 37 covers more details of their application.

height, usually 1/8 inch. By using these symbols, lengthy notes can be replaced and drawing time saved.

20.8 Dimensioning by Computer

Figure 20.17 shows several combinations of dimensioning variables that are available with the *Dim* command (type *Dim*).

You can place text inside the dimension line (*Dimtad: Off*) or above it (*Dimtad: On*). You may place arrowheads at the ends of dimension lines *Dimasz>0* or use tick marks (slashes) instead when *Dimtsz* is set to a value greater than 0, usually about half the letter height. You may select *Units* as architectural (feet and inches), metric (no decimal fractions), decimal inches (two or more decimal fractions), or engineering units (feet and decimal inches).

The text font (letter form) used in dimensioning is the same as the currently-used text font or the font assigned by the *Dimstyle* procedure and a *Style* is selected for dimensioning. In both cases, the text height should be set to 0 (zero) in order for *Dimscale*, (an option with the *Dim* command) to change the text height when invoked (**Figure 20.18**).

Had you set text to a specified height under the *Dimstyle* command, the text would not change with different *Dimscale* values, but would remain at its constant specified height. Text size is the most critical aspect of dimensioning since it must be sufficiently large to be readable (**Figure 20.19**). Using *Dimscale* enables text height, arrows, and extension line offsets to be changed at the same time.

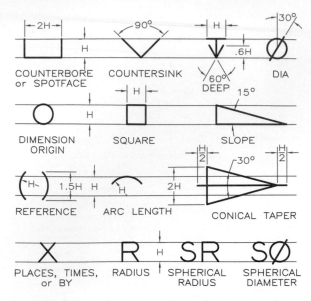

COUNTERBORE or SPOTFACE   COUNTERSINK   60° DEEP   DIA

DIMENSION ORIGIN   SQUARE   SLOPE   15°

REFERENCE   ARC LENGTH   CONICAL TAPER   30°

PLACES, TIMES, or BY   RADIUS   SPHERICAL RADIUS   SPHERICAL DIAMETER

20.16 These symbols can be used instead of words to dimension parts. Their proportions are based on the letter height, H, which usually is 1/8 inch.

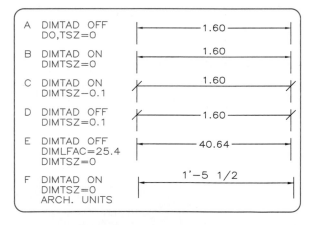

A	DIMTAD OFF DO,TSZ=0	1.60
B	DIMTAD ON DIMTSZ=0	1.60
C	DIMTAD ON DIMTSZ−0.1	1.60
D	DIMTAD OFF DIMTSZ=0.1	1.60
E	DIMTAD OFF DIMLFAC=25.4 DIMTSZ=0	40.64
F	DIMTAD ON DIMTSZ=0 ARCH. UNITS	1'−5 1/2

20.17 These dimension lines illustrate the effects of using the different dimensioning variables in AutoCAD.

20.9 Dimensioning Rules

There are many rules of dimensioning that you should become familiar with in order to place dimensions and notes on drawings most effectively. Each geometrical shape has its own set of rules: prisms, angular surfaces, cylindrical features, pyramids, cones, spheres, and arcs.

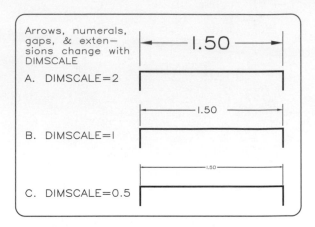

Arrows, numerals, gaps, & extensions change with DIMSCALE

A. DIMSCALE=2   1.50

B. DIMSCALE=1   1.50

C. DIMSCALE=0.5   1.50

20.18 DIMSCALE, a subcommand under DIM, permits changing the sizes of all dimensioning variables by typing a single scale factor. It changes the text size, arrows, offsets, and extensions at the same time.

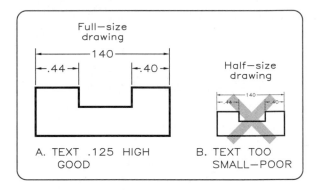

Full−size drawing   140   .44   .40   Half−size drawing   140

A. TEXT .125 HIGH GOOD   B. TEXT TOO SMALL−POOR

20.19 When dimensioning a drawing, be aware of its final plotted size so that you can properly size the dimensioning variables for reduction or enlargement. *Dimscale* is the most efficient command for assigning the proper scale to dimensioning variables.

Dimensioning rules are more guidelines than rules since many drawings make it difficult to apply them rigidly. Quite often rules of dimensioning must be violated or applied in a different manner due to the complexity of the part or the lack of space available.

Prisms

Beginning with **Figures 20.20–20.32**, the fundamental rules of dimensioning prisms are illustrated. The rules are presented in the simplest of examples in order to focus on the

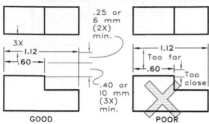

RULE I: First row 3 times letter height from part, minimum. (X=Letter height)

20.20 Place the first row of dimensions at least three times the letter height from the object. Successive rows should be at least two times the letter height apart.

RULE 2: Place dimensions between the views.

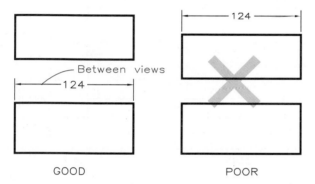

20.21 Place dimensions between the views sharing these dimensions.

RULE 3: Dimension the most descriptive views.

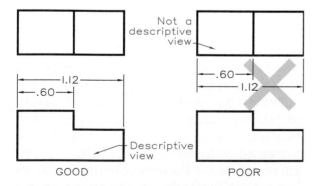

20.22 Place dimensions on the most descriptive views of an object.

RULE 4: Dimension from visible lines, not hidden lines.

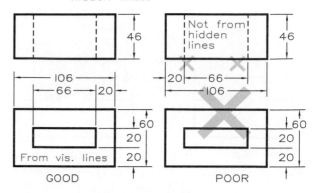

20.23 Dimension visible features, not hidden features.

RULE 5: Give an overall dimension and omit one of the chain dimensions.

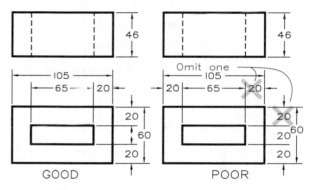

20.24 Leave the last dimension blank in a chain of dimensions and give an overall dimension.

RULE 5 (DEVIATION): When one chain dimension is not omitted, mark one dimension as a reference dimension.

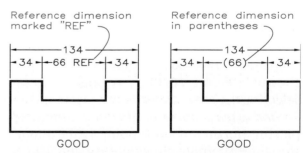

20.25 If you give all dimensions in a chain, mark the reference dimension (the one that would be omitted) with REF or place it in parentheses. Giving a reference dimension is a way of eliminating mathematical calculations in the shop.

RULE 6: Organize and align dimensions for ease of reading.

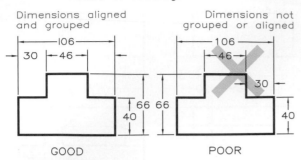

20.26 Place dimensions in well-organized lines for uncluttered drawings.

RULE 7: Do not repeat dimensions.

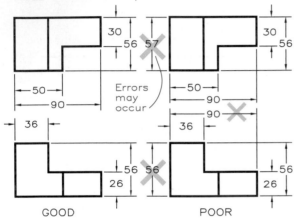

20.27 Do not duplicate dimensions on a drawing to avoid errors or confusion.

RULE 8: Dimension lines should not cross other lines.

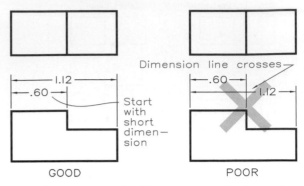

20.28 Dimension lines should not cross any other lines unless absolutely necessary.

RULE 9: Extension lines may cross other lines if they must.

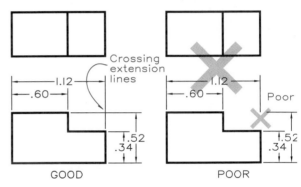

20.29 Extension lines may cross other extension lines or object lines if necessary.

specific rules one point at a time. These generally accepted dimensioning rules were not arbitrarily arrived at; they are based on a logical approach to aid in their application and interpretation.

Computer Method Dimension the part shown in **Figure 20.33** by typing *Dimlinear,* selecting the *First extension line origin* and the *Second extension line origin* when prompted, picking *Horizontal* and a point on the dimension line. Alternatively, extension lines can be found automatically by responding to the prompt, *Select object to dimension,* by selecting the line to be measured, specifying the

direction of the dimension line, and locating the dimension line.

By setting the dimensioning variable *Dimaso* to *ON,* dimensions will be associative dimensions. That is, the measurements in the dimension lines will change as the size of the objects are changed. For example, the *Stretch* command updates the dimensioning measurements as a part's size is modified (**Figure 20.34**). Associative dimensions can be erased as a unit (extension lines, dimension lines, arrows, and number). When *Dimsho* is ON, you will be able to see the dimensioning numerals changing dynamically on the screen as you *Stretch* the part to a new size.

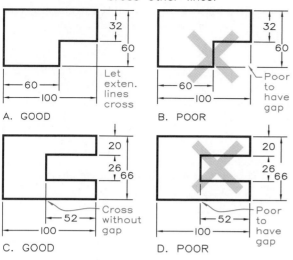

RULE 9 (Cont.): Extension lines may cross other lines.

A. GOOD — Let exten. lines cross

B. POOR — Poor to have gap

C. GOOD — Cross without gap

D. POOR — Poor to have gap

20.30 Leave a small gap from the edges of an object to extension lines that extend from them. Do not leave gaps where extension lines cross object lines or other extension lines.

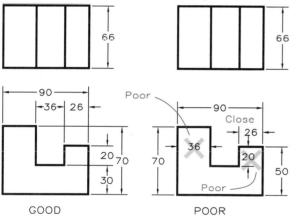

RULE 10: Do not place dimensions within the views unless necessary.

GOOD

POOR — Poor, Close, Poor

20.31 Whenever possible, place dimensions outside objects rather than inside their outlines.

Angles

You may dimension angles either by coordinates locating the ends of sloping surfaces or by angular measurements in degrees (**Figure 20.35**). Fractional angles can be specified in decimal units or in degrees, minutes, and seconds. Recall that there are 60 minutes in a degree and 60 seconds in a minute. It is seldom

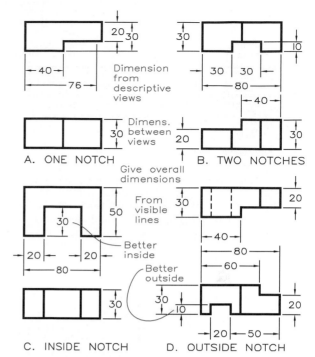

A. ONE NOTCH — Dimension from descriptive views; Dimens. between views

B. TWO NOTCHES

C. INSIDE NOTCH — Give overall dimensions; From visible lines; Better inside; Better outside

D. OUTSIDE NOTCH

20.32 Dimensioning Prisms.

A and B Dimension prisms from descriptive views and between views.

C You may dimension a notch inside the object if doing so improves clarity.

D Dimension visible lines, not hidden lines.

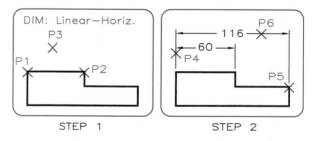

STEP 1 — DIM: Linear—Horiz.

STEP 2

20.33 Linear dimensioning by computer:

Step 1 Command: <u>DIMLINEAR</u> (Enter)

Specify extension line origin or [select object]: <u>P1</u>

Specify second extension line origin: <u>P2</u>

Specify dimension line location or [Mtext/Text/ Angle/ Horizontal/Vertical/Rotated]: <u>P3</u> (Mtext and Text can be used as options for overriding the measured distance.)

Step 2 Command: (Enter) (DIMLINEAR is repeated.)

Specify extension line origin or [select object]: <u>P4</u>

Specify second extension line origin: <u>P5</u>

Specify dimension line location or [Mtext/Text/Angle/ Horizontal/Vertical/Rotated]: <u>P6</u>

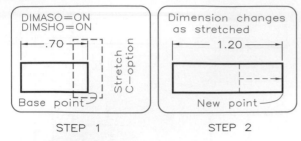

STEP 1   STEP 2

20.34 Associative dimensions:

Step 1 Set dimensioning variable, *Dimaso*, and *Dimsho* to *ON* to associate the dimensions with the size of the part to which they apply. Use the C option of the *Stretch* command to window the ends of an extension line and dimension lines.

Step 2 Select a base point and a new point. Dragging the size of the part recalculates the dimensions dynamically.

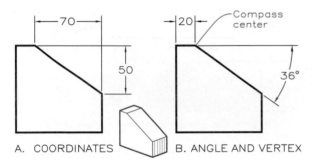

A. COORDINATES   B. ANGLE AND VERTEX

20.35 Dimensioning angles.

A Dimension angular planes by using coordinates.

B Measure angles by locating the vertex and measuring the angle in degrees. When accuracy is essential, specify angles in degrees, minutes, and seconds.

that you will you need to measure angles to the nearest second. **Figures 20.36** and **20.37** illustrate basic rules for dimensioning angles.

Computer Method Type *Dimangular* at the command line to dimension angles (**Figure 20.38**). If room is not available for the arrows between the extension lines, AutoCAD will draw them outside the extension lines.

Cylindrical Parts and Holes

The diameters of cylinders are measured with a micrometer (**Figures 20.39** and **20.40**). Therefore, dimension cylinders in their rectangular views with a diameter as a graphical

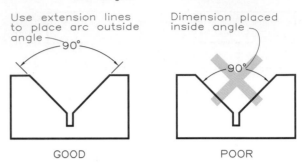

RULE 11: Place angular dimensions outside the angle.

GOOD   POOR

20.36 Place angular dimensions outside the object by using extension lines.

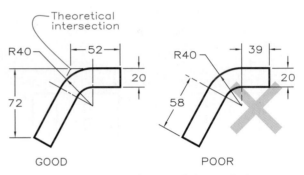

RULE 12: Dimension rounded corners to the theoretical intersection.

GOOD   POOR

20.37 Dimension a bent surface rounded corner by locating its theoretical point of intersection with extension lines.

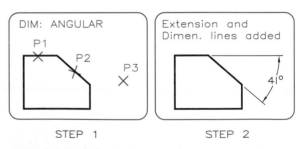

STEP 1   STEP 2

20.38 Angular dimensions by computer:

Step 1 *Command:* <u>DIMANGULAR</u> (Enter)
Select arc, circle, line: <u>P1</u>
Select second line: <u>P2</u>

Step 2 *Specify dimension arc line location or (Mtext/Text/ Angle):* <u>P3</u> (to accept measured angle.)
Dimension text [41]:
Command: (Continue angular dimensioning or enter a new command.)

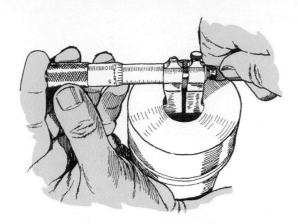

20.39 This internal micrometer caliper measures internal cylindrical diameters (radii cannot be measured).

20.40 This external micrometer caliper measures the outside diameter of a cylinder.

RULE 13: Dimension cylinders in their rectangular views with diameters.

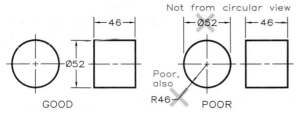

20.41 Dimension the diameter (not the radius) of a cylinder and in the rectangular view.

simulation of how the dimension is obtained (**Figures 20.41** and **20.42**). Place diameter symbols in front of the diametral dimensions to indicate that the dimension is a diameter. You will recall that the diametric symbol is a

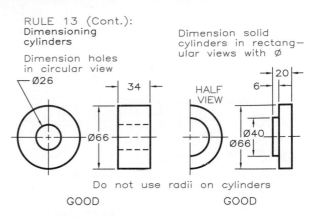

Do not use radii on cylinders

20.42 Dimension holes in their circular views with leaders. Dimension concentric cylinders with a series of diameters.

circle with a slash through it. In the English system, the abbreviation DIA placed after the diametral dimension is still sometimes used, but the metric diameter symbol is preferred.

Space is almost always a problem in dimensioning, and all means of conserving space must be used. Stagger dimensions for concentric cylinders to avoid crowding as shown in **Figure 20.43**. Dimension cylindrical holes in their circular view with leaders (**Figure 20.44**). The circular view is the view that would be used when the hole is located and drilled. Draw leaders specifying hole sizes as shown in **Figure 20.45**. When you must

RULE 14: Stagger dimension numerals to prevent crowding.

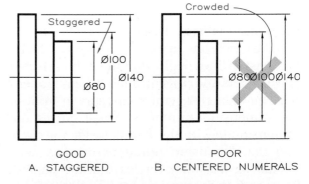

20.43 Dimensions on concentric cylinders are easier to read if they are staggered within their dimension lines.

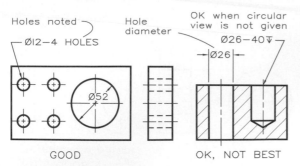

Holes noted — Ø12–4 HOLES

Hole diameter

OK when circular view is not given — Ø26–40

Ø52

Ø26

GOOD

OK, NOT BEST

20.44 Dimension holes in their circular view with leaders whenever possible, but dimension them in their rectangular views if necessary.

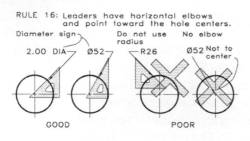

RULE 16: Leaders have horizontal elbows and point toward the hole centers.

Diameter sign

2.00 DIA   Ø52   Do not use radius — R26   No elbow   Ø52 Not to center

GOOD   POOR

20.45 Draw leaders pointing toward the centers of holes.

place diameter notes for holes in the rectangular view instead of the circular view, draw them as shown in **Figure 20.46**. Examples of correctly dimensioned parts with cylindrical features are shown in **Figure 20.47** and **20.48**.

Computer Method   Dimension circles in the circular view as shown in **Figure 20.49**. Dimension lines begin with the point selected on the circle and pass through the circle's center. The diameter symbol appears in front of the dimension numerals. Small circles are dimensioned with the arrows inside the circle and the dimension numerals outside, connected by a leader. Circles that are smaller have both the arrows and dimension outside the circle.

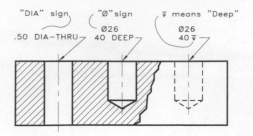

"DIA" sign   "Ø" sign   means "Deep"

.50 DIA–THRU   Ø26 40 DEEP   Ø26 40

20.46 These are examples of holes noted in their rectangular views when they cannot be dimensioned in their circular views.

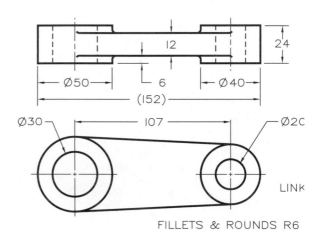

12   24

Ø50   6   Ø40

(152)

Ø30   107   Ø20

LINK

FILLETS & ROUNDS R6

20.47 This drawing illustrates the application of dimensions to cylindrical features. (F&R R6 means that fillets and rounds have a 6-mm radius.)

Pyramids, Cones, and Spheres
Figure 20.50A–C shows three methods of dimensioning pyramids. **Figure 20.50D** and **E** show two acceptable methods of dimensioning cones. Dimension a complete sphere by giving its diameter as shown in **Figure 20.50F**. If the spherical shape is less than a hemisphere (**Figure 20.50G**) use a spherical radius (SR). Only one view is needed to describe a sphere.

Leaders
Used to apply notes and dimensions to features, leaders most often are drawn at standard the angles of triangles (**Figure 20.47**). Leaders should begin at either the first or last

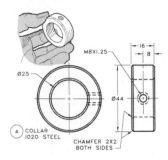

20.48 This cylindrical part, a collar, is shown drawn and dimensioned with properly applied dimensions and notes.

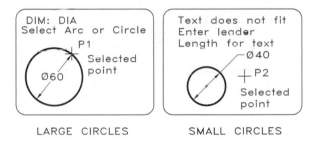

20.49 Dimensioning circles by computer:

Step 1 Type *Dimdiameter* to dimension a circle. Select P1 on the arc (an endpoint of the dimension) to produce the dimension. You have the option of replacing the dimension measured by the computer with a different value.

Step 2 When text does not fit, you can position it with the cursor. Select a point, P2, and the leader and dimension are drawn.

word of a note with a short horizontal line (elbow) from the note and extend to the feature being described as shown in **Figure 20.51**.

Computer Method To produce a leader that begins with an arrow, type *Qleader* or (*Dimension> Leader*) (**Figure 20.52**). To ensure that the arrow touches the circle, use

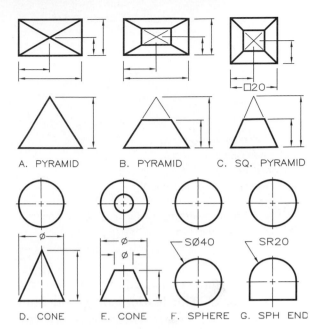

20.50 This drawing shows the proper way to dimension pyramids, cones, and spheres.

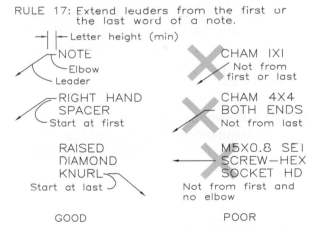

20.51 Extend leaders from the first or the last word of a note with a horizontal elbow.

Osnap and *Nearest* to snap the point of the arrow to the circumference. The program will prompt you to *Specify next point* twice, allowing a leader with an angular bend prior to locating the horizontal elbow at the end of the leader. When prompted, *Enter first line of annotation text*, type the note. Additional

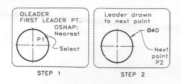

STEP 1   STEP 2

20.52 Leaders by computer:

Step 1 *Command: Qleader* (Enter)
Specify first leader point or [Settings]: <u>P1</u>
Step 2 *Select next point:* <u>P2</u> (Enter)
Specify text width <0.00>: (Enter)
Enter first line of annotation text <Mtext>: <u>%%C40</u> (Enter)
Enter next line of annotation text: (Enter)

prompts will allow you to add more notes in a vertical "stack." The leader command does not measure circles or lines; therefore you must type in the desired values otherwise it applies the last number it used.

If the previous option was *Dimdiametr,* the diameter symbol will precede the dimension. If the previous option was *Dimradius,* the radius symbol R will precede the dimension.

Arcs and Radii

Full circles are dimensioned with diameters by arcs dimensioned with radii (**Figure 20.53**). Current standards specify that radii be dimensioned with an R preceding the dimension (for example, R10). The previous standard was for R to follow the dimension (10R, for example). Thus both methods are seen on drawings.

You may dimension large arcs with a false radius (**Figure 20.54**) by drawing a zigzag to indicate that the line is not the true radius. Where space is not available for radii, dimension small arcs with leaders.

Computer Method Type *Dimradius* and select a point on the arc as the starting point to dimension and arc (**Figure 20.55**). If space permits, the dimension will appear between the center and the arrow. For smaller arcs, the

RULE 18: Dimension arcs (less than 180°) with radii.

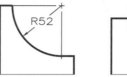

20.53 When space permits, place dimensions and arrows between the center and the arc. When the number will not fit, place it outside and the arrow inside. If the arrow will not fit inside, place both the dimension and arrow outside the arc with a leader.

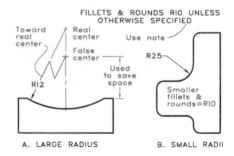

A. LARGE RADIUS   B. SMALL RADII

20.54 Dimensioning radii:

A Show a long radius with a false radius (a line with a zigzag) to indicate that it is not true length. Show its false center on the centerline of the true center.

B Specify fillets and rounds with a note to reduce repetitive dimensioning of small arcs.

arrows will appear inside and the dimensions outside the arc.

When space is not available for arrows inside, both the arrow and dimension will appear outside the arc. Decimal values are preceded with a zero unless you override them by typing in R and the value without a preceding zero. Leading zeros will be omitted if the variable *Dimzin* = <u>4</u>.

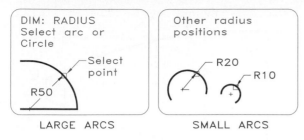

LARGE ARCS   SMALL ARCS

20.55 Type *Dimradius* to select a point on the arc. The program generates the dimension with its arrow at this point and precedes the dimension with an R. The program dimensions smaller arcs by positioning the dimension outside the arc or, because of even more limited space, by placing both the arrows and dimension outside the arc.

Fillets and Rounds

When all fillets and rounds are equal in size, you may place a note on the drawing stating that condition or use separate notes (**Figure 20.56**). If most, but not all, of the fillets and rounds have equal radii, the note may read ALL FILLETS AND ROUNDS R6 UNLESS OTHERWISE SPECIFIED (or abbreviated as F&R R6), with the fillets and rounds of different radii dimensioned separately.

You may note repetitive features as shown in **Figure 20.57** by using the notes TYPICAL, or TYP, which means that the

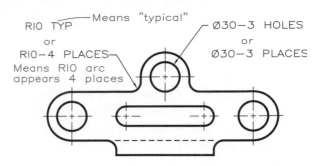

20.57 Use notes to indicate that identical features and dimensions are repeated to simplify dimensioning.

dimensioned feature is typical of those not dimensioned. You may use the note PLACES, or PL, to specify the number of places that identical features appear, although only one is dimensioned.

Figure 20.58 shows a pulley that is drawn and dimensioned. The drawing demonstrates proper application of many of the rules discussed in this section.

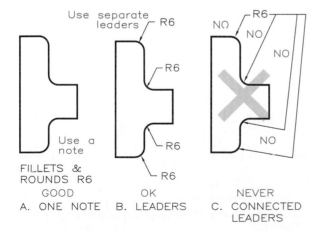

20.56 Indicate fillets and rounds by (A) notes or (B) separate leaders and dimensions. Never use confusing leaders (C).

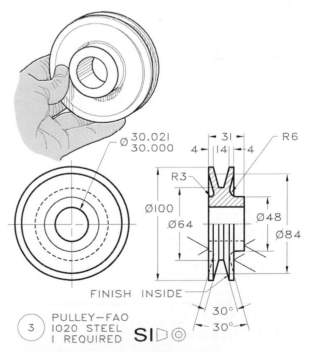

20.58 This dimensioned pulley illustrates the application of many of the rules covered in this section.

20.10 Curved and Symmetrical Parts

Curved Parts

You may dimension an irregular shape comprised of tangent arcs of varying sizes by using a series of radii (**Figure 20.59**). Irregular curves may be dimensioned by using coordinates to locate a series of points along the curve (**Figure 20.60**). You must use your judgment in determining how many located points are necessary to define the curve. Placing extension lines at an angle provides additional space for showing dimensions.

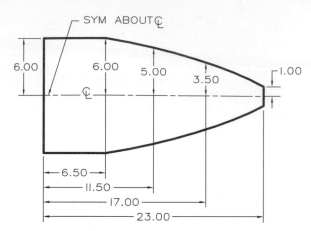

20.61 Use coordinates to dimension points along curves of a symmetrical part.

Symmetrical Parts

Dimension an irregular symmetrical curve with coordinates (**Figure 20.61**). Note the use of dimension lines as extension lines, a permissible violation of dimensioning rules in this case.

Dimension symmetrical objects by using coordinates to imply that the dimensions are symmetrical about the centerline (abbreviated CL) as shown in **Figure 20.62A**. **Figure 20.62B** shows a better method of dimensioning this type of object, where symmetry is dimensioned with no interpretation required on the part of the reader.

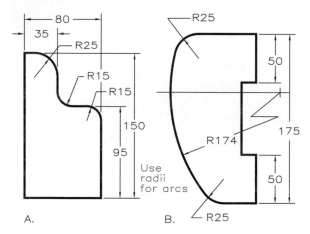

20.59 Use radii to dimension parts composed of arcs and partial circles.

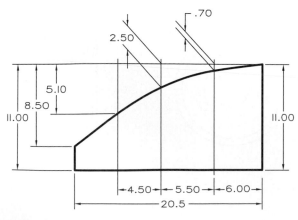

20.60 Use coordinates to dimension points along an irregular curve on a part.

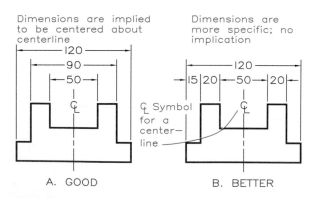

20.62 Symmetrical parts:

A You may dimension symmetrical parts implicitly about their centerlines.

B The better way to dimension symmetrical parts is explicitly about their centerlines.

20.11 Finished Surfaces

Parts formed in molds, called **castings**, have rough exterior surfaces. If these parts are to assemble with and move against other parts, they will not function well unless their contact surfaces are machined to a smooth finish by grinding, shaping, lapping, or a similar process.

To indicate that a surface is to be finished, finish marks are drawn on edge views of surfaces to be finished (**Figure 20.63**). Finish marks should be shown in every view where finished surfaces appear as edges even if they are hidden lines.

The preferred finish mark for the general cases is the uneven mark shown in **Figure 20.63B**. When an object is finished on all surfaces, the note FINISHED ALL OVER (abbreviated FAO) is placed on the drawing.

20.12 Location Dimensions

Location dimensions give the positions, not the sizes, of geometric shapes (**Figure 20.64**). Locate rectangular shapes by using coordinates of their corners and cylindrical shapes by using coordinates of their centerlines. In each case, dimension the view that shows

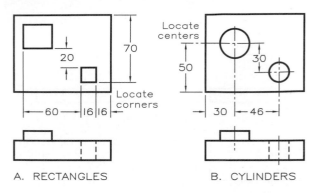

A. RECTANGLES   B. CYLINDERS

20.64 Location dimensions give the positions of geometric features with respect to other geometric features, but not their sizes.

both measurements. Always extend coordinates from any finished surfaces (even if the finished surface is a hidden line) because smooth machined surfaces allow the most accurate measurements. Locate and dimension single holes as shown in **Figure 20.65** and multiple holes as shown in **Figure 20.66**. **Figure 20.67** shows the application of location dimensions to a typical part with size dimensions omitted.

Baseline dimensions extend from two baselines in a single view (**Figure 20.68**). The use of baselines eliminates the possible accumulation of errors in size that can occur from chain dimensioning.

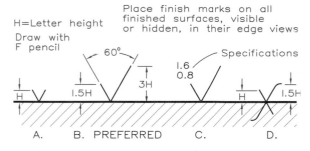

H=Letter height
Draw with
F pencil

Place finish marks on all finished surfaces, visible or hidden, in their edge views

20.63 Finish marks indicate that a surface is to be machined to a smooth surface.

A The traditional V can be used for general applications.

B The unequal finish mark is the best for general applications.

C Where surface texture must be specified, this finish mark is used with texture values.

D The f-mark is the oldest and least used symbol.

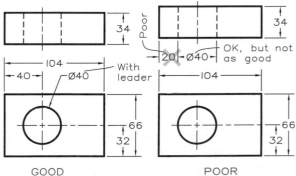

RULE 19: Locate holes in circular views and dimension diameters with a leader.

GOOD   POOR

20.65 Locate cylindrical holes in their circular views by coordinates to their centers.

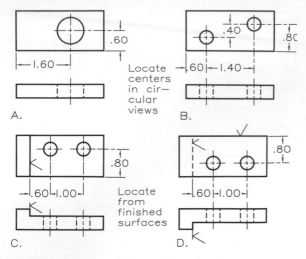

A.

B.

Locate
centers
in cir-
cular
views

C.

Locate
from
finished
surfaces

D.

20.66 Rules in summary:

A Locate cylindrical holes in their circular views from two surfaces of the object.

B Locate multiple holes from center to center.

C Locate holes from finished surfaces, even if the finished surfaces are hidden as in (D).

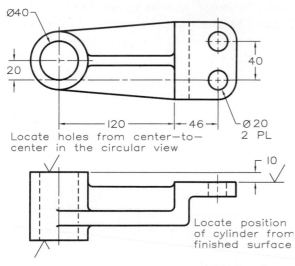

Locate holes from center–to–
center in the circular view

Locate position
of cylinder from
finished surface

20.67 This example of a dimensioned shaft arm shows the application of location dimensions, with size dimensions omitted.

Holes through circular plates can be located by using coordinates or a note as shown in **Figure 20.69**. Dimension the diameter of the imaginary circle passing through the centers of the holes in the circular view as a reference dimension and locate the holes by

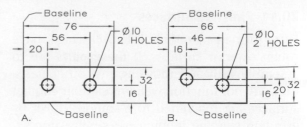

A.

B.

20.68 Measuring holes from two datum planes is the most accurate way to locate them and reduces the accumulation of errors possible in chain dimensioning.

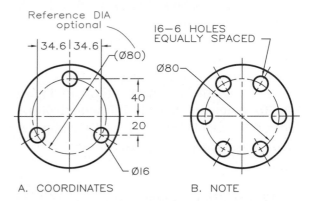

A. COORDINATES

B. NOTE

20.69 Locate holes on a bolt circle by using (A) coordinates or (B) notes.

using coordinates (**Figure 20.69A**) or a note (**Figure 20.69B**). This imaginary circle is called the **bolt circle** or **circle of centers**.

You may also locate holes with radial dimensions and their angular positions in degrees (**Figure 20.70**). Holes may be located on their bolt circle even if the shape of the object is not circular.

20.13 Objects with Rounded Ends

Dimension objects with rounded ends from one rounded end to the other (**Figure 20.71A**) and show their radii as R without dimensions to specify that the ends are arcs. Obviously, the end radius is half the height of the part. If you dimension the object from center to center (**Figure 20.71B**), you must give the radius size. You may specify the overall width as a reference dimension (116) to eliminate the need for calculations.

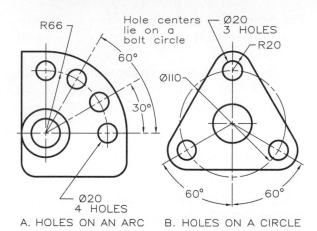

20.70 Locate centers of holes by using (A) a combination of radii and degrees or (B) a circle of centers.

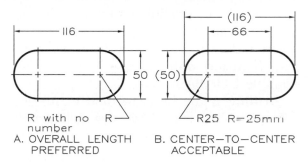

20.71 Rounded ends:

A Dimension objects having rounded ends from end-to-end and give the height.

B A less desirable choice is to dimension the rounded ends from center-to-center and give the radius.

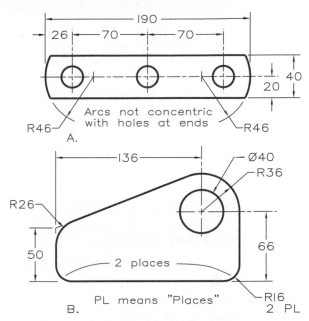

20.72 These drawings show how to dimension parts having (A) rounded ends that are not concentric with the holes and (B) rounds and cylinders.

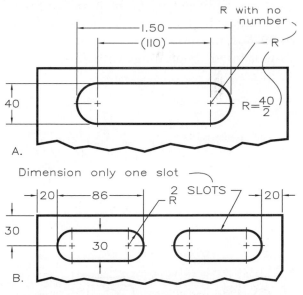

20.73 These drawings illustrate methods of dimensioning parts that have (A) one slot and (B) more than one slot.

Dimension parts with partially rounded ends as shown in **Figure 20.72A**. Dimension objects with rounded ends that are smaller than a semicircle with a radius and locate the arc's center (**Figure 20.72B**).

Dimension a single slot with its overall width and height (**Figure 20.73A**). When there are two or more slots, dimension one slot and use a note to indicate that there are other identical slots (**Figure 20.73B**).

The tool holder table shown in **Figure 20.74** illustrates dimensioning of arcs and slots. To prevent dimension lines from crossing, several dimensions are placed on a less descriptive

view. Notice that the diameters of the semicircular features are given; therefore, notes of R are given to indicate radii but the radius size is unnecessary. Radii are half the diameters.

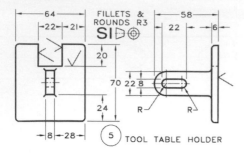

20.74 This dimensioned part has both slots and arcs.

20.14 Outline Dimensioning

Now that you are familiar with most of the rules of dimensioning, you can better understand outline dimensioning, which is a way of applying dimensions to a part's outline (silhouette). By taking this approach, you have little choice but to place dimensions in the most descriptive views of the part. For example, imagine that the T-block shown in **Figure 20.75** has no lines inside its outlines. It is dimensioned beginning with its location

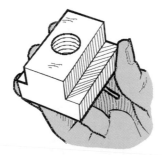

20.75 Outline dimensioning is the placement of dimensions on views as if they had no internal lines. Practice in applying this concept will help you place dimensions on their most descriptive views.

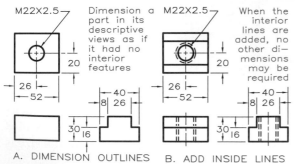

A. DIMENSION OUTLINES B. ADD INSIDE LINES

dimensions. When the inside lines are considered, additional dimensions are seldom needed.

Figure 20.76 shows an example of outline dimensioning. Note the extension of all dimensions from the outlines of well-defined features.

20.15 Machined Holes

Machined holes are formed by machine operations, such as drilling, boring, or reaming (**Figure 20.77**). Occasionally a machining operation is specified in the note, such as 32 DRILL, but it is preferred to omit the specific machining operation. Give the diameter of the hole with its symbol in front of its dimension (for example, 032) with a leader extending from the circular view. You may also note hole diameters with DIA after their size (for example, 2.00 DIA).

Drilling is the basic method of making holes. Dimension the size of a drilled hole with a leader extending from its circular view. You may give its depth in the note from the circular view or dimension it in the rectangular view (**Figure 20.77B**). Dimension the depth of a drilled hole to the usable part of the hole, not to the conical point.

Counterdrilling involves drilling a large hole inside a smaller hole to enlarge it (**Figure**

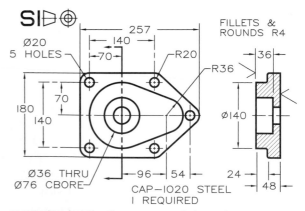

20.76 This drawing illustrates use of the outline method to dimension the cap in its most descriptive views.

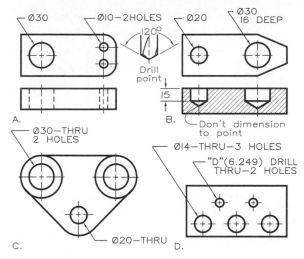

20.77 Cylindrical holes.

A and **B** Dimension cylindrical holes by either of these methods.

C and **D** When you use only one view, you have to note the THRU holes or specify their depths.

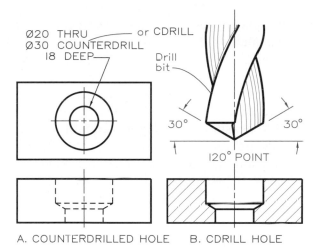

A. COUNTERDRILLED HOLE   B. CDRILL HOLE

20.78 Counterdrilling notes give the specifications for drilling a larger hole inside a smaller hole. Do not dimension the 120° angle because it is a byproduct of the drill point. Noting the counterdrill with a leader from the circular view is preferable.

20.78). The drill point leaves a 120° conical shoulder as a byproduct of counterdrilling.

Countersinking is the process of forming conical holes for receiving screw heads (**Figure 20.79**). Give the diameter of a countersunk hole (the maximum diameter on the surface) and the angle of the countersink in a

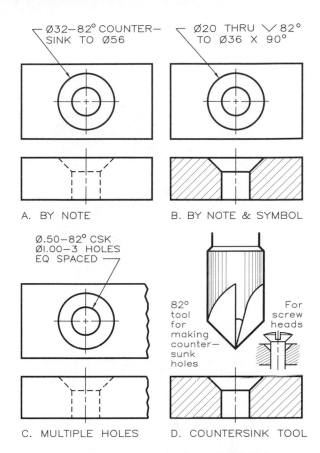

20.79 These illustrations show methods of noting and specifying countersunk holes for receiving screw heads.

note. Countersunk holes also are used as guides in shafts, spindles, and other cylindrical parts held between the centers of a lathe.

Spotfacing is the process of finishing the surface around holes to provide bearing surfaces for washers or bolt heads (**Figure 20.80A**). **Figure 20.80B** shows the method of spotfacing a boss (a raised cylindrical element).

Boring is used to make large holes and it is usually done on a lathe with a bore or a boring bar (**Figure 20.81**).

Counterboring is the process of enlarging the diameter of a drilled hole (**Figure 20.82**) to give a flat hole bottom without the taper as in counterdrilled holes.

Reaming is the operation of finishing or slightly enlarging drilled or bored holes within

20.80 This spotfacing tool is used to finish the cylindrical boss to provide a smooth seat for a bolt head. Spotfacing is the process of smoothing the surface where it will contact a washer, nut, or bolt and is noted as shown below.

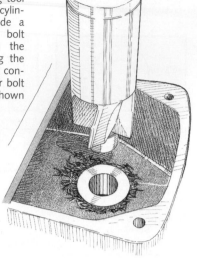

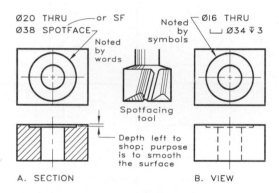

A. SECTION   B. VIEW

Ø20 THRU — or SF
Ø38 SPOTFACE

Noted by words

Spotfacing tool

Depth left to shop; purpose is to smooth the surface

Noted by symbols

Ø16 THRU
⌴ Ø34 ▽ 3

20.81 This photo shows the use of a lathe to bore a large hole with a boring bar. (Courtesy of Clausing Corporation.)

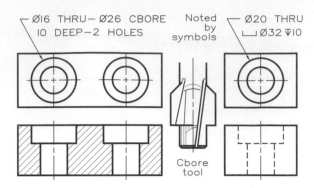

Ø16 THRU— Ø26 CBORE
10 DEEP—2 HOLES

Noted by symbols

Ø20 THRU
⌴ Ø32 ▽10

Cbore tool

20.82 Counterbored holes are similar to counterdrilled holes but have flat bottoms instead of tapered sides. Dimension them as shown.

their prescribed tolerances. A ream is similar to a drill bit.

20.16 Chamfers

Chamfers are beveled edges cut on cylindrical parts, such as shafts and threaded fasteners, to eliminate sharp edges and to make them easier to assemble. When the chamfer angle is 45°, use a note in either of the forms shown in **Figure 20.83A**. Dimension chamfers of other angles as shown in **Figure 20.83B**. When inside openings of holes are chamfered, dimension them as shown in **Figure 20.84**.

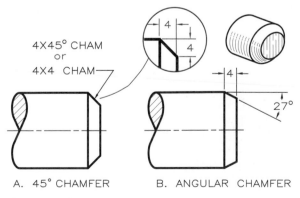

4X45° CHAM
or
4X4 CHAM

A. 45° CHAMFER   B. ANGULAR CHAMFER

20.83 Chamfers.
A Dimension 45° chamfers by one of the methods shown.
B Dimension chamfers of all other angles as shown.

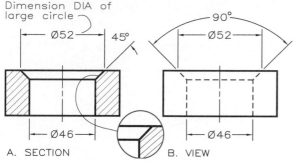

20.84 Dimension chamfers on the insides of cylinders in either of the methods shown.

20.17 Keyseats

A keyseat is a slot cut into a shaft for aligning and holding a pulley or a collar on a shaft. **Figure 20.85** shows how to dimension keyways and keyseats with dimensions taken from the tables in Appendices 24 and 25. The double dimensions on the diameter are tolerances (discussed in Chapter 21).

20.18 Knurling

Knurling is the operation of cutting diamond-shaped or parallel patterns on cylindrical surfaces for gripping, decoration, or press fits between mating parts that are permanently assembled. Draw and dimension diamond knurls and straight knurls as shown in **Figure**

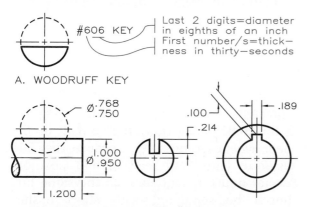

20.85 These drawings show methods of dimensioning (A) Woodruff keys and (B) keyways used to hold a part on a shaft. Appendix 24 gives their tables of sizes.

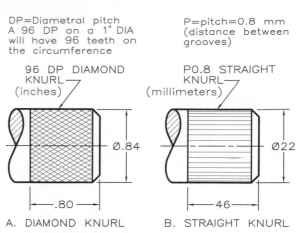

20.86 The diamond knurl has a diametral pitch, DP, of 96 and the straight knurl has a linear pitch, P, of 0.8 mm. Pitch is the distance between the grooves on the circumference.

20.86, with notes that specify type, pitch, and diameter.

The abbreviation DP means diametral pitch, or the ratio of the number of grooves on the circumference (N) to the diameter (D), expressed as $DP = N/D$. The preferred diametral pitches for knurling are 64 DP, 96 DP, 128 DP, and 160 DP.

For diameters of 1 inch, knurling of 64 DP, 96 DP, 128 DP, and 160 DP will have 64, 96, 128, and 160 teeth, respectively, on the circumference. The note P0.8 means that the knurling grooves are 0.8 mm apart. Make knurling calculations in inches and then convert them to millimeters. Specify knurls for press fits with the diameter size before knurling and with the minimum diameter size after knurling.

20.19 Necks and Undercuts

A **neck** is a groove cut around the circumference of a cylindrical part. If cut where cylinders of different diameters join (**Figure 20.87**), a neck ensures that the assembled parts fit flush at the shoulder of the larger cylinder and allows trash that would cause binding to drop out of the way, into the neck.

An **undercut** is a recessed neck inside a cylindrical hole (**Figure 20.88A**). A thread

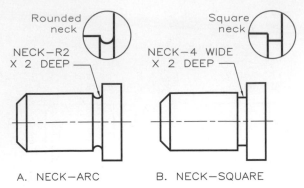

NECK—R2 X 2 DEEP

NECK—4 WIDE X 2 DEEP

A. NECK—ARC   B. NECK—SQUARE

20.87 Necks are recesses cut in cylinders, with rounded or square bottoms, usually at the intersections of concentric cylinders. Dimension necks as shown.

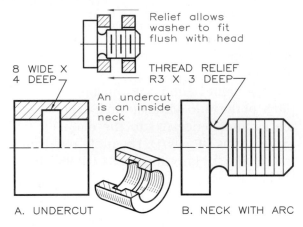

Relief allows washer to fit flush with head

8 WIDE X 4 DEEP

THREAD RELIEF R3 X 3 DEEP

An undercut is an inside neck

A. UNDERCUT   B. NECK WITH ARC

20.88 Undercuts and necks.

A An undercut is a groove cut inside a cylinder.

B A thread relief is a groove cut at the end of a thread to improve the screw's assembly. Dimension both types of necks as shown.

relief is a neck that has been cut at the end of a thread to ensure that the head of the threaded part will fit flush against the part it screws into (**Figure 20.88B**).

20.20 Tapers

Tapers for both flat planes and conical surfaces may be specified with either notes or symbols. Flat taper is the ratio of the difference in the heights at each end of a surface to its length (**Figure 20.89A** and **B**). Tapers on flat surfaces may be expressed as inches per inch

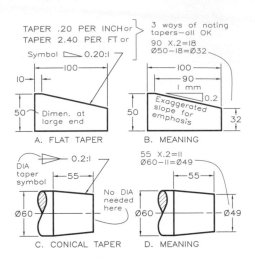

A. FLAT TAPER   B. MEANING

C. CONICAL TAPER   D. MEANING

20.89 Tapers may be specified for either flat or conical surfaces: dimensioning and interpretation for (A and B) a flat taper and (C and D) for a conical taper.

(.20 per inch), inches per foot (2.40 per foot), or millimeters per millimeter (0.20:1).

Conical taper is the ratio of the difference in the diameters at each end of a cone to its length (**Figure 20.89C** and **D**). Tapers on conical surfaces may be expressed as inches per inch (.25 per inch), inches per foot (3.00 per foot), or millimeters per millimeter (0.25:1).

20.21 Miscellaneous Notes

Notes on detail drawings provide information and specifications that would be difficult to represent by drawings alone (**Figures 20.90–20.92**). Place notes horizontally on the sheet whenever possible, because they are easier to letter and read in that position. Several notes in sequence on the same line should be separated with short dashes between them (for example, 15 DIA–30 DIA SPOTFACE). Use standard abbreviations in notes to save space and time.

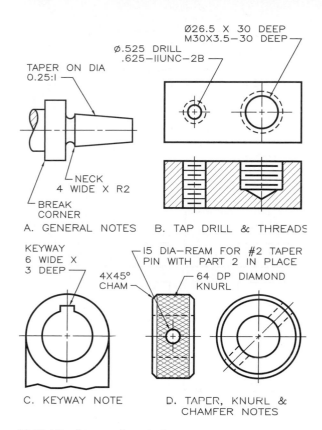

20.90 Miscellaneous dimensioning.

A The notes for this part indicate a neck, a taper, and a break corner, which is a slight round to remove corner sharpness.

B Threaded holes are sometimes dimensioned by giving the tap drill size in addition to the thread specifications, but selection of the tap drill size usually is left to the shop.

C This note is used for dimensioning a keyway.

D The notes for this collar call for knurling, chamfering, and drilling for a #2 taper pin.

Problems

1–48. (Figures 20.93 and **20.94)** Solve these problems on size A paper, one per sheet, if you draw them full size. If you draw them double size, use size B paper. The views are drawn on a 0.20-in. (5 mm) grid. You will need to vary the spacing between views to provide adequate room for the dimensions. Sketching the views and dimensions to determine the required spacing before laying out the solu-

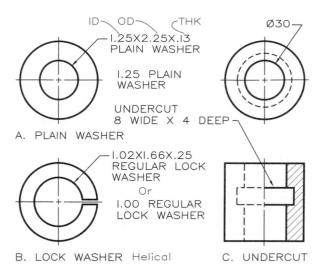

20.91 Washers and undercuts.

A and **B** Dimension washers and lock washers as shown by taking sizes from the tables in the Appendix 2.3.

C Dimension an undercut with a note.

tions with instruments would be helpful. Supply lines that may be missing in all views.

Supplementary Problems:
Problems at the ends of Chapters 13 and 14 can be used as dimensioning exercises to practice the application of the principles covered in this chapter.

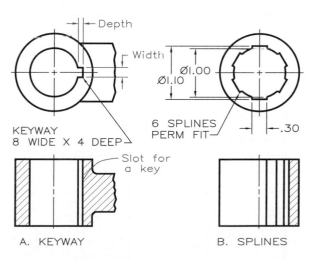

20.92 These drawings illustrate how to dimension (A) keyways and (B) splines.

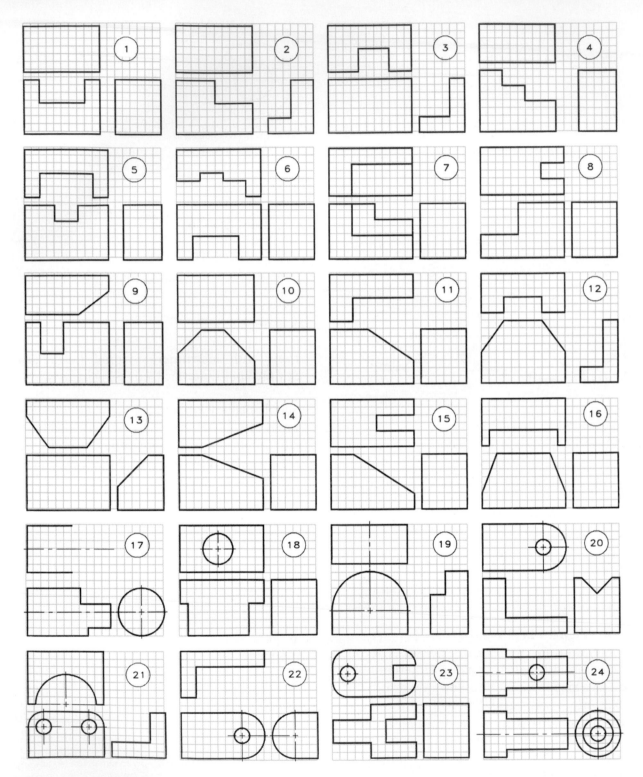

20.93

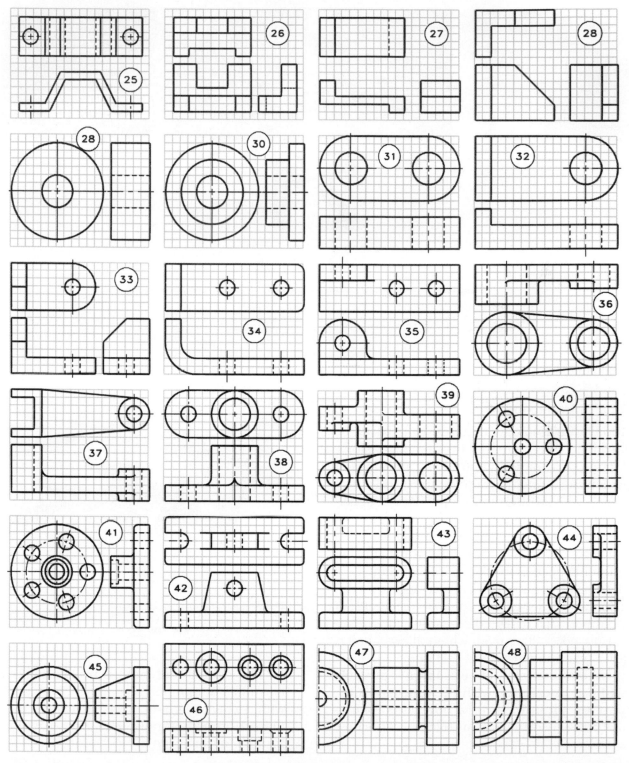

20.94

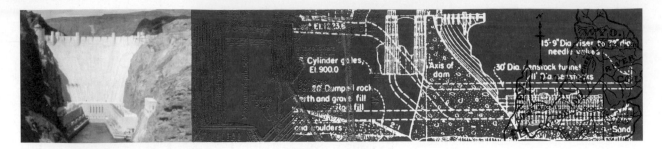

21

Tolerances

21.1 Introduction

Today's technology requires that parts be specified with increasingly exact dimensions. Many parts made by different companies at widely separated locations must be interchangeable, which requires precise size specifications and production.

The technique of dimensioning parts within a required range of variation to ensure interchangeability is called **tolerancing**. Each dimension is allowed a certain degree of variation within a specified zone, or tolerance. For example, a part's dimension might be expressed as 20 ± 0.50, which allows a tolerance (variation in size) of 1.00 mm.

A tolerance should be as large as possible without interfering with the function of the part to minimize production costs. Manufacturing costs increase as tolerances become smaller.

21.1 This cutting-tool holder would not function precisely while being used to make cuts on a lathe unless tolerances were used. All of its parts have been finished within close tolerances.

The cutting-tool holder in **Figure 21.1** illustrates a number of parts that must fit within a high degree of precision in order for it to work properly. It must hold the tool exactly in order to work properly with a lathe.

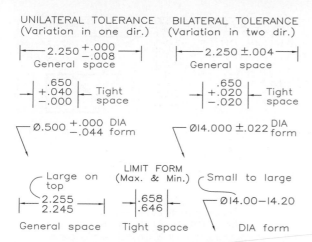

21.2 These are examples of properly applied tolerances in unilateral, bilateral, and limit forms for general and tight spaces and dimensions on leaders.

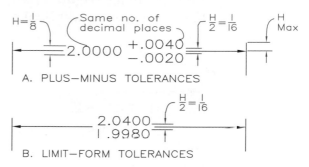

21.3 Place upper limits either above or to the right of lower limits. In plus-and-minus tolerancing, place the plus limits above the minus limits.

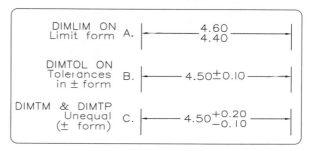

21.4 The spacing and proportions of numerals used to specify tolerances on dimensions.

21.2 Tolerance Dimensions

Three methods of specifying tolerances on dimensions, **unilateral**, **bilateral**, and **limit** forms, are shown in **Figure 21.2**. When plus-or-minus tolerancing is used, it is applied to a theoretical dimension called the **basic dimension**. When dimensions can vary in only one direction from the basic dimension (either larger or smaller), tolerancing is **unilateral**. Tolerancing that permits variation in both directions from the basic dimension (larger and smaller) is **bilateral**.

Tolerances may be given in limit form, with dimensions representing the largest and smallest sizes for a feature. When tolerances are shown in limit form, the basic dimension will be unknown.

The customary methods of applying tolerance values on dimension lines are shown in **Figure 21.3**. The spacing and proportions of the tolerance dimensions are shown in **Figure 21.4**.

Tolerances by Computer

AutoCAD provides an automatic means of showing toleranced dimensions in limit form or

DIMLIM ON Limit form	A.	4.60 / 4.40
DIMTOL ON Tolerances in ± form	B.	4.50±0.10
DIMTM & DIMTP Unequal (± form)	C.	4.50 +0.20 / -0.10

21.5 AutoCAD gives toleranced dimensions in these forms. When *Dimlim* is *On*, tolerances will be applied in limit form; when *Dimtol* is *On*, tolerances will be in plus-or-minus form.

plus-and-minus form (**Figure 21.5**). *Dimlim* must be turned *On* to obtain dimensions in limit form. Assign tolerances to the *Dimtm* and *Dimtp* modes under the *Dim* command.

In addition to linear dimensions, diametral and radial dimensions are automatically given with either a diameter symbol or an R preceding the dimensions (**Figures 21.6** and **21.7**). Angular measurements can be toleranced in the plus-and-minus form or the limit form (**Figure 21.8**). The program measures the angle and automatically computes the upper and lower limits.

21.3 Mating Parts

Mating parts must be toleranced to fit within a prescribed degree of accuracy (**Figure 21.9**). The upper part is dimensioned with limits indicating its maximum and minimum sizes. The slot in the lower part is toleranced to be

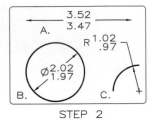

21.6 Limit tolerances by computer:

Step 1 Set the dimensioning variables as shown for tolerances to be given in *Limit* form.

Step 2 Linear dimensions will be in limit form, diameters preceded by a diameter sign, and radii preceded by an R.

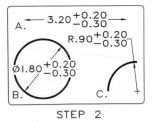

21.7 Plus-or-minus tolerances by computer:

Step 1 Set *Dimtol* to *On*, *Dimtp* (tolerance plus) to 0.2, and *Dimtm* (tolerance minus) to 0.3.

Step 2 Linear dimensions will give a basic diameter followed by plus-and-minus tolerances (deviation form), diameters as linear dimensions preceded by a diameter sign, and radii dimensions preceded by an R.

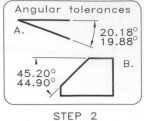

21.8 Angular tolerances by computer:

Step 1 Set the dimensioning variables as shown; *Dimdec > 2* assigns two decimal places.

Step 2 Select the lines forming the angles and the location of the arc, and the toleranced measurements will be given in limit form. Using the UNITS command, you may obtain angular measurements as decimal degrees, minutes and seconds, grads, radians, or surveyor's units.

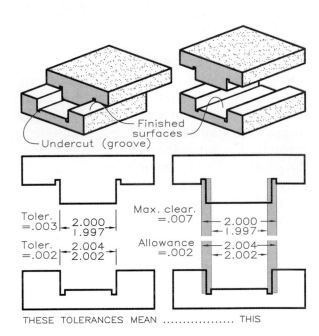

21.9 These mating parts have tolerances (variations in size) of 0.003″ and 0.002″ respectively. The allowance (tightest fit) between the assembled parts is 0.002″.

slightly larger, allowing the parts to assemble with a clearance fit that allows freedom of movement.

Mating parts may be cylindrical forms, such as a pulley, bushing, and shaft (**Figure 21.10**). The bushing should force fit inside the pulley to provide a good bearing surface for

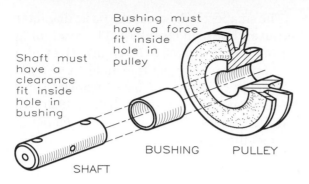

Bushing must have a force fit inside hole in pulley

Shaft must have a clearance fit inside hole in bushing

BUSHING   PULLEY

SHAFT

21.10 These parts must be assembled with cylindrical fits that give a clearance and an interference fit.

the rotating shaft. At the same time, the shaft and the bushing should mate so that the pulley and bushing will rotate on the shaft with a free running fit.

ANSI tables (see Appendices 29–38) prescribe cylindrical-fit tolerances for different applications. Familiarity with the terminology of cylindrical tolerancing is essential in order to apply the data from these tables.

21.4 Tolerancing Terms: English Units

The following terminology and definitions of tolerancing are illustrated in **Figure 21.11**, which shows two mating cylindrical parts.

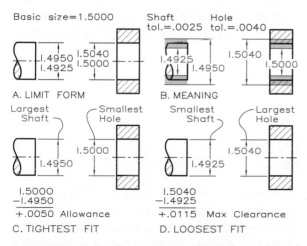

Basic size=1.5000   Shaft tol.=.0025   Hole tol.=.0040

A. LIMIT FORM   B. MEANING

C. TIGHTEST FIT   D. LOOSEST FIT

21.11 The allowance (tightest fit) between these assembled parts is +0.005″. The maximum clearance is +0.0115″.

Tolerance: The difference between the limits of size prescribed for a single feature, or 0.0025 in. for the shaft and 0.0040 for the hole in **Figure 21.11A**.

Limits of tolerance: The maximum and minimum sizes of a feature, or 1.4925 and 1.4950 for the shaft and 1.5000 and 1.5040 for the hole in **Figure 21.11B**.

Allowance: The tightest fit between two mating parts, or +0.0050 in **Figure 21.11C**. Allowance is negative for an interference fit.

Nominal size: A general size of a shaft or hole, usually expressed with common fractions, 1–1/2 in. or 1.50 in. as in **Figure 21.11**.

Basic size: The size to which plus-and-minus tolerance is applied to obtain the limits of size, 1.5000 in **Figure 21.11**. The basic diameter cannot be determined from tolerances that are expressed in limit form.

Actual size: The measured size of the finished part.

Fit: The degree of tightness or looseness between two assembled parts, which can be one of the following: **clearance**, **interference**, **transition,** and **line**.

Clearance fit gives a clearance between two assembled mating parts. The shaft and the hole in **Figure 21.11C** and **D** have a minimum clearance of 0.0050″ and a maximum clearance of 0.0115″.

Interference fit results in a binding fit that requires the parts to be forced together, much as if they are welded (**Figure 21.12A**).

Transition fit may range from an interference to a clearance fit between the assembled parts. The shaft may be either smaller or larger than the hole and still be within the prescribed tolerances, as in **Figure 21.12B**.

Line fit results in surface contact or clearance when the limits are reached (**Figure 21.12C**).

Selective assembly: A method of selecting and assembling parts by hand by trial and

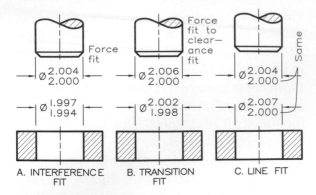

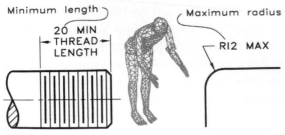

21.12 Three types of fits between mating parts are shown here in addition to the clearance fit shown in Figure 21.11.

error that allows parts to be made with larger tolerances at less cost as a compromise between a high manufacturing accuracy and ease of assembly.

Single-limits: Dimensions designated by either minimum (MIN) or maximum (MAX) as shown in **Figure 21.13**. Depths of holes, lengths, threads, corner radii, and chamfers are sometimes dimensioned in this manner.

21.5 Basic Hole System

The **basic hole system** uses the smallest hole size as the **basic diameter** for calculating tolerances and allowances. The basic hole system is best when drills, reamers, and machine tools are used to give precise hole sizes.

21.13 Single tolerances in maximum (MAX) or minimum (MIN) form can be given in applications of this type.

The smallest hole size is the basic diameter because a hole can be enlarged by machining but not reduced in size. In **Figure 21.11**, the smallest diameter of the hole is 1.500″. Subtract the allowance, 0.0050″, from it to find the diameter of the largest shaft, 1.4950″. To find the smallest limit for the shaft diameter, subtract the tolerance from 1.4950″.

21.6 Basic Shaft System

The **basic shaft system** uses the largest diameter as the **basic diameter** to which the tolerances are applied. This system is applicable when shafts are available in uniform standard sizes.

The largest shaft size is used as the basic diameter because shafts can be machined to a smaller size but not enlarged. For example, if the largest permissible shaft size is 1.500″, add the allowance to this dimension to obtain the smallest hole diameter into which the shaft fits. If the parts are to have an allowance of 0.0040″, the smallest hole would have a diameter of 1.5040″.

21.7 Cylindrical Fits

The *ANSI B4.1* standard gives a series of fits between cylindrical features in inches for the basic hole system. The types of fit covered in this standard are:

RC: running or sliding clearance fits

LC: clearance locational fits

LT: transition locational fits

LN: interference locational fits

FN: force and shrink fits

Appendices 29–33 list these five types of fit, each of which has several classes.

Running or sliding clearance fits (RC) provide a similar running performance, with suitable lubrication allowance. The clearance for the first two classes (RC 1 and RC 2), which are used chiefly as slide fits, increases more slowly with diameter size than other classes to maintain an accurate location even at the expense of free relative motion.

Locational fits (LC, LT, LN) determine only the location of mating parts; they may provide non-moving rigid locations (interference fits) or permit some freedom of location (clearance fits). The three locational fits are: **clearance fits** (**LC**), **transition fits** (**LT**), and **interference fits** (**LN**).

Force fits (FN) are interference fits characterized by a constant bore pressure throughout the range of sizes. There are five types of force fits: FN1 through FN5 varying from a light drive to heavier drives, respectively.

The method of applying tolerance values from the tables in Appendix 29 for an RC9 fit (basic hole system) is shown in **Figure 21.14**.

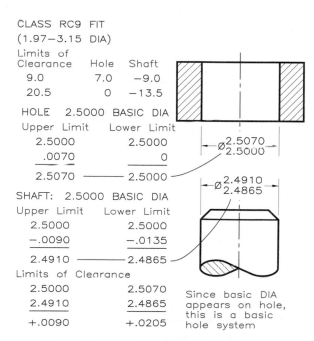

CLASS RC9 FIT
(1.97–3.15 DIA)

Limits of Clearance	Hole	Shaft
9.0	7.0	−9.0
20.5	0	−13.5

HOLE   2.5000 BASIC DIA

Upper Limit	Lower Limit
2.5000	2.5000
.0070	0
2.5070	2.5000

SHAFT: 2.5000 BASIC DIA

Upper Limit	Lower Limit
2.5000	2.5000
−.0090	−.0135
2.4910	2.4865

Limits of Clearance

2.5000	2.5070
2.4910	2.4865
+.0090	+.0205

Since basic DIA appears on hole, this is a basic hole system

21.14 This example shows how to calculate limits and allowances for an RC9 fit between a shaft and hole with a basic diameter of 2.5000″. Values are taken from Appendix 29.

The basic diameter of 2.5000″ falls between 1.97″ and 3.15″ in the size column of the table. Limits are in thousandths, which requires that the decimal point be moved three places to the left. For example, +7 is +0.0070″.

Add the limits to the basic diameter when a **plus sign** precedes the values and **subtract** when a **minus sign** is given. Add the limits (+0.0070″ and 0.0000″) to the basic diameter to find the upper and lower limits of the hole (2.5070″ and 2.5000″). Subtract the limits (−0.0090″ and −.0135″) from the basic diameter to find the limits of the shaft (2.4910″ and 2.4865″).

The tightest fit (the allowance) between the assembled parts (+0.0090″) is the difference between the largest shaft and the smallest hole. The loosest fit (+0.0205″) is the difference between the smallest shaft and the largest hole. These values appear in the *Limit* column of the table (Appendix 29).

This method of using tables of fits is applied to other types of fits by using their respective tables: force fit, interference fit, transition fit, and locational fit. Subtract negative limits from the basic diameter and add positive limits to it. A **minus sign** preceding limits of clearance in the tables indicates an **interference fit** between the assembled features, and a **positive sign** preceding limit of clearance indicates a **clearance fit**.

21.8 Tolerancing: Metric Units

The system recommended by the *International Standards Organization* (*ISO*) in *ANSI B4.2* for metric measurements are fits that usually apply to cylinders—holes and shafts—but these tables can be used to specify fits between parallel contact surfaces such as a key in a slot.

Basic size: The size is usually a diameter from which limits or deviations are calculated

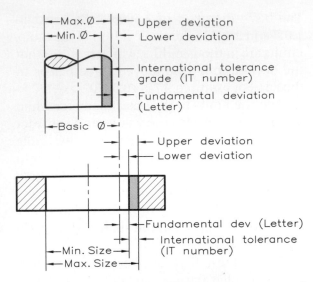

21.15 The terminology and definitions of the metric system of cylindrical fits are given here.

First Choice	Second Choice	First Choice	Second Choice	First Choice	Second Choice
1		10		100	
	1.1		11		110
1.2		12		120	
	1.4		14		140
1.6		16		160	
	1.8		18		180
2		20		200	
	2.2		22		220
2.5		25		250	
	2.8		28		280
3		30		300	
	3.5		35		350
4		40		400	
	4.5		45		450
5		50		500	
	5.5		55		550
6		60		600	
	7		70		700
8		80		800	
	9		90		900
				1000	

21.16 Basic sizes for metric fits selected first from the first-choice column are preferred over those in the second-choice column.

(**Figure 21.15**). Select basic sizes from the *First Choice* column in the table in **Figure 21.16**.

Deviation: The difference between the hole or shaft size and the basic size.

Upper deviation: The difference between the maximum permissible size of a part and its basic size (**Figure 21.15**).

Lower deviation: The difference between the minimum permissible size of a part and its basic size (**Figure 21.15**).

Fundamental deviation: The deviation closest to the basic size (**Figure 21.15**). In the note 40 H8 in **Figure 21.17**, H represents the fundamental deviation for a hole, and in the note 40 f7, the f represents the fundamental deviation for a shaft.

Tolerance: The difference between the maximum and minimum allowable sizes of a part.

International tolerance (IT) grade: A series of tolerances that vary with basic size to provide a uniform level of accuracy within a given grade (**Figure 21.15**). In the note 40 H8 in **Figure 21.17**, the 8 represents the IT grade. There are 18 IT grades: IT01, IT0, IT1, ... , IT16.

Tolerance zone: A combination of the fundamental deviation and the tolerance grade. The H8 portion of the 40 H8 note in **Figure 21.17** is the tolerance zone.

Hole basis: A system of fits based on the minimum hole size as the basic diameter. The fundamental deviation for a hole-basis system is the uppercase "H." Appendices 35 and 36 give hole-basis data for tolerances.

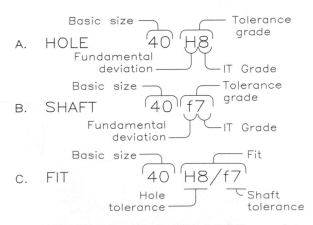

21.17 These tolerance symbols and their definitions apply to holes and shafts.

Shaft basis: A system of fits based on the maximum shaft size as the basic diameter. The fundamental deviation for a shaft-basis system is the lowercase "h." Appendices 37 and 38 give shaft-basis data for tolerances.

Clearance fit: A fit resulting in a clearance between two assembled parts under all tolerance conditions.

Interference fit: A force fit between two parts, requiring that they be driven together.

Transition fit: A fit that may result in either a clearance or an interference between assembled parts.

Tolerance symbols: Notes giving the specifications of tolerances and fits (**Figure 21.17**). The basic size is a number, followed by the fundamental deviation letter and the IT number, which give the tolerance zone. **Uppercase letters** (H) indicate the fundamental deviations for **holes**, and **lowercase letters** (h) indicate fundamental deviations for **shafts**.

Preferred Sizes and Fits

The table in **Figure 21.16** shows the preferred basic sizes for computing tolerances. Under the *First Choice* heading, each number increases by about 25% from the preceding value. Each number in the *Second Choice* column increases by about 12%. To minimize cost, select basic diameters from the first column because they correspond to standard stock sizes for round, square, and hexagonal metal products.

Figure 21.18 shows preferred clearance, transition, and interference fits for the hole-basis and shaft-basis systems. Appendices 35–38 contain the complete tables.

Preferred Fits: Hole-Basis System

The preferred fits for the hole-basis system, in which the smallest hole is the basic diameter, are

Hole Basis	Shaft Basis	Description
H11/c11	C11/h11	Loose Running Fit for wide commerical tolerances on external members
H9/d9	D9/h9	Free Running Fit for large temperature variations, high running speeds, or high journal pressures
H8/f7	F8/h7	Close Running Fit for accurate location and moderate speeds and journal pressures
H7/g6	G7/h6	Sliding Fit for accurate fit and location and free moving and turning, not free running
H7/h6	H7/h6	Locational Clearance for snug fits for parts that can be freely assembled
H7/k6	K7/h6	Locational Transition Fit for accurate locations
H7/n6	N7/h6	Locational Transition Fit for more accurate locations and greater interference
H7/p6	P7/h6	Locational Interference Fit for rigidity and alignment without special bore pressures
H7/s6	S7/h6	Medium Drive Fit for shrink fits on light sections; tightest fit usable for cast iron
H7/u6	U7/h6	Force Fit for parts that can be highly stressed and for shrink fits.

Clearance Fits span H11/c11 through H7/g6. *Transition Fits* span H7/h6 through H7/n6. *Interference Fits* span H7/p6 through H7/u6.

21.18 This list gives the preferred hole-basis and shaft-basis fits for the metric system.

shown in **Figure 21.19**. **Clearance, transition,** and **interference fits** are options of the hole-basis system. **Figure 21.20** illustrates the preferred fits for a hole-basis system where the lower deviation of the hole is zero, which means that the smallest hole is the basic size. Variations in fit between parts range from a clearance fit of H11/c11 to an interference fit of H7/u6 (see **Figure 21.18**).

Preferred Fits: Shaft-Basis System The preferred fits of the shaft-basis system, in which the largest shaft is the basic diameter, are shown in **Figure 21.20**. Variations in fit range from a clearance fit of C11/h11 to an interference fit of U7/h6 (see **Figure 21.17**).

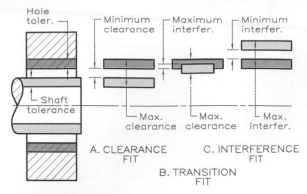

21.19 Types of fits: (A) clearance fit, where clearance is between the parts, (B) transition fit, where there can be either interference or clearance, and (C) interference fit, where the parts must be forced together.

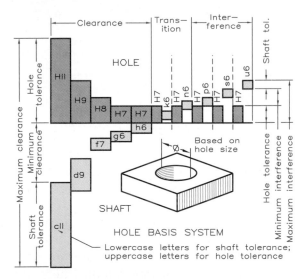

21.20 This diagram illustrates the preferred fits for the hole-basis system listed in Figure 21.18. Appendices 35 and 36 give values for these fits.

Standard Cylindrical Fits

The following examples demonstrate how to calculate and apply tolerances to cylindrical parts. You must use **Figure 21.16**, **Figure 21.18**, and data from Appendix 35.

Example 1 (Figure 21.21)

Required: Use the hole-basis system, a close running fit, and a basic diameter of 49 mm.

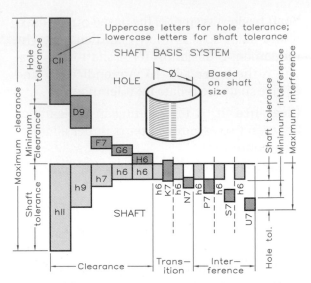

21.21 This diagram illustrates the preferred fits for a shaft-basis system listed in Figure 21.18. Appendices 37 and 38 give values for these fits.

Solution: Use a preferred basic diameter of 50 mm (**Figure 21.18**) and fit of H8/f7 (**Figure 21.20**).

Hole: Find the upper and lower limits of the hole in Appendix 35 under H8 and across from 50 mm. These limits are 50.000 and 50.039 mm.

Shaft: Find the upper and lower limits of the shaft under f7 and across from 50 mm in Appendix 35. These limits are 49.950 and 49.975 mm.

Symbols: **Figure 21.22** shows how to apply toleranced dimensions to the hole and shaft.

Example 2 (Figure 21.23)

Required: Use the hole-basis system, a location transition fit, and a basic diameter of 57 mm.

Solution: Use a preferred basic diameter of 60 mm (**Figure 21.18**) and a fit of H7/k6 (**Figure 21.20**).

Hole: Find the upper and lower limits of the hole in Appendix 36 under H7 and across from 60 mm. These limits are 60.000 and 60.030 mm.

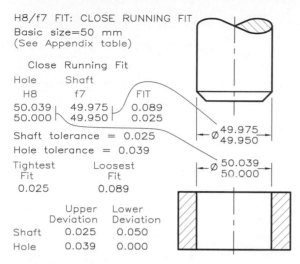

H8/f7 FIT: CLOSE RUNNING FIT
Basic size=50 mm
(See Appendix table)

Close Running Fit

Hole	Shaft	
H8	f7	FIT
50.039	49.975	0.089
50.000	49.950	0.025

Shaft tolerance = 0.025
Hole tolerance = 0.039

Tightest Fit	Loosest Fit
0.025	0.089

	Upper Deviation	Lower Deviation
Shaft	0.025	0.050
Hole	0.039	0.000

Ø 49.975 / 49.950

Ø 50.039 / 50.000

21.22 This drawing shows how to calculate and apply metric limits and fits to a shaft and hole (Appendix 35).

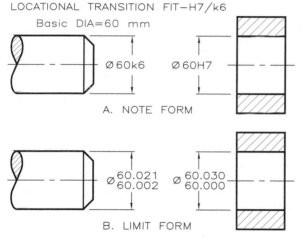

LOCATIONAL TRANSITION FIT—H7/k6

Basic DIA=60 mm

Ø60k6   Ø60H7

A. NOTE FORM

Ø 60.021 / 60.002   Ø 60.030 / 60.000

B. LIMIT FORM

21.23 These methods are used to apply metric tolerances to a hole and a shaft with a locational transition fit (Appendix 36).

Shaft: Find the upper and lower limits of the shaft under k6 and across from 60 mm in Appendix 36. These limits are 60.021 and 60.002 mm.

Symbols: **Figure 21.23** shows two methods of applying the tolerance symbols to a drawing.

Example 3 (Figure 21.24)

Required: Use the hole-basis system, a medium drive fit, and a basic diameter of 96 mm.

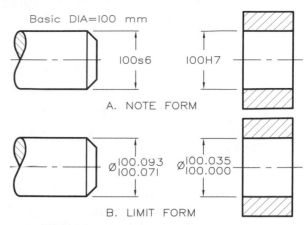

Basic DIA=100 mm

100s6   100H7

A. NOTE FORM

Ø 100.093 / 100.071   Ø 100.035 / 100.000

B. LIMIT FORM

21.24 Either of these formats can be used to apply metric tolerances to a hole and shaft that have an interference fit (Appendix 41).

Solution: Use a preferred basic diameter of 100 mm (**Figure 21.16**) and a fit of H7/s6 (**Figure 21.18**).

Hole: Find the upper and lower limits of the hole in Appendix 36 under H7 and across from 100 mm. These limits are 100.035 and 100.000 mm.

Shaft: Find the upper and lower limits of the shaft under s6 and across from 100 mm in Appendix 36. These limits are 100.093 and 100.071 mm. Appendix 36 gives the tightest fit as an interference of −0.093 mm, and the loosest fit as an interference of −0.036 mm. Minus signs in front of these numbers indicate an interference fit.

Symbols: **Figure 21.24** shows two methods of applying toleranced dimensions to the hole and shaft.

Nonstandard Fits: Nonpreferred Sizes

Limits of tolerances for non-standard sizes can be calculated for any of the preferred fits shown in **Figure 21.18** for non-standard sizes that do not appear in Appendixes 35–38. Limits of tolerances for nonstandard hole sizes are in Appendix 39, and limits of tolerances for nonstandard shaft sizes are in Appendix 40.

Figure 21.25 shows the hole and shaft limits for an H8/f7 fit and a 45-mm DIA. The tolerance

CALCULATION OF NONSTANDARD LIMITS

FIT: H8/f7   Ø45 BASIC DIA

From Appendix				
Hole	Shaft	Hole Limits		45.039
H8	f7			45.000
0.039	−0.025	Shaft Limits		44.975
0.000	−0.050			44.950

21.25 This calculation is for an H8/f7 non-standard diameter of 45 mm (Appendices 39 and 40).

limits of 0.000 and 0.039 mm for an H8 hole are from Appendix 39, across from the size range of 40–50 mm. The tolerance limits of −0.025 and −0.050 mm for the shaft are from Appendix 40. Calculate the hole limits by adding the positive tolerances to the 45-mm basic diameter and the shaft limits by subtracting the negative tolerances from the 45-mm basic diameter.

21.9 Chain versus Datum Dimensions

When parts are dimensioned to locate surfaces or geometric features by a chain of dimensions laid end-to-end (**Figure 21.26A**), variations may accumulate in excess of the specified tolerance. For example, the tolerance between surfaces A and B is 0.02, between A and C it is 0.04, and between A and D it is 0.06.

Tolerance accumulation can be eliminated by measuring from a single plane called a **datum plane** or **baseline**. A datum plane is usually on the object, but it can also be on the machine used to make the part. Because each plane in **Figure 21.26B** is located with respect to a datum plane, the tolerances between the intermediate planes do not exceed the maximum tolerance of 0.02. Always base the application of tolerances on the function of a part in relationship to its mating parts.

Origin Selection

When the shorter of two surfaces is to be the mounting surface, it should be specified as

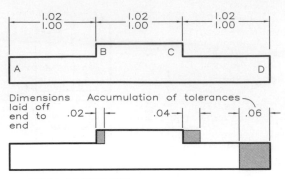

A. CHAIN DIMENSIONS

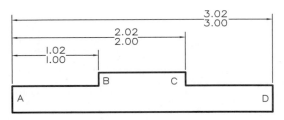

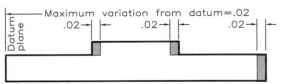

B. DATUM PLANE (BASELINE) DIMENSIONS

21.26 Chain versus datum dimensioning.

A Dimensions given end to end in a chain fashion may result in an accumulation of tolerances of up to 0.06″ at D instead of the specified 0.02″.

B When dimensioned from a single datum, the variations of B, C, and D cannot deviate more than the specified 0.02″ from the datum.

the origin for locating a parallel surface. Had the longer surface been chosen as the origin surface, the shorter surface would have had a greater angular variation from the same tolerance zone (**Figure 21.27**).

21.10 Tolerance Notes

You should tolerance all dimensions on a drawing either by using the rules previously discussed or by placing a note in or near the title block. For example, the note

TOLERANCE ±1/64

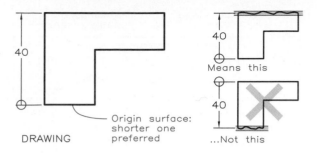

DRAWING

Origin surface: shorter one preferred

40

Means this

40

...Not this

21.27 Selection of the shorter surface as the origin surface for locating a longer parallel surface gives the greatest accuracy.

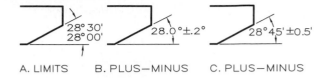

A. LIMITS   B. PLUS—MINUS   C. PLUS—MINUS

21.28 Tolerances on angles can be specified by one of these methods.

might be given on a drawing for less critical dimensions.

Some industries give dimensions in inches with two, three, and four decimal place fractions. A note for dimensions with two and three decimal places might be given on the drawing as

TOLERANCES XX.XX ±0.10

XX.XXX ±0.005.

Tolerances of four places would be given directly on the dimension lines.

The most common method of noting tolerances is to give as large a tolerance as feasible in a note, such as

TOLERANCES ±0.05

and to give tolerances on the dimension lines for dimensions requiring closer tolerances. Give angular tolerances in a general note in or near the title block, such as

ANGULAR TOLERANCES ±0.50° or 30′.

Use one of the formats shown in **Figure 21.28** to give specific angular tolerances directly on angular dimensions.

21.11 General Tolerances: Metric Units

All dimensions on a drawing must be specified within certain tolerance ranges when they are not shown on dimension lines. Tolerances not shown on dimension lines

should be specified by a general tolerance note on the drawing.

Linear Dimensions: Tolerance linear dimensions by indicating plus and minus (±) one half of an international tolerance (IT) grade as given in Appendix 34. You may select the IT grade from the chart in **Figure 21.29**, where IT grades for mass-produced items range from IT12 through IT16. IT grades can be selected from **Figure 21.30** for a particular machining process.

General tolerances using IT grades may be expressed in a note as follows:

UNLESS OTHERWISE SPECIFIED

ALL UNTOLERANCED

DIMENSIONS ARE IT14.

This note means that a tolerance of ±0.700 mm is allowed for a dimension between 315 and 400 mm. The value of the tolerance 1.400 mm is extracted from Appendix 34.

Figure 21.31 shows recommended tolerances for fine, medium, and coarse series for ranges of size. A medium tolerance, for example, can be specified by the following note:

GENERAL TOLERANCES SPECIFIED

IN ANSI B4.3 MEDIUM SERIES APPLY.

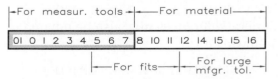

21.29 International tolerance (IT) grades and their applications are shown here. See Appendix 34 to obtain IT tolerance grade numerical values.

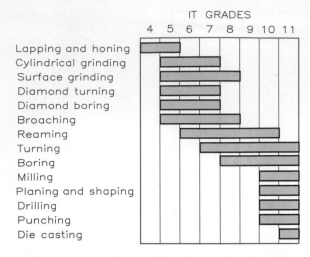

IT GRADES
4 5 6 7 8 9 10 11

Lapping and honing
Cylindrical grinding
Surface grinding
Diamond turning
Diamond boring
Broaching
Reaming
Turning
Boring
Milling
Planing and shaping
Drilling
Punching
Die casting

21.30 International tolerance (IT) values may be selected from this table, which is based on the general capabilities of various machining processes.

GENERAL TOLERANCES:
LINEAR DIMENSIONS (mm)

Basic Dimensions	Fine Series	Medium Series	Coarse Series
0.5 to 3	± 0.05	± 0.1	— —
Over 3 to 6	± 0.05	± 0.1	± 0.2
Over 6 to 30	± 0.1	± 0.2	± 0.5
Over 30 to 120	± 0.15	± 0.3	± 0.8
Over 120 to 315	± 0.2	± 0.5	± 1.2
Over 315 to 1000	± 0.3	± 0.8	± 2
Over 1000 to 2000	± 0.5	± 1.2	± 3

21.31 You may select general tolerance values from this table for fine, medium, and coarse series. Tolerances vary with sizes of dimensions.

Equivalent tolerances may be given in table form (**Figure 21.32**) on the drawing, with the grade (medium in this example) selected from **Figure 21.31**. General tolerances may be given in a table for dimensions expressed with one or no decimal places (**Figure 21.33**).

General tolerances may also be notated in the following form:

UNLESS OTHERWISE SPECIFIED

ALL UNTOLERANCED DIMENSIONS

ARE ±0.8 mm.

Get values from previous table   Specifies a medium series

GENERAL TOLERANCES (mm)
UNLESS OTHERWISE SPECIFIED, THE FOLLOWING TOLERANCES ARE APPLICABLE

LINEAR	Over to	0.5 6	6 30	30 120	120 315	315 1000	1000 2000
TOL.	±	0.1	0.2	0.3	0.5	0.8	1.2

21.32 This table for a medium series of values was extracted from Figure 21.31 for insertion on a working drawing to provide the tolerances for a medium series of sizes.

Medium series for numbers with one decimal place   Coarse series for numbers with no decimal places

GENERAL TOLERANCES (mm)
UNLESS OTHERWISE SPECIFIED, THE FOLLOWING TOLERANCES ARE APPLICABLE

LINEAR	OVER TO	— 120	120 315	315 1000	1000 —
TOL.	ONE DECIMAL ±	0.3	0.5	0.8	1.2
	NO DECIMALS ±	0.8	1.2	2	3

21.33 Placed on a drawing, this table of tolerances would indicate the tolerances for dimensions having one or no decimal places, such as 24.0 and 24, denoting medium and coarse series.

Length of shorter leg (mm)	Up to 10	Over 10 to 50	Over 50 to 120	Over 120 to 400
Degrees	± 1°	± 0° 30'	± 0°20'	± 0°10'
mm per 100	± 1.8	± 0.9	± 0.6	± 0.3

21.34 General tolerances for angular and taper dimensions may be taken from this table of values.

Use this method only when the dimensions on a drawing are similar in size.

Angular Tolerances: Express angular tolerances as (1) an angle in decimal degrees or in degrees and minutes, (2) a taper expressed in percentage (mm per 100 mm), or (3) milliradians. (To find milliradians, multiply the degrees of an angle by 17.45.) **Figure 21.34** shows the suggested tolerances for decimal degrees and taper, based on the length of the shorter leg of the angle. General angular tolerances may be notated on the drawing as follows:

UNLESS OTHERWISE SPECIFIED

THE GENERAL TOLERANCES

IN ANSI B4.3 APPLY.

A second method involves showing a portion of the table from **Figure 21.34** as a table of tolerances on the drawing (**Figure 21.35**). A third method is a note with a single tolerance such as:

UNLESS OTHERWISE SPECIFIED

ANGULAR TOLERANCES ARE ±0°30′

21.12 Geometric Tolerances

Geometric tolerancing is a system that specifies tolerances that control location **form**, **profile**, **orientation**, **location**, and **runout** on a dimensioned part as covered by the *ANSI Y14.5M Standards* and the *Military Standards* (Mil-Std) of the U.S. Department of Defense. Before discussing those types of tolerancing, however, we need to introduce you to symbols, size limits, rules, three-datum-plane concepts, and applications.

Symbols

The most commonly used symbols for representing geometric characteristics of dimensioned drawings are shown in **Figure 21.36**. The proportions of feature control symbols in relation to their feature control frames, based

ANGULAR TOLERANCES				
LENGTH OF SHORTER LEG (mm)	UP TO 10	OVER 10 TO 50	OVER 50 TO 120	OVER 120 TO 400
TOLERANCE	±1°	±0°30′	±0°20′	±0°10′

Values in degrees and minutes taken from previous table

21.35 This table, extracted from Figure 21.34, is placed on the drawing to indicate the general tolerance for angles in degrees and minutes.

GEOMETRIC SYMBOLS

Tolerance		Characteristic	Symbol
INDIVIDUAL FEATURES	Form	Straightness	—
		Flatness	▱
		Circularity	○
		Cylindricity	⌭
BOTH	Profile	Profile: Line	⌒
		Profile: Surface	⌓
RELATED FEATURES	Orientation	Angularity	∠
		Perpendicularity	⊥
		Parallelism	//
	Location	Position	⊕
		Concentricity	◎
		Symmetry	≡
	Runout	Runout: Circular	↗
		Runout: Total	↗↗

21.36 These symbols specify the geometric characteristics of a part's features.

on the letter height, are shown in **Figure 21.37**. On most drawings, a 1/8-in. or 3-mm letter height is recommended. Examples of feature control frames and their proportions are shown in **Figure 21.38**.

Size Limits

Three conditions of size are used when geometric tolerances are applied: **maximum material condition**, **least material condition**, and **regardless of feature size**.

Maximum material condition (MMC) indicates that a feature contains the maximum amount of material. For example, the shaft shown in **Figure 21.39** is at MMC when it has the largest permitted diameter of 24.6 mm. The hole is at MMC when it has the most material, or the smallest diameter of 25.0 mm.

Least material condition (LMC) indicates that a feature contains the least amount of

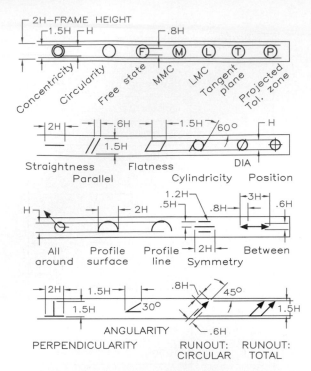

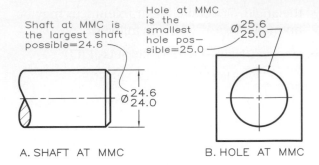

A. SHAFT AT MMC   B. HOLE AT MMC

21.39 A shaft is at maximum material condition (MMC) when it is at the largest size permitted by its tolerance. A hole is at MMC when it is at its smallest size.

21.13 Rules for Tolerancing

Two general rules of tolerancing geometric features should be followed.

Rule 1 (Individual Feature of Size): When only a tolerance of size is specified on a feature, the limits of size control the variation in its geometric form. The forms of the shaft and hole shown in **Figure 21.40** are permitted to vary within the tolerance ranges of the dimensions.

Rule 2 (All Applicable Geometric Tolerances): Where no modifying symbol is specified, RFS (regardless of features size) applies with respect

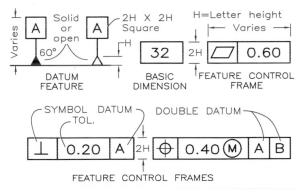

21.37 The proportions of these feature control symbols are based on the letter height, usually 1/8 inch high.

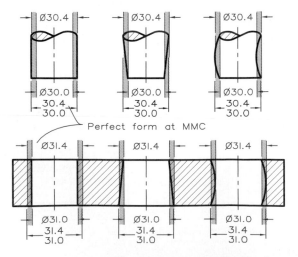

21.40 When only a tolerance of size is specified on a feature, the limits prescribe the form of the features, as shown for these shafts and holes having identical limits.

21.38 These are examples of geometric tolerancing frames and feature control symbols used to indicate geometric tolerances. H is the letter height.

material. The shaft in **Figure 21.39** is at LMC when it has the smallest diameter of 24.0 mm. The hole is at LMC when it has the least material, or the largest diameter of 25.6 mm.

Regardless of feature size (RFS) indicates that tolerances apply to a geometric feature regardless of its size ranging from MMC to LMC.

to the individual tolerance, datum reference, or both, Where required on a drawing, the modifiers MMC or LMC must be specified.

Alternate Practice: For a tolerance of position, RFS may be specified on the drawing with respect to the individual tolerance, datum reference, or both.

The specification of symmetry of the part in **Figure 21.41** is based on a tolerance at RFS from a datum at RFS.

Three-Datum-Plane Concept

A datum plane is used as the origin of a part's features that have been toleranced. Datum planes usually relate to manufacturing equipment, such as machine tables or locating pins.

Three mutually perpendicular datum planes are required to dimension a part accurately. For example, the part shown in **Figure 21.42** sits on the primary datum plane, with at least three points of its base in contact with the datum. The part is related to the secondary plane by at least two contact points. The third (tertiary) datum is in contact with at least one point on the object.

The priority of datum planes is presented in sequence in feature control frames. For example in **Figure 21.43**, the primary datum is surface A, the secondary datum is surface B,

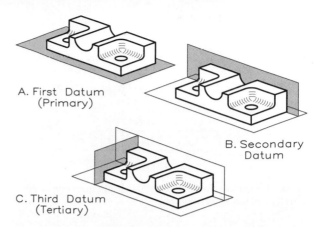

A. First Datum (Primary)

B. Secondary Datum

C. Third Datum (Tertiary)

21.42 When an object is referenced to a primary datum plane, it contacts the datum at at least three points. The vertical surface contacts the secondary datum plane at at least two points. The third surface contacts the third datum at at least one point. Datum planes are listed in order of priority in the feature control frame.

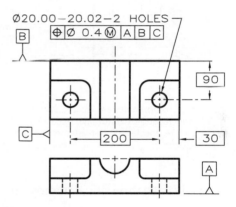

21.43 The three planes of the reference system are noted where they appear as edges. The primary datum plane (A) is given first in the feature control frame; the secondary plane (B), second; and the tertiary plane (C), third. Single numbers in frames are basic dimensions.

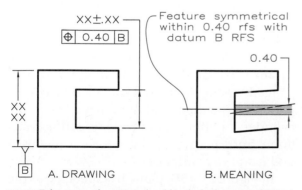

A. DRAWING   B. MEANING

21.41 Tolerances of position should include the note of M or L to indicate maximum material condition or least material condition. When dimensioned in this manner, tolerances of position of 0.40 will be applied regardless of feature size (RFS).

and the tertiary datum is surface C. **Figure 21.44** lists the order of priority of datum planes A–C sequentially in the feature control frames.

21.14 Cylindrical Datum Features

A part with a cylindrical datum feature that is the axis of a cylinder is illustrated in **Figure 21.45**. Datum K is the primary datum. Datum M is

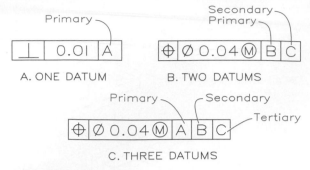

A. ONE DATUM

B. TWO DATUMS

C. THREE DATUMS

21.44 Use feature control frames to indicate from one to three datum planes in order of priority.

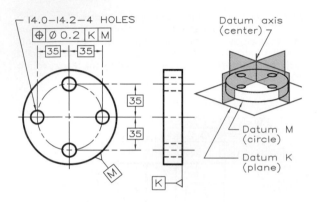

21.45 These true-position holes are located with respect to primary datum K and secondary datum M. Because datum M is a circle, the holes are located about two intersecting datum planes at the crossing centerlines in the circular view, satisfying the three-plane concept.

associated with two theoretical planes—the second and third in a three-plane relationship.

The two theoretical planes are represented in the circular view by perpendicular centerlines that intersect at the point view of the datum axis. All dimensions originate from the datum axis perpendicular to datum K; the other two intersecting datum planes are used for measurements in the x and y directions.

The priority of the datum planes in the feature control frame is significant in the manufacturing and inspection processes. The part shown in **Figure 21.46** is dimensioned in three ways to show the effects of datum-plane selection and material condition on the location of the hole pattern.

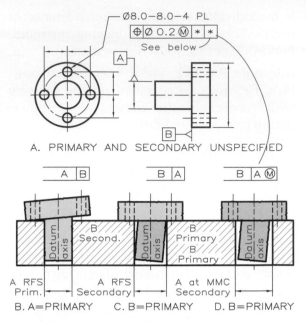

A. PRIMARY AND SECONDARY UNSPECIFIED

B. A=PRIMARY   C. B=PRIMARY   D. B=PRIMARY

21.46 For the unspecified datum planes in (A), examples (B)–(D) illustrate the effects of selecting the datum planes in order of priority and of RFS and MMC.

The effect of specifying diameter A at RFS as the primary datum plane and surface B as the secondary datum plane is illustrated in **Figure 21.46B**. During production, the part is centered on cylinder A. The part is mounted in a chuck, mandrel, or centering device on the processing equipment, which centers the part at RFS. Any variation from perpendicular in surfaces A and B will affect the degree of contact of surface B with its datum plane.

If surface B were specified as the primary datum feature, it would contact datum plane B at no fewer than three points (**Figure 21.46C**). The axis of datum cylinder A will be gauged by the smallest cylinder that is perpendicular to the first datum that will contact cylinder A at RFS. This cylinder identifies variation from perpendicular between planes A and B and size variations.

In **Figure 21.46D**, plane B is specified as the primary datum feature and cylinder A as the secondary datum feature at MMC. The part is mounted on the processing equipment so that at least three points on feature B come into contact with datum B. The datum axis is the axis of a circumscribed cylinder of a fixed size that is perpendicular to datum B. Using the modifier to specify MMC gives a more liberal tolerance zone than when RFS is specified.

Datum Features at RFS

When size dimensions are applied to a feature at RFS, the processing equipment that comes into contact with surfaces of the part establishes the datum. Variable machine elements, such as chucks or center devices, are adjusted to fit the external or internal features and establish datums.

Primary Diameter Datums: For an external cylinder (shaft) at RFS, the datum axis is the axis of the smallest circumscribed cylinder that contacts the cylindrical feature (**Figure 21.47A**). That is, the largest diameter of the part making contact with the smallest cylinder of the machine element holding the part is the datum axis.

For an internal cylinder (hole) at RFS, the datum axis is the axis of the largest inscribed cylinder making contact with the hole. That is,

the smallest diameter of the hole making contact with the largest cylinder of the machine element inserted in the hole is the datum axis (**Figure 21.47B**).

Primary External Parallel Datums: The datum for external features at RFS is the center plane between two parallel planes, at minimum separation, that contact the planes of the object (**Figure 21.48A**). These are planes of a viselike device at minimum separation that hold the part.

Primary Internal Parallel Datums: The datum for internal features is the center plane between two parallel planes, at their maximum separation, that contact the inside planes of the object (**Figure 21.48B**).

Secondary Datums: The secondary datum (axis or center plane) for both external and internal diameters (or distances between parallel planes) has the additional requirement that the cylinder in contact with the parallel elements of the hole be perpendicular to the primary datum (**Figure 21.49**). Datum axis B is the axis of cylinder B.

Tertiary Datums: The third datum (axis or center plane) for both external and internal

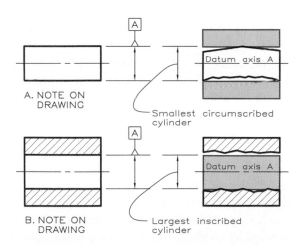

21.47 The datum axis of a shaft is the smallest circumscribed cylinder in contact with the shaft. The datum axis of a hole is the centerline of the largest inscribed cylinder in contact with the hole.

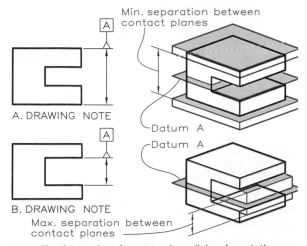

21.48 The datum plane for external parallel surfaces is the center plane between two contact parallel planes at their minimum separation. The datum plane for internal parallel surfaces is the center plane between two contact parallel surfaces at their maximum separation.

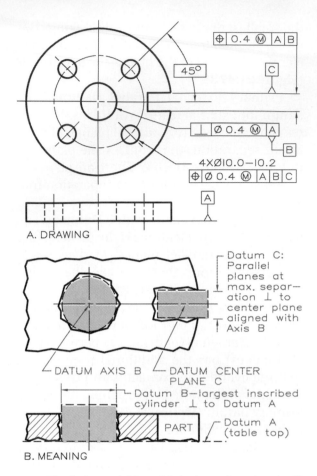

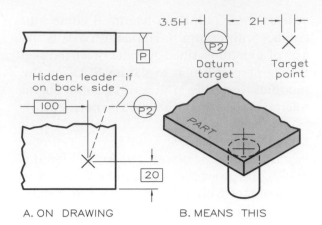

21.50 Target points from which a datum point is established are located with an X and a target symbol. The leader to the target is a hidden line since the target is on the invisible side.

A. DRAWING

Datum C:
Parallel
planes at
max. separ-
ation ⊥ to
center plane
aligned with
Axis B

DATUM AXIS B — DATUM CENTER PLANE C

Datum B—largest inscribed cylinder ⊥ to Datum A

PART

Datum A (table top)

B. MEANING

21.49 The features of this part have been dimensioned with respect to primary, secondary, and tertiary datum planes.

features has the further requirement that either the cylinder or parallel planes be oriented angularly to the secondary datum. Datum C in **Figure 21.49** is the tertiary datum plane.

21.15 Datum Targets

Instead of using a plane surface as a datum, special datum targets can be indicated on a surface of a part where the part is supported by spherical or pointed locating pins. The symbol X is used to indicated target points, which are the points of support by the pins (**Figure 21.50**). Datum target symbols are placed outside the outline of the part with a leader from the target.

When the target is on the near (visible) surface, the leaders are solid lines. When the target is on the far side (invisible) surface, the leaders are drawn as hidden lines as shown in **Figure 21.50**.

Three target points are required to establish the primary datum plane, two for the secondary and one for the tertiary (third). Notice that the target symbol in **Figure 21.50** is labeled P2 to match the designation of the primary datum. The other two points (not shown) would be labeled P2 and P3 to establish the primary datum.

A datum target line is specified in **Figure 21.51** for a part that is supported on a datum line instead of a datum point. An X and a center line are used to locate the line of support.

Target areas are specified for cases where spherical or pointed locating pins are inadequate to support a part. The diameters of the targets are specified with crosshatched circular areas surrounded by dashed lines (**Figure 21.52**). Target symbols give both the diameter of the targets and their number designations. The X symbol and a specified diameter could be used as an alternative method of indicating targets with areas.

The part is located on its datum plane by placing it on the three locating pins with 8-mm

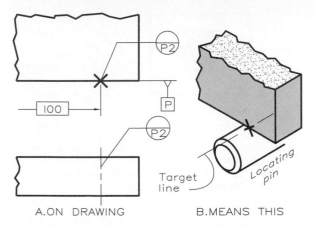

21.51 An X and leader from a datum symbol locate the position of a target line.

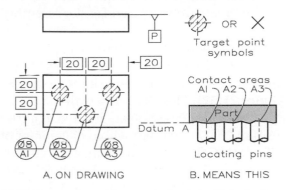

21.52 Target points with areas are located with basic dimensions and target symbols that give the diameters of the target area. Visible leaders indicated that the target areas are on the visible side of the plane.

diameters as shown in **Figure 21.52**. The leaders from the target areas to the target symbols are solid lines to indicate that the targets are on the visible side of the part.

21.16 Location Tolerancing

Tolerances of location deal with **position**, **concentricity**, and **symmetry**.

Position Location dimensions that are toleranced result in a square (or rectangular) tolerance zone for locating the center of a hole (**Figure 21.53A**). In contrast, untoleranced location dimensions, called **basic dimen-**

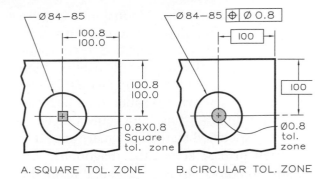

A. SQUARE TOL. ZONE   B. CIRCULAR TOL. ZONE

21.53 Toleranced and untoleranced dimensions.

A Toleranced location dimensions give a square tolerance zone for the axis of the hole.

B Untoleranced basic dimensions (in frames) locate the true position about which a circular tolerance zone of 0.8 mm is specified.

sions, locate the **true position** of a hole's center, about which a circular tolerance zone is specified (**Figure 21.53B**).

In both methods, the size of the hole's diameter is toleranced by identical notes. In the true-position method, a feature control frame specifies the diameter of the circular tolerance zone inside which the hole's center must lie. A circular position zone gives a more precise tolerance of the hole's true position than a square.

Figure 21.54 shows an enlargement of the square tolerance zone resulting from the use

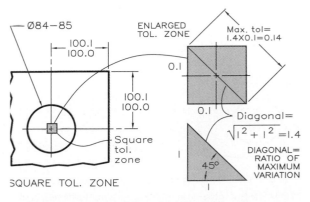

21.54 Toleranced coordinates give a square tolerance zone with a diagonal that exceeds the specified tolerance by a factor of 1.4.

of a toleranced location dimension to locate a hole's center. The diagonal across the square zone is greater than the specified tolerance by a factor of 1.4. Therefore the true-position method, shown enlarged in **Figure 21.55**, can have a larger circular tolerance zone by a factor of 1.4 and still have the same degree of accuracy specified by the 0.1 square zone. If a variation of 0.14 across the diagonal of the square tolerance zone is acceptable in the coordinate method, a circular tolerance zone of 0.14, which is greater than the 0.1 tolerance permitted by the square zone, should be acceptable in the true-position tolerance method.

The circular tolerance zone specified in the circular view of a hole extends the full depth of the hole. Therefore, the tolerance zone for the centerline of the hole is a cylindrical zone inside which the axis must lie. Because both the size of the hole and its position are toleranced, these two tolerances establish the diameter of a gauge cylinder for checking conformance of hole sizes and their locations against specifications (**Figure 21.56**).

Subtracting the true-position tolerance from the hole at MMC (the smallest permissible hole) yields the circle that represents the

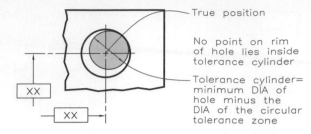

21.56 When a hole at MMC is located at true position, no element of the hole will be inside the imaginary cylinder obtained by subtracting the circular tolerance zone from the minimum diameter of the hole.

least favorable condition when the part is gauged or assembled with a mating part. When the hole is not at MMC, it is larger and permits greater tolerance and easier assembly.

Gauging a Two Hole Pattern Gauging is a technique of checking dimensions to determine whether they meet specified tolerances (**Figure 21.57**). The two holes with diametral size limits of 12.60–12.90 are located at a true position 26.00 mm apart within a diameter of 0.20 at MMC. The gauge pin diameter is calculated to be 12.40 mm (the smallest hole's size, 12.60, minus the true-position tolerance,

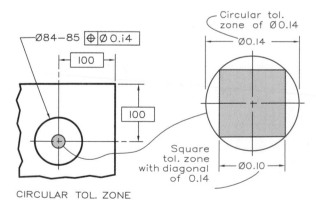

CIRCULAR TOL. ZONE

21.55 The true-position method of locating holes results in a circular tolerance zone. The circular tolerance zone can be 1.4 times greater than the square tolerance zone and still be as accurate.

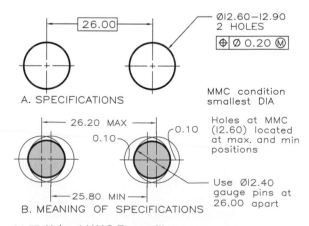

21.57 Holes at MMC: True position.

A These two holes at MMC are to be located at true position as specified.

B The two holes may be gauged with pins 12.52 mm in diameter located 26.00 mm apart.

0.20), as **Figure 21.57B** shows. Thus two pins with diameters of 12.40 mm spaced exactly 26.00 mm apart could be used to check the diameters and positions of the holes at MMC, the most critical size. If the pins can be inserted into the holes, the holes are properly sized and located.

When the holes are not at MMC, or larger than the minimum size, these gauge pins permit a greater range of variation (**Figure 21.58**). When the holes are at their maximum size of 12.90 mm, they can be located as close as 25.50 mm from center to center or as far apart as 26.50 mm from center to center.

Concentricity Concentricity (closely related to a new term, coaxiality) is a feature of location because it specifies the location of two cylinders that share the same axis. In **Figure 21.59**, the large cylinder is labeled as datum A to be used as the datum for locating the small cylinder's axis.

Feature control frames of the type shown in **Figure 21.60** are used to specify concentricity and other geometric characteristics throughout the remainder of this chapter.

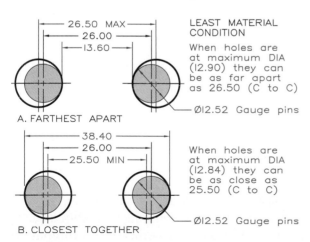

26.50 MAX 26.00 13.60 Ø12.52 Gauge pins	LEAST MATERIAL CONDITION When holes are at maximum DIA (12.90) they can be as far apart as 26.50 (C to C)
A. FARTHEST APART	
38.40 26.00 25.50 MIN Ø12.52 Gauge pins	When holes are at maximum DIA (12.84) they can be as close as 25.50 (C to C)
B. CLOSEST TOGETHER	

21.58 Holes at MMC: Max. and min. spacing.

A These holes at MMC can have their centers spaced as far apart as 26.50 mm apart and still be acceptable.

B The holes may be placed as close as 25.50 mm apart when they are at maximum size.

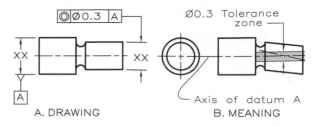

A. DRAWING   B. MEANING

21.59 Concentricity (related to coaxiality) is a tolerance of location. This feature control frame specifies that the axis of the small cylinder be concentric to datum cylinder A, within a tolerance of 0.3 mm diameter.

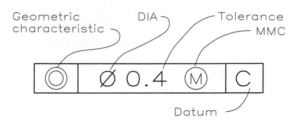

21.60 This typical feature control frame indicates that a surface is concentric to datum C within a cylindrical diameter of 0.4 mm at MMC.

Symmetry Symmetry also is a feature of location in which a feature is symmetrical with the same contour and size on opposite sides of a central plane. **Figure 21.61A** shows how to apply a symmetry feature symbol to the notch that is symmetrical about the part's central datum plane B for a zone of 0.6 mm (**Figure 21.61B**).

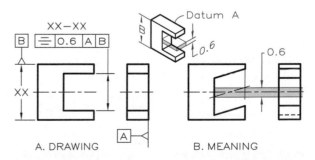

A. DRAWING   B. MEANING

21.61 Symmetry is a tolerance of location that specifies that a part's features be symmetrical about the center plane between parallel surfaces of the part.

21.17 Form Tolerancing

Flatness A surface is flat when all its elements are in one plane. A feature control frame specifies flatness within a 0.4 mm tolerance zone in **Figure 21.62** where no point on the surface may vary more than 0.40 from the highest to the lowest point.

Straightness A surface is straight if all its elements are straight lines within a specified tolerance zone. The feature control frame shown in **Figure 21.63** specifies that the elements of a cylinder must be straight within 0.12 mm. On flat surfaces, straightness is measured in a plane passing through control-line elements, and it may be specified in two directions (usually perpendicular) if desired.

Circularity (Roundness) A surface of revolution (a cylinder, cone, or sphere) is circular when all points on the surface intersected by a plane perpendicular to its axis are equidistant from the axis. In **Figure 21.64** the feature

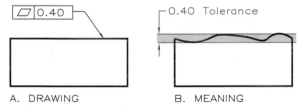

21.62 Flatness is a tolerance of form that specifies a tolerance zone within which a surface must lie.

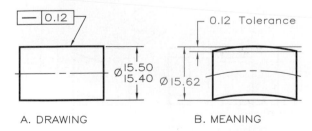

21.63 Straightness is a tolerance of form that indicates that elements of a surface are straight lines. The tolerance frame is applied to the views in which elements appear as straight lines, not points.

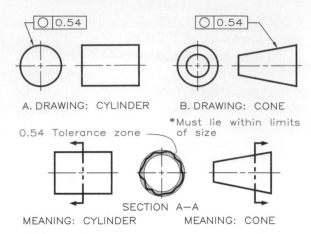

21.64 Circularity (roundness) is a tolerance of form. It indicates that a cross-section through a surface of revolution is round and lies within two concentric circles.

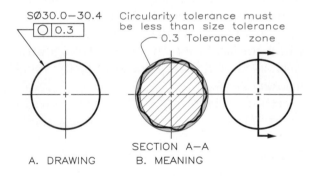

21.65 Circularity of a sphere is a tolerance of form. It means that any cross-section through the sphere is round within the specified tolerance.

control frame specifies circularity of a cone and cylinder, permitting a tolerance of 0.54 mm on the radius. **Figure 21.65** specifies a 0.30 mm tolerance zone for the roundness of a sphere.

Cylindricity A surface of revolution is cylindrical when all its elements lie within a cylindrical tolerance zone, which is a combination of tolerances of roundness and straightness (**Figure 21.66**). Here, a cylindricity tolerance zone of 0.54 mm on the radius of the cylinder is specified.

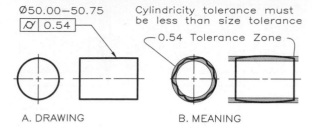

21.66 Cylindricity is a tolerance of form that is a combination of roundness and straightness. It indicates that the surface of a cylinder lies within a tolerance zone formed by two concentric cylinders that are 0.54 apart.

21.18 Profile Tolerancing

Profile tolerancing involves specifying tolerances for a contoured shape formed by arcs or irregular curves, and it can apply to a surface or a single line. The surface with the unilateral profile tolerance shown in **Figure 21.67A** is defined by coordinates. **Figure 21.67B** shows how to specify bilateral and unilateral tolerance zones.

A profile tolerance for a single line is specified as shown in **Figure 21.68**. The curve is formed by tangent arcs whose radii are given as basic dimensions. The radii are permitted to vary 0.20 mm from the basic radii.

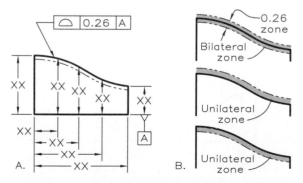

21.67 Profile is a tolerance of form for irregular curving planes. (A) The curving plane is located by coordinates and is toleranced unidirectionally. (B) The tolerance may be applied by any of these methods.

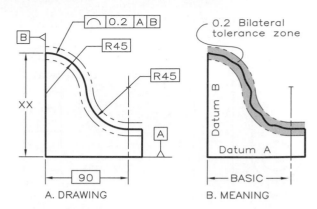

21.68 The profile of a line is a tolerance that specifies the variation allowed from the path of a line. Here, the line is formed by tangent arcs. The tolerance zone may be either bilateral or unilateral, as shown in Figure 21.67.

21.19 Orientation Tolerancing

Tolerances of orientation include parallelism, perpendicularity, and angularity.

Parallelism A surface or line is parallel when all its points are equidistant from a datum plane or axis. Two types of parallelism tolerance zones are:

1. A planar tolerance zone parallel to a datum plane within which the axis or surface of the feature must lie (**Figure 21.69**). This tolerance of orientation also controls flatness.

2. A cylindrical tolerance zone parallel to a datum feature within which the axis of a feature must lie (**Figure 21.70**).

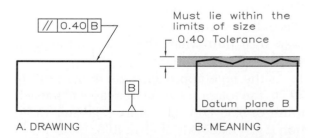

21.69 Parallelism is a tolerance of orientation. It indicates that a plane is parallel to a datum plane within specified limits. Here, plane B is the datum plane.

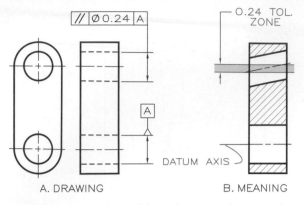

A. DRAWING　　　　　　B. MEANING

21.70 You may specify parallelism of one centerline to another by using the diameter of one of the holes as the datum.

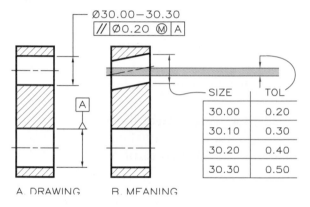

A. DRAWING　　　B. MEANING

SIZE	TOL
30.00	0.20
30.10	0.30
30.20	0.40
30.30	0.50

21.71 The critical tolerance exists when features are at MMC. (A) The upper hole must be parallel to the hole used as datum A within a 0.20 DIA. (B) As the hole approaches its maximum size of 30.30 mm, the tolerance zone approaches 0.50 mm.

Figure 21.71 shows the effect of specifying parallelism at MMC, where the modifier M is given in the feature control frame. Tolerances of form apply at RFS when not specified. Specifying parallelism at MMC means that the axis of the cylindrical hole must vary no more than 0.20 mm when the holes are at their smallest permissible size.

As the hole approaches its upper limit of 30.30, the tolerance zone increases to a maximum of 0.50 DIA. Therefore a greater variation is given at MMC than at RFS.

Perpendicularity The specifications for the perpendicularity of a plane to a datum are

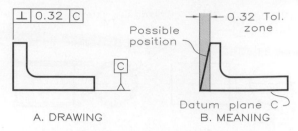

A. DRAWING　　　　　B. MEANING

21.72 Perpendicularity is a tolerance of orientation that gives a tolerance zone of .32 for a plane perpendicular to a specified datum plane.

shown in **Figure 21.72**. The feature control frame shows that the surface perpendicular to datum plane C has a tolerance of 0.32 in. In **Figure 21.73**, a hole is specified as perpendicular to datum plane A.

Angularity　A surface or line is angular when it is at an angle (other than 90°) from a datum or an axis. The angularity of the surface shown in **Figure 21.74** is dimensioned with a basic angle (exact angle) of 30° and an angularity

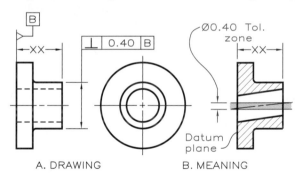

A. DRAWING　　　　　B. MEANING

21.73 Perpendicularity is a tolerance of orientation. Perpendicularity it can apply to the axis of a feature, such as the centerline of a cylinder.

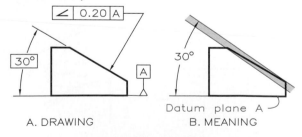

A. DRAWING　　　　　B. MEANING

21.74 Angularity is a tolerance of orientation specifying a tolerance zone for an angular surface with respect to a datum plane. Here, the 30° angle is a true, or basic, angle to which a tolerance of 0.20 mm is applied.

tolerance zone of 0.20 mm inside of which the plane must lie.

21.20 Runout Tolerancing

Runout tolerancing is a way of controlling multiple features by relating them to a common datum axis. Features so controlled are surfaces of revolution about an axis and surfaces perpendicular to the axis.

The datum axis, such as diameter B in **Figure 21.75**, is established by a circular feature that rotates about the axis. When the part is rotated about this axis, the features of rotation must fall within the prescribed tolerance at full indicator movement (FIM).

The two types of runout are **circular runout** and **total runout**. One arrow in the feature control frame indicates circular runout and two arrows indicate total runout.

Circular Runout   Rotating an object about its axis 360° determines whether a circular cross section exceeds the permissible runout tolerance at any point (**Figure 21.76**). This same technique is used to measure the amount of

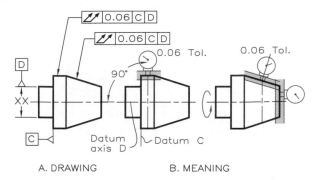

21.76 Here, runout tolerance is measured by mounting the object on the primary datum plane C and the secondary datum cylinder D. The cylinder and conical surface are gauged to check their conformity to a tolerance zone of 0.06 mm. The runout at the end of the cone could have been noted.

wobble in surfaces perpendicular to the axis of rotation.

Total Runout   Used to specify cumulative variations of circularity, straightness, concentricity, angularity, taper, and profile of a surface (**Figure 21.76**), total runout tolerances are measured for all circular and profile positions as the part is rotated 360°. When applied to surfaces perpendicular to the axis, total runout tolerances control variations in perpendicularity and flatness.

Conclusion

The dimensioned part shown in **Figure 21.77** illustrates several of the techniques of geometric tolerancing described in this and previous sections.

21.21 Surface Texture

Because the surface texture of a part affects its function, it must be precisely specified instead of giving an unspecified finished mark such as a V. **Figure 21.78** illustrates most of the terms that apply to surface texture (surface control).

Surface texture: The variation in a surface, including roughness, waviness, lay, and flaws.

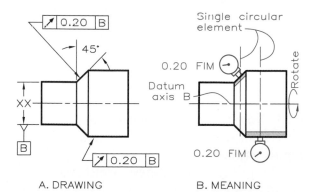

21.75 Runout tolerance, a composite of several tolerance of form characteristics, is used to specify concentric cylindrical parts. The part is mounted on the datum axis and is gauged as it is rotated.

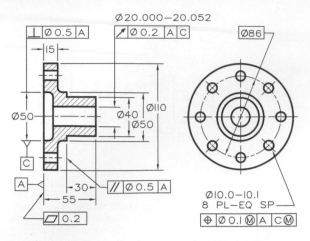

21.77 A combination of notes and symbols describe this part's geometric features.

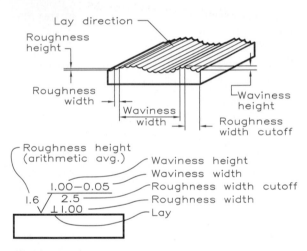

21.78 These are the definitions and terminology of surface texture for a finished surface.

Roughness: The finest of the irregularities in the surface caused by the manufacturing process used to smooth the surface.

Roughness height: The average deviation from the mean plane of the surface measured in microinches (min.) or micrometers (mm), or millionths of an inch and a meter, respectively.

Roughness width: The width between successive peaks and valleys forming the roughness measured in microinches or micrometers.

Roughness width cutoff: The largest spacing of repetitive irregularities that includes average roughness height (measured in inches or millimeters). When not specified, a value of 0.8 mm (0.030 in.) is assumed.

Waviness: A widely spaced variation that exceeds the roughness width cutoff measured in inches or millimeters. Roughness may be regarded as a surface variation superimposed on a wavy surface.

Waviness height: The peak-to-valley distance between waves measured in inches or millimeters.

Waviness width: The spacing between wave peaks or wave valleys measured in inches or millimeters.

Lay: The direction of the surface pattern caused by the production method used.

Flaws: Irregularities or defects occurring infrequently or at widely varying intervals on a surface, including cracks, blow holes, checks, ridges, scratches, and the like. The effect of flaws is usually omitted in roughness height measurements.

Contact area: The surface that will make contact with a mating surface.

Symbols for specifying surface texture are shown in **Figure 21.79**. The point of the V must touch the edge view of the surface, an extension line from it, or a leader pointing to the surface. **Figure 21.80** shows how to specify values as a part of surface texture symbols.

Roughness height values are related to the processes used to finish surfaces and may be taken from the table in **Figure 21.81**. The preferred values of roughness height are listed in **Figure 21.82**.

The preferred roughness width cutoff values in **Figure 21.83** are for specifying the sampling width used to measure roughness height. A value of 0.80 mm is assumed if no

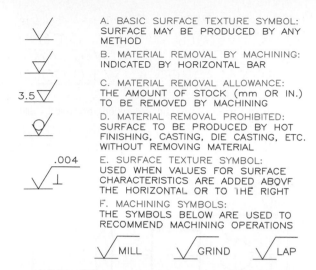

A. BASIC SURFACE TEXTURE SYMBOL: SURFACE MAY BE PRODUCED BY ANY METHOD

B. MATERIAL REMOVAL BY MACHINING: INDICATED BY HORIZONTAL BAR

C. MATERIAL REMOVAL ALLOWANCE: THE AMOUNT OF STOCK (mm OR IN.) TO BE REMOVED BY MACHINING

D. MATERIAL REMOVAL PROHIBITED: SURFACE TO BE PRODUCED BY HOT FINISHING, CASTING, DIE CASTING, ETC. WITHOUT REMOVING MATERIAL

E. SURFACE TEXTURE SYMBOL: USED WHEN VALUES FOR SURFACE CHARACTERISTICS ARE ADDED ABOVE THE HORIZONTAL OR TO THE RIGHT

F. MACHINING SYMBOLS: THE SYMBOLS BELOW ARE USED TO RECOMMEND MACHINING OPERATIONS

MILL   GRIND   LAP

21.79 Use surface texture symbols to specify surface finish on the edge views of finished surfaces.

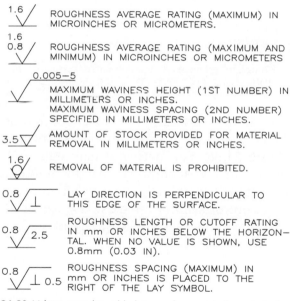

1.6 — ROUGHNESS AVERAGE RATING (MAXIMUM) IN MICROINCHES OR MICROMETERS.

1.6 / 0.8 — ROUGHNESS AVERAGE RATING (MAXIMUM AND MINIMUM) IN MICROINCHES OR MICROMETERS

0.005-5 — MAXIMUM WAVINESS HEIGHT (1ST NUMBER) IN MILLIMETERS OR INCHES. MAXIMUM WAVINESS SPACING (2ND NUMBER) SPECIFIED IN MILLIMETERS OR INCHES.

3.5 — AMOUNT OF STOCK PROVIDED FOR MATERIAL REMOVAL IN MILLIMETERS OR INCHES.

1.6 — REMOVAL OF MATERIAL IS PROHIBITED.

0.8 ⊥ — LAY DIRECTION IS PERPENDICULAR TO THIS EDGE OF THE SURFACE.

0.8 / 2.5 — ROUGHNESS LENGTH OR CUTOFF RATING IN mm OR INCHES BELOW THE HORIZONTAL. WHEN NO VALUE IS SHOWN, USE 0.8mm (0.03 IN).

0.8 ⊥ 0.5 — ROUGHNESS SPACING (MAXIMUM) IN mm OR INCHES IS PLACED TO THE RIGHT OF THE LAY SYMBOL.

21.80 Values may be added to surface control symbols for more precise specifications.

value is given. When required, maximum waviness height values may be selected from the recommended values shown in **Figure 21.84**.

Lay symbols indicating the direction of texture (markings made by the machining

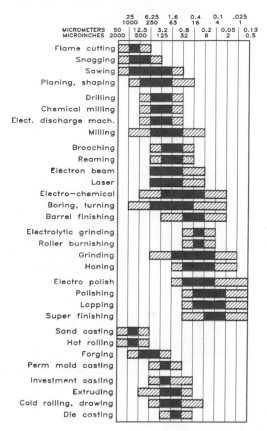

21.81 Various types of production methods result in the surface roughness heights shown in micrometers and microinches (millionths of a meter or an inch respectively).

PREFERRED ROUGHNESS AVERAGE VALUES

Micro— meters μm	Micro— inches μin.		Micro— meters μm	Micro— inches μin.
0.025	1		1.6	63
0.050	2		3.2	125
0.10	4		6.3	250
0.20	8		12.5	500
0.40	16		25	1000
0.80	32		Micrometers=0.001 mm	

21.82 This range of roughness heights is recommended by the ANSI Y14.36 standards for metric and English units.

MILLIMETERS	0.08	0.25	0.80	2.5	8.0	25
INCHES	.003	.010	.030	.1	.3	1

21.83 This range of roughness width cutoff values is recommended in the ANSI Y14.36 standards. When unspecified, assume a value of 0.8 mm or 0.030 inches.

MAXIMUM WAVINESS HEIGHT VALUES

mm	in.	mm	in.
0.0005	.00002	0.025	.001
0.0008	.00003	0.05	.002
0.0012	.00005	0.08	.003
0.0020	.00008	0.12	.005
0.0025	.0001	0.20	.008
0.005	.0002	0.25	.010
0.008	.0003	0.38	.015
0.012	.0005	0.50	.020
0.020	.0008	0.80	.030

21.84 This range of maximum waviness height values is recommended in the ANSI Y14.36 standards.

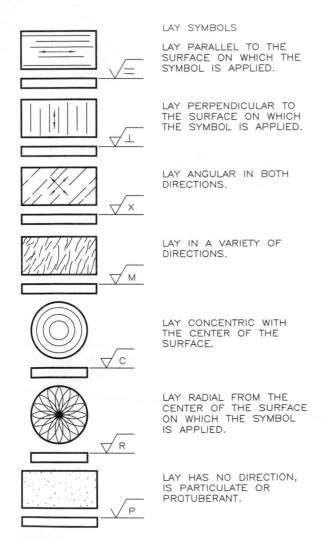

21.85 These symbols are used to indicate the direction of lay with respect to the surface where the control symbol is placed.

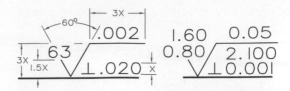

21.86 These are examples and proportions of typical, fully specified surface texture symbols.

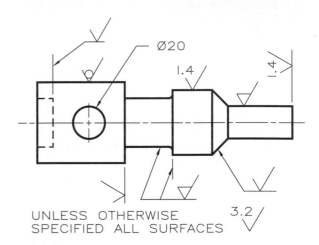

21.87 Various techniques of applying surface texture symbols to a part are illustrated here.

operation) on a surface (**Figure 21.85**) may be added to surface texture symbols as shown in **Figure 21.86**. The perpendicular sign indicates that lay is perpendicular to the edge view of the surface in this view (where the surface control symbol appears). **Figure 21.87** illustrates how to apply a variety of surface texture symbols to a part.

Problems

Solve the following problems on size A sheets laid out on a grid of 0.20 in. or 5 mm.

Cylindrical Fits

1. (**Sheet 1**) Draw the shaft and hole shown (it need not be to scale), give the limits for each diameter, and complete the table of values.

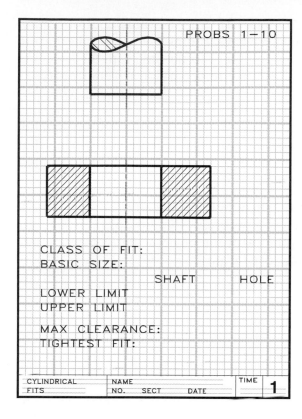

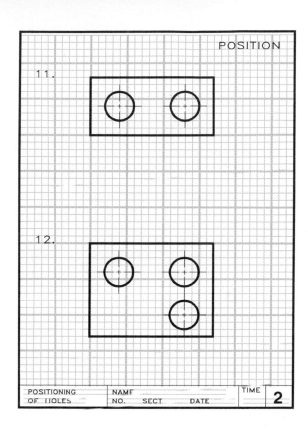

Use a basic diameter of 1.00 in. (25 mm) and a class RC 1 fit or a metric fit of H8/f7.

2. Repeat Problem 1, but use a basic diameter of 1.75 in. (45 mm) and a class RC 9 fit or a metric fit of H11/c11.

3. Repeat Problem 1, but use a basic diameter of 2.00 in. (51 mm) and a class RC 5 fit or a metric fit of H9/d9.

4. Repeat Problem 1, but use a basic diameter of 12.00 in. (305 mm) and a class LC 11 fit or a metric fit of H7/h6.

5. Repeat Problem 1, but use a basic diameter of 3.00 in. (76 mm) and a class LC 1 fit or a metric fit of H7/h6.

6. Repeat Problem 1, but use a basic diameter of 8.00 in. (203 mm) and a class LC 1 fit or a metric fit of H7/k6.

7. Repeat Problem 1, but use a basic diameter of 102 in. (2591 mm) and a class LN 3 fit or a metric fit of H7/n6.

8. Repeat Problem 1, but use a basic diameter of 11.00 in. (279 mm) and a class LN 2 fit or a metric fit of H7/p6.

9. Repeat Problem 1, but use a basic diameter of 6.00 in. (152 mm) and a class FN 5 fit or a metric fit of H7/s6.

10. Repeat Problem 1, but use a basic diameter of 2.60 in. (66 mm) and a class FN 1 fit or a metric fit of H7/u6.

Tolerances of Location
11. (**Sheet 2**) Make an instrument drawing of the part shown. Locate the two holes with a size tolerance of 1.00 mm and a position tolerance of 0.50 DIA. Insert the proper symbols and dimensions.

12. Repeat Problem 11, but locate three holes using the same tolerances for size and position.

13. Give the specifications for a two-pin gauge that can be used to measure the correctness of the two holes specified in Problem 11. Make a sketch of the gauge and show the proper dimensions on it.

14. (**Sheet 3**) Using positioning tolerances, locate the holes and properly note them to provide a size tolerance of 1.50 mm and a locational tolerance of 0.60 DIA.

15. Repeat Problem 14, but locate six equally spaced, equally sized holes using the same tolerances of position.

16. (**Sheet 4**) Using a feature control symbol and the necessary dimensions, indicate that the notch is symmetrical to the left-hand end of the part within 0.60 mm.

17. (**Sheet 4**) Using a feature control symbol and the necessary dimensions, indicate that the small cylinder is concentric with the large one (the datum cylinder) within a tolerance of 0.80.

Tolerances of Form

18. (**Sheet 5**) Using a feature control symbol and the necessary dimensions, indicate that the elements of the cylinder are straight within a tolerance of 0.20 mm.

19. (**Sheet 5**) Using a feature control symbol and the necessary dimensions, indicate that surface A of the object is flat within a tolerance of 0.08 mm.

20–22. (**Sheet 6**) Using feature control symbols and the necessary dimensions, indicate that the cross sections of the cylinder, cone, and sphere are round within a tolerance of 0.40 mm.

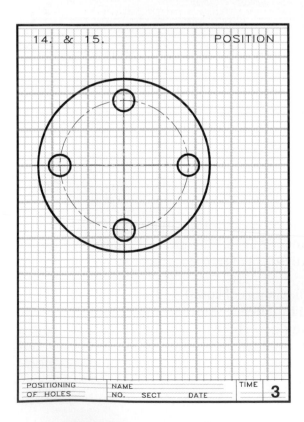

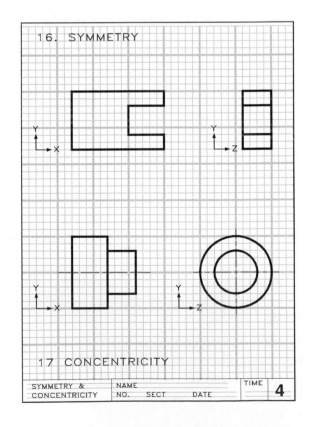

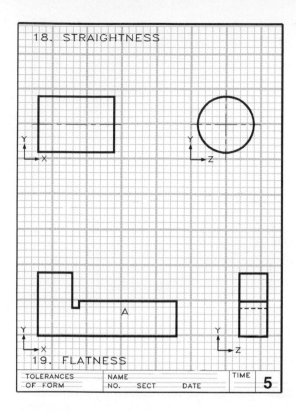

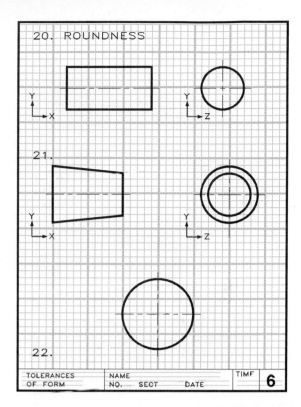

23. (**Sheet 7**) Using a feature control symbol and the necessary dimensions, indicate that the cylindricity of the cylinder is 0.90 mm.

Tolerances of Profile

24. (**Sheet 7**) Using a feature control symbol and the necessary dimensions, indicate that the profile of the irregular surface of the object lies within a bilateral or unilateral tolerance zone of 0.40 mm.

25. (**Sheet 8**) Using a feature control symbol and the necessary dimensions, indicate that the profile of the line formed by tangent arcs lies within a bilateral or unilateral tolerance zone of 0.40 mm.

Tolerances of Orientation

26. (**Sheet 8**) Using a feature control symbol and the necessary dimensions, indicate that the angularity tolerance of the inclined plane

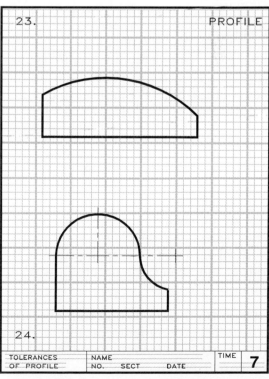

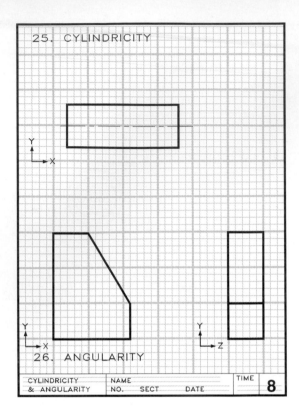

25. CYLINDRICITY

26. ANGULARITY

CYLINDRICITY & ANGULARITY	NAME			TIME	
	NO.	SECT	DATE		8

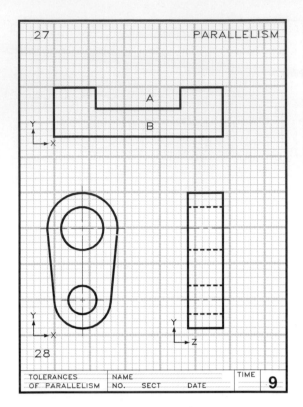

27 PARALLELISM

28

TOLERANCES OF PARALLELISM	NAME			TIME	
	NO.	SECT	DATE		9

is 0.7 mm from the bottom of the object, the datum plane.

27. (**Sheet 9**) Using a feature control symbol and the necessary dimensions, indicate that surface A of the object is parallel to datum B within 0.30 mm.

28. (**Sheet 9**) Using a feature control symbol and the necessary dimensions, indicate that the small hole is parallel to the large hole, the datum, within a tolerance of 0.80 mm.

29. (**Sheet 10**) Using a feature control symbol and the necessary dimensions, indicate that the vertical surface B is perpendicular to the bottom of the object, the datum C, within a tolerance of 0.20 mm.

30. (**Sheet 10**) Using a feature control symbol and the necessary dimensions, indicate that the hole is perpendicular to datum A within a tolerance of 0.08 mm.

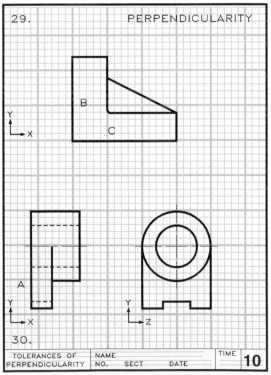

29. PERPENDICULARITY

30.

TOLERANCES OF PERPENDICULARITY	NAME			TIME	
	NO.	SECT	DATE		10

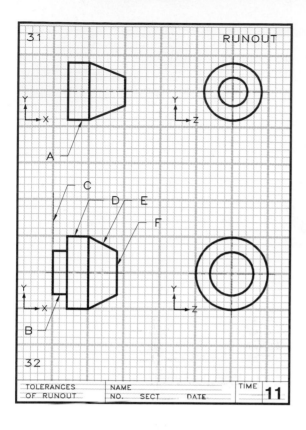

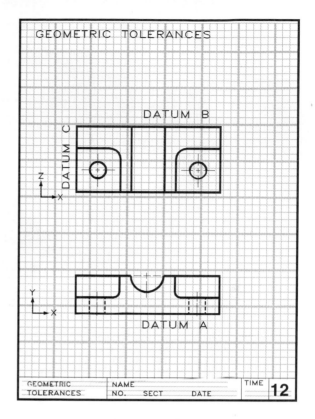

Tolerances of Runout

31. (**Sheet 11**) Using the appropriate geometric tolerancing symbols and cylinder A as the datum, indicate that the conical feature has a runout of 0.80 mm.

32. (**Sheet 11**) Using a feature control symbol with cylinder B as the primary datum and surface C as the secondary datum, indicate that surfaces D, E, and F have a runout of 0.60 mm.

33. (**Sheet 12**) Draw the journal base as it is shown, which is a half-size, two-view drawing. Using metric units as the three datum planes, locate the two holes (size tolerance of 0.20) to lie within a tolerance zone of 1.6 DIA. Indicate that datum A is flat within 0.8. Indicate that datum B is perpendicular to datum A within 1.2. (Refer to **Figure 21.88**.)

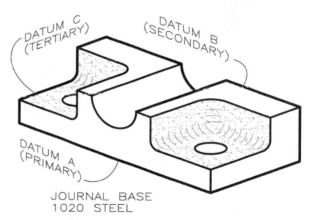

21.88 Problems 33 and 34.

34. Same as Problem 33, but in addition to these specifications, give complete dimensions necessary to fully describe the part. Also indicate that the upper surface in the front view is parallel to the datum A within 1.4.m.

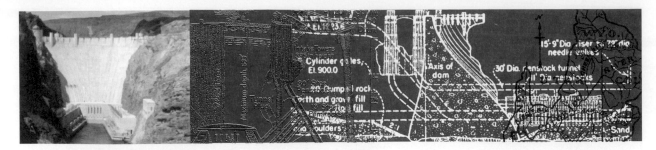

22

Welding

22.1 Introduction

Welding is the process of permanently joining metal by heating a joint to a suitable temperature with or without applying pressure and with or without using filler material. The welding practices described in this chapter comply with the standards developed by the American Welding Society and the American National Standards Institute (ANSI).

Welding is done in shops, on assembly lines, or in the field as shown in **Figure 22.1**, where a welder is joining pipes. Welding is a widely used method of fabrication with its own language of notes, specifications, and symbology. You must become familiar with this system of notations in order to make and read drawings containing welding specifications.

Advantages of welding over other methods of fastening include (1) simplified fabrication, (2) economy, (3) increased strength and rigid-

22.1 This welder is joining two pipes in accordance with specifications on a set of drawings. (Courtesy of Texas Eastern; *TE Today;* photo by Bob Thigpen.)

ity, (4) ease of repair, (5) creation of gas- and liquid-tight joints, and (6) reduction in weight and size.

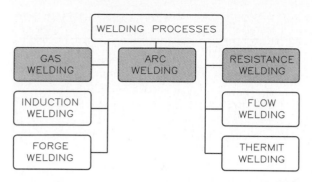

22.2 The three main types of welding processes are gas welding, arc welding, and resistance welding.

22.2 Welding Processes

Figure 22.2 shows various types of welding processes. The three main types are gas welding, arc welding, and resistance welding.

Gas welding involves the use of gas flames to melt and fuse metal joints. Gases such as acetylene or hydrogen are mixed in a welding torch and burned with air or oxygen (**Figure 22.3**). The oxyacetylene method is widely used for repair work and field construction.

Most oxyacetylene welding is done manually with a minimum of equipment. Filler material in the form of welding rods is used to deposit metal at the joint as it is heated. Most metals, except for low- and medium-carbon steels, require fluxes to aid the process of melting and fusing the metals.

Arc welding involves the use of an electric arc to heat and fuse joints, with pressure sometimes required in addition to heat (**Figure 22.4**). The filler material is supplied by a consumable or nonconsumable electrode through which the electric arc is transmitted. Metals well-suited to arc welding are wrought iron, low- and medium-carbon steels, stainless steel, copper, brass, bronze, aluminum, and some nickel alloys. In electric-arc welding, **the flux is a material coated on the electrodes that forms a coating on the metal being welded**. This coating protects the metal from oxidation so that the joint will not be weakened by overheating.

Flash welding is a form of arc welding, but it is similar to resistance welding because both pressure and electric current (**Figure 22.5**) are applied. The pieces to be welded are brought together, and an electric current is passed through them, causing heat to build up between them. As the metal burns, the current is turned off and the pressure between the pieces is increased to fuse them.

Resistance welding comprises several processes by which metals are fused both by the heat produced from the resistance of the parts to an electric current and by pressure. Fluxes and filler materials normally are not used. All resistance welds are either lap- or butt-type welds.

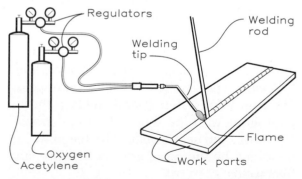

22.3 The gas welding process burns gases such as oxygen and acetylene in a torch to apply heat to a joint. The welding rod supplies the filler material.

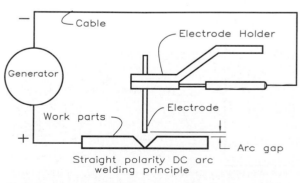

22.4 In arc welding, either AC or DC current is passed through an electrode to heat the joint.

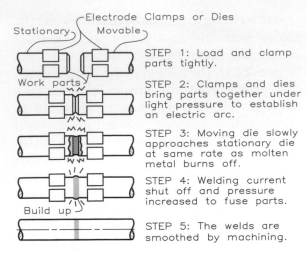

STEP 1: Load and clamp parts tightly.

STEP 2: Clamps and dies bring parts together under light pressure to establish an electric arc.

STEP 3: Moving die slowly approaches stationary die at same rate as molten metal burns off.

STEP 4: Welding current shut off and pressure increased to fuse parts.

STEP 5: The welds are smoothed by machining.

22.5 Flash welding, a type of arc welding, uses a combination of electric current and pressure to fuse two parts.

Material	Spot Welding	Flash Welding
Low—carbon mild steel		
SAE 1010	Rec.	Rec.
SAE 1020	Rec.	Rec.
Medium—carbon steel		
SAE 1030	Rec.	Rec.
SAE 1050	Rec.	Rec.
Wrought alloy steel		
SAE 4130	Rec.	Rec.
SAE 4340	Rec.	Rec.
High—alloy austenitic stainless steel		
SAE 30301—30302	Rec.	Rec.
SAE 30309—30316	Rec.	Rec.
Ferritic and martensistic stainless steel		
SAE 51410—51430	Satis.	Satis.
Wrought heat—resisting alloys		
19—9—DL	Satis.	Satis.
16—25—6	Satis.	Satis.
Cast iron	NA	Not Rec.
Gray iron	NA	Not Rec.
Aluminum & alum. alloys	Rec.	Satis.
Nickel & nickel alloys	Rec.	Satis.

Rec.—Recommended   Satis.—Satisfactory
Not Rec.— Not   NA—Not applicable
    recommended

22.7 Resistance welding processes for various materials are shown here.

Resistance spot welding is performed by pressing the parts together, and an electric current fuses them as illustrated in the lap joint weld in **Figure 22.6**. A series of small welds spaced at intervals, called **spot welds**, secure the parts. **Figure 22.7** lists the recommended materials and processes to be used for resistance welding.

22.3 Weld Joints and Welds

Figure 22.8 shows the five standard weld joints: **butt joint**, **corner joint**, **lap joint**, **edge joint**, and **tee joint**. The **butt joint** can be joined with the square groove, V-groove, bevel groove, U-groove, and J-groove welds. The **corner joint** can be joined with these welds and with the fillet weld. The **lap joint** can be joined with the bevel groove, J-groove, fillet, slot, plug, spot, projection, and seam welds. The **edge joint** uses the same welds as the lap joint along with the square groove, V-groove, U-groove, and seam welds. The **tee joint** can be joined by the bevel groove, J-groove, and fillet welds.

Figure 22.9 depicts commonly used welds and their corresponding ideographs (sym-

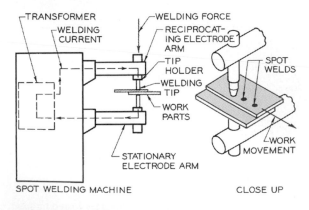

22.6 Resistance spot welding may be used to join lap and butt joints.

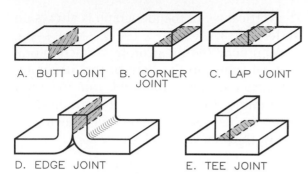

22.8 These diagrams depict the five standard weld joints.

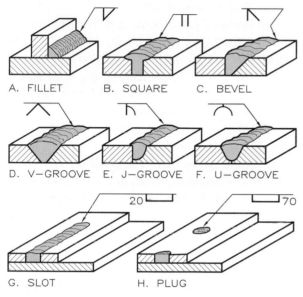

22.9 These views illustrate standard welds and their corresponding ideographs.

bols). The fillet weld is a built-up weld at the intersection (usually 90°) of two surfaces. The square, bevel, V-groove, J-groove, and U-groove welds all have grooves, and the weld is made in these grooves. Slot and plug welds have intermittent holes or openings where the parts are welded. Holes are unnecessary when resistance welding is used.

22.4 Welding Symbols

If a drawing has a general welding note such as ALL JOINTS ARE WELDED THROUGHOUT,

the designer has transferred responsibility to the welder. Welding is too important to be left to chance and should be specified more precisely.

Symbols are used to convey welding specifications on a drawing. The complete welding symbol is shown in **Figure 22.10**, but it usually appears on a drawing in a modified, more general form with less detail. The scale of the welding symbol is based on the letter height used on the drawing, which is the size of the grid on which the symbol is drawn in **Figure 22.11**. The standard height of lettering on a drawing is usually 1/8 in. or 3 mm.

The **ideograph** is the symbol that denotes the type of weld desired, and it generally depicts the cross-section representation of the weld. **Figure 22.12** shows the ideographs used most often. They are drawn to scale on

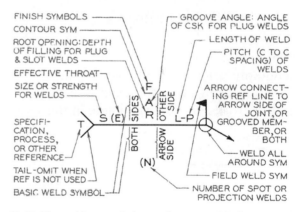

22.10 The welding symbol. Usually it is modified to a simpler form for use on drawings.

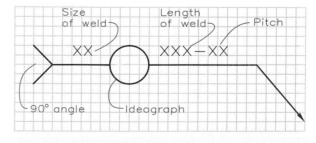

22.11 Welding symbol proportions are based on the letter height used on a drawing, usually 1/8 in. or 3 mm. This grid is equal to the letter height.

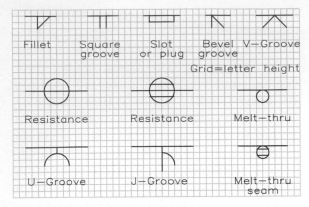

22.12 The sizes of the ideographs shown on the 1/8-in. (3 mm) grid (the letter height) are proportional to the size of the welding symbol (Figure 22.11).

the 1/8-in. (3-mm) grid (equal to the letter height), which represents their full size when added to the welding symbol.

22.5 Application of Symbols

Fillet Welds

In **Figure 22.13A**, placement of the fillet weld ideograph below the horizontal line of the symbol indicates that the weld is at the joint on the arrow side—the right side in this case.

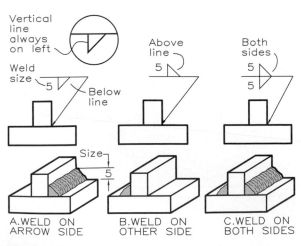

22.13 Fillet welds may be noted with abbreviated symbols. (A) When the ideograph appears below the horizontal line, it specifies a weld on the arrow side. (B) When it is above the line, it specifies a weld on the opposite side. (C) When it is on both sides of the line, it specifies a weld on each side.

The vertical leg of the ideograph is always on the left side.

A numeral (either a common fraction or a decimal value) to the right of the ideograph indicates the size of the weld. You may omit this number from the symbol if you insert a general note elsewhere on the drawing to specify the fillet size, such as:

ALL FILLET WELDS 1/4 IN.

UNLESS OTHERWISE NOTED.

Placing the ideograph above the horizontal line in **Figure 22.13B** indicates that the weld is to be on the other side, that is, the joint on the other side of the part away from the arrow. When the part is to be welded on both sides, use the ideograph shown in **Figure 22.13C**. You may omit the tail and other specifications from the symbol when you provide detailed specifications elsewhere.

A single arrow often is used to specify a weld that is to be made all around two joining parts (**Figure 22.14A**). A circle of 6 mm (twice the letter height) in diameter, drawn at the bend in the leader of the symbol denotes this type of weld. If the welding is to be done in the field rather than in the shop, a solid black triangular "flag" is added also (**Figure 22.14B**).

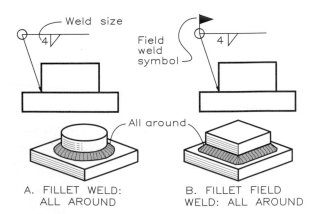

22.14 These symbols indicate fillet welds are to be made all around two types of parts.

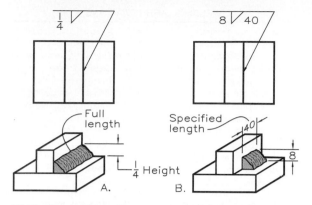

22.15 Fillet weld symbols.

A This symbol indicates full-length fillet welds on the arrow side.

B This symbol indicates fillet welds of specified, but less than, full length of the welded parts.

You may specify a fillet weld that is to run the full length of the two parts as in **Figure 22.15A.** The ideograph is on the lower side of the horizontal line, so the weld is on the arrow side. You may specify a fillet weld that is to run shorter than full length as in **Figure 22.15B,** where 40 represents the weld's length in millimeters.

You may specify fillet welds to run different lengths and be positioned on both sides of a part as in **Figure 22.16A.** The dimension on the lower side of the horizontal gives the length of the weld on the arrow side, and the

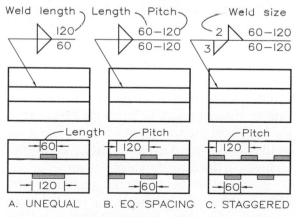

22.16 These symbols specify intermittent welds of varying lengths and alignments.

dimension on the upper side of the horizontal gives the length on the opposite side.

Intermittent welds have a specified length and are spaced uniformly, center to center, at an interval called the pitch. In **Figure 22.16B,** the welds are equally spaced on both sides, are 60 mm long, and have pitches of 120 mm, as indicated by the symbol shown. The symbol shown in **Figure 22.16C** specifies intermittent welds that are staggered in alternate positions on opposite sides.

Groove Welds

The standard types of groove welds are the **V-groove, bevel groove, double V-groove, U-groove,** and **J-groove** (**Figure 22.17**). When you do not give the depth of the grooves, angle of the chamfer, and root openings on a symbol,

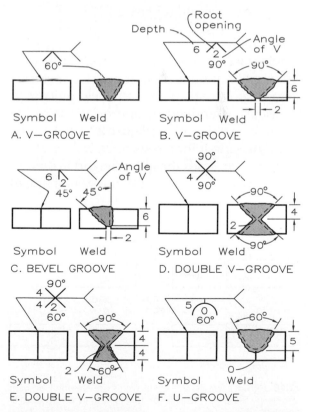

22.17 This drawing shows the various types of groove welds and their general specifications.

you must specify them elsewhere on the drawing or in supporting documents. In **Figure 22.17A** and **B**, the angles of the V-joints are labeled 60° and 90° under the ideographs. In **Figure 22.17B**, the depths of the weld (6) and the root opening (2)—the gap between the two parts—are given.

In a bevel groove weld, only one of the parts is beveled. The symbol's leader is bent and pointed toward the beveled part to call attention to it (**Figs. 22.17C** and **22.18B**). This practice also applies to J-groove welds, where one side is grooved and the other is not (**Figure 22.18A**).

Notate double V-groove welds by weld size, bevel angle, and root opening (**Figure 22.17D** and **E**). Omit root opening sizes or show a zero on the symbol when parts fit flush. Give the angle and depth of the groove in the symbol for a U-groove weld (**Figure 22.17F**).

Seam Welds

A seam weld joins two lapping parts with either a continuous weld or a series of closely spaced spot welds. The seam weld process to be used is identified by abbreviations in the tail of the weld symbol (**Figure 22.19**). The circular ideograph for a resistance weld is about 12 mm (four times the letter height) in diameter and is centered over the horizontal line of

CAW	Carbon—arc w.	IB	Induction brazing
CW	Cold welding	IRB	Infrared brazing
DB	Dip brazing	OAW	Oxyacetylene w.
DFW	Diffusion welding	OHW	Oxyhydrogen w.
EBW	Electric beam w.	PGW	Pressure gas w.
ESW	Electroslag welding	RB	Resist. brazing
EXW	Explosion welding	RPW	Projection weld.
FB	Furnace brazing	RSEW	Resist. seam w.
FOW	Forge welding	RSW	Resist. spot w.
FRW	Friction welding	RW	Resist. welding
FW	Flash welding	TB	Torch brazing
GMAW	Gas metal arc w.	UW	Upset welding
GTAW	Gas tungsten w.	*w.=welding	

22.19 These abbreviations represent the various types of welding processes and are used in welding symbols.

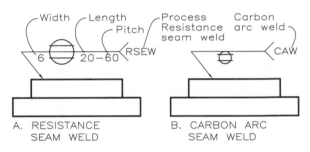

22.20 The process used for (A) resistance seam welds and (B) arc-seam welds is indicated in the tail of the symbol. For the arc weld, the symbol must specify the arrow side or the other side of the piece.

the symbol (**Figure 22.20A**). The weld's width, length, and pitch are given.

When the seam weld is to be made by carbon arc welding (CAW), the diameter of the ideograph is about 6 mm (twice the letter height) and goes on the upper or lower side of the symbol's horizontal line to indicate whether the seam is to be applied to the arrow side or opposite side (**Figure 22.20B**). When the length of the weld is not shown, the seam weld is understood to extend between abrupt changes in the direction of the seam.

Spot welds are similarly specified with ideographs and specifications by diameter, number of welds, and pitch between the

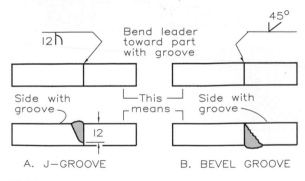

22.18 J-groove welds and bevel welds are specified by bent arrows pointing to the side of the joint that is grooved or beveled.

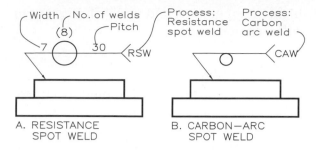

A. RESISTANCE SPOT WELD

B. CARBON—ARC SPOT WELD

22.21 The process to be used for (A) resistance spot welds and (B) carbon arc-spot welds is indicated in the tail of the symbol. For the arc weld, the symbol must specify the arrow side or the other side of the piece.

welds. The process of resistance spot welding (RSW) is noted in the tail of the symbol (**Figure 22.21A**). For carbon arc welding, the arrow side or other side must be indicated by a symbol (**Figure 22.21B**).

Built-Up Welds

When the surface of a part is to be enlarged or built-up by welding, indicate this process with a symbol as shown in **Figure 22.22**. Dimension the width of the built-up weld in the view. Specify the height of the weld above the surface in the symbol to the left of the ideograph. The radius of the circular segment is 6 mm (twice the letter height).

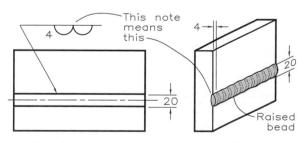

22.22 Use this method to apply a symbol to a built-up weld on a surface.

22.6 Surface Contouring

Contour symbols are used to indicate which of the three types of contours, **flush**, **concave**, or **convex**, is desired on the surface of the weld. Flush contours are smooth with the surface or flat across the hypotenuse of a fillet weld. Concave contours bulge inward with a curve, and convex contours bulge outward with a curve (**Figure 22.23**).

Finishing the weld by an additional process to obtain the desired contour is often necessary. These processes, which may be indicated by their abbreviations, are **chipping** (C), **grinding** (G), **hammering** (H), **machining** (M), **rolling** (R), and **peening** (P), as shown in **Figure 22.24**.

22.7 Brazing

Brazing is a method much like welding for joining pieces of metal. Brazing entails heating joints to more than 800°F and distributing by capillary action a nonferrous filler material, with a melting point below that of the base materials, between the closely fitting parts.

Before brazing, the parts must be cleaned and the joints fluxed. The brazing filler is added before or just as the joints are heated

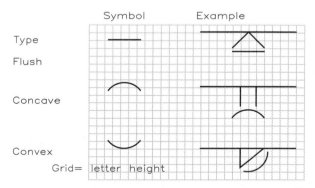

22.23 These contour symbols specify the desired surface finish of a weld.

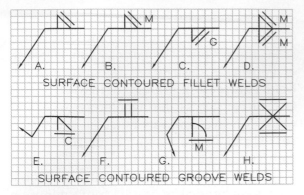

22.24 These examples of contoured weld symbols with letters added indicate the type of finishing to be applied to the weld (M, machining; G, grinding; C, chipping).

beyond the filler's melting point. After the filler material has melted, it is allowed to flow between the parts to form the joint. As **Figure 22.25** shows, there are two basic brazing joints: lap joints and butt joints.

Brazing is used to join parts, to provide gas- and liquid-tight joints, to ensure electrical conductivity, and to aid in repair and salvage. Brazed joints withstand more stress, higher temperature, and more vibration than soft-soldered joints.

22.8 Soldering

Soldering is the process of joining two metal parts with a third metal that melts below the temperature of the metals being joined. Solders are alloys of nonferrous metals that melt below 800°F. Widely used in the automotive and electrical industries, soldering is one of the basic techniques of welding and often is done by hand with a soldering iron like the one depicted in **Figure 22.26**. The iron is placed on the joint to heat it and to melt the solder. Basic soldering is noted on the drawing with a leader simply as SOLDER, and other specifications are noted as needed as shown in **Figure 22.26**.

By necessity, the coverage in this chapter is introductory in nature, but adequate for a basic understanding of how to specify welding on an engineering drawing. More detailed information on welding is available from the American Welding Society, 2501 N.W. 7th Street, Miami, Florida 33125. This society maintains and publishes guidelines and standards for the technology of welding.

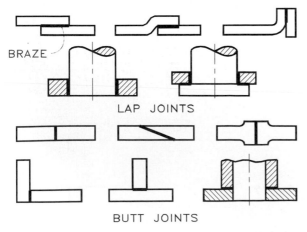

22.25 The two basic types of brazing joints are lap joints and butt joints.

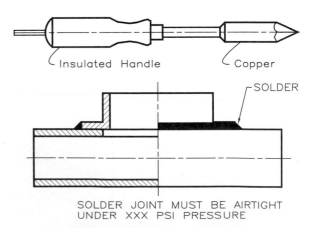

22.26 This typical hand-held soldering iron is used to soft-solder two parts together. The method of notating a drawing for soldering is also shown.

Problems

Solve these problems on size A sheets laid out on a grid of 0.20 inches (5 mm).

1–8. (**Figure 22.27**) Give welding notes to include the information specified for each problem. Omit instructional information from the solution.

9. (**Figure 22.28**) The shaft socket has a base of 4 in. × 4 in. Draw a top and front view of it, approximate its dimensions, and show the appropriate welding notes for its fabrication.

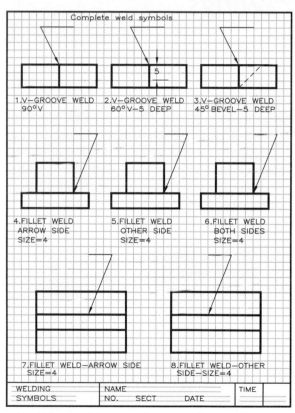

22.27 Problems 1–8.

22.28 Problem 9.

10. (**Figure 22.29**) The angle fixture has faces of 9 × 12 and 10 × 12. Redesign the fixture to have ribs that are welded instead of cast as in this example and show welding symbols. Estimate the unspecified dimensions.

22.29 Problem 10.

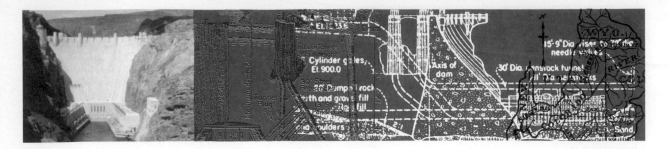

23
Working Drawings

23.1 Introduction

Working drawings are the drawings from which a design is implemented. All principles of orthographic projection and techniques of graphics can be used to communicate the details of a project in working drawings. A **detail drawing** is a working drawing of a single part (or detail) within the set of working drawings.

Specifications are the written instructions that accompany working drawings. When the design can be represented on a few sheets, the specifications are usually written on the drawings to consolidate the information into a single format.

All parts must interact with other parts to some degree to yield the desired function from a design. Before detail drawings of individual parts are made, the designer must thoroughly analyze the working drawing to ensure that the parts fit properly with mating parts, that the correct tolerances are applied, that the contact surfaces are properly finished, and that the proper motion is possible between the parts. Much of the work in preparing working drawings is done by the drafter, but the designer, who is usually an engineer, is responsible for their correctness. It is working drawings that bring products and systems into being.

23.2 Working Drawings as Legal Documents

Working drawings are legal contracts that document the design details and specifications as directed by the engineer. Therefore drawings must be as clear, precise, and thorough as possible. Revisions and modifications of a project at the time of production or construction are much more expensive than when done in the preliminary design stages.

Poorly executed working drawings result in wasted time and resources and increase implementation costs. To be economically competitive, drawings must be as error-free as possible.

Working drawings specify all aspects of the design, reflecting the soundness of engineering and function of the finished product and economy of fabrication. The working drawing is the instrument that is most likely to establish the responsibility for any failure to meet specifications during implementation.

23.3 Dimensions and Units

English System

The inch is the basic unit of the English system, and virtually all shop drawings made with English units are dimensioned in inches.

This practice is followed even when dimensions are several feet in length.

The base flange shown in **Figure 23.1** is an example of a relatively simple working drawing of a part. However, there are many drawings of this simplicity that must be designed, developed, and detailed in order for the overall project to come into being.

The base flange is dimensioned with two-place decimal inches except where four-place decimal inches are used for toleranced dimensions. Inch marks (″) are omitted from dimensions on working drawings because the units are understood to be in inches, and their

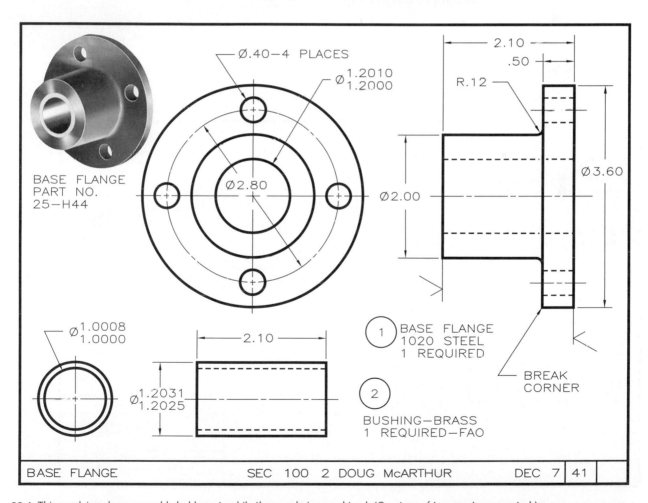

23.1 This revolving clamp assembly holds parts while they are being machined. (Courtesy of Jergens, Incorporated.)

omission saves drafting time. Finish marks are applied to the surfaces that must be machined smooth. Notice that the dimensions are spaced and applied in accordance with the principles covered in Chapters 20 and 21. A photograph showing a three-dimensional view of the flange has been inserted in the corner of the drawing as a raster image by using AutoCAD.

The clamp illustrated in **Figure 23.2** is detailed in three sheets of a working drawing (**Figures 23.3–23.5**), which are also dimensioned in inches. Decimal fractions are preferable to common fractions, although common fractions are still used (although mostly by architects). Arithmetic can be done with

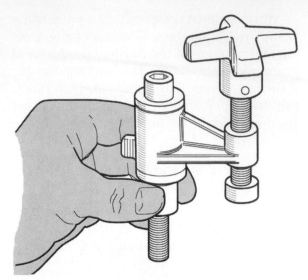

23.2 This revolving clamp assembly holds parts while they are being machined. (Courtesy of Jergens, Incorporated.)

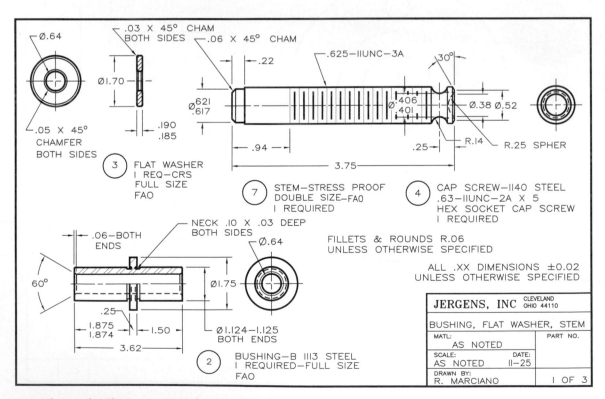

23.3 (Sheet 1 of 3) This is a computer-drawn working drawing of parts of the clamp assembly shown in Figure 23.2. (Courtesy of Jergens, Incorporated.)

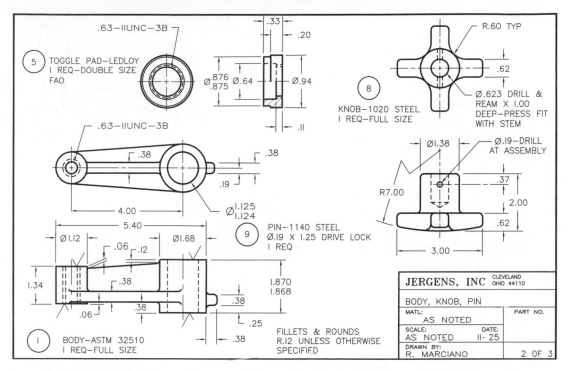

23.4 (Sheet 2 of 3) This continuation of Figure 23.3 shows additional parts of the clamp assembly. (Courtesy of Jergens, Incorporated.)

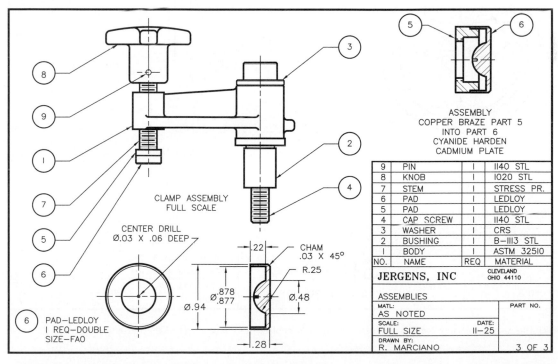

23.5 (Sheet 3 of 3) The pad assembly, an orthographic assembly drawing of the clamp, and a parts list are shown. (Courtesy of Jergens, Incorporated.)

greater ease with decimal fractions than with common fractions.

Usually, several dimensioned orthographic views of parts may be shown on each sheet. However, some companies have policies that views of only one part be drawn on a sheet, even if the part is extremely simple, such as a threaded fastener or the base flange shown in **Figure 23.1**.

The arrangement of views of parts on the sheet need not attempt to show the relationship of the parts when assembled; the views are simply positioned to best fit the available space on the sheet. The views of each part are labeled with a part number, a name for identi-

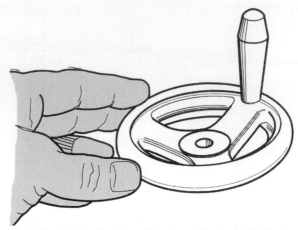

23.6 This photo is of a left-end handcrank that is detailed in the working drawing in Figures 23.7 and 23.8.

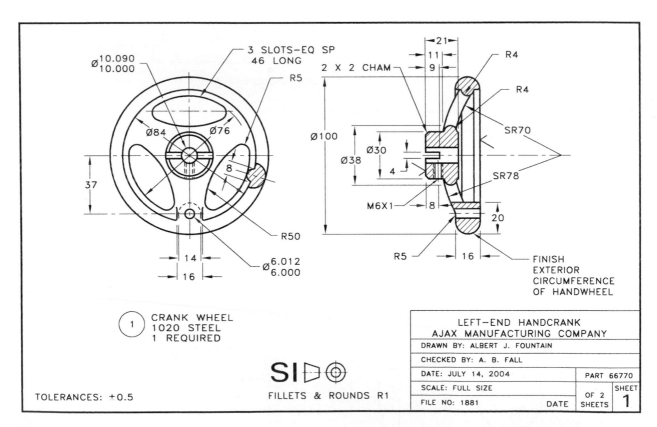

23.7 (Sheet 1 of 2) This set of working drawings (dimensions in mm) depicts the crank wheel of the left-end handcrank shown in Figure 23.6.

fication, the material it is made of, the number of the parts required, and any other notes necessary to explain manufacturing procedures.

The purpose of the orthographic assembly drawing shown on Sheet 3 (**Figure 23.5**) is to illustrate how the parts are to fit together. Each part is numbered and cross-referenced with the part numbers in the parts list that serves as a bill of materials.

Metric System

The millimeter is the basic unit of the metric system, and dimensions usually are given to the nearest whole millimeter without decimal fractions (except to specify tolerances, which may require three-place decimals). Metric abbreviations (mm) after the numerals are omitted from dimensions because the SI symbol near the title block indicates that all units are metric. If you have trouble relating to the length of a millimeter, recall that the fingernail of your index finger is about 10 mm wide.

The left-end handcrank (**Figure 23.6**) is depicted and dimensioned in millimeters in the working drawings shown in **Figure 23.7** and **Figure 23.8** on two size B sheets. Dimensions and notes along with the descriptive views give the information needed to construct the four pieces.

The orthographic, sectioned assembly drawing of the left-end handcrank shown in

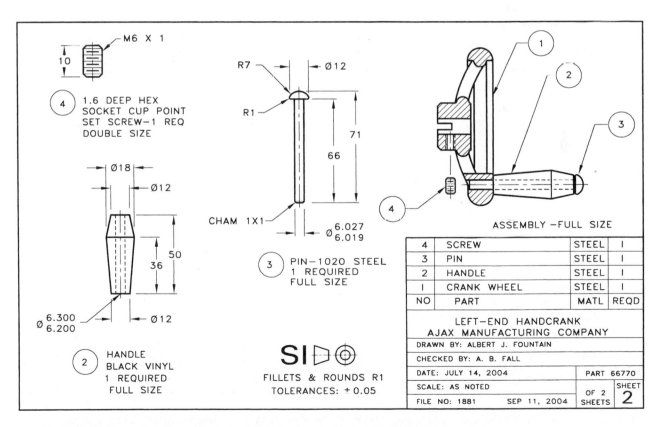

23.8 **(Sheet 2 of 2)** This continuation of Figure 23.7 includes an assembly drawing and parts list.

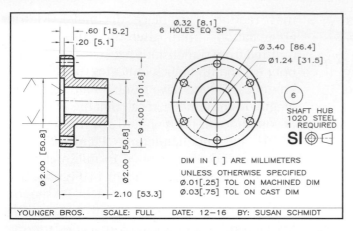

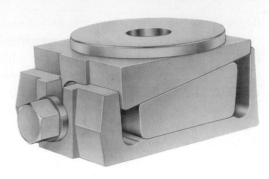

DIM IN [ ] ARE MILLIMETERS
UNLESS OTHERWISE SPECIFIED
Ø.01[.25] TOL ON MACHINED DIM
Ø.03[.75] TOL ON CAST DIM

SI ⊕ ◁

YOUNGER BROS.   SCALE: FULL   DATE: 12—16   BY: SUSAN SCHMIDT

23.9 In this dual-dimensioned drawing, dimensions are shown in millimeters; their equivalents in inches are given in brackets.

23.10 This Lev-L-Line lifting device is used to level heavy machinery. (Courtesy of Unisorb Machinery Installation Systems.)

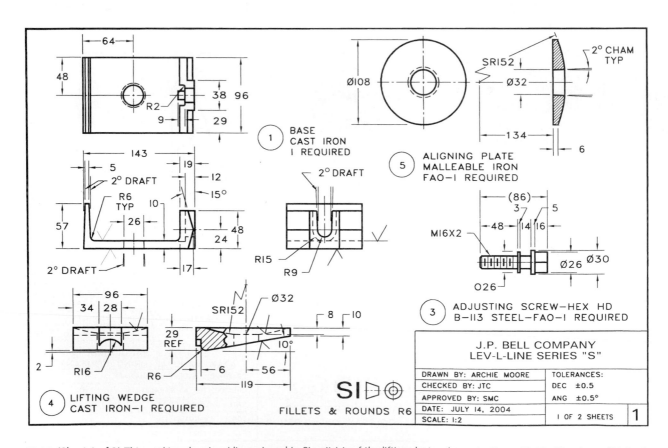

23.11 (Sheet 1 of 2) This working drawing (dimensioned in SI units) is of the lifting device shown in Figure 23.10. (Courtesy of Unisorb Machinery Installation Systems.)

Figure 23.8 illustrates how the parts are to be put together. The numbers in the balloons, the part numbers, provide a cross-reference to the parts list placed just above the title block.

Dual Dimensions

Some working drawings carry both inch and millimeter dimensions as shown in **Figure 23.9**, where the dimensions in parentheses or brackets are millimeters. The units may also appear as millimeters first and then be converted and shown in brackets as inches. Converting from one unit to the other results in fractional round-off errors. An explanation

of the primary unit system for each drawing should be noted in the title block.

Metric Working Drawing Example

Figure 23.10 is a photograph of a lifting device used to level heavy equipment such as lathes and milling machines. The device raises or lowers the machinery when the screw is rotated that slides two wedges together. A two-sheet working drawing that gives the details of the parts of the lifting device is shown in **Figures 23.11** and **23.12**. The SI symbol indicates that the dimensions are in millimeters and the truncated cone indicates

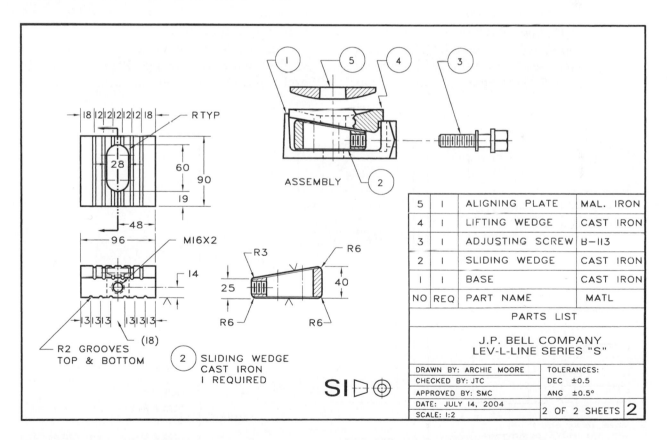

23.12 (Sheet 2 of 2) This continuation of Figure 23.11 is a further continuation of the working drawing and assembly drawing of the lifting device. (Courtesy of Unisorb Machinery Installation Systems.)

that the orthographic views are drawn using third-angle projection. The assembly drawing in **Fig. 23.12** illustrates how the parts are to be assembled after they have been made.

23.4 Laying Out a Detail Drawing

By Computer

You may lay out a simple working drawing of a single part by computer by beginning with the border (**Figure 23.13**). Although this example is elementary, it is useful in describing the procedures for laying out more complex drawings. Allow a margin of at least 0.25 in. (7 mm) between the edge of the sheet and the borders. Outline space for the title block in the lower right-hand corner of the sheet to ensure that the drawing or notes do not occupy this area.

First, make a freehand sketch to determine the necessary views, the number of dimensions, and their placement so that you can select the proper scale. Then use the computer to produce the views close together to make projection from view to view easier. Next, *Move* the views apart to make room for notes and dimensions. Finally, add dimensions, notes, the SI symbol, and the title block to complete the drawing.

By Hand

When making a drawing by hand with instruments on paper or film, first lay out the views and dimensions on a different sheet of paper. Then overlay the drawing with vellum or film and trace it to obtain the final drawing. You must use guidelines for lettering for each dimension and note. Lightly draw the guidelines or underlay the drawing with a sheet containing guidelines.

Figure 23.14 shows the standard sheet sizes for working drawings. Paper, film, cloth, and reproduction materials are available in

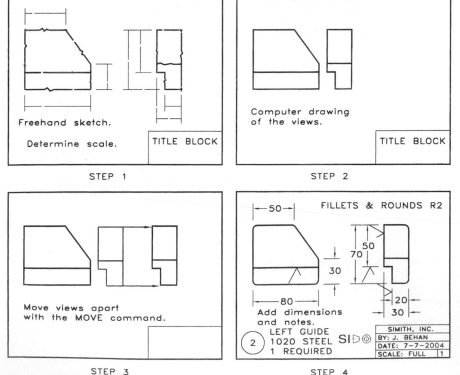

23.13 Laying out a working drawing by computer:

Step 1 Make a freehand sketch to determine the views and dimensions needed.

Step 2 Plot the views close together to make projection easier.

Step 3 Use the *Move* command to move the views apart, making room for the dimensions.

Step 4 Insert dimensions and notes and complete the title block.

ENGLISH SIZES			METRIC SIZES		
A	11 x 8.5		A4	297 X 210	
B	17 X 11		A3	420 X 297	
C	22 X 17		A2	594 X 420	
D	34 X 22		A1	841 X 594	
E	44 X 34		A0	1189 X 841	

23.14 These are the standard sheet sizes for working drawings dimensioned in inches and millimeters.

these modular sizes; good practice requires that you make drawings in one of these standard sizes. Modular sized drawings can be folded to fit standard-sized envelopes, match the sizes of print paper, and fit in standard size filing cabinets.

23.5 Notes and Other Information

Title Blocks and Parts Lists

Figure 23.15 shows a title block and parts list suitable for most student assignments. Title blocks usually are placed in the lower right-hand corner of the drawing sheet against the borders. The parts list (**Figure 23.15**) should be placed directly over the title block (see also **Figures 23.8** and **23.12**).

　　Title Blocks In practice, title blocks usually contain the title or part name, drafter, date,

2	SHAFT	2	1020 STL	.38
1	BASE	1	CAST IRON	
NO	PART NAME	REQ	MATERIAL	
PARTS LIST				

|← 5" (Approximately) →|

TITLE		.38
BY: RED GRANGE	SECT: 500	
DATE: MAY 2, 2004	SHEET 1	
SCALE: FULL SIZE	OF 1 SHEETS	

$\frac{1}{8}$" LETTERS

23.15 This typical title block and parts list is suitable for most student assignments.

REVISIONS	COMPANY NAME
	COMPANY ADDRESS
CHG. HEIGHT	TITLE:　LEFT—END BEARING
FAO	DRAWN BY:　JOHNNY RINGO
	CHECKED BY:　FRED J. DODGE
	DATE:　JULY 14, 2004
	SCALE　HALF SIZE ┃ SHEET 2 OF 3 SHEETS

23.16 This title block, which includes a revision block, is typical of those used in industry.

scale, company, and sheet number. Other information, such as tolerances, checkers, and materials, also may be given. **Figure 23.16** shows another example of a title block, which is typical of those used by various industries. Any modifications or changes added after the first version to improve the design are shown in the revision blocks.

　　Depending on the complexity of the project, a set of working drawings may contain from one to more than a hundred sheets. Therefore, giving the number of each sheet and the total number of sheets in the set on each sheet is important (for example, Sheet 2 of 6, Sheet 3 of 6, and so on).

Computer Method　A computer shortcut for filling in a title block that will be used on several sheets is shown in **Figure 23.17**. By computer,

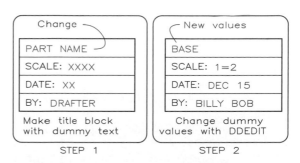

23.17 Producing a title block by computer:

Step 1 Draw the title block and add dummy text values, using the desired text style and size. Make a *Block* of the title block.

Step 2 Position the title block against the lower bottom and right borders. *Explode* the *Block* and use the *Ddedit* command to convert the dummy values to actual values.

you need only draw the title block once, filling it in with dummy values to establish the positions of the text. By using the *Ddedit* command, the dummy entries can be updated with applicable values after the block is inserted or copied into a new drawing.

A similar shortcut in the design of a title block involves the use of *Attributes* (see Chapter 38). In that case, the program will prompt you to *Insert* entries into the title block one at a time.

Parts List   The part numbers and part names in the parts list correspond to those given to each part depicted on the working drawings. In addition, the number of identical parts required are given along with the material used to make each part. Because the exact material (for example, 1020 STEEL) is designated for each part on the drawing, the material in the parts list may be shortened to STEEL, which requires less space.

Patent Rights Note

A note near the title block that names Jack Omohundro as the inventor of the part or process is used to establish ownership of the design (**Figure 23.18**). An associate, J. B. Hickok,

INVENTOR'S SIGNATURE AND DATE

INVENTOR:
JACK OMOHUNDRO
MAY 2, 2004   *Jack Omohundro*

WITNESS:
J. B. HICKOK   *J. B. Hickok*
MAY 2, 2004

WITNESS

DRAFT

DATE:

MATER

FEATU

SHEET

23.18 This note next to the title block names the inventor and is witnessed by an associate to establish ownership of a design for patent purposes.

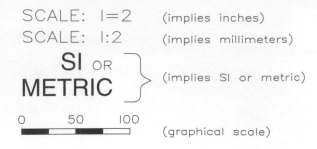

SCALE: 1=2   (implies inches)

SCALE: 1:2   (implies millimeters)

SI OR METRIC }   (implies SI or metric)

0   50   100

(graphical scale)

23.19 Specify scales in English and SI units on working drawings with these methods.

signs and dates the drawing as a witness to the designer's work. This type of note establishes ownership of the ideas and dates of their development to help the inventor obtain a patent. An even better case for design ownership is made if a second witness signs and dates the drawing. As modifications to the design are made, those drawings should receive the same documentation.

Scale Specification

If all working drawings in a set are the same scale, you need to indicate it only once in the title block on each sheet. If several detail drawings on a working drawing are different scales, indicate them on the drawing under each set of views. In this case, indicate AS SHOWN in the title block opposite scale. When a drawing is not to scale, place the abbreviation NTS (not to scale) in the title block.

Figure 23.19 shows several methods of indicating scales. Use of the colon (for example, 1:2) implies the metric system; use of the equal sign (for example, 1=2) implies the English system—but these are not absolute rules. The SI symbol or metric designation on a drawing specifies that millimeters are the units of measurement.

In some cases, you may want to show a graphical scale with calibrations on a drawing to permit the interpretation of linear measurements by transferring them with dividers from the drawing to the scale.

Tolerances

Recall from Chapter 21 that you may use general notes on working drawings to specify the dimension tolerances. **Figure 23.20** shows a table of values with boxes in which you can make a check mark to indicate whether the units are in inches or millimeters. Position plus-and-minus tolerance value under each common or decimal fraction. For example, this table specifies that each dimension with two-place decimals will have a tolerance of ± 0.10 in. You may also give angular tolerances in general notes ($\pm 0.5°$, for example).

Part Labeling

Give each part a name and number, using letters and numbers 1/8-in. (3 mm) high (**Figure 23.21**). Place part numbers inside circles, called balloons, having diameters approximately four times the height of the numerals.

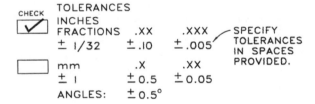

23.20 General tolerance notes on working drawings specify the dimension tolerance values permitted.

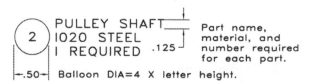

23.21 Name and number each part on a working drawing for use in the parts list, indicate the number of parts of this particular part that are needed and give the part's material.

Place part numbers near the views to which they apply, so their association will be clear. On assembly drawings, balloons are especially important because the same part numbers are used in the parts list. Show the number of parts required near the part name.

23.6 Checking a Drawing

People who check drawings must have special qualifications that enable them to identify errors and to suggest revisions and modifications that will result in a better product at a lower cost. A checker may be a chief drafter experienced in drafting and manufacturing processes, or the engineer or designer who originated the project. In large companies, personnel in the various shops involved in production review the drawings to ensure that the most efficient production methods are specified for each part.

Checkers never check the original drawing; instead, they mark corrections with a colored pencil on a diazo (blue-line) print. They return the marked-up print to the drafter who revises the original and makes another print for final approval.

In **Figure 23.22**, the various modifications made by checkers are labeled with letters that are circled and placed near the revisions. The drafter lists and dates changes in the revision record, which lists the revisions made.

Checkers inspect a working or detail drawing for correctness and soundness of design. In addition, they are responsible for the drawing's completeness, quality, readability, and clarity, which reflect on the lettering and drafting techniques used. However, it is the project engineer who is ultimately responsible for correctness of the drawings and documents. Lettering and text quality is especially important because the shop person is guided by lettered notes and dimensions.

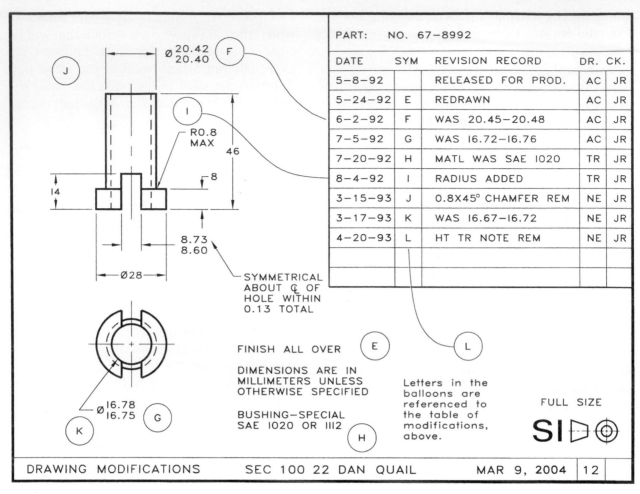

Part: No. 67-8992

DATE	SYM	REVISION RECORD	DR.	CK.
5-8-92		RELEASED FOR PROD.	AC	JR
5-24-92	E	REDRAWN	AC	JR
6-2-92	F	WAS 20.45-20.48	AC	JR
7-5-92	G	WAS 16.72-16.76	AC	JR
7-20-92	H	MATL WAS SAE 1020	TR	JR
8-4-92	I	RADIUS ADDED	TR	JR
3-15-93	J	0.8X45° CHAMFER REM	NE	JR
3-17-93	K	WAS 16.67-16.72	NE	JR
4-20-93	L	HT TR NOTE REM	NE	JR

FINISH ALL OVER

DIMENSIONS ARE IN MILLIMETERS UNLESS OTHERWISE SPECIFIED

BUSHING—SPECIAL SAE 1020 OR 1112

Letters in the balloons are referenced to the table of modifications, above.

FULL SIZE

SI

DRAWING MODIFICATIONS   SEC 100 22 DAN QUAIL   MAR 9, 2004   12

23.22 The modifications to this working drawing are noted near the details revised. The letters in balloons correspond to those in the table of revisions.

Checking Students' Work

The best way for you to check your drawings for adequate dimensions is to rapidly make a scale drawing of the part using the dimensions from the working drawings. You can identify missing dimensions more easily in this way than by reading a drawing by eye. **Figure 23.23** shows a grading scale for checking working drawings prepared by students, with hypothetical grading shown. Use this list as an outline for reviewing working drawings to ensure that you have met the main requirements.

23.7 Drafter's Log

In addition to the individual revision records, drafters should keep a log of all changes made

TITLE BLOCK (5 points)			DRAFTSMANSHIP (19 points)		
Student's name	1	1	Thicknesses of lines	6	5.5
Checker's name	1	1	Lettering	6	5
Date	1	1	Neatness	3	1.5
Scale	1	1	Reproduction quality	4	3
Sheet number	1	1	GENERAL NOTES (5 points)		
DRAWING DETAILS (19 pts)			SI symbol	2	2
Properly drawn views	10	9	Third angle symbol	2	2
Spacing of views	5	4.5	General tolerance note	1	1
Part names and numbers	2	1.5	ASSEMBLY DRAWING (15 points)		
Sections & conventions	2	1.5	Descriptive views	6	5.5
DESIGN INFORMATION (12 pts)			Clarity of assembly	3	2.8
Proper tolerances	5	4	Parts list completeness	4	3.6
Surface texture symbols	2	2	Part numbers in balloons	2	2
Thread notes & symbols	5	4.5	PRESENTATION (5 points)		
DIMEN. PRACTICES (20 pts)			Properly stapled	2	2
Proper arrowheads	3	2.5	Properly trimmed	1	1
Positions of dimensions	3	2.5	Properly folded	1	1
Completeness of dimens	4	3	Grade sheet attached	1	1
Fillet & round notes	2	2	Total 100		88
Machined-hole notes	2	1.5			
Inch/mm marks omitted	2	2			
Dimensions from best views	4	3.5			

23.23 This checklist may be used to evaluate a student's working-drawing assignment.

during a project. As the project progresses, the drafter should record the changes, dates, and people involved. Such a log allows anyone reviewing the project in the future to understand easily and clearly the process used in arriving at the final design.

Calculations often are made during a drawing's preparation. If they are lost or poorly done, they may have to be redone during a later revision; therefore they should be a permanent part of the log in order to preserve previously expended work.

23.8 Assembly Drawings

After parts have been made according to the specifications of the working drawings, they will be assembled (**Figure 23.24**) in accordance with the directions of an assembly drawing. Two general types of assembly drawings are **orthographic assemblies** and **pictorial assemblies.** Dimensions usually are omitted from assembly drawings.

23.24 An assembly drawing explains how the parts of a product, such as this Ford tractor, are to be assembled. (Courtesy of Ford Motor Company.)

The lifting device shown in Fig. 23.10 is depicted in an isometric assembly in **Figure 23.25**. Each part is numbered with a balloon and leader to cross-reference them to the parts list, where more information about each part is given.

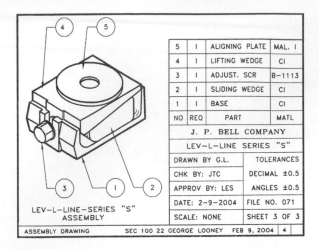

23.25 This isometric assembly drawing depicts the parts of the lifting device shown in Figure 23.10 fully assembled. Dimensions usually are omitted from assembly drawings and a parts list is given.

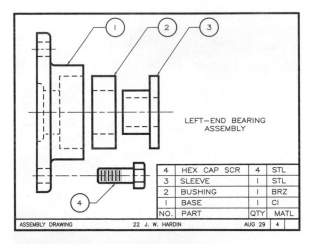

23.26 This exploded orthographic assembly illustrates how the parts shown are to be put together.

Figure 23.26 shows an orthographic exploded assembly drawing. In many applications, the arrangement of parts may be easier to understand when the parts are shown exploded along their centerlines. These views are shown as regular orthographic views, with some lines shown as hidden lines and others omitted.

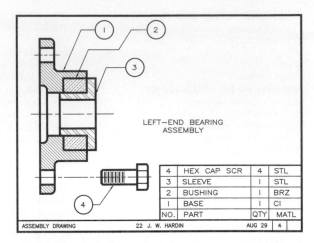

23.27 This sectioned orthographic assembly shows the parts from Figure 23.26 in their assembled positions, except for the exploded bolt.

Assembly of the same part is shown in **Figure 23.27** in an orthographic assembly drawing, in which the parts are depicted in their assembled positions. The views are sectioned to make them easier to understand.

Figure 23.28 shows a pulley assembly in an exploded pictorial assembly drawing, illustrating how the parts fit together, leaving no doubt as to how the parts relate to each other. Part numbers are given in balloons to complete the drawing.

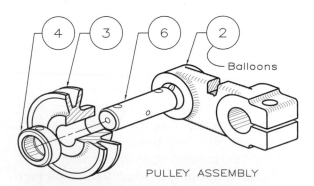

23.28 This exploded pictorial assembly drawing is of a shaft bearing assembly. (Courtesy of Cameron Iron Works, Incorporated.)

23.9 Freehand Working Drawings

A freehand sketch can serve the same purpose as an instrument drawing, provided that the part is sufficiently simple and that the essential dimensions are shown (**Figure 23.29**). Use the same principles of making working drawings with instruments when making working drawings freehand. A sketch can be made quickly, and it can be made in the field, fabrication shop, or other locations where drafting-room instruments are not readily available.

23.10 Working Drawings for Forged Parts and Castings

The two versions of a part shown in **Figure 23.30** illustrate the difference between a part that has been forged into shape (sometimes called a **blank**) and its final state after the forging has been machined. Recall from Chapter 19 that a forging is a rough form made by hammering (forging) the metal into shape or pressing it between two forms (called dies). The forged part is then machined to its specified finished dimensions and tolerances so it will function as intended.

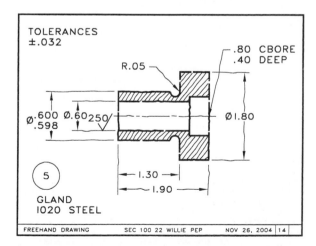

23.29 A freehand working drawing with the essential dimensions can be as adequate as an instrument-drawn detail drawing for simple parts.

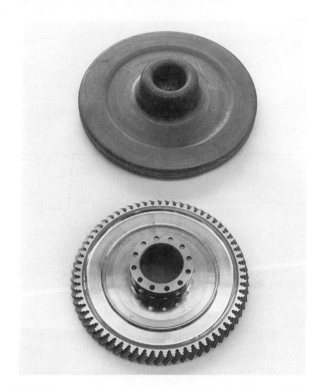

23.30 The top part is a blank that has been forged. It will look like the part on the bottom after it has been machined.

A casting, like a forging, must be machined so that it too will fit and function with other parts when assembled; therefore additional material is added to the areas where metal will be removed by the machining processes. As covered in Chapter 19, a casting is formed by pouring molten metal into a mold formed by a pattern that is slightly larger than the finished part to compensate for metal shrinkage (**Figure 23.31**). For the pattern to be removable from the sand that forms the mold, its sides must be tapered about 5° to 10°. This taper is called **draft**.

Some industries that work extensively with cast and forged parts required that separate working drawing be made for forged and cast parts (**Figure 23.32A**). More often, however, the parts are detailed on the regular working drawing with the understanding that the features

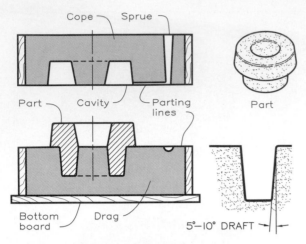

23.31 A two-part sand mold is used to produce a casting. A draft of 5° to 10° is needed to permit withdrawal of the pattern from the sand. Some machining is usually required to finish various features of the casting within specified tolerances.

that are to be machined by operations such as grinding or shaping are made oversize by the fabrication shop.

Problems

Dimensions given on many of the following three-dimensional, pictorial problems (**Figures 23.33–23.40**) do not represent good dimensioning practices because of space limitations and the complexity of the pictorial drawings, but they provide adequate dimensions from which working drawings can be made. In cases where dimensions may be missing, approximate them using your own judgment. Provide all the necessary information, notes, and dimensions to describe the views completely. Use any of the previously covered principles, conventions, and techniques to present the views with the maximum clarity and simplicity.

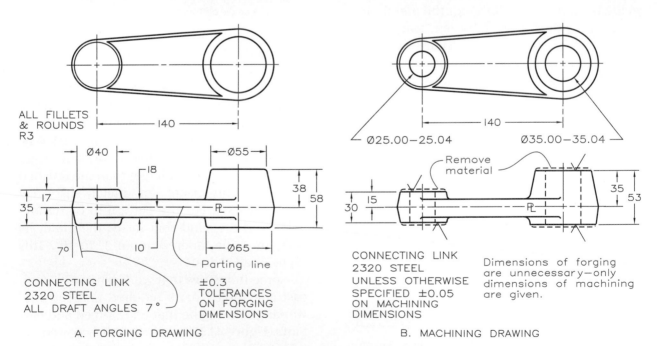

23.32 These separate working drawings, (A) a forging drawing and (B) a machining drawing, give the details of the same part. Often, this information is combined into a single drawing.

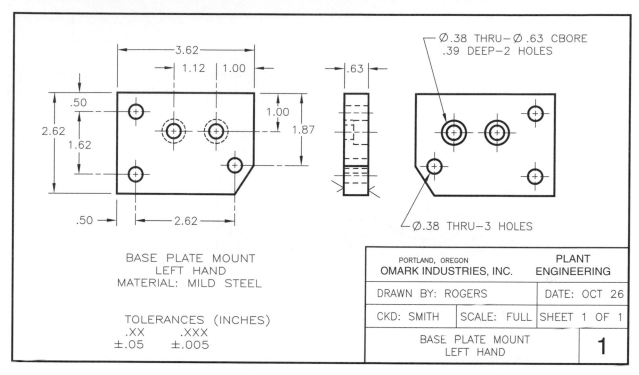

23.33 Duplicate the working drawing of the base plate mount on a size B sheet. (Courtesy of Omark Industries, Inc.)

Working drawings can be made as free-hand sketches, instrument drawings, or computer drawings. The determination of the proper scale, selection of sheet sizes, and the choice and positioning of the views on the drawing sheet will be a major portion of all problem assignments. The making of free-hand, preliminary sketches will be very helpful in making these decisions and saving layout time.

Working Drawing Practice
Reproduce the drawings shown in **Figures 23.41–23.53** as directed. Some have only one sheet and others have more. You may use the dimensions given or convert them to the

other system (from millimeters to inches, for example). The purpose of these assignments is to give you experience in laying out a working drawing and improving your draftsmanship on the board or at the computer.

Working Drawings of Products with Multiple Parts
Make working drawings by hand or by computer, as assigned, of the products consisting of multiple parts shown in **Figures 23.54–23.77** on the sheet sizes specified. Include a title block and the dimensions and notes necessary for manufacturing the part. Draw an assembly that shows how the parts fit together. More than one sheet may be required for a solution.

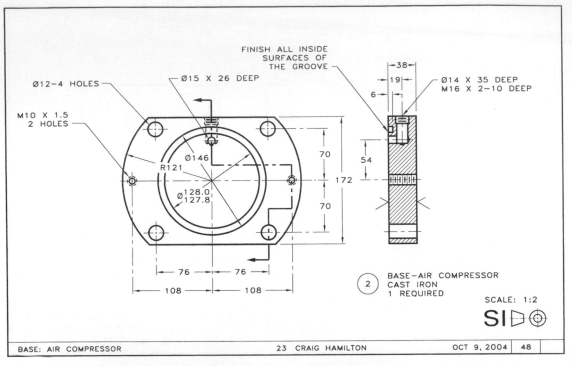

23.34 Duplicate the detail drawing of the air compressor base on a size B sheet.

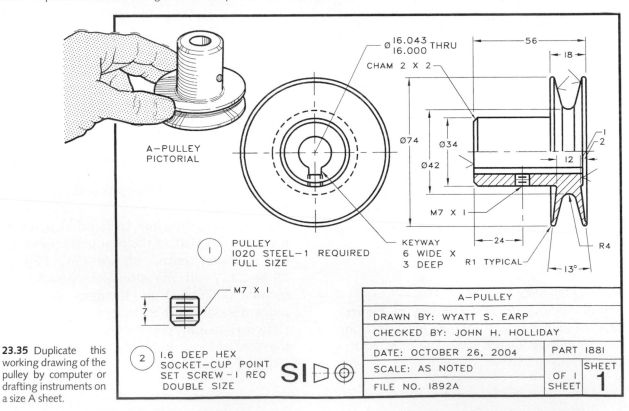

23.35 Duplicate this working drawing of the pulley by computer or drafting instruments on a size A sheet.

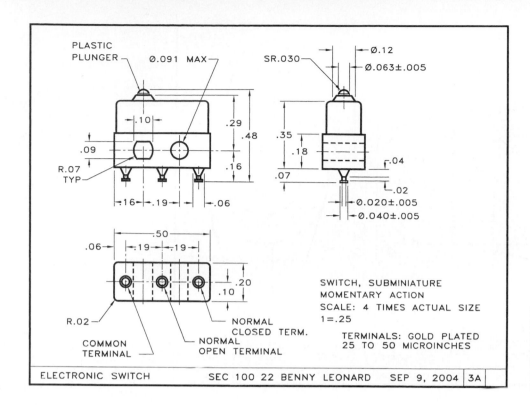

PLASTIC PLUNGER

Ø.091 MAX

SR.030

Ø.12

Ø.063±.005

.10

.29

.48

.09

.35

.18

R.07
TYP

.16

.07

.04

.02

.16 .19 .06

Ø.020±.005

Ø.040±.005

.50

.06 .19 .19

R.02

.20

.10

COMMON
TERMINAL

NORMAL
OPEN TERMINAL

NORMAL
CLOSED TERM.

SWITCH, SUBMINIATURE
MOMENTARY ACTION
SCALE: 4 TIMES ACTUAL SIZE
1=.25

TERMINALS: GOLD PLATED
25 TO 50 MICROINCHES

ELECTRONIC SWITCH   SEC 100 22 BENNY LEONARD   SEP 9, 2004   3A

23.36 Duplicate this working drawing of the switch by computer or drafting instruments on a size A sheet. Note that the views are enlarged by a factor of 4.

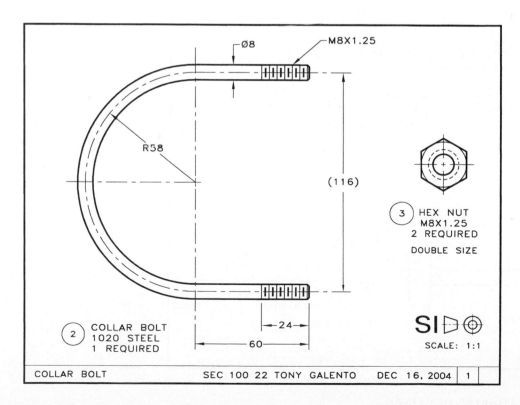

Ø8

M8X1.25

R58

(116)

3 HEX NUT
M8X1.25
2 REQUIRED

DOUBLE SIZE

2 COLLAR BOLT
1020 STEEL
1 REQUIRED

24

60

SI

SCALE: 1:1

COLLAR BOLT   SEC 100 22 TONY GALENTO   DEC 16, 2004   1

23.37 (Sheet 1 of 3) Duplicate this full-size working drawing and assembly (in mm) of the pipe hanger on size A sheets.

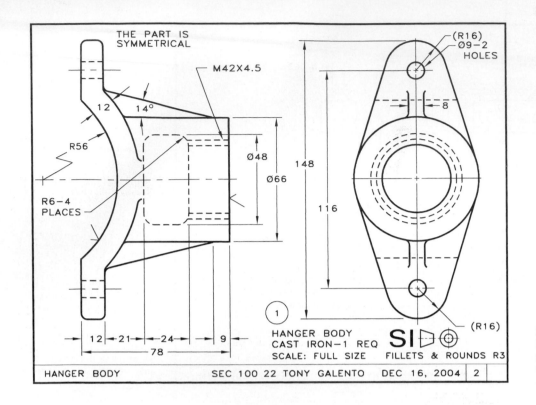

THE PART IS SYMMETRICAL

M42X4.5

12

14°

R56

R6-4 PLACES

Ø48

Ø66

148

116

(R16)
Ø9-2 HOLES

8

(R16)

12 21 24 9

78

1

HANGER BODY
CAST IRON-1 REQ
SCALE: FULL SIZE

SI

FILLETS & ROUNDS R3

HANGER BODY   SEC 100 22 TONY GALENTO   DEC 16, 2004   2

23.38 (Sheet 2 of 3)
Duplicate this second sheet of the pipe hanger working drawing.

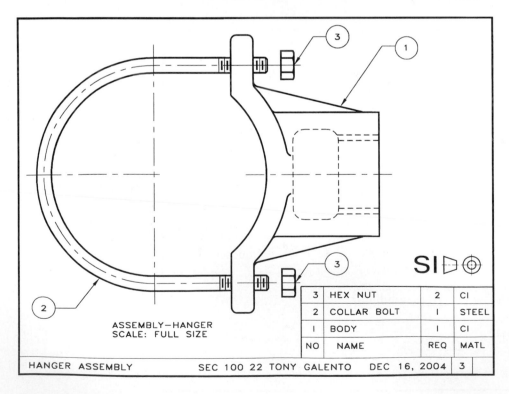

3

1

3

2

ASSEMBLY-HANGER
SCALE: FULL SIZE

SI

3	HEX NUT	2	CI
2	COLLAR BOLT	I	STEEL
I	BODY	I	CI
NO	NAME	REQ	MATL

HANGER ASSEMBLY   SEC 100 22 TONY GALENTO   DEC 16, 2004   3

23.39 (Sheet 3 of 3)
Duplicate this third sheet of the pipe hanger that shows its assembly.

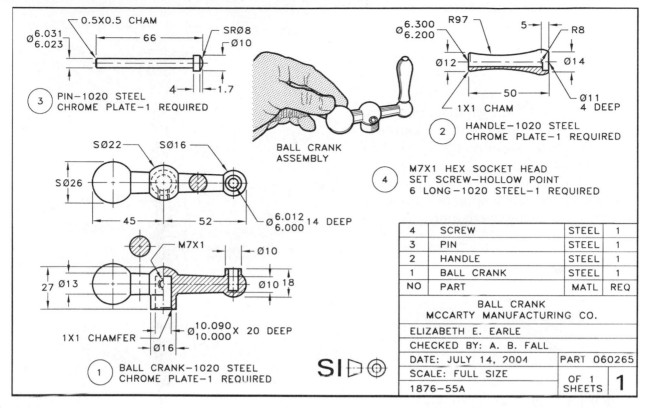

23.40 The ball crank (above) has been detailed in the working drawing. Duplicate the working drawing on a size B sheet. On a second size B sheet, make an assembly drawing of the parts.

Thought Questions

1. How many pages are used to detail the parts of the ball crank (Figure 23.40)?

2. By what means is the handle (part 2) attached to the ball crank (part 1)?

3. What is the allowance between the ball crank (part 1) and the pin (part 3)?

4. What is the allowance between the handle (part 2) and the pin (part 3).

5. By what means is the ball crank (part 1) attached to the shaft (not shown)?

6. One part is not drawn, but specified by a note only. Is this sufficient to describe the part? If so, why?

7. Why does the hole at the bottom of the ball crank (part 1) have an internal chamfer?

8. Why is there a ball at one end of the ball crank?

9. Why are there no toleranced dimensions on the outside of the handle (part 2)?

10. Why were removed sections used to describe the ball crank (part 1)?

Working Drawings: Single Parts

Make working drawings of the assigned parts, providing the necessary information, notes, and dimensions. In cases where dimensions may be missing, approximate them using your own judgment.

The determination of the proper scale, selection of sheet sizes, and the choice and positioning of the views on the drawing sheet will be a major portion of all problem assignments. The making of freehand, preliminary sketches will be very helpful in making these decisions and saving layout time.

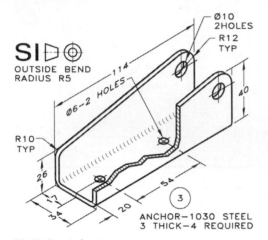

23.41 Size A sheet.

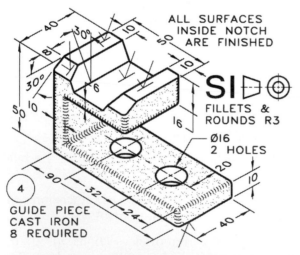

23.42 Size B sheet.

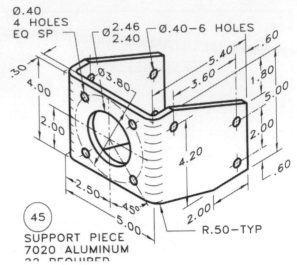

23.43 Size B sheet.

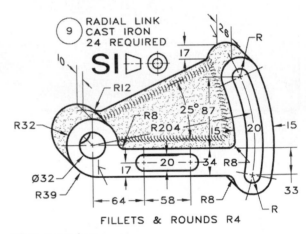

23.44 Size B sheet.

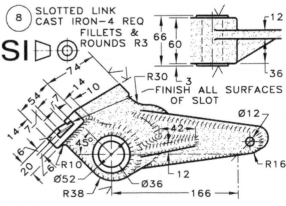

23.45 Size B sheet.

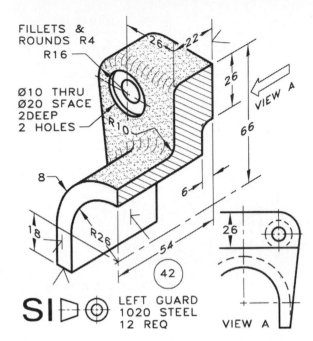

FILLETS &
ROUNDS R4
R16

Ø10 THRU
Ø20 SFACE
2DEEP
2 HOLES

R10

8

18

R26

26

66

6

54

42

26

VIEW A

SI ⬠⊕

LEFT GUARD
1020 STEEL
12 REQ

VIEW A

23.46 Size B sheet.

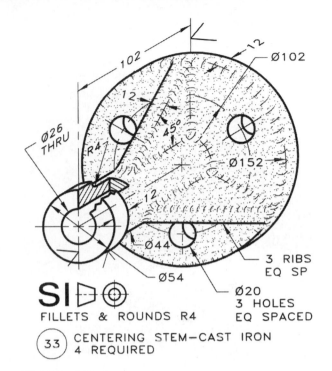

102

Ø26
THRU

R4

12

12

45°

12

Ø54

Ø44

Ø102

Ø152

3 RIBS
EQ SP

Ø20
3 HOLES
EQ SPACED

SI ⬠⊕
FILLETS & ROUNDS R4

33 CENTERING STEM—CAST IRON
4 REQUIRED

23.48 Size B sheet.

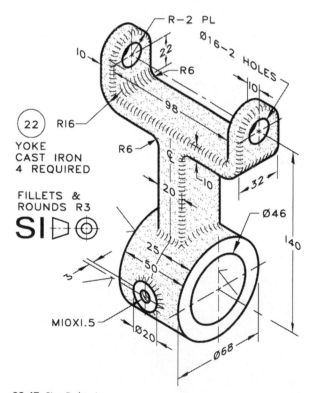

R—2 PL

Ø16—2 HOLES

10

22

R6

10

98

22 R16

R6

YOKE
CAST IRON
4 REQUIRED

FILLETS &
ROUNDS R3

SI ⬠⊕

20

10

32

25

50

3

Ø46

140

M10X1.5

Ø20

Ø68

23.47 Size B sheet.

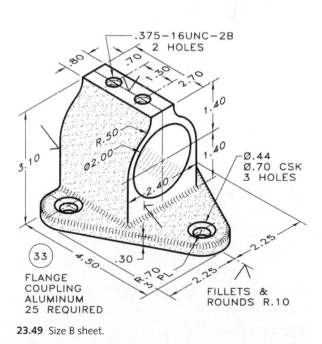

.375—16UNC—2B
2 HOLES

.80

.70

.30

2.70

1.40

1.40

R.50

Ø2.00

3.10

2.40

Ø.44
Ø.70 CSK
3 HOLES

33

FLANGE
COUPLING
ALUMINUM
25 REQUIRED

4.50

.30

R.70
3 PL

2.25

2.25

FILLETS &
ROUNDS R.10

23.49 Size B sheet.

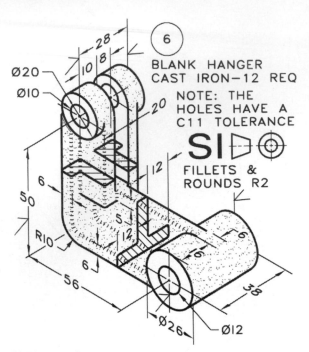

23.50 Size B sheet.

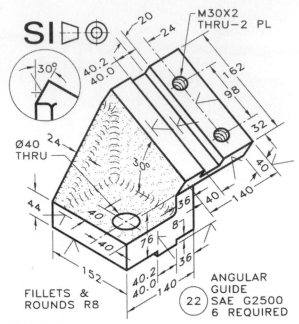

23.52 Size B sheet.

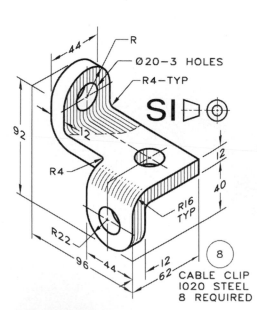

23.51 Size B sheet.

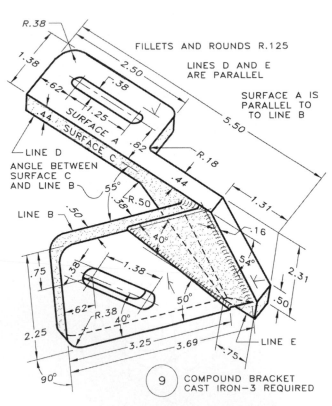

23.53 Size B sheet.

Working Drawings: Multiple Parts

Make working drawings by hand or by computer, as assigned, of the products consisting of the multiple parts shown in **Figures 23.54–23.77** on the suggested sheet sizes. Include a title block, dimensions, and notes necessary for manufacturing the parts. Make an assembly drawing that shows how the parts fit together. More than one sheet may be required.

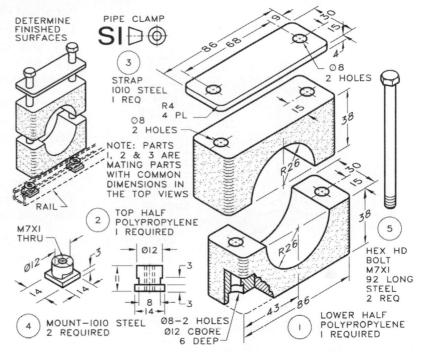

23.55 Size B sheet.

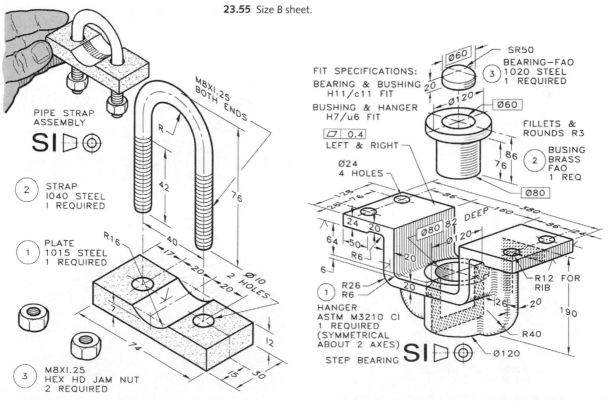

23.54 Size B sheet.

23.56 Size B sheet.

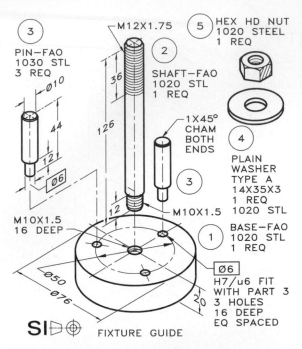

23.57 Make working drawings with an assembly drawing of this adjustable swing stop on size B sheets.

Important Points

1. Fill out the title block completely on all sheets of a working drawing: drafter, checker, company (with address), sheet number, and so forth. This is the beginning step of a working drawing.

2. Prepare the details of each part in a very precise, accurate manner as if you were preparing a legal contract, because that is exactly what working drawing are.

3. Provide adequate space between multiple-view drawings to avoid the crowding of dimensions and notes.

4. The best means of checking your drawing for completeness is by drawing each part using your notes and dimensions.

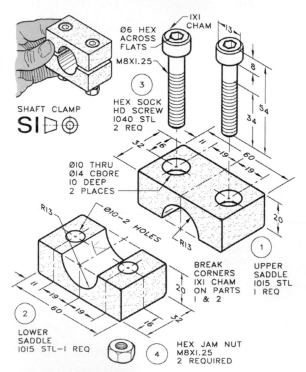

23.58 Make working drawings with an assembly drawing of this flange jig on size B sheets.

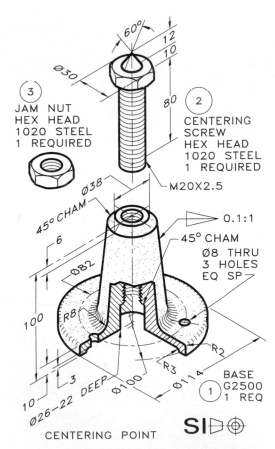

23.59 Make working drawings with an assembly drawing of this pipe clamp on size B sheets.

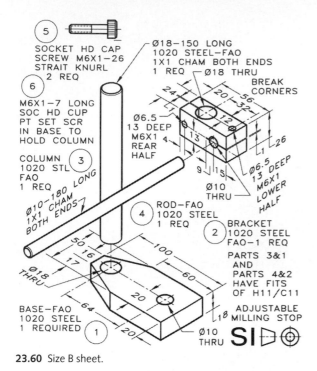

5 SOCKET HD CAP SCREW M6X1-26 STRAIT KNURL 2 REQ

6 M6X1-7 LONG SOC HD CUP PT SET SCR IN BASE TO HOLD COLUMN

COLUMN 3 1020 STL FAO 1 REQ

Ø10-180 LONG 1X1 CHAM BOTH ENDS

Ø18-150 LONG 1020 STEEL-FAO 1X1 CHAM BOTH ENDS 1 REQ — Ø18 THRU

BREAK CORNERS

Ø6.5 13 DEEP M6X1 REAR HALF

Ø6.5 13 DEEP M6X1 LOWER HALF

Ø10 THRU

ROD-FAO 4 1020 STEEL 1 REQ

BRACKET 2 1020 STEEL FAO-1 REQ

PARTS 3&1 AND PARTS 4&2 HAVE FITS OF H11/C11

Ø18 THRU

BASE-FAO 1020 STEEL 1 REQUIRED 1

Ø10 THRU SI

ADJUSTABLE MILLING STOP

23.60 Size B sheet.

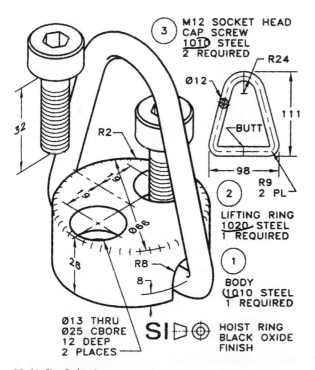

3 M12 SOCKET HEAD CAP SCREW 1010 STEEL 2 REQUIRED

Ø12

R24

R2

BUTT

111

98

R9 2 PL

LIFTING RING 2 1020 STEEL 1 REQUIRED

Ø88

BODY 1 1010 STEEL 1 REQUIRED

R8

28

8

Ø13 THRU Ø25 CBORE 12 DEEP 2 PLACES

SI

HOIST RING BLACK OXIDE FINISH

23.61 Size B sheet.

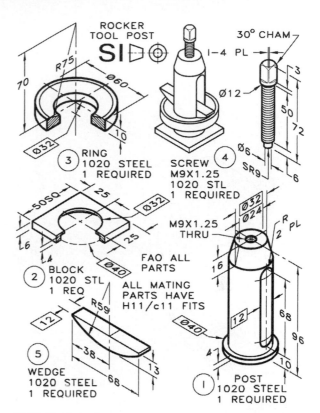

ROCKER TOOL POST SI

R75

70

Ø60

10

Ø32

30° CHAM

1-4 PL

Ø12

3

50

72

Ø6

SR9

6

RING 3 1020 STEEL 1 REQUIRED

SCREW 4 M9X1.25 1020 STL 1 REQUIRED

50 SQ.

25

Ø32

Ø40

25

M9X1.25 THRU

FAO ALL PARTS

BLOCK 2 1020 STL 1 REQ

R59

ALL MATING PARTS HAVE H11/c11 FITS

Ø32 Ø24

R 2 PL

16

12

68

Ø40

96

12

WEDGE 5 1020 STEEL 1 REQUIRED

38

13

68

POST 1 1020 STEEL 1 REQUIRED

4

10

23.62 Size B sheet.

23.60 Make a working drawing with an assembly with a parts list of this adjustable milling stop on size B sheets. The column (part 3) that fits into the base (part 1) must be held in position with a set screw that passes through the end of the base (part 1) and bears against part 3. Select a proper screw for this application and specify it on your working drawing.

23.61 Make a drawing with an assembly of the hoist ring on size B sheets. Give a parts list on the assembly sheet using good drawing and dimensioning practices.

23.62 Make a working drawing with an assembly of the rocker tool post on size B sheets. Give a parts list on the assembly sheet using good drawing and dimensioning practices.

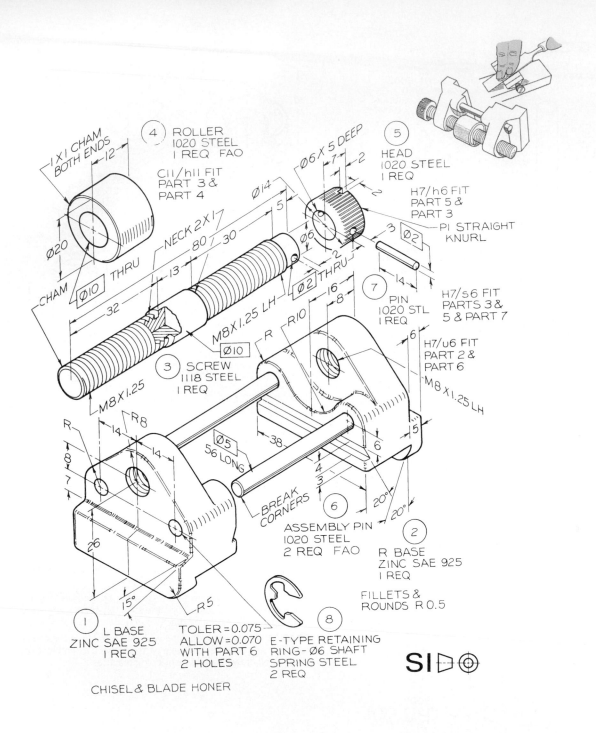

④ ROLLER
1020 STEEL
I REQ FAO

—I X I CHAM
BOTH ENDS—— 12

Ø14

C11/h11 FIT
PART 3 &
PART 4

Ø6 X 5 DEEP

⑤ HEAD
1020 STEEL
I REQ

5

H7/h6 FIT
PART 5 &
PART 3

Ø20

Ø6

NECK 2 X 1   80   30

Ø10   THRU

Ø2

— PI STRAIGHT
KNURL

13

CHAM

Ø10   THRU

32

M8 X 1.25 LH

2

3

Ø2

Ø2 THRU

2

14

⑦ PIN
1020 STL
I REQ

H7/s6 FIT
PARTS 3 &
5 & PART 7

16

8

6

M8 X 1.25

Ø10

③ SCREW
1118 STEEL
I REQ

R   R10

R

6

H7/u6 FIT
PART 2 &
PART 6

M8 X 1.25 LH

R   R8

14

Ø5
56 LONG

38

5

R

8

14

6

7

26

BREAK
CORNERS

4

3

20°

20°

⑥ ASSEMBLY PIN
1020 STEEL
2 REQ   FAO

② R BASE
ZINC SAE 925
I REQ

FILLETS &
ROUNDS R 0.5

15°

R 5

① L BASE
ZINC SAE 925
I REQ

TOLER = 0.075
ALLOW = 0.070
WITH PART 6
2 HOLES

⑧ E-TYPE RETAINING
RING- Ø6 SHAFT
SPRING STEEL
2 REQ

SI ▷⊕

CHISEL & BLADE HONER

23.63 Make a drawing with an assembly of the hoist ring on size B sheets. Give a parts list on the assembly sheet using good drawing and dimensioning practices.

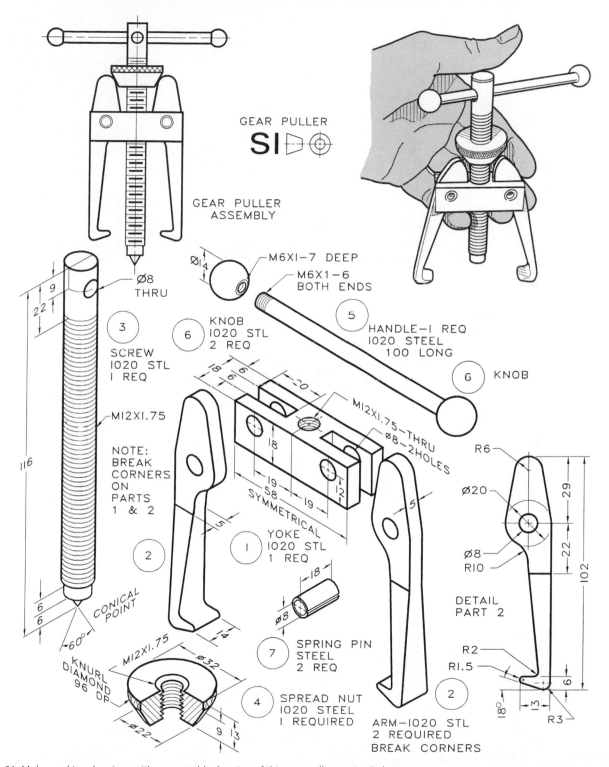

GEAR PULLER
SI ▷⊕

GEAR PULLER
ASSEMBLY

Ø8
THRU

③

SCREW
1020 STL
1 REQ

M12X1.75

NOTE:
BREAK
CORNERS
ON
PARTS
1 & 2

②

9
22
116
6
6

60°

CONICAL
POINT

KNURL
DIAMOND
96 DP

M12X1.75

Ø32
Ø22
9 13

④ SPREAD NUT
1020 STEEL
1 REQUIRED

Ø14
⑥ KNOB
1020 STL
2 REQ

M6X1–7 DEEP
M6X1–6
BOTH ENDS

⑤ HANDLE—1 REQ
1020 STEEL
100 LONG

⑥ KNOB

18
6
7
20

M12X1.75—THRU
Ø8—2HOLES

18

19
58
19

12

5

SYMMETRICAL

① YOKE
1020 STL
1 REQ

18
Ø8

⑦ SPRING PIN
STEEL
2 REQ

5

② ARM—1020 STL
2 REQUIRED
BREAK CORNERS

R6
Ø20
29
22
102
Ø8
R10

DETAIL
PART 2

R2
R1.5
18°
13
6
R3

②

23.64 Make working drawings with an assembly drawing of this gear puller on size B sheets.

PROBLEMS • 409

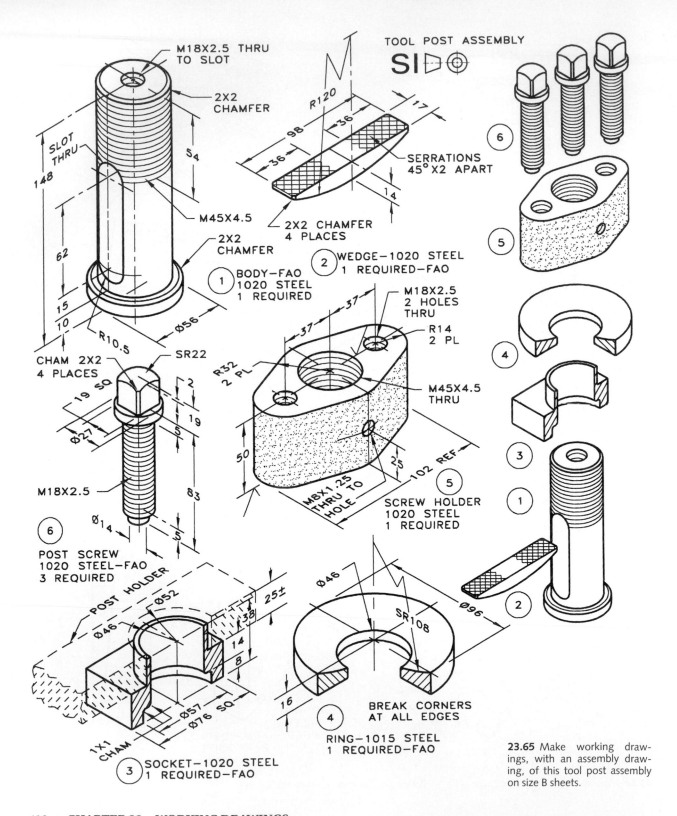

TOOL POST ASSEMBLY

M18X2.5 THRU TO SLOT

2X2 CHAMFER

SLOT THRU

148

54

62

M45X4.5

2X2 CHAMFER

15

10

R10.5

Ø56

① BODY—FAO 1020 STEEL 1 REQUIRED

R120

98

36

36

17

SERRATIONS 45° X2 APART

14

2X2 CHAMFER 4 PLACES

② WEDGE—1020 STEEL 1 REQUIRED—FAO

⑥

⑤

④

③

①

②

CHAM 2X2 4 PLACES

SR22

19 SQ

2

19

Ø27

5

M18X2.5

83

Ø14

5

⑥ POST SCREW 1020 STEEL—FAO 3 REQUIRED

37

37

M18X2.5 2 HOLES THRU

R14 2 PL

R32 2 PL

M45X4.5 THRU

50

M8X1.25 THRU TO HOLE

25

102 REF

⑤ SCREW HOLDER 1020 STEEL 1 REQUIRED

POST HOLDER

Ø52

Ø46

25±

38

14

8

Ø57

Ø76 SQ

1X1 CHAM

③ SOCKET—1020 STEEL 1 REQUIRED—FAO

Ø46

SR108

16

④ RING—1015 STEEL 1 REQUIRED—FAO

BREAK CORNERS AT ALL EDGES

Ø96

23.65 Make working drawings, with an assembly drawing, of this tool post assembly on size B sheets.

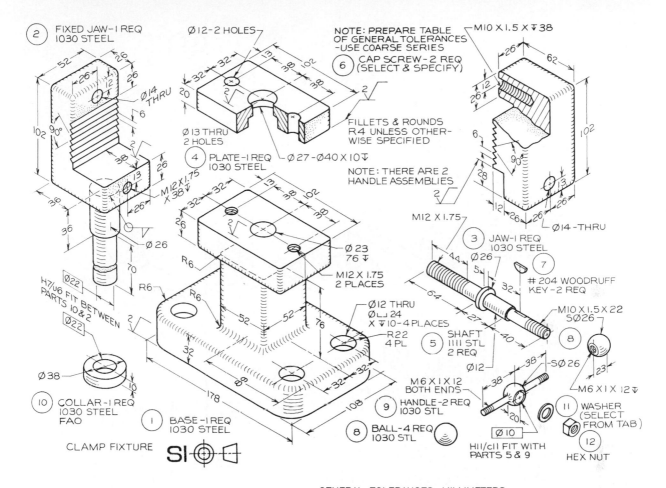

FIXED JAW–1 REQ
1030 STEEL ②

Ø12–2 HOLES

NOTE: PREPARE TABLE
OF GENERAL TOLERANCES
–USE COARSE SERIES

⑥ CAP SCREW–2 REQ
(SELECT & SPECIFY)

M10 X 1.5 X ↧38

Ø14 THRU

FILLETS & ROUNDS
R4 UNLESS OTHER-
WISE SPECIFIED

Ø13 THRU
2 HOLES

④ PLATE–1 REQ
1030 STEEL

Ø27–Ø40 X 10↧

NOTE: THERE ARE 2
HANDLE ASSEMBLIES

M12 X 1.75 X 38 ↧

③ JAW–1 REQ
1030 STEEL

M12 X 1.75

Ø14 –THRU

Ø23
76 ↧

M12 X 1.75
2 PLACES

Ø26

⑦ # 204 WOODRUFF
KEY – 2 REQ

M10 X 1.5 X 22
5 Ø26

⑤ SHAFT
1111 STL
2 REQ

⑧

Ø12 THRU
Ø⌴24
X ↧10–4 PLACES

R22
4 PL

H7/u6 FIT BETWEEN
PARTS 10 & 2

Ø22

Ø22

Ø26

Ø38

Ø26

M6 X 1 X 12
BOTH ENDS

⑨ HANDLE–2 REQ
1030 STL

M6 X 1 X 12 ↧

⑩ COLLAR–1 REQ
1030 STEEL
FAO

① BASE–1 REQ
1030 STEEL

Ø26

⑪ WASHER
(SELECT
FROM TAB)

⑧ BALL–4 REQ
1030 STL

Ø10

⑫ HEX NUT

H11/c11 FIT WITH
PARTS 5 & 9

CLAMP FIXTURE   SI ⌖ ⬚

23.66 Make a half-size working drawing with an assembly drawing of this clamp fixture on size B sheets.

GENERAL TOLERANCES—MILLIMETERS
UNLESS OTHERWISE SPECIFIED

LINEAR TOLERANCE	OVER TO	– 120	120 315	315 1000	1000 –
		±0.3	±0.5	±0.8	±1.2

Thought Questions

1. How deep are the threads in the ball, Part 8?

2. Why are there no fillets or rounds on the collar, Part 10?

3. What is the clearance between the fixed jaw, Part 2, and the base, Part 1?

4. What are the radii of fillets and rounds on the base, Part 1?

5. What is the height of the clamp fixture when it is assembled?

6. What is the weight of the base, Part 1?

7. What is the range of opening sizes of the jaws, Part 2 and Part 3?

8. What is the purpose of the #204 Woodruff key on the shaft, Part 5?

9. How many handle assemblies are there?

10. How is the cylindrical column on the fixed jaw, Part 2, attached to the upper portion of the part?

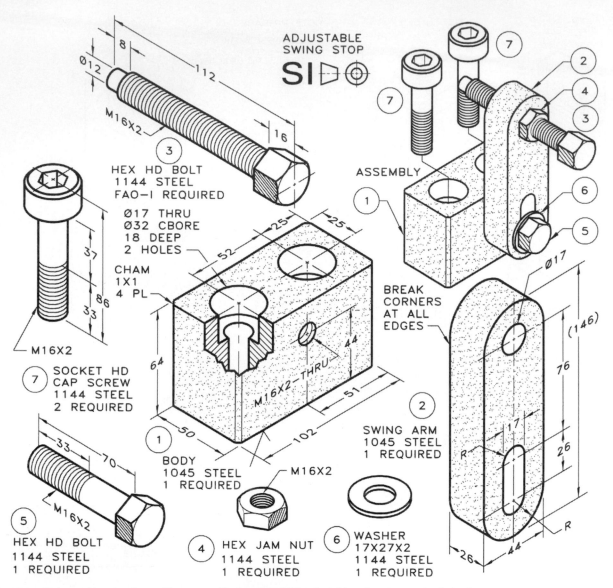

ADJUSTABLE
SWING STOP

SI⊳⊕

Ø12

8

112

16

M16X2

3

HEX HD BOLT
1144 STEEL
FAO-I REQUIRED

37

86

33

M16X2

7

SOCKET HD
CAP SCREW
1144 STEEL
2 REQUIRED

Ø17 THRU
Ø32 CBORE
18 DEEP
2 HOLES

CHAM
1X1
4 PL

25

25

52

64

50

M16X2 THRU

44

51

102

1

BODY
1045 STEEL
1 REQUIRED

M16X2

4

HEX JAM NUT
1144 STEEL
1 REQUIRED

33

70

5

M16X2

HEX HD BOLT
1144 STEEL
1 REQUIRED

6

WASHER
17X27X2
1144 STEEL
1 REQUIRED

ASSEMBLY

7

7

1

2

4

3

6

5

BREAK
CORNERS
AT ALL
EDGES

Ø17

(146)

76

17

26

R

R

26

44

2

SWING ARM
1045 STEEL
1 REQUIRED

23.67 Make working drawings with an assembly drawing of this adjustable swing stop on size B sheets.

Thought Questions

1. What is the purpose of the washer (part 6) in this assembly?

2. Why is there a slot in the swing arm (part 2) instead of circular hole?

3. Why did the designer of this assembly use both socket-head cap screws and hex-head cap screws instead of using one type or the other?

4. When assembled, what is the range of adjustment from the upper surface of the body (part 1) and the centerline of the hex-head bolt (part 3)?

5. What is the maximum distance that the 12-DIA end of the hex-head bolt (part 3) can extend beyond the face of the swing arm (part 2)?

6. Write a paragraph to verbally give a description of either part 1 or part 2.

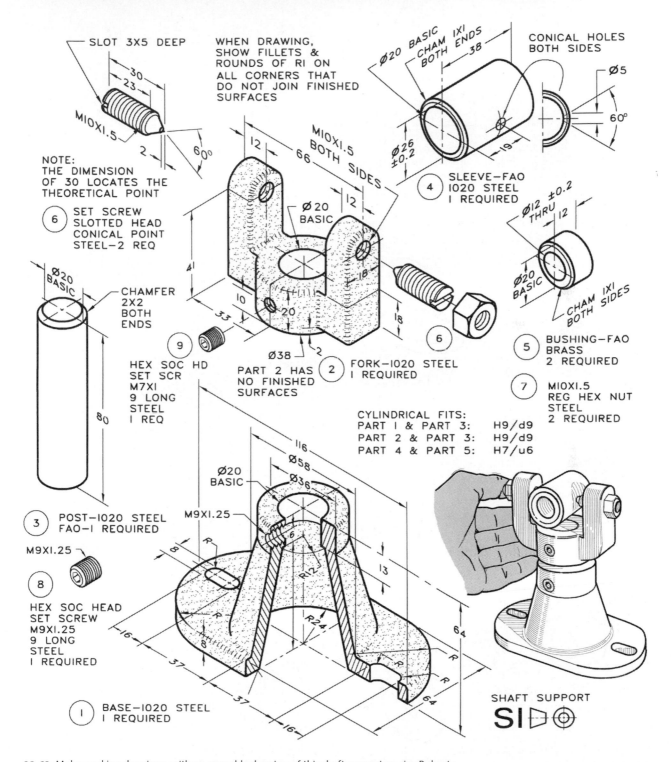

SLOT 3X5 DEEP

30

23

M10X1.5

2

60°

NOTE:
THE DIMENSION
OF 30 LOCATES THE
THEORETICAL POINT

(6) SET SCREW
SLOTTED HEAD
CONICAL POINT
STEEL—2 REQ

WHEN DRAWING,
SHOW FILLETS &
ROUNDS OF R1 ON
ALL CORNERS THAT
DO NOT JOIN FINISHED
SURFACES

M10X1.5
BOTH SIDES

12

66

12

Ø20
BASIC

41

10

20

33

Ø38

2

(2) FORK—1020 STEEL
1 REQUIRED

PART 2 HAS
NO FINISHED
SURFACES

18

18

(6)

Ø20 BASIC
CHAM 1X1
BOTH ENDS

38

CONICAL HOLES
BOTH SIDES

Ø5

60°

Ø26
±0.2

19

(4) SLEEVE—FAO
1020 STEEL
1 REQUIRED

Ø12 ±0.2
THRU

12

Ø20
BASIC

CHAM 1X1
BOTH SIDES

(5) BUSHING—FAO
BRASS
2 REQUIRED

(7) M10X1.5
REG HEX NUT
STEEL
2 REQUIRED

Ø20
BASIC

CHAMFER
2X2
BOTH
ENDS

(9)

HEX SOC HD
SET SCR
M7X1
9 LONG
STEEL
1 REQ

80

(3) POST—1020 STEEL
FAO—1 REQUIRED

M9X1.25

(8)

HEX SOC HEAD
SET SCREW
M9X1.25
9 LONG
STEEL
1 REQUIRED

CYLINDRICAL FITS:
PART 1 & PART 3:   H9/d9
PART 2 & PART 3:   H9/d9
PART 4 & PART 5:   H7/u6

116

Ø58

Ø36

Ø20
BASIC

M9X1.25

R

8

6

R12

13

R

R24

R

R

64

64

16

8

37

37

16

(1) BASE—1020 STEEL
1 REQUIRED

SHAFT SUPPORT

S1

23.68 Make working drawings, with an assembly drawing of this shaft support on size B sheets.

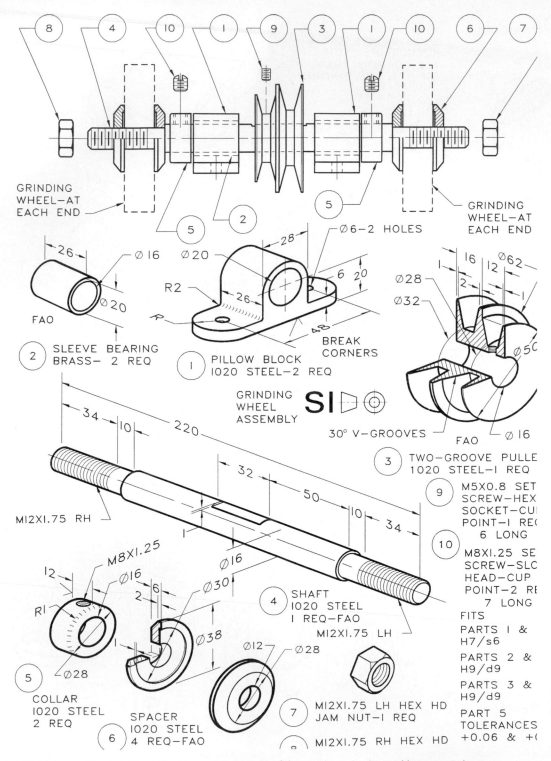

GRINDING WHEEL—AT EACH END

GRINDING WHEEL—AT EACH END

② SLEEVE BEARING BRASS— 2 REQ

① PILLOW BLOCK 1020 STEEL—2 REQ

Ø6—2 HOLES

R2

26

28

Ø20

BREAK CORNERS

GRINDING WHEEL ASSEMBLY

SI

30° V—GROOVES

Ø16

Ø28

Ø32

Ø50

16 Ø62

12

2

1

FAO

③ TWO—GROOVE PULLE 1020 STEEL—1 REQ

④ SHAFT 1020 STEEL 1 REQ—FAO
M12X1.75 LH

M12X1.75 RH

220

34

10

32

50

10

34

⑤ COLLAR 1020 STEEL 2 REQ

M8X1.25

Ø16

Ø16

2 6

Ø30

Ø38

Ø28

R1

⑥ SPACER 1020 STEEL 4 REQ—FAO

Ø12

Ø28

⑦ M12X1.75 LH HEX HD JAM NUT—1 REQ

M12X1.75 RH HEX HD

⑨ M5X0.8 SET SCREW—HEX SOCKET—CUI POINT—1 REC 6 LONG

⑩ M8X1.25 SE SCREW—SLC HEAD—CUP POINT—2 RE 7 LONG

FITS
PARTS 1 &
H7/s6
PARTS 2 &
H9/d9
PARTS 3 &
H9/d9
PART 5
TOLERANCES
+0.06 & +(

23.69 Make working drawings with an assembly drawing of this grinding wheel assembly on size B sheets.

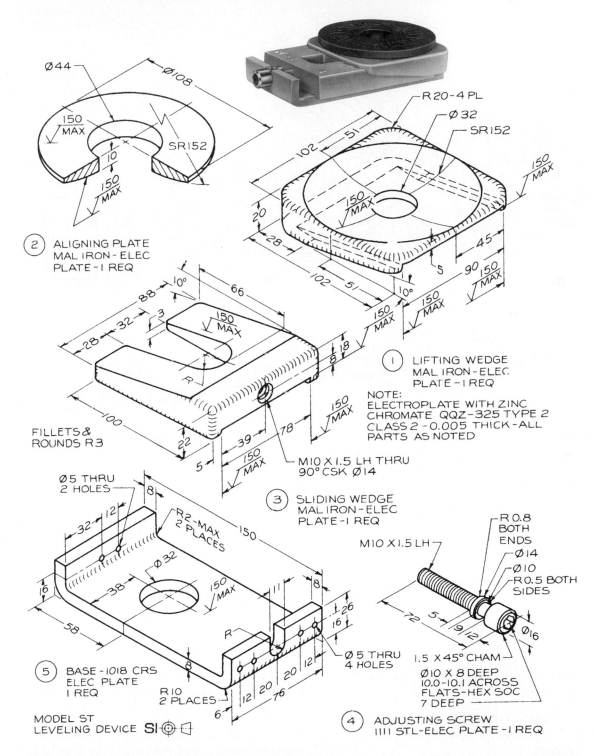

② ALIGNING PLATE
MAL IRON – ELEC
PLATE – 1 REQ

Ø44
Ø108
150 MAX
10
150 MAX
SR152

R 20 – 4 PL
Ø 32
SR152
51
102
150 MAX
150 MAX
20
28
5
45
102
51
90 150 MAX
10°
150 MAX
150 MAX

① LIFTING WEDGE
MAL IRON – ELEC.
PLATE – 1 REQ

NOTE:
ELECTROPLATE WITH ZINC
CHROMATE QQZ–325 TYPE 2
CLASS 2 –0.005 THICK – ALL
PARTS AS NOTED

10°
66
88
3
32
28
150 MAX
8 18
R
100
22
5
39 78
150 MAX
150 MAX
M10 X 1.5 LH THRU
90° CSK Ø14

FILLETS &
ROUNDS R3

③ SLIDING WEDGE
MAL IRON – ELEC
PLATE – 1 REQ

Ø5 THRU
2 HOLES
8
32 12
R2 – MAX
2 PLACES
150
16
Ø 32
38
150 MAX
11
8
58
R
26
16
Ø 5 THRU
4 HOLES
8
R10
2 PLACES
12 20 12
76
6

⑤ BASE – 1018 CRS
ELEC PLATE
1 REQ

MODEL ST
LEVELING DEVICE SI ⊕ ⊟

M10 X 1.5 LH
R 0.8
BOTH
ENDS
Ø14
Ø10
R 0.5 BOTH
SIDES
72
5
9 12
Ø16

1.5 X 45° CHAM
Ø10 X 8 DEEP
10.0 –10.1 ACROSS
FLATS – HEX SOC
7 DEEP

④ ADJUSTING SCREW
1111 STL – ELEC PLATE – 1 REQ

23.70 Make working drawings with an assembly drawing of this levelling device on size B sheets. (Courtesy of Level-Line.)

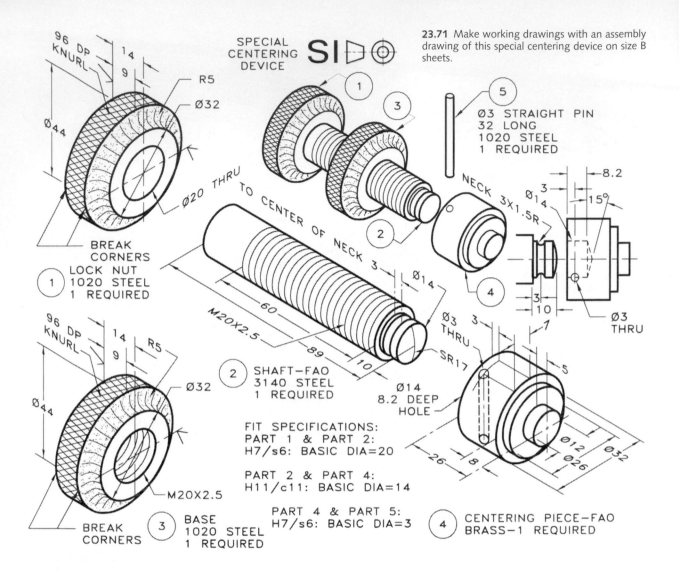

SPECIAL CENTERING DEVICE SI ⊐⊖ ◉

96 DP KNURL
1 4
9
R5
Ø32
Ø44
Ø20 THRU
BREAK CORNERS

(1) **LOCK NUT**
1020 STEEL
1 REQUIRED

96 DP KNURL
1 4
9
R5
Ø32
Ø44
M20X2.5
BREAK CORNERS

(3) **BASE**
1020 STEEL
1 REQUIRED

TO CENTER OF NECK 3
M20X2.5
60
89
10

(2) **SHAFT—FAO**
3140 STEEL
1 REQUIRED

23.71 Make working drawings with an assembly drawing of this special centering device on size B sheets.

5
Ø3 STRAIGHT PIN
32 LONG
1020 STEEL
1 REQUIRED

NECK 3X1.5R
Ø14
Ø14
Ø3 THRU
SR17

8.2
Ø3
14
15°
3
10
Ø3 THRU

(4) Centering piece
Ø14 8.2 DEEP HOLE

3
7
5
26
8
Ø12
Ø26
Ø32

FIT SPECIFICATIONS:
PART 1 & PART 2:
H7/s6: BASIC DIA=20

PART 2 & PART 4:
H11/c11: BASIC DIA=14

PART 4 & PART 5:
H7/s6: BASIC DIA=3

(4) **CENTERING PIECE—FAO**
BRASS—1 REQUIRED

Thought Questions

1. What conversion factor would you use to convert the metric dimensions to English units?

2. Why is knurling given on parts 1 and 3? Why was knurling not given on part 4?

3. Why were part 1 and part 2 not designed with threads for attachment to each other?

4. When assembled, will part 4 rotate about the end of part 2? Determine and explain why it attaches as it does.

5. Why was the fit between parts 4 and 5 selected to be H7/s6 instead of H11/c11?

6. What would be the approximate weight of the total assembly if all materials were assumed to weigh 490 lbs per cu. ft.?

7. Can you explain why parts 1 and 3 were designed with bosses (raised surfaces around the holes) as shown?

8. Which of the parts can be specified on a working drawing by a note without a drawing?

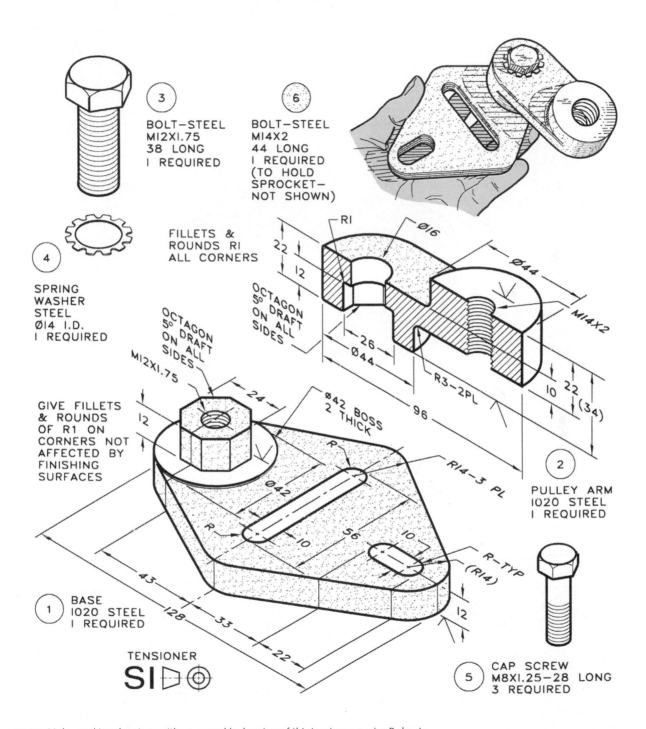

③ BOLT—STEEL
MI2XI.75
38 LONG
I REQUIRED

⑥ BOLT—STEEL
MI4X2
44 LONG
I REQUIRED
(TO HOLD
SPROCKET—
NOT SHOWN)

④ SPRING
WASHER
STEEL
Ø14 I.D.
I REQUIRED

FILLETS &
ROUNDS RI
ALL CORNERS

OCTAGON
5° DRAFT
ON ALL
SIDES

RI

Ø16

Ø44

MI4X2

22

12

26

Ø44

R3—2PL

96

10

22

(34)

② PULLEY ARM
1020 STEEL
I REQUIRED

MI2XI.75

OCTAGON
5° DRAFT
ON ALL
SIDES

GIVE FILLETS
& ROUNDS
OF R1 ON
CORNERS NOT
AFFECTED BY
FINISHING
SURFACES

24

Ø42 BOSS
2 THICK

R

RI4—3 PL

Ø42

R

10

56

10

R—TYP
(RI4)

43

128

33

22

12

① BASE
1020 STEEL
I REQUIRED

TENSIONER
SI ⬚ ◉

⑤ CAP SCREW
M8XI.25—28 LONG
3 REQUIRED

23.72 Make working drawings with an assembly drawing of this tensioner on size B sheets.

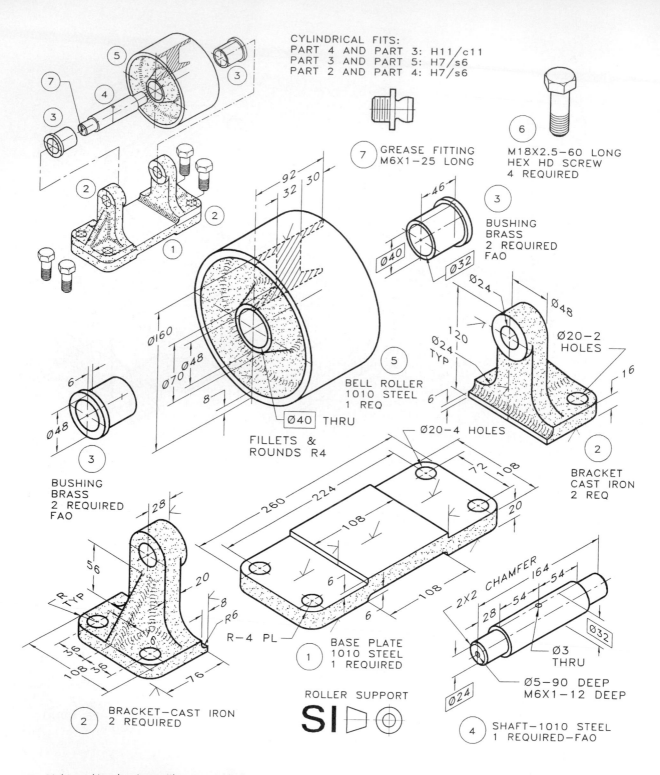

CYLINDRICAL FITS:
PART 4 AND PART 3: H11/c11
PART 3 AND PART 5: H7/s6
PART 2 AND PART 4: H7/s6

7 GREASE FITTING
M6X1-25 LONG

6 M18X2.5-60 LONG
HEX HD SCREW
4 REQUIRED

3 BUSHING
BRASS
2 REQUIRED
FAO
Ø40 Ø32 46

2 BRACKET
CAST IRON
2 REQ
Ø24 Ø48 Ø20-2 HOLES 120 Ø24 TYP 6 16

Ø160 Ø70 Ø48 92 32 30 8
5 BELL ROLLER
1010 STEEL
1 REQ
Ø40 THRU
FILLETS &
ROUNDS R4

3 BUSHING
BRASS
2 REQUIRED
FAO
6 Ø48

Ø20-4 HOLES 72 108 20

2 BRACKET—CAST IRON
2 REQUIRED
28 56 20 8 R6 R TYP R-4 PL 108 36 76

1 BASE PLATE
1010 STEEL
1 REQUIRED
260 224 108 6 6 108

ROLLER SUPPORT
SI ◁ ◉

4 SHAFT-1010 STEEL
1 REQUIRED—FAO
2X2 CHAMFER 164 54 28 54 Ø32 Ø24 Ø3 THRU Ø5-90 DEEP M6X1-12 DEEP

23.73 Make working drawings with an assembly drawing of this pulley support on size B sheets.

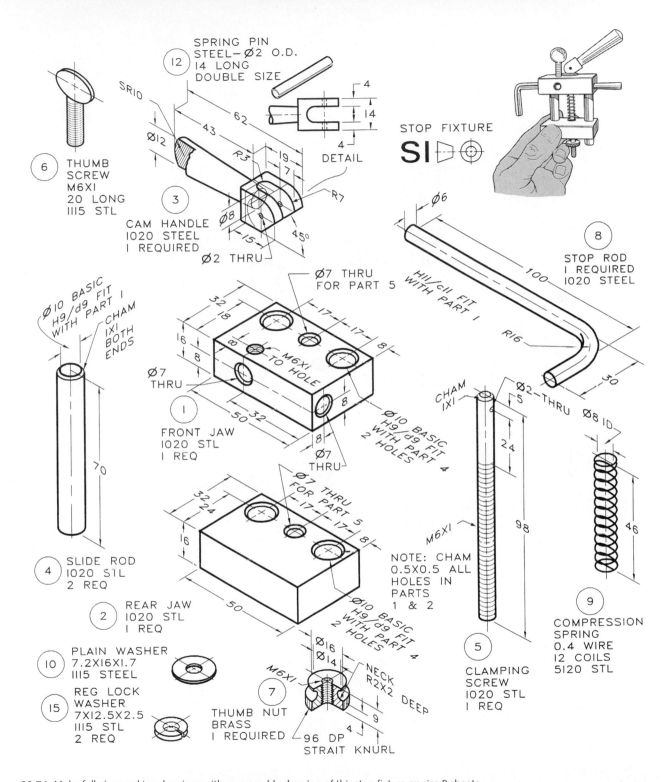

SPRING PIN
STEEL—Ø2 O.D.
(12) 14 LONG
DOUBLE SIZE

4

14

DETAIL

4

STOP FIXTURE

SI ▷ ⊕

SR10

Ø12

43

62

R3

19

7

R7

(3)

Ø8

15

45°

Ø2 THRU

CAM HANDLE
1020 STEEL
1 REQUIRED

(6)
THUMB
SCREW
M6X1
20 LONG
1115 STL

Ø6

H11/c11 FIT
WITH PART 1

(8)
STOP ROD
1 REQUIRED
1020 STEEL

100

R16

30

Ø8 ID

Ø7 THRU
FOR PART 5

Ø10 BASIC
H9/d9 FIT
WITH PART 1

CHAM
1X1
BOTH
ENDS

32

18

17

17

8

16

8

8

M6X1
TO HOLE

Ø7
THRU

(1)
FRONT JAW
1020 STL
1 REQ

50

32

8

Ø7
THRU

Ø10 BASIC
H9/d9 FIT
WITH PART 4
2 HOLES

Ø2 THRU

5

CHAM
1X1

6

24

98

M6X1

(5)
CLAMPING
SCREW
1020 STL
1 REQ

Ø8 ID

(9)
COMPRESSION
SPRING
0.4 WIRE
12 COILS
5120 STL

46

70

(4)
SLIDE ROD
1020 STL
2 REQ

Ø7 THRU
FOR PART 5

32

24

17

17

8

16

(2)
REAR JAW
1020 STL
1 REQ

50

Ø10 BASIC
H9/d9 FIT
WITH PART 4
2 HOLES

NOTE: CHAM
0.5X0.5 ALL
HOLES IN
PARTS
1 & 2

(10)
PLAIN WASHER
7.2X16X1.7
1115 STEEL

(15)
REG LOCK
WASHER
7X12.5X2.5
1115 STL
2 REQ

Ø16

Ø14

M6X1

(7)

THUMB NUT
BRASS
1 REQUIRED

NECK
R2X2
DEEP

9

4

96 DP
STRAIT KNURL

23.74 Make full-size working drawings with an assembly drawing of this stop fixture on size B sheets.

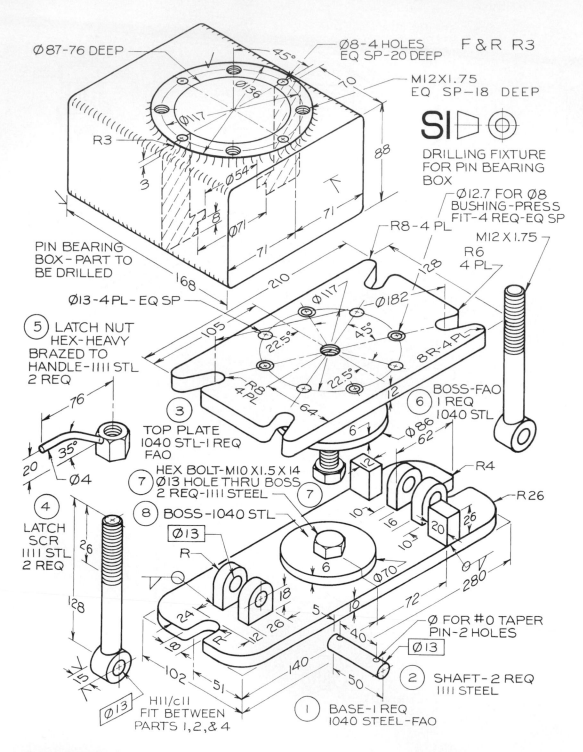

Ø87–76 DEEP

F & R R3

Ø8–4 HOLES
EQ SP–20 DEEP

45°

Ø136

Ø117

R3

3

Ø54

Ø71

8

M12X1.75
EQ SP–18 DEEP

SI ▷ ⊙

DRILLING FIXTURE
FOR PIN BEARING
BOX

Ø12.7 FOR Ø8
BUSHING-PRESS
FIT-4 REQ-EQ SP

M12 X 1.75

R6
4 PL

88

71

71

PIN BEARING
BOX – PART TO
BE DRILLED

168

210

R8–4 PL

128

Ø13–4 PL–EQ SP

Ø117

Ø182

⑤ LATCH NUT
HEX–HEAVY
BRAZED TO
HANDLE–1111 STL
2 REQ

105

22.5°

45°

8R–4 PL

③

TOP PLATE
1040 STL–1 REQ
FAO

R8
4 PL

22.5°

12

⑥ BOSS–FAO
1 REQ
1040 STL

64

6

76

20

35°

Ø4

④

LATCH
SCR
1111 STL
2 REQ

26

128

15

Ø13

HEX BOLT–M10 X1.5 X 14
⑦ Ø13 HOLE THRU BOSS
2 REQ–1111 STEEL

⑦

⑧ BOSS–1040 STL

Ø13

R

Ø86

62

2

R4

10

16

20

26

R26

Ø70

6

10

10

Ø V

280

24

18

R

12

26

18

5

10

72

40

Ø FOR #0 TAPER
PIN–2 HOLES

Ø13

H11/c11
FIT BETWEEN
PARTS 1, 2, & 4

102

51

140

50

② SHAFT–2 REQ
1111 STEEL

① BASE–1 REQ
1040 STEEL–FAO

23.75 Make full-size working drawings with an assembly drawing of this drilling fixture on size B sheets.

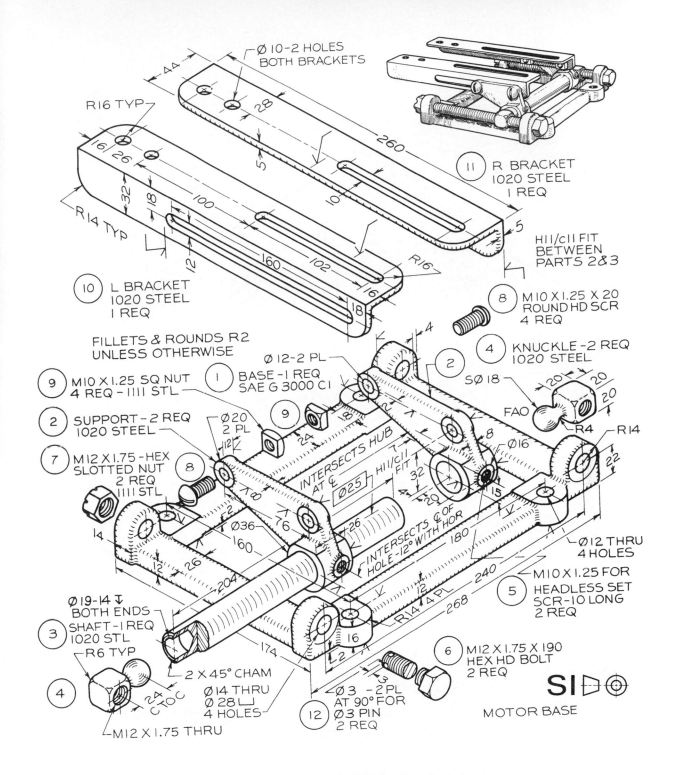

Ø 10-2 HOLES
BOTH BRACKETS

R16 TYP

44

28

260

R14 TYP

16

26

32

18

100

5

10

11 R BRACKET
    1020 STEEL
    1 REQ

H11/C11 FIT
BETWEEN
PARTS 2&3

5

160

102

R16

18

16

12

10 L BRACKET
    1020 STEEL
    1 REQ

FILLETS & ROUNDS R2
UNLESS OTHERWISE

8 M10 X1.25 X 20
   ROUND HD SCR
   4 REQ

4

4 KNUCKLE -2 REQ
   1020 STEEL

9 M10 X1.25 SQ NUT
   4 REQ - 1111 STL

1 BASE -1 REQ
   SAE G 3000 CI

Ø 12-2 PL

2

SØ 18

FAO

20 20

20

R4 R14

2 SUPPORT - 2 REQ
   1020 STEEL

Ø20
2 PL

9

24

12

INTERSECTS HUB
AT ℄

Ø25 H11/C11 FIT

32

Ø16

8

22

7 M12 X1.75 -HEX
   SLOTTED NUT
   2 REQ
   1111 STL

8

8

Ø36

76

2

26

INTERSECTS ℄ OF
HOLE -12° WITH HOR

4

20

15

14

160

12

26

204

180

240

Ø 12 THRU
4 HOLES

M10 X1.25 FOR
HEADLESS SET
SCR-10 LONG
2 REQ

5

3 SHAFT -1 REQ
   1020 STL

Ø 19-14 ↓
BOTH ENDS

R6 TYP

174

R14 4 PL

12

268

16

2

6 M12 X1.75 X 190
   HEX HD BOLT
   2 REQ

4

2 X 45° CHAM

24
C TO C

Ø 14 THRU
Ø 28 ⊔
4 HOLES

M12 X 1.75 THRU

Ø 3 -2 PL
AT 90° FOR
Ø3 PIN
2 REQ

12

3

SI ⊳⊙

MOTOR BASE

23.76 Make full-size working drawings with an assembly drawing of this motor base on size B sheets.

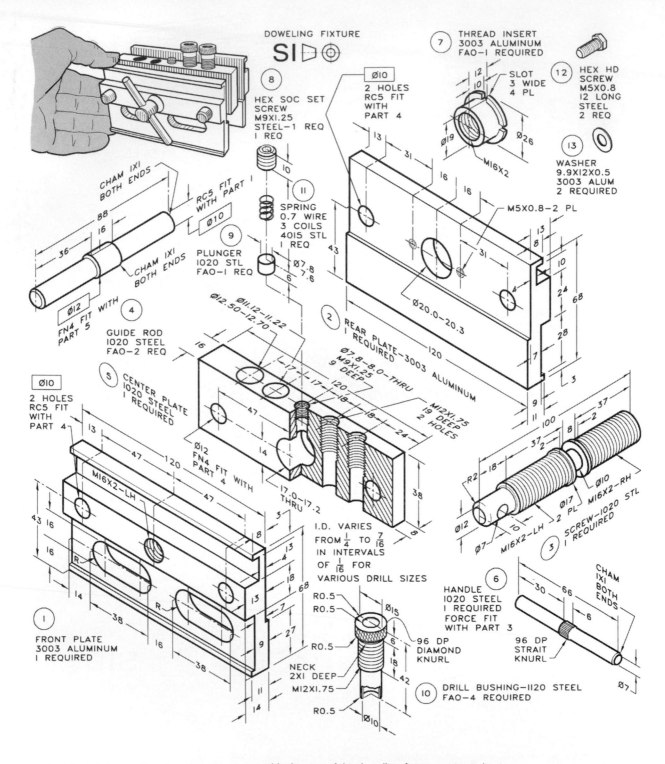

DOWELING FIXTURE

SI ⬭⊙

8 HEX SOC SET SCREW M9X1.25 STEEL–1 REQ 1 REQ

Ø10 2 HOLES RC5 FIT WITH PART 4

7 THREAD INSERT 3003 ALUMINUM FAO–1 REQUIRED

SLOT 3 WIDE 4 PL

12 2 10 Ø19 Ø26 M16X2

12 HEX HD SCREW M5X0.8 12 LONG STEEL 2 REQ

13 WASHER 9.9X12X0.5 3003 ALUM 2 REQUIRED

13 31 16 16

M5X0.8–2 PL

13 8 31 10 24 68 28 7 3

CHAM 1X1 BOTH ENDS

88 16 36

RC5 FIT WITH PART 1

Ø10

CHAM 1X1 BOTH ENDS

11 SPRING 0.7 WIRE 3 COILS 4015 STL 1 REQ

9 PLUNGER 1020 STL FAO–1 REQ

10

43

Ø7.8 7.6 6

2 REAR PLATE–3003 ALUMINUM 1 REQUIRED

Ø20.0–20.3

120

9 11 100

Ø12 FN4 FIT WITH PART 5

4 GUIDE ROD 1020 STEEL FAO–2 REQ

Ø12.50–12.70

Ø11.12–11.22

16

Ø7.8–8.0–THRU M9X1.25 9 DEEP

M12X1.75 19 DEEP 2 HOLES

Ø10 2 HOLES RC5 FIT WITH PART 4

S CENTER PLATE 1020 STEEL 1 REQUIRED

17 17

120

18

18 24

47

14

Ø12 FN4 FIT WITH PART 4

17.0–17.2 THRU

38

8

37 2 8 2 37 R2 18

Ø17 Ø10 M16X2–RH 2 PL

Ø12 Ø7 M16X2–LH 10

3 SCREW–1020 STL 1 REQUIRED

Ø10 2 HOLES RC5 FIT WITH PART 4

47 120

M16X2–LH

47

43 16 16

R

R

14

38

13 4 18 13 7 27 9 11 14

3 8 68 16

1 FRONT PLATE 3003 ALUMINUM 1 REQUIRED

I.D. VARIES FROM ¼ TO 7/16 IN INTERVALS OF 1/16 FOR VARIOUS DRILL SIZES

R0.5 R0.5 Ø15

R0.5

6 96 DP DIAMOND KNURL 18 42

NECK 2X1 DEEP M12X1.75

R0.5 Ø10

6 HANDLE 1020 STEEL 1 REQUIRED FORCE FIT WITH PART 3

30 66 6

96 DP STRAIT KNURL

CHAM 1X1 BOTH ENDS

Ø7

10 DRILL BUSHING–1120 STEEL FAO–4 REQUIRED

23.77 Make full-size working drawings with an assembly drawing of this dowelling fixture on size B sheets.

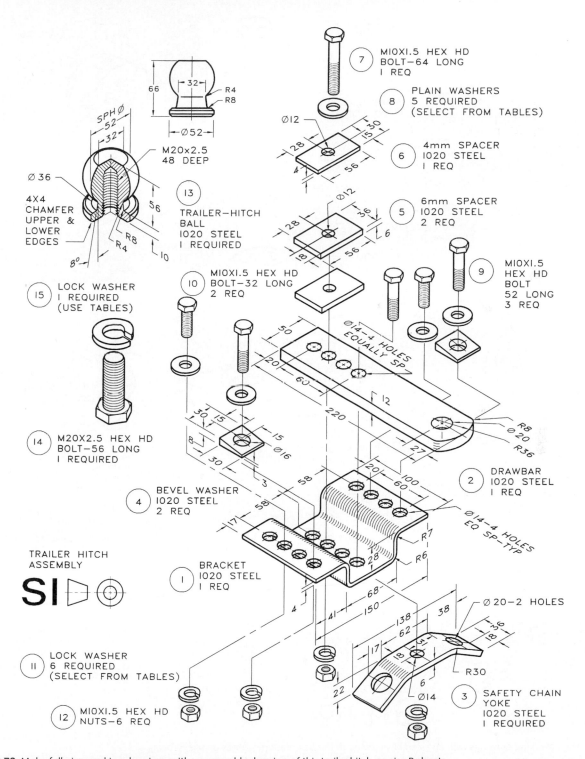

66
┌─32─┐
R4
R8
SPHØ
52
32
Ø52
M20x2.5
48 DEEP
Ø36
4X4
CHAMFER
UPPER &
LOWER
EDGES
8°
R8
R4
56
10

⑬ TRAILER—HITCH
BALL
1020 STEEL
1 REQUIRED

⑦ M10X1.5 HEX HD
BOLT—64 LONG
1 REQ

⑧ PLAIN WASHERS
5 REQUIRED
(SELECT FROM TABLES)

Ø12
28
30
15
4
56

⑥ 4mm SPACER
1020 STEEL
1 REQ

Ø12
28
36
18
6
56

⑤ 6mm SPACER
1020 STEEL
2 REQ

⑨ M10X1.5
HEX HD
BOLT
52 LONG
3 REQ

⑩ M10X1.5 HEX HD
BOLT—32 LONG
2 REQ

⑮ LOCK WASHER
1 REQUIRED
(USE TABLES)

50
20
60
Ø14—4 HOLES
EQUALLY SP
12
220

⑭ M20X2.5 HEX HD
BOLT—56 LONG
1 REQUIRED

R8
Ø20
R36

② DRAWBAR
1020 STEEL
1 REQ

④ BEVEL WASHER
1020 STEEL
2 REQ

30
15
15
8
30
3
Ø16
58
58
17

20
60
100
27
Ø14—4 HOLES
EQ SP-TYP

R7
R6
28

TRAILER HITCH
ASSEMBLY

SI ▷ ◎

① BRACKET
1020 STEEL
1 REQ

4
68
150
41

138
62
17
18
38
Ø20—2 HOLES
16
18

R30

⑪ LOCK WASHER
6 REQUIRED
(SELECT FROM TABLES)

22
6
Ø14

③ SAFETY CHAIN
YOKE
1020 STEEL
1 REQUIRED

⑫ M10X1.5 HEX HD
NUTS—6 REQ

23.78 Make full-size working drawings with an assembly drawing of this trailer hitch on size B sheets.

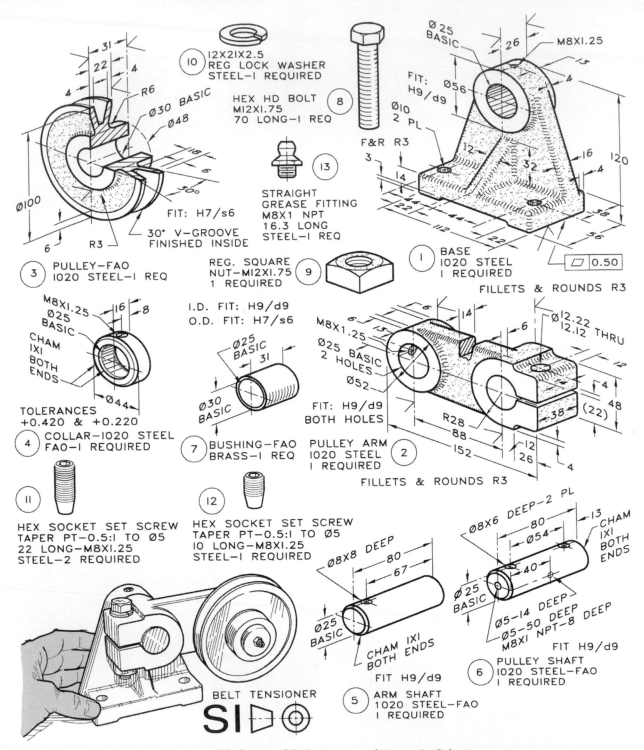

⑩ 12X21X2.5
REG LOCK WASHER
STEEL-1 REQUIRED

HEX HD BOLT
MI2XI.75 ⑧
70 LONG-1 REQ
F&R R3

⑬

STRAIGHT
GREASE FITTING
M8X1 NPT
16.3 LONG
STEEL-1 REQ

③ PULLEY-FAO
1020 STEEL-1 REQ

REG. SQUARE
NUT-MI2XI.75 ⑨
1 REQUIRED

R6
Ø30 BASIC
Ø48
Ø100
Ø30 BASIC
30°
FIT: H7/s6
30° V-GROOVE
FINISHED INSIDE
R3
31 22 4 4 4 18 6 6

① BASE
1020 STEEL
1 REQUIRED

Ø25
BASIC 26 M8X1.25
FIT:
H9/d9 Ø56
12
3
14
Ø10
2 PL
13
4
16
32
4
120
22 44 112 22 36 56
□ 0.50
FILLETS & ROUNDS R3

M8X1.25
Ø25
BASIC 16 8
CHAM
1X1
BOTH
ENDS
Ø44
TOLERANCES
+0.420 & +0.220

④ COLLAR-1020 STEEL
FAO-1 REQUIRED

I.D. FIT: H9/d9
O.D. FIT: H7/s6
Ø25
BASIC 31
Ø30
BASIC

⑦ BUSHING-FAO
BRASS-1 REQ

M8X1.25
Ø25 BASIC
2 HOLES
Ø52
FIT: H9/d9
BOTH HOLES
PULLEY ARM
1020 STEEL
1 REQUIRED
②
6 6 14 6 Ø12.22 THRU 12.12
15
R28
88
152
4 48 (22) 38 12 26 4 12
FILLETS & ROUNDS R3

⑪
HEX SOCKET SET SCREW
TAPER PT-0.5:1 TO Ø5
22 LONG-M8X1.25
STEEL-2 REQUIRED

⑫
HEX SOCKET SET SCREW
TAPER PT-0.5:1 TO Ø5
10 LONG-M8X1.25
STEEL-1 REQUIRED

Ø8X8 DEEP
80
67
Ø25
BASIC
CHAM 1X1
BOTH ENDS
FIT H9/d9

⑤ ARM SHAFT
1020 STEEL-FAO
1 REQUIRED

Ø8X6 DEEP-2 PL
80
Ø54 13 CHAM 1X1 BOTH ENDS
Ø25
BASIC 40
Ø5-14 DEEP
Ø5-50 DEEP
M8X1 NPT-8 DEEP
FIT H9/d9

⑥ PULLEY SHAFT
1020 STEEL-FAO
1 REQUIRED

BELT TENSIONER
SI ◁ ◎

23.79 Make working drawings with an assembly drawing of this belt tensioner device on size B sheets.

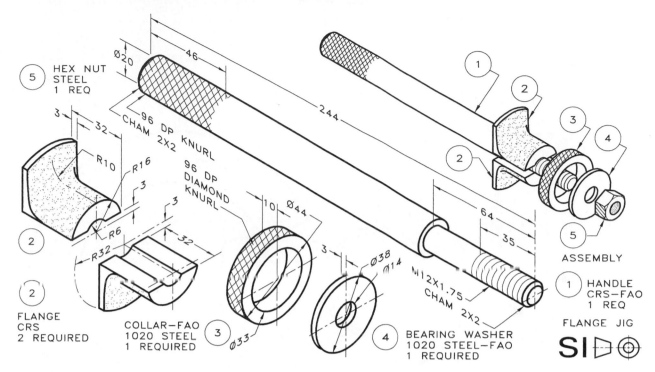

23.80 Size B sheets.

Thought Questions

1. How many parts comprise the flange jig (Figure 23.80)?

2. What are the thread specifications for the hexagon nut (part 5)?

3. What is the pitch of threads on the handle (part 1)?

4. Why are no tolerances given on any of the parts?

5. Why is knurling specified for the collar (part 3)?

6. What is the approximate weight of the handle (part 1)?

7. Are the features that need to be finished have finish marks added? Why?

Working Drawings: Multiple Parts With Design Applications

Make dimensioned working drawings of the multiple parts shown in **Figures 23.81–23.85** on a sheet size of your choice with the necessary dimensions and notes to fabricate the parts. Each part is given in a general format, with partial dimensions, which requires some design effort on your part. You must consider the addition of fillets and rounds, the application of finish marks, and the modification of features of the parts to make them functional and practical. Apply the tolerances to the parts in limit form by using the tables of cylindrical fits in the Appendices. Make an assembly drawing and parts list to show how the parts are to be put together.

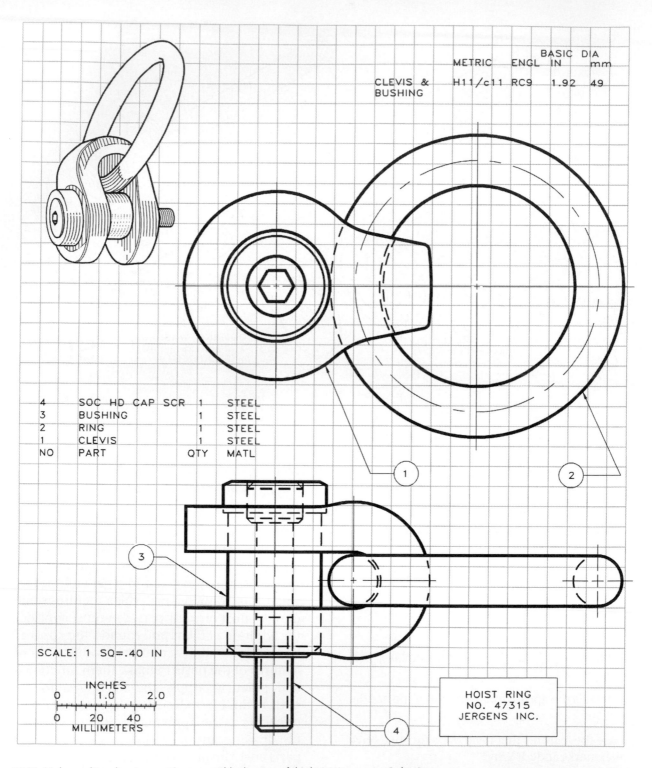

			BASIC DIA	
	METRIC	ENGL	IN	mm
CLEVIS & BUSHING	H11/c11	RC9	1.92	49

4	SOC HD CAP SCR	1	STEEL
3	BUSHING	1	STEEL
2	RING	1	STEEL
1	CLEVIS	1	STEEL
NO	PART	QTY	MATL

SCALE: 1 SQ=.40 IN

INCHES
0   1.0   2.0
0   20   40
MILLIMETERS

HOIST RING
NO. 47315
JERGENS INC.

23.81 Make working drawings, with an assembly drawing of this hoist ring on size B sheets.

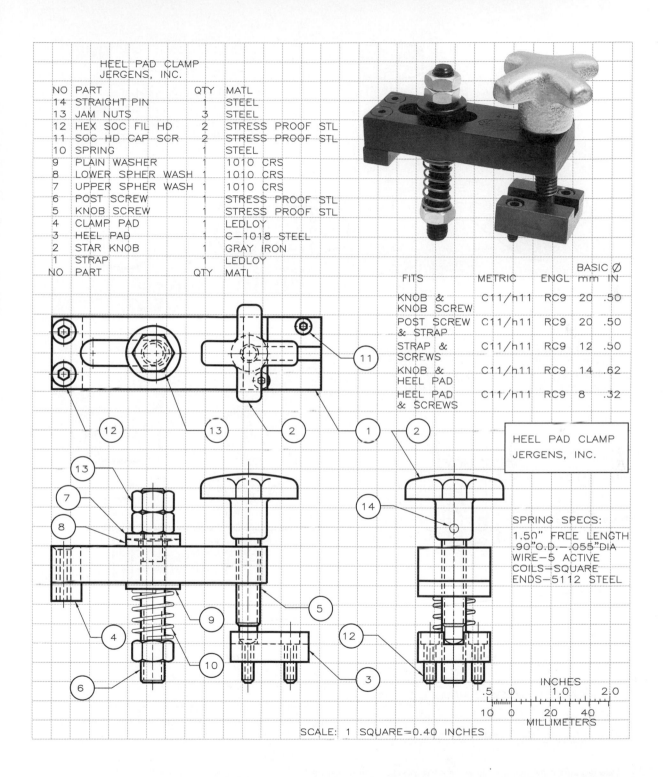

HEEL PAD CLAMP
JERGENS, INC.

NO	PART	QTY	MATL
14	STRAIGHT PIN	1	STEEL
13	JAM NUTS	3	STEEL
12	HEX SOC FIL HD	2	STRESS PROOF STL
11	SOC HD CAP SCR	2	STRESS PROOF STL
10	SPRING	1	STEEL
9	PLAIN WASHER	1	1010 CRS
8	LOWER SPHER WASH	1	1010 CRS
7	UPPER SPHER WASH	1	1010 CRS
6	POST SCREW	1	STRESS PROOF STL
5	KNOB SCREW	1	STRESS PROOF STL
4	CLAMP PAD	1	LEDLOY
3	HEEL PAD	1	C—1018 STEEL
2	STAR KNOB	1	GRAY IRON
1	STRAP	1	LEDLOY
NO	PART	QTY	MATL

FITS	METRIC	ENGL	BASIC Ø mm	IN
KNOB & KNOB SCREW	C11/h11	RC9	20	.50
POST SCREW & STRAP	C11/h11	RC9	20	.50
STRAP & SCREWS	C11/h11	RC9	12	.50
KNOB & HEEL PAD	C11/h11	RC9	14	.62
HEEL PAD & SCREWS	C11/h11	RC9	8	.32

HEEL PAD CLAMP
JERGENS, INC.

SPRING SPECS:
1.50" FREE LENGTH
.90"O.D.—.055"DIA
WIRE—5 ACTIVE
COILS—SQUARE
ENDS—5112 STEEL

INCHES
.5 0   1.0   2.0
10 0   20   40
MILLIMETERS

SCALE: 1 SQUARE=0.40 INCHES

23.82 Make working drawings, with an assembly drawing of this heel pad clamp on size B sheets.

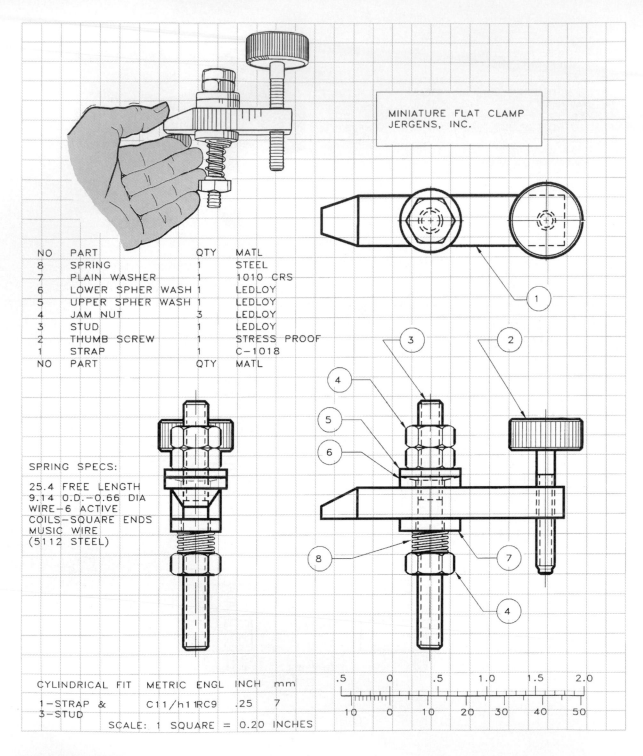

MINIATURE FLAT CLAMP
JERGENS, INC.

NO	PART	QTY	MATL
8	SPRING	1	STEEL
7	PLAIN WASHER	1	1010 CRS
6	LOWER SPHER WASH	1	LEDLOY
5	UPPER SPHER WASH	1	LEDLOY
4	JAM NUT	3	LEDLOY
3	STUD	1	LEDLOY
2	THUMB SCREW	1	STRESS PROOF
1	STRAP	1	C-1018
NO	PART	QTY	MATL

SPRING SPECS:

25.4 FREE LENGTH
9.14 O.D.—0.66 DIA
WIRE—6 ACTIVE
COILS—SQUARE ENDS
MUSIC WIRE
(5112 STEEL)

CYLINDRICAL FIT	METRIC	ENGL	INCH	mm
1—STRAP &	C11/h11	RC9	.25	7
3—STUD				

SCALE: 1 SQUARE = 0.20 INCHES

.5   0   .5   1.0   1.5   2.0
10   0   10   20   30   40   50

23.83 Make working drawings with an assembly drawing, of this miniature
flat clamp on size B sheets. (Courtesy of Jergens Incorporated.)

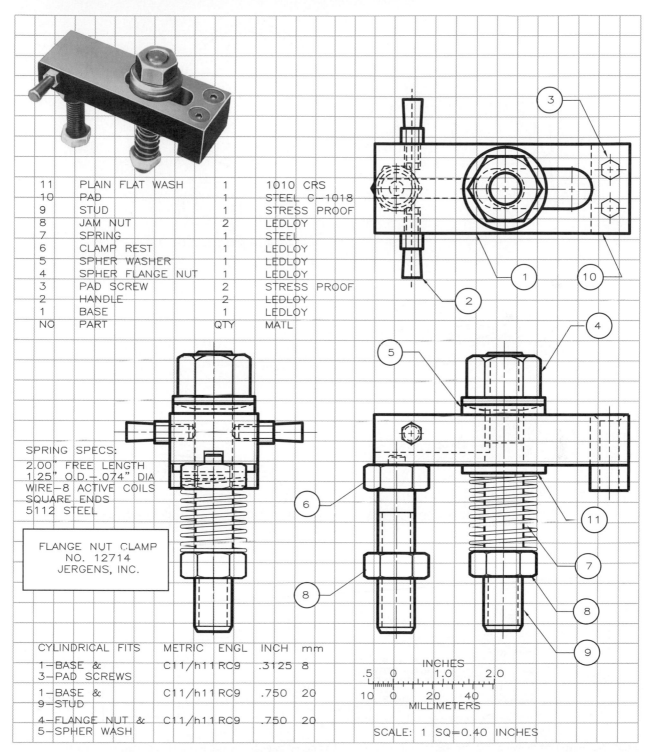

11	PLAIN FLAT WASH	1	1010 CRS
10	PAD	1	STEEL C-1018
9	STUD	1	STRESS PROOF
8	JAM NUT	2	LEDLOY
7	SPRING	1	STEEL
6	CLAMP REST	1	LEDLOY
5	SPHER WASHER	1	LEDLOY
4	SPHER FLANGE NUT	1	LEDLOY
3	PAD SCREW	2	STRESS PROOF
2	HANDLE	2	LEDLOY
1	BASE	1	LEDLOY
NO	PART	QTY	MATL

SPRING SPECS:
2.00" FREE LENGTH
1.25" O.D.-.074" DIA
WIRE-8 ACTIVE COILS
SQUARE ENDS
5112 STEEL

FLANGE NUT CLAMP
NO. 12714
JERGENS, INC.

CYLINDRICAL FITS	METRIC	ENGL	INCH	mm
1-BASE & 3-PAD SCREWS	C11/h11	RC9	.3125	8
1-BASE & 9-STUD	C11/h11	RC9	.750	20
4-FLANGE NUT & 5-SPHER WASH	C11/h11	RC9	.750	20

INCHES
.5   0   1.0   2.0
10 0   20   40
MILLIMETERS

SCALE: 1 SQ=0.40 INCHES

23.84 Make working drawings, with an assembly drawing, of this flange nut clamp on size B sheets. (Courtesy of Jergens Incorporated.)

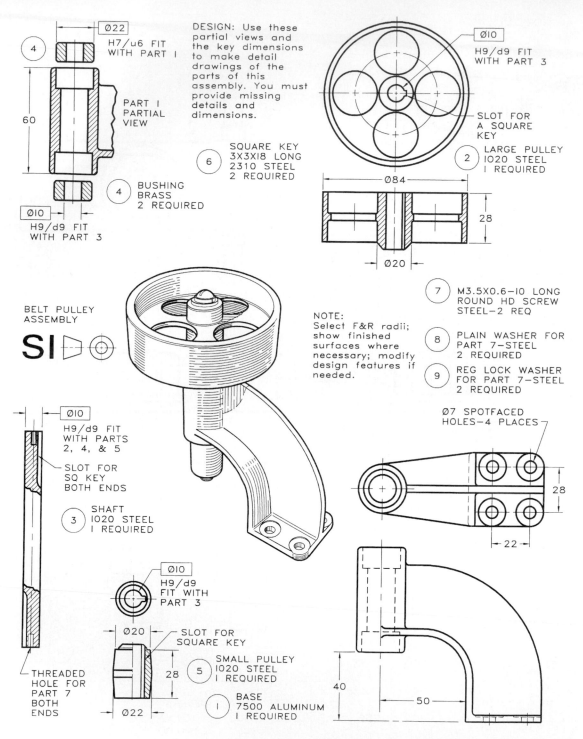

Ø22

H7/u6 FIT
WITH PART 1

4

PART 1
PARTIAL
VIEW

60

DESIGN: Use these
partial views and
the key dimensions
to make detail
drawings of the
parts of this
assembly. You must
provide missing
details and
dimensions.

Ø10

H9/d9 FIT
WITH PART 3

6 SQUARE KEY
3X3X18 LONG
2310 STEEL
2 REQUIRED

4 BUSHING
BRASS
2 REQUIRED

Ø10

H9/d9 FIT
WITH PART 3

H9/d9 FIT
WITH PART 3

SLOT FOR
A SQUARE
KEY

2 LARGE PULLEY
1020 STEEL
1 REQUIRED

Ø84

28

Ø20

7 M3.5X0.6-10 LONG
ROUND HD SCREW
STEEL-2 REQ

8 PLAIN WASHER FOR
PART 7-STEEL
2 REQUIRED

9 REG LOCK WASHER
FOR PART 7-STEEL
2 REQUIRED

BELT PULLEY
ASSEMBLY

SI

NOTE:
Select F&R radii;
show finished
surfaces where
necessary; modify
design features if
needed.

Ø7 SPOTFACED
HOLES-4 PLACES

Ø10

H9/d9 FIT
WITH PARTS
2, 4, & 5

SLOT FOR
SQ KEY
BOTH ENDS

3 SHAFT
1020 STEEL
1 REQUIRED

28

22

Ø10

H9/d9
FIT WITH
PART 3

Ø20

SLOT FOR
SQUARE KEY

5 SMALL PULLEY
1020 STEEL
1 REQUIRED

28

THREADED
HOLE FOR
PART 7
BOTH
ENDS

Ø22

1 BASE
7500 ALUMINUM
1 REQUIRED

40

50

23.85 Make working drawings with an assembly drawing of this belt pulley assembly on size B sheets. (Courtesy of Jergens Incorporated.)

Working Drawings: Design

The following problems require the application of working drawing principles, creative skills, and judgment. You must determine many of the dimensions, tolerances, and standard features of the parts. Make orthographic, dimensioned working drawings of the parts and assemblies shown in **Figure 23.86** through **Figure 23.98** on size A or size B sheets incorporating the design features where specified. Include a title block, dimensions, and notes necessary for making the part.

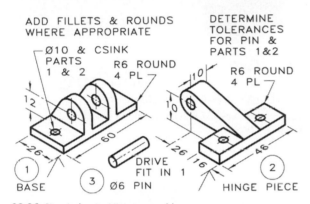

23.86 Size A sheets. Hinge assembly.

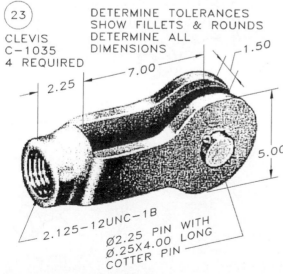

23.87 Size B sheets. Swing hanger.

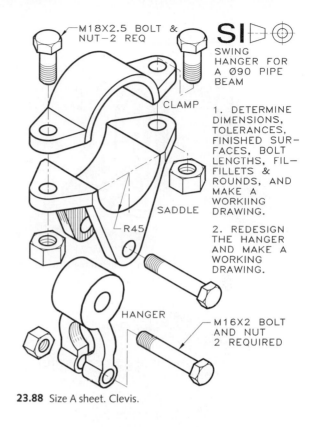

1. DETERMINE DIMENSIONS, TOLERANCES, FINISHED SURFACES, BOLT LENGTHS, FILLETS & ROUNDS, AND MAKE A WORKIING DRAWING.

2. REDESIGN THE HANGER AND MAKE A WORKING DRAWING.

23.88 Size A sheet. Clevis.

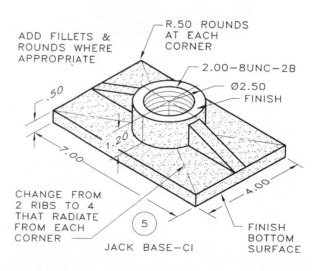

23.89 Size B sheet. Jack base.

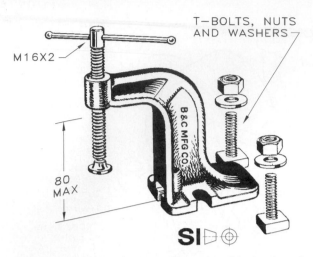

M16X2

T-BOLTS, NUTS
AND WASHERS

80
MAX

B & C MFG CO.

SI

23.90 Make working drawings with an assembly drawing of this hold-down clamp on size B sheets.

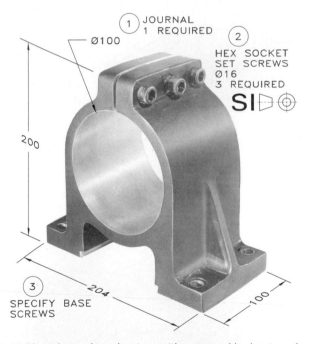

① JOURNAL
1 REQUIRED

Ø100

② HEX SOCKET
SET SCREWS
Ø16
3 REQUIRED

SI

200

204

100

③
SPECIFY BASE
SCREWS

23.91 Make working drawings with an assembly drawing of this journal on size B sheets.

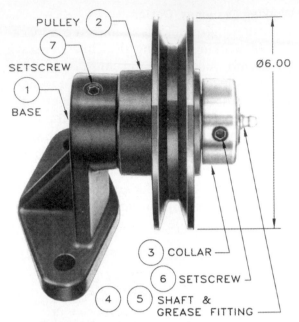

PULLEY ②

⑦
SETSCREW

① BASE

Ø6.00

③ COLLAR

⑥ SETSCREW

④ ⑤ SHAFT &
GREASE FITTING

PULLEY BRACKET ASSEMBLY

23.92 Make working drawings with an assembly drawing of this pulley bracket on size B sheets.

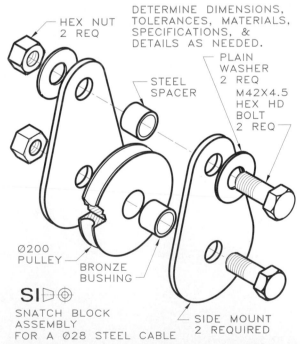

HEX NUT
2 REQ

DETERMINE DIMENSIONS,
TOLERANCES, MATERIALS,
SPECIFICATIONS, &
DETAILS AS NEEDED.

STEEL
SPACER

PLAIN
WASHER
2 REQ
M42X4.5
HEX HD
BOLT
2 REQ

Ø200
PULLEY

BRONZE
BUSHING

SIDE MOUNT
2 REQUIRED

SI

SNATCH BLOCK
ASSEMBLY
FOR A Ø28 STEEL CABLE

23.93 Make working drawings with an assembly drawing of this snatch-block assembly on size B sheets.

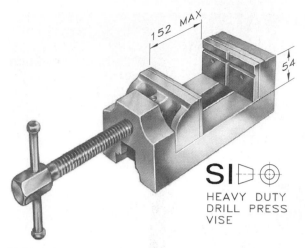

SI⊳⊖

HEAVY DUTY
DRILL PRESS
VISE

152 MAX

54

23.94 Make working drawings with an assembly drawing of this drill press vise on size B sheets.

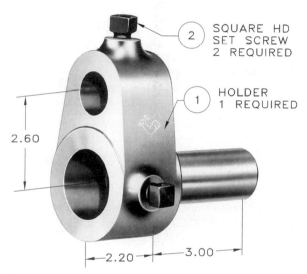

2 SQUARE HD
SET SCREW
2 REQUIRED

1 HOLDER
1 REQUIRED

2.60

2.20

3.00

23.95 Make working drawings with an assembly drawing of this tool holder on size B sheets.

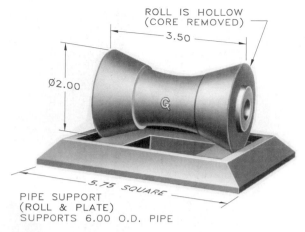

ROLL IS HOLLOW
(CORE REMOVED)

3.50

Ø2.00

5.75 SQUARE

PIPE SUPPORT
(ROLL & PLATE)
SUPPORTS 6.00 O.D. PIPE

23.96 Make working drawings with an assembly drawing of this pipe support on size B sheets.

Ø1.50 CYLINDRICAL
STOCK

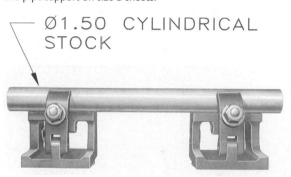

23.97 Make working drawings with an assembly drawing of the milling fixtures that support the cylindrical stock on size B sheets.

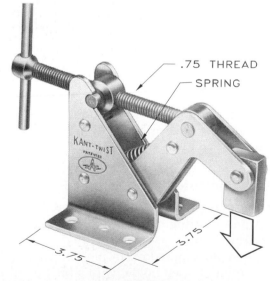

.75 THREAD

SPRING

KANT-TWIST
PATENTED

3.75

3.75

23.98 Make working drawings with an assembly drawing of this roller chain puller on size B sheets.

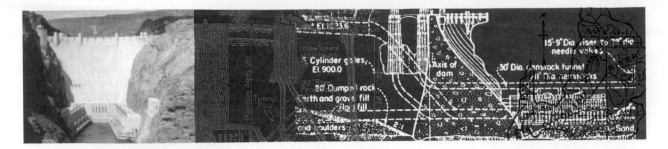

24

Reproduction of Drawings

24.1 Introduction

So far we have discussed the preparation of drawings and specifications through the working-drawing stage, where detailed drawings are completed on tracing film or paper. Now the drawings must be reproduced, folded, and prepared for transmittal to those who will use them to prepare bids or to fabricate the parts.

Several methods of reproduction are available to engineers and technologists for making copies of their drawings. However, most reproduction methods require strong, well-executed line work on the originals in order to produce good copies.

24.2 Computer Reproduction

Three major types of computer reproduction are (A) **pen plotting**, (B) **ink jet printing**, and (C) **laser printing**.

Pen plotting is done by plotter with a single- or a multiple-pen holder with a fiber point that "draws" on the paper or film in ink by moving the pen in x and y directions. Multiple strokes of the pen will give various thicknesses of lines.

Ink jet printing is the process of spraying ink from tiny holes in a flat, disposable printhead onto the drawing surface as it passes through the printer. Prints can be obtained in color and in black and white. Ink jet printers vary in size from 8-1/2 × 11 output (**Figure 24.1**) to large engineering print sizes (**Figure 24.2**).

Laser printing is an electrophotographic process that uses a laser beam to draw an image on a photosensitive drum, where it is electrostatically charged to attract the toner. The electrostatically charged paper is rolled against the drum, the image is transferred, and toner is fused to the paper by heat (**Figure 24.3**). Laser printers make sharp drawings of the highest quality in color or black and white.

Figure 24.4 shows the LaserJet 1200, which is a favorite of offices whose needs do not exceed A-size sheets for both text and graphics. It prints with the highest laser quality of 1200 dots per inch.

24.1 The DesignJet 990c printer provides quiet high-speed operation and high print quality. Its letter- and legal-size format produces excellent color plots of text and graphics. (Courtesy of Hewlett-Packard Company.)

24.2 The DesignJet 1050c printer provides quiet high-speed operation and high print quality. This large-format color plotter can print a D-size color line drawing in less than one minute. (Courtesy of Hewlett-Packard Company.)

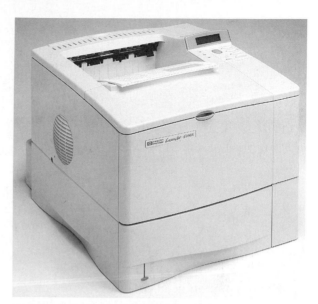

24.3 This network laser printer, the HP LaserJet 4100n, has a speed of 24 pages per minute in order to accommodate more users and higher print volumes. (Courtesy of Hewlett-Packard Company.)

24.4 This HP LaserJet 1200 has a speed of 15 pages per minute and a print quality of 1200 dots per inch for black and white prints. (Courtesy of Hewlett-Packard Company.)

24.3 Types of Reproduction

Drawings made by a drafter are of little use in their original form. If original drawings were handled by checkers and by workers in the field or shop, they would quickly be soiled and damaged, and no copy would be available as a permanent record of the job. Therefore, the reproduction of drawings is necessary for making inexpensive, expendable copies for use by the people who need to use them.

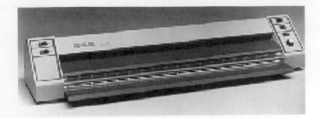

24.5 This typical whiteprinter operates on the diazo process. (Courtesy of Blu-Ray, Incorporated, Essex, CT.)

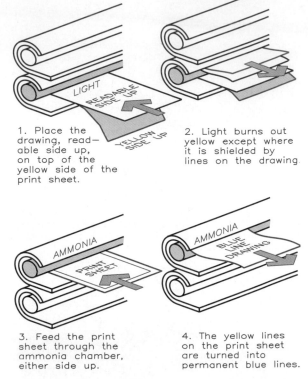

1. Place the drawing, read—able side up, on top of the yellow side of the print sheet.

2. Light burns out yellow except where it is shielded by lines on the drawing.

3. Feed the print sheet through the ammonia chamber, either side up.

4. The yellow lines on the print sheet are turned into permanent blue lines.

24.6 Diazo (blue-line) prints are made by placing the original readable side up and on top of the yellow side of the diazo paper and feeding them under the light, as shown in the steps above.

The most often used processes of reproducing engineering drawings are (1) **diazo printing**, (2) **microfilming**, (3) **xerography**, and (4) **photostating**.

Diazo Printing

The **diazo print** more correctly is called a **whiteprint** or **blue-line print** rather than a **blueprint**, because it has a white background and blue lines. Other colors of lines are available, depending on the type of diazo paper used. (Blueprinting, which creates a print with white lines and a blue background, is a wet process that is almost obsolete at the present.) **Figure 24.5** shows a typical diazo printer.

Diazo printing requires that original drawings be made on semitransparent tracing paper, cloth, or film that light can pass through except where lines have been drawn. The diazo paper on which the blue-line print is copied is chemically treated, giving it a yellow tint on one side. Diazo paper must be stored away from heat and light to prevent spoilage.

The steps of making a diazo print are shown in sequence in **Figure 24.6**. The drawing is placed face up on the yellow side of the diazo paper and then fed through the diazo-process machine, which exposes the drawing to a built-in light. Light rays pass through the tracing paper

and burn away the yellow tint on the diazo paper except where the drawing lines have shielded the paper from the light, similar to how a photographic negative is used. (It is important that your lines be adequately dense to shield the diazo paper enough to make a good print.) The exposed diazo paper becomes a duplicate of the original drawing except that the lines are light yellow and are not permanent.

When the diazo paper is passed through the developing unit of the diazo machine, ammonia fumes develop the yellow lines on it into permanent blue lines. The speed at which the drawing passes under the light determines the darkness of the blue-line copy; the faster the speed, the darker the print is. A slow speed burns out more of the yellow and produces a clear white background, but some of the lighter lines of the drawing may be lost. Most diazo copies are made at a speed fast enough to give a light tint of blue in the background in order to obtain the darkest lines on the copy. Ink drawings, whether made by hand or by computer, give the best reproductions.

Diazo printing has been enhanced by the advent of the computer since computer drawings are made in ink. Thus the print quality is much better than pencil drawings. Also, drawings made by different drafters are more uniform in line weight, lettering, and technique than drawings made by hand.

Microfilming

Microfilming is a photographic process that converts large drawings into film copies—either aperture cards or roll film. Drawings are placed on a copy table and photographed on either 16-mm or 35-mm film.

The roll film or aperture cards are placed in a microfilm enlarger-printer, where the individual drawings can be viewed on a built-in screen. The selected drawings can be printed from the film in standard sizes. Microfilm copies are usually made smaller than the original drawings to save paper and make the drawings easier to use.

Microfilming eliminates the need for large, bulky files of drawings because hundreds of drawings can be stored in permanent archives in miniature on a small amount of film. This is the same process used to preserve newspapers and other large materials by libraries and archives.

Xerography

Xerography is an electrostatic process of duplicating drawings on ordinary, unsensitized paper. Originally developed for business and clerical uses, xerography more recently is currently used for the reproduction of engineering drawings. The xerographic process can also be used to reduce the sizes of the drawings being copied to more convenient and easier-to-use

sizes. The Xerox 2080 can reduce a 24 × 36 inch drawing to 8 × 10 inches.

Photostating

Photostating is a method of enlarging or reducing drawings photographically. The drawing is placed under the glass of the exposure table, which is lit by built-in lamps. The image appears on a glass plate inside the darkroom, where it is exposed on photographically sensitive paper. The exposed negative paper is placed in contact with receiver paper, and the two are fed through the developing solution to obtain a photostatic copy. Photostating also can be used to make reproductions on transparent films and for reproducing halftones (photographs with tones of gray).

24.4 Assembling Drawing Sets

After the original drawings have been copied, they should be stored flat and unfolded in a flat file for future use and updating. Prints made from the originals, however, usually are folded or rolled for ease of transmittal from office to office. The methods of folding size B, C, D, and E sheets so that the image will appear on the outside of the fold are shown in **Figure 24.7**. Drawings should be folded to show the title block always

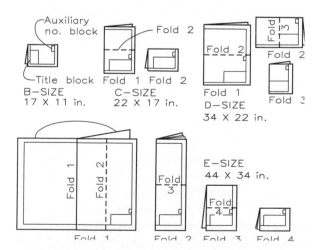

24.7 All standard drawing sheets can be folded to 8-1/2″ × 11″ size for filing and storage.

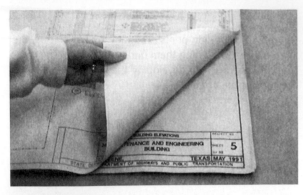

24.8 The title block should appear at the right, usually in the lower right-hand corner of the sheet.

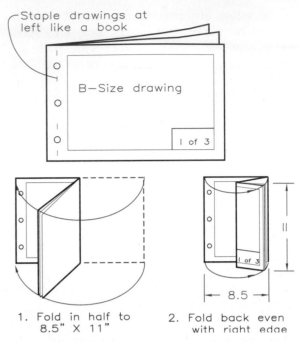

24.9 A set of size B drawings can be assembled by stapling, punching, and folding, as shown here, for safekeeping in a three-ring notebook with the title block visible on top.

on the outside at the right, usually in the lower right-hand corner of the page (**Figure 24.7**). The final size after folding is 8-1/2 × 11 inches (or 9 × 12 inches).

An alternative method of folding and stapling size B sheets often is used for student assignments so that they can be kept in a three-ring notebook (**Figure 24.9**). The basic rules of assembling drawings are listed in **Figure 24.10**.

24.5 Transmittal of Drawings

Prints of drawings are delivered to contractors, manufacturers, fabricators, and others who must use the drawings for implementing the project. Prints usually are placed in standard 9 × 12-inch envelopes for delivery by hand or mail. Sets of large drawings, which may be 30 × 40 inches in size and contain four or more sheets, usually are rolled and sent in a mailing tube when folding becomes impractical. It is not uncommon for a set of drawings to have forty or fifty sheets.

An advanced method of transmitting drawings is by use of large fax machines. Within minutes, large documents can be scanned and transmitted to their destination sites.

Computer drawings can be transmitted on disk by mailing them to their destination, where hard copies can be plotted and reproduced. This procedure offers substantial savings in shipping charges.

Computer drawings can also be transmitted over the Internet in the form of data that is downloaded at its destination. The downloaded data is then printed in the form of a drawing and it is maintained in the database of the computer. In the future, more drawings, documents, and photographs will be sent electronically as data and as scanned images over telephone wires, making them available instantaneously at the desired location.

Only a few years ago, transmission of information and data across the state or nation was time-consuming and carried the risk of loss. Today, any document can be transmitted overnight with certainty of delivery, and most can be transmitted to the receiver within minutes. The OmniShare conferencer (**Figure 24.11**) lets people in two locations collaborate on the same document at the same time over a single phone line.

WORKING DRAWING CHECKLIST

1. Staple along left edge, like a book. Use several staples, never just one.

2. Fold with drawing on outside.

3. Fold drawings as a set, not one at a time separately.

4. Fold to an 8.5"X 11" modular size.

5. The title block must be visible after folding.

6. Sheets of a set should be uniform in size.

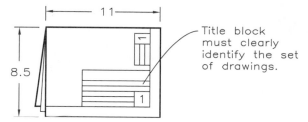

24.10 Follow these basic rules for assembling sets of working drawing prints.

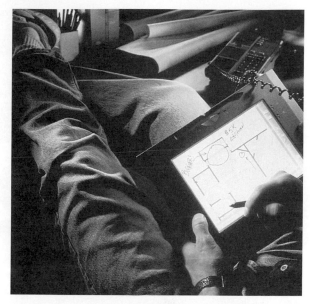

24.11 Hewlett Packard's OmniShare conferencer enables people in two locations to "meet" and collaborate on the same document, at the same time, over a single phone line. (Courtesy of Hewlett-Packard Company.)

24.12 Fast, high-quality output and paper-handling flexibility required of today's business users can be found in the HP LaserJet 4100 printer . In addition, users can add copy, fax, file, and read capabilities by adding the optional LaserJet Companion printer accessory. (Courtesy of Hewlett-Packard Company.)

Hewlett Packard's LaserJet printers have accessories available for fax, copy, file, and read capabilities. Today, the communication of engineering data can be done instantaneously and easily, contributing to an increased productivity (**Figure 24.12**).

Numerically-controlled manufacturing systems can be actuated directly from engineering data once the designs have been digitized. Such systems can be controlled from remote sites to produce products that previously required a high intensity of work hours by individuals. The future will hold many unique innovations in which business, manufacturing, and construction is done.

25

Three-Dimensional Pictorials

25.1 Introduction

A three-dimensional pictorial is a drawing that shows an object's three principal planes, much as they would be captured by a camera. This type of pictorial is an effective means of illustrating a part that is difficult to visualize when only orthographic views are given. Pictorials are especially helpful when a design is complex and when the reader of the drawings is unfamiliar with orthographic drawings.

Sometimes called **technical illustrations**, pictorials are widely used to describe products in catalogs, parts manuals, and maintenance publications (**Figure 25.1**). The ability to sketch pictorials rapidly to explain a detail to an associate in the field is an important communication skill.

The four commonly used types of pictorials are (**1**) **obliques**, (**2**) **isometrics**, (**3**) **axonometrics**, and (**4**) **perspectives** (**Figure 25.2**).

25.1 Many objects cannot be seen as well in real life as they can in a drawing, as shown in this pen set. (Courtesy of Keuffel & Esser Co.)

Oblique pictorials: Three-dimensional drawings made by projecting from the object with parallel projectors that are oblique to the picture plane (**Figure 25.2A**).

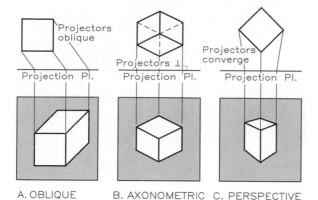

A. OBLIQUE   B. AXONOMETRIC   C. PERSPECTIVE

25.2 The three pictorial projection systems are: (A) oblique pictorials, with parallel projectors oblique to the projection plane; (B) axonometric (including isometric) pictorials, with parallel projectors perpendicular to the projection plane; and (C) perspectives, with converging projectors that make varying angles with the projection plane.

Isometric and axonometric pictorials: Three-dimensional drawings made by projecting from the object with parallel projectors that are perpendicular to the picture plane (**Figure 25.2B**).

Perspective pictorials: Three-dimensional drawings made with projectors that converge at the viewer's eye and make varying angles with the picture plane (**Figure 25.2C**).

25.2 Oblique Drawings

The pulley arm shown in **Figure 25.3** is illustrated by orthographic views and an oblique pictorial. Because most parts are drawn before they are made, photographs cannot be taken; therefore the next best option is to draw a three-dimensional pictorial of the part. Details can usually be drawn with more clarity than can be shown in a photograph.

Oblique pictorials are easy to draw. If you can drawn an orthographic view of a part, you are but one step away from drawing an oblique. For example, **Figure 25.4** shows that drawing a front view of a box twice and connecting its corners yields an oblique drawing.

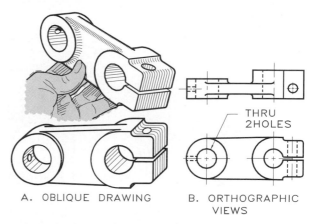

A. OBLIQUE DRAWING   B. ORTHOGRAPHIC VIEWS

25.3 The oblique drawing of this part makes it easier to visualize it than do its orthographic views.

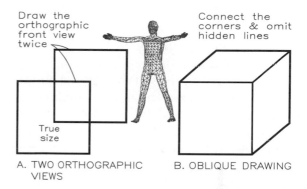

A. TWO ORTHOGRAPHIC VIEWS   B. OBLIQUE DRAWING

25.4 Draw two true-size surfaces of the box, connect them at the corners, and you have an oblique drawing.

Thus an oblique is no more than an orthographic view with a receding axis, drawn at an angle to show the depth of the object. An oblique is a pictorial that does not exist in reality (a camera cannot give an oblique). This type of pictorial is called an oblique because its parallel projectors from the object are oblique to the picture plane. The underlying principles of projection are covered in Section 25.3.

Types of Obliques

The three basic types of oblique drawings are: (**1**) **cavalier**, (**2**) **cabinet**, and (**3**) **general** (**Figure 25.5**). For each type, the angle of the receding axis with the horizontal can be at any angle between 0° and 90°. Measurements

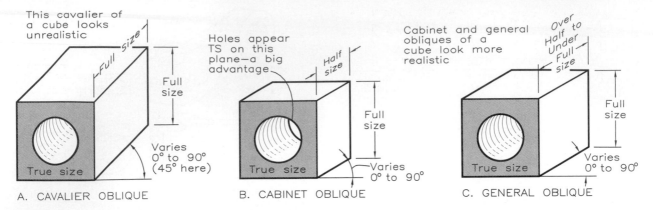

A. CAVALIER OBLIQUE B. CABINET OBLIQUE C. GENERAL OBLIQUE

25.5 The three types of obliques:

A The cavalier oblique has a receding axis at any angle and true-length measurements on the receding axis.

B The cabinet oblique has a receding axis at any angle and half-size measurements along the receding axis.

C The general oblique has a receding axis at any angle and measurements along the receding axis larger than half size and less than full size.

along the receding axes of the cavalier oblique are laid off true length, and measurements along the receding axes of the cabinet oblique are laid off half-size. The general oblique has measurements along the receding axes that are greater than half size and less than full size.

Figure 25.6 shows three examples of cavalier obliques of a cube. The receding axes for each is drawn at a different angle, but the receding axes are drawn true length. **Figure 25.7** compares cavalier with cabinet obliques.

Constructing Obliques

You can easily begin a cavalier oblique by drawing a box using the overall dimensions of height, width, and depth with light construction lines. As demonstrated in **Figure 25.8**, first

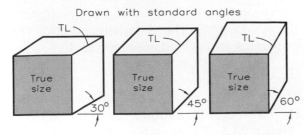

25.6 A cavalier oblique usually has its receding axis at one of the standard angles of drafting triangles. Each gives a different view of a cube.

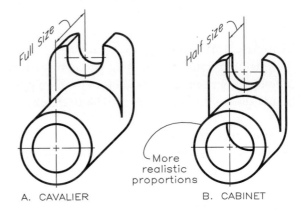

A. CAVALIER B. CABINET

25.7 Measurements along the receding axis of a cavalier oblique are full size and those in a cabinet oblique are half size.

draw the front view as a true-size orthographic view. True measurements must be made parallel to the three axes and transferred from the orthographic views with your dividers. Then remove the notch from the blocked-in construction box to complete the oblique.

Angles

Angular measurements can be made on the true-size plane of an oblique, but not on the other two planes. Note in **Figure 25.9** that a true angle can be measured on a true-size surface, but in **Figure 25.10**, angles along reced-

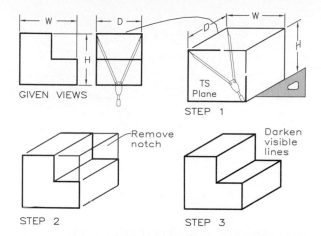

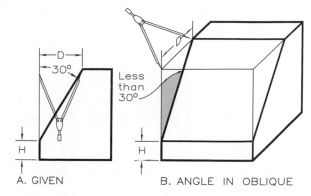

25.10 Angles that do not lie in a true-size plane of an oblique must be located with coordinates.

25.8 Constructing a cavalier oblique:

Step 1 Draw the front surface of the object as a true-size plane. Draw the receding axis at a convenient angle and transfer the true distance D from the side view to it with your dividers.

Step 2 Draw the notch on the front plane and project it to the rear plane.

Step 3 Darken the lines to complete the drawing.

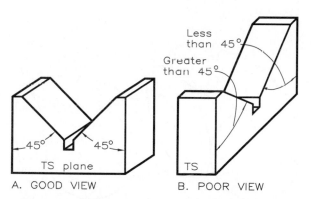

25.9 Objects with angular features should be drawn in oblique so that the angles appear true size. This results in a better pictorial and one that is easier to draw.

ing planes are either smaller or larger than their true sizes. A better, easier-to-draw oblique is obtained when angles are drawn to appear true size.

To construct an angle in an oblique on one of the receding planes, you must use coordinates, as shown in **Figure 25.10**. To find the surface that slopes 30° from the front surface, locate the vertex of the angle, H distance from the bottom. To find the upper end of the sloping

plane, measure the distance D along the receding axis. Transfer H and D to the oblique with your dividers. The angle in the oblique is not equal to the 30° angle in the orthographic view.

Cylinders

The **major advantage** of an **oblique is that circular features can be drawn as true circles** on its frontal plane (**Figure 25.11**). Draw the centerlines of the circular end at A and construct

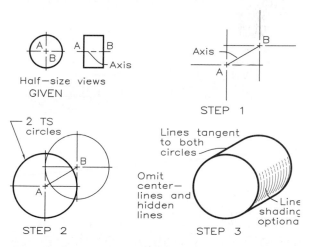

25.11 Drawing a cylinder in oblique:

Step 1 Draw axis AB and locate the centers of the circular ends of the cylinder at A and B. Because the axis is true length, this will be a cavalier oblique.

Step 2 Draw a true-size circle with its center at A by using a compass or computer-graphics techniques.

Step 3 Draw the other circular end with its center at B and connect the circles with tangent lines parallel to axis AB.

the receding axis at the desired angle. Locate the end at B by measuring along the axis, draw circles at each end at centers A and B, and draw tangents to both circles.

These same principles apply to construction of the object having semicircular features shown in **Figure 25.12**. Position the oblique so that the semicircular features are true size. Locate centers A, B, and C and the two semicircles. Then complete the cavalier oblique.

Circles

Circular features drawn as true circles on a true-size plane of an oblique pictorial appear on the receding planes as ellipses.

The four-center ellipse method is a technique of constructing an approximate ellipse with a compass and four centers (**Figure 25.13**). The ellipse is tangent to the inside of a rhombus drawn with sides equal to the circle's diameter. Drawing the four arcs produces the ellipse.

The four-center ellipse method will not work for the cabinet or general oblique, but coordinates must be used. **Figure 25.14** illus-

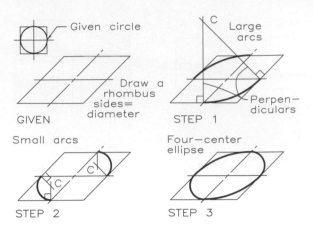

25.13 Constructing a four-center ellipse in oblique:

Given Block in the circle to be drawn in oblique with a square tangent to the circle. This square becomes a rhombus on the oblique plane.

Step 1 Draw construction lines perpendicular at the points of tangency to locate the centers for drawing two segments of the ellipse.

Step 2 Locate the centers for the two remaining arcs with perpendiculars drawn from adjacent tangent points.

Step 3 Draw the four arcs, which yield an approximate ellipse.

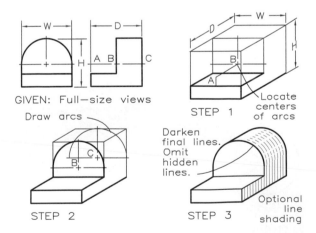

25.12 Drawing semicircular features in oblique:

Step 1 Block in the overall dimensions of the cavalier oblique with light construction lines, ignoring the semicircular feature.

Step 2 Locate centers B and C and draw arcs with a compass or by computer tangent to the sides of the construction boxes.

Step 3 Connect the arcs with lines tangent to each arc and parallel to axis BC and darken the lines.

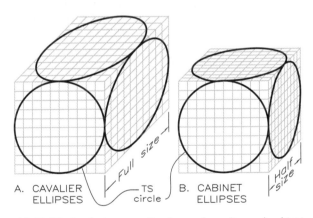

25.14 Circular features on the faces of cavalier and cabinet obliques are compared here. Ellipses on the receding planes of cabinet obliques must be plotted by coordinates. The spacing of the coordinates along the receding axis of cabinet obliques is half size.

trates the method of locating coordinates on the planes of cavalier and cabinet obliques. For the cabinet oblique, the coordinates along the receding axis are half size, and the coordinates along the horizontal axis (true-size axis) are full size. Draw the ellipse with an irregular curve or an ellipse template that approximates the plotted points.

Whenever possible, oblique drawings of objects with circular features should be positioned so circles can be drawn as true circles instead of ellipses. The view in **Figure 25.15A** is better than the one in **Figure 25.15B** because it gives a more descriptive view of the part and is easier to draw.

Curves

Irregular curves in oblique pictorials must be plotted point by point with coordinates (**Figure 25.16**). Transfer the coordinates from the orthographic to the oblique view and draw the curve through the plotted points with an irregular curve. If the object has a uniform thickness, plot the points for the lower curve by projecting vertically downward from the upper points a distance equal to the object's height.

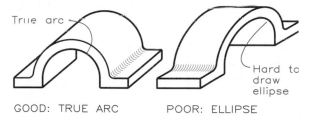

25.15 An oblique should be positioned so that circular and curving features can be drawn most easily.

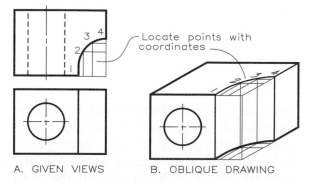

25.16 Coordinates are used to find points along irregular curves in oblique. Projecting the points downward a distance equal to the height of the object yields the lower curve.

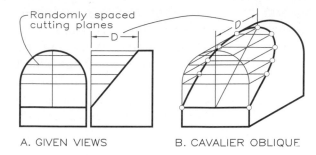

25.17 Construction of an elliptical feature on an inclined surface in oblique requires the use of three-dimensional coordinates to locate points on the curve.

To obtain the elliptical feature on the inclined surface shown in **Figure 25.17**, use a series of coordinates to locate points along its curve. Connect the plotted points by using an irregular curve or ellipse template.

Sketching

Understanding the principles of oblique construction is essential for sketching obliques freehand. The sketch of the part shown in **Figure 25.18** is based on the principles dis-

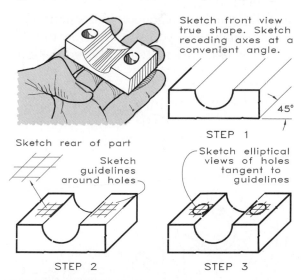

25.18 Sketching obliques:

Step 1 Sketch the front of the object as a true-size surface and draw a receding axis from each corner.

Step 2 Lay off the depth, D, along the receding axes to locate the rear of the part. Lightly sketch pictorial boxes as guidelines for drawing the holes.

Step 3 Sketch the holes inside the boxes and darken all lines.

cussed, but its proportions were determined by eye instead of with scales and dividers.

Lightly drawn guidelines need not be erased when you darken the final lines. When sketching on tracing vellum, you can place a printed grid under the sheet to provide guidelines. Refer to Chapter 13 to review sketching techniques if needed.

Dimensioned Obliques

Dimensioned sectional views of obliques provide excellent, easily understood depictions of objects (**Figure 25.19**). Apply numerals and lettering in oblique pictorials by using either the **aligned** method (with numerals aligned with the dimension lines) or the **unidirectional** method (with numerals positioned horizontally regardless of the direction of the dimension lines), as shown in **Figure 25.20**. Notes connected with leaders are positioned horizontally in both methods.

25.3 Oblique Projection Theory

Now that you have a general understanding of oblique pictorials, you should know the theory on which this system is based. Oblique

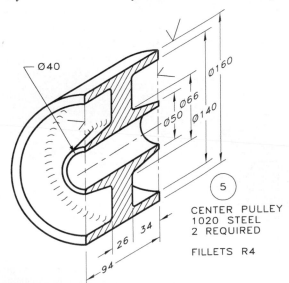

25.19 Oblique pictorials can be drawn as sections and dimensioned to serve as working drawings.

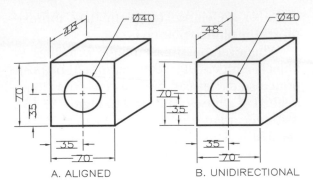

25.20 Either of these methods of lettering, aligned or unidirectional, is acceptable for dimensioning obliques.

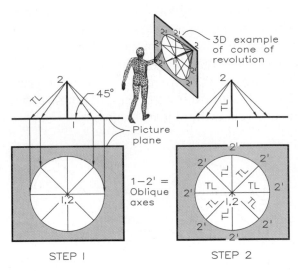

25.21 This drawing demonstrates the underlying principle of the cavalier oblique by using a series of projectors to form a cone:

Step 1 Each element from point 2 makes a 45° angle with the picture plane.

Step 2 The projected lengths of 1–2' are equal in length to line 1–2, which is perpendicular to the picture plane. Thus the receding axis of a cavalier oblique is true length and can be drawn at any angle.

projection, shown in **Figure 25.21**, is the basis of oblique drawings. Receding axis 1–2 is perpendicular to the frontal projection plane. Projectors drawn from point 2 at 45° to the projection plane yield lengths on the front surface that are the same length as 1–2 (true length, in other words). Infinitely many 45°

projectors form a cone of projectors with its apex at 2.

The true-length projections of lines 1–2′ represent receding axes that can be used for cavalier obliques, which by definition have true-length dimensions along their receding axes. Do not use a vertical or a horizontal receding axis, but one between those limits.

To distinguish **oblique projection** from **oblique drawing**, as described in this chapter so far, observe the top and side views of a part and the picture planes shown in **Figure 25.22**. In an oblique projection, projectors from the top and side views are oblique to the edge views of the projection planes, hence the name oblique.

Your line of sight can yield obliques with receding axes longer than true length (which should be avoided). Because of this shortcoming and the complexity of construction, oblique pictorials usually are **oblique drawings** rather than **oblique projections**.

25.4 Isometric Pictorials

In **Figure 25.23** the pulley arm is drawn in orthographic views and as a three-dimensional pictorial drawing. The pictorial is an isometric drawing in which the three planes

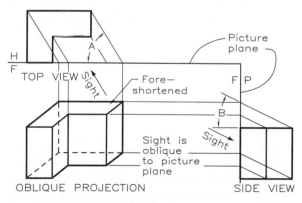

25.22 An oblique projection may be drawn at varying angles of sight. However, a line of sight making an angle of less than 45° with the picture plane would result in a receding axis longer than its true length, thereby distorting the pictorial.

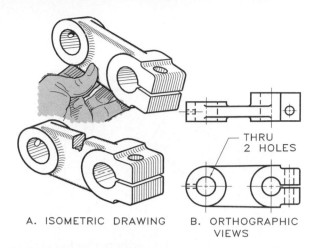

A. ISOMETRIC DRAWING   B. ORTHOGRAPHIC VIEWS

25.23 An isometric drawing gives a more realistic view of a part than an oblique drawing.

of the object are equally foreshortened, representing the object more realistically than an oblique drawing can.

With more realism comes more difficulty of construction. In particular, circles and curves do not appear true shape on any of the three isometric planes.

Isometric Projection versus Drawing

In isometric projection, parallel projectors are perpendicular to the imaginary projection (picture) plane in which the diagonal of a cube appears as a point (**Figure 25.24**). An isometric pictorial constructed by projection is called an **isometric projection**, with the three axes foreshortened to 82% of their true lengths and 120° apart. The name isometric, which means equal measurement, aptly describes this type of projection because the planes are equally foreshortened.

An **isometric drawing** is a convenient approximate isometric pictorial in which the measurements are shown full size along the three axes rather than at 82% as in **isometric projection** (**Figure 25.25**). Thus the isometric drawing method allows you to measure true dimensions with standard scales and lay them off with dividers along the three axes. The only

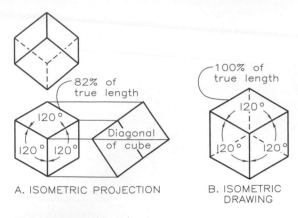

A. ISOMETRIC PROJECTION

B. ISOMETRIC DRAWING

25.24 Projection versus drawing:

A A true isometric projection is found by constructing a view in which the diagonal of a cube appears as a point and the axes are foreshortened.

B An isometric drawing is not a true projection because the dimensions are true size rather than foreshortened.

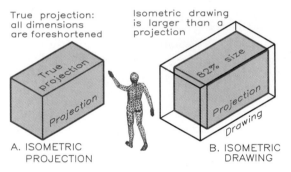

A. ISOMETRIC PROJECTION

B. ISOMETRIC DRAWING

25.25 The true isometric projection is foreshortened to 82% of full size. The isometric drawing is drawn full size for convenience.

difference between the two is the larger size of the drawing. Consequently, isometric drawings are used much more often than isometric projections.

The axes of isometric drawings are separated by 120° (**Figure 25.26**), but more often than not, one of the axes selected is vertical, since most objects have vertical lines. However, isometrics without a vertical axis are still isometrics.

25.5 Isometric Drawings

An isometric drawing is begun by drawing three axes 120° apart. Lines parallel to these

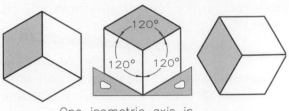

One isometric axis is usually vertical, but they can be at any angle.

25.26 Isometric axes are spaced 120° apart, but they can be revolved into any position. Usually, one axis is vertical, but it can be at any angle with axis spacing remaining the same.

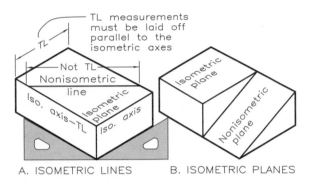

A. ISOMETRIC LINES

B. ISOMETRIC PLANES

25.27 Isometric lines and planes:

A Isometric lines (parallel to the three axes) give true measurements, but nonisometric lines do not.

B Here, the three isometric planes are equally foreshortened, and the nonisometric plane is inclined at an angle to one of the isometric planes.

axes are called **isometric lines** (**Figure 25.27A**). You can make true measurements along isometric lines but not along nonisometric lines. The three surfaces of a cube in an isometric drawing are called **isometric planes** (**Figure 25.27B**). Planes parallel to those planes also are isometric planes.

To draw an isometric pictorial, you need a scale, dividers, and a 30°–60° triangle (**Figure 25.28**). Begin by selecting the three axes and then constructing a plane of the isometric from the dimensions of height, H, and depth, D. Add the third dimension of width, W, and complete the isometric drawing.

Use light construction lines to block in all isometric drawings (**Figure 25.29**) and the

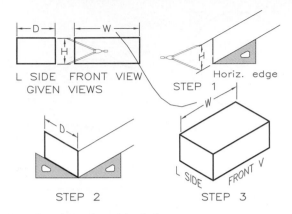

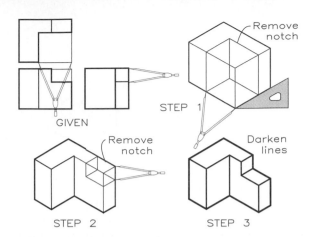

25.28 Drawing an isometric of a box:

Step 1 Use a 30°–60° triangle and a horizontal straight edge to construct a vertical line equal to the height, H, and draw two isometric lines through each end.

Step 2 Draw two 30° lines and locate the depth, D, by transferring depth from the given views with dividers.

Step 3 Locate the width, W, of the object, complete the surfaces of the isometric box, and darken the lines.

25.30 Laying out an isometric drawing:

Step 1 Use the overall dimensions given to block in the object with light lines and remove the large notch.

Step 2 Remove the small notch.

Step 3 Darken the lines to complete the drawing.

notches. The object was blocked in by using the H, W, and D dimensions. Remove the notches in the block to complete the drawing.

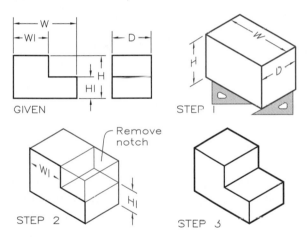

25.29 Constructing an isometric of a simple part:

Step 1 Construct an isometric drawing of a box with the overall dimensions W, D, and H from the given views.

Step 2 Locate the notch by transferring dimensions W1 and H1 from the given views with your dividers.

Step 3 Darken the lines to complete the drawing.

overall dimensions W, D, and H. Take other dimensions from the given views with dividers and measure along their isometric lines to locate notches in the blocked-in drawing.

Figure 25.30 shows an isometric drawing of a slightly more complex object, with two

Angles

You cannot measure an angle's true size in an isometric drawing because the surfaces of an isometric are not true size. Instead, you must locate angles with isometric coordinates measured parallel to the axes (**Figure 25.31**). Lines AD and BC are equal in length in the orthographic view, but they are shorter and longer than true length in the isometric drawing. **Figure 25.32** shows a similar situation,

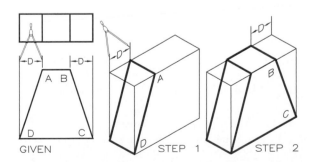

25.31 Use coordinates measured along the isometric axes to obtain inclined surfaces. Angular lines are not true length in isometric.

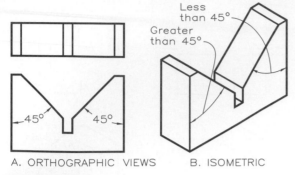

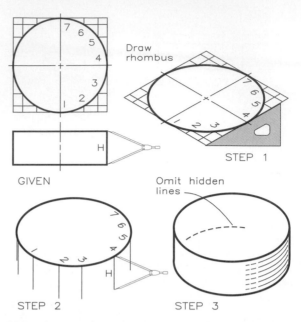

25.32 Angles in isometric may appear larger or smaller than they actually are.

where two angles drawn in isometric are less than and greater than their true dimensions in the orthographic view.

Figure 25.33 shows how to construct an isometric drawing of an object with inclined surfaces. Blocking in the object with its overall dimensions with light construction lines is followed by removal of the inclined portions.

Circles

Three methods of constructing circles in isometric drawings are (**1**) **point plotting**, (**2**) **four-center ellipse construction**, and (**3**) **ellipse template usage**.

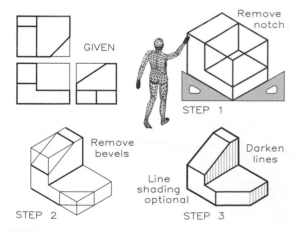

25.33 Drawing inclined planes in isometric:

Step 1 Block in the object with light lines, using the overall dimensions, and remove the notch.

Step 2 Locate the ends of the inclined planes by using measurements parallel to the isometric axes.

Step 3 Darken the lines to complete the drawing.

25.34 Plotting circles in isometric:

Step 1 Block in the circle by using its overall dimensions. Transfer the coordinates that locate points on the circle to the isometric plane and connect them with a smooth curve.

Step 2 Project each point a distance equal to the height of the cylinder to obtain the lower ellipse.

Step 3 Connect the two ellipses with tangent lines and darken all lines.

Point plotting is a method of using a series of x and y coordinates to locate points on a circle in the given orthographic views. The coordinates are then transferred with dividers to the isometric drawing to locate the points on the ellipse one at a time (**Figure 25.34**).

Block in the cylinder with light construction lines and show the centerlines. Draw coordinates on the upper plane and use the height dimension to locate the points on the lower plane. Draw the ellipses with an irregular curve or an ellipse template.

A plotted ellipse is a true ellipse and is equivalent to a 35° ellipse drawn on an isometric plane. An example of a design composed of circular features drawn in isometric is the handwheel shown in **Figure 25.35**.

Four-center ellipse construction is the method of producing an approximate ellipse

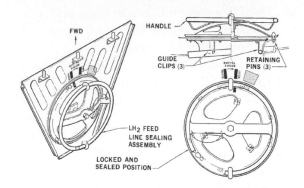

25.35 This handwheel assembly proposed for use in an orbital workshop is an example of parts with circular features drawn as ellipses in isometric. (Courtesy of NASA.)

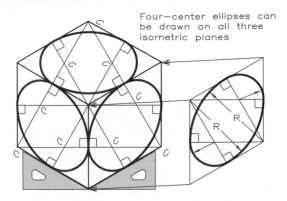

Four-center ellipses can be drawn on all three isometric planes

25.37 Four-center ellipses may be drawn on all three surfaces of an isometric drawing.

(**Figure 25.36**) by using four arcs drawn with a compass. Draw an isometric rhombus with its sides equal to the diameter of the circle to be represented. Find the four centers by constructing perpendiculars to the sides of the rhombus at the midpoints of each side, and draw the four arcs to complete the ellipse. You may draw four-center ellipses on all three isometric planes because each plane is equally foreshortened (**Figure 25.37**). Although it is

only an approximate ellipse, the four-center ellipse technique is acceptable for drawing large ellipses and as a way to draw ellipses when an ellipse template is unavailable.

Isometric ellipse templates are specially designed for drawing ellipses in isometric (**Figure 25.38**). The numerals on the templates represent the isometric diameters of the ellipses because diameters are measured parallel to the isometric axes of an isometric drawing (**Figure 25.39**). Recall that the maximum diameter across the ellipse is its major diameter, which is a true diameter. Thus the size of the diameter marked on the template is less than the ellipse's major diameter. You may use the isometric

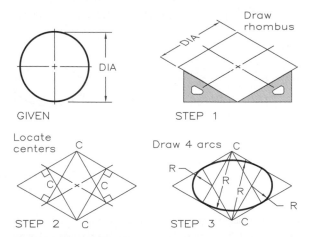

25.36 The four-center ellipse method:

Step 1 Use the diameter of the given circle to draw an isometric rhombus and the centerlines.

Step 2 Draw light construction lines perpendicularly from the midpoints of each side to locate four centers.

Step 3 Draw four arcs from the centers to represent an ellipse tangent to the rhombus.

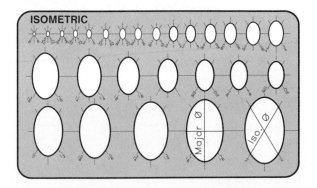

25.38 The isometric template (a 35° ellipse angle) is designed for drawing elliptical features in isometric. The isometric diameters of the ellipses are not major diameters of the ellipses but are diameters that are parallel to the isometric axes.

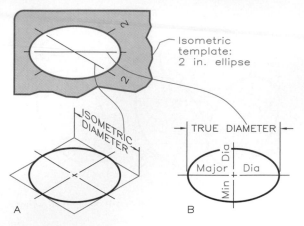

Isometric
template:
2 in. ellipse

25.39 Ellipse terminology:

A Measure the diameter of a circle along the isometric axes. The major diameter of an isometric ellipse thus is larger than the measured diameter.

B The minor diameter is perpendicular to the major diameter.

ellipse template to draw an ellipse by constructing centerlines of the ellipse in isometric and aligning the ellipse template with those isometric lines (**Figure 25.39**).

Cylinders

A cylinder may be drawn in isometric by using the four-center ellipse method (**Figure 25.40**). Use the isometric axes and centerline axis to construct a rhombus at each end of the cylin-

der. Then draw the ellipses at each end, connect them with tangent lines, and darken the lines to complete the drawing.

An easier way to draw a cylinder is to use an isometric ellipse template (**Figure 25.41**). Draw the axis of the cylinder and construct perpendiculars at each end. Because the axis of a right cylinder is perpendicular to the major diameter of its elliptical ends, position the ellipse template with its major diameter perpendicular to the axis. Draw the ellipses at each end, connect them with tangent lines, and darken the visible lines to complete the drawing.

To construct a cylindrical hole in a block (**Figure 25.42**), begin by locating the center of the hole on the isometric plane. Draw the axis of the cylinder parallel to the isometric axis that is perpendicular to the plane of the ellipse through its center. Align the ellipse template with the major diameter, which makes a 90° angle with the cylindrical axis, and complete the elliptical view of the cylindrical hole.

The isometric ellipse template can be used to draw ellipses on all three planes of an iso-

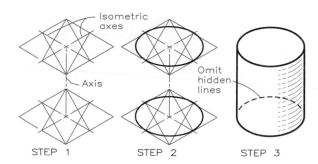

25.40 A cylinder drawn with the four-center method:

Step 1 Draw an isometric rhombus at each end of the cylinder's axis.

Step 2 Draw a four-center ellipse within each rhombus.

Step 3 Draw lines tangent to each rhombus to complete the drawing.

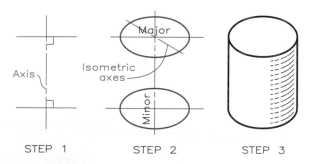

25.41 A cylinder using ellipse template method:

Step 1 Establish the length of the axis of the cylinder and draw perpendiculars at each end.

Step 2 Draw the elliptical ends by aligning the major diameter of the ellipse template with the perpendiculars at the ends of the axis. The isometric diameters of the isometric ellipse template will align with two isometric axes.

Step 3 Connect the ellipses with tangent lines to complete the drawing and omit hidden lines.

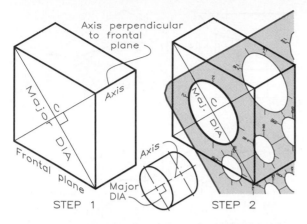

25.42 Constructing cylindrical holes through a block:

Step 1 Locate the center of the hole on a face of the isometric drawing. Draw the axis of the cylinder from the center parallel to the isometric axis perpendicular to the plane of the circle. The major diameter is perpendicular to this axis.

Step 2 Use the 2-in. ellipse template to draw the ellipse by aligning guidelines on the template with the major and minor diameters drawn on the front surface.

metric drawing. On each plane, the major diameter is perpendicular to the isometric axis of the adjacent perpendicular plane. The isometric diameters marked on the template align with the isometric axes. All ellipses drawn on isometric planes must align in the directions shown in **Figure 25.43**.

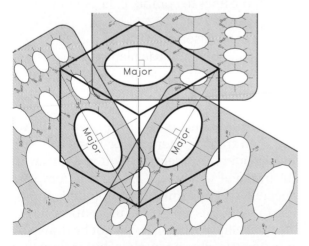

25.43 Position the isometric ellipse template as shown for drawing ellipses of various sizes on the three isometric planes.

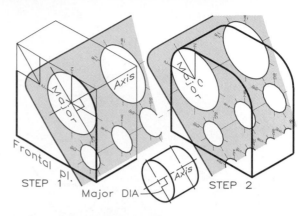

25.44 Drawing rounded corners:

Step 1 Draw the centerlines and isometric axes at the corners. Align the ellipse template with these guidelines and draw one quarter of the ellipse.

Step 2 Draw the other elliptical corner in the same manner with the same size ellipse.

Rounded Corners

The rounded corners of an object can be drawn with an ellipse template (**Figure 25.44**). Block in each corner with light construction lines, draw centerlines, draw the major diameter, and construct ellipses at each corner by positioning the template as shown. The rounded corners may also be constructed by using the four-center ellipse method (see Figure 25.40) or by plotting points with coordinates (see Figure 25.33).

A similar drawing involving the construction of ellipses is the conical shape shown in **Figure 25.45**. Block in the ellipses on the upper and lower surfaces. Then draw the circular features by using a template or the four-center method, and draw lines tangent to each ellipse.

Inclined Planes

Inclined planes in isometric may be located by coordinates, but they cannot be measured with a protractor because they do not appear true size. **Figure 25.46** illustrates the coordinate method. Use horizontal and vertical coordinates (in the x and y directions) to

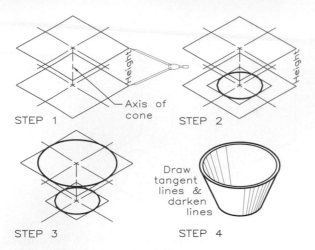

STEP 1

STEP 2

STEP 3

Draw
tangent
lines &
darken
lines

STEP 4

25.45 Constructing a cone in isometric:

Step 1 Draw the axis of the cone and block in the larger end at both ends.

Step 2 Block in the smaller end of the cone.

Step 3 Connect the ellipses with tangents, draw the cone's wall thickness, and darken the lines to complete the drawing.

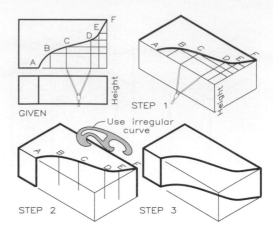

GIVEN

STEP 1

Use irregular curve

STEP 2

STEP 3

25.47 Plotting irregular curves:

Step 1 Block in the shape by using the overall dimensions. Locate points on the irregular curve with coordinates transferred from the orthographic views.

Step 2 Project these points downward the distance H (height) from the upper points to obtain the lower curve.

Step 3 Connect the points and darken the lines.

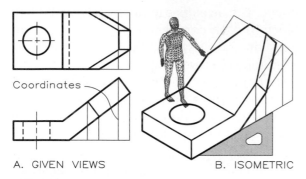

Coordinates

A. GIVEN VIEWS

B. ISOMETRIC

25.46 Inclined surfaces in isometric must be located with three-dimensional coordinates parallel to the isometric axes. True angles cannot be measured in isometric drawings.

locate key points on the orthographic views. Transfer these coordinates to the isometric drawing with dividers to show the features of the inclined surface.

Curves

Irregular curves in isometric must be plotted point by point, with coordinates locating each point. Locate points A through F in the orthographic view with coordinates of width and

depth (**Figure 25.47**). Then transfer them to the isometric view of the blocked-in part and connect them with an irregular curve.

Project points on the upper curve downward a distance of H, the height of the part, to locate points on the lower curve. Connect these points with an irregular curve and darken the lines to complete the isometric.

Ellipses on Nonisometric Planes

Ellipses on nonisometric planes in an isometric drawing, such as the one shown in **Figure 25.48**, must be found by locating a series of points on the curve. Locate three-dimensional coordinates in the orthographic views and then transfer them to the isometric with your dividers. Connect the plotted points with an irregular curve or an ellipse template selected to approximate the plotted points. The more points you select, the more accurate the final ellipse will be. It will not be drawn by an isometric ellipse template, but by one that fits the plotted points.

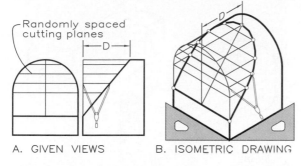

A. GIVEN VIEWS B. ISOMETRIC DRAWING

25.48 To construct ellipses on inclined planes, draw coordinates to locate points in the orthographic views. Then transfer the three-dimensional coordinates to the isometric drawing and connect them with a smooth curve.

25.6 Technical Illustration

Orthographic and isometric views of a spotface, countersink, and boss are shown in **Figure 25.49**. These features may be drawn in isometric by point-by-point plotting of the circular features, the four-center, or ellipse template method (the easiest method of the three).

A threaded shaft may be drawn in isometric as shown in **Figure 25.50**. First draw the cylinder in isometric. Draw the major diameters of the crest lines equally separated by distance P, the pitch of the thread. Then draw ellipses by aligning the major diameter of the ellipse template with the perpendiculars to the cylinder's axis. Use a smaller ellipse at the end for the 45° chamfered end.

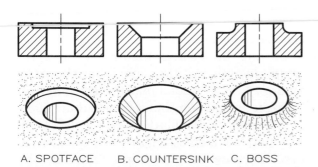

A. SPOTFACE B. COUNTERSINK C. BOSS

25.49 These examples of circular features in isometric may be drawn by using ellipse templates.

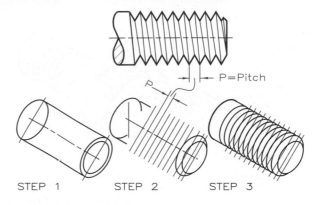

STEP 1 STEP 2 STEP 3

25.50 Threads in isometric:

Step 1 Using an ellipse template, draw the cylinder to be threaded.

Step 2 Lay off perpendiculars, spacing them apart at a distance equal to the pitch of the thread, P.

Step 3 Draw a series of ellipses to represent the threads. Draw the chamfered end by using an ellipse whose major diameter is equal to the root diameter of the threads.

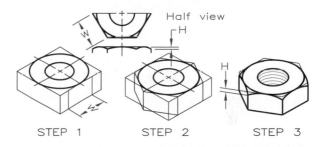

STEP 1 STEP 2 STEP 3

25.51 Constructing a nut:

Step 1 Use the overall dimensions of the nut to block in the nut.

Step 2 Construct the hexagonal sides at the top and bottom.

Step 3 Draw the chamfer with an irregular curve. Draw the threads to complete the drawing.

Figure 25.51 shows how to draw a hexagon-head nut with an ellipse template. Block in the nut and draw an ellipse tangent to the rhombus. Construct the hexagon by locating distance W across a flat parallel to the isometric axes. To find the other sides of the hexagon, draw lines tangent to the ellipse. Lay off distance H at each corner to establish the chamfers.

Figure 25.52 depicts a hexagon-head bolt in two positions. The washer face is on the

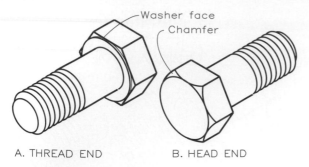

25.52 Isometric drawings of the lower and upper sides of a hexagon-head bolt.

A. THREAD END   B. HEAD END

lower side of the head, and the chamfer is on the upper side.

A portion of a sphere is drawn to represent a round-head screw in **Figure 25.53**. Construct a hemisphere and locate the centerline of the slot along one of the isometric planes. Measure the head's thickness, E, from the highest point on the sphere.

Sections

A full section drawn in isometric can clarify internal details that might otherwise be overlooked (**Figure 25.54**). Half sections also may be used advantageously.

Dimensioned Isometrics

When you dimension isometric drawings, place numerals on the dimension lines, using either **aligned** or **unidirectional** numerals

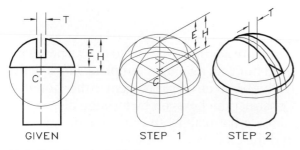

25.53 Drawing spherical features:

Step 1 Use an isometric ellipse template to draw the elliptical features of a round-head screw.

Step 2 Draw the slot in the head and darken the lines to complete the drawing.

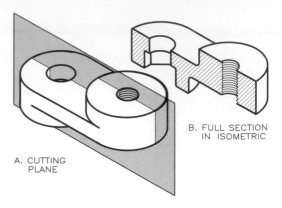

A. CUTTING PLANE   B. FULL SECTION IN ISOMETRIC

25.54 Isometric sections can be used to clarify the internal features of a part.

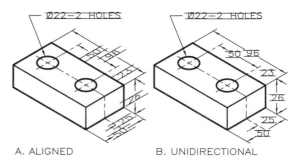

A. ALIGNED   B. UNIDIRECTIONAL

25.55 Either of the techniques shown—aligned or unidirectional—is acceptable for placing dimensions on isometric drawings. Guidelines should always be used for lettering.

(**Figure 25.55**). In both cases, notes connected with leaders usually are positioned horizontally, but drawing them to lie in an isometric plane is permissible. Always use guidelines for your lettering and numerals.

Fillets and Rounds

Fillets and rounds in isometric may be represented by either of the techniques shown in **Figure 25.56** for added realism. The enlarged detail in the balloon shows how to draw fillets and rounds with elliptical segments (A) or with parallel lines (B). Elliptical segments are best drawn by using an ellipse template. The stipple shading was applied by using an adhesive overlay film.

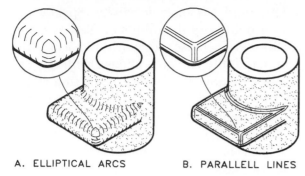

A. ELLIPTICAL ARCS   B. PARALLELL LINES

25.56 Either of these two methods can be used to represent fillets and rounds on the pictorial view of a part.

When fillets and rounds are shown in a three-dimensional drawing of a part they are much more readily understand than when they are depicted in two-dimensional orthographic views (**Figure 25.57**).

Assemblies

Assembly drawings illustrate how to put parts together. **Figure 25.58A** shows common mistakes in applying leaders and balloons to an assembly, and **Figure 25.58B** shows the cor-

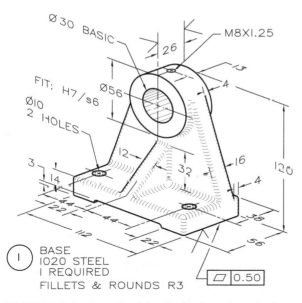

BASE
1020 STEEL
1 REQUIRED
FILLETS & ROUNDS R3

25.57 This three-dimensional drawing has been drawn to show fillets and rounds, dimensions, and notes in order for it to be used as a working drawing.

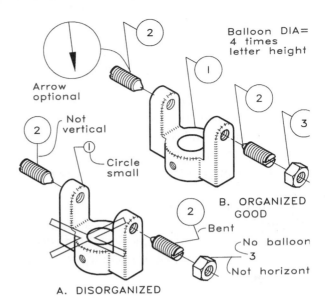

25.58 This drawing shows (A) common mistakes in applying leaders and part numbers in balloons of an assembly, and (B) acceptable techniques of applying leaders and part numbers to an assembly.

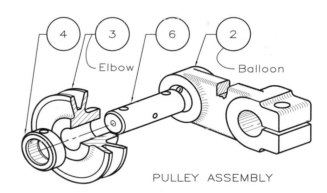

PULLEY ASSEMBLY

25.59 This exploded isometric assembly shows how parts are to be put together.

rect method of applying them. The numbers in the balloons correspond to the part numbers in the parts list. **Figure 25.59** shows an exploded assembly that illustrates the relationship of four mating parts. Illustrations of this type are excellent for inclusion in parts catalogs and maintenance manuals.

25.7 Isometrics by Computer

AutoCAD provides an *Isometric* grid for drawing isometrics. The *Style* option of the *Snap* command allows changing the rectangular *Grid*, called *Standard* (S), to *Isometric* (I) with dots shown vertically and at 30° to the horizontal (**Figure 25.60**). In this mode, you can make the cursor's cross hairs *Snap* to the grid points and align with the axes of isometric drawings.

Isometric drawings made with this system (**Figure 25.61**) are not true three-dimensional drawings. Instead, they are two-dimensional isometrics that cannot be rotated to show other views.

Circles that will appear as ellipses in isometric can be drawn when *Snap* has been set to Isometric. From the *Draw* menu, choose *Ellipse*, and I (isometric circle). Specify the center point and the radius or diameter and the isometric ellipse is drawn. When using this command, the cursor is aligned with each of the three isometric planes by pressing *Ctrl-E* on the keyboard (**Figure 25.62**). When the cursor is aligned with the proper axes of an isometric plane, you may select the center of the isometric ellipse or its diameter's endpoints (**Figure 25.63**).

The *Isoplane* command changes the position of the cursor in the same way *Ctrl-E* does.

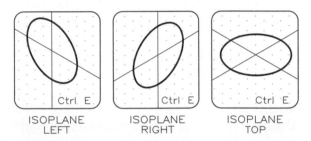

ISOPLANE LEFT   ISOPLANE RIGHT   ISOPLANE TOP

25.62 Use the *Ellipse* command and the *Isocircle* option to draw circles in isometric. By pressing <u>Ctrl-E</u>, you may alternatively rotate the isometric ellipses 120° to fit the three isometric planes.

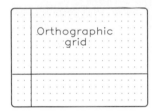

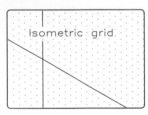

A. ORTHOGRAPHIC GRID   B. ISOMETRIC GRID

25.60 The *Snap* command permits you to use the orthographic grid (Standard) or the isometric grid option (I) for drawing isometric pictorials.

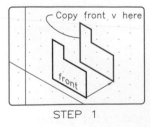

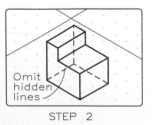

STEP 1   STEP 2

25.61 Isometrics by computer:

Step 1 Set the isometric grid on the screen (SNAP and I), and set *Snap* to the grid. Draw the front view as an isometric and copy it to the backside with the *Copy* command.

Step 2 Connect the visible corner points and ERASE hidden lines to complete the drawing.

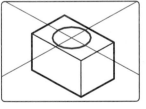

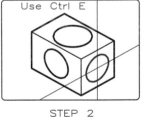

STEP 1   STEP 2

25.63 Isometric ellipses by computer:

Step 1 Use *Snap's* isometric-grid mode to draw isometric ellipses.

Command: <u>ELLIPSE</u> (Enter)

Specify axis endpoint of ellipse or [Arc/ Center/ Isocircle]: I (Enter).

Specify center of isocircle: (Select with cursor.)

Specify radius of isocircle or [Diameter]: (Size the radius with cursor.)

Step 2 Change the orientation of the cursor for drawing isometric ellipses on the other two planes by pressing <u>Ctrl-E</u>. Repeat the process in Step 1.

Isoplane will prompt you to select from Left/Top/Right/[Toggle]: options. To use the Toggle option, press (Enter) to successively move the cursor position from plane to plane.

25.8 Axonometric Projection

An axonometric projection is a type of orthographic projection in which the pictorial view is projected perpendicularly onto the picture plane with parallel projectors. The object is positioned at an angle to the picture plane so that its pictorial projection will be a three-dimensional view. The three types of axonometric projections are: (**1**) **isometric**, (**2**) **dimetric**, or (**3**) **trimetric** (**Figure 25.64**).

Recall that the **isometric projection** is the type of pictorial in which the diagonal of a cube is seen as a point, the three axes and planes of the cube are equally foreshortened, and the axes are equally spaced 120° apart. Measurements along the three axes will be equal but less than true length because the isometric projection is true projection.

A **dimetric projection** is a pictorial in which two planes are equally foreshortened and two of the axes are separated by equal angles. Measurements along two axes of the cube are equal. A **trimetric projection** is a pictorial in which all three planes are

unequally foreshortened. The lengths of the axes are unequal, and the angles between them are different.

25.9 Perspective Pictorials

A perspective pictorial most closely resembles the view seen by the eye or camera and is the most realistic form of pictorial. In a perspective, parallel lines converge at vanishing points (VPs) as the lines recede from the observer. The three basic types of perspectives are (**1**) **one point**, (**2**) **two point**, and (**3**) **three point**, depending on the number of vanishing points used in their construction (**Figure 25.65**).

One-point perspectives have one surface of the objective that is parallel to the picture plane, making it a true shape. The other sides vanish to a single vanishing point on the horizon.

Two-point perspectives are positioned with two sides at an angle to the picture plane, requiring two vanishing points. All horizontal lines converge at the vanishing points on the horizon, but vertical lines remain vertical and have no vanishing point.

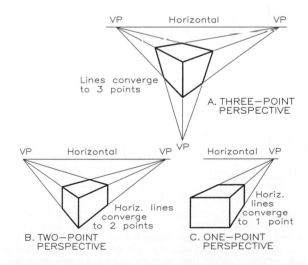

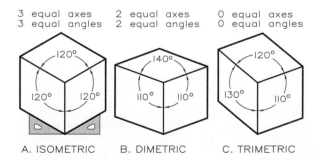

25.64 This drawing illustrates the three types of axonometric projection.

25.65 This drawing compares three-point, two-point, and one-point perspectives.

Three-point perspectives have three vanishing points because the object is positioned so that all of its sides make an angle with the picture plane. Three-point perspectives are used for drawing large objects, such as tall buildings. They are the most realistic perspectives and the most complex to draw. Because of their complexity, we do not show how to construct them in this section.

Construction of One-Point Perspectives

Figure 25.66 shows how to draw a one-point perspective. It shows the top and side views of the object, picture plane, station point, horizon, and ground line. The picture plane (PP) appears as an edge in the top view and is the plane onto which the perspective is projected.

The station point (SP) is the location of the observer's eye in the top view and lies on the horizon in the front view. The horizon is a horizontal line in the front view that represents an infinite horizontal, such as the surface of the ocean, and is aligned with the viewer's eye. The ground line (GL) is an infinite horizontal line parallel to the horizon from which vertical measurements are made.

Constructing Two-Point Perspectives

If two surfaces of an object are positioned at angles to the picture plane, two vanishing points are required to draw it as a perspective. Placing the horizon above the ground line and the height of the object in the front view yields an aerial view (**Figure 25.67**). Placing the ground line and horizon on top of each other in the front view gives a ground-level view (worm's-eye view). Placing the horizon above the ground line and through the object, usually at a person's height for large objects such as buildings, results in a general view.

Figure 25.67 shows how to construct a two-point perspective. Because line AB lies in the picture plane, it will be true length in the perspective. All height dimensions originate at this vertical line because it is the only true-length line of the object.

When drawing any perspective, you should position the station point far enough away from the object that the perspective can be contained in a cone of vision of 30° or less (**Figure 25.68**). A larger cone of vision will distort the perspective.

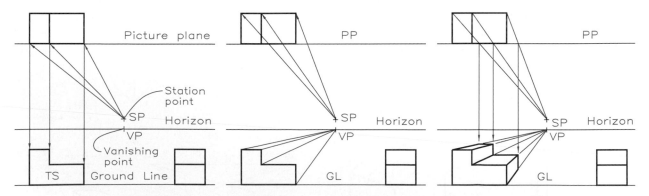

25.66 One-point perspective:

Step 1 Since the object is parallel to the picture plane, there will be only one vanishing point, located on the horizon below the station point. Projections from the top and side views establish the true-size front plane, which lies in the picture plane.

Step 2 Draw projectors from the station point to the rear points of the object in the top view and from the front view to the vanishing point on the horizon. In a one-point perspective, the vanishing point is the front view of the station point on the horizon.

Step 3 Construct vertical projectors from the top view to the front view from the points where the projectors cross the picture plane. These projectors intersect the lines that are drawn to the single vanishing point VP, which located the back side of the perspective.

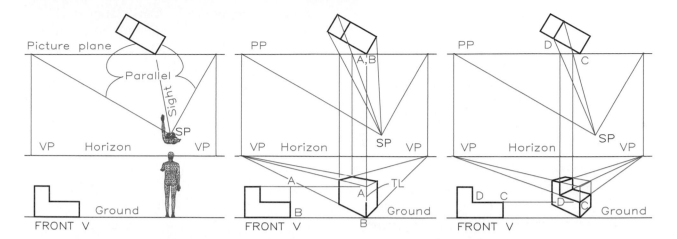

25.67 A two-point perspective:

Step 1 Extend projectors from the top view of the station point to the picture plane parallel to the forward edges of the object. Project these points vertically to the horizon in the front view to locate vanishing points. Draw the ground line below the horizon and construct the side view on the ground line.

Step 2 Lines in the picture plane are true length, so AB is true length. Project AB from the side view to determine its height. Project each end of AB to the vanishing points. Draw projectors from the station point to the exterior edges of the top view and project the intersections of these projectors with the picture plane to the front view.

Step 3 Find point C in the front view by projecting from the side view to AB. Draw a projector from point C to the left vanishing point. Point D lies on this projector beneath the point where a projector from the station point to the top view of point D crosses the picture plane. Draw the notch by projecting to the respective vanishing points.

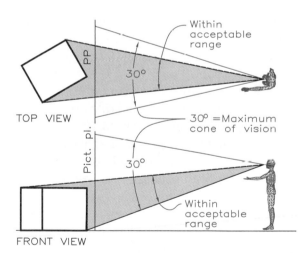

25.68 The station point should be far enough away from the object to permit the cone of vision to be less than 30° to reduce distortion.

The object shown in **Figure 25.69** does not come into contact with the picture plane in the top view as it does in **Figure 25.67**. To draw a perspective of this object, the planes of the object are extended to the picture plane. Measure the height on this line and draw an infinite plane to the right vanishing point. Locate the corner of the object on this infinite plane by projecting the object's right corner to the picture plane in the top view with a projector from the station point and then projecting this point downward to the infinite plane.

Arcs in Perspective

Draw arcs in perspective by using coordinates to locate points along the curves (**Figure 25.70**). Transfer points 1–7 from the semicircular arc in the orthographic view to the perspective by projecting coordinates from the top and side views. All heights are projected to the TL line from the orthographic view. These points do not form a true ellipse but an egg-shaped oval. Connect the points with an irregular curve.

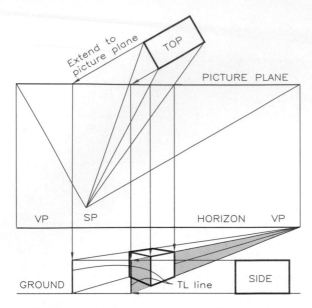

25.69 This two-point perspective is of an object that does not come into contact with the picture plane.

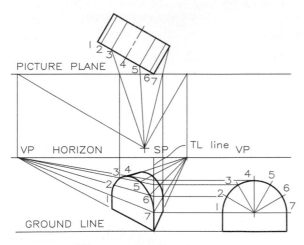

25.70 An object with a semicircular feature is drawn in this two-point perspective.

25.10 Three-Dimensional Modeling

Objects drawn with AutoCAD as true three-dimensional solids can be rotated and viewed from any angle as if they were held in your hand. The object in **Figure 27.71** is an example of a simple object represented by two orthographic views, a wire-frame drawing, a

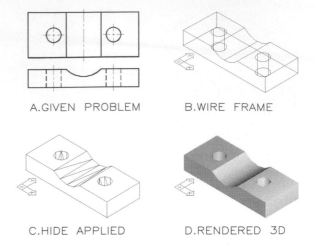

A.GIVEN PROBLEM   B.WIRE FRAME

C.HIDE APPLIED   D.RENDERED 3D

25.71 Modeling a simple part.

A Two orthographic views (top and front) of the part are given.

B A 3-D wire frame drawing of the part is made.

C The hidden lines are suppressed to give a three-dimensional model.

D The model is rendered to give it a realistic look.

hidden-line wire frame drawing, and a rendered solid. The capability to depict objects as rendered solids is a powerful design and communications tool.

Another example of a three-dimensional part that would be difficult to draw by hand is the pulley shown in **Figure 25.72**. A typical section through the pulley and its axis are drawn, the section is revolved about the axis, and the wire frame diagram is rendered. In addition to being able to select various views of the pulley, different lighting combinations and materials can be applied to it in infinite combinations of effects.

An example of an industrial application is given in **Figure 25.73**, which shows an apparatus of a higher degree of complexity that would be a rigorous assignment if drawn by hand. Although it is no easy chore to draw it as a series of solids by computer, the computer drawing enables you to obtain many different views of the parts, and to replicate

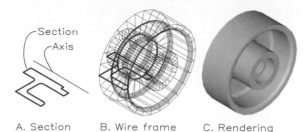

A. Section   B. Wire frame   C. Rendering

25.72 A model by revolution:

A A typical section of the pulley and its axis are drawn.

B The section is rotated about the axis to obtain a wire-frame drawing.

C The wire frame is rendered to obtain a realistic view of the pulley.

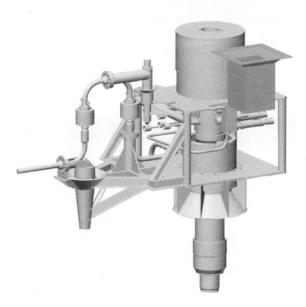

25.73 This apparatus is an example of a rendered three-dimensional model of a moderately complex application. (Courtesy of Cameron.)

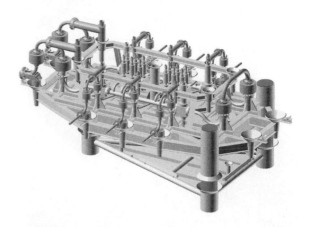

25.74 The apparatus in Figure 25.73 is replicated a number of times in this equipment assembly used in subset production. (Courtesy of Cameron.)

drawings in combination. For example, the apparatus in **Figure 25.73** is applied repetitively in the subset production-equipment assembly in **Figure 25.74**. The savings in time and effort becomes highly significant, and the final rendering greatly improves the understanding of the unit as a whole.

An introduction to three-dimensional modeling by several methods is given in Chapter 38. You will find that solid modeling begins with an understanding of the underlying fundamentals covered in this chapter. The ability to sketch three-dimensional drawings is an invaluable skill that you will use to develop and communicate design applications.

25.11 The Human Figure

An ultimate aspiration of the illustrator has always been the ability to represent the human form in a realistic manner. In addition to determining the interactions between parts, assemblies, and equipment, it is equally important to study the relationship of personnel to their working environment. An example of this type of application in **Figure 25.75** shows workers performing maintenance on a spacecraft. The figures can be moved about the work area and placed in an infinite variety of poses.

Several software packages have been developed that can be used with AutoCAD and other programs that adapt well with computer graphics. The software *Poser®*, by *Fractal Design*, offers many options for representing the human body from stick figures to formally-dressed figures (**Figure 25.76**).

25.75 This computer-drawn scene was at the Kennedy Space Center illustrating the interaction between people and equipment. (Courtesy McDonnell Douglas Space & Defense System—Kennedy Space Center.)

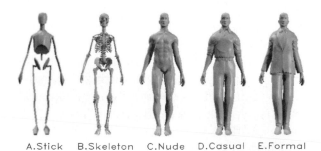

A.Stick   B.Skeleton   C.Nude   D.Casual   E.Formal

25.76 These are several of the rendering options that are available as part of *Poser 2*.

Choices of body styles can be made from many categories, a few of which are age, sex, weight, and pose. Bodies can be positioned and controlled to fit almost any application. Figures can be rotated to obtain orthographic, axonometric, or perspective views of them. Clothing options range from casual to a formal dress, for males and females. Since all designs and projects are to fulfill the needs of people, it is important that the human body interacts with design concepts at all stages of their development.

The Future

The future of 3D graphics is truly exciting. What is available today for the microcomputer was not possible even on much larger and more expensive computers just a few years ago. The capabilities of 3D programs will continue to become more powerful and easier to use. Graphics in the future will include more solid modeling, animation, and sound effects. Get ready for an exciting trip!

Problems

Draw your solutions to the following problems (**Figure 25.77**) on size A or B sheets, as assigned. Select an appropriate scale to take advantage of the space available on each sheet. By letting each square represent 0.20 in. (5 mm), you can draw two solutions on each size A sheet. By setting each square to 0.40 inch (10 mm), you can draw one solution on each size B sheet.

Oblique Pictorials

1–24. Construct cavalier, cabinet, or general obliques of the parts assigned.

Isometric Pictorials

1–24. Construct isometrics of the parts assigned.

Perspective Pictorials

1–24. On size B sheets, lay out perspective views of the parts assigned.

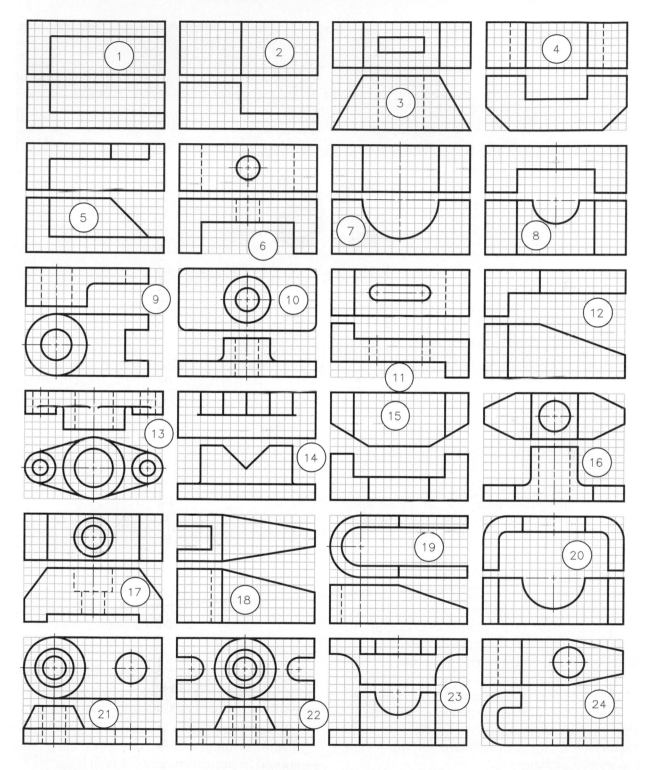

25.77 Problems 1–24.

26

Points, Lines, and Planes

26.1 Introduction

Points, lines, and planes are the basic geometric elements used in three-dimensional spatial geometry, called descriptive geometry. You need to understand how to locate and manipulate these elements in their simplest form because they will be applied to 3D spatial problems in Chapters 26 through 31.

The huge antenna in **Figure 26.1** is composed of many points, lines, and planes that represent its structural members and shapes. Its geometry had to be established one point at a time with great precision in order for it to function and to be properly supported.

The labeling of points, lines, and planes is an essential part of 3D projection because it is your means of analyzing their spatial relationships. **Figure 26.2** illustrates the fundamental requirements for properly labeling these elements in a drawing:

26.1 It is easy to see the numerous applications of points, lines, and planes that were encountered by the team of designers who created this massive antenna.

Lettering: use 1/8-inch letters with guidelines for labels; label lines at each end and planes at each corner with either letters or numbers.

Points: mark with two short perpendicular dashes forming a cross, not a dot; each dash should be approximately 1/8 inch long.

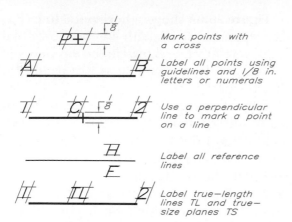

Mark points with a cross

Label all points using guidelines and 1/8 in. letters or numerals

Use a perpendicular line to mark a point on a line

Label all reference lines

Label true—length lines TL and true—size planes TS

26.2 These are standard practices for labeling points, lines, and planes.

Points on lines: mark with a short perpendicular dash crossing the line, not a dot.

Reference lines: label these thin, dark lines as described in Chapter 14.

Object lines: draw these lines used to represent points, lines, and planes twice as thick as hidden lines with an F or HB pencil; draw hidden lines twice as thick as reference lines.

True-length lines: label true length or TL.

True-size planes: label true size or TS.

Projection lines: draw precisely with a 2H or 4H pencil as thin lines, just dark enough to be visible so they need not be erased.

26.2 Projection of Points

A point is a theoretical location in space having no dimensions other than its location. However, a series of points establishes lengths, areas, and volumes of complex shapes.

A point must be located in at least two adjacent orthographic views to establish its position in 3D space (**Figure 26.3**). When the planes of the projection box (**Figure 26.3A**) are opened onto the plane of the drawing surface (**Figure 26.3C**), the projectors from each view of point 2 are perpendicular to the reference

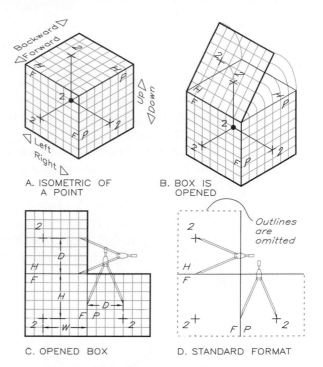

A. ISOMETRIC OF A POINT   B. BOX IS OPENED

C. OPENED BOX   D. STANDARD FORMAT

26.3 Three views of a point:

A The point is projected to three projection planes.

B The projection planes are opened into a single plane.

C In the opened box, point 2 is 5 units to the left of the profile, 5 units below the horizontal, and 4 units behind the frontal plane.

D The outlines of the projection planes are omitted in orthographic projection.

lines between the views. Letters **H**, **F**, and **P** represent the **horizontal**, **frontal**, and **profile planes**, the three principal projection planes.

A point may be located from verbal descriptions with respect to the principal planes. For example, point 2 in **Figure 26.3** may be described as being (1) 5 units left of the profile plane, (2) 5 units below the horizontal plane, and (3) 4 units behind the frontal plane.

When you look at the front view of the box, the horizontal and profile planes appear as edges. In the top view, the frontal and profile planes appear as edges. In the side view, the frontal and horizontal planes appear as edges.

26.3 Lines

A **line** is the straight path between two points in 3D space. A line may appear as (1) **foreshortened**, (2) **true-length**, or (3) **a point** (**Figure 26.4**). Oblique lines are neither parallel nor perpendicular to a principal projection plane (**Figure 26.5**). When line 1–2 is projected onto the horizontal, frontal, and profile planes, it appears foreshortened in each view.

Principal lines are parallel to at least one of the principal projection planes. A principal line is true length in the view where the principal plane to which it is parallel appears true size. The three types of principal lines are **horizontal**, **frontal**, and **profile lines**.

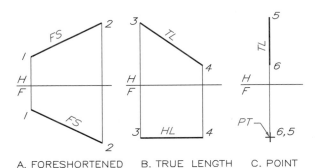

A. FORESHORTENED   B. TRUE LENGTH   C. POINT

26.4 A line in orthographic projection can appear as foreshortened (FS), true length (TL), or a point (PT).

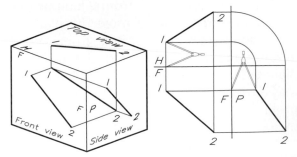

A. THREE–DIMENSIONAL VIEW   B. ORTHOGRAPHIC VIEWS

26.5 A line in space:

A Three views of a line are projected onto the three principal planes.

B These are the standard three orthographic views of a line.

Figure 26.6A shows a **horizontal line** (HL) that appears true length in the horizontal (top) view. Any line shown in the top view will appear true length as long as it is parallel to the horizontal plane.

When looking at the top view, you cannot tell whether the line is horizontal. You must look at the front or side views to do so. In those views, an HL will be parallel to the edge view of the horizontal, the HF fold line (**Figure 26.7**). A line that projects as a point in the front view is a combination horizontal and profile line.

A **frontal line** (FL) is parallel to the frontal projection plane. It appears true length in the front view because your line of sight is perpendicular to it in this view. In **Figure 26.6B**, line 3–4 is an FL because it is parallel to the edge of the frontal plane in the top and side views.

A **profile line** (PL) is parallel to the profile projection planes and appears true length in the side (profile) views. To tell whether a line is a PL, you must look at a view adjacent to the profile view, the top or front view. In **Figure 26.6C**, line 5–6 is parallel to the edge view of the profile plane in both the top and side views.

Locating a Point on a Line

Figure 26.8 shows the top and front views of a line 1–2 with point O located at its midpoint. To find the front view of the point, recall that, in orthographic projection, the projector between the views is perpendicular to the HF fold line. Use that projector to project point O to line 1–2 in the front view. A point located at a line's midpoint will be at the line's midpoint in all orthographic views of the line.

Intersecting and Nonintersecting Lines

Lines that intersect have a common point of intersection lying on both lines. Point O in **Figure 26.9A** is a point of intersection because it projects to a common crossing point in all three views. However, the crossing point of

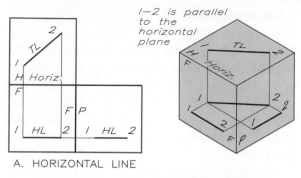

1-2 is parallel to the horizontal plane

A. HORIZONTAL LINE

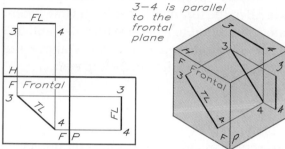

3-4 is parallel to the frontal plane

B. FRONTAL LINE

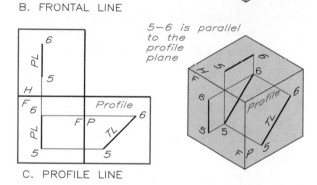

5-6 is parallel to the profile plane

C. PROFILE LINE

26.6 Principal lines:

A A horizontal line is true length in the horizontal (top) view. It is parallel to the edge view of the horizontal plane in the front and side views.

B The frontal line is true length in the front view. It is parallel to the edge view of the frontal plane in the top and side views.

C The profile line is true length in the profile (side) view. It is parallel to the edge view of the profile plane in the top and front views.

the lines in **Figure 26.9B** in the top and front views is not a point of intersection. Point O does not project to a common crossing point in the top and front views, so the lines do not intersect; they simply cross, as shown in the profile view.

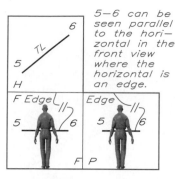

5-6 can be seen parallel to the horizontal in the front view where the horizontal is an edge.

5-6 is seen parallel to horizontal.

A. ORTHOGRAPHIC VIEWS   B. PICTORIAL VIEW

26.7 In order to determine that a line is horizontal, you must look at the front or side views in which the horizontal projection plane is an edge. Line 5-6 is seen parallel to the horizontal edge and is a horizontal line, too.

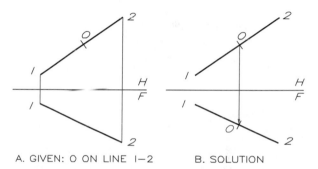

A. GIVEN: O ON LINE 1-2   B. SOLUTION

26.8 Point O on the top view of line 1-2 can be found in the front view by projection. The projector is perpendicular to the HF reference line between the views.

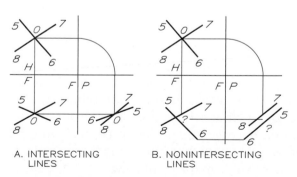

A. INTERSECTING LINES   B. NONINTERSECTING LINES

26.9 Crossing lines:

A These lines intersect because O, the point of intersection, projects as a common point of intersection in all views.

B The lines cross in the top and front views, but they do not intersect because there is no common point of intersection in all views.

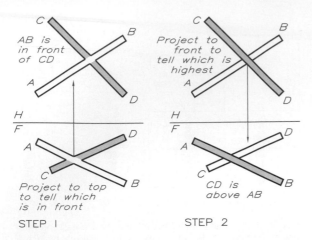

C

B

AB is
in front
of CD

A

D

H
F

A

D

C

B

Project to top
to tell which
is in front

STEP 1

C

B

Project to
front to
tell which is
highest

A

D

H
F

A

D

C

B

CD is
above AB

STEP 2

26.10 Determining visibility of lines:

Step 1 Project the crossing point from the front to the top view. This projector strikes line AB before it strikes line CD, indicating that line AB is in front and thus is visible in the front view.

Step 2 Project the crossing point from the top view to the front view. This projector strikes line CD before it strikes line AB, indicating that line CD is above line AB and thus is visible in the top view.

26.4 Visibility

Crossing Lines

In **Figure 26.10**, nonintersecting lines AB and CD cross in certain views. Therefore portions of the lines are visible or hidden at the crossing points (here, line thickness is exaggerated for purposes of illustration). Determining which line is above or in front of the other is referred to as finding a line's visibility, a requirement of many 3D problems.

You have to determine line visibility by analysis. For example, select a crossing point in the front view and project it to the top view to determine which line is in front of the other. Because the projector contacts line AB first, you know that line AB is in front of CD and is visible in the front view.

Repeat this process by projecting downward from the intersection in the top view to find that line CD is above line AB and is visible in the top view. If only one view were available, visibility would be impossible to determine.

A Line and a Plane

The principles of visibility analysis also apply to determining visibility for a line and a plane (**Figure 26.11**). First, project the intersections of line AB with lines 4–5 and 5–6 to the top view to determine that the lines of the plane (4–5 and 5–6) lie in front of line AB in the front view. Therefore line AB is a hidden line in the front view.

Similarly, project the two intersections of line AB in the top view to the front view, where line AB is found to lie above lines 4–5 and 5–6 of the plane. Because line AB is above the plane, it is a visible line in the top view.

26.5 Planes

A plane may be represented in orthographic projection by any of the four combinations shown in **Figure 26.12**. In orthographic projection, a plane may appear as (1) **an edge**, (2) a **true-size plane**, or (3) a **foreshortened plane** (**Figure 26.13**).

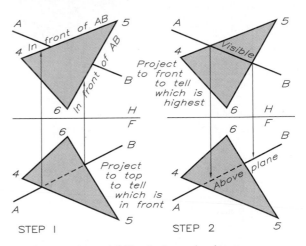

STEP 1

STEP 2

26.11 Determining visibility of a line and a plane:

Step 1 Project the points where line AB crosses the plane from the front view to the top view. These projectors intersect lines 4–6 and 5–6 of the plane first, indicating that the plane is in front of the line and making line AB hidden in the front view.

Step 2 Project the points where line AB crosses the plane in the top view to the front view. These projectors encounter line AB first, indicating that line AB is higher than the plane, thus the line is visible in the top view.

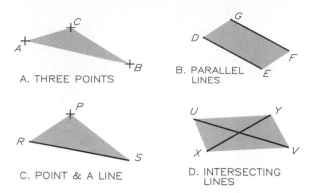

26.12 A plane can be represented as (A) three points not on a straight line, (B) two parallel lines, (C) a line and a point not on the line or its extension, and (D) two intersecting lines.

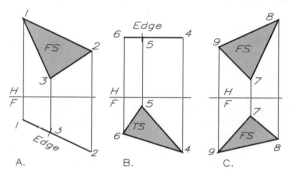

26.13 A plane in orthographic projection can appear as (A) an edge, (B) true size (TS), or (C) foreshortened (FS). A plane that is foreshortened in all principal views is an oblique plane.

Oblique planes (the general case) are not parallel to principal projection planes in any view (**Figure 26.14**). Principal planes are parallel to principal projection planes (**Figure 26.15**). The three types of principal planes are horizontal, frontal, and profile planes.

A **horizontal plane** is parallel to the horizontal projection plane and is true size in the top view (**Figure 26.15A**). To determine that the plane is horizontal, you must observe the front or profile views, where you can see its parallelism to the edge view of the horizontal plane.

A **frontal plane** is parallel to the frontal projection plane and appears true size in the front view (**Figure 26.15B**). To determine that the plane is frontal, you must look at the top

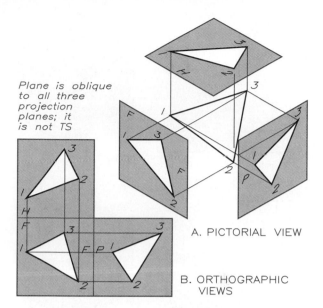

Plane is oblique to all three projection planes; it is not TS

A. PICTORIAL VIEW

B. ORTHOGRAPHIC VIEWS

26.14 An oblique plane is neither parallel nor perpendicular to a projection plane. It is the general-case plane.

or profile views, where you can see its parallelism to the edge view of the frontal plane.

A **profile plane** is parallel to the profile projection plane and is true size in the side view (**Figure 26.15C**). To determine that the plane is profile, you must observe the top or front views, where you can see its parallelism to the edge view of the profile plane.

A Point on a Plane

Point O on the front view of plane 4–5–6 in **Figure 26.16** is to be located on the plane in the top view. First, draw a line in any direction (except vertical) through the point to establish a line on the plane. Then project this line to the top view and project point O from the front view to the top view of the line.

Principal Lines on a Plane

Principal lines may be found in any view of a plane when at least two orthographic views of the plane are given. Any number of principal lines can be drawn on any plane.

Figure 26.17A shows a horizontal line parallel to the edge view of the horizontal projection

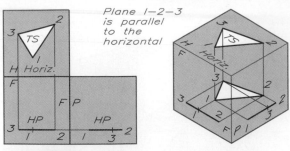

Plane 1–2–3 is parallel to the horizontal

A. HORIZONTAL PLANE

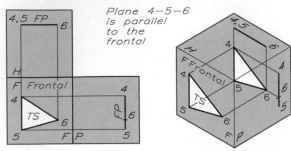

Plane 4–5–6 is parallel to the frontal

B. FRONTAL PLANE

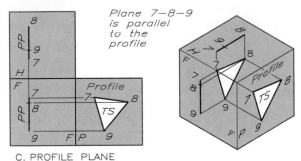

Plane 7–8–9 is parallel to the profile

C. PROFILE PLANE

26.15 Principal planes:

A The horizontal plane is true size in the horizontal (top) view. It is parallel to the edge view of the horizontal plane in the front and profile views.

B The frontal plane is true size in the front view. It is parallel to the edge view of the frontal plane in the top and profile views.

C The profile plane is true size in the profile view. It is parallel to the edge view of the profile plane in the top and front views.

plane in the front view. When projected to the top view, this line is true length.

Figure 26.17B shows a frontal line parallel to the edge view of the frontal projection plane in the top view. When projected to the front view, this line is true length.

Figure 26.17C shows a profile line parallel to the edge view of the profile projection plane

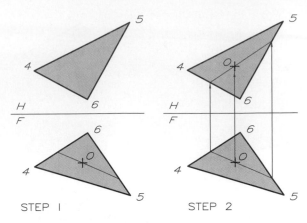

STEP 1   STEP 2

26.16 Locating a point on a plane:

Step 1 In the front view, draw a line through point O in any convenient direction except vertical.

Step 2 Project the ends of the line to the top view and draw the line. Project point O to this line.

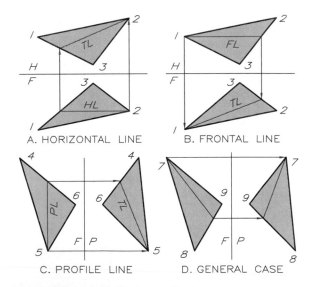

A. HORIZONTAL LINE   B. FRONTAL LINE

C. PROFILE LINE   D. GENERAL CASE

26.17 Finding principal lines on a plane:

A First, draw a horizontal line in the front view parallel to the edge view of the horizontal plane. Then, project it to the top view, where it is true length.

B First, draw a frontal line in the top view parallel to the edge view of the frontal plane. Then, project it to the front view, where it is true length.

C First, draw a profile line in the front view parallel to the edge view of the profile plane. Then project it to the profile view, where it is true length.

D A general-case line is not parallel to the frontal, horizontal, or profile planes and is not true length in any principal view.

in the front. When projected to the profile view, this line is true length.

In the general case (oblique), a line is not parallel to the edge view of any principal projection plane (**Figure 26.17D**). Therefore it is not true length in any principal view.

26.6 Parallelism

Lines

Two parallel lines appear parallel in all views, except in views where both appear as points. Parallelism of lines in 3D space cannot be determined without at least two adjacent orthographic views. In **Figure 26.18**, line AB was drawn parallel to the horizontal view of line 3–4 and through point O, which is the midpoint. Projecting points A and B to the front view with projectors perpendicular to the HF reference plane yields the length of line AB, which is parallel to line 3–4 through point O.

A Line and a Plane

A line is parallel to a plane when it is parallel to any line in the plane. In **Figure 26.19**, a line with its midpoint at point O is to be drawn parallel to plane 1–2–3. In this case, line AB was drawn parallel to a line L 1–3 in the plane

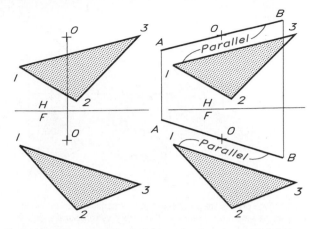

26.19 A line may be drawn through point O parallel to plane 1–2–3 if the line is parallel to any line in the plane. Draw line AB parallel to line 1–3 of the plane in the front and top views, making it parallel to the plane.

in the top and front views. The line could have been drawn parallel to any line in the plane, making infinite solutions possible.

Figure 26.20 shows a similar example. Here, a line parallel to the plane with its midpoint at O was drawn. In this case, the plane is represented by two intersecting lines instead of an outlined area.

Planes

Two planes are parallel when intersecting lines in one plane are parallel to intersecting lines in the other (**Figure 26.21**).

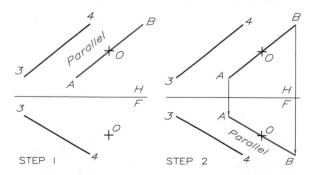

26.18 Constructing a line parallel to a line:

Step 1 Draw line AB parallel to the top view of line 3–4 with its midpoint at O.

Step 2 Draw the front view of line AB parallel to the front view of 3–4 through point O.

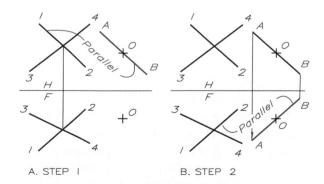

26.20 Constructing a line parallel to plane:

Step 1 Draw line AB parallel to line 1–2 through point O.

Step 2 Draw line AB parallel to the same line, line 1–2, in the front view, which makes line AB parallel to the plane.

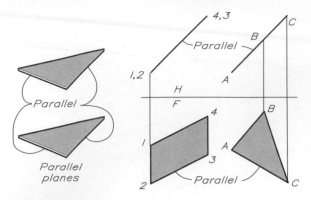

26.21 Two planes are parallel when intersecting lines in one are parallel to intersecting lines in the other. When parallel planes appear as edges, their edges are parallel.

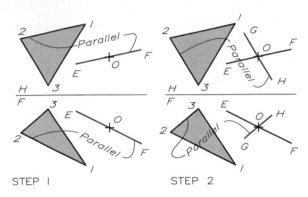

STEP 1   STEP 2

26.22 A plane through a point parallel to a plane:

Step 1 Draw EF parallel to any line in a plane (1–2 in this case). Show the line in both views.

Step 2 Draw a second line parallel to 2–3 in the top and front views. These intersecting lines passing through O represent a plane parallel to 1–2–3.

Determining whether planes are parallel is easy when both appear as edges in a view.

In **Figure 26.22**, a plane is to be drawn through point O parallel to plane 1–2–3. First, draw line EF through point O parallel to line 1–2 in the top and front views. Then draw a second line through point O parallel to line 2–3 of the plane in the front and top views. These two intersecting lines form a plane parallel to plane 1–2–3, as intersecting lines on

one plane are parallel to intersecting lines on the other.

26.7 Perpendicularity

Lines

When two lines are perpendicular, draw them with a true 90° angle of intersection in views where one or both of them appears true length (**Figure 26.23**). In a view where neither of two perpendicular lines is true length, the angle between is not a true 90° angle.

In **Figure 26.23**, the axis is true length in the front view; therefore any spoke of the circular wheel is perpendicular to the axis in the front view. Spokes OA and OB are examples of true length foreshortened axes, respectively, in the front view.

A Line Perpendicular to a Principal Line In **Figure 26.24**, a line is to be constructed through point O perpendicular to frontal line 5–6, which is true length in the front view. First, draw OP perpendicular to line 5–6 because it is true length. Then, project point P to the top view of line 5–6. In the top view, line OP is not perpendicular to line 5–6 because neither of the lines is true length in this view.

A Line Perpendicular to an Oblique Line In **Figure 26.25**, a line is to be constructed from point O perpendicular to oblique line 1–2. First, draw a horizontal line from O to some convenient length in the front view, say, to E.

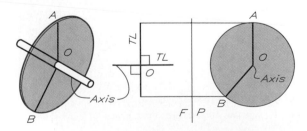

26.23 Perpendicular lines have a true angle of 90° between them in a view where one or both of them appear true length.

474 • CHAPTER 26   POINTS, LINES, AND PLANES

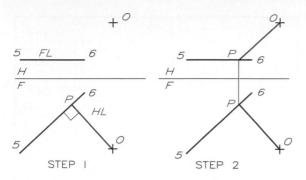

26.24 A line perpendicular to a principal line:

Step 1 Line 5–6 is a frontal line and is true length in the front view, so a perpendicular from point O makes a true 90° angle with it in the front view.

Step 2 Project point P to the top view and connect it to point O. As neither line is true length in the top view, they do not intersect at 90° in this view.

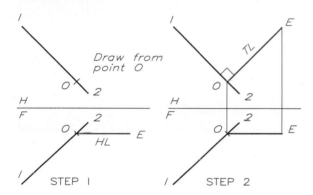

26.25 A line perpendicular to an oblique line:

Step 1 Draw a horizontal line (OE) from O in the front view.

Step 2 Horizontal line OE is true length in the top view, so draw it perpendicular to line 1–2 in this view.

Locate point O in the top view by projection and draw line OE to make a 90° angle with the top view of 1–2. Line OE is true length in the top view, so it makes a true 90° angle with line 1–2.

Planes

A line is perpendicular to a plane when it is perpendicular to any two intersecting lines in the plane (**Figure 26.26A**). A plane is perpendicular to another plane when a line in one plane is perpendicular to the other plane (**Figure 26.26B**).

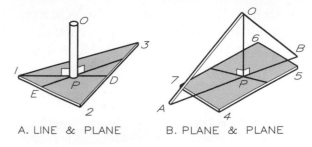

A. LINE & PLANE   B. PLANE & PLANE

26.26 Perpendicularity of lines and plane.

A A line is perpendicular to a plane when it is perpendicular to two intersecting lines on the plane.

B A plane is perpendicular to another plane if it contains a line that is perpendicular to the other plane.

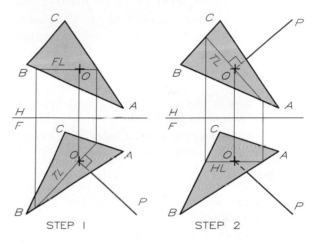

26.27 A line perpendicular to a plane:

Step 1 Draw a frontal line on the plane through O in the top view. This line is true length in the front view, so draw line OP perpendicular to this true-length line.

Step 2 Construct a horizontal line through point O in the front view. This line is true length in the top view, so draw line OP perpendicular to it.

A Line Perpendicular to a Plane

In **Figure 26.27**, a line is to be drawn perpendicular to the plane from point O on the plane. First, draw a frontal line on the plane in the top view through O. Project the line to the front view, where it is true length. Draw line OP at a convenient length perpendicular to the true-length line.

Then, draw a horizontal line through point O in the front view and project it to the top view of the plane, perpendicular to the true-length line. This construction results in a line perpendicular to the plane because the line is perpendicular to two intersecting lines, a horizontal and a frontal line, in the plane.

Problems

Use size A sheets for the following problems and lay out your solutions with instruments. Each square on the grid is equal to 0.20 in. or 5 mm. Use either grid paper or plain paper. Label all reference planes and points in each problem with 1/8-in. letters and numbers using guidelines.

1. (Sheet 1)

(A–D) Draw three views (top, front, and right-side views) of the given points.

(E–F) Draw the three views of the points and connect them to form lines.

2. (Sheet 2)

(A–B) Draw the right-side view of line 1–2 and plane 3–4–5.

(C–E) Draw the missing views of the planes so that 6–7–8 is a frontal plane, 1–2–3 is a horizontal plane, and 4–5–6 is a profile plane.

(F) Complete the top and side views of the plane that appears as an edge in the front view.

3. (Sheet 3)

(A) Draw the side view of line 1–2–3 and and locate point A on it in all views.

(B) Draw the right-side view of the plane and draw two horizontal lines on it in all views.

(C) Draw the right-side view of the plane and draw two frontal lines on it in all views.

(D) Draw the right-side view of the plane and draw two profile lines on it in all views.

(E) Draw three views of the given line and a line through B that is parallel to the plane.

(F) Draw the three views of a plane formed by intersecting lines with A at the end-point of one of the lines.

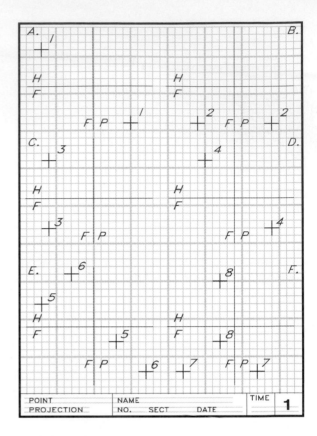

4. (Sheet 4)

(A–B) Draw 1.50-in. lines that pass through point 0 and are parallel to their respective planes.

(C–D) Through point 0, draw the top and front views of lines that are perpendicular to their respective lines.

5. (Sheet 5)

(A–E) The missing views of the planes are to be drawn as three-view projections with top, front, and right-side views. The missing views are to be drawn in the areas of the question marks.

6. (Sheet 6)

(A) The view of four spokes are shown in the circular view of four wheels. Draw these spokes in their side views from their given end points to the axle AB.

(B) Perpendicular spokes IK and JK are given in the front view. Draw them in their proper positions in the side view.

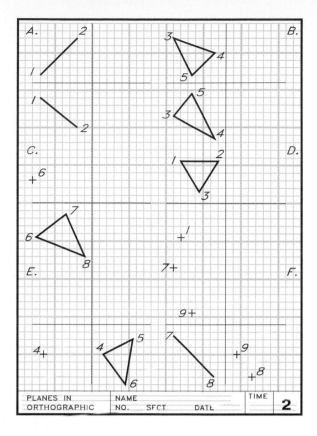

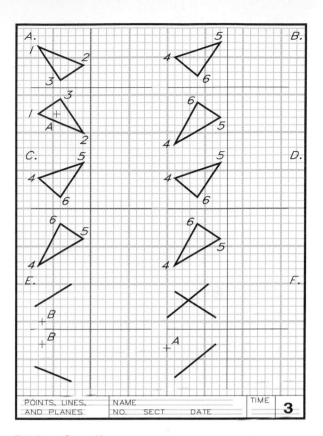

7. (Sheet 7)

  (A) Draw a line from 0 that is perpendicular to 1–2. Show it in both views.

  (B) Draw a 2-in. long line that is perpendicular to 3–4. Show it in both views.

8. (Sheet 8)

  (A) Draw a 1.5 in. long line from 0 on 1-2-3 that is perpendicular to the plane. Show it in both views.

  (B) Draw a 1.5 in. (approximately) long line from 0 on 4-5-6 that is perpendicular to the plane. Show it in both views.

9. (Sheet 9)

  (A) Draw and name the type of each line that has been constructed.

  (B) Draw and name the type of each plane that has been constructed.

  (C) Draw the missing views of the two planes that are given in the front and profile views.

Review Questions

1. How many principal lines are there and what are their names?

2. What are the ways in which a line can appear on a drawing?

3. The three orthographic dimensions are height, width, and depth. Which of these dimensions are necessary to locate a line in the side view? The front view? The top view?

4. What are the three ways in which a plane can appear on a drawing?

5. When is a line parallel to a plane?

6. When is a plane parallel to a plane?

7. When can two perpendicular lines be drawn as perpendicular on a drawing?

8. If a plane appears as an edge in the front view, what can be said about its top view?

9. A line that is vertical appears how in the top view? The side view?

10. A line that is true length in the top view is what type of principal line? How will it appear in the front view?

11. A line that is perpendicular to a plane in the top view makes what angle with a front line on the plane?

12. How many principal lines can be drawn on any view of a plane?

13. A plane that is vertical is true size in which views.

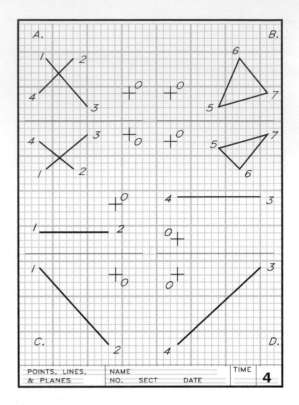

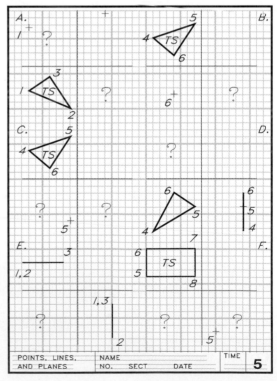

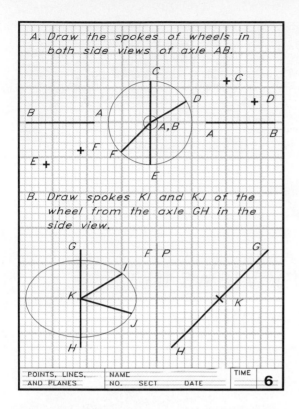

A. Draw the spokes of wheels in both side views of axle AB.

B. Draw spokes KI and KJ of the wheel from the axle GH in the side view.

POINTS, LINES, AND PLANES | NAME NO. SECT DATE | TIME | **6**

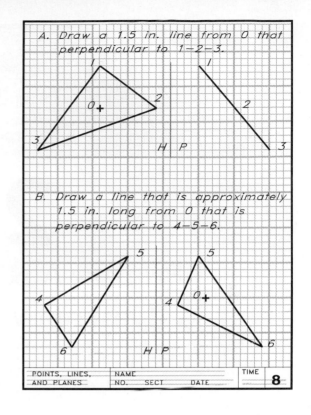

A. Draw a 1.5 in. line from 0 that perpendicular to 1–2–3.

B. Draw a line that is approximately 1.5 in. long from 0 that is perpendicular to 4–5–6.

POINTS, LINES, AND PLANES | NAME NO. SECT DATE | TIME | **8**

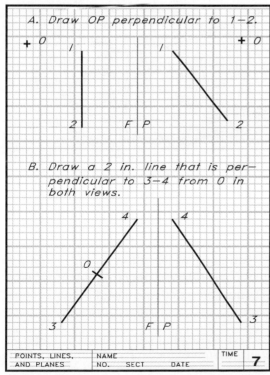

A. Draw OP perpendicular to 1–2.

B. Draw a 2 in. line that is per-pendicular to 3–4 from 0 in both views.

POINTS, LINES, AND PLANES | NAME NO. SECT DATE | TIME | **7**

PROBLEMS • 479

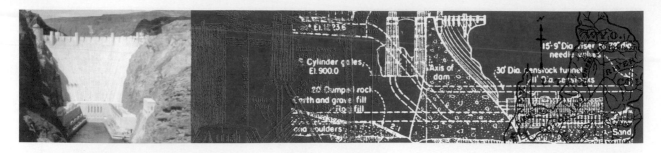

27

Primary Auxiliary Views in Descriptive Geometry

27.1 Introduction

Descriptive geometry is the projection of three-dimensional (3D) orthographic views onto a two-dimensional (2D) plane of paper to allow graphical determination of lengths, angles, shapes, and other geometric information. Orthographic projection is the basis for laying out and solving problems by descriptive geometry.

The primary auxiliary view, which permits analysis of 3D geometry, is essential to descriptive geometry. For example, the design of the helicopter frame shown in **Figure 27.1** contains many complex geometric elements (lines, angles, and surfaces) that were analyzed by descriptive geometry prior to its fabrication.

27.2 Geometry by Computer

Five useful computer routines for solving descriptive geometry problems are covered in this section and documented in Appendix 41.

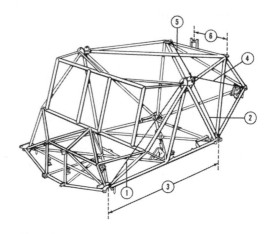

27.1 This helicopter frame was designed using the principles of descriptive geometry to determine lengths, angles, and areas. (Courtesy of Bell Helicopter.)

The commands are *Perpline, Parallel, Transfer, Copydist,* and *Bisect.*[*] This *Lisp* program, called ACAD, must be typed and saved in the *Support*

---

[*]These LISP commands were written by Professor Leendert Kersten of the University of Nebraska at Lincoln.

directory of AutoCAD. To open ACAD, type (Load "ACAD"), being sure to include the parentheses and quote marks. Access the individual commands by typing their names one at a time; *Perpline*, for example.

Figure 27.2 shows how to draw line 3–4 perpendicular to line 1–2 with *Perpline*. Locate the starting point, select the line to which the constructed line is to be perpendicular, and pick a third point in the general area of the line's endpoint.

In **Figure 27.3**, line 3–4 is drawn parallel to line 1–2 with *Parallel*. Locate the starting point of the parallel (point 3) and the general

location of its endpoint (point 4). Select the endpoints of line 1–2 in the same order (from 1 to 2), and the line is drawn.

The *Transfer* command (**Figure 27.4**) transfers distances from reference lines in the same way in which you use your dividers. Select endpoint 2 in the front view and then the reference line. Select endpoint 2 in the top view, then the auxiliary reference line and a circle appears to locate the endpoint at its center in the auxiliary view.

A distance may be copied from one position to another with *Copydist* as shown in **Figure 27.5**. Select the endpoints of the line to

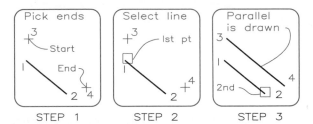

27.2 The *Perpline* command:
Step 1 *Command:* <u>PERPLINE</u> (Enter)
Select START point of perpendicular line: (Select pt. 3.)
Step 2 *Select ANY point on line to which perp'lr:*
(Select point on line.)
Step 3 *Select END point of desired perpendicular* (for length only): (Select pt. 5.) (3–4 is drawn.)

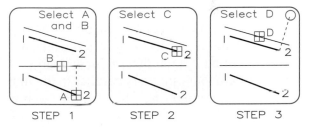

27.4 The *Transfer* command:
Step 1 *Command:* <u>TRANSFER</u> (Enter)
Select start of transfer distance: (Select pt. A.)
Select the reference plane: (Select pt. B.)
Step 2 *Select point to be projected:* (Select pt. C.)
Step 3 *Select other reference plane:* (Select pt. D.) (Point is located at the center of the circle.)

27.3 The *Parallel* command:
Step 1 *Command:* <u>PARALLEL</u> (Enter)
Select START point of parallel line: (Select pt. 3.)
Select END point of parallel line: (Select pt. 4.)
Step 2 *Select 1st point on line for parallelism:* (Select pt. 1.)
Step 3 *Select 2nd point on line for parallelism:* (Select pt. 2.)
(3–4 is drawn.)

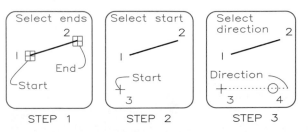

27.5 The *Copydist* command:
Step 1 *Command:* <u>COPYDIST</u> (Enter)
Select start point of line distance to be copied: (Select end 1.)
End point?: (Select end 2.)
Step 2 *Start point of new distance location:* (Select pt. 3.)
Step 3 *Which direction?:* (Select with cursor, and endpoint 4 is located at the center of the circle.)

be copied; locate its beginning point in the new position, and locate the direction of the line to be copied. A circle will appear to locate the endpoint of the line at its center for the exact distance being copied.

The *Bisect* command bisects angles. It is sufficiently self-explanatory in its prompts on the screen for coverage here to be necessary.

27.3 True-Length Lines

Primary Auxiliary View

Figure 27.6 shows the top and front views of line 1–2 pictorially and orthographically. Line 1–2 is not a principal line, so it is not true length in a principal view. Therefore a primary auxiliary view is required to find its true-length view. In **Figure 27.6A**, the line of sight is perpendicular to the front view of the line and reference line F1 is parallel to the line's frontal view. The auxiliary plane is parallel to the line and perpendicular to the frontal plane, accounting for its label, F1, where F and 1 are abbreviations for frontal and primary planes, respectively.

Projecting parallel to the line of sight and perpendicular to the F1 reference line yields the auxiliary view (**Figure 27.6B**). Transferring distance D with dividers to the auxiliary view locates point 2 because the frontal plane appears as an edge in both the top and auxiliary views. Point 1 is located in the same manner, and the points are connected to find the true-length view of the line.

Figure 27.7 summarizes the steps of finding the true-length view of an oblique line. Letter all reference planes using the notation suggested in Chapter 26 and as shown in the examples throughout this chapter, with the exception of noted dimensions such as D. Use your dividers to transfer dimensions.

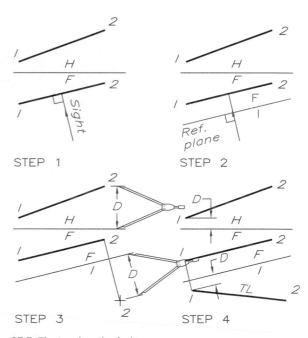

27.7 The true length of a line:

Step 1 To find the true length of line 1–2, the line of sight must be perpendicular to one of its views, the front view here.

Step 2 Draw the F1 reference line parallel to the line and perpendicular to the line of sight.

Step 3 Project point 2 perpendicularly from the front view. Transfer distance D from the top view to locate point 2 in the auxiliary view.

Step 4 Locate point 1 in the same manner to find line 1–2 true length in the auxiliary view.

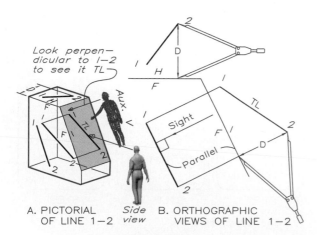

27.6 True-length line by auxiliary view:

A A pictorial of line 1–2 is shown inside a projection box where an auxiliary plane is parallel to the line and perpendicular to the frontal plane.

B The auxiliary view is projected from the front orthographic view to find 1–2 true length.

Computer Method The method of using the *Parallel* and *Transfer* commands to find a line's true length by an auxiliary view, are illustrated in **Figure 27.8**. Draw a line parallel to the top view of line 1–2, and *Transfer* the endpoints of the line. Use the *Center* option of the *Osnap* command to draw line 1–2 from the centers of the circles. Erase the circles afterward.

True Length by Analytical Geometry

The method of finding the true length of frontal line 3–4 mathematically is shown in **Figure 27.9**. The Pythagorean theorem states that the hypotenuse of a right triangle is equal to the square root of the sum of the squares of the other two sides. Because the line is true length

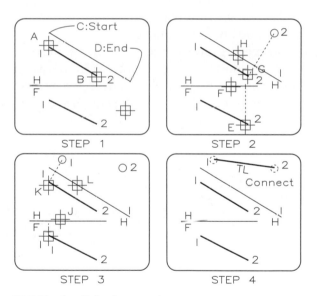

STEP 1   STEP 2

STEP 3   STEP 4

27.8 True length line by computer:

Step 1 *Command:* <u>PARALLEL</u> (Enter) (Follow the prompts in the caption of Figure 27.3 to produce the reference line parallel to line 1–2 with *Parallel.*)

Step 2 *Command:* <u>TRANSFER</u> (Enter) (Follow the prompts in the caption of Figure 27.4 to locate point 2 in the auxiliary view with *Transfer.*)

Step 3 Locate points 1 and 2 in the auxiliary view with *Transfer.*

Step 4 Draw a line from the centers of the circles obtained in the preceding steps by using the *Center* option of the *Osnap* command. *Erase* the circles after this step.

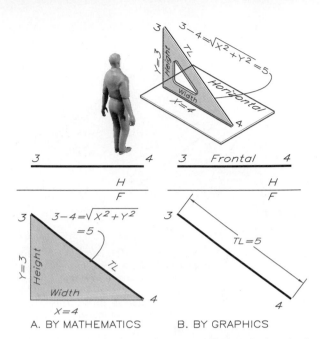

A. BY MATHEMATICS   B. BY GRAPHICS

27.9 Apply the Pythagorean theorem to calculate the length of a line that appears true length in a view, the front view here. Because line 3–4 is true length in the front view, it can be measured to find its length.

in the front view, measuring that length provides a check on the mathematical solution.

The true length of a line shown pictorially in **Figure 27.10A** (line 1–2) is determined by analytical geometry from its length in the front view where the X and Y distances form a right triangle. **Figure 27.10B** shows a second right triangle, 1–0–2, whose hypotenuse is the true length of line 1–2. Thus the true length of an oblique line is the square root of the sum of the squares of the X, Y, and Z distances that correspond to the width, height, and depth of the triangles.

True-Length Diagram

A true-length diagram is two perpendicular lines used to find a line of true length (**Figure 27.11**). The two measurements laid out on the true-length diagram may be transferred from any two adjacent orthographic views. One measurement is the distance between the end-

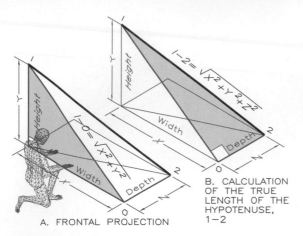

A. FRONTAL PROJECTION

B. CALCULATION OF THE TRUE LENGTH OF THE HYPOTENUSE, 1–2

27.10 To calculate the true length of a 3D line that is not true length in the principal view, find (A) the frontal projection, line 1–O, by using the X and Y distances, and (B) the hypotenuse of the right triangle 1–O–2 by using the length of line 1–O and the Z distance. Then apply the Pythagorean theorem to find its length of 5. Orthographic views are shown in (C).

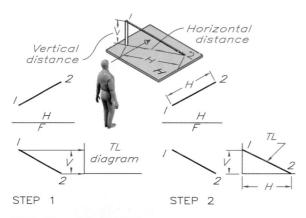

STEP 1   STEP 2

27.11 Using a true-length diagram:

Step 1 Transfer the vertical distance between the ends of line 1–2 to the vertical leg of the TL diagram.

Step 2 Transfer the horizontal length of the line in the top view to the horizontal leg of the TL diagram. The diagonal is the true length of line 1–2.

points in one of the views. The other measurement, from the adjacent view, is the distance between the endpoints perpendicular to the reference line between the two views. Here, these dimensions are vertical, V, and horizontal, H, between points 1 and 2. This method does not give the line's direction, only its true length.

27.4 Angles between Lines and Principal Planes

To measure the angle between a line and a plane, the line must appear true length and the plane as an edge in the same view (Figure 27.12). A principal plane appears as an edge in a primary auxiliary view projected from it, so the angle a line makes with this principal plane can be measured if the line is true length in this auxiliary view.

27.5 Sloping Lines

Slope is the angle that a line makes with the horizontal plane when the line is true length and the plane is an edge. Figure 27.13 shows the three methods for specifying slope: slope angle, percent grade, and slope ratio.

The **slope** angle of line AB in **Figure 27.14** is 31°. It can be measured in the front view where the line is true length.

Percent grade is the ratio of the vertical (rise) divided by the horizontal (run) between the ends of a line, expressed as a percentage. The percent grade of line AB is determined in the front view of **Figure 27.14** where the line is

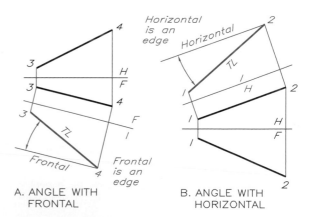

A. ANGLE WITH FRONTAL

B. ANGLE WITH HORIZONTAL

27.12 Angles between lines and principal planes:

A An auxiliary view projected from the front view that shows the line true length will show the frontal plane as an edge, where its angle with the frontal plane can be measured.

B An auxiliary view projected from the top that shows the line true length will show the horizontal plane as an edge, where its angle with the horizontal plane can be measured.

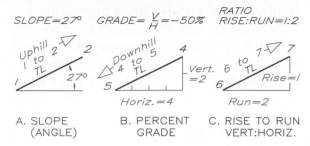

A. SLOPE (ANGLE)   B. PERCENT GRADE   C. RISE TO RUN VERT:HORIZ.

27.13 The inclination of a line with the horizontal may be measured and expressed as: (A) slope angle, (B) percent grade, and (C) slope ratio.

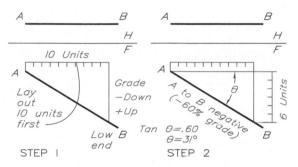

STEP 1   STEP 2

27.14 Percent grade of a line:

Step 1 The percent grade of a line can be measured in the view where the horizontal appears as an edge and the line is true length (here, the front view). Lay off ten units parallel to the horizontal from the end of the line.

Step 2 A vertical distance from the end of the 10 units to the line measures 6 units. The percent grade is 6 divided by 10, or 60%. This is a negative grade from A to B because the line slopes downward from A. The tangent of this slope angle is 6/10, or 0.60, which can be used to verify the slope of 31° from trigonometric tables.

true length and the horizontal plane is an edge. Line AB has a -60% grade from A to B because the line slopes downward; it would be positive (upward) from B to A. Trigonometric tables verify that an angle whose tangent is 0.60 (6/10) is 31°.

The **slope ratio** is the ratio of a rise of 1 to the run. The rise is always written as 1, followed by a colon and the run (for example, 1:10, 1:200). **Figure 27.15** illustrates the graphical method of finding the slope ratio. The rise of 1 unit is laid off on the true-length view of CD. The corresponding run measures 2 units, for a slope ratio of 1:2.

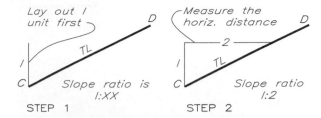

STEP 1   STEP 2

27.15 Slope ratio:

Step 1 Slope ratio always begins with 1, so lay out a vertical distance of 1 from end C.

Step 2 Lay off a horizontal distance from the end of the vertical line and measure it. It is 2, so the slope ratio (always expressed as 1:XX) of this line is 1:2.

The slope of oblique lines are found true length in an auxiliary view projected from the top view so that the horizontal reference plane will appear as an edge (**Figure 27.16A**). The slope is expressed as an angle, or 26°.

To find the **percent grade** of an oblique line (**Figure 27.16B**), lay off 10 units horizontally, parallel to the H1 reference line. The corresponding vertical distance measures 4.5 units, for a -45% grade from point 3 to point 4.

The principles of true-line length and angles between lines and planes are useful in applications such as the design of aggregate conveyors (**Figure 27.17**) where slope is crucial to optimal operation of the equipment.

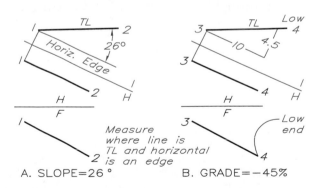

A. SLOPE=26°   B. GRADE=-45%

27.16 Slope of an oblique line:

A Find the slope angle of an oblique line: (26° in this case) in a view where the horizontal appears as an edge and the line is true length.

B Find the percent grade in an auxiliary view projected from the top view where line 3–4 is true length (-45% from 3 to 4, the low end, in this case).

27.17 The design of these aggregate conveyors required the application of sloping-line principles in order to obtain their optimal slopes. (Courtesy of Link-Belt.)

27.6 Bearings and Azimuths of Lines

Two types of bearings of a line's direction are **compass bearings** and **azimuths**. Compass bearings are angular measurements from north or south. The line in **Figure 27.18A** that makes a 30° angle with north has a bearing of N 30° W. The line making a 60° angle with south toward the east has a bearing of south 60° east, or S 60° E. Because a compass can be read only when held level, bearings of a line must be found in the top, or horizontal, view.

Azimuths are measured clockwise from north through 360° (**Figure 27.18B**). Azimuth

bearings are written N 120°, N 210°, and so on, indicating that they are measured from north.

The bearing of a line is toward the low end of the line unless otherwise specified. For example, line 2–3 in **Figure 27.19** has a bearing of N 45° E because the line's low end is point 3 in the front view.

Figure 27.20 shows how to find the bearing and slope of a line. This information may be used verbally to describe the line as having a bearing of S 65° E and a slope of 26° from point

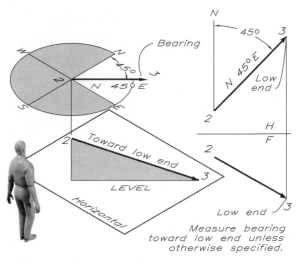

27.19 Measure the compass bearing of a line in the top view toward its low end (unless otherwise specified). Line 2–3 has a bearing of N 45° E from 2 to 3 toward the low end at point 3.

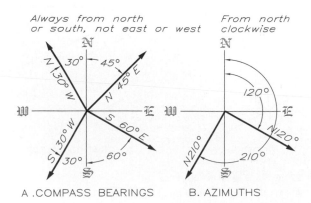

27.18 Compass directions:

A Compass bearings are measured with respect to north and south.

B Azimuths are measured clockwise from north up to 360°.

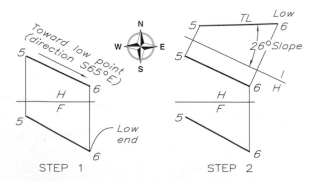

27.20 The slope and bearing of a line:

Step 1 Measure the bearing in the top view toward its low end, or S 65° E in this case.

Step 2 Measure the slope angle of 26° from the H1 reference line in an auxiliary view projected from the top view where the line is true length.

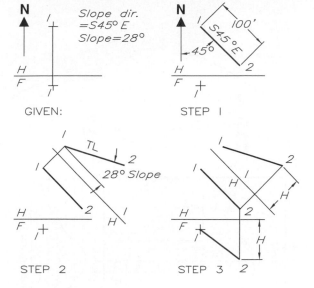

GIVEN: STEP 1

STEP 2   STEP 3

27.21 A line from slope specifications:

Given Draw a line through point 1 that bears S 45° E for 100 ft horizontally and slopes 28°.

Step 1 Draw the bearing and the horizontal distance in the top view.

Step 2 Project an auxiliary view from the top view and draw the line at a slope of 28°.

Step 3 Find the front view of line 1–2 by locating point 2 in the front view.

5 to point 6. This information and the location of one point in the top and front views is sufficient to complete a 3D drawing of a line as illustrated in **Figure 27.21**.

27.7 Application: Plot Plans

A typical plot plan for a tract of land is shown in **Figure 27.22**. The boundary lines of the tract, their bearings, and the interior angles are used to legally define the property. AutoCAD provides an option called *Surveyor's units* that is an excellent means of drawing and labeling a plot plan.

Computer Method To obtain the *Surveyor's units* option, type *Units* (Enter) to get the *Text Window* on the screen. You will be given the following prompts:

```
System of angle measure: (Example)
1. Decimal degrees 45.0000
```

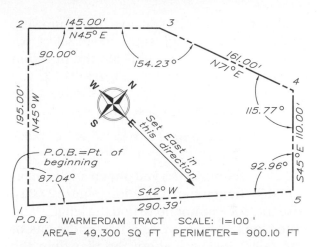

27.22 This typical plot plan shows the lengths and bearings of each side of a tract of land, the interior angles, and the north arrow by using AutoCAD's *Surveyor's Units* option.

```
2. Degree/minutes/seconds 45d0'0"
3. Grads 50.0000g
4. Radians 0.7854r
5. Surveyor's units N45d0'0"E
Enter choice, 1 to 5 <default>: 5
(Enter)
```

By entering 5, you obtain surveyor's units that give directional angles with respect to north or south in east or west directions, such as N 30d45'10", where d = degrees, ' = minutes, and " = seconds. Interior angles measured with the *Dim* command give the angles in the same form, but without reference to compass directions, such as 152d34'17".

Next, the *Units* command gives a prompt for setting the direction of east to establish the relationship of the drawing to the directional north arrow:

```
Direction for angle 0:
East    3 o'clock  =      0
North  12 o'clock  =     90
West    9 o'clock  =    180
South   6 o'clock  =    270
Enter direction for angle E <cur-
rent>:
```

With the cursor, select an area on the drawing and the *Text Window* is suppressed. Select two west-to-east points to indicate the direction of east in your drawing (**Figure 27.23**). By picking these two points, you establish the direction of north. The last prompt reads:

```
Do you want angles measured clock-
wise? <N>: N (Enter)
```

By selecting *No* (N) you obtain angles measured in the standard direction (**Figure 27.22**). Use the *Dist* command to find the lengths and directions of lines by selecting endpoints of the lines clockwise about the plot from the point of beginning (P.O.B.).

27.8 Contour Maps and Profiles

A contour map depicts variations in elevation of the earth in two dimensions (**Figure 27.24**). Three-dimensional representations involve (1) conventional orthographic views of the contour map combined with profiles and (2) contoured surface views (often in the form of models).

Contour lines are horizontal (level) lines that represent constant elevations from a

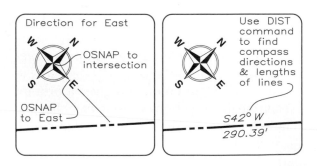

27.23 Compass direction by computer:

Step 1 Insert the compass arrow on the plot plan. Then use the *Units* command and select 5, *Surveyor's units*. It will prompt you to give the direction for east. Select a point on the drawing with the cursor and select two points (west to east) to indicate the direction of east.

Step 2 Using the *Dist* command, select the endpoints of each side clockwise about the plot from the point of beginning (P.O.B.). Label the sides with direction inside and lengths outside the plot.

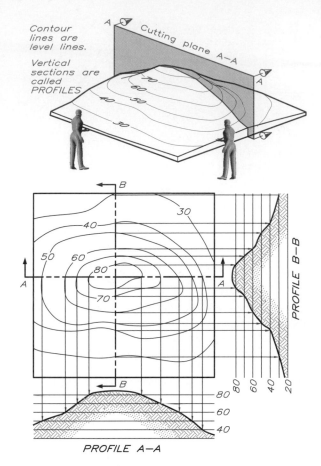

27.24 A contour map shows variations in elevation on a surface. A profile is a vertical section through the contour map. To construct a profile, draw elevation lines parallel to the cutting planes, spacing them equally to show the difference in elevations of the contours (10 ft in this case). Then project crossing points of contours and the cutting plane to their respective elevations in the profile and connect them.

horizontal datum such as sea level. The vertical interval of spacing between the contours shown in **Figure 27.24** is 10 feet. Contour lines may be thought of as the intersection of horizontal planes with the surface of the earth.

Contour maps contain contour lines that connect points of equal elevation on the earth's surface and therefore are continuous (**Figure 27.24**). The closer the contour lines are to each other, the steeper the terrain is.

Profiles are vertical sections through a contour map that show the earth's surface at any desired location (**Figure 27.24**). Contour lines represent edge views of equally spaced horizontal planes in profiles. True representation of a profile involves the use of a vertical scale equal to the scale of the contour map; however, the vertical scale usually is drawn larger to emphasize changes in elevation that often are slight compared to horizontal dimensions.

Contoured surfaces also are depicted in drawings with contour lines (**Figure 27.24**) or on models. When applied to objects other than the earth's surface—such as airfoils, automobile bodies, ship hulls, and household appliances—this technique of showing contours is called **lofting**.

Station numbers identify distances on a contour map. Surveyors use a chain (metal tape) 100 feet long, so primary stations are located 100 feet apart (**Figure 27.25**). For example, station 7 is 700 feet from the beginning point, station 0; a point 33 feet beyond station 7 is station 7 + 33.

A **plan-profile** combines a section of a contour map (a plan view) and a vertical section (a profile view). Engineers use plan-profile drawings extensively for construction projects such as pipelines, roadways, and waterways.

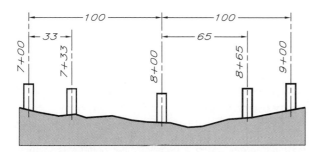

27.25 Primary station points are located 100 ft apart; for example, station 7 is 700 ft from station 0 (not shown). A point 33 ft beyond station 7 is labeled station 7 + 33. A point 865 ft from the origin is labeled station 8 + 65.

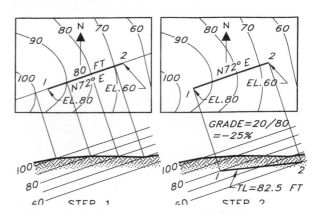

27.26 Drawing profiles (vertical sections):

Step 1 An underground pipe has elevations of 80 ft and 60 ft at its ends. Project an auxiliary view perpendicularly from the top view and draw contours at 10-ft intervals corresponding to their elevations in the plan view. Locate the ground surface by projecting from the contour lines in the plan view.

Step 2 Locate points 1 and 2 at elevations of 80 ft and 60 ft in the profile. Line 1–2 is TL in the section, so measure its slope (percent grade) here and label its bearing and slope in the top view.

Application: Vertical Sections

In **Figure 27.26**, a vertical section passed through the top view of an underground pipe gives a profile view. The pipe is known to have elevations of 80 feet at point 1 and 60 feet at point 2. Project an auxiliary view perpendicularly from the top view, locate contour lines, and draw the top of the earth over the pipe in profile. To measure the true lengths and angles of slope in the profile, use the same scale for both the contour map and the profile.

Pipeline installation (**Figure 27.27**) requires major outlays for engineering design and construction. The use of profiles, found graphically, is the best way to make cost estimations for constructing ditches for laying underground pipe.

27.9 Plan-Profiles

A plan-profile drawing shows an underground drainage system from manhole 1 to manhole 3 in **Figures 27.28** and **27.29**. The

27.27 Pipeline construction applies the principles of descriptive geometry, true-length lines, and slopes of 3D lines. (Courtesy of Consumers Power Company.)

cross the top view of the pipe to their respective elevations in the profile with your dividers to show the surface of the ground over the pipe.

Figure 27.29 shows the manholes, their elevations, and the bottom line of the pipe. Find the drop from manhole 1 to manhole 2 (4.40 ft) by multiplying the horizontal distance of 220.00 ft by a −2.00% grade. The pipes intersect at manhole 2 at an angle, so the flow of the drainage is disrupted at the turn. A drop of 0.20 ft (2.4 in.) across the bottom of the manhole compensates for the loss of pressure (head) through the manhole.

The true lengths of the pipes in the profile view cannot be measured when the vertical scale is different from the horizontal scale. Instead, you must use trigonometry to calculate them.

profile has a larger vertical scale to emphasize variations in the earth's surface and the grade of the pipe, although the vertical scale may be drawn at the same scale as the plan if desired.

The location of manhole 1 is projected to the profile orthographically, but the remaining points are not (**Figure 27.28**). Instead, transfer the distances where the contour lines

27.10 Edge Views of Planes

The edge view of a plane appears in a view where any line on the plane appears as a point. Recall that you can find a line as a point

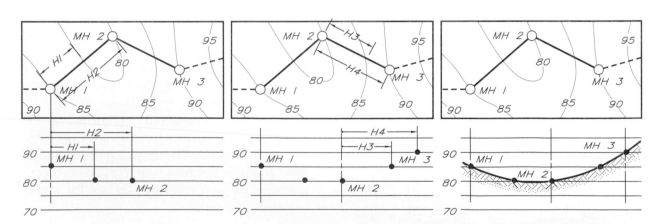

27.28 Plan-profile (vertical section):

Required Find the profile of the earth's surface over the pipeline.

Step 1 Transfer distances *H1* and *H2* from MH 1 in the plan to their respective elevations in the profile view.

Step 2 Measure distances H3 and H4 from manhole 2 in the plan and transfer them to their respective elevations in profile. These points represent elevations of points on the earth above the pipe.

Step 3 Connect these points with a freehand line and crosshatch the drawing to represent the earth's surface. Draw centerlines to show the locations of the three manholes.

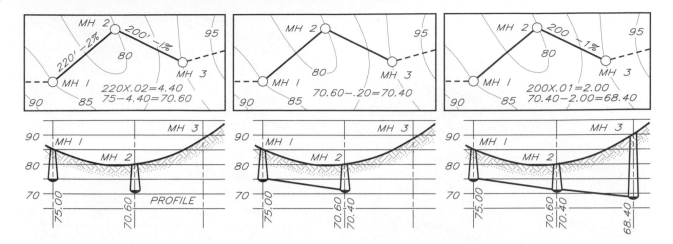

27.29 Plan-profile (manhole location):

Step 1 Multiply the horizontal distance from *MH1* to *MH2* by −2%. Find the elevation of the bottom of *MH2* by subtracting the amount of fall from the elevation of *MH1* (70.60′).

Step 2 The lower side of *MH2* is 0.20 ft lower than the inlet side to compensate for loss of head (pressure) because of the turn in the pipeline. Find the elevation on the lower side (70.40′) and label it.

Step 3 Calculate the elevation of *MH3* (200 ft) at a −1% grade from *MH2* (68.40′). Draw the flow line of the pipeline from manhole to manhole and label the elevations at each manhole.

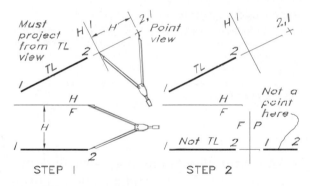

27.30 The point view of a line:

Step 1 Line 4–5 is horizontal in the front view and therefore is true length in the top view.

Step 2 Find the point view of line 4–5 by projecting parallel to its true length to the auxiliary view.

by projecting from its true-length view (**Figure 27.30**). You may obtain a true-length line on any plane by drawing a line parallel to one of the principal planes and projecting it to the adjacent view (**Figure 27.31**). You then get the edge view of the plane in an auxiliary view by finding the point view of line 3–4 on its surface.

Dihedral Angle

The angle between two planes, a dihedral angle, is found in a view where the line of intersection between two planes appears as a point. In this view, both planes appear as edges and the angle between them is true size. The line of intersection, line 1–2 between the two planes shown in **Figure 27.32**, is true length in the top view. Project an auxiliary view from the top view to find the point view of line 1–2 and the edge views of both planes.

27.11 Planes and Lines

Piercing Points

By Projection Finding the piercing point of line 1–2 passing through the plane by projection is shown in **Figure 27.33**. Pass cutting planes through the line and plane in the top view. Then project the trace of this cutting plane, line *DE*, to the front view to find piercing point *P*. Locate the top view of *P* and determine the visibility of the line.

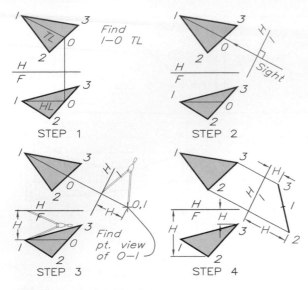

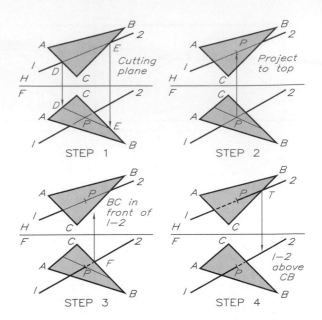

27.31 The edge view of a plane:

Step 1 To find the edge view of plane 1–2–3, draw horizontal line 1–O on its front view of the plane and project it to the top view, where it is true length.

Step 2 Draw a line of sight parallel to the true-length line 1–0. Draw H1 perpendicular to the line of sight.

Step 3 Find the point view of 1–0 in the auxiliary view by transferring height (H) from the front view.

Step 4 Locate points 2 and 3 in the same manner to find the edge view of the plane.

27.33 A piercing point by projection:

Step 1 Pass a vertical cutting plane through the top view of line 1–2, which cuts the plane along line DE. Project line DE to the front view to locate piercing point P.

Step 2 Project point P to the top view of line 1–2.

Step 3 Determine visibility in the front view by projecting the crossing point of lines CB and 1–2 to the top view. Because CB is encountered first, it is in front of 1–2, making segment PF hidden in the front view.

Step 4 Determine visibility in the top view by projecting the crossing point of lines CB and 1–2 to the front view. Because line 1–2 is encountered first, it is above line CB, making TP visible in the top view.

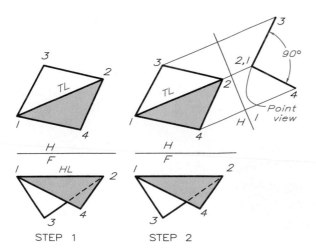

27.32 A dihedral angle:

Step 1 The line of intersection between the planes, line 1–2, is true length in the top view.

Step 2 The angle between the planes (the dihedral angle) is found in the auxiliary view where the line of intersection appears as a point and both planes are edges.

By Auxiliary View You may also find the piercing point of a line and a plane by auxiliary view in which the plane is an edge (**Figure 27.34**). The location of piercing point P in Step 2 is where line *AB* crosses the edge view of the plane. Project point *P* to *AB* in the top view from the auxiliary view, and then to the front view. To verify the location of point P in the front view, transfer dimension *H* from the auxiliary view with dividers.

You can easily determine visibility for the top view because you see in the auxiliary view that line *AP* is higher than the plane and therefore is visible in the top view. Similarly, the top view shows that endpoint *A* is the for-

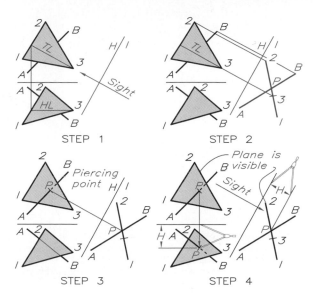

STEP 1   STEP 2

STEP 3   STEP 4

27.34 A piercing point by auxiliary view:

Step 1 Draw a horizontal line on the plane in the front view and project it to the top view where it is true length on the plane.

Step 2 Find the edge view of the plane in an auxiliary view and project AB to this view. P is the piercing point.

Step 3 Project point P to line AB in the top view. Line *AP* is nearest the *H1* reference line, so it is the highest end of the line and is visible in the top view.

Step 4 Project P to line *AB* in the front view. *AP* is visible in the front view because line *AP* is in front of 1–2.

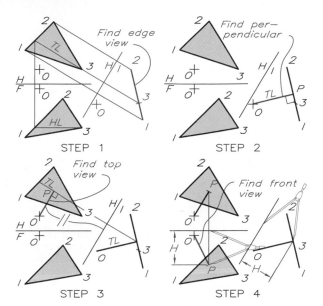

STEP 1   STEP 2

STEP 3   STEP 4

27.35 A line perpendicular to a plane:

Step 1 Find the edge view of the plane by finding the point view of a line on it in an auxiliary view. Project point O to this view, also.

Step 2 Draw line *OP* perpendicular to the edge view of the plane, which is true length in this view.

Step 3 Because line *OP* is true length in the auxiliary view, it must be parallel to the *H1* reference line in the preceding view. Line *OP* is visible in the top view because it appears above the plane in the auxiliary view.

Step 4 Project point *P* to the front view and locate it by transferring height H from the auxiliary view with dividers.

ward-most point and line AP therefore is visible in the front view.

Perpendicular to a Plane

A perpendicular line appears true length and perpendicular to a plane where the plane appears as an edge. In **Figure 27.35**, a line is to be drawn from point *O* perpendicular to the plane. Obtain an edge view of the plane and draw the true-length perpendicular to locate piercing point *P*. Locate point P in the top view by drawing line *OP* parallel to the *H1* reference line. (This principle is reviewed in **Figure 27.36**.) Line *OP* also is perpendicular to a true-length line in the top view of the plane. Obtain the front view of point P, along with its visibility, by projection.

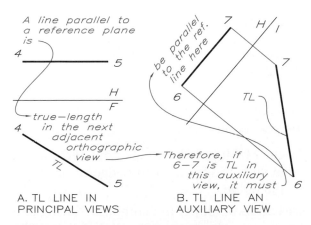

A. TL LINE IN PRINCIPAL VIEWS   B. TL LINE AN AUXILIARY VIEW

27.36 In the top view of Figure 27.35, line *OP* is parallel to the *H1* reference line because it is true length in the auxiliary view. Here, lines 4–5 and 6–7 are examples of this principle: Both are true length in one view and parallel to the reference line in the preceding view.

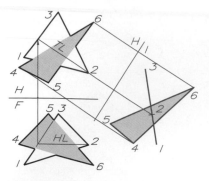

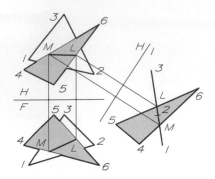

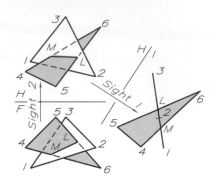

27.37 The intersection of planes by auxiliary view:

Step 1 Locate the edge view of one of the planes in an auxiliary view and project the other plane to this view.

Step 2 Piercing points *L* and *M* are found on the edge view of the plane in the auxiliary view. Project the line of intersection, *LM*, back to the top and front views.

Step 3 The line of sight from the top view strikes *L*–5 first in the auxiliary view, indicating that *L*–5 is visible in the top view. 4–5 is farthest forward in the top view and is visible in the front.

Intersection Between Planes

To find the intersection between planes, find the edge view of one of the planes (**Figure 27.37**). Then project piercing points *L* and *M* from the auxiliary view to their respective lines, 5–6 and 4–6, in the top view. Plane 4–5–*L*–*M* is visible in the top view because sight line 1 has an unobstructed view of the 4–5–*L*–*M* portion of the plane in the auxiliary view. Plane 4–5–*L*–*M* is visible in the front view because sight line 2 has an unobstructed view of the top view of this portion of the plane.

27.12 Sloping Planes

Slope and Direction of Slope

The slope of a plane is described using the following definitions:

Angle of Slope: the angle that the plane's edge view makes with the edge of the horizontal plane.

Direction of Slope: the compass bearing of a line perpendicular to a true-length line in the top view of a plane toward its low side (the direction in which a ball would roll on the plane).

As **Figure 27.38** shows, a ball would roll perpendicular to all horizontal lines on the roof toward the low side. This direction is the slope direction and it can be measured in the top view as a compass bearing.

The steps of determining the slope and direction of slope of a plane are shown in **Figure 27.39**. Conversely, a plane can be drawn in three-dimensional space by working from slope and direction specifications as shown in **Figure 27.40**. Draw the direction of slope in the top view to locate a perpendicular true-length line on the plane. Find the edge view of the plane by locating point 1 and constructing a slope of 30° through it in an auxiliary view. Transfer points 3 and 2 to the front view from the auxiliary view.

Application: Cut and Fill

A level roadway through irregular terrain and the embankment for an earthen dam (**Figure 27.41**) involve the principles of cut and fill. Cut and fill is the process of cutting away high ground and filling low areas, generally of equal volumes.

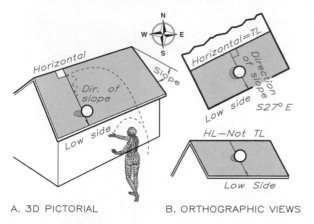

| A. 3D PICTORIAL | B. ORTHOGRAPHIC VIEWS |

27.38 Slope definition:

A The direction of slope of a plane is the compass bearing of the direction in which a ball on the plane will roll.

B Slope direction is measured in the top view toward the low side of the plane and perpendicular to a horizontal line on the plane.

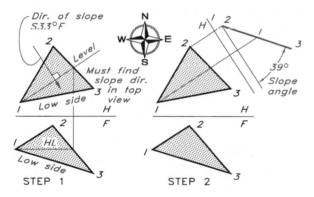

27.39 The slope and bearing of a plane:

Step 1 Slope direction is perpendicular to a true-length, level line in the top view toward the low side of the plane, or S 33° E in this case.

Step 2 Find the slope in an auxiliary view where the horizontal is an edge and the plane is an edge, or 39° in this case.

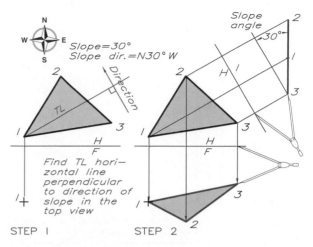

27.40 A plane from slope specifications:

Step 1 If the top view of a plane, the front view of point 1, and slope specifications are given, you can complete the front view. Draw the direction of slope in the top view and a true-length horizontal line on the plane perpendicular to the slope direction.

Step 2 Find a point view of the *TL* line in the auxiliary view to locate point 1. Draw the edge view of the plane through point 1 at a slope of 30°, according to the specifications. Find the front view by transferring height dimensions from the auxiliary view to the front view.

27.41 This dam was built by applying the principles of cut and fill. (Courtesy of the Bureau of Reclamation, U.S. Department of the Interior.)

In **Figure 27.42**, a level roadway at an elevation of 60 feet is to be constructed along a specified centerline with specified angles of cut and fill. First, draw the roadway in the top view. Use contour intervals in the profile view of 10 feet to match those in the top view.

Next, measure and draw the cut angles on both sides of the roadway. Project the points where the cut angles cross each elevation line to the respective contour lines in the plan view to find the limits of cut.

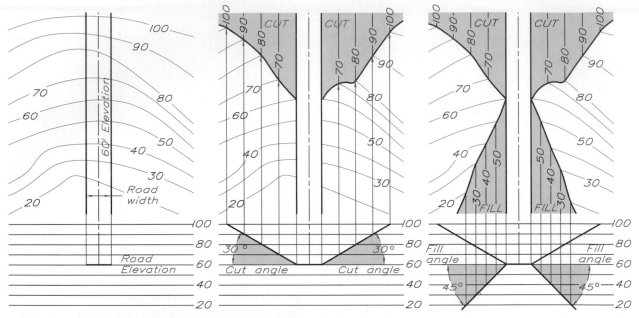

27.42 Cut and fill for a level roadway:

Step 1 Draw and label a series of elevation planes in the front view at the same scale as the contour map. Draw the width and elevation (60 ft. in this case) of the roadway in the top and front views.

Step 2 Draw the cut angles on the higher sides of the road in the front view. Project the points of intersection between the cut angles and the contour planes in the front view to their respective contour lines in the top view to determine the limits of cut.

Step 3 Draw the fill angles on the lower sides of the road in the front. Project the points in the front where the fill angles cross the contour lines to these respective contour lines in the top view. Draw new contours parallel to the centerline in the cut-and-fill areas.

Then, measure and draw the fill angles in the profile view. Project the points where the fill angles cross each elevation line to their respective contour lines in the plan view to find the limits of fill. Finally, draw new contour lines inside the areas of cut and fill parallel the centerline.

Some of the terminology associated with the design of a dam are (**1**) **crest**, or the top of the dam; (**2**) **water level**; and (**3**) **freeboard**, or the height of the crest above the water level. These terms are illustrated in **Figure 27.43**.

Strike and Dip

Strike and dip are terms used in geology and mining engineering to describe the location of strata of ore under the surface of the earth:

Strike: the compass bearings (two are possible) of a level line in the top view of a plane.

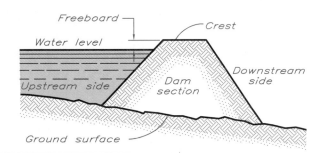

27.43 These terms and symbols are used in the design of a dam.

Dip: the angle that the edge view of a plane makes with the horizontal and its general compass direction, such as NW or SW.

The dip angle lies in the primary auxiliary view projected from the top view. The dip direction is perpendicular to the strike and toward its low side.

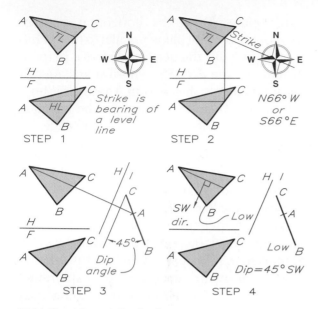

STEP 1

STEP 2

STEP 3

STEP 4

27.44 The strike and dip of a plane:

Step 1 Draw a horizontal line on the plane in the front view and project it to the top view, where it is true length.

Step 2 Strike is the compass direction of a level line on the plane in the top view, either N 66° W or S 66° E.

Step 3 Find the edge view of the plane in the auxiliary view. The dip angle of 45° is the angle between the H1 reference line and the edge view of the plane.

Step 4 The general compass direction of dip is toward the low side, and perpendicular to a strike in the top view or SW in this case. Dip direction is written as 45° SW.

Figure 27.44 demonstrates how to find the strike and dip of a plane. Here, the true-length line in the top view of the plane has a strike of N 66° W or S 66° E. The dip angle appears in an auxiliary view projected from the top view that shows the horizontal (*H1*) and the sloping plane as edges.

You can construct a plane from strike and dip specifications shown in **Figure 27.45**. First, draw the strike as a true-length horizontal line on the plane's top view and the dip direction perpendicular to the strike. Then find the edge view of the plane in the auxiliary view through point 1 at a dip of 30°. Locate points 2 and 3 in the front view by transferring them from the auxiliary view.

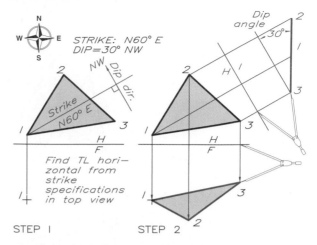

STEP 1

STEP 2

27.45 Strike and dip specifications:

Step 1 Draw the strike in the top view of the plane as a true-length horizontal line. Draw the direction of dip perpendicular to the strike toward the NW as specified.

Step 2 Find the point view of strike in the auxiliary view to locate point 1 where the edge view of the plane passes through it at a 30° dip, as specified. Complete the front view by transferring height (*H*) dimensions from the auxiliary to the front view.

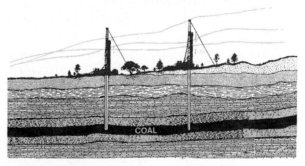

27.46 Test wells are drilled into coal zones to determine the elevations of coal seams that may contribute to the exploration for gas. (Courtesy of Texas Eastern *News*.)

27.13 Ore Vein Applications

The principles of descriptive geometry can be applied to find the distance from a point to a plane. Techniques of finding such distances often are used to solve mining and geological problems. For example, test wells are drilled into coal seams to learn more about them (**Figure 27.46**).

Application: Underground Ore Veins Geologists and mining engineers usually assume

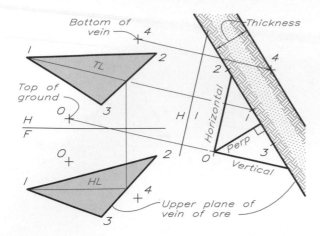

27.47 To find the vertical, horizontal, and perpendicular distances from a point to an ore vein, project an auxiliary view from the top view, where the vein appears as an edge. The thickness of an ore vein is perpendicular to the upper and lower planes of the vein.

that strata of ore veins have upper and lower planes that are parallel. In **Figure 27.47**, point *O* is on the upper surface of the earth and plane 1–2–3 is an underground ore vein. Point 4 is on the lower plane of the vein.

Find the edge view of plane 1–2–3 by projecting from the top view and then draw the lower plane through point 4 parallel to the

upper plane. Draw the horizontal distance from point *O* to the plane parallel to the H1 reference line and the vertical distance perpendicular to line *H1*. The shortest distance is perpendicular to the ore vein. These three lines from point *O* are true length in the auxiliary view where the ore vein appears as an edge.

Application: Ore Vein Outcrop The same assumption is made in **Figure 27.48** that underground ore veins are defined by parallel planes that will outcrop on the Earth's surface if they are inclined (non-horizontal). When ore veins outcrop, open-pit mining can be employed to reduce costs. To find the outcrop of an ore vein, locations of exploratory drillings given on a contour map. Points *A*, *B*, and *C* are located on the upper plane of the ore vein, and point *D* is located on the lower plane of the vein. Draw these points in the front view at their determined elevations.

Find the edge view of the ore vein in an auxiliary view projected from the top view. Then project points on the upper surface where the vein crosses elevation lines back to their respective contour lines in the top view.

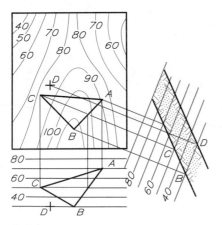
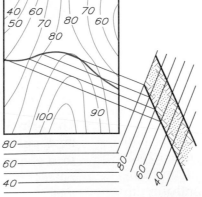
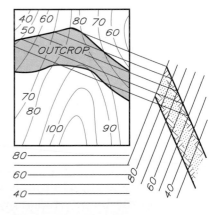

27.48 Locating an ore vein outcrop:

Step 1 Use points *A*, *B*, and *C* on the upper surface of the ore vein to find its edge view by projecting an auxiliary from the top view. Draw the lower surface of the vein parallel to the upper plane through point *D*.

Step 2 Project points of intersection between the upper plane of the vein and their elevation lines in the auxiliary view to their respective contours in the top view to find a line of the outcrop.

Step 3 Project points from the lower plane in the auxiliary view to their respective contours in the top view to find the second line of outcrop. Crosshatch the area between the lines to depict the outcrop of the vein.

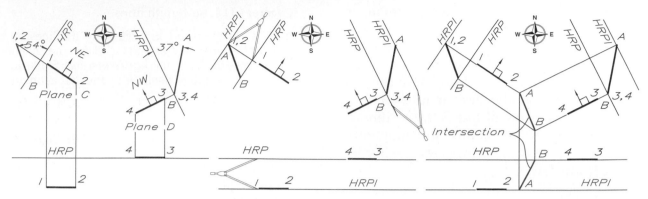

27.49 Intersection of planes: Strike and dip method:

Step 1 Lines 1–2 and 3–4 are strike lines and are true length in the top view. Use a common reference plane, *HRP*, to find the point view of each strike line by auxiliary views. Find the edge views by drawing the dip angles with the *HRP* line through the point views.

Step 2 Draw a supplementary horizontal plane, *HRP1*, at a convenient location in the front view. This plane, shown in both auxiliary views, is located H distance from *HRP*. The *HRP1* cuts through each edge in both auxiliary views, locating A and B.

Step 3 Project points A from both auxiliary views on *HRP1* to the top view of their intersection at *A*. Project points *B* and *HRP* to their intersection in the top view. Project points *A* and *B* to their respective planes in the front view. *AB* is the line of intersection.

Also project points on the lower surface of the vein (through point *D*) to the top. If the ore vein extends uniformly at its angle of inclination to the earth's surface, the area between these two lines will be the outcrop of the vein.

27.14 Intersections between Planes

Strike and Dip Method   The method of locating the intersection of two planes located with strike and dip specifications is shown in **Figure 27.49**. The given strike lines are true-length level lines in the top view, so the edge view of the planes appear in the auxiliary views where the strikes appear as points. Draw the edge views using the given dip angles and directions.

Use the additional horizontal datum plane *HRP1* to find lines on each plane at equal elevations that intersect when projected to the top view from their auxiliary views. Connect points *A* and *B* as the line of intersection between the two planes in the top view and project it to the front view to establish line *AB* in three dimensions.

Cutting Plane Method   In **Figure 27.50**, top and front views of two planes are given and

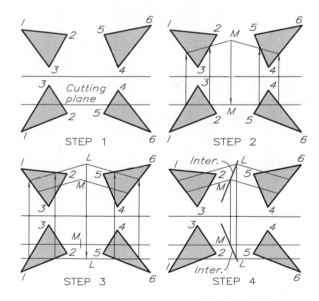

27.50 Intersection of planes by the cutting plane method:

Step 1 Draw a cutting plane that passes through both planes in the front view in any convenient direction.

Step 2 Project the intersections of the cutting plane to the top views of the planes. Find intersection point *M* in the top view and project it to the front view.

Step 3 Draw a second cutting plane through the front view of the planes and project it to the top view. Find point *L* in the top view and project it to the front view.

Step 4 Connect *L* and *M* in the top and front views to represent the line of intersection of the extended planes.

the line of intersection between them, if they were extended, is to be determined. Draw cutting planes through both planes in either view at any angle and project them to the top view. Find points *L* and *M* in the top view to establish the line of intersection. Find the front view of line *LM*, the line of intersection, by projecting its endpoints from the top view to their respective planes in the front view.

Problems

Use size A sheets for the following problems; lay out the solutions using instruments. Each square on the grid is equal to 0.20 inch or 5 mm. Use either grid or plain paper. Label all reference planes and points with 1/8-in. or 3 mm letters or numbers with guidelines.

1. **(Sheet 1)** True-length lines.

(A–D) Find the true-length views of the lines by auxiliary view as indicated by the given lines of sight. *Alternative method:* Find the true-length of the lines by the Pythagorean theorem.

2. **(Sheet 2)** True-length lines.

(A–D) Find the true-length views of the lines by auxiliary view as indicated by the lines of sight. *Alternative method:* Find the true length of the lines by the Pythagorean theorem.

3. **(Sheet 3)** True-length diagram.

(A–B) Find the true length of the lines by using a true-length diagram.

(C–D) Find the point views of the lines.

4. **(Sheet 4)** Sloping lines.

(A–D) Find the slope angle, tangent of the slope angle, and the percent grade of the four lines.

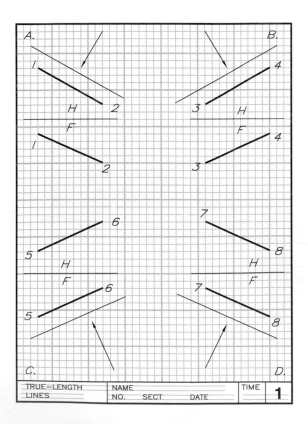

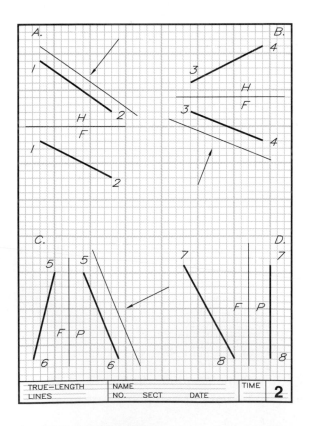

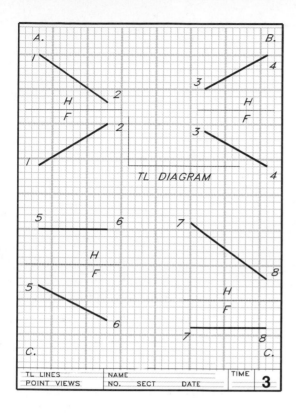

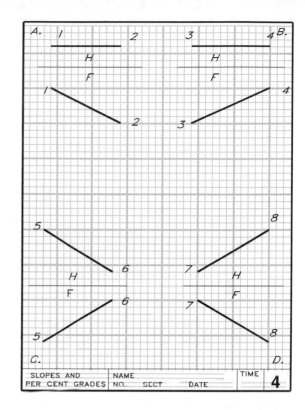

5. **(Sheet 5)** Edge views of planes.

 (A–B) Find the edge views of the planes.

6. **(Sheet 6)** Intersections: lines and planes.

 (A) Find the angle between the planes.

 (B) By projection, find the point of intersection between the line and plane, and show visibility.

 (C) By the auxiliary view method, find the point of intersection between the line and plane and show visibility.

7. **(Sheet 7)** Perpendiculars to planes.

 (A) Construct a 1 in. line perpendicular from point *O* on the plane and show it in all views.

 (B) Draw a line perpendicular to the plane from *O*, find the piercing point, and show visibility.

8. **(Sheet 8)** Dihedral angles.

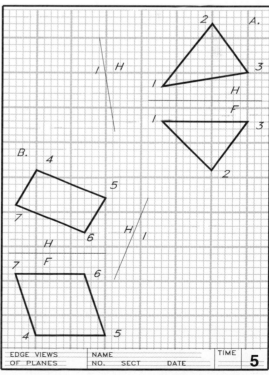

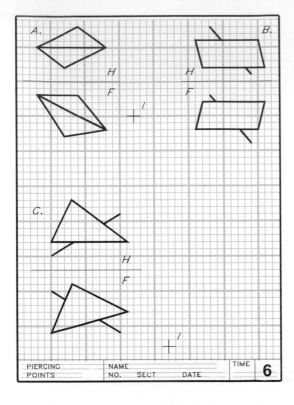

A.

B.

C.

PIERCING POINTS	NAME			TIME	**6**
	NO.	SECT	DATE		

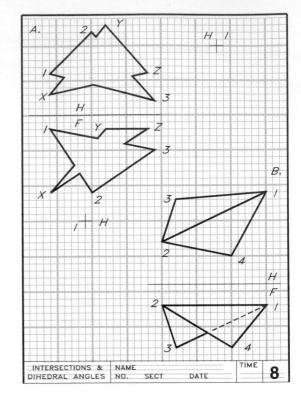

A.

B.

INTERSECTIONS & DIHEDRAL ANGLES	NAME			TIME	**8**
	NO.	SECT	DATE		

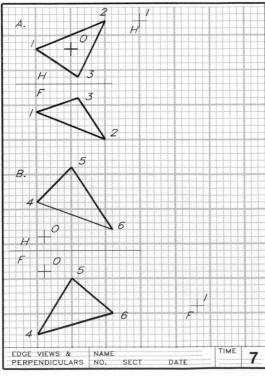

A.

B.

EDGE VIEWS & PERPENDICULARS	NAME			TIME	**7**
	NO.	SECT	DATE		

By using auxiliary views, find:

  (A) The line of intersection between the planes.

  (B) The angle between the planes.

9. (Sheet 9) Slope and direction of slope.

 (A–B) Find the direction of slope and the slope angle of the planes. *Alternative solution:* Find the strike and dip of the planes.

10. (Sheet 10) Distances to a plane.

Find the shortest distance, the horizontal distance, and the vertical distance from point *O* to the underground ore vein represented by plane 1–2–3. Point *B* is on the lower plane of the vein. Find the thickness of the vein. Label your solutions. SCALE: 1=10FT

11. (Sheet 11) Intersecting planes.

  (A) Find the line of intersection between the two planes by the cutting plane method.

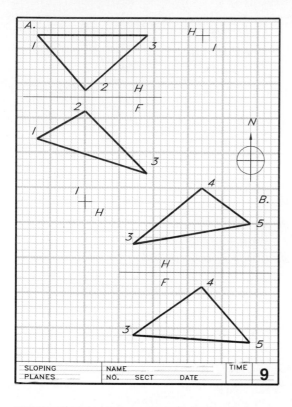

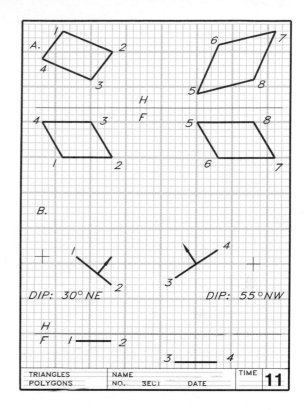

DIP: 30° NE

DIP: 55° NW

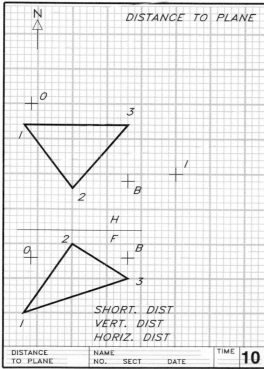

DISTANCE TO PLANE

SHORT. DIST
VERT. DIST
HORIZ. DIST

(B) Find the line of intersection between the two planes indicated by strike lines 1–2 and 3–4. The plane with strike line 1–2 has a dip of 30°, and the plane with strike line 3–4 has a dip of 55°.

12. (Sheet 12) Cut and Fill.

Find the limits of cut and fill in the plan view of the roadway. Use a cut angle of 35° and fill angle of 40°.

13. (Sheet 13) Outcrop.

Find the outcrop of the ore vein represented by plane 1–2–3 on its upper surface. Point B is on the lower surface. SCALE: 1=20FT

14. (Sheet 14) Plan-profile.

Complete the plan-profile drawing of the drainage system from manhole 1 through manhole 2 to manhole 3, using the grades indicated. Allow a drop of 0.20 ft across each manhole to compensate for loss of pressure.

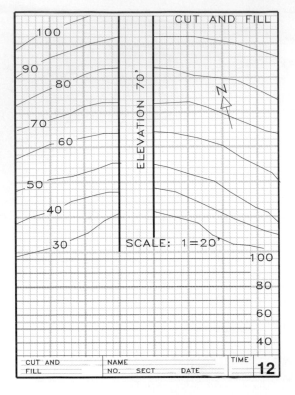

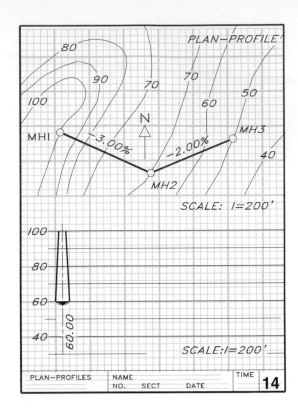

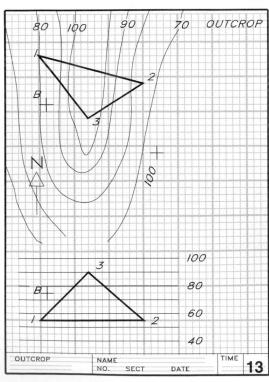

15. (Sheet 15) Contour map.

Draw the contour map with its contour lines. Give the lengths of the sides, their compass directions, interior angles, scale, and north arrow.

16. (Sheet 16) Vertical section.

Draw the contour map and construct the profile (vertical section) as indicated by the cutting plane line in the plan view. Note that the profile scale is different from the plan scale.

Design Problems

Lay out the necessary orthographic views in order to solve the following problems on size A or size B sheets. (Courtesy of Boeing Company.)

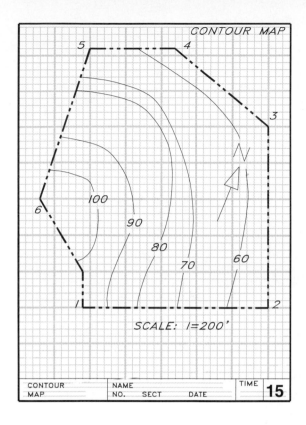

CONTOUR MAP

5   4
3
100
6
90
80
70   60
1   2

SCALE: 1=200'

CONTOUR MAP	NAME			TIME	**15**
	NO.	SECT	DATE		

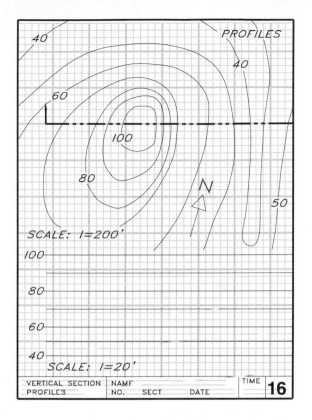

PROFILES

40
40
60
100
80
50

SCALE: 1=200'

100
80
60
40

SCALE: 1=20'

VERTICAL SECTION PROFILES	NAME			TIME	**16**
	NO.	SECT	DATE		

Design 1: Bolt Support

(A) Use the given dimensions and lay out the views necessary to find the angle the tapered washer makes with the inclined block that supports the aircraft seat.

(B) Make a working drawing with dimensions to describe the tapered circular washer.

Design 2: Bulkhead Clearance

Lay out the orthographic views (Design 2) on a size-B sheet that are necessary to find the clearance between the bulkhead and the electrical connector box. If necessary, relocate the connector so it will have a 0.50 clearance with the bulkhead.

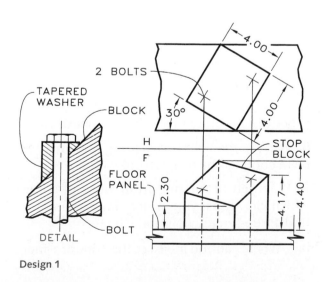

Design 1

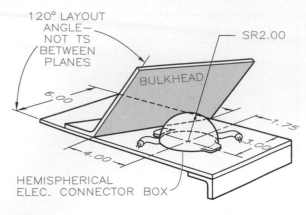

Design 2

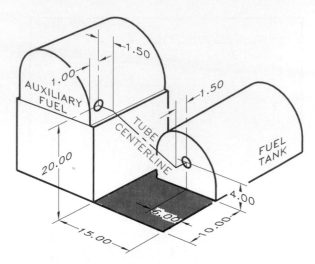

Design 3

Design 3: Fuel-Tank Tube

Layout the orthographic views on a size-*B* sheet that are necessary to find (A) the true length of the tube, and (B) the angles the tube makes with the 1.50 stubs extending from both fuel thanks.

Review Questions

1. What conditions must exist in order for you to measure the angle between a line and plane?

2. In which view can you find the compass bearing of a sloping line?

3. In what direction is a sloping line assumed to slope if not specified?

4. What is the percent grade of a line that is 100 ft. long in the top view, and has a 5 ft. vertical distance between its ends in the front view? What is the rise to run ratio of this line? What is its angle of slope?

5. Which view must an auxiliary view be projected from in order to obtain the slope of a line?

6. What is the azimuth bearing of a line that has a compass bearing of S 30° W?

7. When given top and front views of a line, which view must you project from in order to find a true-length view of the line?

8. Where can you measure the slope of a line?

9. What is the difference between a level line and a contour line?

10. What can be said about the terrain where contour lines are closer together? Farther apart?

11. What is the distance between two points when one has a station number of 8 + 20 and the other has a station number of 9 + 10?

12. Why does a manhole usually have a drop of about 0.2 ft between pipes coming into and leaving it? How many inches is 0.2 ft?

13. Describe the view in which a dihedral angle can be found.

14. In what type of view can the piercing point between a line and a plane be easily seen?

15. In what type of view can a line that is perpendicular to a plane be drawn at a true right angle to it?

16. When will a plane appear as an edge?

17. How many frontal lines can be drawn on a given plane? Horizontal lines?

18. What is the slope of a plane?

19. In what view can the direction of slope be measured and specified. What is the definition of slope direction?

20. If a line in the top view is parallel to the HF reference line, what do you know about its front view?

21. What is the definition of the strike of a plane.

22. In what view must strike be measured?

23. What is the definition of the dip of a plane.

24. What is the difference between a level line and a horizontal line?

25. What is the freeboard of a dam?

26. If a level line on a plane has an azimuth bearing of N 60° E, what is the strike of the plane?

27. What is the area called where an underground ore vein extends to the surface of the earth?

28. What is the slope of a level line?

29. Can you find the point view of a line by projecting from the front view of a horizontal line? Explain.

30. How many orthographic views of a line are required in order to find it true length by the Pythagorean theorem? Explain.

Thinking with a Pencil:
31. Sketch a TL vertical line in the front view about 2 in. tall. Sketch its top and side views. How will it appear in the top? In the side? What type of line is it?

32. Sketch a line AB that is 2 in. long in the top view and has a direction of N 45° E. Draw its front view with B 0.5 in. lower than A. What is its percent grade? In which view is it TL?

33. Sketch line AB (Problem 32) in a TL diagram. Approximately how long is it?

34. Sketch a right cone with a horizontal base approximately 3 in. in diameter with a height of 4 in. Sketch contour lines in both views at 0.5 in. intervals to describe the cone.

35. Sketch a plot plan (4 in. × 4 in.) using contour lines spaced 10 ft. apart vertically to represent a hilltop located in the center of the plot. The highest point is 80 ft. and the lowest point is 20 ft. Multiple solutions are possible.

36. Sketch a profile of your plot made in Problem 35.

37. Follow Problem 35, but modify the hilltop to have a vertical face on one side from an elevation of 30 ft. to 80 ft.

38. Using the plot plan sketched in Problem 35, sketch the path of a ball rolling from its peak in four directions: north, south, east, and west.

39. Sketch top and front views of a horizontal plane that is triangular in shape. Draw top and front views of a line that is higher than the line and show visibility.

40. Sketch top and front views of a three-legged stool with a 12 in. circular seat. The top of the seat is to be 12 in. from the floor; each leg makes a 10° angle with the vertical. Sketch a view that will show one of the legs TL. What angle must the drilled holes make with the seat and in what direction?

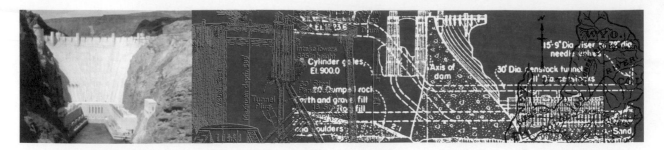

28

Successive Auxiliary Views

28.1 Introduction

A detailed drawing and specifications for a design cannot be completed without determining its geometry, which usually requires the application of descriptive geometry. The structural supports for the shopping mall shown in **Figure 28.1** are examples of complex spatial geometry problems in which lengths must be determined, angles between lines and planes calculated, and three-dimensional connectors designed.

The process of determining the 3D geometry of a design requires the use of secondary and successive auxiliary views. **Secondary auxiliary views are views projected from primary auxiliary views, and successive auxiliary views are views projected from secondary auxiliary views.**

28.2 Point View of a Line

Recall that, when a line appears true length, you can find its point view in a primary auxil-

28.1 This structural support of the roof system of a shopping mall was designed and fabricated through the application of descriptive geometry. (Courtesy of Lorna Stuckgold, Kaiser Engineers.)

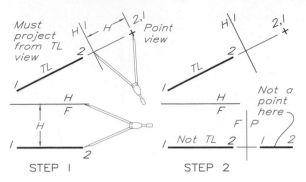

28.2 To find the point of view of a line, project an auxiliary view from the true-length view of the line.

iary view projected parallel from it. In **Figure 28.2**, line 1–2 is true length in the top view because it is horizontal in the front view. To find its point view in the primary auxiliary view, first construct reference line *H1* perpendicular to the true-length line. Transfer the height dimension, *H*, to the auxiliary view to locate the point view of 1–2.

Line 3–4 in **Figure 28.3** is not true length in either view. Finding the line's true length by a primary auxiliary view enables you to find its point view. To obtain a true-length view of line 3–4, project an auxiliary view from the front

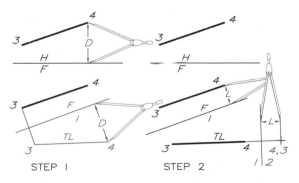

28.3 A point view of an oblique line:

Step 1 Draw a line of sight perpendicular to one of the views, the front view in this case. Line 3–4 is found true length in an auxiliary view projected perpendicularly from the front view.

Step 2 Draw a secondary reference line, 1–2, perpendicular to the true-length view of line 3–4. Find the point view by transferring dimension *L* from the front view to the secondary auxiliary view.

view (or from the top view). Projecting parallel from the true-length view to a secondary auxiliary view gives the point view of line 3–4. Label the line 4–3 because you see point 4 first in the secondary auxiliary view. Label the reference line between the primary and secondary planes 1–2 to represent the primary (1) and secondary (2) planes.

28.3 Dihedral Angles

Recall that the angle between two planes is called a **dihedral angle and can be found in a view where the line of intersection appears as a point.** The line of intersection lies on both planes, so both appear as edges when the intersection is a point view.

The planes shown in **Figure 28.4** represent a special case because their line of intersection, line 1–2, is true length in the top view. This condition permits you to find the line's point view in a primary auxiliary view and measure the true angle between the planes.

Figure 28.5 presents a more typical case. Here, the line of intersection between the two planes is not true length in either view. The line of intersection, line 1–2, is true length in a

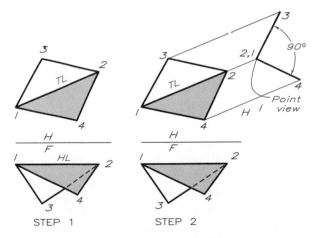

28.4 The angle between planes (the dihedral angle) appears in the view where their line of intersection projects as a point. The line of intersection, 1–2, is true length in the top view, so it can be found as a point in a view projected from the top view.

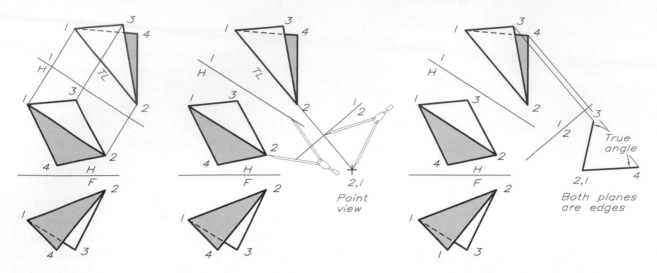

28.5 Angle between two planes:

Step 1 The angle between two planes is found in a view where the line of intersection (1–2) appears as a point. Find the *TL* view of the intersection in an auxiliary view.

Step 2 Obtain the point view of the line of intersection in the secondary auxiliary view by projecting parallel to the true-length view of line 1–2 in the primary auxiliary view.

Step 3 Complete the edge views of the planes in the secondary auxiliary view by locating points 3 and 4. Measure the angle between the planes (the dihedral angle) in this view.

primary auxiliary view, and the point view of the line appears in the secondary auxiliary view, where you measure the dihedral angle.

This principle was applied to determine the angles between wall panels of the control tower shown in **Figure 28.6**. That allowed the corner braces to be designed and the structure to be assembled correctly.

28.4 True Size of a Plane

A plane can be found true size in a view projected perpendicularly from an edge view of a plane. The front view of plane 1–2–3 in **Figure 28.7** appears as an edge in the front view as a special case. The plane's true size is in a primary auxiliary view projected perpendicularly from the edge view.

Figure 28.8 depicts a general case in which you can find the true-size view of plane 1–2–3 by finding the edge view of the plane and constructing a secondary auxiliary view projected perpendicularly from the edge view to find the plane's true size.

This principle can be applied to find the angle between lines such as bends in an automobile exhaust pipe (**Figure 28.9**). The method of solving a problem of this type is shown in **Figure 28.10**. The top and front views of intersecting centerlines are given, and the angles of bend and the radii of curvature must be found. Angle 1–2–3 is an edge in the primary auxiliary view and true size in the secondary view, where it can be measured and the radius of curvature drawn. The support frame of the materials conveyor (**Figure 28.11**) is an example of a design involving planes and lines in geometric combinations that was designed with applications of auxiliary views.

28.5 Shortest Distance from a Point to a Line: Line Method

The shortest distance from a point to a line can be measured in the view where the line appears as a point. The shortest distance from point 3 to line 1–2 that appears in a

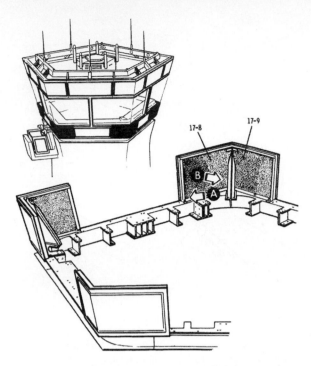

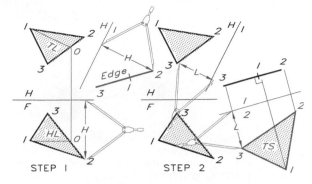

28.8 The true size of a plane (general case):

Step 1 Find the edge view of plane 1–2–3 by obtaining the point view of true-length 1–O in an auxiliary view.

Step 2 Find a true-size view by projecting a secondary auxiliary view perpendicularly from the edge view of the plane found in the primary auxiliary view.

28.6 These control tower wall panels that intersect at compound angles illustrate the need to determine and measure dihedral angles. The designer had to determine the angles between the wall panels in order to design connectors for securing the panels at their joints.

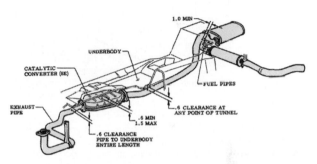

28.9 The designer determined the angles of bend in this automobile exhaust pipe by applying the principles of the angle between two lines. (Courtesy of General Motors Corporation.)

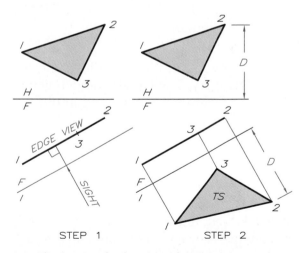

28.7 The true size of a plane (special case):

Step 1 Because plane 1–2–3 appears as an edge in the front view, it is a special case. Draw the sight line perpendicular to its edge and the F1 parallel to the edge.

Step 2 Find the true size of plane 1–2–3 in the primary auxiliary view by locating the vertex points with the depth *D* dimension.

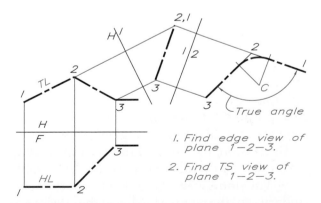

1. Find edge view of plane 1–2–3.

2. Find TS view of plane 1–2–3.

28.10 The angle between two lines is obtained by finding a true-size view of the plane formed by 1–2 and 2–3, the two lines.

28.11 The support frame of this materials conveyor is an example of the application of determining lengths and angles during its design stages. It is used on construction sites to rapidly move cement, aggregate, and sand. (Courtesy of Speed King Manufacturing Co.)

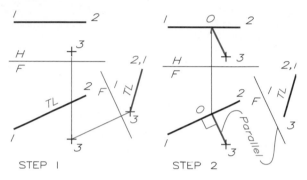

28.12 The shortest distance from a point to a line:

Step 1 The shortest distance from a point to a line is the true length where the line (1–2) appears as a point. The true-length view of the connecting line appears in the primary auxiliary view.

Step 2 When the connecting line is projected back to the front view, it must be parallel to the *F1* reference line in the front view. Project line 3–O back to the top view.

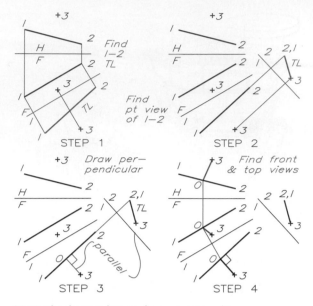

28.13 The shortest distance from a point to a line:

Step 1 The shortest distance from a point to a line is found in the view where the line appears as a point. Find the true length of line 1–2 by projecting from the front view.

Step 2 Line 1–2 is a point in a secondary auxiliary view projected from the true-length view of line 1–2. The shortest distance to it is true length in this view.

Step 3 Since 3–O is true length in the secondary auxiliary view, it is parallel to the 1–2 reference line in the primary auxiliary view and perpendicular to the line.

Step 4 Find the front and top views of 3–O by projecting from the primary auxiliary view in sequence.

primary auxiliary view in **Figure 28.12** (Step 1) is a special case. The distance from point 3 to the line is true length in the auxiliary view where the line is a point, so it is parallel to reference line *F1* in the front view.

Figure 28.13 shows how to solve a general-case problem of this type, where neither line appears true length in the given views. Line 1–2 is true length in the primary auxiliary view projected perpendicularly from the front view. The point view of line 1–2 lies in the secondary auxiliary view, where the distance from point 3 is true length. Because line O–3 is true length in this view, it will be parallel to

reference line 1–2 in the preceding view, the primary auxiliary view. It is also perpendicular to the true-length view of line 1–2 in the primary auxiliary view.

28.6 Shortest Distance between Skewed Lines: Line Method

Randomly positioned (nonparallel) lines are called skewed lines. The shortest distance between two skewed lines is found in the view where one of the lines appears as a point.

The shortest distance between two lines is a line perpendicular to both lines. The location of the shortest distance between lines is both functional and economical. **Figure 28.14**

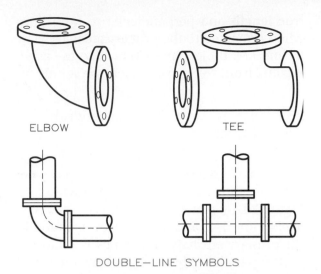

ELBOW   TEE

DOUBLE—LINE SYMBOLS

28.14 The shortest distance between two lines, the perpendicular distance, is the most economical connector between them. This also permits the use of standard fittings, 90° tees and elbows.

shows standard 90° pipe connectors (tees and elbows) that are used to make the shortest connections between skewed pipes.

Figure 28.15 illustrates how to find the shortest distance between skewed lines with the line method. Find the true length of line

3–4 and then its point view in the secondary auxiliary view, where the shortest distance is perpendicular to line 1–2. Because the distance between the lines is true length in the secondary auxiliary view, it is parallel to reference line 1–2 in the primary auxiliary view. Find point O by projection and draw OP perpendicular to line 3–4. Project the line back to the given principal views.

28.7 Shortest Distance between Skewed Lines: Plane Method

You may also determine the shortest distance between skewed lines by the plane method, which requires construction of a plane through one of the lines parallel to the other (**Figure 28.16**). The top and front views of line O–2 are parallel to their respective views of line 3–4. Therefore plane 1–2–O is parallel to line 3–4. Both lines will appear parallel in an auxiliary view, where plane 1–2–O appears as an edge.

Figure 28.17 demonstrates this principle. First construct plane 3–4–O. When its edge view is found in a primary auxiliary view, the

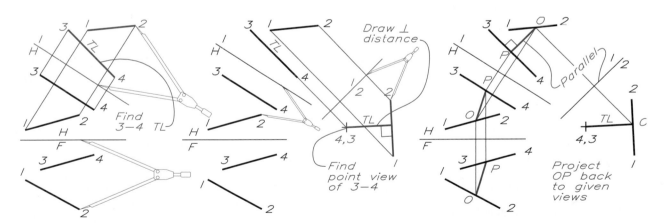

28.15 Shortest distance between skewed lines: Line method

Step 1 The shortest distance between skewed lines appears where one of the lines is a point. Find the 3–4 *TL* by projecting from the top view.

Step 2 Find the point view of line 3–4 in a secondary auxiliary view projected from the true-length view of line 3–4. The shortest distance between the lines is perpendicular to line 1–2.

Step 3 The shortest distance is true length in the secondary auxiliary view, so it must be parallel to the 1–2 reference line in the preceding view. Project line *OP* back to the given views.

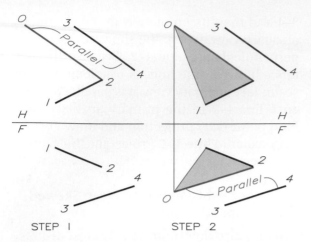

28.16 A plane through a line parallel to another line:

Step 1 Draw line O–2 parallel to line 3–4 to a convenient length.

Step 2 Draw the front view of line O–2 parallel to the front view of line 3–4. Find the length of line O–2 in the front view by projecting from the top view of O. Plane 1–2–O is parallel to line 3–4.

lines appear parallel. To find the secondary auxiliary view where both lines are true length and cross, project a secondary auxiliary view perpendicularly from these parallel lines. The crossing point is the point view of the shortest distance between the lines. That distance is

true length and perpendicular to both lines when projected to the primary auxiliary view as line *LM*. Project line *LM* back to the given top and front views to complete the solution.

28.8 Shortest Level Distance between Skewed Lines

The shortest level (horizontal) distance between two skewed lines can be found by the plane method but not by the line method. In **Figure 28.18**, plane 3–4–O is constructed parallel to line 1–2, and its edge view is found in the primary auxiliary view. Lines 1–2 and 3–4 appear parallel in this view, and the horizontal reference plane *H1* appears as an edge.

A line of sight parallel to *H1* is used, and the secondary reference line, 1–2, is drawn perpendicular to *H1*. The crossing point of the lines in the secondary auxiliary view locates the point view of the shortest horizontal distance between the lines. This line, *LM*, is true length in the primary auxiliary view and parallel to the *H1* plane. Line *LM* is projected back to the given views. As a check on construction, *LM* must be parallel to the *HF* line

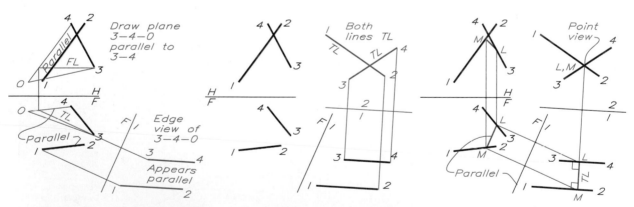

28.17 Shortest distance between skewed lines: Plane method.

Step 1 Construct a plane through 3–4 parallel to 1–2. Find plane 3–4–O as an edge by projecting it from the front view. The lines will appear parallel.

Step 2 The shortest distance is true length in the primary auxiliary view and perpendicular to both lines. Project the secondary auxiliary view perpendicularly from the lines in the primary auxiliary view to find both lines' true length.

Step 3 The crossing point of the two lines is the point view of the perpendicular distance (*LM*) between them. Project *LM* to the primary auxiliary view, where it is true length, and back to the given views.

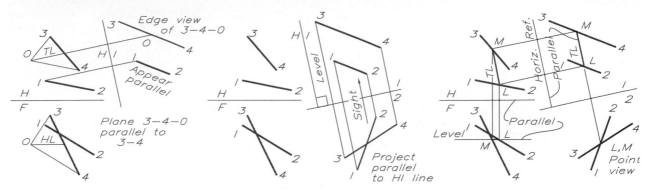

28.18 Shortest level distance between skewed lines: Plane method.

Step 1 Construct plane O–3–4 parallel to 1–2 by drawing O–4 parallel to 1–2. Find the edge view of plane O–3–4 by projecting off the top view; the lines appear parallel. Project the auxiliary view from the top view where the horizontal plane is an edge.

Step 2 Infinitely many horizontal (level) lines may be drawn parallel to reference line H1 in the auxiliary view, but the shortest one appears true length. Construct the secondary auxiliary view by projecting parallel to H1 to find the point view of the shortest level line.

Step 3 The crossing point of the lines in the secondary auxiliary view is the point view of the level connector, LM. Project LM back to the given views. LM is parallel to the horizontal reference plane in the front view, verifying that it is a level line.

in the front view, verifying that it is a level or horizontal line.

28.9 Shortest Grade Distance between Skewed Lines

Features of many applications (such as highways, power lines, or conveyors) are connected to other features at specified grades other than horizontal or perpendicular. For example, the design of the refinery installation shown in **Figure 28.19** involved the application of slopes and grades of conveyor chutes that were critical to their optimum operation.

If you need to find a 40% grade connector between two lines (**Figure 28.20**), use the plane method. To obtain an edge view of the horizontal plane from which the 40% grade is constructed, you must project the primary auxiliary view from the top view. Construct a view in which the lines appear parallel and draw a 40% grade line from the edge view of the horizontal (*H1*) by laying off rise and run units of 4 and 10, respectively. The grade line

28.19 This massive complex with numerous conveyors involved many applications of skewed lines at specified percent grades.

may be constructed in two directions from the H1 reference line, but the shortest distance is the direction most nearly perpendicular to both lines.

Project the secondary auxiliary view parallel to this 40% grade line to find the crossing point of the lines to locate the shortest con-

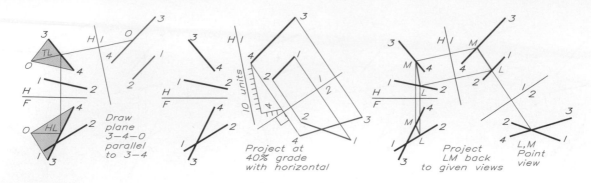

28.20 Grade distance between skewed lines:

Step 1 To find a level line or a line on a grade between two skewed lines, the primary auxiliary must be projected from the top view. Construct plane 3–4–O parallel to 1–2. Find the edge view of the plane; the lines appear parallel.

Step 2 Construct a 40% grade line from the edge view of the H1 reference line in the primary auxiliary view that is most nearly perpendicular to the lines. Project secondary auxiliary view parallel to the grade line. The shortest grade distance appears true length in the primary auxiliary.

Step 3 The point of crossing of the two lines in the secondary auxiliary view establishes the point view of the 40% grade line, *LM*. Project *LM* back to the primary auxiliary view to find its true length. Project *LM* back to the top and front views to complete the problem.

nector, *LM*. Project line *LM* back to all views; it is true length in the primary auxiliary view, where the given lines appear parallel.

The shortest distances between skewed lines—perpendicular, horizontal, and perpendicular—are true length in the view where the lines appear parallel. The plane method is the general-case method that can be used to find any shortest connector between two lines.

28.10 Angular Distance to a Line

Standard connectors used to connect pipes and structural members are available in standard angles of 90° and 45° (see Figure 28.14). Specifying these standard connectors in a design is far more economical than calling for the fabrication of specially made connectors.

In **Figure 28.21**, a line from point *O* that makes an angle of 45° with line 1–2 is to be found. Connect point *O* with the line's endpoints, 1 and 2, to find plane 1–2–*O* in the top and front views, and find its edge view in a primary auxiliary view. Find the true-size view of plane 1–2–*O* by projecting perpendicularly

from its edge view. Measure the angle of the line from point *O* in this view, where the plane of the line and point is true size.

Draw the 45° connector from point *O* toward point 2 (the low point) if it slopes downward or toward point 1 if it slopes upward. Determine the upper and lower ends of line 1–2 by referring to the front view, where the height is easily seen. Project the 45° line, *OP*, back to the given views.

28.11 Angle between a Line and a Plane: Plane Method

The angle between a line and a plane can be measured in the view where the plane appears as an edge and the line appears true length. In **Figure 28.22**, the edge view of plane 1–2–3 lies in a primary auxiliary view projected from the top view and is true size (Step 2) where the line appears foreshortened. Line *AB* is true length in a third successive auxiliary view projected perpendicularly from the secondary auxiliary view of line *AB*. The line appears true length and the plane appears as an edge in the third successive auxiliary view.

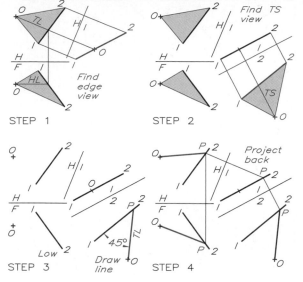

STEP 1   STEP 2

STEP 3   Draw   STEP 4
          line

28.21 A line through a point with a given angle to a line:

Step 1 Connect O to each end of the line to form plane 1–2–O in both views. Draw a horizontal line in the front view of the plane and project it to the top view where it is *TL*. Find the point view of AO and the edge view of the plane.

Step 2 Find the true size of plane 1–2–O in the primary auxiliary view by projecting perpendicularly from its edge view. Omit the outline of the plane in this view and only show line 1–2 and point O.

Step 3 Construct line OP at an angle of 45° with line 1–2. If you draw the angle toward point 2 (the low end), the line slopes downward; if toward point 1, it slopes upward.

Step 4 Project line OP back to the previous views in sequence to complete the solution.

Therefore, the angle between the line and plane can be measured here.

A tracking antenna that sends signals into outerspace illustrates examples of the need for finding the angles between lines and planes as well as other applications covered in this chapter.

28.12 Angle between a Line and a Plane: Line Method

An alternative method (not illustrated here) of finding the angle between a line and a plane is the line method. In that method, the line, rather than the plane, is the primary geometric element that is projected. The line is found as true length in a primary auxiliary view; it is found as a point in the secondary auxiliary view; and the plane is found as an edge in the third successive auxiliary view.

Because this last view was projected from a point view of the line, the line appears true length in this view where the plane appears as an edge. Project the piercing point back to the secondary, primary, top, and front views in sequence to complete the problem.

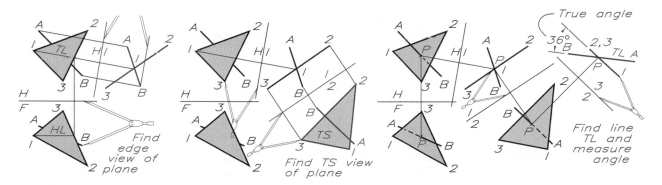

28.22 Angle between a line and plane: Plane method.

Step 1 The angle between a line and a plane is found in the view where the plane is an edge and the line is *TL*. Find the plane as an edge by projecting it from the top view.

Step 2 To find the plane's true size, project a secondary auxiliary view perpendicularly from the edge view of the plane. A view projected in any direction from a true-size plane will show the plane as an edge.

Step 3 Project a third successive auxiliary perpendicularly from AB. The line is *TL*, the plane is an edge, and the angle is *TS*. Project AB back in sequence to the given views; find the piercing points and visibility of the views.

Use size A sheets for the following problems and lay out your solutions with instruments on grid or plain paper. Each square on the grid is equal to 0.20 in. (5 mm). Label all reference planes and points in each problem with 1/8-in. letters or numbers, using guidelines.

Use the crosses marked "1" and "2" for positioning the primary and secondary reference lines. Primary reference lines should pass through "1" and secondary reference lines through "2."

1. (**Sheet 1**) Lines and planes.

(**A–B**) Find the point views of the lines.

(**C–D**) Find the angles between the planes.

2. (**Sheet 2**) True-size planes.

(**A–B**) Find the true-size views of the planes.

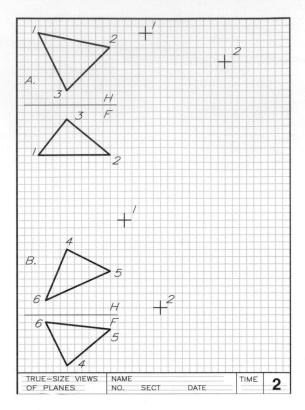

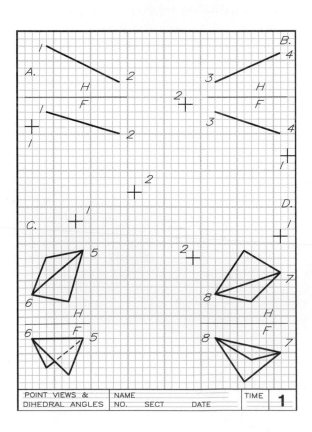

3. (**Sheet 3**) Angles between lines.

(**A–B**) Find the angles between the lines.

4. (**Sheet 4**) Distance to lines.

(**A–B**) Find the shortest distances from the points to the lines. Show this distance in all views.

5. (**Sheet 5**) Distances between lines.

(**A–B**) Find the shortest distances between the lines by the line method. Show this distance in all views.

6. (**Sheet 6**) Skewed line problems.

Find the shortest distance between the lines by the plane method. Show the line in all views. *Alternative problem*: Find the shortest horizontal distance between the two lines and show the distance in all views.

7. (**Sheet 7**) Grade distance between lines.

Find the shortest 20% grade distance between the two lines. Show this distance in all views.

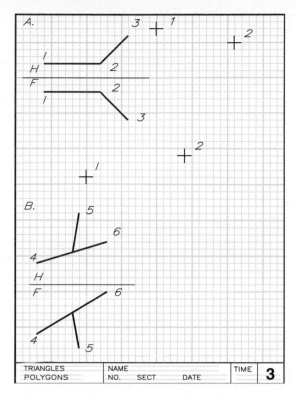

A.

B.

TRIANGLES POLYGONS	NAME		TIME	**3**
	NO. SECT DATE			

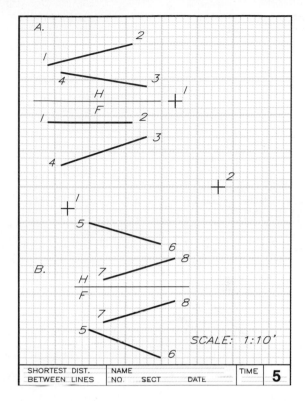

A.

B.

SCALE: 1:10'

SHORTEST DIST. BETWEEN LINES	NAME		TIME	**5**
	NO. SECT DATE			

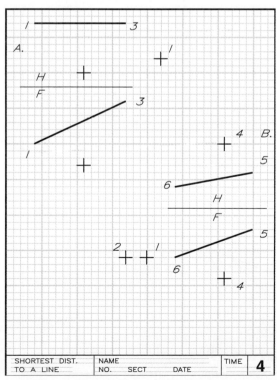

A.

B.

SHORTEST DIST. TO A LINE	NAME		TIME	**4**
	NO. SECT DATE			

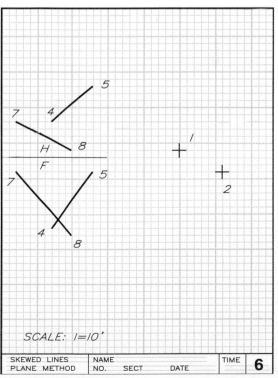

SCALE: 1=10'

SKEWED LINES PLANE METHOD	NAME		TIME	**6**
	NO. SECT DATE			

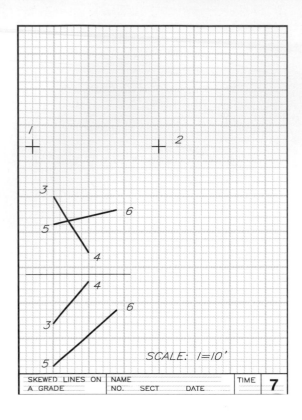

SCALE: 1=10'

| SKEWED LINES ON | NAME | | TIME | **7** |
| A GRADE | NO. SECT DATE | | | |

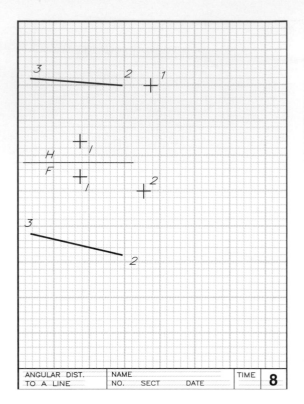

| ANGULAR DIST. | NAME | | TIME | **8** |
| TO A LINE | NO. SECT DATE | | | |

8. (**Sheet 8**) Angular distance to a line.

Find the connector from point 1 that intersects line 1–2 at 60°. Show this line in all views. Project from the top view. Scale: full size.

9. (**Sheet 9**) Angle between a line and plane. Find the angle between the line and the plane by the plane method. Show visibility in all views.

Design Problems

The following problems must be laid out with the necessary orthographic views in order to determine the specified design information. Problems can be presented on either A size or B size sheets.

(*These problems have been adapted from applications encountered at the Boeing Company.*)

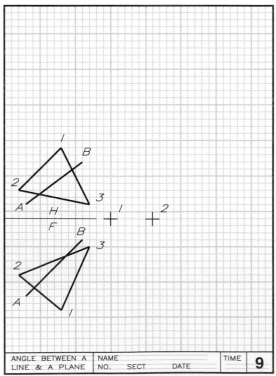

| ANGLE BETWEEN A | NAME | | TIME | **9** |
| LINE & A PLANE | NO. SECT DATE | | | |

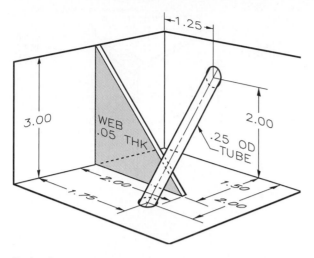

Design 1

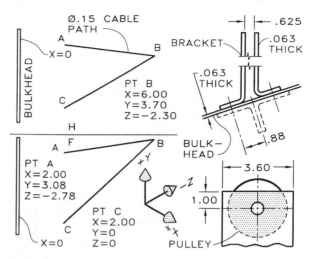

Design 3

Design 1 Lay out the necessary views in order to determine the clearance (if any) between the web and the aircraft tube. Indicate the points on the web and tube where they are closest.

Design 2 Determine the true angle between the oblique web and the proposed structural strut. Provide the necessary details to explain the design.

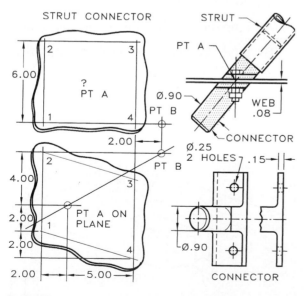

Design 2

Design 3 Points A, B, and C define the points on a control cable. Two brackets are to be designed to attach to the bulkhead (with 4–0.25 DIA bolts) and support a pulley that lies in the plane of the cable path. Lay out the necessary views to determine the following:

(A) The angle the brackets make with the bulkhead.

(B) The size of the pulley that will give the cable a clearance of .25 in. with the bulkhead.

(C) Make a flat-pattern development of both brackets.

(D) Make working drawings of the pulley, brackets, and pulley axle.

Design 4 (For only the best students!) Design a bracket so the strut clears the deck, bulkhead, and the brace by 0.10 in. The lower edge of the bracket flanges should clear the deck by 0.10. Make a developed flat pattern of your bracket design. Suggested scale: 1 = 5.

Design 5 The beam must pass through an opening to be installed in an airplane. The gauge of the door and structure is 0.125 in.

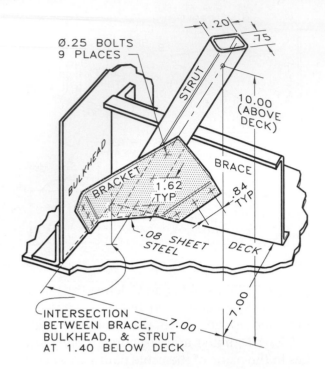

Ø.25 BOLTS
9 PLACES

1.20

.75

STRUT

10.00
(ABOVE
DECK)

BULKHEAD

BRACKET

1.62
TYP

.84
TYP

BRACE

.08 SHEET
STEEL

DECK

7.00

7.00

INTERSECTION
BETWEEN BRACE,
BULKHEAD, & STRUT
AT 1.40 BELOW DECK

Design 4

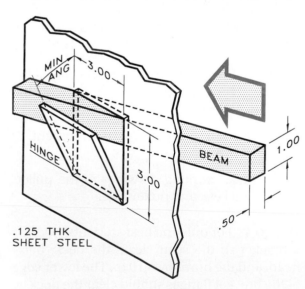

MIN
ANG

3.00

HINGE

BEAM

3.00

1.00

.50

.125 THK
SHEET STEEL

Design 5

Determine the minimum opening of the door to permit the beam to pass through the opening.

Design 6 The 0.50 OD overflow tube extends through an access hole in a bulkhead.

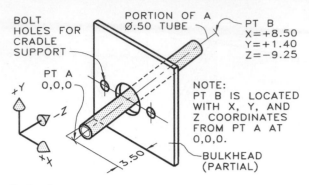

BOLT
HOLES FOR
CRADLE
SUPPORT

PORTION OF A
Ø.50 TUBE

PT B
X=+8.50
Y=+1.40
Z=−9.25

PT A
0,0,0

+Y

−Z

+X

NOTE:
PT B IS LOCATED
WITH X, Y, AND
Z COORDINATES
FROM PT A AT
0,0,0.

3.50

BULKHEAD
(PARTIAL)

Design 6

(A) Find the angle between the tube and the bulkhead and the minimum diameter of the hole.

(B) Design a support to cradle the tube and hold it to the bulkhead. Make a working drawing of your design.

Design 7 Determine the angles in the tube at B and C and make a working drawing of the tube. What is the tube's length?

Design 8 Design details of a proposed section of a duct for a nose pressurization system. The following must be determined:

(A) Find the lengths from point *A* to point *F*.

(B) Find the angles at point *C* and *E* to determine the miter for the joining ducts.

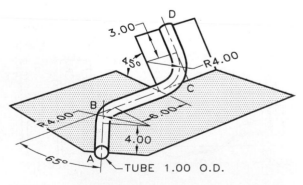

3.00

D

45°

R4.00

R4.00

B

C

6.00

4.00

65°

A

TUBE 1.00 O.D.

Design 7

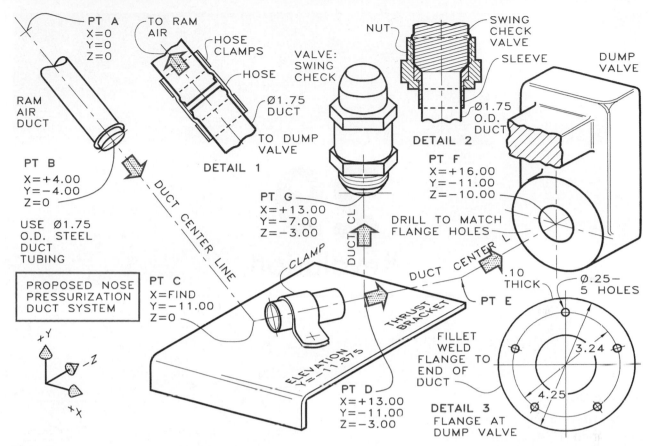

Design 8

Alternate solution: Use double-line symbols to represent the ducts and draw the connectors in the views where the ducts appear true length.

Design 9 Connector Bracket. The bracket is made of 3.00-in. wide by 0.16-in. thick flat stock steel that is bent to form the shape as dimensioned. The true angle of bend at the left side is 60°, but it cannot be measured as 60° in the front view.

(A) Construct the front and top views of the bracket.

(B) Construct a developed flat view of the bracket.

(C) Calculate the weight of the bracket. (Steel weighs 490 lbs per in.)

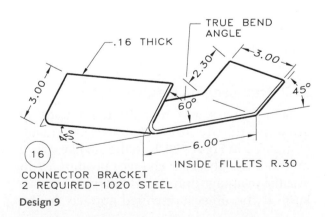

CONNECTOR BRACKET
2 REQUIRED—1020 STEEL

Design 9

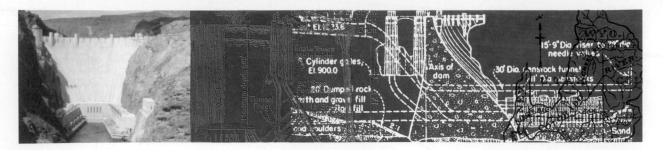

29

Revolution

29.1 Introduction

Many products and systems involve revolution, such as an automobile's front suspension, which was designed to revolve about several axes at each wheel. This design is just one of many based on the principles of revolution. Revolution is an alternative technique for solving spatial problems in orthographic views to yield a true-size view of a surface, the angle between planes or a line, and many others. Revolution was used to solve descriptive geometry problems before the introduction of the auxiliary-view method.

29.2 True-Length Lines: Front View

Auxiliary views and revolution methods of obtaining the true size of inclined surfaces are compared in **Figure 29.1**. In the auxiliary view method, the observer changes position to an auxiliary vantage point and looks perpendicularly at the object's inclined surface. In the

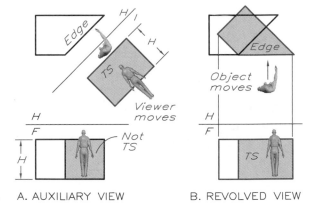

A. AUXILIARY VIEW   B. REVOLVED VIEW

29.1 Auxiliary views versus revolved views:
A The surface is found true size in an auxiliary view.
B The surface is revolved to be seen true size in the front view.

revolution method, the top view of the object is revolved about the axis until the edge view of the inclined plane is parallel to the frontal plane and perpendicular to the standard line of sight from the front view. In other words, the observer's line of sight does not change, but the object is revolved until the plane appears true size in the observer's normal line of sight.

To find a true-length line in the front view by revolution (**Figure 29.2**), revolve line AB into the frontal plane. The top view represents the circular base of a right cone, and the front view is the triangular view of a cone. Line AB′ is the outside element of the cone's frontal line and is true length in the front view.

Figure 29.3 illustrates the technique of finding line 1–2 true length in the front view. The observer's line of sight is not perpendicular to the triangle containing line 1–2 in its first position, and line 1–2 is not seen true length. When the triangle is revolved into the frontal plane, however, the observer's line of sight is perpendicular to the plane, and line 1–2 is seen true length.

In the Top View

A surface that appears as an edge in the front view may be found true size in the top view by a primary auxiliary view or by a single revolution (**Figure 29.4**). The axis of revolution is a point in the front view and true length in the

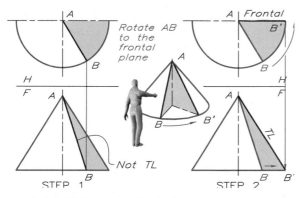

29.2 Determining true length in the front view:

Step 1 Use the top view of line *AB* as a radius to draw the base of a cone with point *A* as the apex. Draw the front view of the cone with a horizontal base through point *B*.

Step 2 Revolve the top view of line *AB* to be parallel to the frontal plane. When projected to the front view, frontal line AB′ is the outside element of the cone and is *TL*.

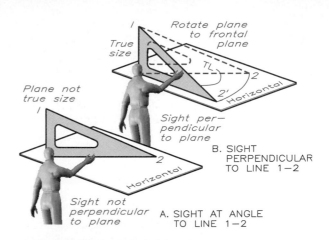

29.3 Line true length by revolution:

A Line 1–2 does not appear true length in the front view because the observer's line of sight is not perpendicular to it.

B When the triangle is revolved into the frontal plane, his line of sight is perpendicular to it and line 1–2′ is seen true length.

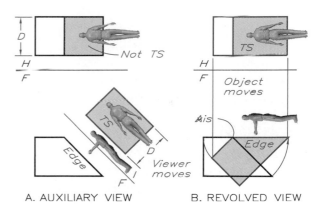

29.4 Auxiliary views versus revolved views:

A By looking perpendicular to the edge view of the plane, it appears true size in the primary auxiliary view.

B When the edge view of the plane is revolved to become horizontal, it appears true size in the top view.

top view. Revolving the edge view of the plane into the horizontal in the front view and projecting it to the top view yields the surface's true size. As in the auxiliary view method, the depth dimension, D, does not change.

In **Figure 29.5**, revolving line *CD* into the horizontal gives its true length in the top view. The arc of revolution in the front view represents the base of the cone of revolution. Line

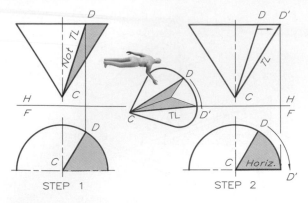

29.5 Finding the true length of a line in the top view:

Step 1 Use the front view of line *CD* as a radius to draw the base of a cone with *C* as the apex. Draw the top view of the cone with the base as a frontal plane.

Step 2 Revolve the front view of line *CD* into a horizontal position, *CD'*. When projected to the top view, *CD'* is the outside element of the cone and is true length.

CD' is true length in the top view because it is both horizontal as well as an outside element of the cone. Note that the depth in the top view does not change.

In the Profile View

In **Figure 29.6**, revolving the front view of line EF into the profile plane gives a true-length view of it. Projecting the circular view of the

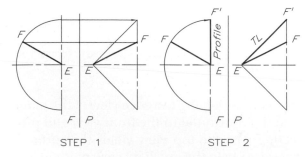

29.6 Finding true length of a line in the side view:

Step 1 Use the front view of line *EF* as a radius to draw the circular view of the base of a cone. Draw the side view of the cone with its base through *F*.

Step 2 Revolve *EF* in the frontal view to position *EF'* where it is a profile line, the outside element of the cone, and true length in the side view.

cone to the side view gives a triangular view of the cone. Because line *EF'* is a profile line, it is true length in the side view where it is the outside element of the cone.

Alternative Points of Revolution

In the preceding examples, each line is revolved about one of its ends. However, a line may be revolved about any point on its length. **Figure 29.7** shows how to find line 5–6 true length by revolving it about point *O*.

29.3 True Size of a Plane

When a plane appears as an edge in a principal view (the top view in **Figure 29.8**), it can be revolved to be parallel to the frontal reference plane. The new front view is true size when projected horizontally across from its original front view.

The combination of an auxiliary view and a single revolution finds the plane in **Figure 29.9** true size. After finding the plane as an edge by finding the point of view of a true-length line in the plane, revolve the edge view to be parallel to the *F1* reference line. To find the true size of the plane, project the original points (1, 2, and 3) in the front view parallel to the F1 line to intersect the projectors from 1′

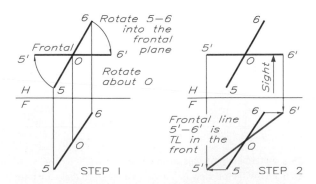

29.7 True-length views of lines may be obtained by revolving them about any point on the lines, not just their endpoints. Line 5–6 is revolved about its midpoint in the top view until it's parallel to the frontal plane and is true length in the front view.

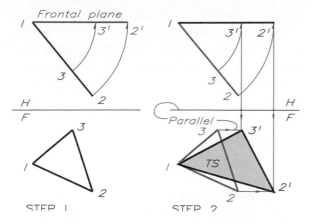

STEP 1   STEP 2

29.8 Determining true size of a plane:

Step 1 Revolve the edge view of the plane until it is parallel to the frontal plane.

Step 2 Project points 2' and 3' to the horizontal projectors from points 2 and 3 in the front view.

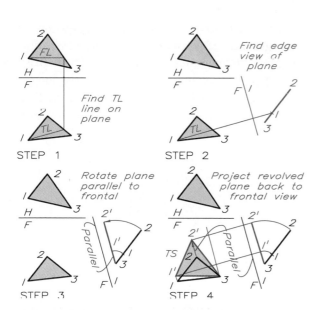

STEP 1   STEP 2

STEP 3   STEP 4

29.9 Obtaining true size of a plane by revolution:

Step 1 To find the edge view of the plane by revolution, draw a frontal line on the plane that is true length in the front view.

Step 2 Find the edge view of the plane by finding the point view of the frontal line.

Step 3 Revolve the edge view of the plane until it is parallel to the F1 reference line.

Step 4 Project the revolved points 1' and 2' to the front view to the projectors from points 1 and 2 that are parallel to the F1 reference line.

and 2'. The true size of the plane may also be found by projecting from the top view to find the edge view.

By Double Revolution

The edge view of a plane can be found by revolution without using auxiliary views (**Figure 29.10**). Draw a frontal line on plane 1–2–3, and project it to the front view where it is true length. Revolve the plane until the true-length line is vertical in the front view. The true-length line projects as a point in the top view; therefore the plane appears as an edge in this view. Projectors from points 2 and 3 from the top view are parallel to the *HF* reference line.

A second revolution, called a double revolution, positions this edge view of the plane parallel to the frontal plane as shown in Step 1 of **Figure 29.11**. Projecting the top views of points 1″ and 2″ to the front view gives a true-size plane 1″–2″–3″ (Step 2). We could have shown this second revolution in **Figure 29.10**, but it would have resulted in overlapping views, making observation of the separate steps difficult.

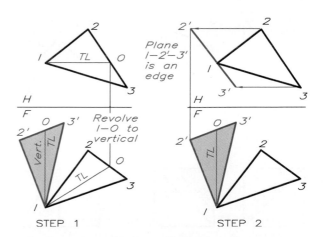

STEP 1   STEP 2

29.10 Finding the edge view of a plane:

Step 1 Draw a true-length frontal line on the plane. Revolve the front view until the *TL* line is vertical.

Step 2 The true-length line, 1–0, is vertical and appears as a point in the top view, and the plane appears as edge 1–2'–3'.

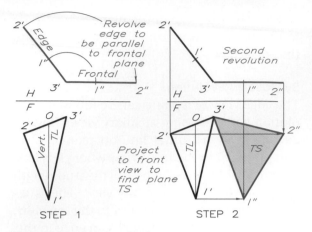

29.11 Finding true size of a plane:

Step 1 The plane in the top view of Figure 29.10 is revolved to a position parallel to the frontal line.

Step 2 Project points 1″ and 2″ to the front view to intersect with the horizontal projectors from the original points 1′ and 2′. Plane 1″–2″–3″ is true size in this view.

Figure 29.12 shows how to use double revolution to find the true size of the oblique plane (1–2–3) of the object. Revolve the true-length line 1–2 on the plane in the top view until it is perpendicular to the frontal plane. Line 1–2 appears as a point in the front view, and the plane appears as an edge. This revolution changes the width and depth, but not the

height. Then, revolve the edge view of the plane into a vertical position parallel to the profile plane. To find the plane in true size, project to the profile view, where the depth remains unchanged but the height is greater.

29.4 Angle between Planes

The angle between the planes of the nuclear detection satellite (**Figure 29.13**) may be found by revolution instead of by auxiliary views. The application of this geometry was an essential principle in designing the satellite.

In **Figure 29.14**, finding the dihedral angle involves drawing its edge view perpendicular to the line of intersection and projecting the plane of the angle to the front view. When the edge view of the angle is revolved into a frontal plane and projected to the front view, the angle appears true size and can be measured.

Figure 29.15 shows how to solve a similar problem. Here, the line of intersection does not appear true length in the given views; therefore an auxiliary view is needed to find its true length. Draw the plane of the dihedral angle as an edge perpendicular to the true-

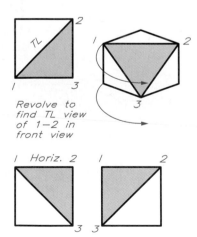

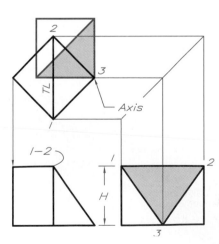

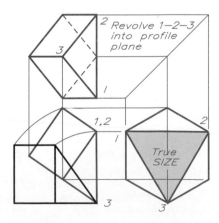

29.12 True size of a plane by double revolution:

Required Find the true size of the plane by revolution.

Step 1 Line 1–2 is horizontal in the frontal view and true length in the top view. Revolve the top view so that line 1–2 appears as a point in the front view.

Step 2 Plane 1–2–3 is an edge; revolve it into a vertical position in the front view to find its true size in the side view. The depth does not change.

29.13 The principles of revolution can be used to determine the angles between the planes of the Saturn S-IVB during its design. (Courtesy of National Aeronautics and Space Administration.)

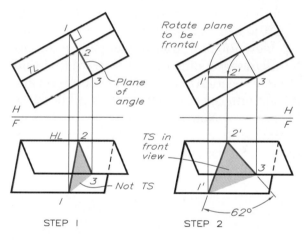

29.14 Finding the angle between planes:

Step 1 Draw the edge of the angle perpendicular to the TL line of intersection between the planes in the top view; project it to the front view.

Step 2 Revolve the edge view of the plane of the angle to position angle 1′–2′–3 in the top view parallel to the frontal plane. Project this angle to the frontal view, where it is true size.

length line of intersection. Project the edge view of plane 1–2–3 to the top view. Then revolve the edge view of plane 1–2–3 in the primary auxiliary view until it is parallel to the *H1* reference line. Project the revolved edge view of the angle back to the top view, where it is true size.

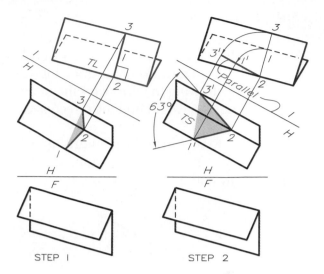

29.15 Determining the angle between oblique planes:

Step 1 Find the true-length view of the line of intersection by projecting perpendicularly from its top view. Draw the edge of the angle perpendicular to the true length of the line of intersection and project it to the top view.

Step 2 Revolve the edge view of the plane of the angle (plane 2–1′–3′) until it is parallel to the *H1* reference line so that it appears true size in the top view.

29.5 Determining Direction

To solve more advanced problems of revolution, you must be able to locate the basic directions of up, down, forward, and backward in any given view. In **Figure 29.16A**, directional arrows in the top and front views identify the directions of backward and forward.

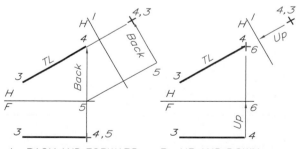

A. BACK AND FORWARD   B. UP AND DOWN

29.16 The directions of backward, forward, up, and down can be identified in the given views with arrows pointing in these directions. Directional arrows can be projected to successive auxiliary views. This drawing shows the directions of (A) backward and (B) up.

Pointing backward in the top view, line 4–5 appears as a point in the front view. Projecting arrow 4–5 to the auxiliary view, as you would any other line, determines the direction of backward. By drawing the arrow on the other end of the line, you would find the direction of forward.

Locate the direction of up in **Figure 29.16B** by drawing line 4–6 in the direction of up in the front view and as a point in the top view.

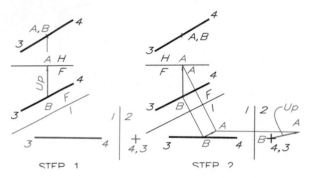

29.17 Direction in a secondary auxiliary view:

Step 1 To find the direction of up in the secondary auxiliary view, draw arrow AB pointing up in the front view. It appears as a point in the top view.

Step 2 Project arrow AB to the primary and secondary auxiliary views to show the direction of up.

Then find the arrow in the primary auxiliary by the usual projection method. The direction of down is in the opposite direction.

You find the location of directions in secondary auxiliary views in the same way. To determine the direction of up in **Figure 29.17,** begin with an arrow that points up in the front view and appears as a point in the top view. Project the arrow *AB* from the front view to the primary auxiliary view and then to a secondary auxiliary view to show the direction of up. Identify the other directions in the same way by beginning with the two principal views of a known direction.

29.6 Revolution: Point about an Axis

In **Figure 29.18**, point *O* is to be revolved about axis 3–4 to its most forward position. Find the axis as a point in the primary auxiliary view and draw the circular path of revolution. Draw the direction of forward and find the new location of point *O* at *O′*. Project back through the successive views to find point *O′* in each view. Note that point *O′* lies on the line in the front view, verifying that point *O′* is in its most forward position.

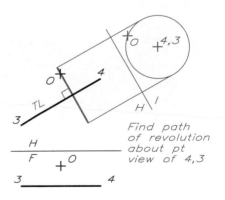
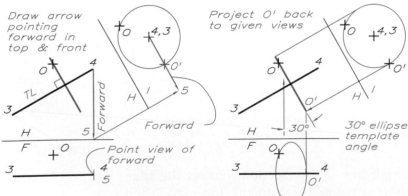

29.18 A point about an axis:

Step 1 To rotate *O* about axis 3–4 to its most forward position, find the point view of 3–4. The path of revolution is a circle in the auxiliary view and an edge perpendicular to 3–4 in the top view.

Step 2 Locate the most forward position of point *O* by drawing an arrow pointing forward in the top view that appears as a point in the front view. Find the arrow, 4–5, in the auxiliary view to locate point *O′* on the circular path of revolution.

Step 3 Project *O′* back to the given views. The path of revolution appears as an ellipse in the front view because the axis is not true length in this view. Draw a 30° ellipse because this is the angle of your line of sight with the circular path in the front view.

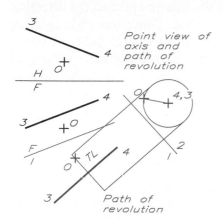

29.19 A point about an oblique axis:

Step 1 To rotate *O* about axis 3–4 to its highest position, find the point view of 3–4 and draw the circular path. The path of revolution is perpendicular to 3–4 in the primary auxiliary views.

Step 2 To locate the highest position on the path of revolution, draw arrow 3–5 pointing up in the front view and project it to the secondary auxiliary view to find *O'*.

Step 3 Project point *O'* back to the given views by transferring the dimensions J and D with your dividers. The highest point lies over the line in the top view. The path of revolution is elliptical where the axis is not true length.

In **Figure 29.19**, an additional auxiliary view is needed in order to rotate a point about an axis because axis 3–4 is not true length in the given views. You must find the true length of the axis before you can find it as a point in the secondary auxiliary view, where the path of revolution appears as a circle. Revolve point *O* into its highest position, *O'*, and locate the up arrow, 3–5, in the secondary auxiliary view. Project back to the given views to locate *O'* in each view. Its position in the top view is over the axis, which verifies that the point is at its highest position.

The paths of revolution appear as edges when their axes are true length and as ellipses when their axes are not true length. The angle of the ellipse template for drawing the ellipse in the front view is the angle the projectors from the front view make with the edge view of the revolution in the primary auxiliary view. To find the ellipse in the top view, project an auxiliary view from the top view to obtain the path of revolution as an edge perpendicular to the true-length axis.

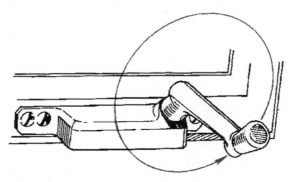

29.20 The handcrank on a casement window is an example of a problem solved by applying revolution principles. The handle must be properly positioned so as not to interfere with the window sill or wall.

The handcrank of a casement window (**Figure 29.20**) is an example of the application of revolution techniques. The designer must determine the clearances between the sill and the window frame when designing the crank in order for it to operate properly.

A Right Prism

The coal chute shown in **Figure 29.21** conveys coal continuously between two buildings. The sides of the enclosed chute must be vertical

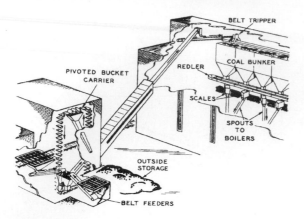

29.21 A conveyor chute must be installed so that two edges of its right section are vertical for the conveyors to function properly. (Courtesy of Stephens-Adamson Manufacturing Company.)

point view of the axis and project the direction of up to this view. Draw the right section about the axis so that two of its sides are parallel to the up arrow. Find the right section in the other views. Then construct the sides of the chute parallel to the axis. The bottom of the chute's right section will be horizontal and properly positioned for conveying coal.

29.7 A Line at Specified Angles

In **Figure 29.23**, a line is to be drawn through point O that makes angles of 35° with the frontal plane and 44° with the horizontal plane and slopes forward and down. First, draw the cone containing elements making 35° with the frontal plane and then the cone with elements making 44° with the horizontal plane. The length of the elements of both cones must be equal so that the cones will intersect with equal elements. Finally, find lines O–1 and O–2, which are elements that lie on each cone and make the specified angles with the principal planes.

and the bottom of the chute's right section must be horizontal. Design of the chute required application of the technique of revolving a prism about its axis.

In **Figure 29.22**, the right section is to be positioned about centerline AB so that two of its sides will be vertical. To do so, find the

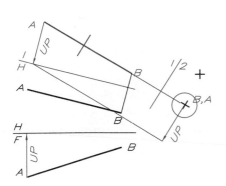 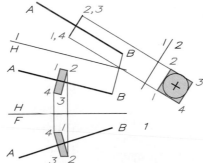 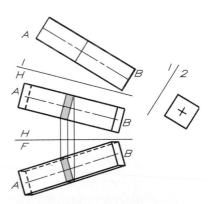

29.22 A prism about its axis:

Step 1 To draw a square chute with two of its sides vertical, find the point view of centerline AB in the secondary auxiliary view. Draw a circle about the axis with a diameter equal to the square section. Draw a vertical arrow in the front and top views; project it to the secondary auxiliary view to show the direction of vertical.

Step 2 Draw the right section, 1–2–3–4, in the secondary auxiliary view with two sides parallel to the vertical directional arrow. Project this section back to the previous views by transferring measurements with dividers. Locate the edge view of the section anywhere along AB in the primary auxiliary view.

Step 3 Draw the edges of the prism through the corners of the right section parallel to AB in all views. Terminate the ends of the prism in the primary auxiliary view where they appear as edges perpendicular to the center line. Project the corner points of the ends to the top and front views to find the ends.

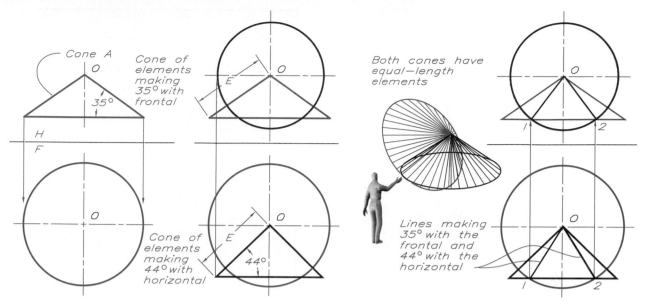

29.23 A line at specified angles:

Step 1 To draw lines at angles of 35° and 44° with the frontal and horizontal, respectively, draw a cone in the top view with outside elements at 35° with the frontal plane. Draw the circular view of the cone in the front with *O* as the apex. All elements of the cone are 35° with the frontal plane.

Step 2 Draw a second cone in the front view with outside elements that make an angle of 44° with the horizontal plane. Draw the elements of this cone equal in length to element *E* of cone *A*. All elements of cone B are 44° with the horizontal plane.

Step 3 Because the elements are equal in length, two elements lie on the surface of each cone: lines *O*–1 and *O*–2. Locate points 1 and 2 at the point where the bases of the cones intersect in both views. These lines slope forward and down from point *O* at the specified angles.

Problems

Lay out the problems on size A sheets with instruments. Each grid is equal to 0.20 in. (5mm). Label all reference planes and points in each problem with 1/8-in. letters or numbers.

Primary and secondary reference lines should pass through the crosses marked "1" and "2", respectively.

1. (**Sheet 1**) True-length lines.

  (**A–B**) Find the true-length views of the lines in their front views by revolution.

  (**C–D**) Find the true-length views of the lines in their top views by revolution.

2. (**Sheet 2**) True-size planes.

  (**A–B**) By revolution, find the true-size views of

plane 1–2–3 in the front view and plane 4–5–6 in the top view.

  (**C**) Use an auxiliary view projected from the top view and one revolution to find the true-size view of plane 7–8–9.

3. (**Sheet 3**) Angles between planes.

  (**A–B**) Find the angles between the planes by revolution. Show construction.

4. (**Sheet 4**) Revolution of a point.

  (**A**) Show point *O* revolved about the axis into its highest position.

  (**B**) Show point *O* revolved about the axis into its most forward position.

5. (**Sheet 5**) Chute design.

Construct a chute from *A* to *B* with the longer sides of its cross-section being vertical.

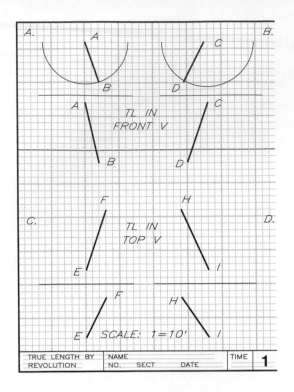

A.

A

B

TL IN
FRONT V

A

B

B.

C

D

C

D

C.

F

H

TL IN
TOP V

E

I

D.

F

H

E SCALE: 1=10' I

TRUE LENGTH BY REVOLUTION	NAME		TIME	**1**
	NO. SECT DATE			

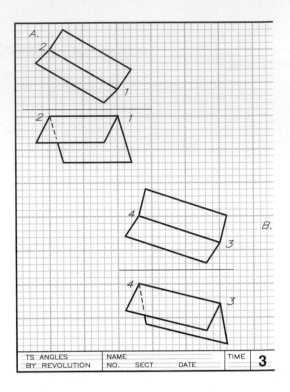

A.

2

1

3

1

2

C.

8

7

9

8

7

9

B.

4

6

5

5

6

TRUE SIZE BY REVOLUTION	NAME		TIME	**2**
	NO. SECT DATE			

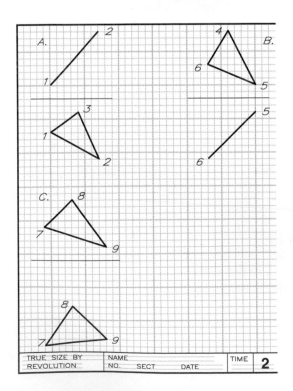

A.

2

1

2

1

4

3

4

3

B.

TS ANGLES BY REVOLUTION	NAME		TIME	**3**
	NO. SECT DATE			

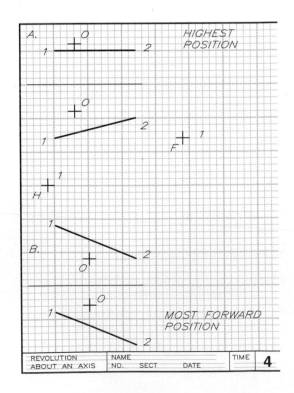

A.

O

1

2

HIGHEST POSITION

O

1

2

F

1

H

1

1

B.

2

O

O

1

2

MOST FORWARD POSITION

REVOLUTION ABOUT AN AXIS	NAME		TIME	**4**
	NO. SECT DATE			

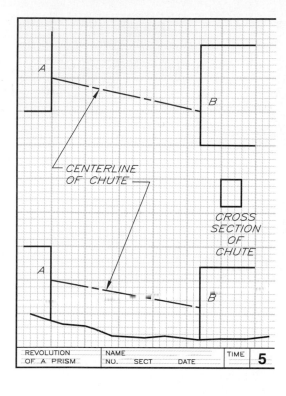

CENTERLINE
OF CHUTE

CROSS
SECTION
OF
CHUTE

A

B

A

B

REVOLUTION OF A PRISM	NAME			TIME	5
	NO.	SECT	DATE		

6. (**Sheet 6**) Lines at Specified Angles. Draw the views of the line that is 3.2 inches long and makes a 30° angle with the frontal plane and a 52° angle with the horizontal plane.

7. (**Design 1**) Aircraft Valve. The shut-off valve must be located in the pipe at the position shown, but it must be revolved in order for path of its handle to clear the bulkhead by 1.00 inch. Lay out the necessary views to solve the problem and show your construction. What is the valve's rotation from its lowest position and the distance B that estabilishes the limit of the handle's path?

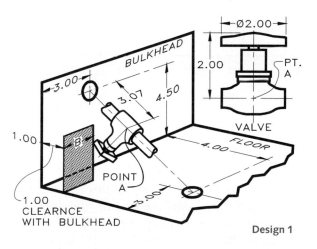

Design 1

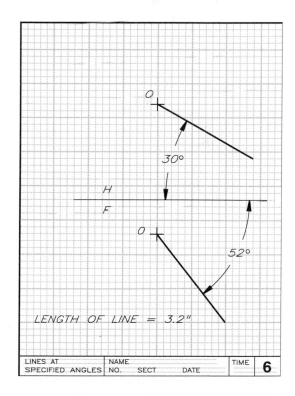

LINES AT SPECIFIED ANGLES	NAME			TIME	6
	NO.	SECT	DATE		

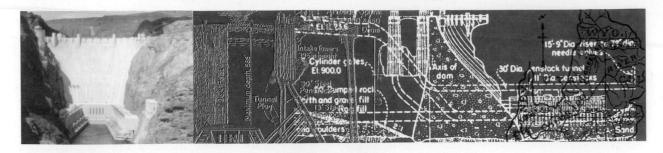

30

Vector Graphics

30.1 Introduction

Design of a structural system requires analysis of each member to determine the loads they must support and whether those loads are in tension or compression. Forces may be represented graphically by vectors and their magnitudes and directions determined in 3D space. Graphical methods are useful in the solution of vector problems as alternatives to conventional trigonometric and algebraic methods. Quantities such as distance, velocity, and electrical properties also may be represented as vectors for a graphical solution.

30.2 Definitions

To help you more easily understand the discussion of vectors in this chapter, the following terms are defined:

Force: a push or pull tending to produce motion. All forces have (1) magnitude, (2) direction, and (3) a point of application. The

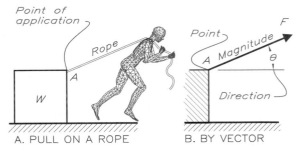

30.1 A force applied to an object (A) may be represented by vector depicting the magnitude and direction of the force (B).

person shown pulling the rope in **Figure 30.1A** is applying a force to the weight W.

Vector: a graphical representation of a force drawn to scale and depicting magnitude, direction, and point of application. The vector in **Figure 30.1B** represents the force applied through the rope to pull the weight W.

Magnitude: the amount of push or pull represented by the length of the vector line, usually measured in pounds or kilograms.

Direction: the inclination of a force (with respect to a reference coordinate system) indicated by a line with an arrow at one end.

Point of application: the point through which the force is applied on the object or member (point A in **Figure 30.1A**).

Compression: the state created in a member by forces that tend to shorten it. Compression is represented by the letter C or a plus sign (+).

Tension: the state created in a member by pulling forces that tend to stretch it. Tension is represented by the letter T or a minus sign (−).

System of forces: the combination of all forces acting on an object as shown (forces A, B, and C in **Figure 30.2**).

Resultant: a single force that can replace all the forces of a force system and have the same effect (force R in **Figure 30.2**).

Equilibrant: the opposite of a resultant; the single force that can be used to counterbalance all forces of a force system.

Components: separate forces that, if combined, would result in a single force; forces A and B are components of resultant $R1$ in **Figure 30.2**.

Space diagram: a diagram depicting the physical relationship between structural members, as given in **Figure 30.2**.

Vector diagram: a diagram of vectors representing the forces in a system and used to solve for unknown vectors in the system.

Metric units: standard units of weights and measures. The kilogram (kg) is the unit of mass (load), and one kilogram is approximately 2.2 pounds.

30.3 Coplanar, Concurrent Forces

When several forces, represented by vectors, act through a common point of application, the system is **concurrent**. In **Figure 30.2**, vectors A, B, and C act through a single point; therefore this system is concurrent. When all vectors lie in the same plane, the system is **coplanar** and only one view is necessary to show them true length.

The **resultant** is the single vector that can replace all forces acting on the point of application. Resultants may be found graphically by (1) the **parallelogram method** and (2) the **polygon method**.

An **equilibrant** has the same magnitude, orientation, and point of application as the resultant in a system of forces, but in the opposite direction. The resultant of the system of forces shown in **Figure 30.3** is balanced by the equilibrant applied at point O, thereby causing the system to be in equilibrium.

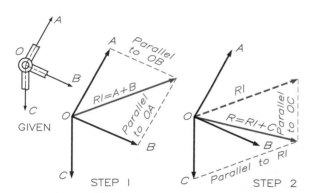

30.2 Resultant by the parallelogram method:

Step 1 Draw a parallelogram with its sides parallel to vectors A and B. The diagonal $R1$ is the resultant of forces A and B.

Step 2 Draw a parallelogram using vectors $R1$ and C to find diagonal R, or the overall resultant that can replace forces A, B, and C.

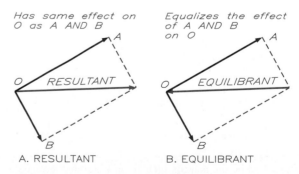

30.3 The (A) resultant and (B) equilibrant are equal in all respects except in direction (shown by arrowheads).

Resultant: Parallelogram Method

In **Figure 30.2**, the vectors lie in the same plane, act through a common point, and are scaled to their known magnitudes. Use of the parallelogram method to determine resultants requires that the vectors be drawn to scale. Vectors A and B form two sides of a parallelogram. Constructing parallels to these vectors completes the parallelogram. Its diagonal, $R1$, is the resultant of forces A and B; that is, resultant $R1$ is the vector sum of vectors A and B.

Replaced by $R1$, vectors A and B now may be disregarded. Resultant $R1$ and vector C are two sides of a second parallelogram. Its diagonal, R, is the vector sum of $R1$ and C and the resultant of the entire system. Resultant R may be thought of as the only force acting on the point, thereby simplifying further analysis.

Resultant: Polygon Method

Figure 30.5 shows the same system of forces, but here the resultant is determined by the polygon method. Again, the vectors are drawn to scale but in this case head-to-tail, in their true directions, to form the polygon. The vectors are laid out in a clockwise sequence beginning with vector A. The polygon does not close, so the system is not in equilibrium but tends to be in motion. The resultant (from the tail of vector A to the head of vector C) closes the polygon.

30.4 Noncoplanar, Concurrent Forces

When vectors lie in more than one plane of projection, they are noncoplanar, requiring 3D views for analysis of their spatial relationships. The resultant of a system of noncoplanar forces may be obtained by the parallelogram method if their projections are given in two adjacent orthographic views.

Resultant: Parallelogram Method

In **Figure 30.5**, vectors 1 and 2 were used to construct the top and front views of a parallelogram and its diagonal $R1$ in both views. The front view of $R1$ must be an orthographic projection of its top view.

Then resultant $R1$ and vector 3 are resolved to form the overall resultant in both views. The top and front views of the resultant must project orthographically. The overall resultant

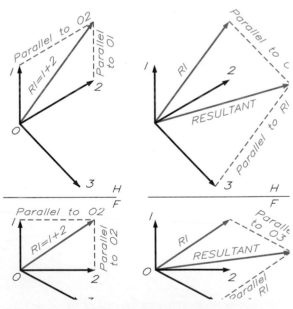

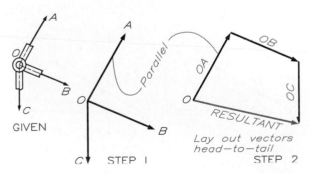

30.4 The resultant of a coplanar, concurrent system may be determined by the polygon method, in which the vectors are drawn head-to-tail. The vector that closes the polygon is the resultant.

30.5 Resultant by the parallelogram method:

Step 1 Use vectors 1 and 2 to construct a parallelogram in the top and front views. Diagonal $R1$ is the resultant of vectors 1 and 2.

Step 2 Use vectors 3 and $R1$ to construct a second parallelogram to find the overall resultant, R.

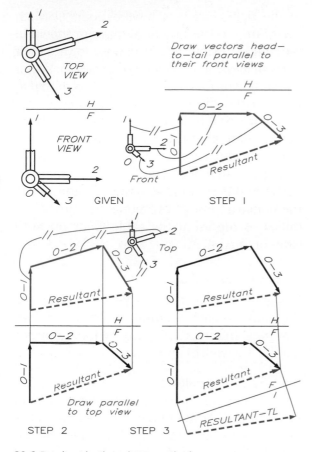

30.7 The loads in the members of this crane can be determined by vector graphics as a coplanar system in equilibrium. (Courtesy of Pacific Hoist Company.)

30.6 Resultant by the polygon method:

Step 1 Lay off each vector head-to-tail parallel to the given view and find the front view of the resultant.

Step 2 Draw the same vectors head-to-tail in the top view, a 3D polygon projected above the front view.

Step 3 The resultant is found true length in an auxiliary view projected from the front view.

replaces vectors 1, 2, and 3. However, it is an oblique line, so an auxiliary view (**Figure 30.6**) or revolution must be used to obtain its true length.

Resultant: Polygon Method

Figure 30.6 shows the solution of the same system of forces for the resultant by the polygon method. **Each vector is laid off head-to-tail clockwise, beginning with vector 1 in the front view.**

Then the vectors are projected orthographically from the front view to the top view of the vector polygon. The vector polygon does not close, so the system is not in equilibrium. In both views, the resultant (from the tail of vector 1 to the head of vector 3) closes the polygon. However, the resultant is an oblique line, requiring an auxiliary view to obtain its true length.

30.5 Forces in Equilibrium

The manufacturing hoist shown in **Figure 30.7** can be analyzed graphically to determine the loads carried by each member and cable since it is a coplanar, concurrent structure in equilibrium. **A structure in equilibrium is one that is static with no motion taking place; they balance each other.**

The coplanar, concurrent structure depicted in **Figure 30.8** is designed to support a load of $W = 165$ kg. The maximum loading of each structural member determines the material and size of the members to be used in the design.

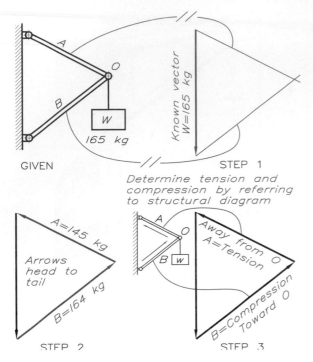

GIVEN

Known vector W=165 kg

STEP 1

Determine tension and compression by referring to structural diagram

STEP 2

Arrows head to tail

A=145 kg

B=164 kg

STEP 3

Away from O
A=Tension

B=Compression Toward O

30.8 Coplanar forces in equilibrium:

Find the forces in the structural members supporting the 165 kg load.

Step 1 Draw the load of 165 kg as a vector. Draw vectors A and B parallel to their directions from the ends of the 165 kg load and extend them to their intersection.

Step 2 The direction of the 165 kg vector is known. Draw the arrows on the polygon head-to-tail to find the directions of A and B.

Step 3 Vector A points away from point O when transferred to the structural diagram and thus is in tension. Vector B points toward point O and is in compression.

A single view of a vector polygon in equilibrium allows you to find only two unknown values. (Later, we show how to solve for three unknowns by using descriptive geometry.) Lay off the only known force, $W = 165$ kg, parallel to its given direction (here pointing vertically downward). Then draw the unknown forces A and B parallel to the supports to form the vector polygon and scale, or calculate, the magnitude of these forces.

Analyze vectors A and B to determine whether they are in tension or compression and thus find their direction. Vector B points upward

to the right, which is toward point O when transferred to the structural diagram shown in the small drawing. Vectors that act toward their point of application are in compression. Vector A points away from point O when transferred to the structural diagram and is in tension.

Figure 30.9 is a similar example involving determination of the loads in the structural members caused by the weight of 110 pounds acting through a pulley. The only difference between this solution and the previous one is the construction of two equal vectors at the outset to represent the cable loads on both sides of the pulley.

30.6 Coplanar Truss Analysis

Designers use vector polygons to determine the loads in each member of a truss by two graphical methods: (1) joint-by-joint analysis and (2) Maxwell diagrams.

Joint-by-Joint Analysis

In the **Fink truss** shown in **Figure 30.10**, 3000-lb loads are applied at its joints. The exterior forces on the truss are labeled with letters placed between them, and numerals are placed between the interior members. Each vector is referred to by the number on each of its sides clockwise about its joint. For example, the vertical load at the left is denoted *AB*, with *A* at the tail and *B* at the head of the vector. This method of designating forces is called **Bow's notation**.

First analyze the joint at the left end with a reaction of 4500 lb. When you read clockwise about the joint, the force is *EA*, where *E* is the tail and *A* is the head of the vector. Continuing clockwise, the next forces are *A*–1 and 1–*E*, which close the polygon at *E*, the beginning letter. Place arrowheads in a head-to-tail sequence beginning with the known vector *EA*.

Determine tension and compression by relating the directions of each vector to the original joint. For example, *A*–1 points toward the joint and is in compression, whereas 1–*E*

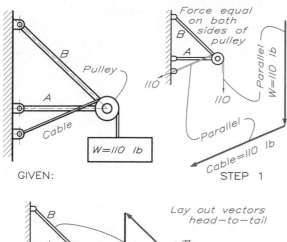

GIVEN:

Force equal on both sides of pulley

STEP 1

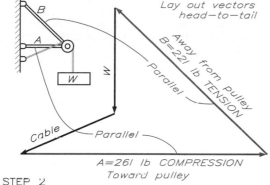

STEP 2

Lay out vectors head-to-tail

B=221 lb TENSION Away from pulley Parallel

A=261 lb COMPRESSION Toward pulley

30.9 Forces in equilibrium: (pulley application).

Find the forces in the members.

Step 1 The force in the cable is equal to 110 lb on both sides of the pulley. Draw these two forces as vectors head-to-tail and parallel to their directions in the space diagram.

Step 2 Draw A and B head-to-tail to close the polygon. Vector A points toward the point of application and thus is in compression. Vector B points away from the point and is in tension.

points away and is in tension. The truss is symmetrical and equally loaded, so the loads in the members on the right will be equal to those on the left.

Analyze the other joints in the same way. The directions of the vectors are opposite at each end. For example, vector A–1 is toward the left in Step 1 and toward the right in Step 2.

Maxwell Diagrams

The **Maxwell diagram** is virtually the same as the joint-by-joint analysis, with the exception that the polygons overlap, with some vectors common to more than one polygon. In **Figure 30.11** (Step 1) the exterior loads are laid out head-to-tail in clockwise sequence—*AB, BC,*

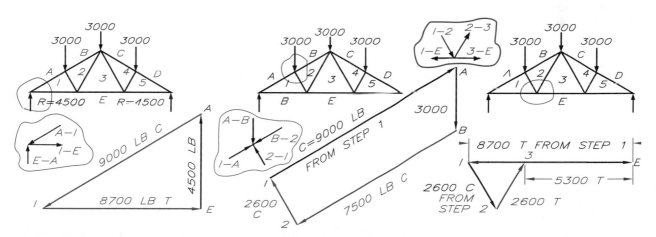

30.10 Truss analysis (joint-by-joint):

Step 1 Label the truss with letters between the exterior loads and numbers between interior members. Analyze the left joint with two unknowns, *A–1* and *1–E*. Draw vectors *A–1* and *1–E* parallel to their directions from both ends of *EA* in a head-to-tail sequence.

Step 2 Use vector 1–*A* and load *AB* from Step 1 to find *B–2* and 2–1. Draw 1–*A* first, then *AB*, and draw *B–2* and 2–1 to close the polygon, moving clockwise about the joint. A vector pointing toward the point of application is in compression. A vector pointing away from the point of application is in tension.

Step 3 Lay out vectors *E–1* and *1–2* from the preceding steps. Vectors 2–3 and 3–*E* close the polygon and are parallel to their directions in the space diagram. Vectors 2–3 and 3–*E* point away from the point of application and thus are in tension.

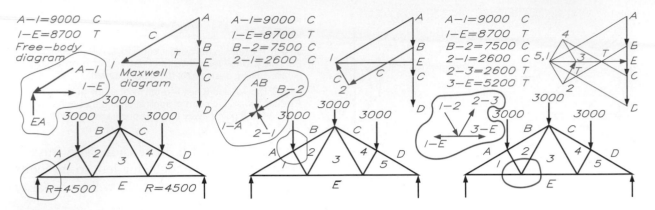

30.11 Truss analysis: Maxwell diagram.

Step 1 Label the outer spaces with letters and the internal spaces with numbers. Draw the loads head-to-tail in a Maxwell diagram; sketch a free-body diagram of the first joint. Use *EA*, *A*–1, and 1–*E* (head-to-tail) to draw a vector diagram. *A*–1 is in compression because it points toward the joint; 1–*E* is in tension because it points away from it.

Step 2 Sketch the next joint to be analyzed. Because *AB* and *A*–1 are known, only 2–1 and *B*–2 are unknown. Draw them parallel to their direction (head-to-tail) in the Maxwell diagram using the previously found vector. Vectors *B*–2 and 2–1 are in compression, since each points toward the joint. Vector *A*–1 becomes vector 1–*A* when read in a clockwise direction.

Step 3 Sketch a free-body diagram of the next joint to be analyzed, where the unknowns are 2–3 and 3–*E*. Draw their vectors in the Maxwell diagram parallel to their given members to find point 3. Vectors 2–3 and 3–*E* are in tension because they point away from the joint. Repeat this process to find the vectors on the opposite side.

CD, *DE*, and *EA*—with a letter placed at each end of each vector. The forces are parallel so this force diagram is a vertical line.

Vector analysis begins at the left end where the force *EA* of 4500 lb is known. A free-body diagram is sketched to isolate this joint. The two unknowns, *A*–1 and 1–*E*, are drawn parallel to their directions in the truss, with *A*–1 beginning at point *A*, 1–*E* beginning at point E, and both extended to point 1.

Because resultant *EA* points upward, *A*–1 must have its tail at *A* and its direction toward point 1. The free-body diagram shows that the direction is toward the point of application, which means that *A*–1 is in compression. Vector 1–*E* points away from the joint, which means that it is in tension. The vectors are coplanar and may be scaled to determine their magnitudes.

In Step 2, where vectors 1–*A* and *AB* are known, the unknown vectors, *B*–2 and 2–1,

may be determined. Vector *B*–2 is drawn parallel to its structural member through point *B* in the Maxwell diagram, and the line of vector 2–1 is extended from point 1 to intersect with *B*–2 at point 2. The arrows of each vector are drawn head-to-tail. Vectors *B*–2 and 2–1 point toward the joint in the free-body diagram, and therefore are in compression.

In Step 3, the next joint is analyzed to find the forces in 2–3 and 3–*E*. The truss and its loading are symmetrical, so the Maxwell diagram will be symmetrical when completed.

If the last force polygon in the series does not close perfectly, an error in construction has occurred. A slight error may be disregarded, as a rounding error may be disregarded in mathematics. Arrowheads are unnecessary and usually are omitted on Maxwell diagrams because each vector will have the opposite direction when applied to a different joint.

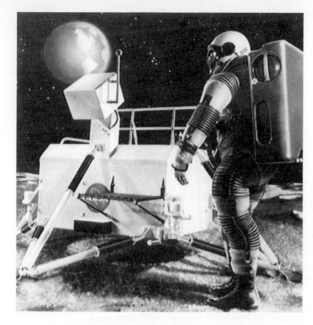

30.12 The structural members of this tripod support for a moon vehicle may be analyzed graphically to determine design load requirement. (Courtesy of NASA.)

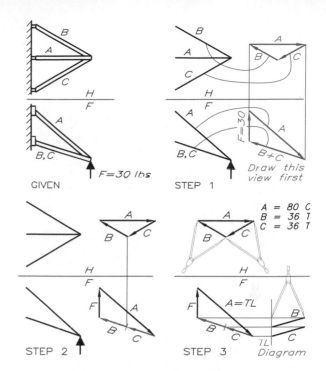

30.13 Noncoplanar structural analysis (special case):

Step 1 Forces B and C coincide in the front view, resulting in only two unknowns. Draw vector F (30 lb) and the two unknown forces parallel to their front view in the front view of the vector polygon. Find the top view of A by projecting from the front. Draw vectors B and C parallel to their top views.

Step 2 Project the point of intersection of vectors B and C in the top view to the front view to separate the head-to-tail vectors.

Step 3 Vectors B and C are in tension because they point away from the point of application in the space diagram. Vector A is in compression because it points toward the point of application.

30.7 Noncoplanar Vectors: Special Cases

The solution of 3D vector systems requires the use of descriptive geometry because the system must be analyzed in 3D space. An example is the manned flying system (MFS) shown in **Figure 30.12**, which was analyzed to determine the loads on its support members. Weight on the moon is 0.165 of earth weight. Thus a tripod that must support 182 lb on earth needs to support only 30 lb on the moon.

In general, only two unknown vectors can be determined in a single view of a vector polygon that is in equilibrium. However, the system shown in **Figure 30.13** is a special case because members B and C lie in the same edge view of the plane in the front view. Therefore solving for three unknowns is possible in this case.

Construct a vector polygon in the front view by drawing force F as a vector and using the other vectors as the sides of the polygon. Draw the top view using vectors B and C to form the

polygon that closes at each end of vector A. Then find the front view of vectors B and C.

A true-length diagram gives the lengths of the vectors; measure them to determine their magnitudes. Vector A is in compression because it points toward the point of application. Vectors B and C are in tension because they point away from the point.

General Case
The structural frame shown in **Figure 30.14** is attached to a vertical wall to support a load of $W = 1200$ lb. There are three unknowns in each

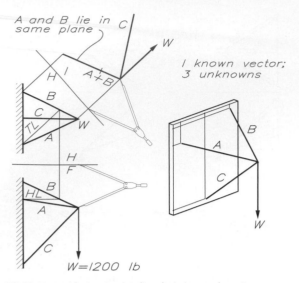

30.14 Noncoplanar structural analysis (general case):

Step 1 Draw an auxiliary view to show the edge view of the plane containing vectors *A* and *B* which are both unknowns. The load of *W* = 1200 lb is true-length and the only known vector in this view.

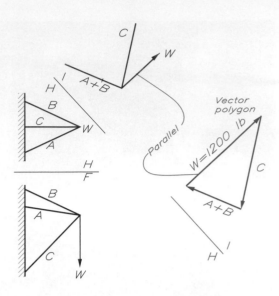

Step 2 Construct a vector polygon with its vectors parallel to the members found in the auxiliary view in step 1. Beginning with true-length vector W, draw the vector polygon head-to-tail using the two unknown vectors (Refer to the solution in Figure 30.13.).

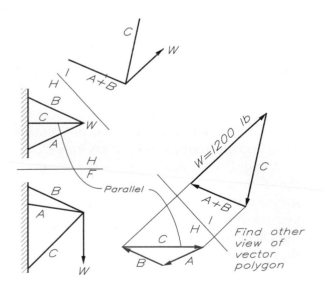

Step 3 Draw the adjacent orthographic view of the vector polygon by projecting perpendicularly to the *H1* reference line. Load *W* appears as a point in this view and vectors *A* and *B* are parallel to their corresponding members in top view of the space diagram.

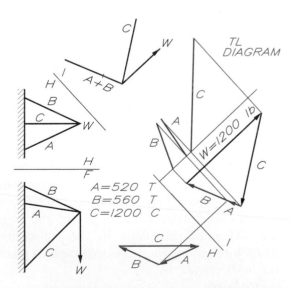

Step 4 Project the intersection of *A* and *B* to the adjacent vector polygon. *A* and *B* are in tension since they point toward the application point when transferred to the space diagram. The magnitudes of vectora A, B, and C are found a TL diagram and are listed in a table.

of the views, so begin by projecting an auxiliary view from the top view to obtain the edge view of a plane containing vectors A and B, thereby reducing the number of unknowns to two. You no longer need to refer to the front view.

Draw a vector polygon with vectors parallel to their members in the auxiliary view. Then draw an adjacent orthographic view of the vector polygon with vectors parallel to their members in the top view. Use a true-length diagram to find the true length of the vectors and measure their magnitudes.

30.8 Resultant of Parallel, Nonconcurrent Forces

The beam in **Figure 30.15** supports the three loads shown. It is necessary to determine the magnitude of supports R1 and R2, the magnitude of the resultant of the loads, and the resultant's location. Begin by labeling the spaces between all vectors clockwise with Bow's notation and draw a vector diagram.

Extend the lines of force in the space diagram and draw the strings from the vector

diagram in their respective spaces, parallel to their original directions. For example, string oa is parallel to string oA in space A between forces EA and AB, and string ob is in space B, beginning at the intersection of oa with vector AB. The last string, oe, closes the diagram, called a **funicular diagram**.

Transfer the direction of string oe to the vector diagram, and lay it off through point O to intersect the load line at E (Step 2). Vector DE represents R2 (refer to Bow's notation as it was applied in Step 1), and vector EA represents R1. It is easy to see that DE and EA are equal to the sum of the downward loads represented by AD.

To find the location of the resultant from R1, extend the outside strings of the funicular diagram, oa and od, to their intersection. The resultant will pass downward through this point of intersection. The resultant has a magnitude of 500 lb, a vertical downward direction, and a point of application at X = 6.1 ft. Notice that two scales are used in this problem: one for the vectors in pounds, and one in feet for the space diagram.

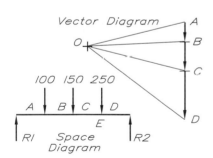 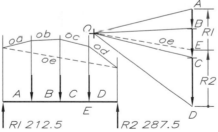 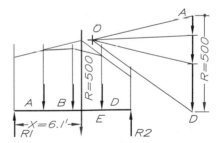

30.15 Parallel, nonconcurrent loads:

Step 1 Letter the spaces between the loads using Bow's notation. Sum the vertical loads by drawing them head-to-tail in a vector diagram. Locate pole point O at a convenient location and draw strings from O to the end of each vector.

Step 2 Extend the lines of the vertical loads and draw a funicular diagram with string oa in the A space, ob in the B space, oc in the C space, and so on. The last string, oe, closes the diagram. Transfer oe to the vector diagram to locate E, thus establishing R1 and R2, which are EA and DE, respectively.

Step 3 The resultant of the three downward forces equals their graphical summation, line AD. Locate the resultant by extending strings oa and od in the funicular diagram to a point of intersection. The resultant, R = 500 lb, acts through this point in a downward direction at distance X = 6.1 ft. from the left end.

Problems

Draw your solutions to these problems with instruments on size A grid or plain sheets. Each grid represents 0.20 in. (5mm). Letter written matter legibly, using 1/8-in. letters with guidelines.

1. (Sheet 1) Resultants.

(A–B) Find the resultants of the force systems by the parallelogram and polygon methods. Scale: 1″ = 100 lb.

2. (Sheet 2) Resultants.

(A–B) Find the resultants of the force systems by the parallelogram and polygon methods. Scale: 1″ = 100 lb.

3. (Sheet 3) Concurrent, coplanar.

(A–B) Find the forces in the coplanar force systems and label your construction properly.

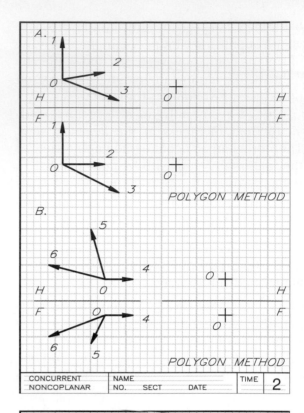

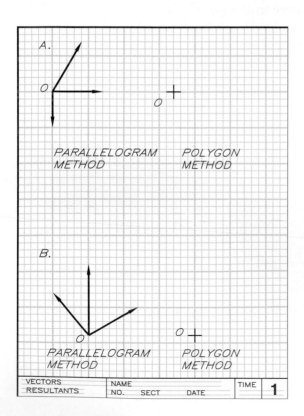

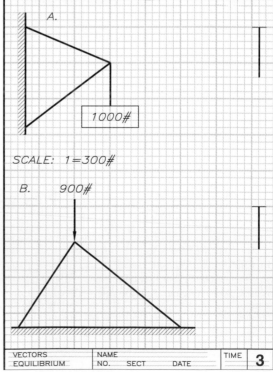

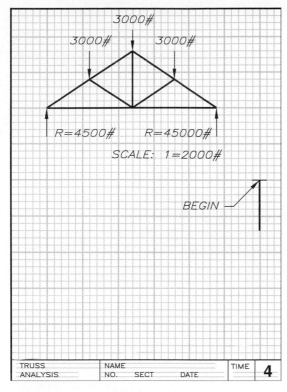

SCALE: 1=2000#

BEGIN

TRUSS ANALYSIS	NAME			TIME	4
	NO.	SECT	DATE		

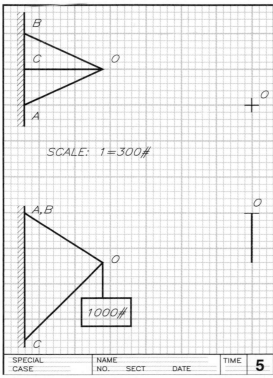

SCALE: 1=300#

1000#

SPECIAL CASE	NAME			TIME	5
	NO.	SECT	DATE		

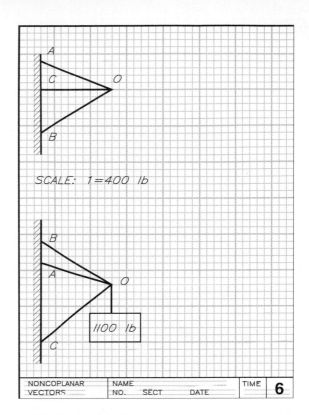

SCALE: 1=400 lb

1100 lb

NONCOPLANAR VECTORS	NAME			TIME	6
	NO.	SECT	DATE		

4. (Sheet 4) Truss analysis.

Find the loads in each member of the truss by using a Maxwell diagram. Make a table of forces and indicate compression and tension.

5. (Sheet 5) Non-coplanar, special case.

Find the forces in the members of the concurrent noncoplanar system. Make a table of forces and indicate compression and tension.

6. (Sheet 6) Non-coplanar, general case.

Find the forces in the members of the concurrent noncoplanar system. Make a table of forces and indicate compression and tension.

7. (Sheet 7) Beam Analysis

   (A) Find forces $R1$ and $R2$ necessary to equalize the loads applied to the beam.

   (B) Find the value and location of the single support that could replace both $R1$ and $R2$.

8. (Sheet 8) Beam Analysis

Repeat Problem 7 for this configuration.

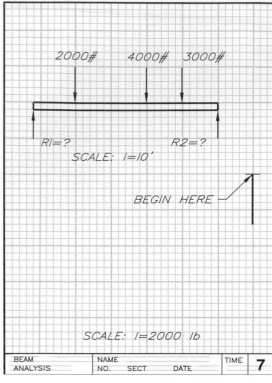

2000#   4000#   3000#

R1=?   R2=?

SCALE: 1=10'

BEGIN HERE

SCALE: 1=2000 lb

| BEAM | NAME | | TIME | 7 |
| ANALYSIS | NO.   SECT   DATE | | | |

9. (**Sheet 9**) Concurrent, non-coplanar forces.

Find the forces in the support members. Make a table of forces; indicate compression or tension.

10. (**Sheet 10**) Concurrent, coplanar.

Find the forces in the coplanar system. Make a table of forces; indicate compression or tension.

11. Repeat Problem 4, but use the joint-by-joint analysis instead of the Maxwell diagram method.

12. (**Sheet 11**) Find the loads in the structural members of the truss, list them in a table, and show whether they are in tension (T) or compression (C).

13. (**Sheet 12**) Determine the forces in the three members of the tripod used to lift concrete slabs into a vertical position when a pull of 480 lbs is applied at O.

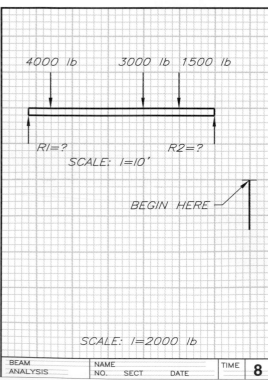

4000 lb   3000 lb 1500 lb

R1=?   R2=?

SCALE: 1=10'

BEGIN HERE

SCALE: 1=2000 lb

| BEAM | NAME | | TIME | 8 |
| ANALYSIS | NO.   SECT   DATE | | | |

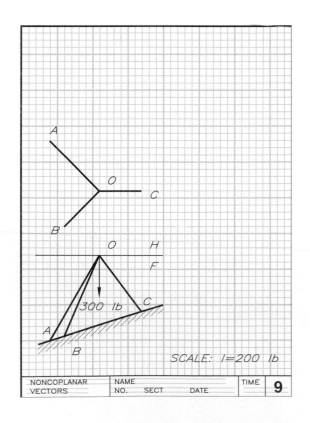

300 lb

SCALE: 1=200 lb

| NONCOPLANAR | NAME | | TIME | 9 |
| VECTORS | NO.   SECT   DATE | | | |

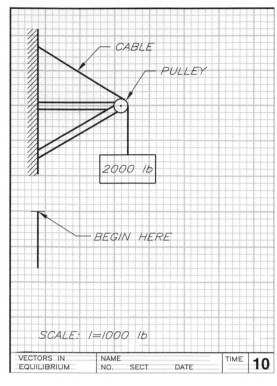

CABLE

PULLEY

2000 lb

BEGIN HERE

SCALE: 1=1000 lb

VECTORS IN EQUILIBRIUM	NAME			TIME	**10**
	NO.	SECT	DATE		

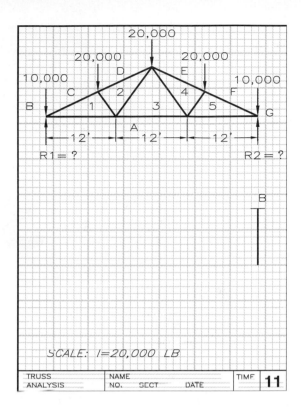

20,000

20,000

20,000

10,000

10,000

D
C
E
B
1
2
3
4
5
F
G
A

12'   12'   12'

R1 = ?   R2 = ?

B

SCALE: 1=20,000 LB

TRUSS ANALYSIS	NAME			TIME	**11**
	NO.	SECT	DATE		

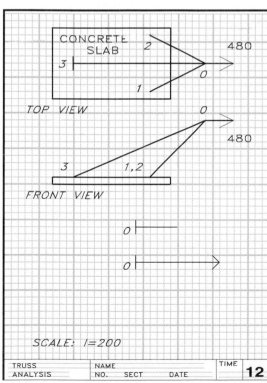

CONCRETE SLAB

2

480

3

0

1

TOP VIEW

0

480

3   1,2

FRONT VIEW

0 |——|

0 |——|→

SCALE: 1=200

TRUSS ANALYSIS	NAME			TIME	**12**
	NO.	SECT	DATE		

Thought Questions

1. Can a force in a leg of a tripod be greater than the magnitude of the load that is supporting? Explain.

2. What is the sum of the forces in the *x*- and *y*-directions when the structure is in equilibrium? Explain.

3. What is the notational method used to label the structural members of a truss when the loads in its members are being analyzed?

4. What do the following definitions of vectors mean: Direction, magnitude, equilibrant, point of application, vector diagram?

5. What is meant when a structural member is described as being in compression? In tension?

6. What can you say about a vecto diagram in which the vectors do close? Explain.

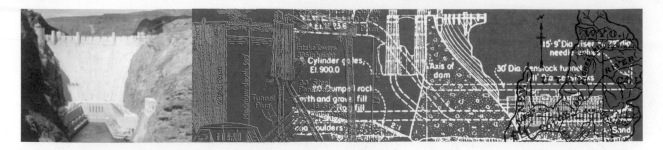

31

Intersections and Developments

31.1 Introduction

Several methods may be used to find lines of intersection between parts that join. Usually such parts are made of sheet metal, or of plywood if used as forms for concrete. After **intersections** are found, **developments**, or flat patterns, can be laid out on sheet metal and cut to the desired shape. You will see examples of intersections and developments ranging from air-conditioning ducts to massive refineries.

31.2 Intersections of Lines and Planes

Figure 31.1 illustrates the fundamental principle of finding the intersection between a line and a plane. This example is a special case in which the point of intersection clearly shows in the view where the plane appears as an edge. Projecting the piercing point P to the front view completes the visibility of the line.

This same principle is applied to finding the line of intersection between two planes (**Figure 31.2**). By locating the piercing points

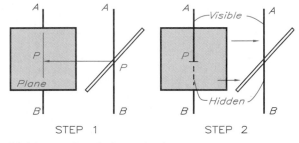

31.1 Intersection of a line and a plane:

Step 1 Find the point of intersection in the view where the plane appears as an edge, the side view in this case, and project it to the front view.

Step 2 Determine visibility in the front view by looking from the front view to the right-side view.

of lines *AB* and *DC* and connecting these points, the line of intersection is found.

The angular intersection of two planes at a corner gives a line of intersection that bends around the corner (**Figure 31.3**). First, find piercing points 2′ and 1′. Then project corner point 3 from the side view where the vertical corner pierces the plane to the front view of the corner. Point 2′ is hidden in the front view because it is on the back side.

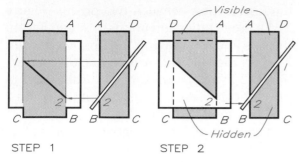

STEP 1   STEP 2

31.2 Intersection of planes:

Step 1 Find the piercing points of lines *AB* and *DC* with the plane where the plane appears as an edge and project them to the front view points 1 and 2.

Step 2 Line 1–2 is the line of intersection. Determine visibility by looking from the front view to the right-side view.

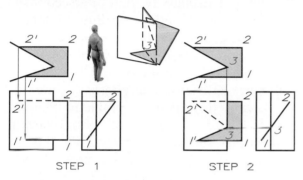

STEP 1   STEP 2

31.3 The intersection of a plane at a corner:

Step 1 The intersecting plane appears as an edge in the side view. Project intersection points 1′ and 2′ from the top and side views to the front view.

Step 2 The line of intersection from 1′ to 2′ must bend around the vertical corner at 3 in the top and side views. Project point 3 to the front view to locate line 1′–3–2′.

Figure 31.4 shows how to find the intersection between a plane and prism where the plane appears as an edge. Obtain the piercing points for each corner line and connect them to form the line of intersection. Show visibility to complete the intersection.

Figure 31.5 depicts a more general case of an intersection between a plane and prism. Passing vertical cutting planes through the planes of the prism in the top view yields traces (cut lines) on the front view of the oblique plane on which the piercing points of the vertical corner lines lie. Connect the

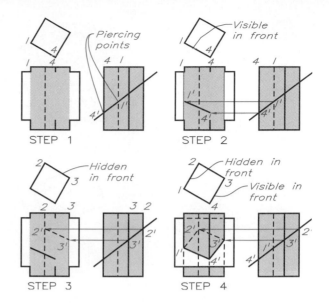

STEP 1   STEP 2

STEP 3   STEP 4

31.4 The intersection of a plane and a prism:

Step 1 Vertical corners 1 and 4 intersect the edge view of the plane in the side view at points 1′ and 4′.

Step 2 Project points 1′ and 4′ from the side view to lines 1 and 4 in the front view. Connect them to form a visible line of intersection.

Step 3 Vertical corners 2 and 3 intersect the edge view of the plane at points 2′ and 3′ in the side view. Project points 2′ and 3′ to the front view to form a hidden line of intersection.

Step 4 Connect points 1′, –2′, –3′, and –4′ and determine visibility by analyzing the top and side views.

points and determine visibility to complete the solution.

In **Figure 31.6**, finding the intersection between a foreshortened plane and an oblique prism involves finding an auxiliary view to obtain the edge view of the plane and simplify the problem. The piercing points of the corner lines of the prism lie in the auxiliary view and project back to the given views. Points 1, 2, and 3, projected from the auxiliary view to the given views, are shown as examples. An analysis of crossing lines determines visibility to complete the line of intersection in the top and front views.

31.3 Intersections between Prisms

The techniques used to find the intersections between planes and lines also apply to finding

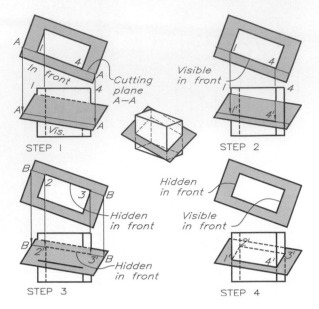

31.5 The intersection of an oblique plane and a prism:

Step 1 Pass vertical cutting plane A–A through corners 1 and 4 in the top view and project endpoints to the front view.

Step 2 Locate piercing points 1′ and 4′ in the front view where line A–A crosses lines 1 and 4.

Step 3 Pass vertical cutting plane B–B through corners 2 and 3 in the top view and project them to the front view to locate piercing points 2′ and 3′.

Step 4 Connect the four piercing points and determine visibility by analysis of the top view.

the intersection between two prisms (**Figure 31.7**). Project piercing points 1, 2, and 3 from the side and top views to the front view. Point X lies in the side view where line of intersection 1–2 bends around the vertical corner of the vertical prism. Connect points 1, X, and 2 and determine visibility.

Figure 31.8 illustrates how to find the line of intersection between an inclined prism and a vertical prism. An auxiliary view reveals the end view of the inclined prism where its planes appear as edges. In the auxiliary view, plane 1–2 bends around corner *AB* at point *P*. Project points of intersection 1′ and 2′ from the top and auxiliary to their intersections in the front view. Then draw the line of intersection 1′–*P*–2′ for this portion of the line of intersection. Connect the remaining lines, 1′–3′ and 2′–3′ to complete the solution.

Figure 31.9 shows an alternative method of solving this type of problem. Piercing points 1′ and 2′ appear in the front view as projections from the top view. Point 5 is the point where line 1′–5–2′ bends around vertical corner *AB*. To find point 5 in the front view,

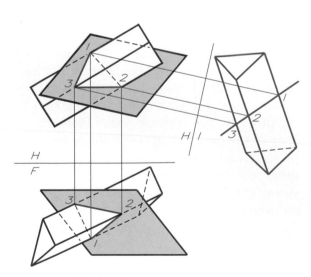

31.6 To find the intersection between a plane and a prism, construct a view in which the plane appears as an edge. Project piercing points 1, 2, and 3 back to the top and front views.

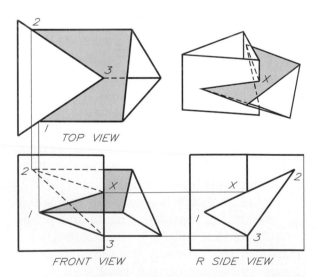

31.7 These are three views of intersecting prisms. The points of intersection are best found where intersecting planes appear as edges.

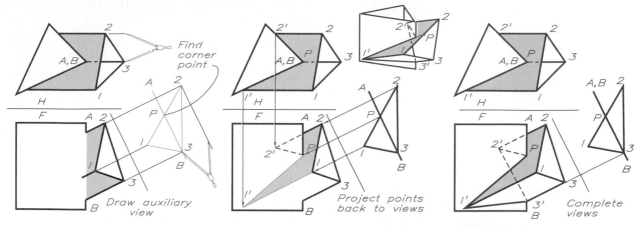

31.8 The intersection of prisms by auxiliary view:

Step 1 Draw the end view of the inclined prism in an auxiliary view from the front view. Show line *AB* of the vertical prism in the auxiliary view.

Step 2 Locate piercing points 1' and 2' in the top and front views. Intersection line 1'–2' bends around corner *AB* at *P* projected from the auxiliary view.

Step 3 The intersection lines from 2' and 1' to 3' do not bend around the corner, but are straight lines. Line 1'–3' is visible, and line 2'–3' is invisible.

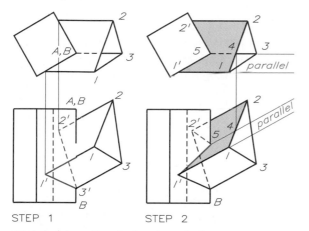

31.9 The intersection of prisms by projection:

Step 1 Project the piercing points of lines 1, 2, and 3 from the top view to the front view to locate piercing points 1', 2', and 3'.

Step 2 Pass a frontal plane through corner *AB* in the top view to locate point 5, where intersection line 1'–2' bends around the vertical prism. Find point 5 in the front view and draw line 1'–5–2'.

pass a frontal plane through corner *AB* in the top view and project its trace (4, 5) to the front view. Draw the lines of intersection, 1'–5–2'.

Some applications of intersections and developments are massive in size, as illustrated by the huge blast furnace under construction in **Figure 31.10**. The workers who are constructing it are dwarfed by its enormity.

31.10 The design of this huge blast furnace used applications of large-scale intersections and developments. (Courtesy of the Jones & Laughlin Steel Corporation.)

31.4 Intersections between Planes and Cylinders

The standard sheet-metal vent pipe that is common to all homes is an example of a cylinder intersecting a plane. **Figure 31.11** shows how to find the intersection between a plane and a cylinder. Cutting planes passed vertically through the top view of the cylinder establish pairs of elements on the cylinder and their piercing points. Space the cutting planes conveniently apart by eye. Then project the piercing points to each view and draw the elliptical line of intersection.

Figure 31.12 shows the solution of a more general problem. Here, the cylinder is vertical and the plane is oblique and does not appear as an edge. Passing vertical cutting planes through the cylinder and the plane in the top view gives elements on the cylinder and their piercing points on the plane. Projecting these points to the front view completes the elliptical line of intersection (Step 2). The more cutting

planes used, the more accurate the line of intersection will be.

Figure 31.13 demonstrates the general case of the intersection between a plane and cylinder, where both the plane and cylinder are oblique in the given views. An auxiliary view is used to show the edge view of the plane. Cutting planes passed through the cylinder parallel to its axis in the auxiliary view establish elements on the cylinder and their piercing points. The points are projected back to the given views and connected to give an elliptical line of intersection in the front and side views.

31.5 Intersections between Cylinders and Prisms

An inclined prism intersects a vertical cylinder in **Figure 31.14**. A primary auxiliary view is drawn to show the end view of the inclined prism where its planes appear as edges. A series of vertical cutting planes in the top view

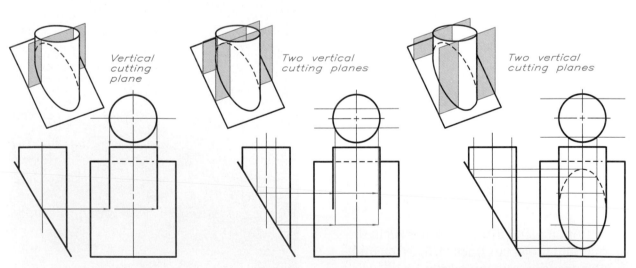

31.11 The intersection between a cylinder and a plane:

Step 1 Pass a vertical cutting plane through the cylinder parallel to its axis to find two points of intersection. On the plane in the front view.

Step 2 Use two more cutting planes to find four additional points in the top and left side views. Project these points to the front view.

Step 3 Use additional cutting planes to find more points. Determine visibility and connect these points to form the elliptical line of intersection.

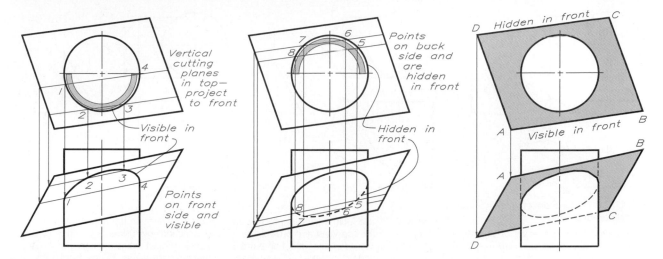

31.12 Intersection of a cylinder and an oblique plane:

Step 1 Pass vertical cutting planes through the cylinder in the top view to find elements on it and the plane. Project points 1, 2, 3, and 4 to the front view of their respective lines and connect them with a visible line.

Step 2 Use additional cutting planes to find other piercing points—5, 6, 7, and 8; project them to the front view of their respective lines on the oblique plane. The points are on the back side and are hidden; connect them with a hidden line.

Step 3 Determine visibility of the plane and cylinder in the front view. Line *AB* is visible by inspection of the top view, since it is the farthest out in front. Line *CD* is the farthest back in the top view and is therefore hidden in the front view.

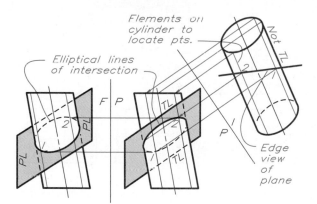

31.13 To find the intersection between an oblique cylinder and an oblique plane, construct a view that shows the plane as an edge. Cutting planes passed through the cylinder locate points on the line of intersection.

establish lines lying on the surfaces of the cylinder and prism. The cutting planes, also shown in the auxiliary view, are the same distance apart as in the top view.

Projecting the line of intersection from 1 to 3 from the auxiliary view to the front view

yields an elliptical line of intersection. The visibility of this line changes from visible to hidden at point X, which appears in the auxiliary view and is projected to the front view. Continuing this process gives the lines of intersection of the other two planes of the prism.

31.6 Intersections between Cylinders

To find the line of intersection between two perpendicular cylinders, pass cutting planes through them parallel to their centerlines (**Figure 31.15**). Each cutting plane locates a pair of elements on both cylinders that intersect at a piercing point. Connecting the points and determining visibility completes the solution. An example of a transmission pipe fabricated with intersecting cylinders is shown in **Figure 31.16**. Each cylinder had to be precisely cut to form accurate intersections for tight joints before welding them together.

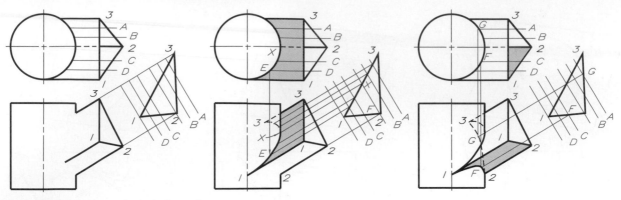

31.14 The intersection of a cylinder and an inclined-prism:

Step 1 Find the edge views of the planes of the triangular prism in an auxiliary view. Project from the front view. Draw frontal cutting planes through the top view and locate them in the auxiliary view with dividers.

Step 2 Locate points along intersection line 1–3 in the top view and project them to the front view. For example, find point E on cutting plane *D* in the top and auxiliary views and project it to the front view where the projectors intersect. Visibility changes in the front view at point *X*.

Step 3 Determine the remaining points of intersection by using the other cutting planes. Project point *F*, shown in the top and auxiliary views, to the front view of line 1–2. Connect the points and determine visibility.

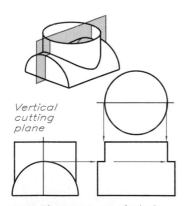
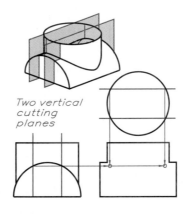
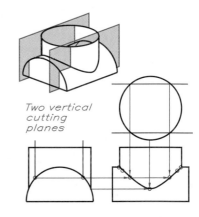

Vertical cutting plane

Two vertical cutting planes

Two vertical cutting planes

31.15 The intersection of cylinders:

Step 1 Pass a cutting plane through the cylinders parallel to their axes, locating two points of intersection.

Step 2 Use two more cutting planes to find four more additional points of intersection.

Step 3 Use two more cutting planes to locate four more points. Connect the points with a smooth curve to complete the line of intersection.

Figure 31.17 illustrates how to find the intersection between nonperpendicular cylinders. This method involves passing a series of vertical cutting planes through the cylinders parallel to their centerlines. Points 1 and 2, labeled on cutting plane *D,* are typical of points on the line of intersection. Other points may be found in the same manner. Although the auxiliary view is not essential to the solution, it is an aid in visualizing the problem. Projecting points 1 and 2 on cutting plane *D* in the auxiliary view to the front view provides a check on the projections from the top view.

31.16 Intersections between cylinders have been welded in shop fabrication of a transmission pipe, as shown in these two views.

31.7 Intersections between Planes and Cones

To find points of intersection on a cone, use cutting planes that are (1) perpendicular to the cone's axis or (2) parallel to the cone's axis. The vertical planes in the top view of **Figure 31.18A** cut radial lines on the cone and establish elements on its surface. The horizontal planes in **Figure 31.18B** cut circular sections that appear true size in the top view of a right cone.

A series of radial cutting planes define elements on a cone (**Figure 31.19**). These elements cross the edge view of the plane in the front view to locate piercing points of each element that, when projected to the top view of the same elements, lie on the line of intersection.

A series of horizontal cutting planes may be used to determine the line of intersection

31.17 To find this intersection, find the end view of the inclined cylinder in an auxiliary view. Use vertical cutting planes to find the piercing points of the cylindrical elements and the line of intersection.

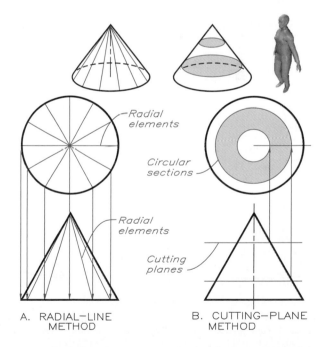

31.18 To find intersections on conical surfaces, use (A) radial cutting planes that pass through the cone's centerline and are perpendicular to its base, or (B) cutting planes that are parallel to the cone's base.

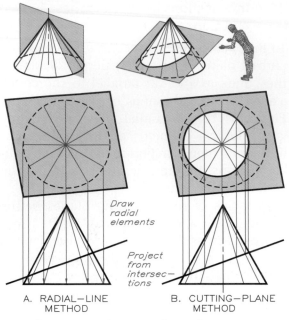

Draw
radial
elements

Project
from
intersec-
tions

A. RADIAL—LINE
METHOD

B. CUTTING—PLANE
METHOD

31.19 The intersection between a plane and a cone.

A Divide the base evenly in the top view and connect these points with the apex to establish elements on the cone. Project these elements to the front view.

B Project the piercing point of each element on the edge view of the plane to the top view of the same elements, and connect them to form the line of intersection.

between a cone and an oblique plane (**Figure 31.20**). The sections cut by these imaginary planes are circles in the top view. The cutting planes also locate lines on the oblique plane that intersect the circular sections cut by each respective cutting plane. The points of intersection found in the top view project to the front view. We could have used the radial-line method shown in **Figure 31.19** to obtain the same results.

31.8 Intersections between Cones and Prisms

A primary auxiliary view gives the end view of the inclined prism that intersects the cone in **Figure 31.21**. Cutting planes that radiate from the apex of the cone in the top view locate elements on the cone's surface that intersect the edge view of the prism in the auxiliary view. These elements are projected to the front view.

Wherever the edge view of plane 1–3 intersects an element in the auxiliary view, the piercing points project to the same element in

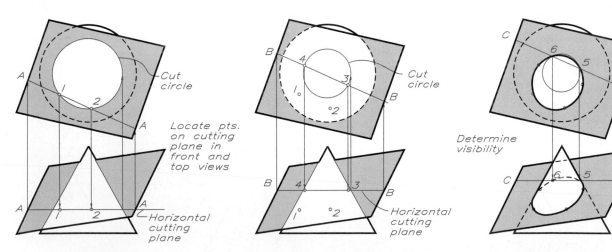

31.20 The intersection between an oblique plane and a cone:

Step 1 Pass a horizontal cutting plane through the front view to find a circular section on the cone and a line on the plane in the top view. The piercing points of this line are on the circle. Project points 1 and 2 to the front view.

Step 2 Pass horizontal cutting plane B–B through the front view in the same manner to locate piercing points 3 and 4 in the top view. Project these points to the horizontal plane in the front view from the top view.

Step 3 Use additional horizontal planes to find a sufficient number of points to complete the line of intersection in the same manner as covered in the previous steps. Draw the intersection and determine visibility.

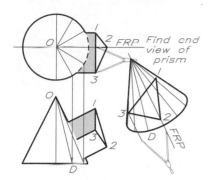

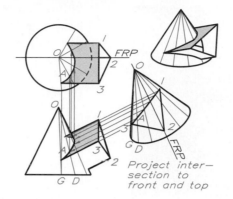

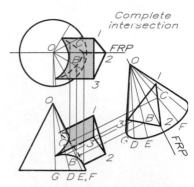

31.21 Intersection of a cone and prism:

Step 1 Draw an auxiliary view to obtain the edge views of the planes of the prism. In the top view, pass vertical cutting planes through the cone through the apex 0. Project these elements to the front and auxiliary views.

Step 2 Find the piercing points of the cone's elements with the edge view of plane 1–3 in the auxiliary view and project them to the front and top views. For example, point A lies on element OD in the auxiliary view, so project it to the front and top views of OD.

Step 3 Locate the piercing points where the conical elements intersect the edge views of the planes of the prism in the auxiliary view. For example, find point B on QF in the primary auxiliary view and project it to the front and top views of OE.

the front and top views. Passing an extra cutting plane through point 3 in the auxiliary view locates an element that projects to the front and top views. Piercing point 3 projects to this element in sequence from the auxiliary view to the top view.

This same procedure yields the piercing points of the other two planes of the prism. All projections of points of intersection originate in the auxiliary view, where the planes of the prism appear as edges.

In **Figure 31.22**, horizontal cutting planes passed through the front view of the cone and cylinder give a series of circular sections in the top view. Points 1 and 2, shown on cutting plane C in the top view, are typical and project to the front view. The same method produces other points.

This method is feasible only when the centerline of the cylinder is perpendicular to the axis of the cone, producing circular sections in the top view (rather than elliptical sections, which would be difficult to draw). A series of intersecting cylinders can be seen in this gas-powered turbine in **Figure 31.23**.

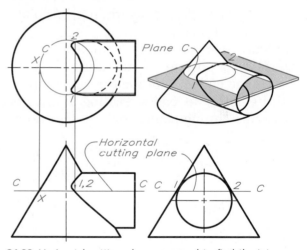

31.22 Horizontal cutting planes are used to find the intersection between the cone and the cylinder. The cutting planes cut circles in the top view. Only one cutting plane is shown here as an example.

31.9 Intersections Between Pyramids

Figure 31.24 shows how to find the intersection of an inclined prism with a pyramid. An auxiliary view shows the end view of the inclined prism and the pyramid. The radial lines OB and OA drawn through corners 1 and 2 in the auxiliary view project back to the front

31.23 Examples of intersecting cylinders are shown in this gas turbine power plant in Hong Kong. (Courtesy of General Electric Company.)

and top views. Projection locates intersecting points 1 and 2 on lines *OB* and *OA* in each view. Point P is the point where line 1–2 bends around corner *OC*. Finding lines of intersection 1–3 and 4–2 and determining visibility completes the solution.

Figure 31.25 shows a horizontal prism that intersects a pyramid. An auxiliary view depicts the end view of the horizontal prism with its planes as edges. Passing a series of horizontal cutting planes through the corner

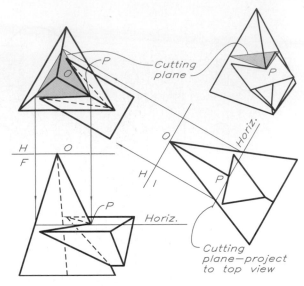

31.25 The intersection of this pyramid and prism is found by obtaining the end view of the prism in an auxiliary view. Horizontal cutting planes are passed through the fold lines of the prism to find the piercing points on the line of intersection. One cutting plane used to find corner point *P* is shown.

points of the horizontal prism and the pyramid in the auxiliary view gives the lines of intersection, which form triangular sections in the top view.

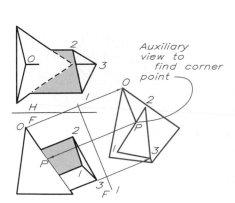

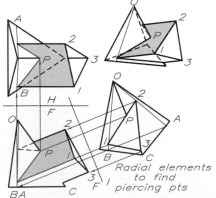

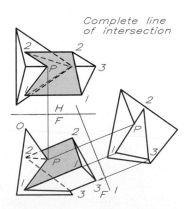

31.24 Intersection of a prism and pyramid:

Step 1 Find the edge views of the planes of the prism in an auxiliary view. Project the pyramid into this view also, showing only the visible surfaces.

Step 2 Pass planes *A* and *B* through *O* and points 1 and 3 in the auxiliary view. Project *OA* and *OB* to the front and top views; project 1 and 3 to them. Point 2 lies on *OC*. Connect 1, 2, and 3 for the intersection of this plane.

Step 3 Point 3 lies on *OC* in the auxiliary view. Project this point to the principal views. Connect 3 to points 1 and 2 to complete the intersections; show visibility. Assume that these shapes are constructed of sheet metal.

The cutting plane through corner point *P* in the auxiliary view is an example of a typical cutting plane. At point P, the line of intersection of this plane bends around the corner of the pyramid. Other cutting planes are passed through the corner lines of the prism in the auxiliary and front views. Each corner line of the prism extends in the top view to intersect the triangular section formed by the cutting plane, as shown in the same matter P was found.

31.10 Principles of Development

The bodies of aircraft (**Figure 31.26**) are designed to be fabricated from sheet metal stock formed into shapes. Although aircraft design is one of the most advanced applications of developments, the principles are the same as for the design of a garbage can.

31.26 Essentially the entire body of an aircraft, such as the Super Hornet shown in the foreground, is a series of applications of intersections and developments. (Courtesy of Northrup Grumman Corporation.)

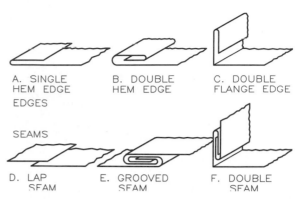

31.27 These are examples of several types of edges and seams used to join sheet-metal developments. Other seams are joined by riveting and welding.

Figure 31.27 illustrates some of the standard edges and joints for sheet metal. The application determines the type of seam that is used.

The development of patterns for four typical shapes is shown in **Figure 31.28**. The sides of a box are unfolded into a common plane. The cylinder is rolled out along a stretch-out line equal in length to its circumference. The pattern of a right cone and right pyramid are developed with the length of an element serving as a radius for drawing the base arc.

The construction of patterns for geometric shapes with parallel elements, such as the prisms and cylinders shown in **Figure 31.29A** and **B**, begins with drawing stretch-out lines parallel to the edge views of the shapes' right sections. The distance around the right section becomes the length of the stretch-out line. The

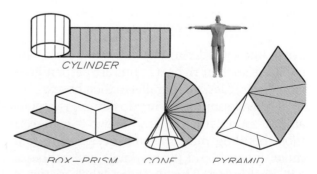

31.28 Four standard types of developments are the box, cylinder, cone, and pyramid.

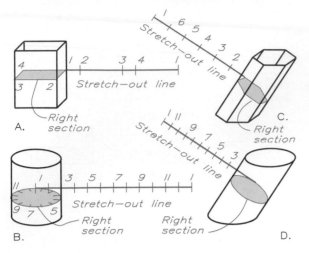

31.29 Stretch-out lines.

A and **B** To obtain the developments of right prisms and right cylinders, roll out the right sections along a stretch-out line.

C and **D** Draw stretch-out lines parallel to the edge views of the right section of cylinders and prisms, or perpendicular to their true-length elements.

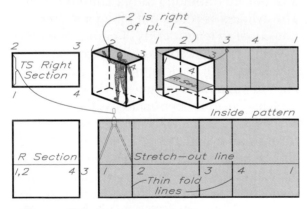

31.30 To develop a rectangular prism for an inside pattern, draw the stretch-out line parallel to the edge view of the right section. Transfer the distances between the fold lines from the true-size right section to the stretch-out line.

prism and cylinder in **Figure 31.30C** and **D** are inclined, so their right sections are perpendicular to their sides, not parallel to their bases.

In development, an inside pattern is preferable to an outside pattern for two reasons: (1) most bending machines are designed to fold metal inward, and (2) markings and scribings will be hidden. The designer labels patterns with a series of lettered or numbered points on

the layouts. **All lines on developments must be true length. Patterns should be laid out so that the seam line (a line where the pattern is joined) is the shortest line in order to reduce the expense of riveting or welding the seams.**

31.11 Development of Rectangular Prisms

The development of a flat pattern for a rectangular prism is illustrated in **Figure 31.30**. The edges of the prism are vertical and true length in the front view. The right section is perpendicular to these sides and the right section is true size in the top view. The stretch-out line begins with point 1 and is drawn parallel to the edge view of the right section.

If an inside pattern is to be laid out to the right, you must determine which point is to the right of the beginning point, point 1. Let's assume that you are standing inside the top view and are looking at point 1: You will see point 2 to the right of point 1.

To locate the fold lines of the pattern, transfer lines 2–3, 3–4, and 4–1 with your dividers from the right section in the top view to the stretch-out line. The length of each fold line is its projected true length from the front view. Connect the ends of the fold lines to form the boundary of the developed surface. Draw the fold lines as thin dark lines and the outside lines as thicker, visible object lines.

The fuselage of the Premier in **Figure 31.31** was designed using the principles of intersections and developments. Development of the prism depicted in **Figure 31.32** is similar to that shown in **Figure 31.33**. Here, though, one of its ends is beveled (truncated) rather than square. The stretch-out line is parallel to the edge view of the right section in the front view. Lay off the true-length distances around the right section along the stretch-out line (beginning with the shortest one) and locate the fold lines. Find the lengths of the fold lines by projecting from the front view of these lines.

31.31 The assembly of Premier is an example of a fuselage designed by applying principles of intersections and developments. (Courtesy of Raytheon Aircraft.)

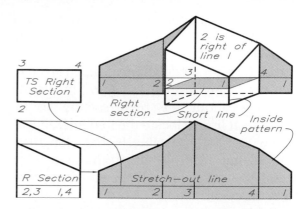

31.32 To develop an inside pattern of a rectangular prism with a beveled end, draw the stretch-out line parallel to the right section. Then find the fold lines by transferring distances between the fold lines from the true-size right section to the stretch-out line.

31.12 Development of Oblique Prisms

The prism shown in **Figure 31.33** is inclined to the horizontal plane, but its fold lines are true length in the front view. The right section is an edge perpendicular to these TL fold lines, and the stretch-out line is parallel to the edge of the right section. A true-size view of the right section is found in the auxiliary view.

Transfer the distances between the fold lines from the true-size right section to the stretch-out line. Find the lengths of the fold lines by projecting from the front view. Determine the ends of the prism and attach them to the pattern so that they can be folded into position.

In **Figure 31.34**, the fold lines of the prism are true length in the top view, and the edge

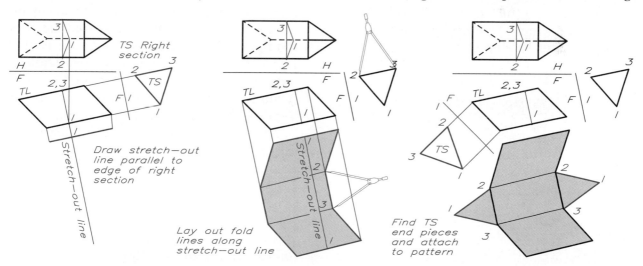

31.33 Development of an oblique prism:

Step 1 Draw the edge view of the right section perpendicular to the true-length axis in the front view. Find the true-size view of the right section in the auxiliary view. Draw the stretch-out line parallel to the edge of the right section. The line through point 1 is the first line of the development.

Step 2 Because the pattern is to be laid out to the right from line 1, the next point is line 2 (from the auxiliary view). Transfer true-length lines 1–2, 2–3, and 3–1 from the right section to the stretch-out line to locate fold lines. Determine the lengths of bend lines by projection.

Step 3 Find true-size views of the end pieces by projecting auxiliary views from the front view. Connect these ends to the development to form the completed pattern. Draw fold lines as thin dark lines and outside lines as thicker, visible object lines.

view of the right section is perpendicular to them. The stretch-out line is parallel to the edge view of the right section, and the true size of the right section appears in an auxiliary view projected from the top view. Transfer the distances about the right section to the stretch-out line to locate the fold lines, beginning with the shortest line. Find the lengths of the fold lines by projecting from the top view. Attach the end portions to the pattern to complete the construction.

A prism that does not project true length in either view may be developed as shown in **Figure 31.35**. The fold lines are true length in a primary auxiliary view projected from the front view. The right section appears as an edge perpendicular to the fold lines in the auxiliary view and true size in a secondary auxiliary view.

Draw the stretch-out line parallel to the edge view of the right section. Locate the fold lines on the stretch-out line by measuring around the right section in the secondary auxiliary view, beginning with the shortest one. Then project the lengths of the fold lines to the development from the primary auxiliary view to the development.

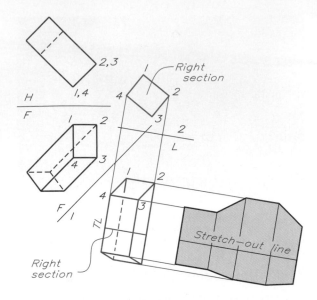

31.35 To develop an oblique prism, draw a primary auxiliary view in which the fold lines are true length and a secondary auxiliary view in which the right section appears true size. Use these views to develop the pattern the same way as in Figure 31.34.

31.13 Development of Cylinders

Figure 31.36 illustrates how to develop a flat pattern of a right cylinder. The elements of the cylinder are true length in the front view, so the right section appears as an edge in this view and true size in the top view. The stretch-out line is parallel to the edge view of the right section, and point 1 is the beginning point because it lies on the shortest element.

Let's assume that you are standing inside the cylinder in the top view and are looking at point 1; you will see that point 2 is to the right of point 1. Therefore, lay off point 2 to the right of point 1 for developing an inside pattern.

By drawing radial lines at 15° or 30° intervals you can equally space the elements in the top view and conveniently lay them out along the stretch-out line as equal measurements. To complete the pattern, find the lengths of the elements by projecting from the front view. An application of a large developed cylinder joined with a transition piece is shown in **Figure 31.37**.

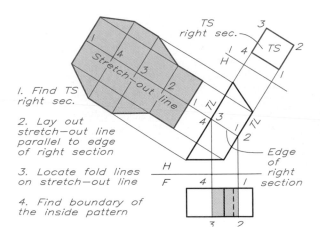

31.34 To develop this oblique chute, locate the right section true size in the auxiliary view. Draw the stretch-out line parallel to its right section. Find fold lines by transferring their spacing from the true-size right section to the stretch-out line.

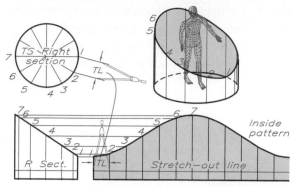

31.36 To develop an inside pattern for a truncated right cylinder, draw the stretch-out line parallel to the right section. Transfer points 1–7 from the top view to the stretch-out line that is parallel to the right section. Point 2 is to the right of point 1 for an inside pattern.

31.14 Development of Oblique Cylinders

The pattern for an oblique cylinder (**Figure 31.38**) involves the same determinations as the preceding cases, but with the additional step of finding a true-size view of the right section in an auxiliary view. First, locate a series of equally spaced elements around the

31.37 This large transition piece makes a 90° turn enabling the cylinder to join with a square-end duct.

right section in the auxiliary view and project them back to the true-length view. Draw the stretch-out line parallel to the edge view of the right section in the front view. Lay out the spacing between the elements along the stretch-out line, and draw the elements

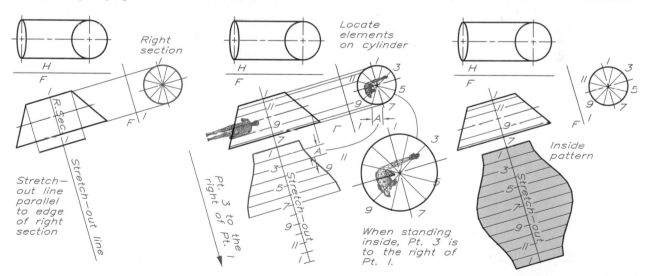

31.38 Development of an oblique cylinder:

Step 1 Draw the right section perpendicular to the true-length axis in the front view. Draw an auxiliary view to find the right section true size; divide it into equal chords. Draw a stretch-out line parallel to the edge of the right section. Locate the shortest line at 1.

Step 2 Project elements from the right section to the front view. Transfer the chordal measurements in the auxiliary view to the stretch-out line to locate cylindrical elements and determine their lengths by projection.

Step 3 Locate the remaining elements to complete the construction as begun in Step 2. Connect the ends of the elements with a smooth curve. This is an inside pattern with a seam along its shortest element.

through these points perpendicular to the stretch-out line. Find the lengths of the elements by projecting from the front view and complete the pattern.

A more general case is the oblique cylinder shown in **Figure 31.39**, where the elements are not true length in the given views. A primary auxiliary view gives the elements true length, and a secondary auxiliary view yields a true-size view of the right section. Draw the stretch-out line parallel to the edge view of the right section in the primary auxiliary view. Transfer the elements to the stretch-out line from the true-size right section.

Draw the elements perpendicular to the stretch-out line and find their lengths by projecting from the primary auxiliary view. Connect the endpoints with a smooth curve to complete the pattern.

31.15 Development of Pyramids

All lines used to draw patterns must be true length, but pyramids have few lines that are true length in the given views. For this reason

you must find the sloping corner lines true length before drawing a development.

Figure 31.40 shows the method of finding the corner lines of a pyramid true length by revolution. Revolve line O–5 into the frontal plane to line O–5′ in the top view so that it will be true length in the front view. An application of the development of a pyramid is the sheet metal hopper shown in **Figure 31.41**.

Figure 31.42 shows the development of a right pyramid. Line O–1 is revolved into the frontal plane in the top view to find its true length in the front view. Because it is a right pyramid, all corner lines are equal in length. Line O–1′ is the radius for the base circle of the development. When you transfer distance 1–2 from the base in the top view to the development, it forms a chord on the base circle. Find lines 2–3, 3–4, and 4–1 in the same manner and in sequence. Draw the fold lines as thin lines from the base to the apex, point O.

A variation of this case is the truncated pyramid (**Figure 31.43**). Development of the inside pattern proceeds as in the preceding case, but establishing the upper lines of the development requires an additional step. Revolution yields

31.39 To develop an oblique cylinder, construct a primary auxiliary view in which the elements appear true length. Find the right section true size in a secondary auxiliary view and complete the construction as shown in Figure 31.38.

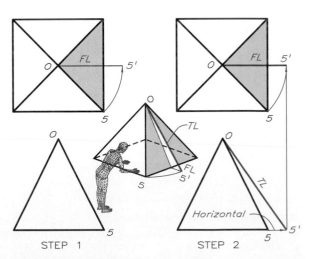

31.40 True length (Pyramid) by revolution:

Step 1 To find the true length of corner line O–5 of a pyramid, revolve it into the frontal plane in the top view, to O–5′.

Step 2 Project point 5′ to the front view, where frontal line O–5′ is true length.

31.41 This sheet-metal hopper is an application of a design involving development of a pyramid. (Courtesy of Gar-Bro.)

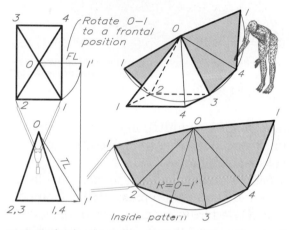

31.42 To develop this right pyramid, lay out an arc by using the true length of corner line O–1' as the radius. Transfer true-length distances around the base in the top view to the arc by triangulation and darken the lines.

the true-length lines from the apex to points 1', 2', 3', and 4'. Lay off these distances along their respective lines on the pattern to find the upper boundary of the pattern.

31.16 Development of Cones

All elements of a right cone are equal in length (**Figure 31.44**). Revolving element O–6 into its frontal position at O–6' gives its true length when projected to the front view. Line O–6' is true length and is the outside element of the cone. Projecting point 7 horizontally to element O–6' locates point 7'. Line O–7' is TL.

To develop the right cone depicted in **Figure 31.45**, divide the base into equally spaced elements in the top view and project

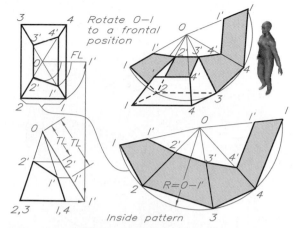

31.43 To develop an inside pattern of a truncated right pyramid, use the method shown in Figure 31.42. Find the true lengths of elements O–1', O–2', O–3' and O–4' in the front view by revolution. Lay them off along their respective elements to find the upper boundary of the pattern.

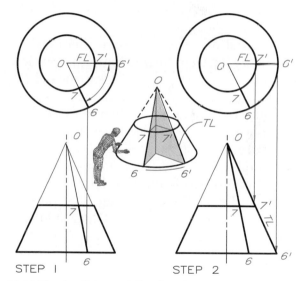

31.44 True length by revolution (cone):

Step 1 Revolve an element of a cone, O–6, into a frontal plane in the top view, O–6'.

Step 2 Project point 6' to the front view, where it is a true-length outside element of the cone. Find the true length of line O–7' by projecting point 7 to the outside element in the front view, 7'.

them to the front view, where they radiate to the apex at O. The outside elements in the front view, O–10 and O–4, are true length.

Using element O–10 as a radius, draw the base arc of the development. The spacing of the elements along the base circle is equal to

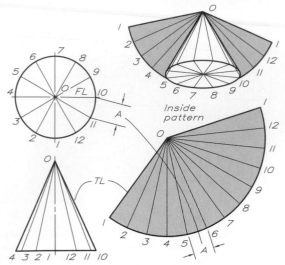

31.45 To develop an inside pattern of a right cone, use a true-length element (O–4 or O–10 in the front view) as the radius. Transfer chordal distances from the true-size base in the top view and mark them off along the arc.

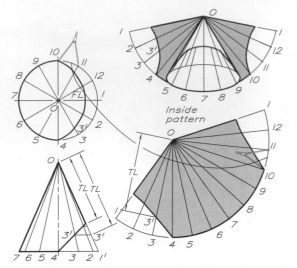

31.46 To develop a conical surface with a side opening, begin by laying it out as in Figure 31.45. Find true-length elements by revolution in the front view and transfer them to their respective elements in the pattern.

the chordal distances between them on the base in the top view. Inspection of the top view from the inside, where point 2 is to the right of point 1, indicates that this is an inside pattern. The nose of the Premier in **Figure 31.28** is an example of cone developed in sheet metal.

Figure 31.46 shows the development of a truncated cone. To find its pattern, lay out the entire cone by using the true-length element O–1 as the radius, ignoring the portion removed from it. Locate the hyperbolic section formed by the inclined plane through the front view of the cone in the top view by projecting points on each element of the cone to the top view of these elements. For example, determine the true length of line O–3′ by projecting point 3′ horizontally to the true-length element O–1 in the front view. Lay off these distances, and others, along their respective elements to establish a smooth curve.

31.17 Development of Transition Pieces

A transition piece changes the shape of a section at one end to a different shape at the other end. In **Figure 31.47**, you can see examples of transition pieces that convert one cross-sec-

tional shape to another. In industrial applications, transition pieces may be huge or relatively small (**Figure 31.48**).

Figure 31.49 shows the steps in the development of a transition piece. Radial elements are extended from each corner to the equally spaced points on the circular end of the piece. Revolution is used to find the true length of each line. True-length lines 2–D, 3–D, and 2–3 yield the inside pattern of 2–3–D.

The true-length radial lines, used in combination with the true-length chordal distances in the top view, give a series of abutting trian-

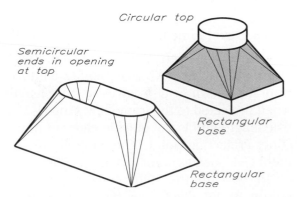

31.47 These are examples of transition pieces that connect parts having different cross sections.

gles to form the pattern beginning with element *D2*. Adding the triangles *A*–1–2 and *G*–3–4 at each end of the pattern completes the development of a half-pattern. Only a half pattern is shown in this example.

Problems

The scale of these problems allows you to fit two solutions on a size A sheet for a grid size of 0.20 in. (5 mm). For a grid size of 0.40 in. (10 mm), you can fit only one solution on a size A sheet.

31.48 This transition piece joins the circular shape at the upper end of the hopper with its rectangular section at the lower end.

Intersections
1–24. (Figure 31.50) Lay out the given views of the problems and find the intersections between the shapes that are necessary to complete the views.

Developments
25–48. (Figure 31.51) Lay out the problems and draw their developments. Orient the long side of the sheet horizontally to allow space at the right of the given views for the development.

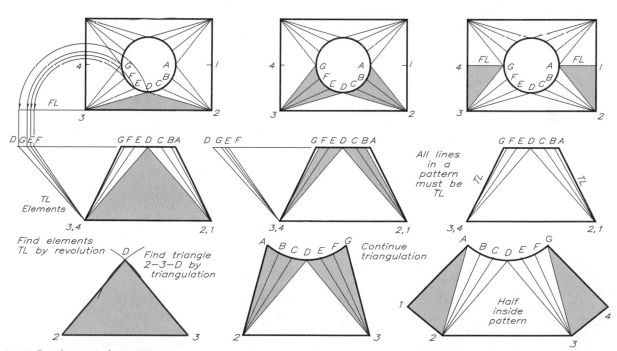

31.49 Development of a transition piece:

Step 1 Divide the circular end into equal parts in the top view and connect these points with lines to corner points 2 and 3. Find the true length of these lines by revolving and projecting them to the front view. Using the true-length lines, draw triangle 2–3–*D*.

Step 2 Using other true-length lines and the chord distances on the circular end in the top view, draw a series of triangles joined at common sides. For example, draw arcs 2–*C* from point 2. To find point *C*, draw arc *DC* from *D*. Chord *DC* is true length in the top view.

Step 3 Construct the remaining planes, *A*–1–2 and *G*–3–4, by triangulation to complete the inside half-pattern of the transition piece. Draw the fold lines, where the surface is to be bent, as thin lines. The seam line for the pattern is line *A*–1, the shortest line.

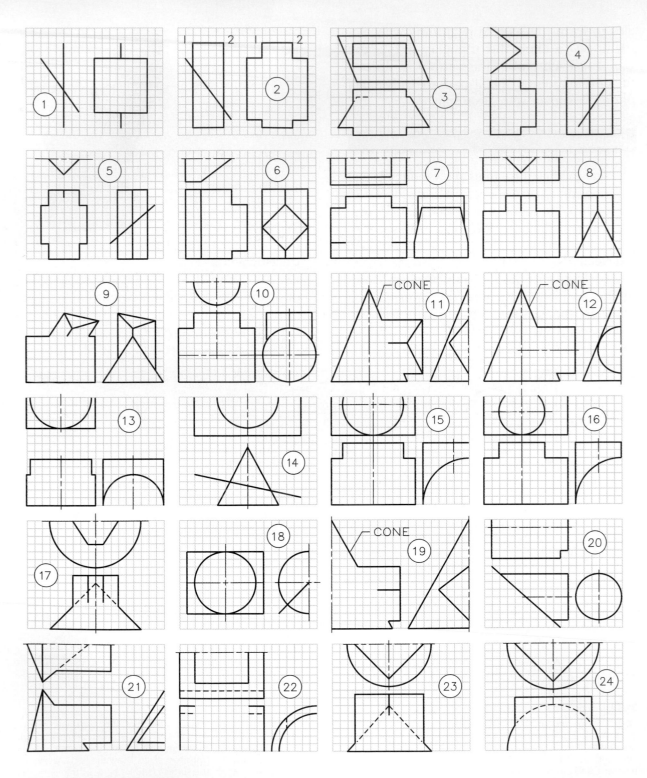

31.50 **(Problems 1–24)** Intersections.

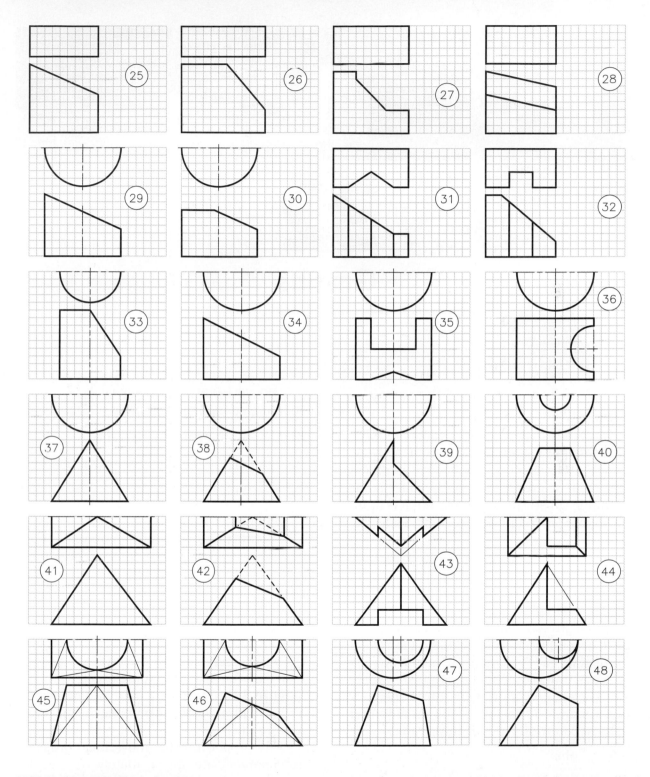

31.51 (Problems 25–48) Developments.

32

Graphs

32.1 Introduction

Data and information expressed as numbers and words are usually difficult to analyze or evaluate unless transcribed into graphical form, or as a **graph**. The term **chart** is an acceptable substitute for graph, but it is more appropriate when applied to maps, a specialized form of graphs.

Graphs are especially useful in presenting data at briefings where the data must be interpreted and communicated quickly to those in attendance. Graphs are convenient ways to condense and present data visually, allowing the data to be grasped much more easily than when presented verbally or as tables of numbers.

Several different types of graphs are widely used. Their application depends on the data and the nature of the presentation required. The most common types of graphs are:

1. Pie graphs

2. Bar graphs

3. Linear coordinate graphs

4. Logarithmic coordinate graphs

5. Semilogarithmic coordinate graphs

6. Schematics and diagrams

Proportions

Graphs are used on large display boards and in technical reports, as slides for a projector, or as transparencies for an overhead projector. Consequently, the proportion of the graph must be determined before it is constructed to match the page, slide, or transparency.

A graph that is to be photographed with a 35-mm camera must be drawn to the proportions of the film, or approximately 3×2 (**Figure 32.1**). This area may be enlarged or reduced proportionally by using the diagonal-line method.

The proportions of an overhead projector transparency are approximately 10×8. The image size should not exceed 9.5 inches × 7.5

572

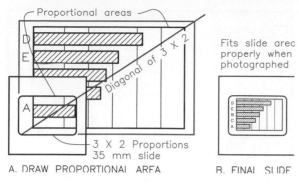

A. DRAW PROPORTIONAL AREA   B. FINAL SLIDE

32.1 This diagonal-line method may be used to lay out drawings that are proportional to the area of a 35-mm slide.

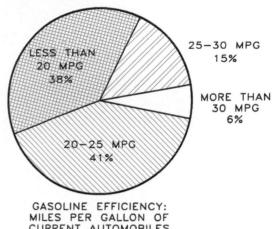

GASOLINE EFFICIENCY:
MILES PER GALLON OF
CURRENT AUTOMOBILES

32.2 A pie graph shows the relationship of parts to a whole. It is most effective when there are only a few parts.

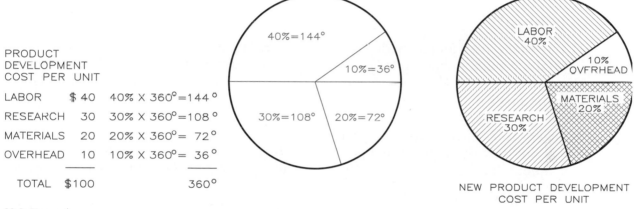

PRODUCT
DEVELOPMENT
COST PER UNIT

LABOR	$ 40	40% X 360° = 144°
RESEARCH	30	30% X 360° = 108°
MATERIALS	20	20% X 360° = 72°
OVERHEAD	10	10% X 360° = 36°
TOTAL	$100	360°

NEW PRODUCT DEVELOPMENT
COST PER UNIT

32.3 Pie graphs:

Step 1 Find the sum of the parts and the percentage that each is of the total. Multiply each percentage by 360° to obtain the angle of each sector.

Step 2 Draw the circle and construct each sector using the degrees of each from Step 1. Place small sectors as nearly horizontal as possible.

Step 3 Label sectors with their proper names and percentages. Exact numbers also may be included in each sector to add more clarity.

inches to allow an adequate margin for mounting the transparency on a frame of cardboard or plastic.

32.2 Pie Graphs

Pie graphs compare the relationship of parts to a whole. For example, **Figure 32.2** shows a pie graph that compares the energy efficiency (in miles per gallon) of the current models of automobiles on the market.

Figure 32.3 illustrates the steps involved in drawing a pie graph. The data in this example, as simple as it is, are not as easily compared in numerical form as when drawn as a pie graph. Position thin sectors of a pie graph as nearly horizontal as possible to provide more space for labeling. When space is not available within the sectors, place labels outside the pie graph and, if necessary, use leaders (see **Figure 32.2**). Showing the percentage repre-

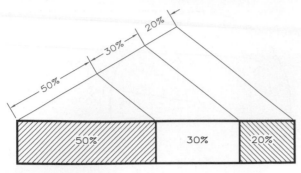

32.4 The diagonal-line method may be used to find the percentages of the parts to the whole, where the total bar represents 100%.

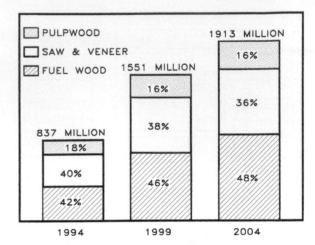

32.5 In this bar graph, each bar represents 100% of the total amount and shows the percentages of the parts to the total.

sented by each sector is important and giving the actual numbers or values as part of the label is also desirable.

32.3 Bar Graphs

Bar graphs are widely used for comparing values because the general public understands them. A bar graph may be a single bar (**Figure 32.4**) where the length of the bar representing 100% is divided into lengths proportional to the percentages of its three parts. In **Figure 32.5** the bars not only show the overall production of timber (the total heights of the

bars), but also the percentages of the total devoted to three uses of the timber.

Figure 32.6 shows how to convert data into a bar graph that can be used in a report or briefing. The axes of the graph carry labels, and its title appears inside the graph where space is available.

The bars of a bar graph should be sorted in ascending or descending order unless there is an overriding reason not to, such as a chrono-

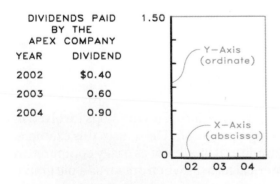

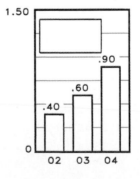

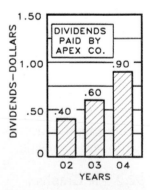

32.6 Drawing a bar graph:

Given These numerical data are to be plotted as a bar graph for better presentation in a report.

Step 1 Scale the vertical and horizontal axes so that the data will fit on the grid. Begin the bars at zero.

Step 2 The width of the bars should be greater than the space between them. Lines should not cross the bars.

Step 3 Strengthen lines, place a title in the graph, label the axes, and cross-hatch the bars.

logical sequence. An arbitrary arrangement of the bars, such as in alphabetical or numerical order, makes a graph difficult to evaluate (**Figure 32.7A**). However, ranking the categories by bar length allows easier comparisons from smallest to largest (**Figure 32.7B**). If the data are sequential and involve time, such as sales per month, a better arrangement of the bars is chronologically, to show the effect of time.

Bars in a bar graph may be horizontal (**Figure 32.8**) or vertical. Data cannot be compared accurately unless each bar is full length and originates at zero. Also, bars should not extend beyond the limits of the graph (giving the impression that the data were "too hot" to hold). Another form of bar graph shows plus and minus changes from a base value (**Figure 32.9**).

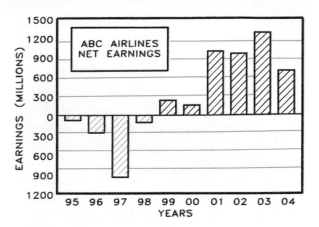

32.9 Bar graphs may be drawn with bars in both negative and positive directions.

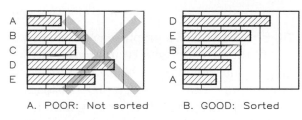

A. POOR: Not sorted B. GOOD: Sorted

32.7 Arranging bars by length.

A When bars are arbitrarily arranged, such as alphabetically, the bar graph is difficult to interpret.

B When the bars are sorted by length, the graph is much easier to interpret.

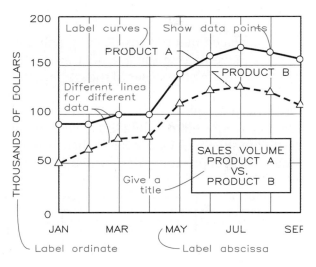

32.10 This basic linear coordinate graph illustrates the important features on a graph.

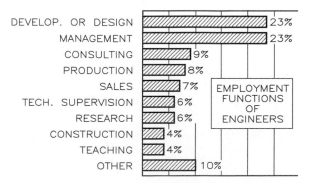

32.8 The horizontal bars of this graph are arranged in descending order to show the employment functions of engineers.

32.4 Linear Coordinate Graphs

Figure 32.10 shows a typical linear coordinate graph, with notes explaining its important features. Divided into equal divisions, the axes are referred to as linear scales. Data points are plotted on the grid by using measurements, called coordinates, along each axis from zero. The plotted points are marked with symbols such as circles or squares that may be easily drawn with a template. The horizontal scale of the graph is called the **abscissa** or *x*-axis. The vertical scale is called the **ordinate** or *y*-axis.

When the points have been plotted, a curve is drawn through them to represent the data. The line drawn to represent data points is called a **curve** regardless of whether it is a straight line, smooth curve, or broken line. The curve should not extend through the plotted points; rather, the points should be left as open circles or other symbols.

The curve is the most important part of the graph, so it should be drawn as the most prominent (thickest) line. If there are two curves in a graph, they should be drawn as different line types and labeled. The title of the graph is placed in a box inside the graph and units are given along the *x*- and *y*-axes with labels identifying the scales of the graph.

Broken-Line Graphs

The steps required to draw a linear coordinate graph are shown in **Figure 32.11**. Because the data points represent sales, which have no predictable pattern, the data do not give a smooth progression from point to point. Therefore the points are connected with a **broken-line curve** drawn as an angular line from point to point.

Again, leave the symbols used to mark the data points open rather than extending grid lines or the data curve through them (**Figure 32.12**). Each circle or symbol used to plot points should be about 1/8 in. (3 mm) in diameter. **Figure 32.13** shows typical data-point symbols and lines.

Computer Method   Data points may be produced as *Circles*, *Donuts* (open and closed), or *Polygons*. The grid lines and curves that pass through the open symbols may be removed easily with the *Trim* command (**Figure 32.14**).

Titles   The title of a graph may be located in any of the positions shown in **Figure 32.15**. A graph's title should never be as meaningless as "graph" or "coordinate graph." Instead, it should identify concisely what the graph shows.

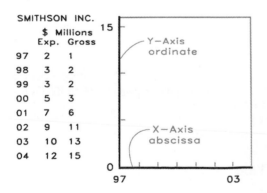
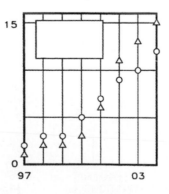
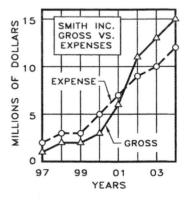

32.11 A broken-line graph:

Given A record of the Smith Company's gross income and expenses.

Step 1 Lay off the vertical (ordinate) and horizontal (abscissa) axes to provide space for the largest values.

Step 2 Draw division lines and plot the data, using different symbols for each set of data.

Step 3 Connect points with straight lines, label the axes, title the graph, darken the lines, and label the curves.

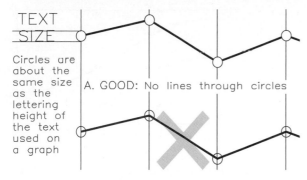

TEXT SIZE

Circles are about the same size as the lettering height of the text used on a graph

A. GOOD: No lines through circles

B. POOR: Lines through circles

32.12 The curve of a graph drawn from point to point should not extend through the symbols used to represent data points.

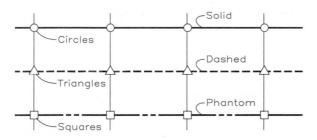

Solid

Circles

Dashed

Triangles

Phantom

Squares

32.13 Symbols and lines such as these may be used to represent different curves on a graph. The data-point symbols should be drawn about the same size as the letter height being used, usually a little less than 1/8 in. (3 mm) in diameter.

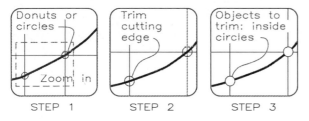

Donuts or circles

Zoom in

Trim cutting edge

Objects to trim: inside circles

STEP 1   STEP 2   STEP 3

32.14 Editing data points:

Step 1 Open data points may be plotted on a graph as *Circle*, *Donuts*, or *Polygons*. To remove lines from inside the open points, *Zoom* in on several points.

Step 2 *Command:* TRIM (Enter)

Select cutting edges (s): ...

Select objects: (Select the circle.) (Enter)

Step 3 *Select object to trim:* (Select the lines inside the circle, and they will be removed.)

Continue this process for all points.

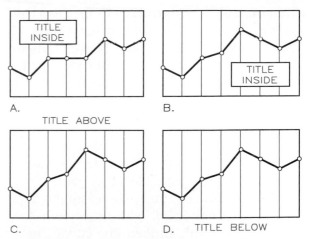

TITLE INSIDE

A.

TITLE ABOVE

TITLE INSIDE

B.

C.

D.   TITLE BELOW

32.15 Placement of titles on a graph.

A and **B** The title of a graph may be placed inside a box within the graph. Box perimeter lines should not coincide with grid lines.

C Titles may be placed over the graph.

D Titles may be placed under the graph.

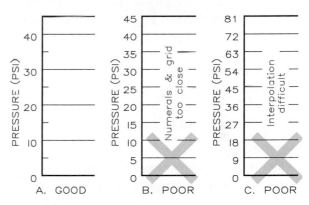

PRESSURE (PSI)

A. GOOD   B. POOR   C. POOR

Numerals & grid too close

Interpolation difficult

32.16 Calibrating graph scales.

A The scale is properly labeled and calibrated. It has about the right number of grid lines and divisions, and the numbers are well spaced and easy to interpolate.

B The numbers are too close together, and there are too many grid lines.

C The increments selected make interpolation difficult.

Scale and Labeling The calibration and labeling of the axes affects the appearance and readability of a graph. **Figure 32.16A** shows a properly calibrated and labeled axis. **Figure 32.16B** and **C** illustrates common mistakes: placing the grid lines too close together and labeling too many divisions along the axis. In **Figure 32.16C**, the choice of the interval between the labeled values (9 units) makes interpolation between them difficult. For example, locating the value 22 by eye is more difficult on this scale than on the one shown in **Figure 32.16A**.

Smooth-Line Graphs

The strength of concrete related to its curing time is plotted in **Figure 32.17**. The strength of concrete changes gradually and continuously in relation to curing time. Therefore the data points are connected with a smooth-line curve rather than a broken-line curve. These relationships are represented by the **best-fit curve**, a smooth curve that is an average representation of the points.

There is a smooth-line curve relationship between miles per gallon and the speed at which a car is driven. **Figure 32.18** compares the results for two engines.

A smooth-line curve on a graph implies that interpolations between data points can be made to estimate other values. Data points connected by a broken-line curve imply that interpolations between the plotted points cannot be made.

Straight-Line Graphs

Some graphs have neither broken-line curves nor smooth-line curves, but straight-line

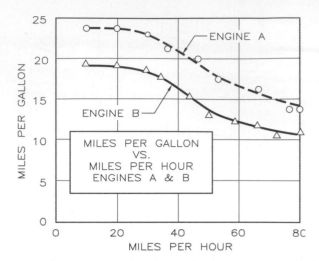

32.18 These data are represented by best-fit curves that approximate the data without necessarily passing through each point. The relationship of these data indicates that this curve should be a smooth-line curve rather than a broken-line curve.

curves (**Figure 32.19**). On this graph, a third value can be determined from the two given values. For example, if you are driving 70 miles per hour and you take 5 seconds to react and apply your brakes, you will have traveled 550 feet in that time.

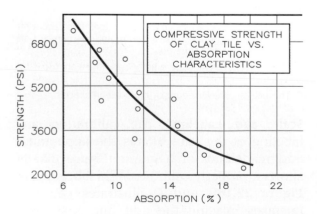

32.17 When the data being graphed involve gradual, continuous changes in relationships, the curve is drawn as a smooth line.

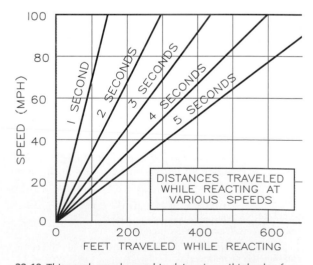

32.19 This graph may be used to determine a third value from the other two variables. For example, select a speed of 70 mph and a time of 5 seconds to find a distance traveled of 550 ft.

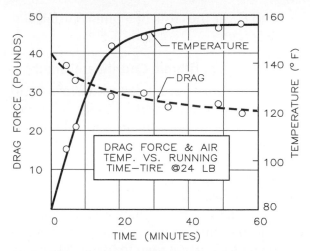

32.20 A two-scale graph has different scales along each y axis, and labels identify which scale applies to which curve.

Two-Scale Coordinate Graphs

Graphs may contain different scales in combination, as shown in **Figure 32.20**, where the vertical scale at the left is in units of pounds and the one at the right is in degrees of temperature. Both curves are drawn with respect to their y axes and each curve is labeled. Two-scale graphs of this type may be confusing unless they are clearly labeled. Two-scale graphs are effective for comparing related variables, as shown here.

Optimization Graphs

Figure 32.21 depicts the optimization of an automobile's depreciation in terms of maintenance cost increases. These two sets of data cross at an x-axis value of slightly more than five years, or the optimum point. At that time the cost of maintenance is equal to the value of the car, indicating that it might be a desirable time to buy a new car.

The steps involved in drawing an optimization graph are illustrated in **Figure 32.22**. Here, the manufacturing cost per unit reduces as more units are made, causing warehousing costs to increase. Adding the two curves to get a third (total) curve indicates that the optimum number to manufacture at a time is

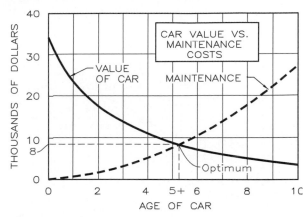

32.21 This graph shows the optimum time to sell a car based on the intersection of curves representing the depreciating value of the car and its increasing maintenance costs.

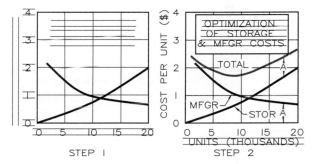

32.22 Constructing an optimization graph:

Step 1 Lay out the graph and plot the curves from the data given.

Step 2 Graphically add the two curves to find a third curve. For example, transfer distance A to locate a point on the third curve. The lowest point of the "total" curve is the optimum point, or 8000 units.

about 8000 units (the low point on the total curve). When more or fewer units are manufactured, the total cost per unit is greater.

Composite Graphs

The graph shown in **Figure 32.23** is a composite (or combination) of an area graph and a coordinate graph. The upper curve is a plot of the company's gross income. The lower curve is a plot of the company's expenses. The difference between the two is the company's profit. Cross-hatching the areas emphasizes the relative values of expenses and profits.

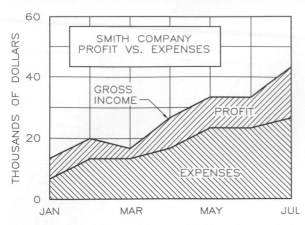

32.23 This composite graph is a combination of a coordinate graph and an area graph. The upper area represents the difference between the two plotted curves.

Break-Even Graphs

Break-even graphs help in evaluating marketing and manufacturing costs to determine the selling price of a product. As **Figure 32.24** shows, if the desired break-even point for a product is 10,000 units, it must sell for $3.50

each to cover the costs of manufacturing and development.

32.5 Semilogarithmic Coordinate Graphs

Semilogarithmic graphs are called **ratio graphs** because they graphically represent ratios. One scale, usually the vertical scale, is logarithmic, and the other is linear (divided into equal divisions).

The same data plotted on a linear grid and on a semilogarithmic grid are compared in **Figure 32.25**. The semilogarithmic graph reveals that the percentage change from 0 to 5 is greater for curve B than for curve A because here, curve B is steeper. The plot on the linear grid appears to show the opposite result.

Figure 32.26 shows the relationship between the linear scale and the logarithmic scale. Equal divisions along the linear scale have unequal ratios, but equal divisions along the log scale have equal ratios.

Log scales may have one or many cycles. Each cycle increases by a factor of 10. For example, the scale shown in **Figure 32.27A** is a

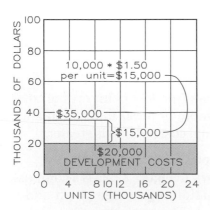

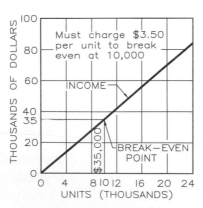

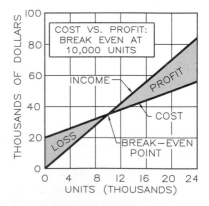

32.24 Drawing a break-even graph:

Step 1 Plot the development cost ($20,000). At $1.50 per unit to make, the total cost would be $35,000 for 10,000 units, the break-even point.

Step 2 To break even at 10,000, the manufacturer must sell each unit for $3.50. Draw a line from zero through the break-even point of $35,000 to represent income.

Step 3 There is a loss of $20,000 at zero units, but progressively less until the break-even point is reached. Profit is the difference between the curves at the right of break-even point.

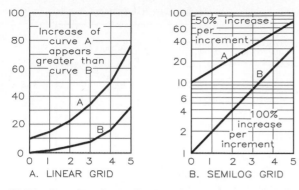

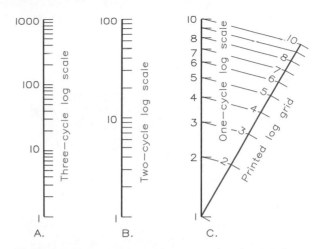

32.25 When plotted on a linear grid, curve A appears to be increasing at a greater rate than curve B. However, plotting the data on a semilogarithmic grid reveals the true rate of change.

32.27 Logarithmic scales may have several cycles: (A) three-cycle scales, (B) two-cycle scales, and (C) one-cycle scales. Calibrations may be projected to a scale of any length from a printed scale as shown here.

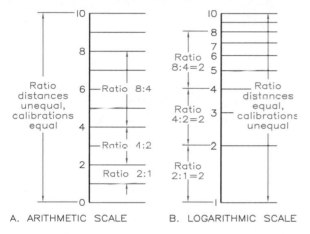

32.26 (A) The divisions on an arithmetic scale are equal and represent unequal ratios between points. (B) The divisions on logarithmic scales are unequal and represent equal ratios.

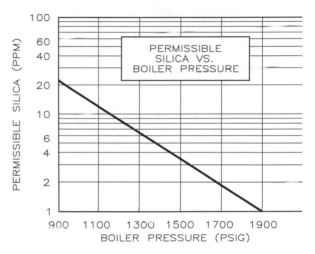

32.28 This semilogarithmic graph relates permissible silica (parts per million) to boiler pressure.

three-cycle scale, and the one shown in **Figure 32.27B** is a two-cycle scale. When scales must be drawn to a certain length, commercially printed log scales may be used to graphically transfer the calibrations to the scale being used (**Figure 32.27C**).

An application of a semilogarithmic graph for presenting industrial data is illustrated in **Figure 32.28**. People who do not realize that semilog graphs are different from linear coordinate graphs may misunderstand them. Also, zero values cannot be shown on log scales.

Percentage Graphs

The percentage that one number is of another, or the percentage increase of one number to a greater number, can be determined on a semilogarithmic graph. Data plotted in **Figure 32.29A** are used to find the percentage that 30 is of 60 (two points on the curve) by arithmetic. The vertical distance between them is the difference between their

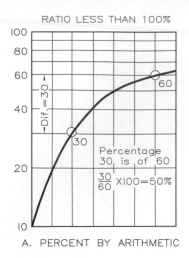
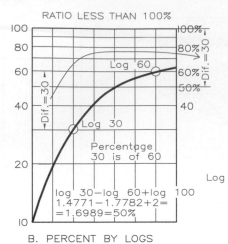
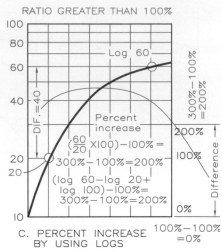

RATIO LESS THAN 100%

A. PERCENT BY ARITHMETIC

RATIO LESS THAN 100%

B. PERCENT BY LOGS

RATIO GREATER THAN 100%

C. PERCENT INCREASE
BY USING LOGS

32.29 Percentage graphs:

A To find the percentage that one data point is of another point (the percentage that 30 is of 60, for example) you may calculate it mathematically: (30/60)(100) = 50%.

B Find the percentage that 30 is of 60 by using the logarithms of the numbers. Or find it graphically by transferring the distance between 30 and 60 to the scale at the right, which shows that 30 is 50% of 60.

C To find a percentage increase greater than 100%, divide the smaller number into the larger number. Find the difference between the logs of 60 and 20 with dividers and measure upward from 100% to find the increase of 200%.

logarithms, so the percentage can be found graphically in **Figure 32.29B**. The distance from 30 to 60 is transferred to the log scale at the right of the graph and subtracted from the log of 100 to find the value of 50% as a direct reading of percentage.

In **Figure 32.29C**, the percentage increase between two points is transferred from the grid to the lower end of the log scale and measured upward because the increase is greater than zero. These methods may be used to find percentage increases or decreases for any set of points on the grid.

32.6 Schematics

The schematic diagram shown in **Figure 32.30** depicts the steps required to complete a construction project. Each step is blocked in and connected with arrows to give the sequence.

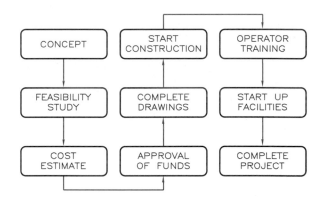

32.30 This block diagram shows the steps required to complete a project.

The organization of a group of people can be represented in an organizational graph (usually called an organization chart), as shown in **Figure 32.31**. This particular example was drawn by a computer program especially designed for drawing these types of graphs. The positions represented by the

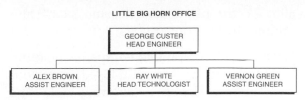

LITTLE BIG HORN OFFICE

32.31 A computer-drawn organizational chart clearly outlines the lines of responsibility.

blocks in the lower part of the graph are responsible to those represented by the blocks above them. Lines of authority connecting the blocks give the routes for communication both upward and downward.

Geographical charts are used to combine maps and other relationships, such as weather (**Figure 32.32**). Here, different hatching symbols represent the annual rainfall in various parts of the nation.

32.7 Graphs by Computer

Many computer programs are available for converting numerical data into various types of graphs to improve comprehension and interpretation. These programs vary from data representation as bar graphs, coordinate graphs, and pie graphs to 3D mathematical models.

Computer-produced graphs are especially useful for preparing visual aids for projection on a screen for a presentation and for technical reports. **Figure 32.33** is an example of a 3D

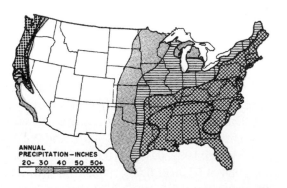

32.32 This map shows the annual rain for various parts of the nation. (Courtesy of the Structural Clay Products Institute.)

bar graph printed from data that was input in tabular form. This same data can be instantaneously plotted in a different 3D form as shown in **Figure 32.34**, and plotted in yet a different format as shown in **Figure 32.35**.

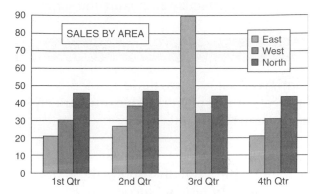

32.33 A computer-drawn three-dimensional bar graph.

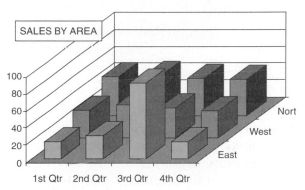

32.34 A three-dimensional plot of the data shown in Figure 32.33, drawn by a computer program.

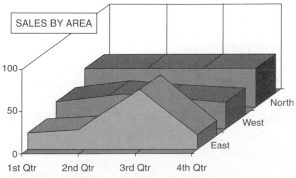

32.35 Another three-dimensional version of the data from Figure 32.33 drawn by a computer program.

It is up to you, the user, to determine which format is best in communicating a particular type of data to your audience, whether by an illustration in a written report or by a visual aid during an oral presentation. Whether drawn by hand or by computer, the principles of preparing effective graphs are the same.

Problems

Draw your solutions to these problems on size A sheets. Apply the techniques and principles covered in this chapter.

Pie Graphs

1. Draw a pie graph that shows the comparative sources of retirees' income: investments, 34%; employment, 24%; social security, 21%; pensions, 19%; other, 2%.

2. Draw a pie graph that shows the number of members of the technological team: engineers, 985,000; technicians, 932,000; scientists, 410,000.

3. Construct a pie graph of the employment status of graduates of two-year technician programs a year after graduation: employed, 63%; continuing full-time study, 23%; considering job offers, 6%; military, 6%; other, 2%.

4. Draw a pie graph showing the types of degrees held by aerospace engineers: bachelor's, 65%; master's, 29%; Ph.D.'s, 6%.

Bar Graphs

5. Draw a bar graph that shows the starting salaries for 1999 engineering graduates. Aerospace $40,662, chemical $46,920, civil $36,076, electrical $46,180, industrial $43,086, mechanical $43,275, metallurgical 43,038, mining $39,587, and petroleum $50,440.

6. Draw a bar graph that represents 100% of a die casting alloy. The proportional parts of the alloy are: tin, 16%; lead, 24%; zinc, 38.8%; aluminum, 16.4%; copper, 4.8%.

7. Draw a bar graph that compares the number of skilled workers employed in various occupations. Use the following data and arrange the graph for ease of comparing occupations: carpenters, 82,000; all-round machinists, 310,000; plumbers, 350,000; bricklayers, 200,000; appliance servicers, 185,000; automotive mechanics, 760,000; electricians, 380,000; painters, 400,000.

8. Draw a bar graph that shows the characteristics of a typical U.S. family's spending: housing, 29.6%; food, 15.1%; transportation, 16.7%; clothing, 5.9%; retirement, 8.6%; entertainment, 4.9%; insurance, 5.2%; health care, 3.0%; charity, 3.1%; other, 7.9%.

9. Draw a bar graph that compares the corrosion resistance of the materials listed in the following table.

	Loss in Weight (%)	
	In Atmosphere	*In Sea Water*
Common steel	100	100
10% nickel steel	70	80
25% nickel steel	20	55

10. Draw a bar graph of the data from Problem 1.

11. Draw a bar graph of the data from Problem 2.

12. Draw a bar graph of the data from Problem 3.

13. Construct a bar graph comparing sales and earnings of Apple Computer from 1980 through 1989. Data are by year for sales and earnings (profit) in billions of dollars: 1980, 0 and 0; 1981, 0.33 and 0.05; 1982, 0.60 and 0.07; 1983, 1.00 and 0.09; 1984, 1.51 and 0.08; 1985, 1.90 and 0.07; 1986, 1.85 and 0.12; 1987, 2.70 and 0.25; 1988, 4.15 and 0.40; 1989, 5.50 and 0.45.

Linear Coordinate Graphs

14. Draw a linear coordinate graph to show world-wide commerce in billions on the internet (W) and USA internet commerce (U) from 1998–2003. 1998: $20 U and $22 W, 1999: $25 U and $35 W, 2000: $45 U and $62 W, 2001: $60 U and $150 W, 2002: $100 U and $250 W, 2003: $147 U and $380 W.

15. Construct a linear coordinate graph that shows the relationship of energy costs (mills per kilowatt-hour) on the y-axis to the percent capacity of a nuclear power plant and a gas-or oil-fired power plant on the x-axis. Gas- or oil-fired plant data: 17 mills, 10%; 12 mills, 20%; 8 mills, 40%; 7 mills, 60%; 6 mills, 80%; 5.8 mills, 100%. Nuclear plant data: 24 mills, 10%; 14 mills, 20%; 7 mills, 40%; 5 mills, 60%; 4.2 mills, 80%; 3.7 mills, 100%.

16. Plot the data from Problem 13 as a linear coordinate graph.

17. Construct a linear coordinate graph to show the relationship between the transverse resilience in inch-pounds (ip) on the y-axis and the single-blow impact in foot-pounds (fp) on the x-axis of gray iron. Data: 21 fp, 375 ip; 22 fp, 350 ip; 23 fp, 380 ip; 30 fp, 400 ip; 32 fp, 420 ip; 33 fp, 410 ip; 38 fp, 510 ip; 45 fp, 615 ip; 50 fp, 585 ip; 60 fp, 785 ip; 70 fp, 900 ip; 75 fp, 920 ip.

18. Draw a linear coordinate graph to show the estimated cost per month of colonizing the moon. The y-axis is systems cost in billions of dollars and x-axis is duration of the project in years. 10 years, $250; 15 years, $125; 20 years, $110; 25 years, $100; 30 years, $95.

19. Draw a linear coordinate graph that shows the voltage characteristics for a generator as given in the following table of values: abscissa, armature current in amperes, (Ia); ordinate, terminal voltage in volts, (Et):

Ia	Et	Ia	Et	Ia	Et
0	288	31.1	181.8	41.5	68.0
5.4	275	35.4	156.0	40.5	42.5
11.8	257	39.7	108.0	39.5	26.5
15.6	247	40.5	97.0	37.8	16.0
22.2	224.5	40.7	90.0	13.0	0
26.2	217	41.4	77.5		

20. Draw a linear coordinate graph for the centrifugal pump test data in the following table. The units along the x-axis are to be gallons per minute. Use four curves to represent the variables given.

Gallons per Minute	Water HP	Electric HP
0	0.00	1.36
75	0.72	2.25
115	1.00	2.54
154	1.00	2.74
185	0.74	2.80
200	0.63	2.83

21. Draw a linear coordinate graph that compares two of the values shown in **Table 32.1**—ultimate strength and elastic limit—with degrees of temperature (x-axis).

Table 32.1

F°	Ultimate Strength	Elastic Limit
400	257,500	208,000
500	247,000	224,500
600	232,500	214,000
700	207,500	193,500
800	180,500	169,000
900	159,500	146,500
1000	142,500	128,500
1100	126,500	114,000
1200	114,500	96,500
1300	108,000	85,500

Break-Even Graphs

22. Draw a break-even graph that shows the earnings for a new product that has a development cost of $12,000. The break-even point is at 8000 units and each costs $0.50 to manufacture. What would be the profit at volumes of 20,000 and 25,000?

23. Repeat Problem 22 except that the development costs are $80,000, the manufacturing cost of the first 10,000 units is $2.30 each, and the desired break-even point is 10,000 units. What is the profit at volumes of 20,000 and 30,000?

Logarithmic Graphs

24. Use the data given in **Table 32.2** to construct a logarithmic graph. Plot the vibration amplitude (A) as the ordinate and the vibration frequency (F) as the abscissa. The data for curve 1 represent the maximum limits of machinery in good condition with no danger from vibration. The data for curve 2 are the lower limits of machinery that is being vibrated excessively to the danger point. The vertical scale is three cycles, and the horizontal scale is two cycles.

25. Plot this data on a two-cycle log graph to show the current in amperes (y axis) versus the voltage in volts (x axis) of precision temperature-sensing resistors. Data: 1 volt, 1.9 amps; 2 volts, 4 amps; 4 volts, 8 amps; 8 volts, 17 amps; 10 volts, 20 amps; 20 volts, 30 amps; 40 volts, 36 amps; 80 volts, 31 amps; 100 volts, 30 amps.

26. Plot the data in Problem 20 as a logarithmic graph.

Semilogarithmic Graphs

27. Construct a semilogarithmic graph with the y axis as a two-cycle log scale from 1 to 100 and the x-axis as a linear scale from 1 to 7 to show the survivability of a shelter at varying distances from the atmospheric detonation of a one-megaton thermonuclear bomb. Plot overpressure in psi along the y-axis and distance from ground zero in miles along the x-axis. The data points represent an 80% chance of survival of the shelter. Data: 1 mile—55 psi; 2 miles—11 psi; 3 miles—4.5 psi; 4 miles—2.5 psi; 5 miles—2.0 psi; 6 miles—1.3 psi.

28. The growth of Division A and Division B of a company is to be plotted on a semilog graph with a one-cycle log scale on the y axis for sales in thousands of dollars and a linear scale on the x axis for years. Data: first year, $A = \$11,700$ and $B = \$44,000$; second year, $A = \$19,500$ and $B = \$50,000$; third year, $A = \$25,000$ and $B = \$55,000$; fourth year, $A = \$32,000$ and $B = \$64,000$; fifth year, $A = \$42,000$ and $B = \$66,000$; sixth year, $A = \$48,000$ and $B = \$75,000$. Which division has the better growth rate?

29. Draw a semilog chart showing probable engineering progress based on the following indices: 40,000 B.C., 21; 30,000 B.C., 21.5; 20,000 B.C., 22; 16,000 B.C., 23; 10,000 B.C., 27; 6000 B.C., 34; 4000 B.C., 39; 2000 B.C., 49; 500 B.C., 60; A.D. 1900, 100. Use a horizontal scale of 1 in. = 10,000 years, a height of about 5 in., and two-cycle printed paper, if available.

30. Plot the data in Problem 19 as a semilogarithmic graph.

31. Plot the data in Problem 21 as a semilogarithmic graph.

Table 32.2

F	100	200	500	1000	2000	5000	10,000
A(1)	0.0028	0.002	0.0015	0.001	0.0006	0.0003	0.00013
A(2)	0.06	0.05	0.04	0.03	0.018	0.005	0.001

Percentage Graphs

32. Using the graph plotted in Problem 28, determine the percentage of increase of Division A and Division B growth from year 1 to year 4. What percentage of sales of Division A are the sales of Division B at the end of year 2? At the end of year 6?

33. Plot the values for water horsepower and electric horsepower from Problem 20 on semilog paper. What percentage of electric horsepower is water horsepower when 75 gallons per minute are being pumped?

Logarithmic Graph

34. Plot the data below that gives the variation in pressure with the increase in water flow through a 1 inch diameter globe valve. The x-axis represents flow in gallons per minute (1 to 10) and the y-axis represents pressure drop in psi (0.1 to 1).

x	1	1.5	2	3	5	10
y	0.1	0.2	0.35	0.8	—	—

Two-Curve Coordinate Graph

35. Plot the data below that gives a comparison between stress and temperature of joints soldered by method A and method B. The x-axis represents temperature °F (0 to 180); the y-axis represents stress in 1000 pounds per square inch (0 to 12,000).

x	0	40	60	80	100	120	140	180
A	1	2.9	4	5	6.5	8.8	10.2	—
B	3.8	0.8	0.5	1.2	2.2	3.2	4.5	8

Semilog graphs

36. Sound is measured in decibels as the distance from the source is increased. Plot the following data on a semilogarithmic grid with decibel reduction on the x-axis (0 to 28) and distance from the source in feet on the y-axis (1 to 100, 2 cycles).

x	0	8	12	16	20	24	28
y	3	7.7	12	18	30	46	75

Two-Scale Coordinate Graph

37. Two curves are to be plotted on a linear coordinate graph to show change in temperature and drag force for a tire with 25 psi pressure over a period of 60 minutes. The x-axis represents time in minutes (0 to 60); the y-axis at the left is drag force in pounds (0 to 50); and the y-axis at the right is temperature °F (80 to 160).

x	0	5	10	20	30	40	50	60
Y-L	42	37	32	28	27	25	25	25
X-R	0	112	130	148	155	156	156	156

Linear Coordinate Graph

38. Plot the data below on a linear coordinate graph to compare the compressive strength of Gray Iron with its tensile strength. The x-axis represents compressive strength in 1000 psi (0 to 250) and the y-axis represents tensile strength in 1000 psi (0 to 100).

x	0	50	100	150	200	250
y	0	15	28	50	72	100

39. The data below yields a learning curve that indicates a production effectiveness of 20 percent when quantities of parts are doubled. For example, 60 units require 80 percent per unit than was required for 30 units. The x-axis represents units produced (0 to 100) and the y-axis represents cost per unit (0 to 40).

x	1	5	10	20	40	60	80	100
y	20	12	10	8	6.2	6	5.5	5

33

Nomography

33.1 Introduction

An additional aid in analyzing data is a graphical computer called a **nomogram**, **nomograph**, or "number chart." Basically, it is any graphical arrangement of calibrated scales and lines that may be used for calculations, usually those of a repetitive nature.

The term *nomogram* frequently denotes a specific type of scale arrangement called an **alignment graph**. Two examples of alignment graphs are shown in **Figure 33.1**. Other types have curved scales or different scale arrangements for use with more complex problems.

An alignment graph usually is constructed to solve for one or more unknowns in a formula or empirical relationship between two or more quantities. For example, such a graph can be used to convert degrees Celsius to degrees Fahrenheit or to find the size of a structural member to sustain a certain load. To read an alignment graph, **place a straightedge or draw a line**, **called an isopleth**, across the scales of

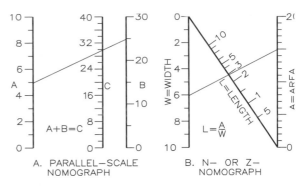

33.1 These two types of alignment graphs are typical nomographs.

the graph and read corresponding values from the scale on this line. **Figure 33.2** shows readings for the formula $W = U + V$.

33.2 Alignment Graph Scales

To construct any alignment graph, you must first determine the graduations of the scales that give the desired relationships. Alignment graph scales, called **functional scales**, are

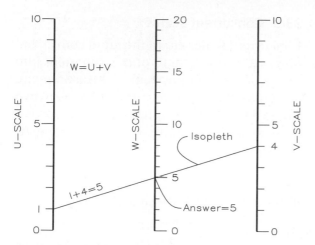

33.2 This graph illustrates the use of an isopleth to solve graphically for unknowns in an equation, $W = U + V$.

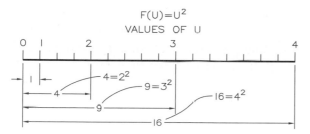

33.3 This functional scale contains units of measurement proportional to $F(U) = U^2$.

graduated according to values of some function of a variable. **Figure 33.3** illustrates a functional scale for $F(U) = U^2$. Substituting a value of $U = 2$ into the equation gives the position of U on the functional scale 4 units $(2^2 = 4)$ from zero. Repeating this procedure for as many values of U as required yields a scale for finding all corresponding values of the function.

The Scale Modulus
Because the graduations on a functional scale are spaced in proportion to values of the function, a proportionality, or scaling factor, is needed. This constant of proportionality is called the scale modulus, m:

$$m = \frac{L}{F(U_2) - F(U_1)} \tag{1}$$

where

m = scale modulus, in inches per functional unit,

L = desired length of the scale in inches,

$F(U_2)$ = functional value at the end of the scale, and

$F(U_1)$ = functional value at the start of the scale.

For example, suppose that you are to construct a functional scale for $F(U) = sin\ U$ from $0°$ to $45°$ and a scale 6 in. in length. Thus, $L = 6$ in., $F(U_2) = \sin 45° = 0.707$, $F(U_1) = \sin 0° = 0$. Substituting these values into Eq. (1) gives

$$m = \frac{6}{0.707 - 0} = 8.49 \text{ in. per (sine) unit.}$$

The Scale Equation
A scale equation makes possible graduation and calibration of functional scales. The general form of this equation is a variation of Eq. (1):

$$X = m[F(U) - F(U_1)] \tag{2}$$

where

X = distance from the measuring point of the scale to any graduation point,

m = scale modulus,

$F(U)$ = functional value at the graduation point, and

$F(U_1)$ = functional value at the measuring point of the scale.

For example, let's construct a functional scale for the equation $F(U) = \sin U$ $(0°$ to $45°)$. We have already determined that $m = 8.49$, $F(U) = \sin U$, and $F(U_1) = \sin 0° = 0$. By substitution, Eq. (2) becomes

$$X = 8.49(\sin U - 0) = 8.49 \sin U.$$

Using this equation, we can substitute values of U and construct a table of positions at $5°$ intervals as shown in **Figure 33.4**. The values

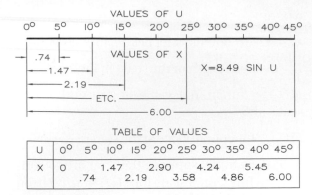

			VALUES OF X				

X=8.49 SIN U

.74
1.47
2.19
ETC.
6.00

TABLE OF VALUES

U	0°	5°	10°	15°	20°	25°	30°	35°	40°	45°
X	0		1.47		2.90		4.24		5.45	
		.74		2.19		3.58		4.86		6.00

33.4 This functional scale is calibrated with values from the table that were derived from the scale equation, X = 8.49 sin U.

of *X* give the positions in inches for the corresponding graduations measured from $U = 0°$. The initial measuring point does not have to be at one end of the scale, but an end is usually the most convenient point, especially if the functional value is zero at that point.

Figure 33.5 shows how to locate functional values along a scale with the proportional-line method. Measure the sine functions along a line at 5° intervals, with the end of the line passing through the 0° end of the scale. Transfer these functions from the inclined line with parallel lines back to the scale and label the functions.

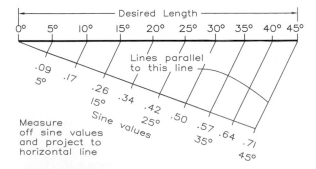

33.5 This functional scale shows the sine of the angles from 0° to 45° drawn by the proportional-line method. Draw the scale to a desired length and lay off the sine values of angles at 5° intervals along a construction line passing through the 0° end of the scale.

33.3 Concurrent Scales

Concurrent scales aid in the rapid conversion of terms in one system of measurement into terms of a second system of measurement. Formulas of the type $F_1 = F_2$, which relate two variables, may be adapted to the concurrent scale format. An example is the Fahrenheit-Celsius temperature relationship:

$$°F = \frac{9}{5}°C + 32$$

Another is the area of a circle:

$$A = \pi r^2$$

Construction of a concurrent scale chart involves determining a functional scale for each side of the mathematical formula so that the position and lengths of each scale coincide. To construct a conversion chart 5 inches long that gives the areas of circles whose radii range from 1 to 10, we first write $F_1(A) = A$; $F_2(r) = \pi r^2$, $r_1 = 1$; and $r_2 = 10$. The scale modulus for *r* is

$$m_R = \frac{L}{F_2(r_2) - F_2(r_1)}$$

$$= \frac{5}{\pi(10)^2 - \pi(1)^2} = 0.0161$$

Thus, the scale equation for *R* becomes

$$X_r = m_R[F_2(r) - F_2(r_1)]$$
$$= 0.0161[\pi r^2 - \pi(1)^2]$$
$$= 0.0161 \, \pi(r^2 - 1)$$
$$= 0.0505 \, (r^2 - 1)$$

Figure 33.6 shows a table of values for X_r and *r*. The *r*-scale values come from this table. From the original formula, $A = \pi \, r^2$, the limits of *A* are found to be $A_1 = \pi = 3.14$ and $A_2 = 100 \, \pi = 314$. The scale modulus for concurrent scales is always the same for equal-length scales; therefore $m_A = m_R = 0.0161$, and the scale equation for *A* becomes

$$X_A = m_A[F_1(A) - F_1(A_1)]$$
$$= 0.0161(A - 3.14)$$

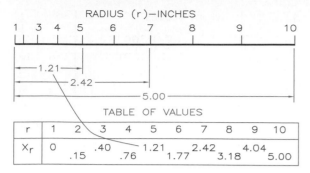

33.6 Calibrate one scale of a concurrent scale chart by using values from the table that have been calculated.

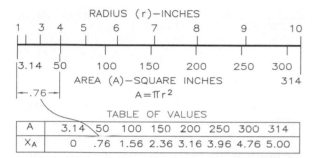

33.7 This completed concurrent scale chart is for the formula $A = \pi R^2$. Values for the A scale are from the table.

We then compute the corresponding table of values for selected values of A, as shown in **Figure 33.7**. We superimposed the A scale on the R scale and placed its calibrations on the other side of the line to facilitate reading.

To expand or contract one of the scales, use the technique shown in **Figure 33.8**. Draw the scales parallel at any distance apart and calibrate them in opposite directions. If they

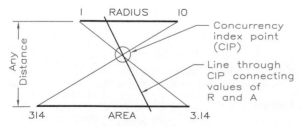

33.8 This construction is used to draw a concurrent graph with unequal scales.

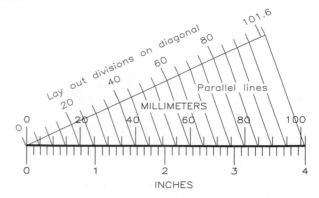

33.9 The proportional-line method can be used to construct an alignment graph that converts inches to millimeters. The units at each end of the scales must be known; 101.6 mm and 4 inches, in this case.

have different lengths, a different scale modulus and scale equation must be calculated for each scale.

To draw concurrent scales, use the proportional line method shown in **Figure 33.9**. There are 101.6 mm in 4 inches. Project millimeters to the upper side of the inch scale with a series of parallel projectors.

33.4 Alignment Graphs: Three Variables

For a formula containing three functions (of one variable each), draw a nomograph by selecting the lengths and positions of two scales according to the size of the graph desired. Calibrate these scales by using the scale equations presented in Section 33.3. Mathematical relationships may be used to locate the third scale, but graphical methods are simpler and less subject to error. Examples of the various forms of nomographs are shown in the following sections.

33.5 Parallel Scale Graphs: Linear Scales

Any formula of the type $F_3 = F_1 + F_2$ *may* be represented as a parallel scale alignment graph (**Figure 33.10**). For addition, the three scales increase (functionally) in the same

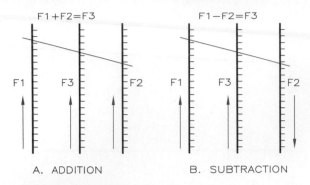

A. ADDITION   B. SUBTRACTION

33.10 These two common forms of parallel-scale alignment nomographs show the directions in which the scales increase for addition and subtraction.

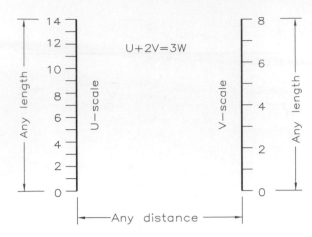

$U + 2V = 3W$

33.12 This calibration is of the outer scales for the formula $U + 2V = 3W$, where $U = 0$ to 12 and $V = 0$ to 8.

direction, and the function of the middle scale represents the sum of the other two. Reversing the direction of any scale changes the sign of its function in the formula, as for $F_1 - F_2 = F_3$.

The formula $Z = X + Y$ is used to illustrate this type of alignment graph (**Figure 33.11**). Draw and calibrate the outer scales for X and Y, then use two sets of data that yield a Z of 8 to locate the parallel Z scale. Divide the Z scale into

16 units. Add various values of X and Y with an isopleth to find their sums along the Z scale.

Figure 33.12 illustrates how to calibrate the outer scales of a parallel-scale nomograph for the equation $U + 2V = 3W$. The scales are placed any distance apart and are divided into

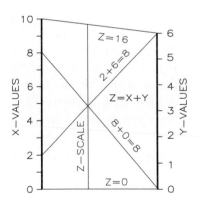
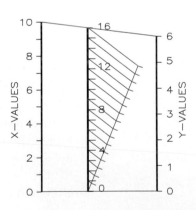
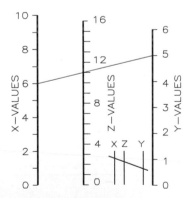

33.11 Constructing a parallel scale nomograph (linear scales):

Step 1 Draw and calibrate two parallel scales of any length. Locate the parallel Z scale by using two sets of values that give the same value (8 in this case). The ends of the Z scale are 0 and 16, the sum of the end values of X and Y.

Step 2 Draw the Z scale through the point located in Step 1 parallel to the other scales. Calibrate the scale from 0 to 16 by using the proportional line method.

Step 3 Calibrate and label the Z scale. Draw a key to show how to use the nomograph. If the Y scale were calibrated with 0 at the upper end instead of at the bottom, a different Z scale could be computed and the nomograph could be used for $Z = X - Y$.

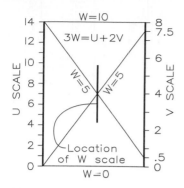
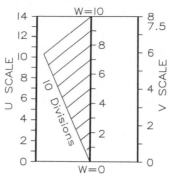
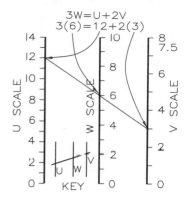

33.13 Constructing parallel scale nomograph (linear scales):

Step 1 Substitute the end values of the U and V scales into the formula to find end values of the W: 0 and 10. Use any two sets of U and V that give the same W (U = 0 and V = 7.5, W = 5, and U = 14 and V = 0.5, W = 5) to locate the W scale.

Step 2 Draw the W scale parallel to the outer scales to the limit lines of W = 10 and W = 0. This scale is 10 linear divisions long, so divide it graphically into 10 units. The W scale is a linear scale; construct it as shown in Figure 33.11.

Step 3 Connect any values of U and V with an isopleth to determine the resulting value of W. Draw a key that shows how to use the nomograph. The example values of U = 12 and V = 3 verify the graph's accuracy.

linear divisions from 0 to 14 for U and 0 to 8 for V. These scales are used to complete the nomograph that is explained in **Figure 33.13**.

Obtain the end calibrations for the middle scale by connecting the endpoints of the outer scales and substituting these values into the formula. W is 0 and 10 at its ends. Select two pairs of corresponding values of U and V that give the same value of W. For example, U = 0 and V = 7.5 give W = 5. Also W = 5 where U = 14 and V = 0.5. Because the W scale is linear (3W is a linear function), it can be subdivided into uniform intervals of equal parts. For a nonlinear scale, find the scale modulus and equation by substituting length and end values into Eq. (1).

Logarithmic Scales

Problems involving equations of the type $F_3 = (F_1)(F_2)$ can be solved in the manner described in **Figure 33.11** by using logarithmic scales instead of linear scales. The first step in drawing a nomograph with logarithmic scales is to

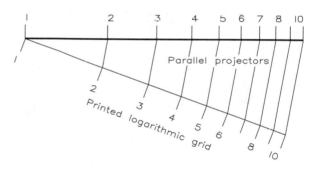

33.14 Here, a scale is calibrated graphically by projection from a printed logarithmic scale.

transfer the logarithmic functions to the scales. **Figure 33.14** shows the graphical method where units are projected from a printed logarithmic scale to the nomographic scale.

Figure 33.15 illustrates the conversion of the formula $Z = XY$ into a nomograph. The desired end values of the X and Y scales are 1 and 10. Sets of values of X and Y that give the same value of Z (10 in this case) are used to locate the Z axis with end values of 1 and 100.

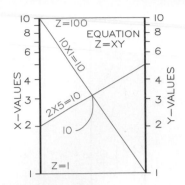 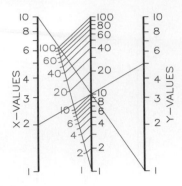 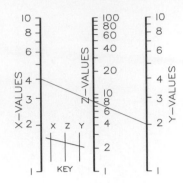

33.15 Parallel-scale nomograph (logarithmic scales):

Step 1 For the equation $Z = XY$ draw parallel log scales. Construct sets of X and Y points that give the same value of Z (10 in this case) to locate the Z scale with end values of 1 and 100.

Step 2 Graphically calibrate the Z axis as a two-cycle logarithmic scale from 1 to 100 by projecting from a printed log scale. The Z scale is parallel to the X and Y scales.

Step 3 Draw a key showing how to use the nomograph. By reversing the Y value scale from 1 to 10 downward and computing a different Z scale, the nomograph could be used for $Z = Y/X$.

The z-axis is drawn and calibrated as a two-cycle log scale. A key explains how to use an isopleth to add the logarithms of X and Y to give the log of Z. The addition of logarithms performs multiplication. Had the Y-axis been calibrated in the opposite direction with 1 at the upper end and 10 at the lower end, a new Z axis could have been calibrated for the formula $Z = Y/X$ for division by subtracting logarithms.

33.6 N or Z Nomographs

Whenever F_2 and F_3 are linear functions, we can avoid using logarithmic scales for formulas of the type

$$F_1 = \frac{F_2}{F_3}$$

We use an N graph (**Figure 33.16**) where the outer scales are functional scales and are linear if F_2 and F_3 are linear. If a parallel-scale graph were used for the same formula, all its scales would be logarithmic. Main features of N graphs are:

1. The outer scales are parallel functional scales of F_2 and F_3.
2. The outer scales increase functionally in opposite directions.

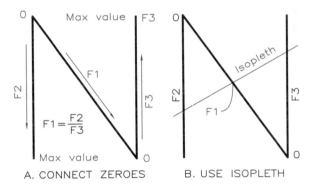

33.16 This N graph solves an equation of the form $F_1 = F_2/F_3$.

3. The diagonal scale connects the functional zeros of the outer scales.

4. The diagonal scale is not a functional scale for the function F_1 and is nonlinear.

Construction of an N nomograph is simplified because locating the middle (diagonal) scale is usually less of a problem than it is for a parallel scale graph. Calibration of the diagonal scale is most easily accomplished graphically.

Figure 33.17 shows how to construct a basic N graph of the equation $Z = Y/X$. Draw the diagonal to connect the zero ends of the

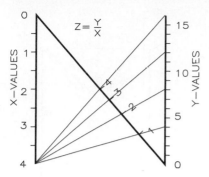 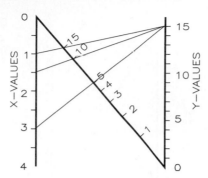 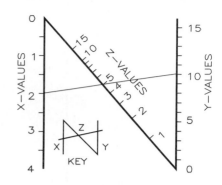

33.17 Constructing an N graph:

Step 1 Draw an N graph for the equation of $Z = Y/X$ and connect the zero ends of each scale with a diagonal scale. Draw isopleths to locate units along the diagonal scale.

Step 2 Draw additional isopleths to locate other units along the diagonal. The units on the diagonal should be whole units to make interpolation between them easy.

Step 3 Label the diagonal scale and draw a key. An isopleth confirms that 10/2 = 5. The accuracy of the N graph is greatest at the 0 end of the diagonal; the other end approaches infinity.

scales, then locate whole values along the diagonal by using combinations of X and Y values. Use whole-value units along the diagonal that are easy to interpolate between. Label the diagonal and give a key explaining how to use the nomograph. A sample isopleth verifies the correctness of the graphical relationship of the scales.

A more advanced N graph can be drawn for the equation

$$A = \frac{B+2}{C+5}$$

where $B=0$ and $C=15$. This equation takes the form

$$F_1 = \frac{F_2}{F_3}$$

where $F1(A) = A$, $F2(B) = B + 2$, and $F3(C) = C + 5$.

Thus the outer scales will represent $B + 2$ and $C + 5$ and the diagonal scale will be for A.

Begin the construction in the same manner as for a parallel scale graph by selecting the layout of the outer scales (**Figure 33.18**). Determine the limits of the diagonal scale by connecting the endpoints on the outer scales, giving $A = 0.1$ for $B = 0$, $C = 15$ and $A = 2.0$ for $B = 8$, and $C = 0$.

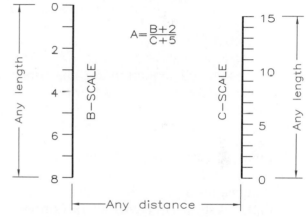

33.18 This calibration of the outer scales of an N graph is for the equation $A = (B + 2)/(C + 5)$.

Figure 33.19 shows these relationships and gives the remainder of the construction.

Locate the diagonal scale by finding the function zeros of the outer scales (that is, the points where $B + 2 = 0$ or $B = -2$ and $C + 5 = 0$ or $C = -5$). Then draw the diagonal scale by connecting these points. Calibrating the diagonal scale is most easily accomplished by substituting into the formula. Selecting the upper limit of an outer scale, say, $B = 8$, gives the formula:

$$A = \frac{10}{C+5}$$

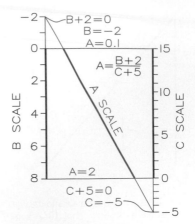

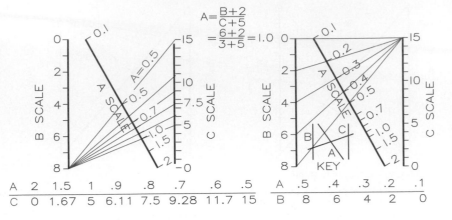

33.19 Constructing an N graph:

Step 1 Locate the diagonal scale by finding the functional zeros of the outer scales. Set $B + 2 = 0$ and $C + 5 = 0$, which gives a zero value for A. Therefore, $B = -2$ and $C = -5$ respectively.

Step 2 Select the upper limit of an outer scale ($B = 8$ in this case), substitute it into the equation, and obtain values of C for whole values of A. Draw isopleths from $B = 8$ to the values of C to calibrate the A scale.

Step 3 Calibrate the rest of the A scale by substituting $C = 15$ into the equation to determine a value on the B scale for whole values on the A scale. Draw isopleths from $C = 15$ to calibrate the A scale.

Then solve this equation for the other outer scale variable:

$$C = \frac{10}{A} - 5$$

Using it as a "scale equation," make a table of values for the desired values of A and corresponding values of C (up to the limit of C in the graph), as shown in **Figure 33.19**. Connect isopleths from $B = 8$ to the tabulated values of C. Their intersections with the diagonal scale give the required calibrations for approximately half the diagonal scale. Calibrate the rest of the diagonal scale by substituting the end value of the other outer scale ($C = 15$) into the formula, giving

$$A = \frac{B + 2}{20}$$

Solving for B yields $B = 20A - 2$.

Construct a table for the desired values of A (**Figure 33.19**) with isopleths connecting $C = 15$ with values of B, calibrating the rest of the A scale.

Problems

Solve the following problems on size A sheets. Show both the construction and calculations in the solutions involving both. If the calculations are extensive use a separate sheet.

Concurrent Scales

Draw concurrent scales for converting one type of unit to the other using the ranges given.

1. Kilometers and miles:

    1.609 km = 1 mile; from 10 to 100 miles.

2. Liters and U.S. gallons:

    1 L = 0.2692 U.S. gal; from 1 to 10 L.

3. Knots and miles per hour:

    1 knot = 1.15 mph; from 0 to 45 knots.

4. Horsepower and British thermal units:

    1 hp = 42.4 Btu; from 0 to 1200 hp.

5. Radius and area of a circle:

$$A = \pi r^2; \text{ from } r = 0 \text{ to } 10.$$

6. Inches and millimeters:

$$1 \text{ in.} = 25.4 \text{ mm}; \text{ from } 0 \text{ to } 5 \text{ in.}$$

7. Numbers and their logarithms:

Use logarithm tables; numbers from 1 to 10.

Addition and Subtraction Nomographs

Construct parallel scale nomographs to solve the following addition and subtraction problems.

8. $A = B + C$, where $B = 0$ to 10 and $C = 0$ to 5.

9. $Z = X + Y$, where $X = 0$ to 8 and $Y = 0$ to 12.

10. $Z = Y - X$, where $X = 0$ to 6 and $Y = 0$ to 24.

11. $A = C - B$, where $C = 0$ to 30 and $B = 0$ to 6.

12. $W = 2V + U$, where $U = 0$ to 12 and $V = 0$ to 9.

13. $W = 3U + V$, where $U = 0$ to 10; $V = 0$ to 10.

Multiplication and Division: Parallel Scales

Construct parallel scale nomographs with logarithmic scales for performing the following multiplication and division operations.

14. Area of a rectangle: $A = H \times W$, where $H = 1$ to 10 and $W = 1$ to 12.

15. Area of a triangle: $A = 1/2B \times H$, where $B = 1$ to 10, and $H = 1$ to 5.

16. Pythagorean theorem:

$$C^2 = A^2 + B^2$$

where C = hypotenuse of a right triangle in (cm), A = one leg of the right triangle, 5 to 50 cm, and B = second leg of the triangle, 20 to 80 cm.

17. Miles per gallon (mpg) an automobile gets: Miles vary from 1 to 500; gallons from 1 to 24.

18. Cost per mile (cpm) of an automobile: Miles vary from 1 to 500; cost varies from $1 to $28.

N Nomographs

Construct N graphs that will solve the following equations.

19. Stress $= P/A$, where P varies from 0 to 1000 psi and A varies from 0 to 15 in².

20. Volume of a cylinder: $V = \pi r^2 h$, where V = volume in in³; r = radius (5 to 10 ft); and h = height (2 to 20 in.).

21. Repeat Problem 14.

22. Repeat Problem 15.

23. Repeat Problem 16.

24. Repeat Problem 17.

25. Repeat Problem 18.

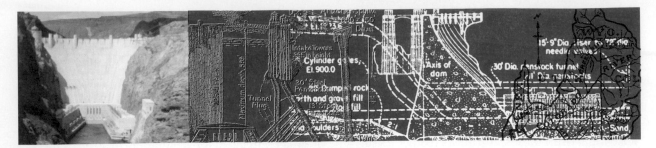

34

Empirical Equations and Calculus

34.1 Introduction

Graphical methods are useful supplements to mathematical techniques of solving problems dealing with experimental data. Graphics can be used to determine the mathematical equation of data obtained from laboratory testing and experimentation. Data from laboratory experiments and field tests are called **empirical data**. Empirical data often are expressed as one of three types of equations: **linear**, **power**, or **exponential**.

Analysis of empirical data begins by plotting the data on three types of standard grids: **rectangular (linear)**, **logarithmic,** or **semilogarithmic**. When the data plots as a straight line on one of these grids, its mathematical equation can be determined by using the characteristics of that particular grid (**Figure 34.1**).

To find the equation of a straight-line plot of data, you will need to know the **slope (M)** of the curve, and its **intercept**, **B**, and apply it to the **slope-intercept** equation of $Y = MX + B$.

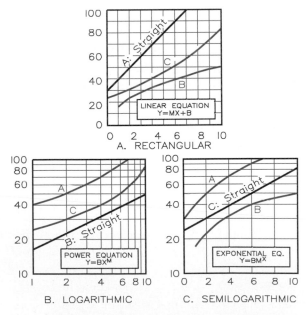

34.1 Plotting empirical data on each type of grid determines which yields a straight-line plot. A straight-line plot on one of these grids means that a mathematical equation can be derived from which to describe the data.

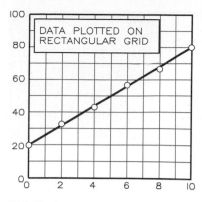

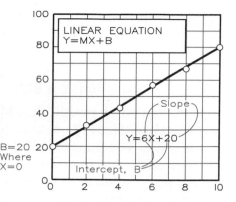

34.2 The linear equation, Y = MX + B:

Step 1 When plotted on a linear grid, the data is a straight line. It will have the equation of Y = MX + B.

Step 2 Two points, (0,20) and (5,50), are selected on the data curve to find the slope, M, which is Y/X. The slope is found to be 6.

Step 3 The intercept, B, where the curve crosses the y-axis is where X = 0 is found to be 20. The equation of the data is found to be Y = 6X + 20.

Slope is found by taking any two widely separated points on the curve and drawing a triangle to find the ratio of the height (ΔY) to the width (ΔX), which is the tangent of the curve's angle with the horizontal. The slope in **Figure 34.2** is found to be 6 by using this construction.

The intercept (B) is the point where the curve intersects the y-axis at X = 0. The intercept in **Figure 34.2** is B = 20. The values of slope and intercept can be combined with variables X and Y to find the equations of the data, as will be shown in the following examples.

34.2 Linear Equation: Y = MX + B

The curve representing empirical data in Figure 34.2 plots as a straight line on a rectangular graph, which identifies the data as linear. Each measurement along the y-axis is directly proportional to the measurement along the x-axis.

By selecting two points on the curve, the slope and the intercept can be found as the first step in determining its equation. The vertical and horizontal differences between the coordinates of each point establish the adjacent sides of a right triangle in Step 2. In the slope-intercept equation, Y = MX + B, the slope, M, is the tangent of the angle between the curve and the

horizontal, B,; is the Y intercept of the curve (where X = 0),; and X and Y are variables. Here, M = 30/5 = 6, and the intercept is at 20.

Substituting these values into the slope-intercept equation, we obtain Y = 6X + 20, from which we may determine values of Y by substituting any value of X into the equation. If the curve had sloped downward to the right, the slope would have been negative.

Figure 34.3 shows the relationship between the transverse strength and impact resistance

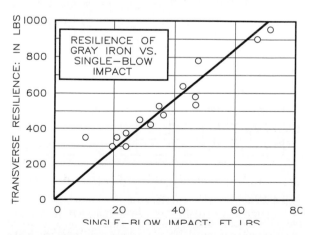

34.3 The relationship between the transverse strength of gray iron and impact resistance plots as a straight line yielding an equation of the form Y = MX + B or Y = 13.3X.

of gray iron obtained from empirical data. The data plot as a straight line on a linear grid, so its equation takes the linear form.

34.3 Power Equation: $Y = BX^M$

Data plotted on a logarithmic grid that yields a straight line (Figure 34.4) may be expressed in the power equation form in which Y is a function of X raised to a power, or $Y = BX^M$. We obtain the equation of the data by using the Y intercept as B and the slope of the curve as M.

Select two points on the curve to form the slope triangle and use an engineer's scale to measure its slope. If you draw the horizontal side of the right triangle as 1 or a multiple of 10, the vertical distance can be read directly. Here, the slope M (tangent of the angle) is 0.54 and the Y intercept, B, is 7; thus the equation is $Y = 7X^{0.54}$, which in logarithmic form is

$$\log Y = \log B + M \log X,$$
$$= \log 7 + 0.54 \log X.$$

We used base–10 logarithms in these examples, but natural logs may be used with e (2.718) as the base. In **Figure 34.5**, the inter-

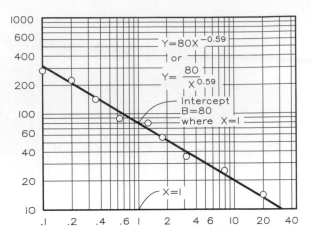

34.5 When using the slope-intercept method on a logarithmic grid, the intercept must be selected where X = 1. In this case the intercept lies at 80 (near the middle of the graph).

cept, B, lies on the y-axis where X = 1 because the curve is plotted on a logarithmic grid. The Y intercept (B = 80) is found where X = 1. Recall that the intercept at this point is analogous to the linear form of the equation because the log of 1 is 0. The curve slopes downward to the right making the slope, M, negative.

Figure 34.6 shows plots of empirical data that relate the specific weight (pounds per

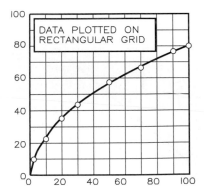

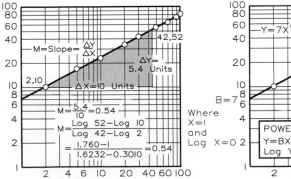

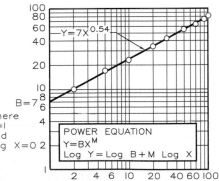

34.4 Power equation, $Y = BX^M$:

Step 1 Data plotted on a rectangular (linear) grid give a parabolic curve. Because the curve is not a straight line on this grid, its equation is not linear.

Step 2 The data plots as a straight line on a logarithmic grid. Find the slope, M, graphically with an engineer's scale by setting ΔX at 10 units and measuring ΔY as 5.4 units. M = 5.4/10, or 0.54.

Step 3 The intercept B = 7 lies at X = 1. Substitute the slope (0.54) and intercept (7) into the equation form of $Y = BX^M$ to obtain the equation of the data: $Y = 7X^{0.54}$.

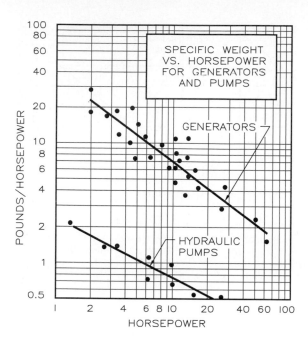

34.6 Empirical data plotted on a logarithmic grid showing the specific weight versus horsepower of electric generators and hydraulic pumps are straight lines. Their equations take the power form, Y = BX^M.

horsepower) of generators and hydraulic pumps to horsepower. The plots of the data are represented by straight lines on a logarithmic grid, which means that their equations take the power form.

34.4 Exponential Equation: Y = BM^X

When data plotted on a semilogarithmic grid (Figure 34.7) yield a straight line, the equation takes the form Y = BM^X, where B is the Y intercept, and M is the slope of the curve. Select two points along the curve, draw a right triangle, and find its slope, M. The slope of the curve is

$$\log M = \frac{\log 40 - \log 6}{8 - 3} = 0.1648,$$

or

$$M = (10)^{0.1648} = 1.46.$$

Substitute this value of M into the exponential equation:

$$Y = BM^X \qquad \text{or} \qquad Y = 2(1.46)^X,$$

$$Y = B(10)^{MX} \qquad \text{or} \qquad Y = 2(10)^{0.1648X},$$

where X is a variable that can be substituted into the equation to give infinitely many values for Y. In logarithmic form, the equation is

$$\log Y = \log B + X \log M,$$

or

$$\log Y = \log 2 + X \log 1.46.$$

The same methods give the negative slope of a curve. The curve shown in **Figure 34.8** slopes downward to the right and has a negative slope. M is the antilog of −0.0277. The intercept of 70 and the slope yield the equation for the curve.

The half-life decay of radioactivity plotted in **Figure 34.9** compares decay to time. The half-life of different isotopes varies, so different values would be assigned along the x-axis for them. However, the curves for all isotopes would be straight lines with the exponential equation form.

34.5 Graphical Calculus

If the equation of a curve is known, calculus may be used to perform various types of calculations. However, experimental data often do not fit standard mathematical equations, making impossible the mathematical application of calculus. In these cases, graphical calculus can be used. The two basic forms of calculus are (1) **differential calculus** and (2) **integral calculus.**

Differential calculus is used to determine the rate of change of one variable with respect to another (**Figure 34.10A**). The rate of change at any instant along the curve is the slope of a line tangent to the curve at that point.

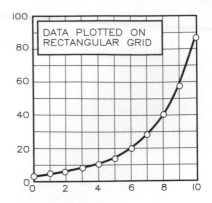

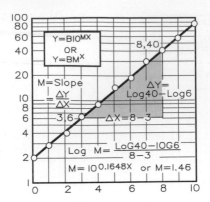

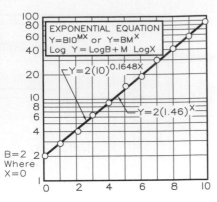

34.7 Exponential equation, $Y = BM^X$:

Step 1 This data (shown plotted on a rectangular grid) plots as a straight line on a semi-logarithmic grid. It takes the equation form of $Y = BM^X$.

Step 2 The slope must be found mathematically, because the X and Y scales are unequal. Write the slope equation in either of the forms shown above. The slope is found to be M = 1.46.

Step 3 Find the intercept, B = 2, on the Y axis, where X = 0. Substitute the values for M (1.46) and B (2) into the equation to get the equation $Y = 2(10)^{0.1648X}$ or $Y = 2(1.46)^X$.

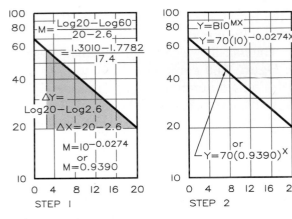

34.8 Negative slope (M).

A When a curve slopes downward to the right, its slope is negative.

B Substitution gives these two forms of the equation (shown above) for the curve.

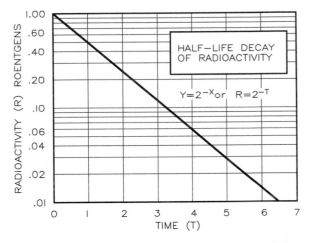

34.9 The decay of radioactivity is represented by a straight line on a semilog grid, indicating that its equation takes the exponential form, $Y = BM^X$ or $R = 2^{-T}$.

Constructing a chord at any interval allows approximation of this slope. The tangent, $\Delta Y/\Delta X$, may represent miles per hour, weight versus length, or various other rates of change important in the analysis of data.

Integral calculus is the reverse of differential calculus. Integration is used to find the area under a curve (the product of the variables plotted on the *x*- and *y*-axes). The area under a curve is approximated by dividing one of the variables into a number of very small rectangular bars under the curve (**Figure 34.10B**). Each bar is drawn so that equal areas lie above and below the curve and the average height of the bar is near its midpoint.

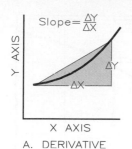

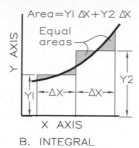

A. DERIVATIVE

B. INTEGRAL

34.10 Derivatives and integrals.

A The derivative of a curve is the rate of change at any point on the curve, or its slope, $\Delta Y/\Delta X$.

B The integral of a curve is the cumulative area enclosed by the curve, or the summation of the incremental areas comprising the whole.

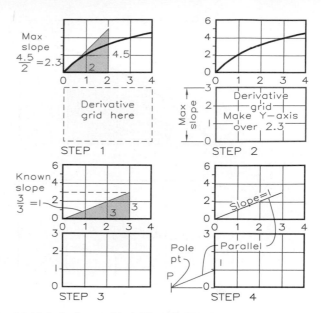

34.11 Scales for graphical differentiation:

Step 1 Estimate the maximum slope of the curve (here, 2.3) by drawing a line tangent to the curve where it is steepest.

Step 2 Draw the derivative grid with a maximum ordinate of 3.0 to accommodate the slope of 2.3.

Step 3 Find a known slope of 1 on the given grid. The slope has no relationship to the data curve.

Step 4 Draw a line from 1 on the y-axis of the derivative grid parallel to the slope of the triangle drawn in the given grid. This locates the pole point on the extension of the x-axis.

34.6 Graphical Differentiation

Graphical differentiation is used to determine the rate of change of two variables with respect to each other at any given point. **Figure 34.11** illustrates the preliminary construction of a derivative scale and the pole point for plotting a derivative curve.

The graphical differentiation process is illustrated in **Figure 34.12**. The maximum slope of the data curve is estimated to be slightly less than 12; an ordinate scale long enough to accommodate the maximum slope is selected. To locate the pole point, a line is drawn from point 4 on the ordinate axis of the derivative grid parallel to the known slope on the given curve grid of 12/3 = 4 and the line is extended to the x-axis.

A series of chords is constructed on the given curve. The interval between 0 and 1, where the curve is steepest, is divided in half to obtain a more accurate plot. After other bars are found, a smooth curve is drawn through the top of them so that the area above and under the top of each bar is equal. The rate of change, $\Delta Y/\Delta X$, can be found for any value of X in the derivative graph.

Applications

The mechanical handling shuttle shown in **Figure 34.13** converts rotational motion into linear motion. The drawing of the linkage shows the end positions of point P, which is the zero point for plotting travel versus degrees of revolution. Rotation is constant at one revolution every three seconds, so the degrees of revolution may be converted to time (**Figure 34.14**). The drive crank, R1, is revolved at 30° intervals, and the distance that point P travels from its end position is plotted on the graph to give the distance-time relationship.

The ordinate scale of the derivative grid is scaled with an end value of 100 in./sec, or slightly larger than the estimated maximum slope of the curve. A slope of 40 is drawn on

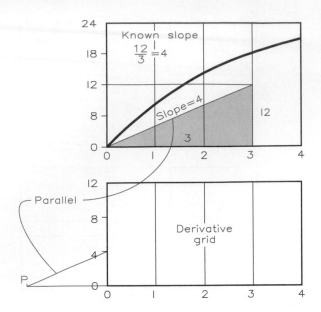

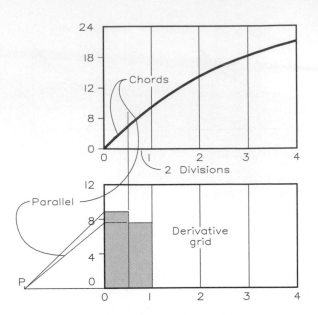

34.12 Graphical differentiation:

Required Find the derivative curve of the given data.

Step 1 Find the derivative grid and the pole point by using the steps introduced in Figure 34.11.

Step 2 Construct chords between intervals on the given curve and draw lines parallel to them through point P on the derivative grid. These lines locate the heights of bars in their respective intervals. Divide the first interval into two bars where the curve is sharpest.

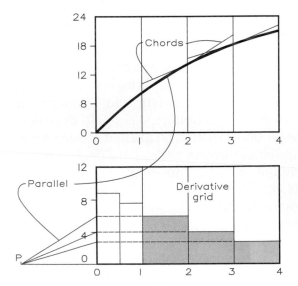

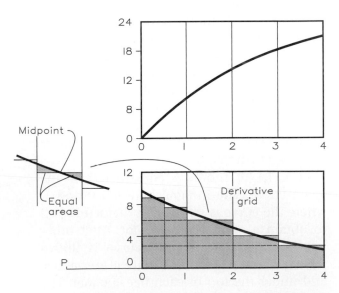

Step 3 Draw additional chords on the curve in the last three intervals. Draw lines parallel to the chords through point P to the y-axis. This construction locates the remainder of the bars needed to draw the derivative curve.

Step 4 The vertical bars represent the slopes of the curve at different intervals. Draw the derivative curve through the midpoints of the bars so that the areas below and above the bars are equal.

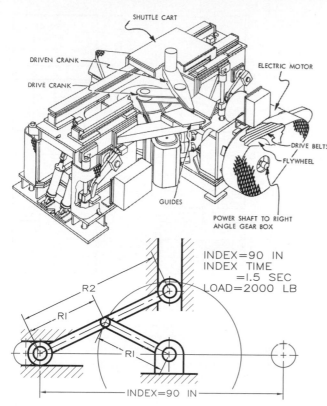

34.13 These are drawings of a mechanical handling shuttle used to move automobile parts on an assembly line. (Courtesy of General Motors Corporation.)

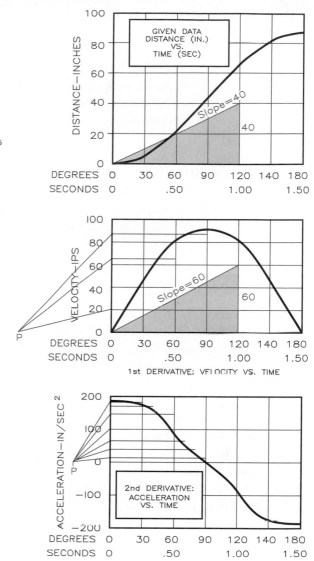

34.14 Velocity and acceleration of the mechanical handling shuttle may be obtained with graphical differential calculus.

the given data grid. Pole point P is found by drawing a line from 40 on the derivative ordinate scale parallel to the slope of 40 in the given data graph to the extension of the *x*-axis of the derivative graph.

Chords drawn on the curve approximate the slope at various points. Draw lines from point P of the derivative scale parallel to the chordal lines to the ordinate axis. These intersections on the *y*-axis are projected horizontally to their respective intervals to form vertical bars. A curve is drawn through the tops of the bars to give an average of the bars. This curve shows the velocity of the shuttle in inches per second at any time interval.

The second derivative curve, acceleration versus time, is drawn in the same manner as the first derivative curve. Inspection of the first derivative curve shows that the maximum slope is about 200 in./sec/sec. An easily measured scale for the ordinate is chosen. A pole, P, is found in the same manner as for the previous pole.

Chords are drawn at intervals on the first derivative curve. Lines drawn from P parallel to these chords intersect the Y axis of the second derivative graph, where they are projected horizontally to their respective intervals, establishing the bars. A smooth curve is drawn through the tops of the bars to give a close approximation of the average areas of the bars. The minus scale indicates deceleration.

These velocity and acceleration graphs show that parts being handled by the shuttle accelerate at a rapid rate until the maximum velocity is attained at 90°, at which time deceleration begins and continues until the parts come to rest.

34.7 Graphical Integration

Graphical **integration** is used to determine the area (product of two variables) under a curve. For example, if the y-axis represented pounds and the x-axis represented feet, the integral curve would give the product of the variables, foot-pounds, at any interval along the x-axis. The steps in determining the pole point and the scales for integration are illustrated in **Figure 34.15**.

In **Figure 34.16**, the total area under the curve is estimated to be less than 80 units. Therefore the maximum height of the y-axis on the integral curve is 80, drawn at a convenient scale. Pole P is found by the steps shown in **Figure 34.15**.

A series of vertical bars is constructed to approximate the areas under the curve. The narrower the bars, the more accurate will be the resulting plotted curve. The interval between 2 and 3 is divided into two bars to obtain more accuracy where the curve is sharpest. The tops of the bars are extended horizontally to the y-axis and are connected to point P.

Lines are drawn parallel to AP, BP, CP, DP, and EP in the integral grid to correspond to

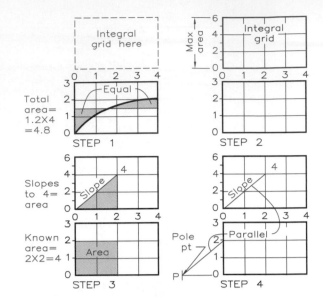

34.15 Scales for graphical integration:

Step 1 To determine the maximum value on the y-axis, draw a line to approximate the area under the given curve (4.8 square units here).

Step 2 Draw the integral graph with a y-axis of 6 to accommodate the maximum area.

Step 3 Find a known area of 4 on the given grid. Draw a slope from 0 to 4 on the integral grid directly above the known area to establish the integral.

Step 4 Draw a line from 2 on the y-axis of the given grid parallel to the slope line in the integral grid. Locate the pole point P on the extension of the x-axis.

the respective intervals in the given grid. The intersection points of the chords are connected by a smooth curve—the integral curve—to give the cumulative product of the X and Y variables along the x-axis.

Problems

Use size A sheets to solve the following problems. Show mathematical calculations either on the same or separate sheet if space is not available on the sheet with the graphical solutions.

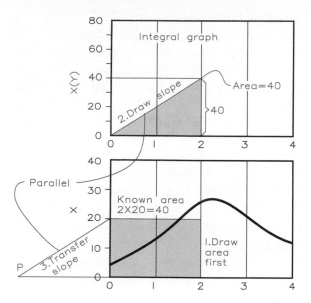

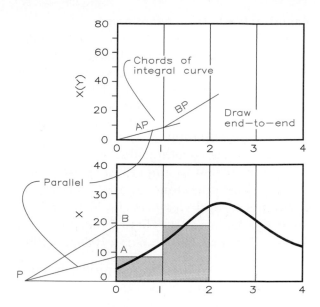

34.16 Graphical integration.

Required Plot the integral curve of the given data.

Step 1 Find pole point P by using the technique described in Figure 34.15.

Step 2 Construct bars in the given graph to approximate the areas under the curve. Project the heights of the bars to the y-axis and draw lines to pole P. Draw sloping lines AP and BP in their respective intervals parallel to the lines drawn to P in the integral graph.

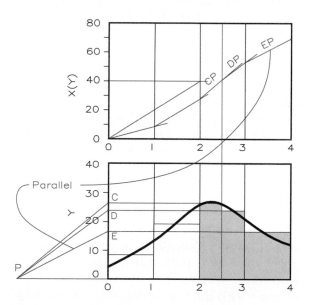

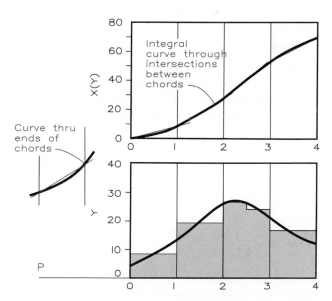

Step 3 Draw additional bars from 2 to 4 on the x-axis on the given graph. Project the heights of the bars to the y-axis and draw rays to pole P. Draw lines CP, DP, and EP at their respective intervals and parallel to their rays in the integral grid.

Step 4 The lines connected in the integral grid are chords of the integral curve. Draw the curve to pass through the intersections of the chords. An ordinate value on the integral curve represents the cumulative area under the given curve from zero to that point on the x-axis.

Empirical Equations: Logarithmic

1. Find the equation of the empirical data below that compares input voltage, V (y-axis), with input current in amperes, I (x-axis), to a heat pump: V = 0.7, I = 8; V = 1.1, I = 20; Y = 1.6, I = 30; V = 1.8, I = 40; V = 2.1, I = 60.

2. Find the equation of the empirical data that gives the relationship between peak allowable current in amperes, I (y-axis), with the overload operating time in cycles at 60 cycles per second, C (x-axis): I = 2000, C = 1; I = 1840, C = 2; I = 1640, C = 5; I = 1480, C = 10; I = 1300, C = 20; I = 1200, C = 50; I = 1000, C = 100.

3. Find the equation of the empirical data of a low-voltage circuit breaker used on a welding machine that gives the maximum loading during welding in amperes, rms (y-axis), for the percentage of duty, pdc (x-axis): rms = 7500, pdc = 3; rms = 5200, pdc = 6; rms = 4400, pdc = 9; rms = 3400, pdc = 15; rms = 2300, pdc = 30; rms = 1700, pdc = 60.

4. Construct a three-cycle by three-cycle logarithmic graph to find the equation of a machine's vibration displacement in mills (y-axis) and vibration frequency in cycles per minute, cpm (x-axis). Data: 100 cpm, 0.80 mills; 400 cpm, 0.22 mills; 1000 cpm, 0.09 mills; 10,000 cpm, 0.009 mills; 50,000 cpm, 0.0017 mills.

5. Find the equation of the data shown below that compares the velocities of air moving over a plane surface in feet per second, V (X-axis), at different heights in inches, Y (Y-axis), above the surface: Y = 0.3, V = 1.0; Y = 0.46, V = 2.0; Y = 0.8, V = 5.0; Y = 1.2, V = 10.0; Y = 2.0, V = 20.0, Y = 3.2, V = 50.0.

6. Find the equation of the data below that shows the distance traveled in feet, S (y-axis), at various times in seconds, T (x-axis), of a test vehicle: T = 1, S = 15.8; T = 2, S = 63.3; T = 3, S = 146; T = 4, S = 264; T = 5, S = 420; T = 6, S = 580.

Empirical Equations: Linear

7. Construct a linear graph to determine the equation for the annual cost of a compressor (y-axis) in relation to the compressor's size in horsepower (x-axis). Data: 0 hp, $0; 50 hp, $2100; 100 hp, $4500; 150 hp, $6700; 200 hp, $9000; 250 hp, $11,400. Write the equation for these data.

8. Construct a linear graph on which the x-axis is the mat depth from 0 to 4 in. and the y-axis is tons per hour per foot of width from 0 to 70 for a conveyor traveling at a rate of 50 feet per minute. The conveyor is a moving shaker that screens particles of coal by size. X = 0, Y = 0; X = 1, Y = 13; X = 2, Y = 25; X = 3, Y = 38; and X = 4, Y = 50.

9. Plot the empirical data on a linear graph and determine its equation which shows the deflection in centimeters of a spring, D (y-axis), when it is loaded with different weights in kilograms, W (x-axis): W = 0, D = 0.45; W = 1, D = 1.10; W = 2, D = 1.45; W = 3, D = 2.03, W = 4, D = 2.38.

10. Plot the empirical data on a linear graph and determine its equation. It shows temperatures on a Fahrenheit thermometer, °F (y-axis), and a Celsius thermometer, °C (x-axis): C = −6.8, F = 20; C = 6 F = 43; C = 16, F = 60.8; C = 32.2, F = 90; C = 52, F = 125.8; C = 76, F = 169.

Empirical Equations: Semilogarithmic

11. Draw a semilog graph of the data to find its equation. The y-axis is a two-cycle log scale for V (voltage) and the x-axis is a 10-unit linear scale for T (time) in sixteenths of a second. This data represents resistor voltage during capacitor charging. Data: 0 sec, 10 V; 2 sec, 6 V; 4 sec, 3.6 V; 6 sec, 2.2 V; 8 sec, 1.4 V; 10 sec, 0.8 V.

12. Construct a semilog graph of the data to find its equation. Make the y-axis a three-cycle log scale for the reduction factor, R, and the x-axis a linear scale from 0 to 250 for mass thickness per square foot, MT, of a nuclear protection barrier. Data: 0MT, 1.0R; 100MT, 0.9R; 150MT, 0.028R; 200MT, 0.009R; 300MT, 0.0011R.

13. An engineering firm considering expansion is reviewing its past income. Their years of operation are represented by X (x-axis), and their annual income (in tens of thousands of dollars) by N (y-axis): X = 1, N = 0.05; X = 2, Y = 0.08; X = 3, N = 012; X = 4, N = 0.2; X = 5, Y = 0.32; X = 6, N = 0.51; X = 7, N = 0.8; X = 8, N = 1.3; X = 9, N = 2.05; X = 10, N = 2.5.

Calculus: Graphical Differentiation

14. Plot the equation $Y = X^2/6$ as a linear graph. Graphically determine the first derivative curve.

15. Plot the equation $Y = 2X^2$; find its derivative curve on a graph placed below the first.

16. Plot the equation $4Y = 8 - X^2$; find its derivative curve on a graph placed below the given grid.

17. Plot the equation $3Y = X^2 + 16$; find its derivative curve on a graph placed below the given grid.

18. Plot the equation $X = 3Y^2 - 5$; find the derivative curve on a graph placed below the given grid.

Calculus: Graphical Integration

19. Plot the equation $Y = X^2$; find its integral curve on a graph placed above the first.

20. Plot the equation $Y = 9 - X^2$; find its integral curve on a graph placed above the first.

21. Plot the equation $Y = X$ on a graph; find its integral curve on a graph placed above the first.

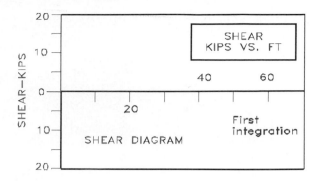

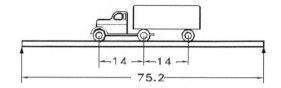

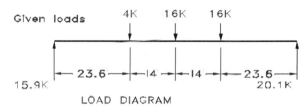

34.18 Shear diagram.

22. A truck applies loads to a bridge as shown in **Figure 34.18**. Plot the integral curves specified below, label the axes, provide titles, and use the correct principles of graphing.

(A) Plot an integral curve for the truck in this position to show the shear in kips (thousands of lbs) along the y-axis and distance in feet along the x-axis. Where is shear greatest and least? What is the shear at 40 feet from the left end?

(B) Plot a second curve as the integral of the shear curve to find the bending moment in foot kips along the y-axis and distance in feet along the x-axis. Where is the bending moment greatest and least? What is the bending moment at 40 feet from the left end?

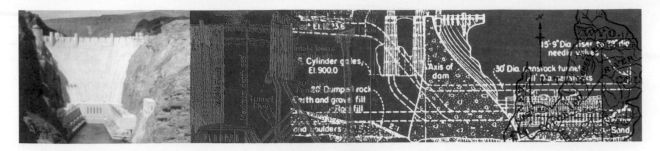

35

Pipe Drafting

35.1 Introduction

An understanding of pipe drafting begins with a familiarity with the types of pipe that are available. The commonly used types of pipe are (1) steel pipe, (2) cast-iron pipe, (3) copper, brass, and bronze pipe and tubing, and (4) plastic pipe. The standards for the grades and weights for pipe and pipe fittings are specified by several organizations to ensure the uniformity of size and strength of interchangeable components. Several of these organizations are the American National Standards Institute (ANSI), the American Society for Test Materials (ASTM), the American Petroleum Institute (API), and the Manufacturers Standardization Society (MSS).

35.2 Welded and Seamless Steel Pipe

Traditionally, steel pipe has been specified in three weights: **standard (STD)**, **extra strong (XS)**, and **double extra strong (XXS)**. These designations and their specifications are listed in the ANSI B 36.10 standards. However, additional designations for pipe, called schedules, have been introduced to provide the pipe designer with a wider selection of pipe to cover more applications.

The ten schedules are: Schedule 10, Schedule 20, Schedule 30, Schedule 40, Schedule 60, Schedule 80, Schedule 100, Schedule 120, Schedule 140, and Schedule 160. The wall thicknesses of the pipes vary from the thinnest, in Schedule 10, to the thickest, in Schedule 160. The outside diameters are of a constant size for pipes of the same nominal size in all schedules. Schedule designations correspond to STD, XS, and XXS specifications in some cases (**Table 35.1**). This table has been abbreviated from the ANSI B 36.10 tables by omitting a number of the pipe sizes and schedules. The most often used schedules are 40, 80, and 120.

Pipes from the smallest size up to and including 12-in. pipes are specified by their inside diameter (ID), which means that the outside diameter (OD) is larger than the specified size. The inside diameters are the same

Table 35.1 Dimensions and Weights of Welded and Seamless Steel Pipe (ANSI B 36.10–1979)

Inch Units				Identification		SI Units		
Inch Nominal Size (in.)	Outside Diameter (in.)	Wall Thickness (in.)	Weight lbs/ft	*STD XS XXS	Schedule Number	Outside Diameter (mm)	Wall Thickness (mm)	Weight kg/m
1/2	0.84	0.11	0.85	STD	40	21.3	2.8	1.3
1	1.32	0.13	1.68	STD	40	33.4	3.4	2.5
1	1.3	0.18	2.17	XS	80	33.4	4.6	3.2
1	1.3	0.36	3.66	XXS		33.4	9.1	5.5
2	2.38	0.22	3.65	STD	40	60.3	3.9	5.4
2	2.38	0.22	5.02	XS	80	60.3	5.5	7.5
2	2.38	0.44	9.03	XXS		60.3	11.1	13.4
4	4.50	0.23	10.79	STD	40	114.3	6.0	16.1
4	4.50	0.34	14.98	XS	80	114.3	8.6	42.6
4	4.50	0.67	27.54	XXS		114.3	17.1	41.0
8	8.63	0.32	28.55	STD	40	219.1	8.2	42.6
8	8.63	0.50	43.39	XS	80	219.1	12.7	64.6
8	8.63	0.88	74.40	XXS		219.1	22.2	107.9
12	12.75	0.38	49.56	STD		323.0	9.5	67.9
12	12.75	0.50	65.42	XS		323.0	12.7	97.5
12	12.75	1.00	125.40	XXS	120	133.9	25.4	187.0
14	†14.00	0.38	54.57	STD	30	355.6	9.5	87.3
14	14.00	0.50	72.08	XS		355.6	12.7	107.4
18	18.00	0.38	70.59	STD		457	9.5	106.2
18	18.00	0.50	93.45	XS		457	12.7	139.2
24	24.00	0.38	94.62	STD	20	610	9.5	141.1
24	24.00	0.50	125.49	XS		610	12.7	187.1
30	30.00	0.38	118.65	STD		762	9.5	176.8
30	30.00	0.50	157.53	XS	20	762	12.7	234.7
40	40.00	0.38	158.70	STD		1016	9.5	236.5
40	40.00	0.50	210.90	XS		1016	12.7	314.2

*Standard (STD)

X-strong (XS)

XX-strong (SSS)

†Beginning with 14-in. DIA pipe, the nominal size represents the outside diameter (O.D.)

This table has been compressed by omitting many of the available pipe sizes. The nominal sizes of pipes that are listed in the complete table are: ⅛″, ¼″, ⅜″, ½″, ¾″, 1″, 1¼″, 1½″, 2″, 2½″, 3″, 3½″, 4″, 5″, 6″, 8″, 10″, 12″, 14″, 16″, 18″, 20″, 22″, . . . (at 2″ increments up to 60″).

size as the nominal sizes of the pipe for STD weight pipe. For XS and XXS pipe, the inside diameters are slightly different in size from the nominal size. Beginning with the 14-in. diameter pipes, the nominal sizes represent the outside diameters of the pipe.

The standard lengths for steel pipe are 20 ft and 40 ft. **Seamless steel (SMLS STL)** pipe is a smooth pipe with no weld seams along its length. Welded pipe is formed into a cylinder and is butt-welded (BW) at the seam, or it is joined with an electric resistance weld (ERW).

35.2 WELDED AND SEAMLESS STEEL PIPE • 611

35.3 Cast-Iron Pipe

Cast-iron pipe is used for the transportation of liquids, water, gas, and sewerage. When used as a sewerage pipe, cast-iron pipe is referred to as "soil pipe." Cast-iron pipe is available in diameter sizes from 3 in. to 60 in.

The standard lengths of cast-iron pipe are 5 ft and 10 ft. Cast iron is more brittle and more subject to cracking when loaded than is steel pipe. Therefore, cast-iron pipe should not be used where high pressures or weights will be applied to it.

35.4 Copper, Brass, and Bronze Piping

Copper, brass, and bronze are used to manufacture piping and tubing for use in applications where there must be a high resistance to corrosive elements, such as acidic soils and chemicals transmitted through the pipes. Copper pipe is used when the pipes are placed within or under concrete slab foundations of buildings to ensure that they will resist corrosion. The standard length of pipes made of these nonferrous materials is 12 ft.

Tubing is a smaller-size pipe that can be easily bent when it is made of copper, brass, or bronze. The term **piping** refers to rigid pipes that are larger than tubes, usually in excess of 2 in. in diameter.

35.5 Miscellaneous Pipes

Other materials that are used to manufacture pipes are aluminum, asbestos-cement, concrete, polyvinyl chloride (PVC), and various other plastics. Each of these materials has its own special characteristics that make it desirable or economical for certain applications. The method of designing and detailing piping systems by the pipe drafter is essentially the same regardless of the piping material used.

35.6 Pipe Joints

The basic connection in a pipe system is the joint where two straight sections of pipe fit together. Three types of joints are illustrated in **Figure 35.1**: **screwed**, **welded**, and **flanged**.

Screwed joints are joined by pipe threads of the type covered in Chapter 17 and Appendix 42. Pipe threads are tapered at a ratio of 1 to 16 along the outside diameter (**Figure 35.2**). As the pipes are screwed together, the threads bind to form a snug, locking fit. A cementing compound is applied to the threads before joining to improve the seal.

Flanged joints, shown in **Figure 35.3**, are welded to the straight sections of pipe, which are then bolted together around the perimeter of the flanges. Flanged joints form strong rigid joints that can withstand high pressure and permit disassembly of the joints when needed. Several types of flange faces are shown in **Figure 35.4** and in Appendix 43.

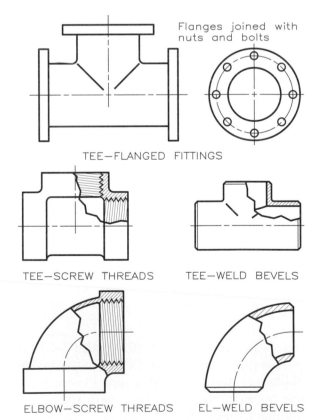

35.1 Examples of the types of fitting that are used to join pipes are shown here.

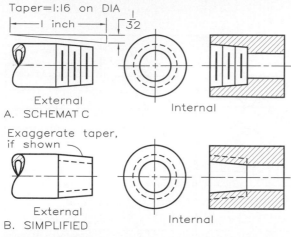

Taper=1:16 on DIA

A. SCHEMATIC   External   Internal

Exaggerate taper, if shown

B. SIMPLIFIED   External   Internal

35.2 Pipe threads have a slight taper and are used to connect screwed pipe fittings.

Welded joints are joined by welded seams around the perimeter of the pipe to form butt welds. Welded joints are used extensively in "big inch" pipelines that are used for transporting petroleum products cross-country.

Bell and spigot (**B&S**) joints are used to join cast-iron pipes (**Figure 35.5**). The spigot is placed inside the bell and the two are sealed with molten lead or a sealing ring that snaps into position to form a sealed joint.

Soldering is used to connect smaller pipes and tubes, especially nonferrous tubing.

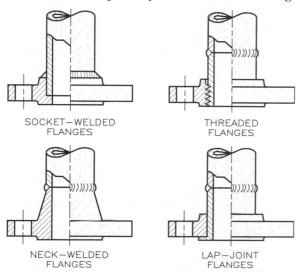

SOCKET—WELDED FLANGES

THREADED FLANGES

NECK—WELDED FLANGES

LAP—JOINT FLANGES

35.3 Types of flanged joints and the methods of attaching the flanges to the pipes are shown here.

Screwed fittings are available to connect tubing as shown in **Figure 35.6**.

35.7 Pipe Fittings

Pipe fittings are placed within a pipe system to join pipes at various angles, to transform the pipe diameter to a different size, or to control the flow and its direction within the system. Fittings are placed in the system using any of the previously covered joints. A pipe system can be drawn with single-line symbols or double-line symbols.

The fittings in **Figure 35.7** are represented as single-line symbols with flanged, screwed, bell

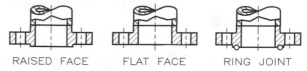

RAISED FACE   FLAT FACE   RING JOINT

35.4 Three types of flange faces are the raised face (RF), the flat face (FF), and the ring joint (RJ).

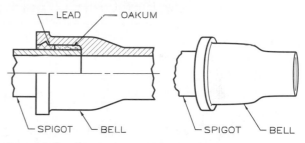

LEAD   OAKUM

SPIGOT   BELL   SPIGOT   BELL

35.5 A bell and spigot joint (B&S) is used to connect cast-iron pipes.

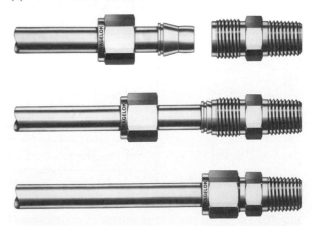

35.6 Screwed joints can be used to join small tubing. (Courtesy of Crawford Fitting Co.)

	FLANGED	SCREWED	BELL & SPIGOT	WELDED	SOLDERED
1. JOINT					
2. 90° ELBOW					
TOP VIEW					
FRONT VIEW					
BOTTOM VIEW					
3. ELBOW—LONG RADIUS					
4. ELBOW— REDUCING					
5. TEE					
TOP VIEW					
FRONT VIEW					
BOTTOM VIEW					
6. 45° ELBOW					
TOP VIEW					
FRONT VIEW					
BOTTOM VIEW					
7. 45° LATERAL					
TOP VIEW					
FRONT VIEW					
8. REDUCER					
9. GATE VALVE					
10. GLOBE VALVE					
11. CHECK VALVE					
12. UNION					

35.7 These single-line pipe fittings symbols are extracted from the ANSI Z32.2 standards.

and spigot, welded, and soldered joints. Most symbols have been shown as they would be drawn to appear in various orthographic view—top, front, and side views. These symbols have been extracted from ANSI Z32.2.3 standards.

35.8 Screwed Fittings

A number of standard fittings are shown in **Figures 35.8** through **35.11**. The two types of graphical symbols that are used to represent fittings and pipe are **double-line symbols** and **single-line symbols**.

Double-line symbols are more descriptive of the fittings and pipes since they are drawn to scale with double lines. Single-line symbols are more symbolic since the pipe and fittings are drawn with single lines and schematic symbols.

Fittings are available in three weights: standard (STD), extra strong (XS), and double extra strong (XXS) to match the standard

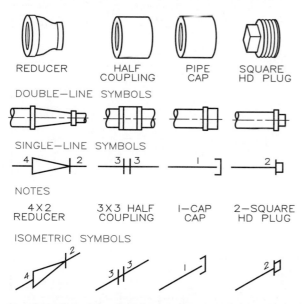

35.8 Examples of standard fittings for screwed connections along with the single-line and double-line symbols that are used to represent them are shown here. Nominal pipe sizes can be indicated by numbers placed near the joints. The major flow direction is labeled first, with the branches labeled second. The large openings are labeled to precede the smaller openings.

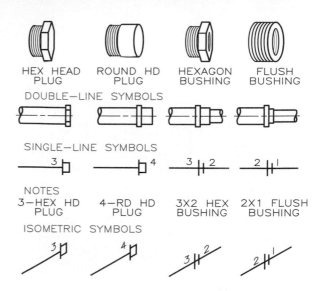

35.9 This is a single-line piping system with the major valves represented as double-line symbols.

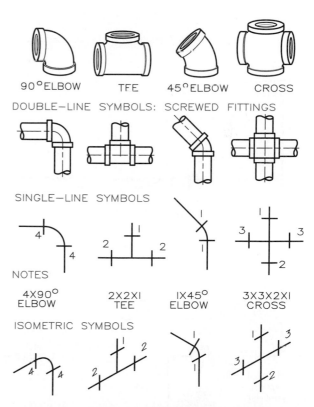

35.10 Further examples of standard fittings for screwed connections are shown here.

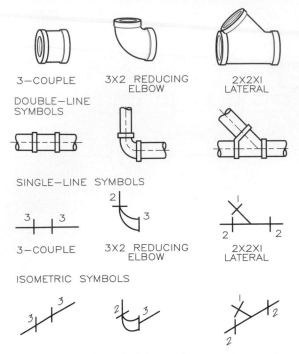

3—COUPLE   3X2 REDUCING   2X2XI
                  ELBOW   LATERAL

DOUBLE—LINE
SYMBOLS

SINGLE—LINE SYMBOLS

3—COUPLE   3X2 REDUCING   2X2XI
                  ELBOW   LATERAL

ISOMETRIC SYMBOLS

35.11 Examples of standard fittings for screwed connections are shown here.

weights of the pipes with which they will be connected. Other weights of fittings are available, but these three weights are stocked by practically all suppliers.

A piping system of screwed fittings is shown in **Figure 35.12** with double-line symbols in a single-line system to call attention to them. These could just as well have been drawn using the single-line symbols.

The most common symbols for representing fittings are shown in **Figure 35.7** and have been extracted from ANSI Z 32.2.3 standards.

35.9 Flanged Fittings

Flanges are used to connect fittings into a piping system when heavy loads are supported in large pipes and where pressures are great. Since flanges are expensive, their usage should be kept to a minimum if other joining methods can be used. Flanges are welded to straight pipe sections so that they can be bolted together.

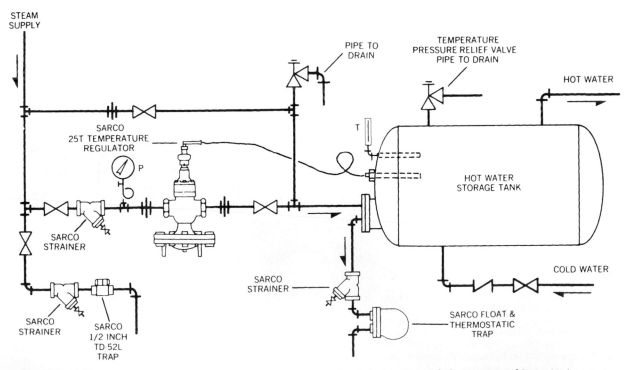

35.12 This is a single-line piping system with the major valves represented with double-line symbols. (Courtesy of Sarco, Inc.)

Examples of several flanged fittings are drawn as double-line and single-line symbols in **Figures 35.13** and **35.14**. The elbow is commonly referred to as an "ell" and it is available in angles of 90° and 45° in both short and long radii. The radius of a short-radius ell is equal to the diameter of the larger end. The long-radius (LR) ells have radii that are approximately 1.5 times the nominal diameter of the large end of the ell. A table of dimensions for 125 LB and 250 LB cast-iron fittings is given in Appendices 42–43.

35.10 Welded Fittings

Welding is used to join pipes and fittings for permanent, pressure-resistant joints. Examples of double-line and single-line fittings connected by welding are shown in **Figures 35.15**

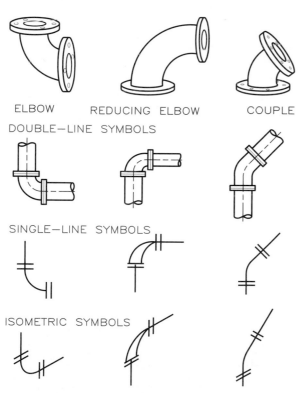

35.13 Examples of standard flanged fittings along with the single-line and double-line symbols that are used to represent them are shown here.

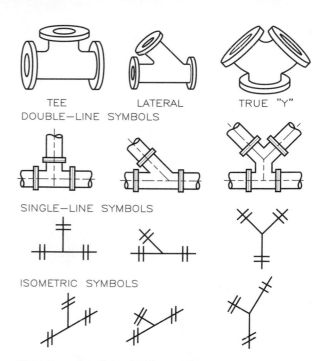

35.14 Examples of standard flanged fittings along with the single-line and double-line symbols that are used to represent them are shown here.

and **35.16**. Fittings are available with beveled edges that are ready for welding.

A piping layout in **Figure 35.17** illustrates a series of welded joints with a double-line drawing. The location of the welded joints has been dimensioned. Several flanged fittings have been welded into the system in order for the flanges to be used.

35.11 Valves

Valves are used to regulate the flow of gas and liquid transported within a pipelines or to turn off the flow completely. Several types of valves are **gate**, **globe**, **angle**, **check**, **safety**, **diaphragm**, **float**, and **relief** valves. The three basic valves—gate, globe, and check—are shown in **Figure 35.18** drawn with single-line symbols.

Gate valves are used to turn the flow within a pipe on or off with the least restriction of flow

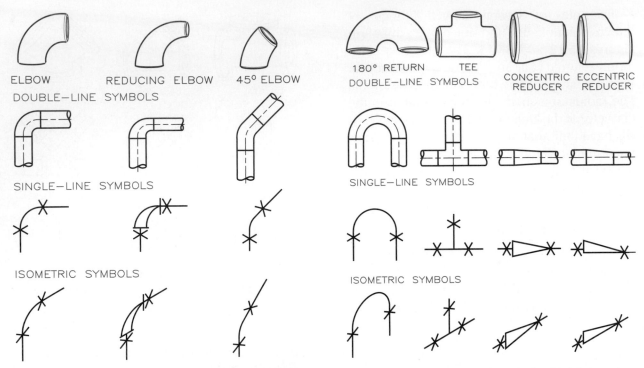

ELBOW   REDUCING ELBOW   45° ELBOW

DOUBLE–LINE SYMBOLS

180° RETURN   TEE   CONCENTRIC   ECCENTRIC
DOUBLE–LINE SYMBOLS   REDUCER   REDUCER

SINGLE–LINE SYMBOLS

SINGLE–LINE SYMBOLS

ISOMETRIC SYMBOLS

ISOMETRIC SYMBOLS

35.15 Examples of welded fittings along with the single-line and double-line symbols that are used to represent them are shown here.

35.16 Additional examples of standard welded fittings along with the single-line and double-line symbols that are used to represent them are shown here.

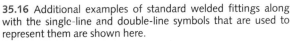

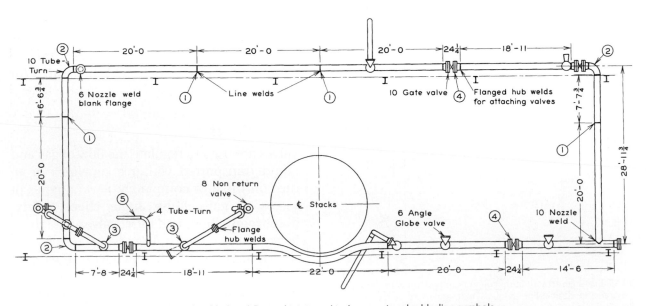

35.17 This piping layout uses a series of welded and flanged joints and is drawn using double-line symbols.

35.18 The three basic types of valves are gate, globe, and check valves. (Photographs courtesy of Walworth/Aloyco.)

through the valve. Gate valves are not meant to be used to regulate the degree of flow.

Globe valves are not only used to turn the flow on and off, but they are also used to regulate the flow to a desired level.

Angle valves are types of globe valves that turn at 90° angles at bends in the piping system. They have the same controlling features as the straight globe valves.

Check valves restrict the flow in the pipe to only one direction. A backward flow is prevented by either a movable piston or a swinging washer activated by a reverse in the flow (**Figure 35.18**).

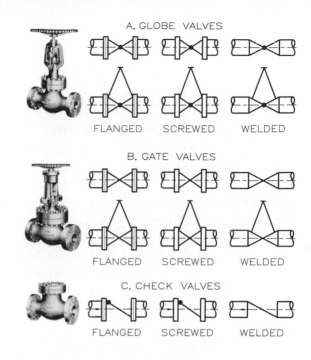

A. GLOBE VALVES

FLANGED   SCREWED   WELDED

B. GATE VALVES

FLANGED   SCREWED   WELDED

C. CHECK VALVES

FLANGED   SCREWED   WELDED

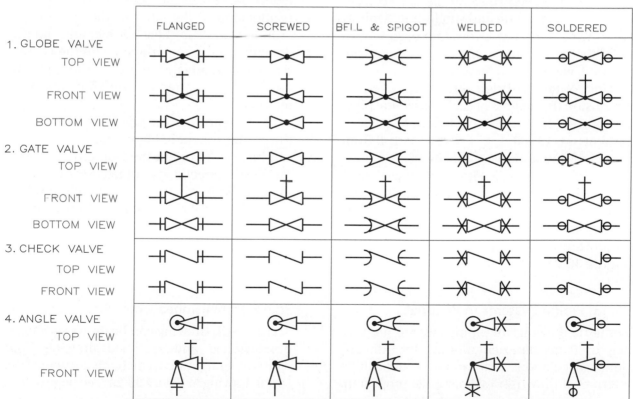

35.19 Examples of orthographic views of valves drawn as single-line symbols.

Orthographic symbols for types of valves are shown in **Figure 35.19**. These symbols can be converted to isometric as shown in **Figure 35.21**.

35.12 Fittings in Orthographic Views

Fittings and valves must be shown from any view in orthographic projection. Two and three views of typical fittings are shown in **Figure 35.7** as single-line screwed fittings. The same general principles are used to represent joints as double-line drawings. Observe the various views of the fittings and notice how the direction of an elbow can be shown by a slight variation in the adjacent views. The same techniques are used to represent tees and laterals.

A piping system is shown in a single orthographic view in **Figure 35.20**, where a combination of double-line and single-line symbols are drawn. Note that arrows are used to give the direction of flow in the system. Joints are screwed, welded, and flanged.

Horizontal elevation lines are given to dimension the heights of each horizontal pipe. Station $5 + 12 - 0 - 1/4''$ represents a distance of 500 feet plus $12' - 0 - 1/4''$, or $512' - 0 - 1/4''$ from the beginning station point of $0 + 00$.

The dimensions in **Figure 35.20** are measured from the centerlines of the pipes; this is indicated by the **CL** symbols. In some cases, the elevations of the pipes are dimensioned to the bottom of the pipe, abbreviated as **BOP** (**Figure 35.24**).

35.13 Piping Systems in Pictorial

Isometric drawings of piping systems are very helpful in the representation of three-dimensional installations that would be difficult to interpret if drawn in orthographic projection.

Isometric and axonometric drawings of piping systems, called "**spool drawings**," can be drawn using either single-line or double-line symbols.

A three-dimensional piping system is drawn orthographically in **Figure 35.21** (see page 622) with top and front views. Although this is a relatively simple three-dimensional system, a thorough understanding of orthographic projection is required to read the drawing.

In **Figure 35.22** (see page 623), the piping system is drawn with all of the pipes revolved into the same horizontal plane. You will notice that the vertical pipes and their fittings are drawn true size in the top view. This is called a **developed pipe drawing**. The fittings and pipe sizes are noted on this preliminary sketch from which the finished drawing will be made in **Figure 35.23** (see page 622).

An axonometric schematic is drawn in **Figure 35.21** to explain the three-dimensional relationship of the parts of the system. The rounded bends in the elbows in an isometric drawing can be constructed with the isometric ellipse template, or the corners can be drawn square to reduce the effort and time required.

A north arrow is drawn on the plan view of the piping system in **Figure 35.21**, which this can be used to orient the isometric pictorial. This north direction is not necessarily related to compass north, but it is a direction that is selected parallel to a major set of pipes within the system. In the isometric drawing, it is preferred that the north arrow point to the upper-left or upper-right corner of the pictorial.

35.14 Dimensioned Isometrics

An isometric drawing can be drawn as a fully-dimensioned and specified drawing from which a piping system can be constructed. The spool drawing in **Figure 35.24** (see page 623) is

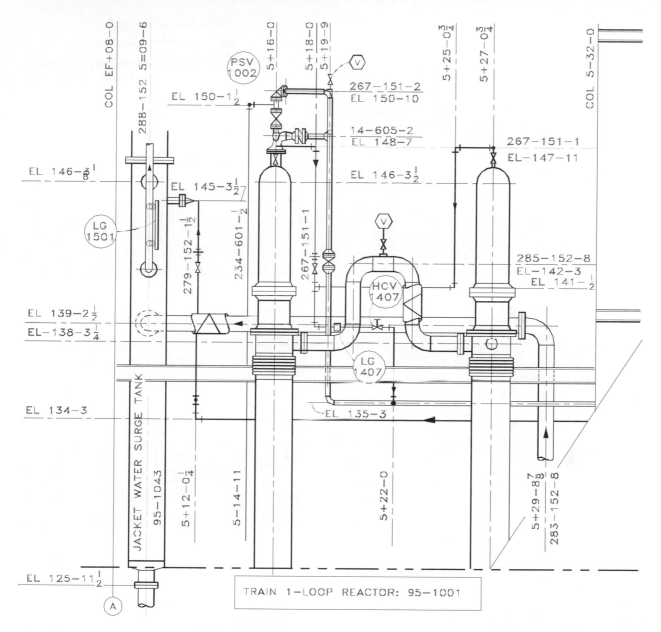

35.20 This piping system is drawn using a combination of single-line and double-line symbols. The connections are shown as screwed, welded, and flanged. (Courtesy of Bechtel Corporation.)

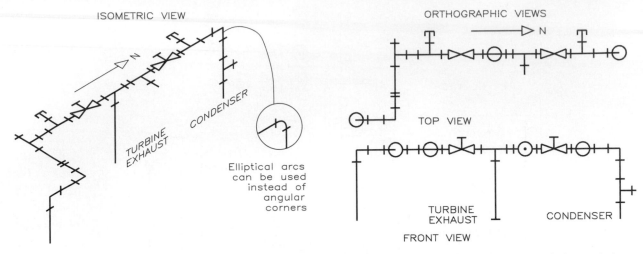

ISOMETRIC VIEW

ORTHOGRAPHIC VIEWS

Elliptical arcs can be used instead of angular corners

TOP VIEW

TURBINE EXHAUST

CONDENSER

FRONT VIEW

35.21 Top and front orthographic views are used as the basis for drawing a three-dimensional piping system using single-line symbols.

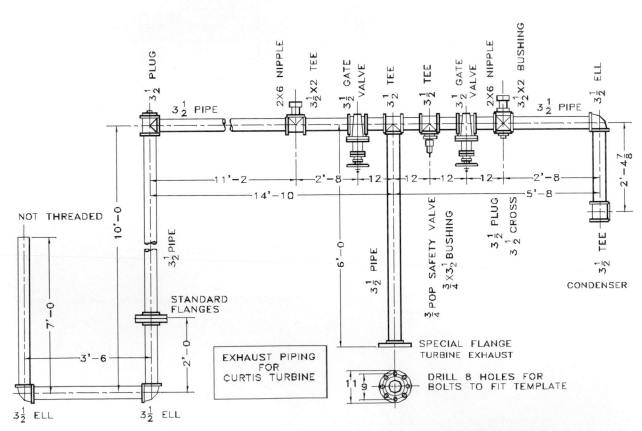

35.23 A finished developed drawing that shows all of the components in the system true size with double-line symbols.

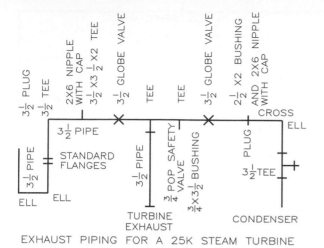

EXHAUST PIPING FOR A 25K STEAM TURBINE

35.22 The vertical pipes shown in Figure 28.21 are revolved into the horizontal plane to form a developed drawing. The fittings and valves are noted on the sketch.

an example where the specifications for the pipe, fittings, flanges, and valves are noted on the drawing and are itemized in the bill of materials.

A number of abbreviations are used to specify piping components and fittings, as you can see by referring to the bill of materials. Many of the standard abbreviations associated with pipe drawings and specifications are given in **Table 35.2**. Part number 1, for example, is an 8-in. diameter pipe of a Schedule 40 weight that is made of seamless steel by the open hearth (OH) process. Instead of OH, the abbreviation EF may be used, which is the abbreviation for electric furnace.

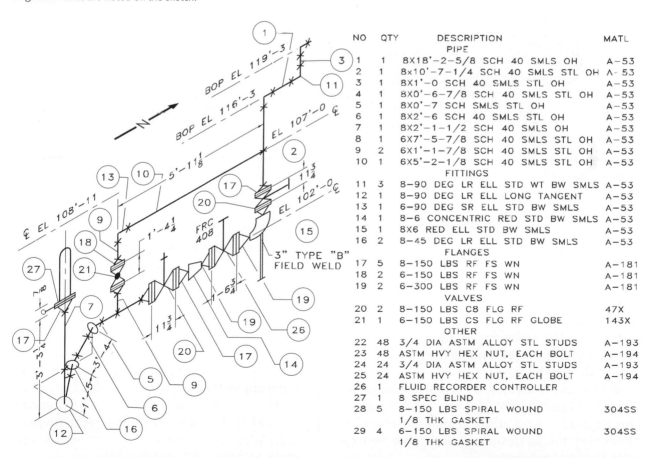

NO	QTY	DESCRIPTION	MATL
		PIPE	
1	1	8X18'-2-5/8 SCH 40 SMLS OH	A-53
2	1	8x10'-7-1/4 SCH 40 SMLS STL OH	A-53
3	1	8X1'-0 SCH 40 SMLS STL OH	A-53
4	1	8X0'-6-7/8 SCH 40 SMLS STL OH	A-53
5	1	8X0'-7 SCH SMLS STL OH	A-53
6	1	8X2'-6 SCH 40 SMLS STL OH	A-53
7	1	8X2'-1-1/2 SCH 40 SMLS OH	A-53
8	1	6X7'-5-7/8 SCH 40 SMLS STL OH	A-53
9	2	6X1'-1-7/8 SCH 40 SMLS STL OH	A-53
10	1	6X5'-2-1/8 SCH 40 SMLS STL OH	A-53
		FITTINGS	
11	3	8-90 DEG LR ELL STD WT BW SMLS	A-53
12	1	8-90 DEG LR ELL LONG TANGENT	A-53
13	1	6-90 DEG SR ELL STD BW SMLS	A-53
14	1	8-6 CONCENTRIC RED STD BW SMLS	A-53
15	1	8X6 RED ELL STD BW SMLS	A-53
16	2	8-45 DEG LR ELL STD BW SMLS	A-53
		FLANGES	
17	5	8-150 LBS RF FS WN	A-181
18	2	6-150 LBS RF FS WN	A-181
19	2	6-300 LBS RF FS WN	A-181
		VALVES	
20	2	8-150 LBS C8 FLG RF	47X
21	1	6-150 LBS CS FLG RF GLOBE	143X
		OTHER	
22	48	3/4 DIA ASTM ALLOY STL STUDS	A-193
23	48	ASTM HVY HEX NUT, EACH BOLT	A-194
24	24	3/4 DIA ASTM ALLOY STL STUDS	A-193
25	24	ASTM HVY HEX NUT, EACH BOLT	A-194
26	1	FLUID RECORDER CONTROLLER	
27	1	8 SPEC BLIND	
28	5	8-150 LBS SPIRAL WOUND 1/8 THK GASKET	304SS
29	4	6-150 LBS SPIRAL WOUND 1/8 THK GASKET	304SS

35.24 This dimensioned isometric pictorial is called a "spool drawing." It is sufficiently complete that it can serve as a working drawing when used with the bill of materials.

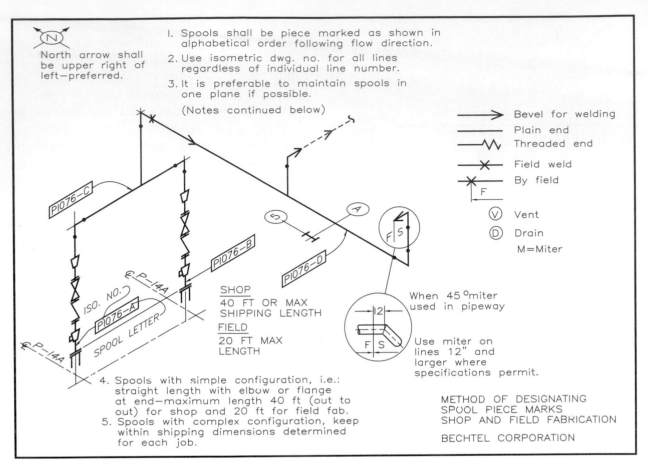

35.25 A suggested format for preparing spool drawings from a company standard. (Courtesy of the Bechtel Corporation.)

Under the column "materials," you will notice a code that begins with the letter A, such as A–53. The letter A is used to represent a grade of carbon steel that is listed in Table A of the ANSI B31.3: *Petroleum Refinery Piping Standards.* The codes for fittings, flanges, and valves are taken from the manufacturers' catalogs of these products.

A suggested format for spool drawings is given in **Figure 35.25**. This format is used by the Bechtel Corporation, a major construction company, in designing and constructing pipelines and refineries.

35.15 Vessel Detailing

Vessels are containers, usually cylindrical in shape, that are used to contain petroleum products and other chemicals. The cylinders can be installed in vertical or horizontal positions. Vessels can also be spherical or ellipsoidal.

A detailed drawing of a cylindrical vessels are drawn and dimensioned with specifications by following the general rules of working drawings. In addition, the types of welded joints required are specified to ensure that the vessel is properly fabricated to withstand the pressures and weights that it will be subjected to.

Table 35.2 Standard Abbreviations Associated with Pipe Specifications

AVG	average	FS	forged steel	SPEC	specification
BC	bolt circle	FSS	forged stainless steel	SR	short radius
BE	beveled ends	FW	geld weld	SS	stainless steel
BF	blind flange	GALV	galvanized	STD	standard
BM	bill of materials	GR	grade	STL	steel
BOP	bottom of pipe	ID	inside diameter	STM	steam
B&S	bell & spigot	INS	insulate	SW	socketweld
BWG	Birmingham wire gauge	IPS	iron pipe size	SWP	standard working pressure
CAS	cast alloy steel	LR	long radius	TC	test connection
CI	cast iron	LW	lap weld	TE	threaded end
CO	clean out	MI	malleable iron	TEMP	temperature
CONC	concentric	MFG	manufacture	T&G	tongue & groove
CPLG	coupling	OD	outside diameter	TOS	top of steel
CS	carbon steel, cast steel	OH	open hearth	TYP	typical
DWG	drawing	PE	plain end—not beveled	VC	vitrified clay
ECC	eccentric	PR	pair	WE	weld end
EF	electric furnace	RED	reducer	WN	weld neck
EF7N	electric fusion weld	RF	raised face	W3	welded bonnet
ELEV	elevation	RTG or RJ	ring type joint	WT	weight
ERW	electric resistance weld	SCH	schedule	XS	extra strong
FF	flat face	SCRD	screwed	XXS	double extra strong
FLG	flange	SMLS	seamless		
FOB	flat on bottom	SO	slip-on		

35.16 Computer Drawings

Pipe drawings range from plumbing drawings of typical homes to offshore oil wells and similar large-scale applications. Due to the broadness of this field, several software packages have been developed for engineers and technicians who are responsible for the preparation of pipe drawings.

Models have been extensively used for the design and layout of complex systems—refineries, processing plants, and similar installations—because of the difficulty in representing the intricate details of three-dimensional piping systems in two-dimensional drawings. Models (by computer and by hand) are used to design, develop, and explain these projects with the supplementation of drawings (**Figure 35.26**).

The ability to use three-dimensional models made by computer has alleviated some of the need for shop-made models. Computer programs are available that enable the piping layout to be designed as a three-dimensional model on the screen, which permits the generation of views from any angle. An example

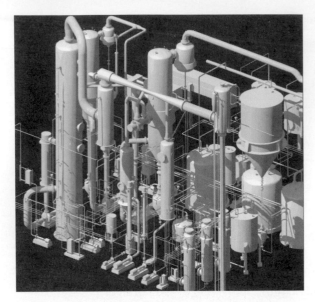

35.26 Models (by computer or by hand) are often necessary in the design, development, and explanation of piping installations. (Courtesy of Coade, Inc.)

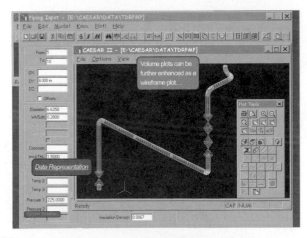

35.27 This computer program is available for drawing piping systems in 2D and 3D. (Courtesy of Coade, Inc.)

of a pipe layout program is shown in **Figure 35.27** where the pipe shown is being drawn as a rendered, three-dimensional model.

The area of pipe drafting is a complex study in graphics and technology worthy of a sizeable textbook on this field alone. The coverage here was limited to a brief introduction to the basics of piping. The standards of pipe drafting vary to a notable degree from company to company.

Problems

1. On a size A sheet, draw five orthographic views of the fittings listed below. The views should include the front, top, bottom, and the left and right views. Draw two fittings per page. Refer to **Figure 35.17** and **Table 35.2** as a guide. Use single-line symbols to draw the following screwed fittings: 90° ell, 45° ell, tee, lateral, cap, reducing ell, cross, concentric reducer, check valve, union, globe valve, gate valve, and bushing.

2. Same as Problem 1, but draw the fittings as flanged fittings.

3. Same as Problem 1, but draw the fittings as welded fittings.

4. Same as Problem 1, but draw the fittings as double-line screwed fittings.

5. Same as Problem 1, but draw the fittings as single-line flanged fittings.

6. Same as Problem 1, but draw the fittings as double-line welded fittings.

7. Convert the single-line sketch in **Figure 35.28** into a two-view, single-line drawing that will fit on a Size A sheet.

8. A. Convert the single-line pipe system in **Figure 35.29** into a double-line drawing that will fit on a Size B sheet, using the graphical scale given in the drawing to select the best scale for the system.
B. Same as A but draw using single-line symbols.

9. Convert the pipe system in **Figure 35.30** into a single-line orthographic (screwed joints) drawing that will fit on a Size B sheet.

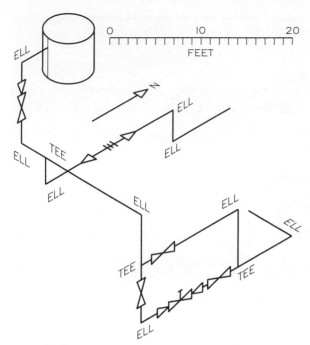

35.28 Problem 7.

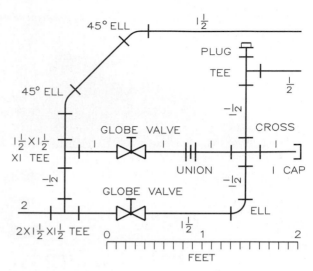

35.29 Problem 8.

10. Convert the pipe system given in **Figure 35.21** into a two-view orthographic drawing using double-line symbols that will fit on a size B sheet.

11. Convert the isometric drawing of the pipe system in **Figure 35.24** into a two-view orthographic, double-line drawing that will fit on a Size B sheet. Take the measurements from the given drawing, and select a convenient scale.

12. Convert the isometric drawing of the pipe system in **Figure 35.25** into a two-view, double-line orthographic drawing that will fit on a Size B sheet.

13. Convert the orthographic pipe system in **Figure 35.12** into a single-line isometric drawing that will fit on a Size B sheet. Estimate the dimensions.

14. Convert the orthographic pipe system in **Figure 35.12** into a double-line orthographic view that will fit on a Size B sheet.

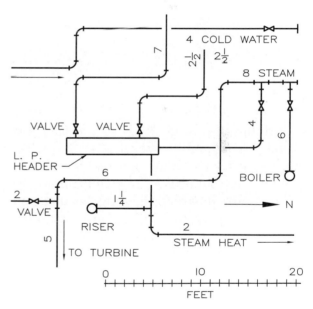

35.30 Problem 9.

36

Electric/Electronic Graphics

36.1 Introduction

Electric/electronics graphics is a specialty area of the field of graphics technology. **Electrical** graphics is related to the transmission of electrical power that is used in large quantities in homes and industry for lighting, heating, and equipment operation. **Electronics** graphics deals with circuits in which transistors and electronic tubes are used, where power is used in much smaller quantities. Examples of electronic equipment are radios, televisions, computers, and similar products.

Electronics drafters are responsible for the preparation of drawings that will be used in fabricating the circuit, and thereby bringing the product into being. They will work from sketches and specifications developed by the engineer or electronics technologist. This chapter will review the drafting practices that are necessary for the preparation of electronic diagrams.

A major portion of this chapter has been adapted from ANSI Y14.15, *Electrical and Electronics Diagrams,* the standards that regulate the graphics techniques used in this area. The symbols used were taken from ANSI Y32.2, *Graphic Symbols for Electrical and Electronics Diagrams.*

36.2 Types of Diagrams

Electronic circuits are classified and drawn in the format of one of the following types of diagrams:

1. Single-line diagrams

2. Schematic diagrams

3. Connection diagrams

The suggested line weights for drawing these types of diagrams are shown in **Figure 36.1**.

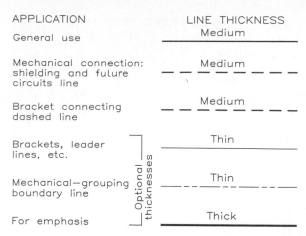

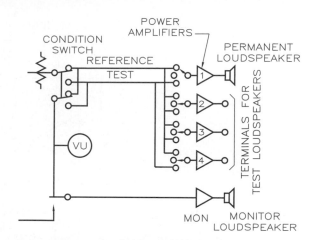

36.1 The recommended line weights for drawing electronics diagrams.

36.3 A typical single-line diagram for illustrating a circuit. Its basic functions are shown, but many of the details, components, and their ratings that are shown in a schematic are omitted.

Single-line Diagrams

Single-line diagrams are drawn with single lines and general symbols that are adequate to trace the flow of current through the circuit and obtain a basic understanding of the parts and devices within it. Descriptions of the circuit components are not specified in detail. Single lines are used to represent both AC and DC systems, as illustrated in **Figure 36.2**. An example of a single-line diagram of an audio system is shown in **Figure 36.3**.

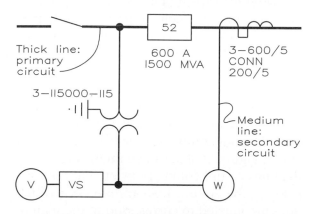

36.2 A portion of a single-line diagram where heavy lines represent the primary circuits, and medium lines represent the connections to the current and potential sources.

Primary circuits are indicated by thick connecting lines, and medium lines are used to represent connections to the current and potential sources.

Single-line diagrams show the connections of meters, major equipment, and instruments. Ratings are often given to supplement the graphic symbols to provide such information as kilowatts, voltages, cycles and revolutions per minute, and generator ratings (**Figure 36.4**).

Schematic Diagrams

Schematic diagrams use graphic symbols to show the electrical connections and functions of a specific circuit arrangement. Schematics provide more information and specifications that are necessary for the composition of a circuit than do single-line diagrams. Although a schematic diagram enables one to trace the circuit and its functions, physical sizes, shapes, and locations of various components are not shown. A schematic diagram in **Figure 36.6** can be referred to for many applications of the principles covered in this chapter. Another example of a schematic is given in **Figure 36.48**.

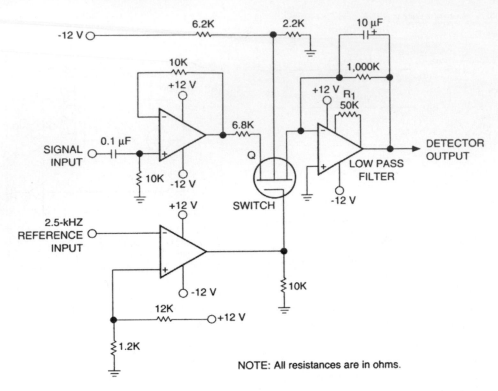

36.4 A single-line diagram illustrates a switching circuit complete with notation of device designations. (Courtesy of NASA.)

Connection Diagrams

Connection diagrams show the connections and installations of the parts and devices of the system. In addition to showing the internal connections, external connections, or both, they show the physical arrangement of the parts. It can be described as an installation diagram such as the one shown in **Figure 36.5**.

36.3 Schematic Diagram Connecting Symbols

The most basic symbols of a circuit are those that are used to represent connections of parts within the circuit. Connections, or junctions, are indicated by using small black dots, as shown in **Figure 36.7A**. The dots distinguish between connecting lines and those that simply pass over each other (Part B). The use of dots to show connections is optional; it is preferable to omit them if clarity is not sacrificed. Also, it is preferred that connecting wires have single junctions wherever possible.

When the layout of a circuit does not permit the use of single junctions, and lines within the circuit must cross, then dots must be used to distinguish between crossing and connecting lines (Parts C and D).

Interrupted paths are breaks in lines within a schematic diagram that are interrupted to conserve space when this can be done without confusion. For example, the circuit in **Figure 36.8** has been interrupted since the lines do not connect the left and the right sides of the circuit. Instead, the ends of the lines are labeled to correspond to the matching notes at the other side of the interrupted circuit.

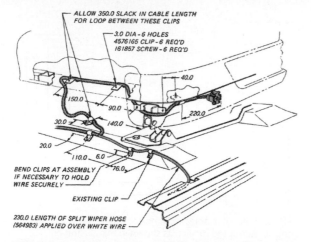

ALLOW 350.0 SLACK IN CABLE LENGTH
FOR LOOP BETWEEN THESE CLIPS

3.0 DIA - 6 HOLES
4576165 CLIP - 6 REQ'D
161857 SCREW - 6 REQ'D

40.0

150.0

90.0

220.0

30.0

140.0

20.0

110.0   6.0

76.0

BEND CLIPS AT ASSEMBLY
IF NECESSARY TO HOLD
WIRE SECURELY

EXISTING CLIP

230.0 LENGTH OF SPLIT WIPER HOSE
(564983) APPLIED OVER WHITE WIRE

36.5 This three-dimensional connections diagram shows the circuit and its components with the necessary details to explain how it is connected or installed. (Courtesy of the General Motors Corporation.)

There will be occasions where sets of lines in a horizontal or vertical direction will be interrupted (**Figure 36.9**). Brackets will be used to interrupt the circuit and notes will be placed outside the brackets to indicate the destinations of the wires or their connections.

In some cases, a dashed line is used to connect brackets that interrupt circuits (**Figure 36.10**). The dashed line should be drawn so that it will not be mistaken as a continuation of one of the lines within the bracket.

Mechanical linkages that are closely related to electronic functions may be shown as part of a schematic diagram (**Figure 36.11**). An arrangement of this type helps clarify the relationship of the electronics circuit with the mechanical components.

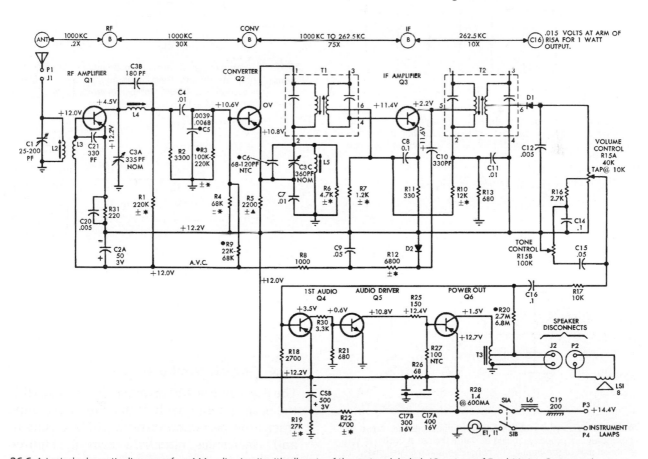

36.6 A typical schematic diagram of an AM radio circuit with all parts of the system labeled. (Courtesy of Ford Motor Company.)

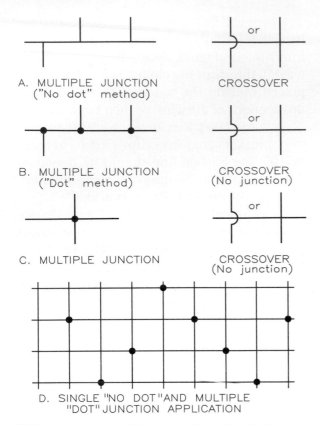

A. MULTIPLE JUNCTION ("No dot" method)

CROSSOVER

B. MULTIPLE JUNCTION ("Dot" method)

CROSSOVER (No junction)

C. MULTIPLE JUNCTION

CROSSOVER (No junction)

D. SINGLE "NO DOT" AND MULTIPLE "DOT" JUNCTION APPLICATION

36.7 Connections should be shown with single-point junctions as shown at (A). Dots may be used to call attention to connections as shown at (B), and dots must be used when there are multiples of the type shown at (C).

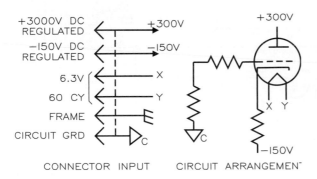

CONNECTOR INPUT   CIRCUIT ARRANGEMENT

36.8 Circuits may be interrupted and connections not shown by lines if they are properly labeled to clarify their relationship to the removed part of the circuit. The connections above are labeled to match those on the left and right sides of the illustration.

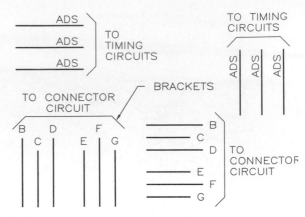

36.9 Brackets and notes may be used to specify the destinations of interrupted circuits, as shown in this illustration.

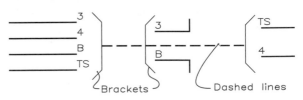

36.10 The connections of interrupted circuits can be indicated by using brackets and a dashed line in addition to labeling the lines. The dashed line should not be drawn to appear as an extension of one of the lines in the circuit.

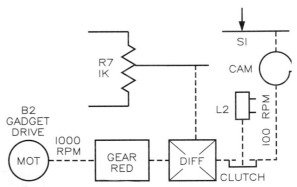

36.11 If mechanical functions are closely related to electrical functions, it may be desirable to link them within the schematic diagram.

36.4 Graphic Symbols

The electronics drafter must be familiar with the basic graphic symbols that are used to represent the parts and devices within electrical and electronics circuits shown in **Figures 36.12–36.16**. These symbols, extracted from

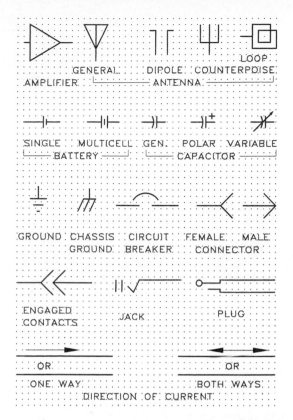

36.12 The proportions of these symbols are drawn on a 3-mm (0.13 in.) grid. When enlarged so the grid is full size, their suggested sizes can be found.

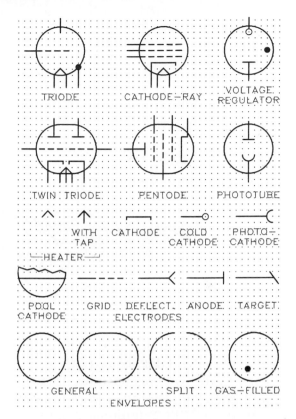

36.13 The upper six symbols are used to represent often-used types of electron tubes drawn on a 3-mm (0.13 in.) grid. An explanation of the parts that make up each symbol is given in the lower half of the figure. (Electron tubes have been mostly replaced by transistors.)

the *ANSI Y32.2* standard, are adequate for practically all diagrams. However, when a highly specialized part needs to be shown and a symbol for it is not provided in these standards, it is permissible for the drafter to develop his or her own symbol provided it is properly labeled and its meaning clearly conveyed.

The symbols shown in **Figures 36.12–36.16** are drawn on a grid of 3-mm (0.13 in.) squares that have been reduced. The size of this grid is equal to the letter height used on the final drawing. It is general practice to size graphic symbols based on letter height since text and numerals cannot be enlarged or reduced as easily as graphic symbols without affecting readability. Symbols may be drawn larger or smaller to fit the size of your layout, provided the relative proportions of the symbols are kept about the same.

The symbols in **Figures 36.12–36.16** are but a few of the more commonly used symbols. There are between five and six hundred different symbols in the ANSI standards for variations of the basic electrical/electronics symbols.

The preparation of a schematic diagram begins with drawing a freehand sketch to show the circuit and the placement of its components (**Figure 36.17A**). Using a printed grid makes sketching easier. When the sketch is completed, an instrument drawing can be made by hand or by computer to show the

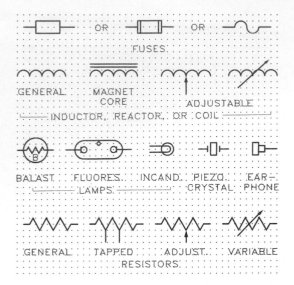

36.14 Graphics symbols of standard circuit components drawn on a 3-mm (0.13 in.) grid.

components at a proper scale and reference designations added (**Figure 36.17B**).

Some of the symbols have been noted to provide designations of their sizes or ratings. The need for this additional information depends on the requirements and the usage of the schematic diagrams.

36.5 Terminals

Terminals are the ends of the circuit where devices are attached with connecting wires. Examples of devices with terminals that are specified in circuit diagrams are switches, relays, and transformers. The graphic symbol for a terminal is an open circle that is the same size as the solid circle used to indicate a connection.

Switches are used to turn a circuit on or off, or else to actuate a certain part of it while turning another off. Examples of labeling switches are shown in **Figure 36.6**. At the lower right of this schematic diagram the switches are labeled S1A and S1B.

When a group of parts is enclosed or shielded (drawn enclosed with dashed lines) and the terminal circles have been omitted,

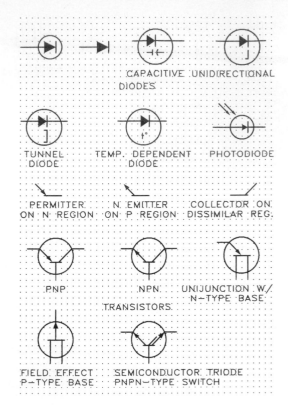

36.15 Graphics symbols for semiconductor devices and transistors drawn on a 3 mm (0.13 in.) grid. The arrows in the middle of the figure illustrate the meanings of the arrows used in the transistor symbols shown below them.

the terminal markings should be placed immediately outside the enclosure, as shown in **Figure 36.5** at T1 and T2. The terminal identifications should be added to the graphic symbols that correspond to the actual physical markings that appear on or near the terminals of the part (not given in this diagram). Several examples of notes and symbols that explain the parts of a circuit diagram are shown in **Figures 36.18**, **36.19**, and **36.20**.

When colored wires, numbers, or geometric symbols are used to identify the various leads or terminals of multilead parts, show this identification near the connecting line adjacent to the symbol. Colored wires can be identified on a diagram with color abbreviations to denote their color.

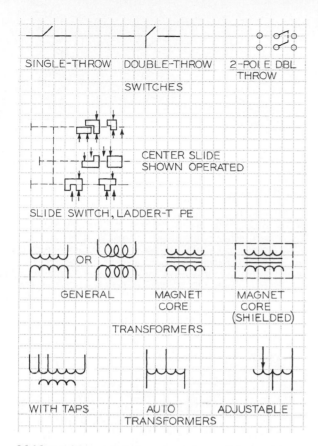

36.16 Graphics symbols for representing switches and transformers drawn on a 3-mm (0.13 in.) grid.

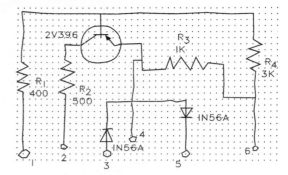

A. A RAPID FREEHAND SKETCH

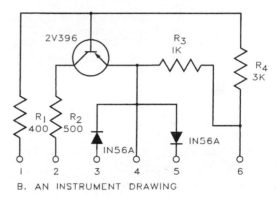

B. AN INSTRUMENT DRAWING

36.17 Drawing a diagram from a sketch.

A The circuit designer can make a freehand sketch of a circuit as a preliminary drawing.

B The final drawing of the circuit is made using the proper symbols, notes, and lines.

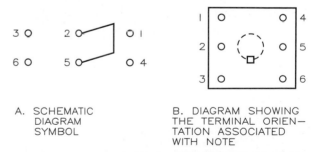

A. SCHEMATIC DIAGRAM SYMBOL

B. DIAGRAM SHOWING THE TERMINAL ORIEN—TATION ASSOCIATED WITH NOTE

36.18 An example of a method of labeling the terminals of a toggle switch on a schematic diagram (A) and a diagram that illustrates the toggle switch when it is viewed from its rear (B).

Rotary terminals are used to regulate the resistance in some circuits, and the direction of rotation of the dial is indicated on the schematic diagram. The abbreviations CW (clockwise) or CCW (counterclockwise) are placed adjacent to the movable contact when it is in its extreme clockwise position, as shown in **Figure 36.21A**. The movable contact can be identified by an arrow at its end.

If the device terminals are not marked, numbers may be used with the resistor symbol and the number 2 assigned to the adjustable contact (**Figure 36.21B**). Other fixed taps may be sequentially numbered and added, as shown in Part C.

The position of a switch as it relates to the function of a circuit should be indicated on a schematic diagram. A method of showing functions of a variable switch is shown in

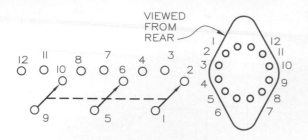

A. SYMBOL ON
SCHEMATIC DIAGRAM

B. TERMINAL ORIENTATION
DIAGRAM ASSOCIATED
WITH NOTE

36.19 An example of a rotary switch as it would appear on a schematic diagram (A) and a diagram that shows the numbered terminals of the switch when viewed from its rear (B).

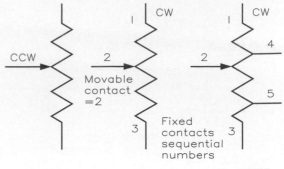

A. COUNTER—
CLOCKWISE

B. MOVABLE
CONTACT

C. FIXED
CONTACTS

36.21 To indicate the direction of rotation of rotary switches on a schematic diagram, the abbreviations CW (clockwise) and CCW (counterclockwise) are placed near the moveable contact (A). If the device terminals are not marked, numbers may be used with the resistor symbols and the number 2 assigned to the adjustable contact (B). Additional contacts may be labeled, as shown at C.

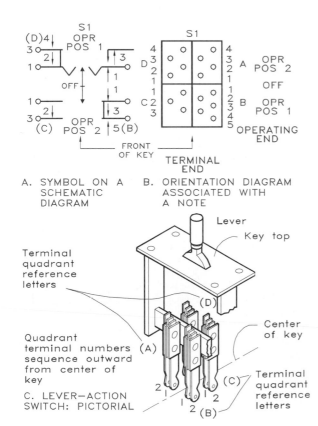

A. SYMBOL ON A
SCHEMATIC
DIAGRAM

B. ORIENTATION DIAGRAM
ASSOCIATED WITH
A NOTE

C. LEVER—ACTION
SWITCH: PICTORIAL

36.20 (A) An example of a typical lever switch as it would appear on a schematic diagram. (B) An orientation diagram that shows terminal end and the numbered terminals of the switch when viewed from its operating end. (C) A pictorial of the lever switch and its four quadrants.

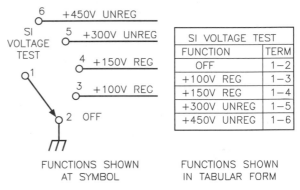

FUNCTIONS SHOWN
AT SYMBOL

FUNCTIONS SHOWN
IN TABULAR FORM

SI VOLTAGE TEST	
FUNCTION	TERM
OFF	1—2
+100V REG	1—3
+150V REG	1—4
+300V UNREG	1—5
+450V UNREG	1—6

36.22 For more complex switches, position-to-position function relations may be shown using symbols on the schematic diagram, or by a table of values located elsewhere on the diagram.

Figure 36.22. The arrow, representing the movable end of the switch, can be positioned to connect with several circuits. The different functional positions of the rotary switch are shown both by symbol and by table.

Another method of representing a rotary switch is shown in **Figure 36.23** by symbol and table. The tabular form is preferred due to the complexity of this particular switch. The dashes between the numbers in the table indicate that the numbers have been connected. For example, when the switch is in

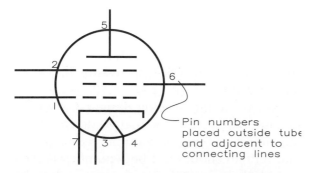

A. SYMBOL ON SCHEMATIC DIAGRAM

(SWITCH VIEWED FROM FRONT)

B. FUNCTIONS SHOWN IN TABULAR FORM

POS	FUNCTION	TERMINALS
SI REAR		
I	OFF (SHOWN)	I—2, 5—6, 9—I0
2	STANDBY	I—3, 5—7, 9—II
3	OPERATE	I—4, 5—8, 9—I2

36.23 A rotary switch may be shown on a schematic diagram with its terminals labeled, as shown at the left, or its functions can be given in a table placed elsewhere on the drawing, as shown at the right. Dashes are used to indicate the linkage of the numbered terminals. For example, 1–2 means that terminals 1 and 2 are connected in the "off" position.

Position 2, the following terminals are connected: 1 and 3, 5 and 7, and 9 and 11. A table of this type should be placed at the bottom of a schematic diagram, if applicable.

Electron tubes have pins that fit into sockets that have terminals connecting into circuits. Pins are labeled with numbers placed outside the symbol used to represent the tube, as shown in **Figure 36.24**, and are numbered in a clockwise direction with the tube viewed from its bottom.

36.6 Separation of Parts

In complex circuits, it is often advantageous to separate elements of a multi-element part with portions of the graphic symbols drawn in different locations on the drawing. An example of this method of separation is the switch labeled S1A and S1B in **Figure 36.5**. The switch is labeled S1 and the letter that follows, called a suffix, is used to designate different parts of the same switch. Suffix letters may also be used to label subdivisions of an enclosed unit that is made up of a series of internal parts, such as the crystal unit shown in **Figure 36.25**. These crystals are referred to as Y1A and Y1B.

Rotary switches of the type shown in **Figure 36.26** are designated as S1A, S1B, etc. The suffix letters A, B, etc., are labeled in sequence beginning with the knob and working away from it. Each end of the various sections of the switch should be viewed from the same end. When the rear and front of the switches need to be used, the words FRONT and REAR are added to the designations.

Portions of items such as terminal boards, connectors, or rotary switches may be separated on a diagram. The words PART OF may precede the identification of the portion of the circuit of which it is a part, as shown in **Figure 36.27A**. A second method of showing a

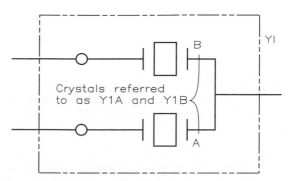

Crystals referred to as Y1A and Y1B

36.24 Tube pin numbers should be placed outside the tube envelope and adjacent to the connecting lines. (Courtesy of ANSI.)

Pin numbers placed outside tube and adjacent to connecting lines

36.25 As subdivisions within the complete part, crystals A and B are referred to as Y1A and Y1B.

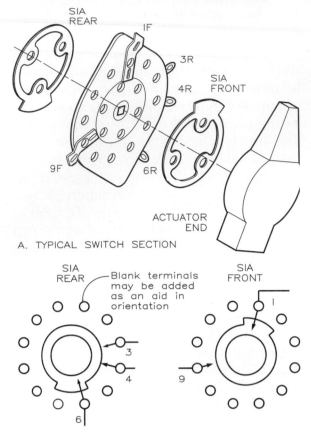

A. TYPICAL SWITCH SECTION

B. GRAPHICAL SYMBOL

36.26 Parts of rotary switches are designated with suffix letters A, B, C, etc., and are referred to as S1A, S1B, S1C, etc. The words FRONT and REAR are added to these designations when both sides of the switch are used. (Courtesy of ANSI.)

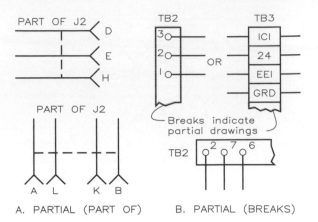

A. PARTIAL (PART OF)   B. PARTIAL (BREAKS)

36.27 The portions of connectors or terminal boards are functionally separated on a diagram; the words PART OF may precede the reference designation of the entire portion. Or, conventional breaks can be used to indicate graphically that the part is only a portion of the whole.

part of a system is by using conventional break lines that make the note, PART OF, unnecessary.

36.7 Reference Designations

A combination of letters and numbers that identify items on a schematic diagram are called reference designations. These designations are used to identify the components not only on the drawing, but in the related documents that refer to them. Reference designations should be placed close to the symbols that represent the replaceable items of a cir-

cuit on a drawing. Items that are not separately replaceable may be identified if this is considered necessary. Mounting devices for electron tubes, lamps, fuses, and so forth, are seldom identified on schematic diagrams.

It is standard practice to begin each reference designation with an uppercase letter that may be followed by a numeral with no hyphen between them. The number usually represents a portion of the part being represented. The lowest number of a designation should be assigned to begin at the upper left of the schematic diagram and proceed consecutively from left to right and top to bottom throughout the drawing.

Some of the standard abbreviations used to designate parts of an assembly are: amplifier-A, battery-BT, capacitor-C, connector-J, piezoelectric crystal-Y, fuse-F, electron tube-V, generator-G, rectifier-CR, resistor-R, transformer-T, and transistor-Q.

As the circuit is being designed, some of the numbered elements may be deleted from the circuit drawing. The numbered elements that remain should not be renumbered even though there is a missing element within the sequence of numbers used to label the parts.

HIGHEST REFERENCE DESIGNATIONS	
R72	C40
REFERENCE DESIGNATIONS NOT USED	
R8, R10, R61	C12, C15, C17
R64, R70	C20, C22

36.28 Reference designations are used to identify parts of a circuit. They are labeled in a numerical sequence from left to right beginning at the upper left of the diagram. If parts are later deleted from the system, the ones deleted should be listed in a table, along with the highest reference number designations.

Instead, a table of the type shown in **Figure 36.28** can be used to list the parts that have been omitted from the circuit. The highest designations are also given in the table as a check to be sure that all parts were considered.

Electron tubes are labeled not only with reference designations but with type designation and circuit function as shown in **Figure 36.29**. This information is labeled in three lines, such as V5/35C5/OUTPUT, which are located adjacent to the symbol.

36.8 Numerical Units of Function

Functional units such as the values of resistance, capacitance, inductance, and voltage should be specified with the fewest number of

zeros by using the multipliers in **Figure 36.30A** as prefixes. Examples using this method of expression are shown in the parts B and C, where units of resistance and capacitance are given. When four-digit numbers are given, omit the commas; write one thousand as 1000, not as 1,000. You should recognize and use the lowercase or uppercase prefixes as indicated in the table of **Figure 36.30**.

A. MULTIPLIERS		SYMBOL	
		METHOD 1	METHOD 2
MULTIPLIER	PREFIX		
10^{12}	TERA	T	T
10^{9}	GIGA	G	G
10^{6} (1,000,000)	MEGA	M	M
10^{3} (1,000)	KILO	k	K
10^{-3} (0.001)	MILLI	m	MILLI
10^{-6} (0.000,001)	MICRO	μ	U
10^{-9}	NANO	n	N
10^{-12}	PICO	p	P
10^{-13}	FEMTO	f	F
10^{-16}	ATTO	a	A

B. RESISTANCE		
RANGE IN OHMS	EXPRESS AS	EXAMPLE
LESS THAN 1,000	OHMS	0.031 470
1,000 TO 99,999	OHMS OR KILOHMS	1800 15,853 10k
100,000 to 999,999	KILOHMS OR MEGOHMS	220k 0.22M
1,000,000 OR MORE	MEGOHMS	3.3M

C. CAPACITANCE		
RANGE IN PICOFARADS	EXPRESS AS	EXAMPLE
LESS THAN 10,000	PICOFARADS	152.4pF 4700pF
10,000 OR MORE	MICROFARADS	0.015μF 30μF

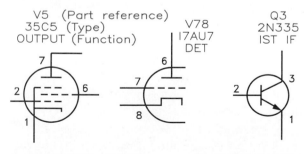

V5 (Part reference)
35C5 (Type)
OUTPUT (Function)

V78
17AU7
DET

Q3
2N335
IST IF

A. ELECTRON TUBES   B. TRANSISTOR

36.29 Three lines of notes can be used with electron tubes and transistors to specify reference designation, type designation, and function. This information should be located adjacent to and preferably above the symbol.

36.30 Multipliers should be used to reduce the number of zeros in a number (A). Examples of units of capacitance and resistance are shown at (B) and (C).

A general note can be used where certain units are repeated on a drawing to reduce time and effort:

UNLESS OTHERWISE SPECIFIED, RESISTANCE VALUES ARE IN OHMS; CAPACITANCE VALUES ARE IN MICROFARADS.

or

CAPACITANCE VALUES ARE IN PICOFARADS.

A note for specifying capacitance values is:

CAPACITANCE VALUES SHOWN AS NUMBERS EQUAL TO OR GREATER THAN UNITS ARE IN pF AND NUMBERS LESS THAN UNITY ARE IN μF.

Examples of the placement of the reference designations and the numerical values of resistors are shown in **Figure 36.31**.

36.9 Functional Identification of Parts

The readability of a circuit is improved if parts are labeled to indicate their functions. Test points are labeled on drawings with the letters "Tp" and their suffix numbers. The sequence of the suffix numbers should be the same as the sequence for troubleshooting the circuit when it is defective. As an alternative, the test function can be indicated on the diagram below the reference designation.

Additional information may be included on a schematic diagram to aid in the maintenance of the system:

DC resistance of windings and coils.

Critical input and output impedance values.

Wave shapes (voltage or current) at significant points. Wiring requirements for critical ground points, shielding, pairing, etc.

Power or voltage ratings of parts.

Caution notation for electrical hazards at maintenance points.

Circuit voltage values at significant points (tube pins, test points, terminal boards, etc.).

Zones (grid system) on complex schematics.

Signal flow direction in main signal paths shall be emphasized.

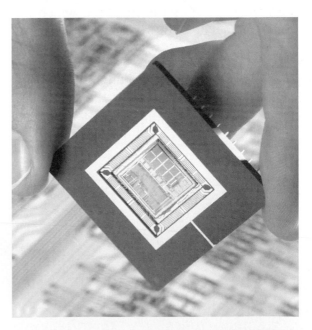

36.32 Jack S. Kilby wrote in his notebook in 1958, "Extreme miniaturization of many electrical circuits could be achieved by making resistors, capacitors and transistors, and diodes on a single slice of silicon." (Courtesy of Texas Instruments.)

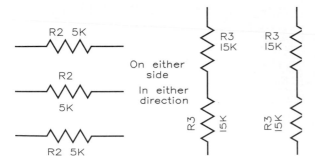

36.31 Methods of labeling the units of resistance on a schematic diagram are shown here.

36.10 Printed Circuits

Printed circuits are universally used for miniature electronic components and computer systems (**Figure 36.32**). For years, the vacuum tube was the best means of controlling electrical current; it was bulky, fragile, unreliable, and consumed large amounts of power. In 1947, the transistor replaced the vacuum tube, eliminated the disadvantages of the tube, and reduced its size to fit on the head of pin.

In 1958, Jack Kilby advanced miniaturization further with his integrated circuit in which resistors, capacitors, transistors, and other components were merged on a slice of silicon called a "chip." With each advancement in electronics has come even more breakthroughs, which have offered more productivity, at greater speeds, and at lower costs. The dramatic degree of miniaturization that has occurred can be seen in **Figure 36.33** where a complex circuit fits on a one-square-centimeter chip.

Printed circuit boards, on which chips and other devices are assembled, are drawn up to four times its final size. The drawings are precisely drawn in ink on a highly stable acetate film and are photographically reduced to the

36.34 A magnified view of a circuit that has been printed and etched on a board and the devices soldered in position. (Courtesy of Bishop Industries Corp.)

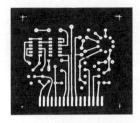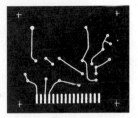

36.35 A printed circuit applied to both sides of the circuit board requires two drawings, one for each side, that are photographically converted to negatives for printing. (Courtesy of Bishop Industries Corp.)

36.33 An integrated circuit on a one-square-centimeter chip was a major breakthrough in electronics, but it has since been surpassed with more advanced technology. (Courtesy of Texas Instruments.)

desired size. The circuit is "printed" onto an insulated board made of plastic or ceramics, and the devices within the circuit are connected and soldered (**Figure 36.34**).

Some printed circuits are printed on both sides of the circuit board, requiring two photographic negatives, as shown in **Figure 36.35**, that were made from positive drawings (black lines on a white background). Each drawing for each side can be made on separate sheets of acetate that are laid over each other when the second diagram is drawn. However, a more efficient method

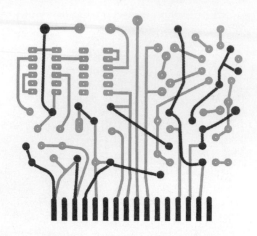

36.36 By using two colors, such as blue and red, one circuit drawing can be made, and two negatives made from the same drawing by using camera filters that screen out one of the colors with each shot. The circuits are then printed on each side of the board. (Courtesy of Bishop Industries Corp.)

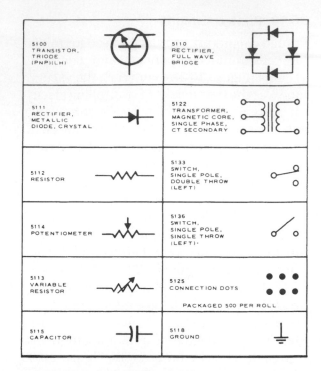

36.37 Stick-on symbols are available for laying out schematic diagrams rather than drawing them. They give a higher contrast and sharpness that improves their reproducibility. (Courtesy of Bishop Industries Corp.)

uses red and blue tape that can be used for making a single drawing (**Figure 36.36**) from which two negatives are photographically made. Filters are used on the process camera to drop out the red for one negative and a different filter for dropping out the blue for the second negative.

Printed circuits are usually coated with silicone varnish to prevent malfunction because of the collection of moisture or dust on the surface. They may also be enclosed in protective shells.

36.11 Shortcut Symbols

Preprinted symbols are available commercially that can be used for "drawing" high quality electronic circuits and printed circuits. The symbols are available on sheets or on tapes that can be burnished onto the surface of the drawing to form a permanent schematic diagram (**Figure 36.37**).

The symbols can be connected with matching tape to represent wires between them instead of drawing the lines. Schematic symbols provided by computer programs have replaced most other techniques of preparing schematics due to ease of making changes and modifications.

36.12 Installation Drawings

Many types of electric/electronics drawings are used to produce the finished installation, from the designer who visualized the system at the outset of the project to the contractor who builds it. Drawings are used to design the circuit, detail its parts for fabrication, specify the arrangement of the devices within the system, and instruct the contractor how to install the project.

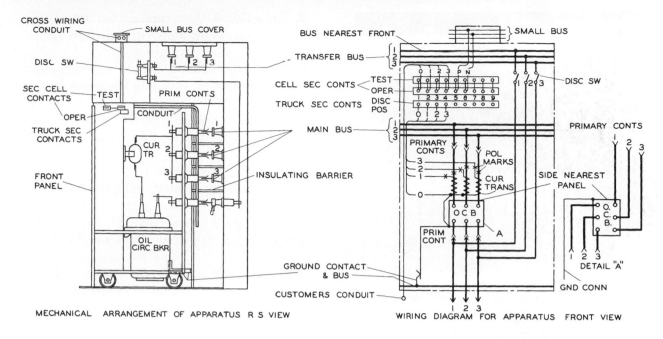

CROSS WIRING CONDUIT — SMALL BUS COVER
DISC SW
SEC CELL CONTACTS — TEST
OPER
TRUCK SEC CONTACTS
FRONT PANEL
PRIM CONTS
CONDUIT
CUR TR
INSULATING BARRIER
OIL CIRC BKR
GROUND CONTACT & BUS
CUSTOMERS CONDUIT

MECHANICAL ARRANGEMENT OF APPARATUS R S VIEW

BUS NEAREST FRONT — SMALL BUS
TRANSFER BUS
CELL SEC CONTS — TEST / OPER
TRUCK SEC CONTS — DISC POS
DISC SW
MAIN BUS
PRIMARY CONTS
POL MARKS
CUR TRANS
SIDE NEAREST PANEL
O C B
A
PRIM CONT
GND CONN
PRIMARY CONTS
DETAIL "A"

WIRING DIAGRAM FOR APPARATUS FRONT VIEW

36.38 This drawing shows views of a metal-enclosed switchgear to describe the arrangement of the apparatus; it also gives the wiring diagram for the unit.

A combination arrangement and wiring diagram drawing is shown in **Figure 36.38**, where the system is shown in a front and right-side view. The wiring diagram explains how the wires and components within the system are connected for the metal-encased switchgear. Bus bars are conductors for the primary circuits.

An installation/circuit diagram in **Figure 36.39** is a drawing used in a maintenance manual to show how to troubleshoot a defective circuit. It is a combination drawing that shows physical arrangement and the circuit as well.

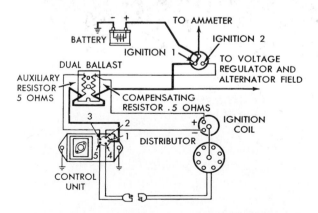

TO AMMETER
BATTERY
IGNITION 2
IGNITION 1
DUAL BALLAST
TO VOLTAGE REGULATOR AND ALTERNATOR FIELD
AUXILIARY RESISTOR 5 OHMS
COMPENSATING RESISTOR .5 OHMS
IGNITION COIL
DISTRIBUTOR
CONTROL UNIT

36.39 An installation/circuit diagram that describes how to troubleshoot a circuit. (Courtesy of Chrysler Corporation.)

1. On a Size A sheet, make a schematic diagram of the circuit shown in **Figure 36.40**.

2. On a Size A sheet, make a schematic diagram of the circuit shown in **Figure 36.41**.

3. On a Size A sheet, make a schematic diagram of the circuit shown in **Figure 36.42**.

4. On a Size A sheet, make a schematic diagram of the circuit shown in **Figure 36.43**.

5. On a Size A sheet, make a schematic diagram of the circuit shown in **Figure 36.44**.

6. On a Size B sheet, make a schematic diagram of the circuit shown in **Figure 36.45**.

7. On a Size B sheet, make a schematic diagram of the circuit shown in **Figure 36.46**.

8. On a Size B sheet, make a schematic diagram of the circuit shown in **Figure 36.47**.

9. On a Size C sheet, draw a schematic diagram of the circuit in **Figure 36.48** and show the parts list.

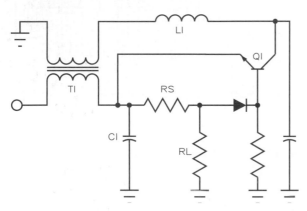

36.40 (Problem 1) A low-pass inductive-input filter. (Courtesy of NASA.)

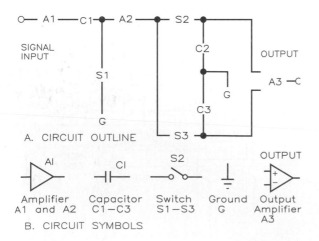

36.41 (Problem 2) A quadruple-sampling processor. (Courtesy of NASA.)

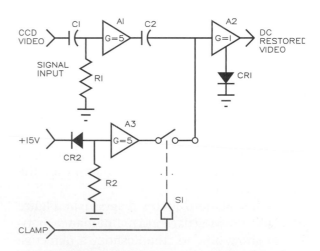

36.42 (Problem 3) A temperature-compensating DC restorer circuit. (Courtesy of NASA.)

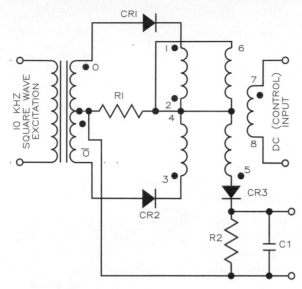

36.43 (Problem 4) A magnetic amplifier DC transducer. (Courtesy of NASA.)

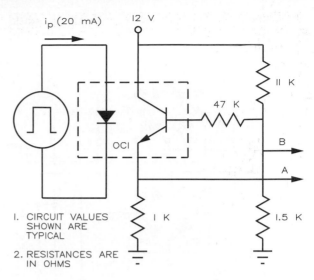

1. CIRCUIT VALUES SHOWN ARE TYPICAL

2. RESISTANCES ARE IN OHMS

36.44 (Problem 5) An improved power-factor controller. (Courtesy of NASA.)

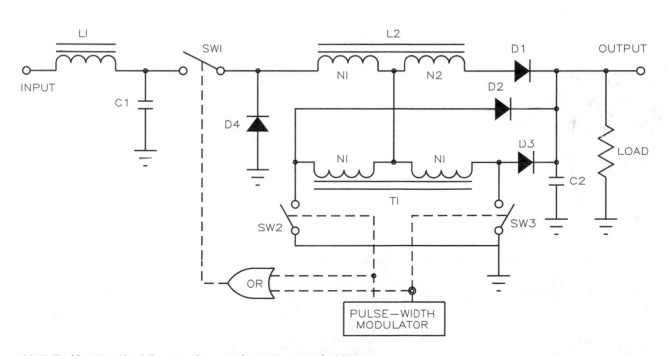

36.45 (Problem 6) A "buck/boost" voltage regulator. (Courtesy of NASA.)

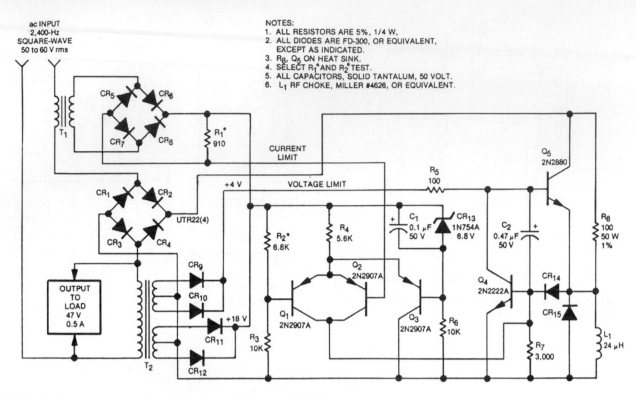

36.46 (Problem 7) An overload protection circuit. (Courtesy of NASA.)

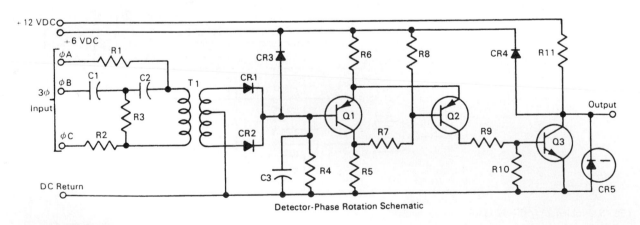

Detector-Phase Rotation Schematic

36.47 (Problem 8) A schematic of a phase detector circuit. (Courtesy of NASA.)

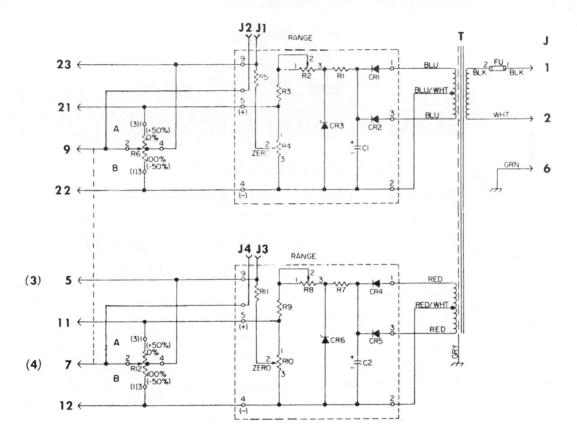

COMPONENT	DESCRIPTION	
CR1, CR2, CR3, CR4	DIODE	1N3193
CR3 & CR6	ZENER DIODE	24V 3W ±.5V TOL
R1 & R7	RESISTOR	560 Ω 5W 5%
R3 & R9	RESISTOR	56 Ω
R5 & R11	RESISTOR	2.7K 1/2W 5%
R2 & R8	TRIM POT	50 Ω
R4 & R10	TRIM POT	5K
R6 & R12	POT	500 Ω 2W DUAL
C1 & C2	CAPACITOR	60 MFD 60V
T	TRANSFORMER	
FU	FUSE 1 AMP	
M	EDGEWISE METER	
J	CONNECTOR	
J1, J2, J3, J4	TEST JACK	

TYPE	R6 SCALE		M SCALE	
	A	B	BOTTOM	TOP
RS1100C	0%	100%	0%	100%
RS1110C	0%	100%	PER ENGINEERING DATA	
RS3100C	+50%	-50%	-50%	+50%
RS4100C	0%	100%	0%	100%
RS2120C	100%	0%	OMIT	
RS2100C	100%	0%	0%	100%

NOTE: OUTPUT WIRED TO TERMINALS 3 AND 4 ON
TYPES RS1100C, RS1110C, AND RS3100C.

OUTPUT WIRED TO TERMINALS 5 AND 7 ON
TYPES RS2100C, RS2120C, AND RS4100C.

36.48 (Problem 9) A schematic diagram of a dual output manual station and its parts list. (Courtesy of NASA.)

37

2D Computer Graphics
AutoCAD 2004

37.1 Introduction

This chapter provides an introduction to computer graphics using AutoCAD 2004 which runs on an Intel 486 or Pentium processor with 32 MB (64 MB preferred) of RAM, at least a 200 Mb of hard disk space, a mouse (or tablet), an A-B plotter, and/or printer. Windows 98, Windows 2000, or Windows NT are recommended as the operating systems. AutoCAD was selected as the software for presenting computer graphics because it is the most widely used computer graphics program.

The coverage of AutoCAD in this book is brief and many operations could not be included because of space limitations. AutoCAD's concisely written *User's Guide* has 856 pages, and other manuals on the market have as many as 1500 pages. However, AutoCAD is covered here sufficiently well enough to guide you through the applications necessary for a typical engineering design graphics course.

You will find that the learning of computer graphics and its successive upgrades will be a career-long experience. We recommend that you begin this self-teaching process by experimenting with the peripheral commands and options that are not covered in this book. Also, reference to *Help* should be made routinely as a means of learning new commands and refreshing your memory when necessary.

37.2 Computer Graphics Overview

The major areas of computer graphics are **CAD** (computer-aided design), **CADD** (computer-aided design and drafting), **CIM** (computer-integrated manufacturing), and **CAD/CAM** (computer-aided manufacturing).

CAD (computer-aided design) is used to solve design problems, analyze design data, and store design information for easy retrieval. Many CAD systems perform these functions in an integrated manner, greatly increasing the designer's productivity.

CADD (computer-aided design drafting) is the computer process of making engineering drawings and technical documents more closely related to drafting than is CAD.

CAD/CAM (computer-aided design/computer-aided manufacturing) is a system that can be used to design a part or product, devise the production steps, and electronically communicate this data to control the operation of manufacturing equipment and robots.

CIM (computer-integrated manufacturing) is a more advanced system of the CAD/CAM that coordinates and operates all stages of manufacturing from design to finished product (Figure 37.1).

Advantages of CAD and CADD

Computer-graphics systems offer the designer and drafter some or all of the following advantages.

37.1 This automatic Chrysler assembly line uses computer-controlled robots for welding body parts together. (Courtesy of Chrysler Corporation.)

1. **Increased accuracy.** CAD systems are capable of producing drawings that are essentially 100% accurate in size, line quality, and uniformity.

2. **Increased drawing speed.** Engineering drawings and documents can be prepared more quickly, especially when standard details from existing libraries are incorporated in new drawings.

3. **Easy to revise.** Drawings can be more easily modified, changed, and revised than is possible by hand techniques.

4. **Better design analysis.** Alternative designs can be analyzed quickly and easily. Software is available to simulate a product's operation and test it under a variety of conditions, which lessens the need for models and prototypes.

5. **Better presentation.** Drawings can be presented in 2D or 3D and rendered as technical illustrations to better communicate designs.

6. **Libraries of drawing aids.** Databases of details, symbols, and figures that are used over-and-over can be archived for immediate used in making drawings.

7. **Improved filing.** Drawings can be conveniently filed, retrieved, and transmitted on disks and tapes.

37.3 Hardware

The hardware of a computer graphics system includes the **computer**, **monitor**, **input device** (keyboard, digitizer, mouse, or light pen), and **output device** (plotters and printers).

Computer

The computer, with an installed **program**, receives input from the user through the keyboard, executes the instructions, and produces output. The part of the computer that follows the program's instructions is the **CPU** (central processing unit). The computer graphics computer should have at least 32 MB or RAM, and its hard disk storage should be large, preferably 6 to 8 gigabytes and larger (**Figure 37.2**).

The monitor is a **CRT (cathode-ray tube)** and has an electron gun that emits a beam that sweeps rows of raster lines onto the screen. Each line consists of dots called **pixels**. Raster-scanned CRT's refresh the picture display many times per second. A measure of monitor quality is **resolution** which is the number of pixels per inch that can be produced on the screen. The greater the number of pixels, the greater will be the clarity of the image on the screen (**Figure 37.3**).

Input Devices

Besides the standard keyboard, the **digitizer** is used to enter graphic data to the computer. The **mouse**, a hand-held device that is moved about the table top to transmit information to the computer, is the most commonly used input device. Variations of the mouse are **thumbwheels** operated by fingertips, **joysticks** that let the user "steer" about the screen by tilting a lever, and **spherical balls** that can

37.3 Current technology enables computers to send 3D graphics over networks for instantaneous communication. (Courtesy of Hewlett Packard Company.)

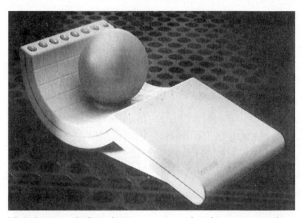

37.4 The Spaceball(R) digitizer is manipulated to move an object in 3D space on the screen by pushing, pulling, and twisting it. (Courtesy of CalCom Corporation.)

be rotated to input 3D data to the screen (**Figure 37.4**).

A **tablet** and digitizer used in combination are an alternative to the mouse. The digitizer (stylus) can be used to select commands from the menu attached to the tablet. Also, drawings can be attached to tablets and "traced" with the stylus to convert it to x- and y- coordinates. The **light pen** enables the user to "draw" on the screen with it to select points and lines.

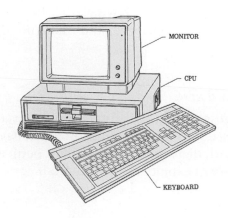

37.2 The basic components of a desktop computer system are the CPU (central processing unit), monitor, and keyboard.

MONITOR

CPU

KEYBOARD

Output Devices

Plotters make drawings on paper or film with a pen in the same manner a drawing is made by hand. Plotter types are **flatbed plotters**, **drum plotters**, and **sheet-fed plotters**. In flatbed plotters, the drawing paper is held stationary while pens are moved about its surface. Drum plotters roll the paper up-and-down over a cylinder while the pen moves left and right to make the drawing. Sheet-fed plotters hold the paper with grit wheels in a flat position as the sheet is moved forward and backward as the pen moves left and right to make the drawing.

Printers are of the impact type (much like typewriters) or nonimpact type where images are formed by sprays, laser beams, photography, or heat (**Figure 37.5**). The laser printer gives an excellent resolution of dense, accurately drawn lines in color, as well as in black and white. Inkjet technology has enabled images to be sprayed onto the drawing surface in color or in black that approaches the quality of the laser. Larger nonimpact printers (24″ × 36″ and larger) are most often inkjet printers since large lasers are much more expensive.

37.5 The DeskJet 990c series of printers produces photo-quality color images that are as close to traditional photographs as desktop printers have ever come. (Courtesy of Hewlett Packard Company.)

37.4 Your First Session

If this is your first session, you are anxious to turn the computer on, make a drawing on the screen, and plot it without reading the instructions. This section is what you're looking for.

Format of Presentation

In this chapter, the progression from one step of a command to the next level will be separated by an angle pointing to the right (>) in order to simplify presentation and reduce explanatory text. Since there are about three different ways of actuating most commands, these methods will be used alternatively in the following examples. The commands and prompts that appear on the screen will be given in italics to distinguish them from supplementary notes of explanation. (Enter) is the keyboard key with this name. Once a command is selected, additional prompts will be given at the *Command* line at the bottom of the screen or in dialogue boxes that must be followed.

Booting Turn on the computer and boot the system by typing ACAD2004 (or the command used by your system) to activate the program (**Figure 37.6**). Experiment by moving the cursor around the screen with your mouse, select items, and try the pull-down menu.

Mouse Most interactions with the computer will be accomplished with a mouse (**Figure 37.7**), but many commands can be entered at the keyboard (maybe more quickly after you learn them). Press the left mouse button to click on, select, or pick a command or object; a double click is needed in some cases. The right button has the same effect as pressing (Enter) on the keyboard.

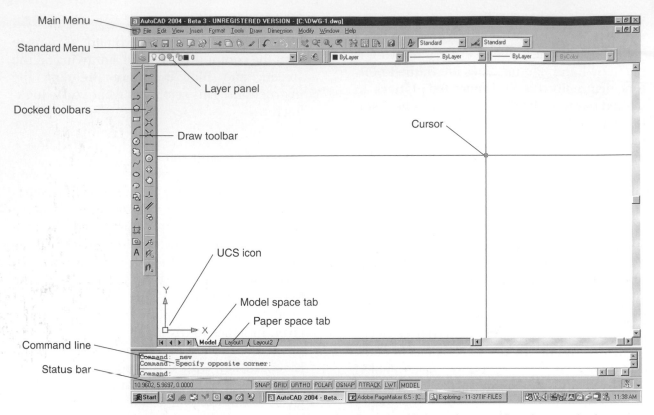

Main Menu

Standard Menu

Layer panel

Docked toolbars

Draw toolbar

Cursor

UCS icon

Model space tab

Paper space tab

Command line

Status bar

37.6 This is a typical view of the AutoCAD 2004 main screen. Additional toolbars can added to and removed from the screen by the user.

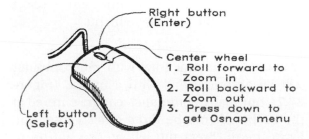

37.7 The left mouse button is used for picking points when drawing and selecting operational buttons. The right mouse button has the same effect as pressing the (Enter) button on the keyboard.

Creating a File To create a new file on a disk, place your formatted disk in its slot, from the *Main menu* bar pick *Files* and *New* from the dialogue box (**Figure 37.8**), and the *Create New Drawing* dialogue box will appear on the screen (**Figure 37.9**). Select the *Use a Wizard* icon button, *Quick setup,* and *OK,* and the *Quick Setup* box will appear where *Size Units* can be specified (**Figure 37.10**); select *Decimal* units and the *Next* button. When the *Area* box appears on the screen, insert the *Width* and *Length* (11″ × 8.5″) and select the *Finish* button (**Figure 37.11**). The program returns to the

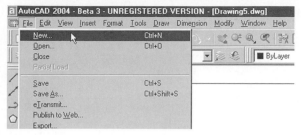

37.8 To begin a new drawing, select *Main Menu > File> New* to obtain the *Create New Drawing* box.

37.9 In the *Create New Drawing* box, select Use a *Wizard icon* button> *Quick Setup>* and *Ok*.

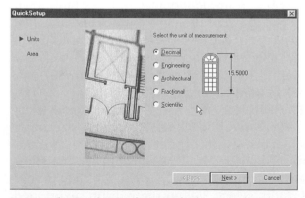

37.10 In the *Quick Setup* box specify the type of *Units* you wish to have (*Decimal* units in this example), and select *Next*.

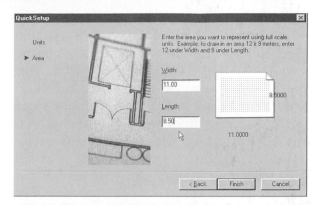

37.11 Type the dimensions for *Width* and *Length* (11 and 8.5 for an A-size sheet) and select *Finish*.

screen, and its command menus are ready for your drawing.

Making a Drawing Since you don't have the menus and toolbars figured out, type L (for *Line*), press (Enter), and draw some lines on the screen with the mouse for the fun of it by selecting endpoints with the left button as shown in **Figure 37.12**. A line on the screen "rubberbands" from point to point. To disengage the rubberband, press the right mouse button which is the same as pressing (Enter). Press the right button again to return to the *Line* command.

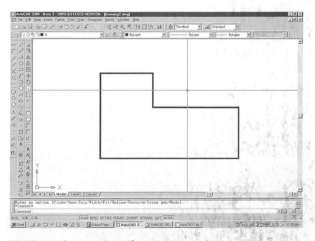

37.12 By *Command*: L (for *Line*), or selecting *Main Menu> Draw> Line*, endpoints of lines can be selected on the screen with the left mouse button to draw a figure.

Instead of typing L to enter the Line command, pick the *Main Menu* bar> *Draw* and select the *Line* option to get the prompt line *Specify first:* in the *Command* line at the bottom of the screen. Now, use your mouse to draw on the screen. The *Line* command can also be selected from the *Draw* toolbar that you will learn about soon. Try drawing circles and other objects on the screen by selecting icons from the *Draw* pull-down menu.

Repeat Commands By pressing (Enter) on the keyboard (or the right mouse button) twice after the previous command, the command can be repeated. For example, *Line* will appear in the Command line at the bottom of the screen after pressing (Enter) twice, if *Line* was the previous command.

Saving Your File. Click on *File* on the *Main Menu* bar and pick *Save As* from the pull-down menu to get the menu box shown in (**Figure 37.13**). Type A:DRW-5 (if your disk is in Drive A) and select the *Save* button; the light over drive A will blink briefly and the drawing named *DRW-5* is saved to the disk in the A drive.

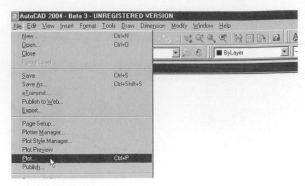

37.14 Select *Main Menu> File> Plot* to pick plotter settings.

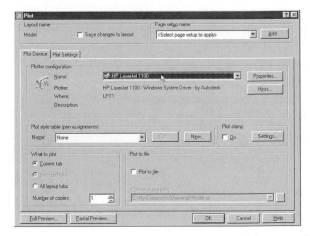

37.15 In the *Plot box*, select the *Plot Device* tab and the name of the plotter or printer that you will use, then select the *Plot Settings* tab.

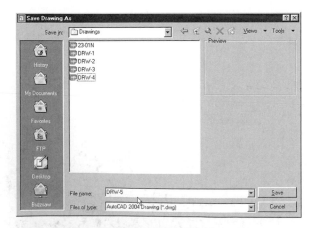

37.13 Select *Main Menu> File> Save As* to obtain the *Save Drawing As* box; type the drawing drive (A:) and the file name *DRW-5* in the File name box and select *Save*. File *A:DRW-5* is saved.

Plotting Your Drawing. Select *File* (**Figure 37.14**) and the *Plot* option from its pull-down to get the *Plot* dialogue box. Select the *Plot Device* tab, click on the pull-down window of the *Plotter configuration* area and pick the plotter that you intend to use (**Figure 37.15**). Select the *Plot Settings* tab of the *Plot* box (**Fig. 37.16**), select *Limits*, set *Scale* to 1:1, and select *Center* under *Plot Offset*. Load the A-size paper sheet in the printer or plotter; pick *Full Preview* to see how the drawing will appear when plotted. If it appears correct, press the right mouse button, select *Plot* on the pop-up menu, and the drawing is plotted.

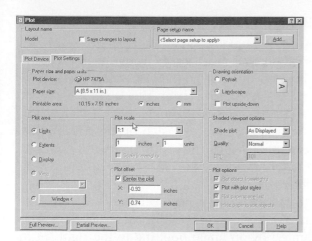

37.16 Specify *Paper size* (8.5 X 11), *Drawing orientation* (Landscape), *Plot area* (*Limits*), *Plot scale* (1:1), *Plot offset* (*Center the plot*), and click on *Full Preview* to obtain a view of what will be plotted. If it looks correct, press the right mouse button, select *Plot*, and the drawing is plotted.

Quitting AutoCAD To be sure that your latest changes have been saved to disk, select *File* from the *Main Menu* and pick *Save* from the pull-down menu to update *A:DRW-5*. To quit AutoCAD, close the active file (*Main Menu> File> Close*) and the file will close and leave the screen if it has been *Saved* in its current form (**Figure 37.17**). If changes have been made to it since its last *Save*, the pop-up box in **Figure 37.18** let's you decide to *Save* it or not (*Yes* or *No*) or *Cancel*. To quit your session *Main Menu> File> Exit*, and AutoCAD closes (**Figure 37.19**). If you intend to continue your drawing session, do not *Exit* now but continue with other drawings. When working from a floppy disk, do not

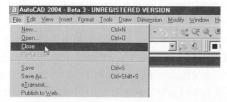

37.17 Select *Main Menu> File> Close* to close the current file; you will be prompted to save it if it has not been saved.

37.18 If the file has not been *Saved*, this dialogue prompts you to decide whether or not you want to save (*Yes*, *No*, or *Cancel*).

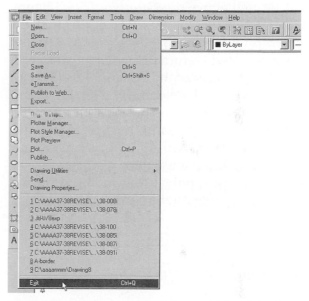

37.19 If you wish to quit AutoCAD completely: *Main Menu> File> Exit*. It is always good practice to save your drawing again before exiting from the program, just to be safe.

remove it from its drive until it has been saved with either the *Save* or *Save As* options.

That's how it works. Now, let's get into the details and learn more.

37.5 Introduction to Windows

The recommended operating system for AutoCAD 2004 for a single user is Windows 98 or higher, which allows several programs to be open and running at the same time. For example, a word-processing program can be running in addition to AutoCAD.

A drawing file can be manipulated with the three buttons in the upper-right corner of the window (**Figure 37.20**). The "overlapping-boxes" button is selected to display the file covering only a portion of the screen and to allow other files or programs to share the screen. The "X" button closes the current file. The dash button minimizes a file to an icon box (that retains the three buttons of the original) located above the *Command* line (**Figure 37.21**). Maximize the minimized file to fill the screen by selecting the "box" button (**Figure 37.22**) and it will return to the screen and cover the other displays on the screen.

Windows can be resized by selecting a border, or the corner of the border, while holding down the left mouse button, and "dragging" the window to size (**Figure 37.23**). When several overlapping program windows appear on the screen, select any point on a window to move it to the front to make it the current program.

37.20 The icons in the upper right of the screen can be used minimize a drawing to an icon (dash), reduce it to partial size (two boxes), or exit from the file (X).

37.21 Minimized files are displayed as an icon box above the *Command* line.

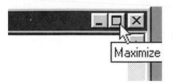

37.22 A partial screen can be enlarged to full-screen size by selecting the box button.

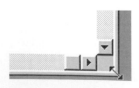

37.23 Select the lower right corner, hold down the left mouse button, and "drag" the border to size the screen as you desire.

37.6 Format of Presentation

AutoCAD has several ways of using a command in almost all cases. For example, a circle can be drawn by typing C at the *Command* line; selecting the *circle icon* from the *Draw* toolbar; or using the *Main Menu* bar> *Draw*> *Circle* on the drop-down menu. In each of these examples, you must select from options— *Center, radius; Center, diameter;* and others—before drawing the circle.

In this chapter, the progression from one step of a command to the next level will be separated by an angle pointing to the right (>) in order to simplify presentation and reduce explanatory text. The commands and prompts that appear on the screen will be given in *italic* type to distinguish them from supplementary notes of explanation. (Enter) is the keyboard key with this name. Once a command is selected, additional prompts given at the *Command* line at the bottom of the screen and/or in dialogue boxes must be responded to.

Command Line When a circle is drawn by typing at the *Command* line it will be presented as follows: *Command:* Circle (or C)> *Center point*> P1> *Radius*> 4 (Enter). Entries that are typed in response to prompts are underlined. The underlined P1 is a point selected on the screen and the underlined 4 is the radius which is typed at the command line in response to the prompt, *Radius.*

Main Menu Bar When the *Main Menu* bar is used to draw a circle, the sequence of steps is presented as: *Main Menu*> *Draw*> *Circle*> *Center, radius*> *Center point*> P1> *Radius*> Drag to a radius of 4.

Draw Toolbar When the *Draw* toolbar is used to draw a circle, the steps are presented as: *Main Menu*> *View*> *Draw* toolbar> *Circle* icon> *Center, radius*> *Center point*> P1> *Radius*> 4 (Enter).

37.7 Using Dialogue Boxes

AutoCAD 2004 has many dialogue boxes with names beginning with *DD* (*DDlmodes*, for example) to interact with the user. The command *Filedia* can be used to turn off (0 = off and 1 = on) the dialogue boxes if you prefer to type the commands without dialogue boxes. When a command on a menu followed by three dots (...) or an arrow (>) is selected, supplemental dialogue boxes will be displayed (**Figure 37.24**) and some of these boxes have subdialogue boxes.

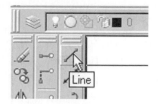

37.25 *Definition* boxes are flyout boxes that explain the functions of the various icons when the cursor is rested on them briefly.

Definition boxes are provided to identify the functions of each box on the screen. By resting the cursor on a box, a *flyout* will appear to define its function as shown in **Figure 37.25**.

Double clicking the mouse (quickly pressing the left button twice) selects a file from a list and displays it. Single clicking followed by *OK* is an alternative for activating a selection. With experimentation you will soon learn where double clicking can be best applied.

Right clicking when the cursor (right mouse key) is placed over an icon will display a set of commands that are related to that particular command.

The *Select File* box in **Figure 37.26** has dialogue boxes, lists, blanks, and buttons that can be selected by the cursor. When a file is selected, it is darkened by a gray bar and a thumbnail illustration of it is shown in the window. To find a file for which you have a name,

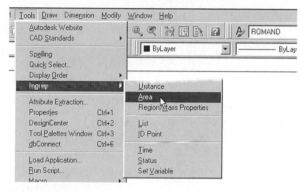

37.24 The *Main Menu* has many pull-down menus. Commands in the pull-down menu followed by black arrows have sub-dialogue boxes. The sequence in this example is *Main Menu> Tools> Inquiry> Area*.

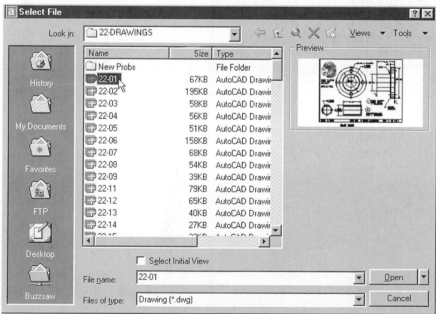

37.26 *Main Menu> File> Open* gives the Select File box that helps you find the file that you wish to open. A thumbnail view of the selected file is previewed in the window before it is opened.

from the *Select File* box select Find to display the *Find* box (**Figure 37.27**). Pick *Name & Location* tab> type file name in *Named* window> select file type from the *Type* window> *Browse* to select drive or folder to search> *Find Now*. The matching files will appear in the window at the bottom of the *Find* box.

In many cases, speed is increased if you type commands at the *Command* line instead of using dialogue boxes. What could be easier than typing L and pressing (Enter) for drawing a *Line*?

37.8 Drawing Aids

Function Keys

Convenient drawing aids are available from the function keys on the keyboard which can be used to turn settings on and off (**Figure 37.28**). *F1* (*Help*) can be pressed to open the help screen for instant troubleshooting. *F2* (*Flip screen*) alternates between the graphics on the screen to its corresponding text mode. *F3* controls *Snap* and *F5* controls *Isoplane*. *F4* (*Tablet*) activates a digitizing tablet if one is attached to your computer. *F6* (*Coordinates*)

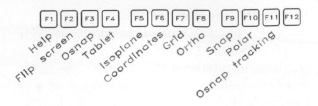

37.28 These function keys on the keyboard control the options as indicated by the notes.

shows numerical coordinates of the cursor in the status bar at the bottom of the screen as the cursor is moved. *F7* (*Grid*) turns the grid on or off and refreshes the screen in the process, removing any blips or erasures. *F8* (*Ortho*) forces all lines to be drawn either in horizontal or vertical directions. *F9* (*Snap*), when on, makes all object points lie on points defined by an invisible grid.

One of the first settings to make when beginning a drawing is that of the area size, called *Limits*, in which the drawing will be made. To set, type <u>Limits</u> and respond to the *Command* line prompts by typing the coordinates of the diagonal across the area. *Limits*

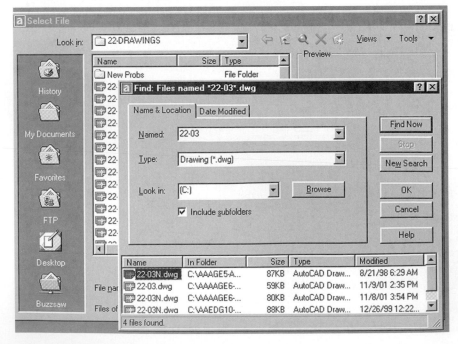

37.27 To locate a file for which you have a name, select *Tools> Find>* get *Find* box> *Name & Location* tab> type file name in *Named* window> select file type in *Type* window> *Browse* to select drive or folder to search> *Find Now* and the matching files appear in window at the bottom of the *Find* box.

command can also be accessed by: *Main Menu> Format> Drawing Limits*. A drawing that fills an A-size sheet (11 × 8.5 inches) has a plotting area of about 10.4 × 7.8 inches or 254 × 198 mm. The drawing area is specified with the *Limits* command as follows:

Command: Limits (Enter)

Specify lower left corner [ON/OFF] <0.00,0.00>:

(Enter) to accept default value of 0,0.

Specify upper right corner <12.00,9.00>: 11, 8.5

(Enter)

Press F7 (*Grid*) to see the dot pattern of the grid fill the *Limits*.

Command: Zoom (Enter)> type All (Enter) and the drawing *Limits* and *Grid* will fill the screen.

Limits can be reset at any time during the drawing session by repeating these steps.

Drafting Settings

The *Drafting Settings* box is found by *Main Menu> Tools> Drafting Settings*, right clicking

on the *Osnap* icon above the *Command* line, and selecting *Settings*, or by typing DDrmodes (**Figure 37.29**). The *Snap* and *Grid* tab gives buttons for selection and blanks for filling in to activate these settings. *Snap* forces the cursor to stop only at points on an imaginary grid of a specified spacing. The *Snap X* and *Snap Y* values can be set by typing values in the blanks. *Snap* can also be set by typing *Snap* at the *Command* line and specifying the interval desired. When *On*, the *Snap* icon in the *Status* line at the bottom of the screen is highlighted. *Snap* can be toggled on and off by clicking on this icon or by pressing *F9*.

Grid of the *Drafting Settings* box fills the *Limits* area with dots spaced apart by typing values in the *Grid X spacing* and *Grid Y spacing* boxes. When *Snap type & style> Grid snap> Rectangular snap* is selected the cursor snaps to the grid if the *Snap* and *Grid* spacings are equal.

The *Object Snap* tab (**Figure 37.30**) gives options for making lines and other geometry of a drawing snap to previously drawn objects at specified points. *Osnap* is covered in Section 37.42.

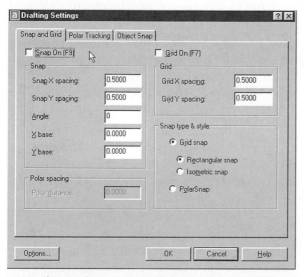

37.29 The *Drafting Settings* dialogue box (or, *Comand:* DDrmodes (Enter)) has three tabs from which to make settings: *Snap* and *Grid*, *Polar Tracking*, and *Object Snap*.

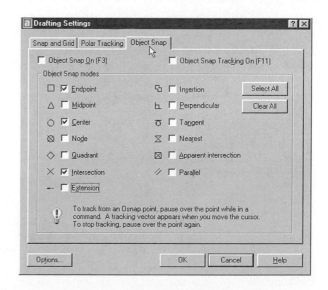

37.30 *Osnap* settings can be made from the *Object Snap* tab of the *Drafting Settings* dialogue box (*DDrmodes*).

37.31 *Drawing Aides* boxes in the *Status* bar beneath the *Command* line at the bottom of the screen display the current settings. Click on these buttons to turn them on or off.

The *Status* window at the bottom of the screen displays several of the settings discussed above when they are turned *On* (**Figure 37.31**). Single clicking on these buttons toggles them off or on.

Blips (*Command:* Blips) are temporary markers made on the screen when selections are made with the mouse. They are removed by refreshing the screen by pressing *F7* (*Grid*).

37.9 Help Commands

Several examples of helpful commands that can be typed at the Command line are shown here.

Help (*Main Menu> Help> AutoCAD 2004 Help*) gives menus *Contents, Index, Search, Favorites,* and *Ask me* to help you with all aspects of AutoCAD. Under *Find,* you will be able to insert key words for commands and steps that you need assistance with as shown

37.32 By selecting *Help* from the *Main Menu* (or pressing F1), the *AutoCAD 2004 Help* box with five tabs appears on the screen. Select *Index* tab, type the keyword in the blank, and you will obtain instructions about your topic.

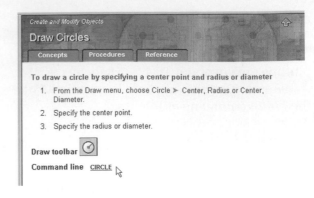

37.33 The help box can be obtained by *Main Menu> Help> AutoCAD 2004 User Help>* type subject (CIRCLE)> Select *CIR-CLE* from list (all caps for a command)> *Display* button> select *CIRCLE> Display* button> Select *To draw a circle by specifying a center point and a radius or diameter* to get the help screen shown here.

in **Figure 37.32**. When a topic, *Circle* for example, is selected and *Display* is picked, a screen of instructions will appear to help you with using the *Circle* command (**Figure 37.33**). Additional options are provided under *Help* that are self-explanatory when you experiment with them.

The Purge command (*Main Menu> File> Drawing Utilities> Purge> All*) can be used at any time to remove unused layers, blocks, and other attributes from files. *Purge* can be activated by using *Command:* Purge (Enter) to obtain the *Purge* box where you can select *All items* or *Blocks/Dimension styles/Layers/ Linetypes/Mline styles/Plot styles/Shapes/Text styles*. By purging unused attributes of a file, clutter is eliminated and disk space is saved.

The *All* option is used to eliminate all unused references one at a time as prompted. The other purge options purge selected features of a drawing.

List (*Main Menu> Tools> Inquiry> List*) asks you to select any object drawn on the screen and gives information about it. For example, when a circle is selected it gives its radius, circumference and area plus the coordinates of its center point.

Copy (*Modify toolbar> Copy* icon) is used to select objects on the screen (single objects or groups of objects) with the cursor, pick a new position, and make a duplicate of the selection. The *Multiple* (*M*) option can be selected for making more than a single copy. *Unlock* is used to select *Files* to *Unlock* and then, pick *OK*.

37.10 Drawing Layers

An almost infinite number of layers can be created, each assigned a *Name, Color, Linetype, Lineweight,* and *Plot Style* on which to draw. For example, a yellow layer named *Hidden* for drawing dashed lines may be created.

Architects use separate copies of the same floor plan for different applications: dimensions, floor finishes, electrical details, and so forth. The same basic plan is used for all of these applications by turning on the needed layers and turning off others.

Working with Layers Layers and their settings are created and manipulated in the *Layer Properties Manager* box which is displayed by selecting the paper-stack icon next to the *Layer Control panel* (**Figure 37.34**). For most working drawings, the layers shown in the *Layer Properties Manager* box in **Figure 37.35** are sufficient. Layers are assigned linetypes, lineweights, and different colors so they can be easily distinguished from each other. The 0 (zero) layer is the default layer which can be turned off or frozen but not deleted.

Layers can be created by selecting the *New* box to obtain *Layer1*—the default name which

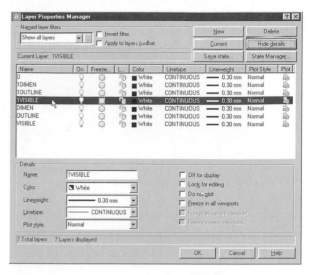

37.35 *Main Menu> Format> Layer> Layer Properties Manager* box (or, *Command:> DDlmodes*) lists the layers and their properties. From this box, layers can be created and deleted, linetypes and colors assigned, and other settings made.

can be replaced by a new name (**Figure 37.35**). The new layer will appear in the listing of layers with a default color of white and a continuous linetype, all of which can be changed to your specifications.

Color for a new layer can be changed from its default color (white) by selecting its current color in color column to obtain the *Select Color* menu. From this menu, pick a color for that selected layer and pick the <u>Ok</u> button to finalize the color change (**Figure 37.36**).

Linetypes are found by clicking on the default linetype of the new layer in the *Layer Properties Manager* box and the *Select Linetype* box appears; select Load and the *Load* or *Reload Linetypes* box appears with a list of linetypes from which to select (**Figure 37.37**). To assign a hidden *Linetype* to a layer named *Hidden*, select the linetype of that layer, select *Hidden* from the *Select Linetype* box, pick *Ok*, and the line is assigned to the layer. All lines drawn on the *Hidden* layer will be dashed lines.

37.34 Select the paper stack next to *Layer Control* panel to obtain the *Layer Properties Manager* box for setting all layer properties.

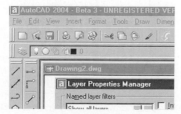

37.36 *Main Menu> Format> Color> Select Color box* (or type DDcolor) to get this box from which colors can be selected and assigned to layers.

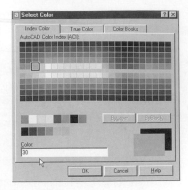

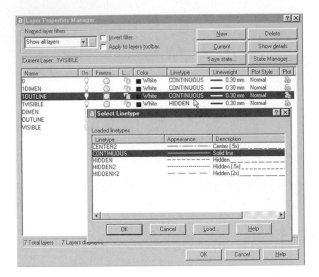

37.37 In *Main Menu> Format> Layers> Layer Properties Manager* box> Select *Linetype* (or, *Command*: DDltype) (Enter) and Load to get this box from which linetypes can be loaded, selected, and assigned to layers.

Lineweights are found by clicking on the default lineweight of a layer in the *Layer Properties Manager* box and the *Lineweight* box appears with a list of linetypes from which to select. Select the desired lineweight, pick *Ok*, and the line is assigned to that layer. Lines drawn on this layer will have this line thickness.

Ltscale, when typed at the *Command* line, modifies the lengths of line segments of *all* hidden lines and other noncontinuous linetypes at one time.

Layer Control Panel

A layer must be selected as the current layer in order to draw on it using the assigned color and linetypes. The *Layer Control Panel* offers the quickest method of setting a layer (**Figure 37.38**). By picking a point anywhere within the *Layer Control Panel*, a listing of the named layers and their properties will be displayed from which a layer can be selected to make it the current layer (**Figure 37.39**). Now you can draw on this current layer.

This same portion of the *Layer Control Panel* can also be used to make other layer assignments: *On* and *Off*, *Freeze* and *Thaw*, *Lock* and *Unlock*, and others. By selecting the *Layers* icon (**Figure 37.38**), the *Layers Properties Manager* box will be displayed from which the previously covered setting can be made. This box can also be obtained by typing LA at the *Command* line and (Enter).

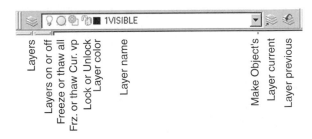

37.38 The *Layer Control* bar offers easy access to layers and their properties by selecting one of the icons shown here.

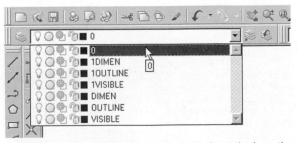

37.39 Select any point on the *Layer Control* panel where the name of the active layer's name appears, or select the down button, and this drop-down list of the layers appears.

Renaming a layer can be done from the *Layers Properties Manager* box by dragging the cursor across the existing name and typing a new one in its place.

On/Off is applied to a layer in the *Layer Control* panel by selecting the lightbulb icon (**Figure 37.39**). Layers can be turned on or off with buttons from the *Layers Properties Manager* box (**Figure 37.37**) (or by typing LA at the *Command* line) and picking *Off.* An *Off* layer that is selected as the current layer can be drawn on, but this is seldom done.

Freeze and *Thaw* options (sun icon) under the *Layer Control Panel* are used like the *On* and *Off* options (**Fig. 37.39**). *Freeze* a layer and it will (unlike an *Off* layer) be ignored by the computer until it has been *Thawed* which makes regeneration faster than when *Off* is used.

37.11 Toolbars

Main Menu> View> Toolbars gives a menu of toolbars that can be selected to suit the current application (**Figure 37.40**). A portion of the *Standard* toolbar (**Figure 37.41**) gives a

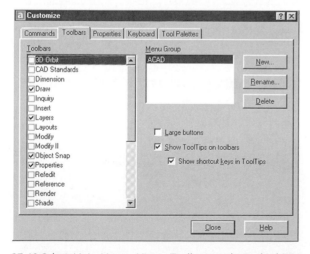

37.40 Select *Main Menu> View> Toolbars* to obtain this listing of toolbars that can be placed on or removed from the screen by checking a toolbar box.

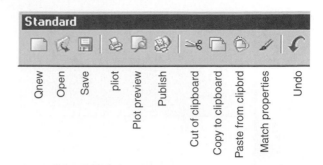

37.41 This left half of the *Standard* toolbar provides many routine operations that can be accessed with the cursor.

sequence of icons for commands from *New* to *Tracking Point.* The remainder of the *Standard* toolbar commands from *UCS* to *Help* are given in **Figure 37.42**.

Toolbars can be moved about the screen by selecting a point on one of their edges, holding down the select button of the mouse, and moving the cursor to a new position. *Toolbars* can be *docked* by moving them into contact with a border on the screen. They can be changed from single strips, to double and triple blocks, by moving the corners of the toolbars, or into vertical or a horizontal strips depending on which border of the screen they are moved to (**Figure 37.43**). When located in the open area of the screen, toolbars will appear as *floating menus*. When a small icon needs explanation, place the pointer on an icon and a flyout box will appear with its definition.

37.42 The right half of the *Standard* toolbar has more helpful commands that you will need for most drawings.

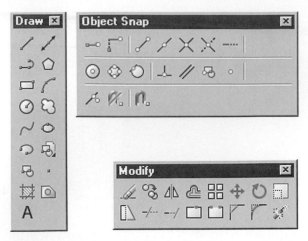

37.43 *Toolbars* can be arranged by dragging their corners to make single, double, or triple strips that are horizontal or vertical. They can be docked at the borders or left "floating" on the screen.

37.12 A New Drawing

Set Up a Title Block

To create a new drawing, select *Main Menu> File> New>* (**Figure 37.44**) and the *Create New Drawing* box appears giving two options: *Advanced Setup* or *Quick Setup* (**Figure 37.45**). Pick *Advanced Setup*, set the *Units* to *Decimal* with a *Precision* (decimal places) of 0.00, and select *Next* (**Figure 37.46**). In the following screen (**Figure 37.47**), set the *Angle* to *Decimal Degrees* with a *Precision* of 0 and pick *Next* (**Figure 37.47**).

In the *Angle Measure* box (**Figure 37.48**), set the angle measurement to *East* and pick *Next*. In the *Angle Direction* box (**Figure 37.49**), set the angle direction to *Counter-Clockwise* and pick *Next*. In the Area box (**Fig. 37.50**), set the

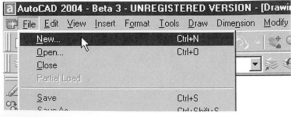

37.44 Begin the creation of a new drawing using these steps: *Main Menu> File> New*.

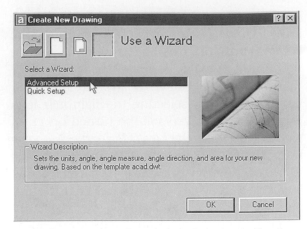

37.45 The *Create New Drawing* box appears and offers three options for a new drawing. Select *Use a Wizard> Advanced Setup>* and *OK*.

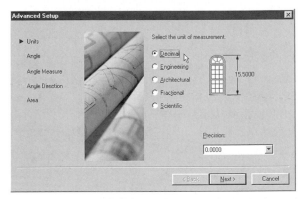

37.46 Set *Units* to *Decimal* and set *Precision* to 0.00 (2 decimal places: X.00 for example).

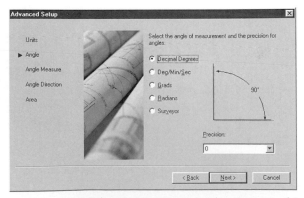

37.47 For *Angle*, select *Decimal Degrees* with a *Precision* of 0 and pick *Next*.

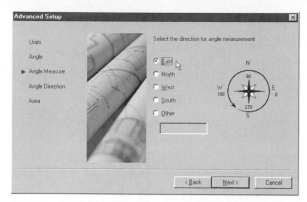

37.48 For *Angle Measure*, select the *East* and *Next* button.

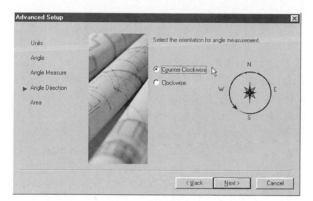

37.49 For *Angle Direction* select *Counter-Clockwise* and *Next*.

values for *Width* (11) and *Length* (8.5) and pick *Finish*.

Prototype Drawing

The drawing screen appears ready for drawing. If the grid dots do not show, press *F7* to show them; the upper right dot having coordinates of about 11 and 8.5, the sheet size that was assigned. Draw a border using *Line*> 0,0> 10.4, 0> 10.4, 7.8> 0, 7.8> 0,0 to get a 10.4 × 7.8 border (**Figure 37.51**). Save this A-size border for future use as a prototype file for making drawings with these same settings (*Main Menu> File> Save As> ABORD-HORIZ* (Enter)). Close this file (*Main Menu> File> Close* and the drawing leaves the screen.

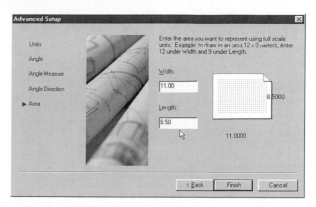

37.50 For the drawing *Area*, type values of 11 and 8.5 for *Width* and *Length*, respecitvely, for an 8.5 X 11 sheet size and pick *Finish*.

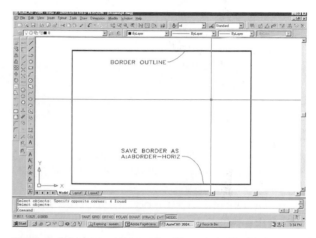

37.51 In the new screen, press *F7* to turn the grid on and *Command*: Zoom> (Enter)> All (Enter) to make the grid fill the drawing area. Draw a border (10.4 X 7.8) and *Save As* A:ABORD-HORIZ and close the file.

Using the Prototype Drawing

So far, you have been working entirely in *Model Space*, which is adequate for the two-dimensional drawing covered in this chapter. The application of *Model Space* and *Paper Space* in combination will be covered in Chapter 38.

To make a drawing using the settings and border made in the previous sequence, open the prototype file (*Main Menu> File> Open> A:ABORD-HORIZ*), and the border and grid appear on the screen as it did at the end of the

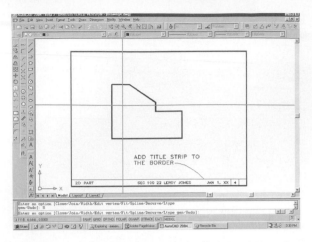

37.52 Open A:ABORD-HORIZ, save the file (*Main Menu> Save As> A:DWG5* (Enter)) and DWG5 become the current drawing. Make your drawing, add a title strip to the border, and save the file (*Main Menu> File> Save*).

last drawing (**Figure 37.52**). Save the file with a new name (*Main Menu> File> Save As> A:DWG5>* (Enter)) and DWG5 becomes the current file and replaces the prototype file on the screen. (This same procedure can be used to create a new file by saving the prototype file to a new name, while preserving the prototype file in its original form.)

Add a title strip to the border for your name, date, and other information required by your instructor. You may consider making this information part of your prototype file so it will be displayed every time it is used: *File> Save As> A:ABORD-HORIZ.* Since you are saving to an existing file, the *Save Drawing As* box in **Figure 37.53** will ask you if you want to replace it; select *Yes.* Prototype file ABORD-HORIZ is

updated so the title strip will be included when it is used next.

Make your drawing within the border, save it (*Main Menu> File> Save As> A:DWG5*), and your drawing is ready to be plotted.

Section 37.66 covers the techniques of customizing your title block for classroom drawings, but first, you must learn more about the operation of AutoCAD. If this section seems a little advanced for you, leave it, and come back when you have learned a few more drawing principles.

37.13 Drawing Scale

It is best and easiest to work with a drawing at a full-size scale where 1 inch is equal to 1 inch. The previous examples of files and title blocks were developed as full-size layouts which permits text size and measurements to be easily handled.

Half-size drawings can be made by creating a new drawing, DWG6, by using the *Advanced Setup* steps covered in the last section. Now, insert (**Figure 37.54**) the prototype drawing, A:ABORD-HORIZ, with the following steps: (*Main Menu> Insert> Block> File> A:ABORD-HORIZ>* Set parameters: check *Uniform Scale*, type 2 in *X-Scale window> Ok>* type 0,0 at the *Command* line as the insertion point (**Figure 37.55**). These commands insert the border and title strip at double size (scale = 2). For a double-size metric drawing, the scale would be $25.4 \times 2 = 50.8$ for millimeters. A full-

37.53 When saving a drawing file with *Save As*, you will be asked by this box if you want to update an existing file. By selecting *Yes* the named file is replaced by the current file.

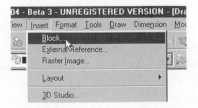

37.54 To insert a file into another file, select *Insert* and *Block* from the *Main Menu* bar.

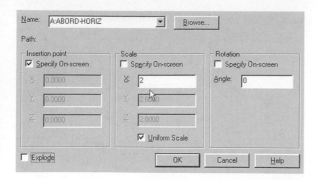

37.55 Type the name of the file to insert, <u>A:ABORD-HORIZ,</u> select *Uniform Scale*, and give a *Scale* factor of <u>2</u>, Select *Ok*, and insert the file at <u>0,0</u> at the *Command* line.

size drawing is made within the double-size border. At plot time the double-size border must be reduced to half size (0.5) so it will fit on its sheet, and the full-size drawing within it is reduced to half size.

Double-size drawings can be made with the same steps as half-size drawings except the *Scale* is <u>0.5</u> (half size) when the prototype file, A:ABORD-HORIZ, is inserted. (The factor would be 25.4 × 0.5 = 12.7 for millimeters.) At plot time, the half-size border must be enlarged by a factor of two so it will fill the sheet, and the full-size drawing within it will be doubled in size.

Using this logic, other combinations of scale factors can be determined for drawings of any scale. The most important point to remember is that you are better off working with full-size drawings and scaling at plot time.

37.14 Saving and Exiting

The pull-down file (*Main Menu> File*) gives options for saving a drawing—*Save*, *Save As*, and *Exit*—which can be selected from the pull-down menu, the *Standard* toolbar, or by typing one of these commands at the *Command* line.

Saving

Use the *Command*: <u>Save</u> (Enter) to "quick save" to the current file's name if it has been previously named and saved. If the file is unnamed, the Save command prompts for a file name by displaying the *Save Drawing As* dialogue box in **Figure 37.56**. Select the directory and name the file (<u>A:NEW7</u>) to save it on the disk in drive A. The new drawing, *A:NEW7*, becomes the current file on the screen.

Ending the Session

To exit the drawing session and close AutoCAD, select *Main Menu> File> Exit*. If you have not saved immediately before selecting *Exit*, the dialogue box shown in **Figure 37.53** asks if you want to save the updated drawing. Select *Yes* and the *Save Drawing As* menu box appears for assigning the drive, directory, file name, and file type. To exit without saving, respond to *Save Changes?* with *No* and the latest changes made since the last *Save* will be discarded.

Use *Command*: <u>Close</u> (Enter) and you will be prompted to *Save* the drawing if has been changed since the last Save. If the drawing has not been previously named, the *Save Drawing As* dialogue box will prompt you for a file

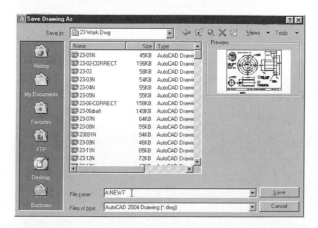

37.56 Name the file, select the drive to save it to, and save it by using this dialogue box. (*Main Menu> File> Save As*)

name. The previous version of the drawing is automatically saved as a backup file with a *.bak* extension and the current drawing is saved with a *.dwg* extension. Several drawing files can be open at the same time during a session.

37.15 Plotting Parameters

To plot a drawing before ending a drawing session select *File> Plot* to obtain the *Plot* box (**Figure 37.57**). Select the *Plot Device* tab to obtain the subdialogue box for selecting *HP 7475 plotter* to make a pen drawing. (A *plotter* has pens that plot a drawing, whereas a *printer* has no pins but sprays the ink onto the drawing.)

Plotter Settings

Main Menu> File> Plot> Plot Device tab> *HP 7475A> Properties> Device and Document Settings* tab> *Custom Properties> Device Options* allow the physical pens to be set for a pen plotter (**Figure 37.58**). Since plotters use physical pens, the pens to be used by each

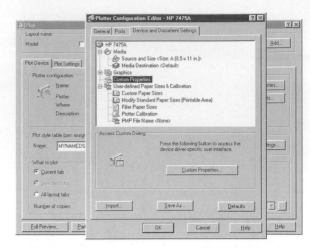

37.58 Pen assignments for a plotter with physical pens are set in successive windows of this figure (not shown), which are found by *Main Menu> File> Plot> Device* and *Document Settings* tab> *Properties> Custom Properties> Physical Pen Configuration> Device Options*.

color on the screen must be specified. The colors of the lines on the paper plot depend upon the colors of the physical pens, which may be all one color, and lineweights will be determined by the lineweights of the physical pens.

A *Color* of red, a pen *Speed* of 9 in. per sec (228 mm per sec), and a *Width* of .01 (.25 mm) is assigned to pen 1 (the slot occupied by the pen in the pen holder). All lines drawn in red on the screen will plot with pen 1 at this speed. Select *Save As> Pen Plotter HP7475* and the settings are saved with a *.pc3* extension for this setup.

Printer Settings

Use *Main Menu> Plot> Plot Device* tab to select the configured printer that you intend to use for plotting your drawing. Select the *Plot Settings* tab, select *Drawing orientation* (*Portrait* or *Landscape*), specify *Plot Area* (*Limits, Extents,* or *Display*), assign a *Scale,* set *Plot offset* to *Center the plot.* Save this *Page setup* by selecting the *Add* button (**Figure 37.59**) and naming it in the setup in the *User*

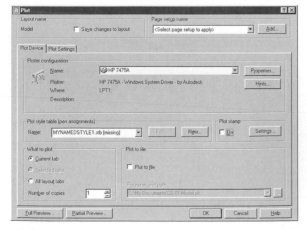

37.57 The printing options for a laser printer (without physical pens) are specified in this box (*Main Menu> Plot> Plot Device* tab). Select the printer to be used from the pull-down window across from *Name:*.

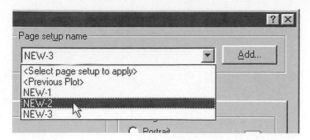

37.59 Settings made in the *Plot* menu can be saved with the *Add...* button for use in future drawings. The *Previous Plot* option enables you to specify settings identical to the last plot.

37.60 From the *Add...* button, the *User Defined Page Setups* box appears where you can name a new page setup and select *Ok* to save it for future drawings.

Defined Page Setups box that appears on the screen (**Figure 37.60**) and pick <u>Ok</u>. These settings will be saved with the drawing when the file is saved.

Plot Settings Definitions

Plot settings (**Figure 37.61**) applicable to both pen plotters and printers are covered below.

Limits is used to plot the portion of the drawing bounded by the grid pattern defined by its *Limits*.

Extents plots a drawing to its extents if the scale selected permits. It is good practice to apply

37.61 The *Plot area* section of *Plot* box is used to specify the portion of the drawing that will be plotted: *Limits*, *Extents*, or *Display*.

Zoom> Extents to ready a drawing for plotting.

Display plots the portion of the drawing shown on the screen.

View lists saved *Views* that can be selected and plotted. This box is gray if no views are saved. *Window* specifies the portion of a drawing to be plotted when the window is sized with the cursor or coordinates typed from the keyboard.

Paper Size lists standard and user-specified plot sizes (**Figure 37.62**).

Drawing Orientation offers *Portrait* and *Landscape* options (**Figure 37.63**).

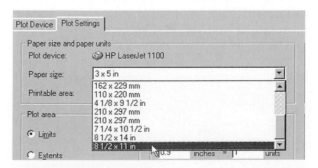

37.62 The *Paper Size* window gives a listing of the standard plot sizes that are available.

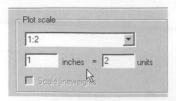

37.63 Select the desired scale (1 = 2, or half size, in this case). If *Scaled to Fit* is selected, the drawing fills the sheet if the placement of the orgin permits.

Plot Scale can be typed in the edit boxes or selected from the pull-down window where 1:2 appears (**Figure 37.63**). The scale of 1 = 1 is full size, 1 = 2 is half size, and 2 = 1 is double size.

Scaled to Fit calculates the scale that makes the drawing's extent fill the plotting area and be as large as possible.

Center the plot is picked to specify that the drawing will be centered on the sheet (**Figure 37.64**).

Plot Offset can be set with x and y coordinates to locate a plot on a sheet (**Figure 37.64**).

Hide paperspace objects removes hidden lines from 3D drawings that are plotted.

Partial Preview shows rectangles representing the paper size and the plotting area (**Figure 37.65**). The part of the drawing area that exceeds the paper size cannot be plotted unless the scale, origin, or both are adjusted. The triangular icon signifying the lower left corner of a drawing is shown in the upper left corner of this partial preview of the drawing.

Full Preview shows the entire drawing on the screen and its relationship to the paper limits when plotted (**Figure 37.66**). Right click

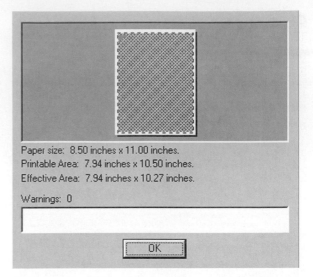

37.65 Select *Partial Preview* to get two rectangles on the screen that represent the paper size and plotting size.

to *Pan* and *Zoom* about the preview drawing. Press (Esc) to return to the *Plot* dialogue box.

To update the *Page setup* settings in *Plot box> add>* Select the named file> *Ok>* and type <u>Yes</u> when asked if you want to replace the existing file.

37.64 Select *Center the plot* box to position the plot at the center of the sheet; x- and y-coordinates of the drawing's origin are given.

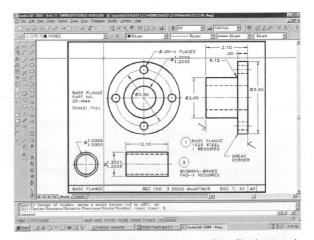

37.66 Select *Full Preview* to get this view of the file that is to be printed when *Plot* is selected.

General

These approaches to making settings for plotters and printers while working from *Model Space* are the simplest for plotting drawings, yet they are sufficient for essentially all two-dimensional drawings required in a beginning graphics course. The application of *Plot Styles* is not covered in this chapter for this reason. Plot styles will be touched on in Chapter 38 where three-dimensional principles are covered.

37.16 Readying the Plotter or Printer

Plotter

Load paper in the plotter, as shown in **Figure 37.67** with the thick pen (P.7) in slot 1 and the thin pen (P.3) in slot 2 as specified by *Pen Assignments* in **Figure 37.58**. Press (Enter), and the plot will begin. Plotting can be cancelled by pressing (Esc), but it may take almost a minute for it to take effect. When completed, select *File> Exit* to close AutoCAD and end the session.

Printer

Load the printer with the necessary sheets of paper, be sure the printer is on, select *Main*

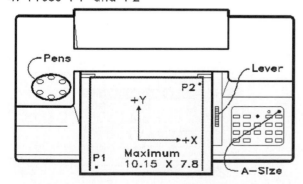

37.67 The Hewlett-Packard 7475A plotter is often the plotter of choice for size A and size B plots with pens.

37.68 The Hewlett-Packard LaserJet 1200 printer has no physical pens; the ink is sprayed onto the paper. (Courtesy Hewlett Packard Company.)

Menu> Plot> OK, and the drawing file is sent to the printer where it is plotted. A laser printer has no physical pens; the ink is sprayed onto the paper (**Figure 37.68**).

Now that we know how to set a few drawing aids, save files, and plot, it is time to learn how to make drawings.

37.17 Lines

Open your prototype drawing, *Main Menu> File> Open>* A:ABORD-HORIZ and use *Save As* to name the drawing as A:NO1, which becomes the current drawing with the same settings of A:ABORD-HORIZ. Load the *Draw* toolbar (*Main Menu> View> Toolbars> Draw*) to obtain the command options shown in (**Figure 37.69**).

A *Line* (called an object) can be drawn by using the keyboard, *Draw* toolbar, or the *Main Menu* (*Main Menu> Draw> Line> select endpoints*). Use *Command:* Line (or L) and respond to the prompts as shown in

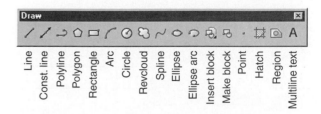

37.69 The *Draw* toolbar makes it easy for you to select commands with the cursor. (*Main Menu> View> Toolbars> Draw*)

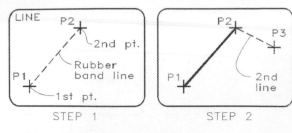

STEP 1   STEP 2

37.70 Line.

Step 1 *Menu Menu> Draw> Line> Specify first point*: <u>P1</u>
Specify next point or [Undo]: <u>P2</u>

Step 2 *Specify next point or [Close/Undo]*: <u>P3</u>. (Enter)

Figure 37.70 to draw lines by picking endpoints with the left button of the mouse. The current line will rubber band from the last point, and lines are drawn in succession until (Enter) or the right button of your mouse is pressed.

The *Draw* toolbar can be used to select four types of lines: *Line*, *Ray*, *Construction*, and *Multilines*. These types of lines can also be selected from *Main Menu> Draw* (**Figure 37.71**). A *Construction line* is drawn totally across the screen, and a *Ray* is drawn from the selected point to the edge of the screen.

A comparison of absolute and polar coordinates is shown in **Figure 37.72**. *Delta coordinates* can be typed as @2,4 to specify the end of a line 2 units in the x-direction and 4 units in the y-direction from the current end.

Polar coordinates are *2D* coordinates that are typed as @3.6 < 56 to draw a 3.6 long line from the current (and active) end of a line at an angle of 56° with the x-axis. *Last coordinates*

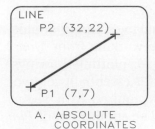
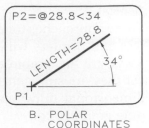

A. ABSOLUTE COORDINATES   B. POLAR COORDINATES

37.72 Lines by coordinates.

A Absolute coordinates can be typed (7,7 and 32,22) at the *Command* line to establish the ends of a line.

B Polar coordinates are relative to the current point and are specified with a length and the angle measured clockwise from the horizontal (@28.8 < 34).

are found by typing @ while in the *Line* command. This causes the cursor to move to the last point.

World coordinates locate points in the *World Coordinate System* regardless of the *User Coordinate System* being used by preceeding the coordinates with an asterisk (*). Examples are *4,3; *90 < 44; and @*1,3.

The *Status* line at the bottom of the screen shows the length of the line and its angle from the last point as it is rubberbanded from point to point. The *Close* command will close a continuous series of lines from the last to first point selected.

37.18 Circles

The *Circle* command (*Main Menu> Draw> Circle*) draws circles when you select a center and radius, a center and diameter, or three points (**Figure 37.73**). Use *Command*: <u>Circle</u>

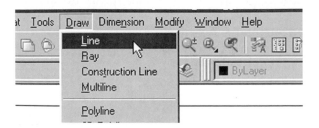

37.71 These types of lines can be selected from the Draw drop-down menu or from the *Main Menu* toolbar.

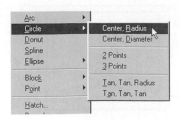

37.73 The **Circle** command (*Main Menu> Draw> Circle*) has a flyout menu with these options for drawing circles.

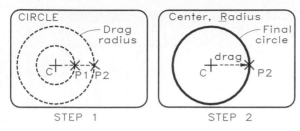

37.74 Circle.

Step 1 *Main Menu> Draw> Circle> Specify center point for circle or [3P/2P/Ttr (tan tan radius)]:* Pick C (Enter)

Specify radius of circle or [Diameter] <0.00>: R (Enter)

Step 2 Drag radius to P1 and P2 to enlarge the circle, click the left mouse button, (Enter), and the final circle is drawn.

(or C) (Enter)) is the fastest means of activating the *Circle* command, but the *Circle* icon on the *Draw* toolbar (**Figure 37.69**) can also be selected. **Figure 37.74** illustrates how a circle is drawn.

The *TTR* (*tangent, tangent, radius*) option of *Circle* draws a circle tangent to a circle and a line, two lines, or two circles. A circle is drawn tangent to a line and a circle by selecting the circle, the line, and giving the radius (**Figure 37.75**). The *tan, tan, tan* option calculates the radius length and draws a circle tangent to three lines (**Figure 37.76**).

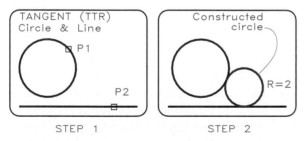

37.75 Circle: Tangent to two objects (TTR).

Step 1 *Main Menu> Draw> Circle> Specify center point for circle or [3P/ 2P/ Ttr (tan tan radius)]:* TTR (Enter)

Specify point on object for first tangent of circle: P1

Step 2 *Specify point on object for second tangent of circle:* P2

Specify radius of circle <0.00>: 2 (Enter)

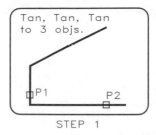 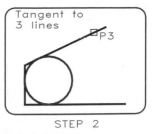

37.76 Circle: Tangent to 3 lines (TTT).

Step 1 *Main Menu> Draw> Circle> Tan, Tan, Tan*

Specify center point for circle or [3P/ 2P/ Ttr (tan tan radius)]: 3P > *Specify first point on circle: tan* to P1

Specify second point on circle: tan to P2

Step 2 *Specify third point on circle: tan* to P3. The tangent circle is drawn.

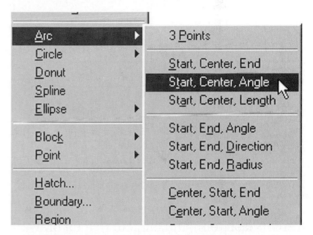

37.77 To draw an arc, select *Main Menu> Draw> Arc>* and these options are given. When using the *Start, Center, Angle,* these elements must be specified in this order on the screen. This selection can be made from the *Draw* toolbar, also.

37.19 Arcs

The *Arc* command (*Main Menu> Draw> Arc>* options) (**Figure 37.77**) or the *Arc* icon in the *Draw* toolbar (**Figure 37.69**) has eleven combinations of variables that use abbreviations for starting point, center, angle, ending point, length of chord, and radius. The *S, C, E* version requires that you locate the starting point S, the center C, and the ending point E (**Figure 37.78**). The arc begins at point S and is drawn counterclockwise by default to a point near E.

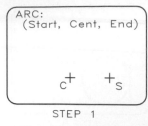
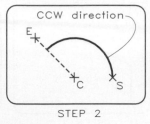

STEP 1 | STEP 2

37.78 Arc: Start, Center, End option (SCE).

Step 1 *Main Menu> Draw> Arc> Start, Center, End> Specify start point of arc or [Center]*: <u>Start</u> (Enter)
Specify center point of arc: <u>Center</u> (Enter)

Step 2 *Specify end point of arc or [Angle/chord Length]*: <u>E</u> Drag the arc to the end of the radial line CE, click the left mouse button (Enter) and the arc is drawn.

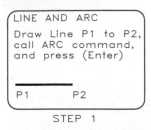
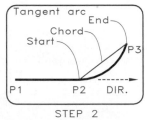

STEP 1 | STEP 2

37.79 Arc: Tangent at the end of a line.

Step 1 *Draw toolbar> Line> Specify first point*: <u>P1</u>
Specify next point or [Undo]: <u>P2</u> (Enter)

Step 2 *Draw toolbar> Arc> Specify start point of arc or [Center]*: (Enter) to grab <u>P2</u> > *Specify end point of arc*: Drag tangent arc to <u>P3</u>.

A line can be continued as an arc drawn from its last point and tangent to it for drawing runouts of fillets and rounds (**Figure 37.79**). It can be used to draw a tangent line from an arc by applying the commands in reverse and dragging the line to its final length.

37.20 Polygons

The *Polygon* and many other objects can be drawn by selecting their icons from the *Draw* toolbar (*View> Toolbars> Draw*) (**Fig. 37.80**). A Polygon is drawn from the *Draw* toolbar in **Fig. 37.81**. An equal-sided polygon can be drawn by using *Main Menu> Draw> Polygon* icon and following the prompts. A polygon is drawn in a counterclockwise direction about the center point. Polygons can have a maximum of 1024 sides.

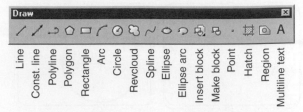

37.80 Select *Main Menu> View> Toolbars> Draw* to obtain the *Draw* toolbar. Select the *Polyline* icon to draw a polygon as shown in **Figure 37.81**.

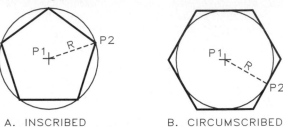

A. INSCRIBED | B. CIRCUMSCRIBED

37.81 Polygons: Inscribed and circumscribed.

A *Draw toolbar> Polygon> Enter number of sides <4>*: <u>5</u> (Enter)> *Specify center of polygon or [Edge]*: <u>P1</u>
Enter an option [Inscribed in circle/Circumscribed about circle] <I>: <u>I</u> (Enter)

B Select *Circumscribed* option. *Specify radius of circle*: Drag radius to <u>P2</u> to size the circumscribed circle.

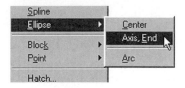

37.82 *Draw toolbar> Ellipse> Axis, End* can be used for drawing ellipses.

37.21 Ellipses

The *Ellipse* command in the *Main Menu> Draw* menu gives icons for three types of ellipses (**Figure 37.82**). An ellipse is drawn by selecting the endpoints of the major axis and a third point <u>P3</u>, to give the length of the minor radius (**Figure 37.83**). <u>P3</u> need not lie on the ellipse; the distance from the midpoint of the axes to <u>P3</u> merely gives the length of the minor axis, which is perpendicular to the major diameter, regardless of the direction in which the distance is specified.

In **Figure 37.84** points are picked at the center of the ellipse at <u>P1</u>, one axis endpoint at <u>P2</u>, and the second axis length from <u>P1</u> to <u>P3</u>. The ellipse is drawn through points <u>P2</u> and the endpoint of the minor diameter specified by <u>P3</u>. The endpoints of the axis can be located

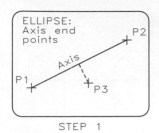
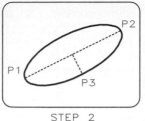

STEP 1   STEP 2

37.83 Ellipse: axis, end option.
Step 1 *Main Menu> Draw> Ellipse> Axis, End>*
Specify axis endpoint of ellipse or [Arc/Center]: P1
Specify other endpoint of axis: P2
Step 2 *Specify distance to other axis or [Rotation]:* P3,
Ellipse is drawn through P1, P2, and point established by P3.

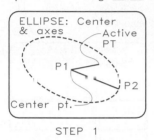
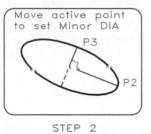

STEP 1   STEP 2

37.84 Ellipse: Center option.
Step 1 *Draw* toolbar> *Ellipse> Center*
Specify axis endpoint of ellipse or [Arc/Center]: C (Enter)
Specify center of ellipse: P1> Specify endpoint of axis: P2
Step 2 *Specify distance to other axis or [Rotation]:* P3
The ellipse is drawn.

and the rotation angle can be specified. An angle of 0° gives an ellipse as a full circle, and an angle of 90° gives an edge view of the circle.

37.22 Fillets

The corners of two lines can be rounded with the *Fillet* command (*Main Menu> Modify> Fillet*) whether or not they intersect. When the fillet is drawn, the lines are either trimmed or extended as shown in **Figure 37.85**. The assigned radius is remembered until it is changed. By setting the radius to 0, lines will be extended to a perfect intersection. Fillets of a specified radius can be drawn tangent to circles or arcs as shown in **Figures 37.86** and **37.87**.

The *Fillet* command can also be selected by typing *Fillet* at the *Command* line. Practically all commands can be entered at the command line in this manner without referring to screen menus.

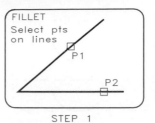
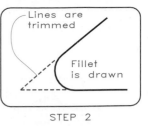

STEP 1   STEP 2

37.85 Fillet.
Step 1 *Command:* Fillet
Current settings: Mode=TRIM, Radius=0.0000
Select first object or [Polyline/Radius/Trim/mUltiple]: R (Enter)>
Specify fillet radius <0.0000>: 1.00 (Enter) (Enter)
Select first object or [Polyline/Radius/Trim/mUltiple]: P1>
Step 2 *Select second object:* P2
The tanget arc is drawn and the lines are trimmed.

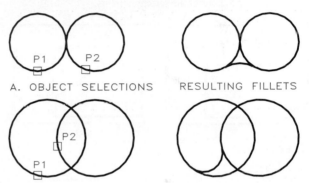

A. OBJECT SELECTIONS   RESULTING FILLETS

B. OBJECT SELECTIONS   RESULTING FILLETS

37.86 Fillets: Tangent arcs.
Step 1 *Main Menu> Modify> Fillet>*
Mode=TRIM, Radius=2.0000,
Select first object of [Polyline/Radius/Trim/mUltiple]: R (Enter)
Specify fillet radius <2.0000>: 1.2 (Enter)
Step 2 (Enter) *FILLET, Current settings: Mode=TRIM, Radius=1.2000> Select first object, or [Polyline/Radius/Trim/mUltiple]:* P1> *Select second object:* P2

37.23 Chamfers

The *Chamfer* command (*Main Menu> Modify> Chamfer*) draws angular bevels at intersections of lines or polylines. Two chamfer distance can be assigned if the bevel has unequal distances. After assigning chamfer distances (D), select two lines and they are trimmed or extended and the *Chamfer* is drawn (**Figure 37.88**). Press (Enter) to repeat this command using the previous settings. By setting the chamfer distance to 0 (zero), this command can be used to trim nonparallel lines to perfect intersections.

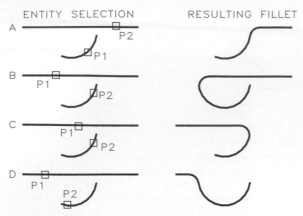

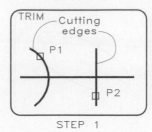 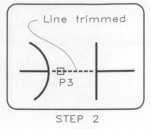

37.87 The applications of fillets between lines and arcs are determined by the positions of the selection points.

37.89 Trim: Cutting edges.

Step 1 *Modify toolbar> Trim> Current settings: Projection=UCS Edge=Extend Select cutting edges...*

Select objects: P1 *1 found*

Select objects: P2 *1 found> Select objects: (Enter)*

Step 2 *Select object to trim or [Project/Edge/Undo]:* P3, line between the cutting edges is removed.

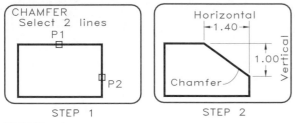

37.88 Chamfer.

Step 1 *Main Menu> Modify> Chamfer>*

(TRIM mode) Current chamfer Dist1=0.50, Dist2=0.50

Select first line or [Polyline/Distance/Angle/Trim/Method/mUltiple]: D> *Specify first chamfer distance <0.50>:* 1.40 *(Enter)*

Specify second chamfer distance <1.40>: 1.00 *(Enter)*

Step 2 *(Enter) Chamfer (Trim mode) Current chamfer Dist1=1.4000, Dist2=1.0000*

Select first line or [Polyline/Distance/Angle/Trim/Method/mUltiple]: P1> *Select second line:* P2

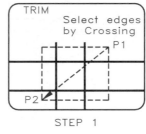 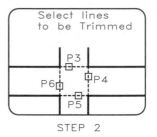

37.90 Trim: Crossing window.

Step 1 *Main Menu> Modify> Trim> Current settings: Projection=UCS, Edge=None, Select cutting edges...*

Select objects: P1> *Specify opposite corner:* P2 *4 found.*

Step 2 *Select object to trim or shift-select or [Project/Edge/Undo]:* P3

Select object to trim . . . or [Project/Edge/Undo]: P4

Select object to trim . . . or [Project/Edge/Undo]: P5

Select object to trim . . . or [Project/Edge/Undo]: P6 *(Enter)*

37.24 Trim

The *Trim* command (*Main Menu> View> Toolbars> Modify> Trim*) selects cutting edges for trimming selected lines, arcs, or circles that cross the edges at their crossing points (**Figure 37.89**). A series of objects can be selected one at a time or selected as a group by a window. A *Crossing* window is used to select four cutting edges for trimming four lines in Figure **37.90**.

37.25 Extend

The *Extend* command lengthens lines, plines, and arcs to intersect a selected boundary (**Figure 37.91**). You are prompted to select the

boundary object and the object to be extended to the boundary. More than one object can be extended at a time. *Extend* will not work on "closed" figure such as polygons.

37.26 Trace

Wide lines (either solid or open) can be drawn with the *Trace* command. For solid lines, select *Command:* Fill *(Enter)* pick On *(Enter)*. To draw trace lines, select *Command:* Trace *(Enter)* as shown in **Figure 37.92A**. When *Fill* is *Off*, the lines will be drawn as open, parallel lines with "mitered" angles (**Figure 37.92B**). Notice that the poor corner joint of the *Trace* lines can be

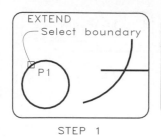

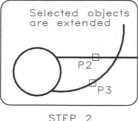

STEP 1   STEP 2

37.91 Extend.

Step 1 *Main Menu> Modify> Extend*
Current settings: Projection=UCS, Edge=None
Select boundary edges... Select objects: <u>P1</u> *1 found*
Step 2 *Select objects: Select object to extend or [Project/Edge/Undo]:* <u>P2</u> *2 found*
Select object to extend or shift-select to trim or[Project/Edge/Undo]: <u>P3</u> *(Enter)*

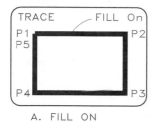

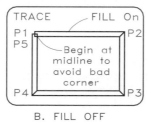

A. FILL ON   B. FILL OFF

37.92 Trace.

A *Command:* <u>Fill</u>, *Enter mode [On/Off]:* <On> *(Enter) Command:* <u>Trace</u> *(Enter), Specify trace width <0.00>:* <u>0.20</u>, *Specify start point:* <u>P1</u>, *Select next point:* <u>P2</u> *Select next point:* Continue point selection; (Enter) to end.
B To obtain a *Trace* with unfilled lines: *Command:* <u>Fill</u>, Set mode to *Off*, (Enter) *Command:* <u>Regen</u> (Enter) Draw the trace as in (A) above.

omitted if the line's beginning point is selected along the line rather than at a corner point.

37.27 Zoom and Pan

Drawings can be enlarged or reduced by the *Zoom* command (*Main Menu> View> Toolbars> Standard>* to get the *Zoom* options shown in **Figure 37.93**. Other options can be selected from the *Zoom* toolbar (*Main Menu> View> Toolbars> Zoom*) shown in **Figure 37.94A**. The same options are available in a pull-down menu (*Main Menu> View> Zoom*). An example of a *Zoom Window* used to enlarge part of a drawing is shown in **Figure 37.95**.

Zoom (*Command:> <u>Z</u>*) gives the following

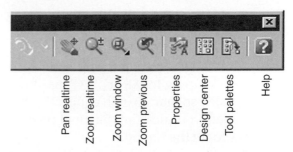

Pan realtime Zoom realtime Zoom window Zoom previous Properties Design center Tool palettes Help

37.93 This part of the *Standard* toolbar contains many options for *Panning* and *Zooming* on the screen.

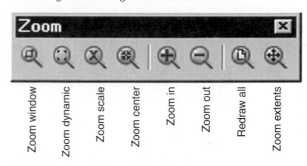

Zoom window Zoom dynamic Zoom scale Zoom center Zoom in Zoom out Redraw all Zoom extents

37.94 The *Zoom* toolbar contains these options for *Zooming* on the screen.

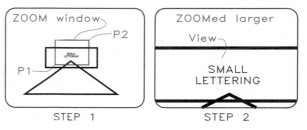

STEP 1   STEP 2

37.95 Zoom.

Step 1 *Standard* toolbar> *Zoom realtime* icon> right-click to activate pop-up menu> *Specify corner of window, enter a scale facor (nX or nXP), or [All/Center/Dynamic/Extents/ previous/ Scale/Window] <real time>:* Hold down left button and drag a window around the zoom area (<u>P1</u> and <u>P2</u>).

Step 2 The area of the window is expanded to fill as much of the screen as possible.

options: *All, Center, Dynamic, Extents, Previous, Scale, Window,* and *Real time.*

All expands the drawing's *Limits* (dot pattern) to fill the screen.

Center is picked to select the center of the *Zoomed* image and to specify its magnification or reduction.

Dynamic lets you *Zoom* and *Pan* by selecting points with the cursor.

Extents enlarges the drawing to its maximum size on the screen.

Previous displays the last *Zoomed* view.

Scale NXP magnifies a drawing relative to paper space. By typing 1/4XP or .25XP the drawing will be scaled so that .25 inch equals 1 inch.

Window lets you pick the diagonal corners of a window to fill the screen.

Real time (the default) lets you drag to the left to make a crossing window, or to the right for a window to specify the area to be enlarged.

The *Pan* (P) command (**Figure 37.96**) is used to pan the view across the screen by selecting two points: a handle point, and its final position. The drawing is not relocated as in the *Move* command; only your viewpoint of it is changed.

37.28 Selecting Objects

A reoccurring prompt, *Select objects:*, asks you to select an object or objects that are to be *Erased*, *Changed*, or modified in some way. Type *Select* when in a current command that requires a selection (*Move*, for example) and the options will be displayed: *Window, Last, Crossing, BOX, ALL, Fence, WPolygon, CPolygon, Group, Add, Remove, Multiple, Previous, Undo, AUto,* and *SIngle*. **Figure 37.97** shows ways of selecting objects for applicable

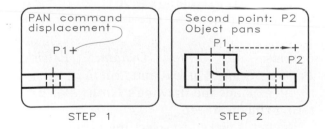

37.96 Pan.

Step 1 *Standard* toolbar> *Pan* icon

Step 2 Pick P1, hold down left button, and drag to P2 to change your view of the screen.

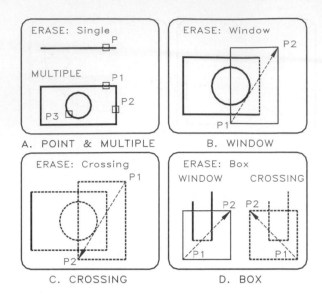

37.97 Selection options.

A *Select objects*: selects the objects one at a time.

B *Window (W)* selects objects lying completely within the window; drag window left to right.

C *Crossing Window (C)* selects objects lying within or crossed by the window; drag window right to left.

D *Box* makes a *Window* or a *Crossing Window* that is determined by the direction dragging.

commands, such as *Copy*, to select objects. The application of the options of *Single* and *Multiple, Window, Crossing,* and *BOX* are shown in parts A, B, C, and D, respectively.

Window (W) lets you select objects lying completely within it.

Crossing (C) lets you select objects lying within or crossed by the window.

Last is used to pick the most recently drawn object.

Box (B) lets you make a window by selecting a point and dragging to the right. A crossing window can be made by dragging to the left.

Other techniques for selecting objects are shown in **Figure 37.98**.

WPolygon (WP) forms a solid-line polygon that has the same effect as a window.

CPolygon (CP) forms a dotted-line polygon that has the same effect as a crossing window.

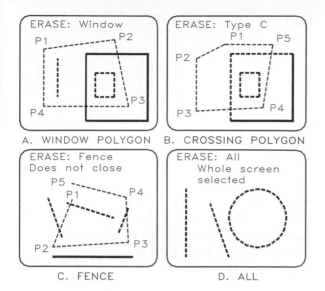

37.98 Selection of objects.

A *Window Polygon* (*WP*) selects all objects inside it.

B *Crossing Polygon* (*CP*) selects objects inside it and crossed by it.

C *Fence* (*F*), a nonclosing polyline selects objects crossed by it.

D *All* selects all objects on the screen.

Fence (*F*) selects corner points of a polyline that will select any object it crosses the same as a crossing window.

ALL selects everything on the screen.

Remove and *Add* are used while selecting objects to remove one by typing R (remove) or to add one by typing A (Add). When finished, press (Enter) to end the *Select/remove* prompt.

Multiple (*M*) selects multiple points (without highlighting) to speed up the selection process.

Previous (*P*) recalls the previously selected set of objects. For example, enter *Move*, and type P, and the last selected objects are recalled.

Undo (*U*) removes objects in reverse order one at a time by typing U (undo) repetitively.

Auto (*A*) selects an object by pointing to it and pointing to a blank area selects the first corner of a box defined by the *Box* option.

Single (*SI*) causes the program to act on the object or sets of objects without pausing for a response.

A *Window* (*W*) or a *Crossing Window* (C) can be obtained automatically by pressing the pick button, holding it down, and selecting the diagonal of a window. By dragging it to the right, a window is obtained; by dragging it to the left, a crossing window is obtained (**Figure 37.99**).

37.29 Erase and Break

The *Erase* command (*Modify* toolbar> *Erase* icon) (**Fig. 37.100**) deletes specified parts of a drawing. The selection techniques described previously can be used to select objects to be erased as shown in **Figures 37.101** and **37.102**. The default of the *Erase* command, *Select Objects*, allows you to pick one or more objects and delete them by pressing (Enter). Type

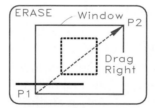 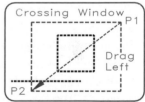

37.99 Selection windows.

A A *Window* is formed by holding down the select button while dragging a diagonal corner to the **right**.

B A *Crossing Window* is formed in the same manner, but dragged to the **left**.

37.100 The *Modify* toolbar, a portion of which is shown here, has options for making changes in a drawing.

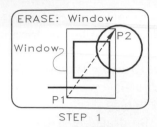
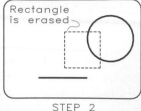

STEP 1 STEP 2

37.101 Erase: Window option.
Step 1 *Modify* toolbar> *Erase* icon
Select objects: P1, *Other corner*: P2 (left to right)
Step 2 *Select objects*: (Enter) The box completely within the erasing window is removed.

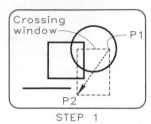
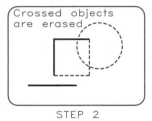

STEP 1 STEP 2

37.102 Erase: Crossing option.
Step 1 *Modify* toolbar> *Erase* icon
Select objects: P1, *Other corner*: P2 (right to left for a crossing window).
Step 2 *Select objects*: (Enter) The objects crossed by or completely within the window are removed.

Oops to restore the last erasure, but only the last one can be removed.

The *Break* command (*Modify* toolbar> *Break* icon) removes part of a line, pline, arc, or circle (**Figure 37.103**). To specify a break at an intersection with another line as shown in **Figure 37.104**, select the line to be broken, select the *F* option, and pick the two endpoints of the line to

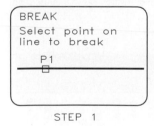
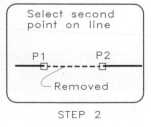

STEP 1 STEP 2

37.103 Break.
Step 1 *Modify* toolbar> *Break* icon
Select object: P1,
Step 2 *Specify second break point or [First point]*: P2, Line between P1 and P2 is removed.

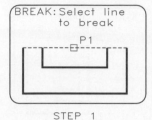
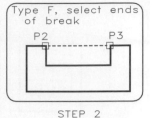

STEP 1 STEP 2

37.104 Break: F option.
Step 1 *Modify* toolbar> *Break* icon
Select object: P1, *Specify second break point or [First point]*: F (Enter)
Step 2 *Specify first break point*: P2, *Specify second break point*: P3, Line P2-P3 is removed.

be removed. Endpoints can be selected without fear of selecting the wrong lines.

37.30 Move and Copy

The *Move* command (*Modify* toolbar> *Move* icon) repositions a drawing (**Figure 36.105**) and the *Copy* command duplicates it, leaving the original in its original position. The *Copy* (*C*) command is applied in the same manner as the *Move* command. The *Copy* command has a *Multiple* option (type M when prompted) for locating more than one copy of the selected drawing in different positions.

37.31 Undo

The *Undo* command (*Command*: Undo, or U) can reverse the previous commands one at a time back to the beginning of a session. The

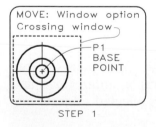
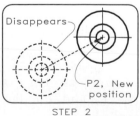

STEP 1 STEP 2

37.105 Move.
Step 1 *Modify* toolbar> *Move* icon
Select objects: W , Select by a Window.
Specify base point or displacement: P1
Step 2 *Specify second point of displacement or <use first point as displacement>*: Drag to new position of P2.

Redo command reverses the last *Undo; Oops* will not work. The *Undo* command has options of: *Auto, Control, BEgin, End, Mark, Back,* and *Number.*

> *Command*: <u>Undo</u> (Enter)
> *Auto/Control/BEgin/End/Mark/Back/*
> *<Number>*: <u>4</u> (Enter)

Entering 4 has the same effect as using the *U* command four separate times.

Mark identifies a point in the drawing process to which subsequent additions can be undone by the *Back* option. Only the part of the drawing added after placing the *Mark* will be undone at the prompt:

> *This will undo everything.*
> *OK? <Y>*: <u>Y</u> (Enter)

By responding <u>Y</u>, the *Mark* will be removed, making it possible for the next *Undo* (*U*) to proceed backward past the mark.

The *BEgin* and *End* options group a sequence of operations until *End* terminates the group. *Undo* treats the group as a single operation. The *Control* subcommand has three options: *All, None,* and *One. All* turns on the full features of the *Undo* command, *None* turns them off, and *One* uses *Undo* commands for single operations and requires the least disk space.

37.32 Change

Use *Command*: <u>Change</u> to modify features: *Lines, Circles, Text, Attribute Definitions, Blocks, Color, Layers, Linetypes,* and *Thickness.* The position of an endpoint of a *Line* is changed by selecting one end and locating a new endpoint (**Figure 37.106**). *Change* varies the size of a circle by picking a point on its arc and dragging to a new size (**Figure 37.107**).

Text can be modified with the *Change* command by pressing (Enter) until the prompts *Insertion point, Style, Height, Rotation Angle,* and *New Text* appear in sequence (**Figure**

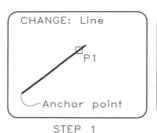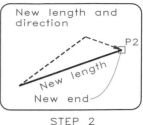

37.106 Change: Line.

Step 1 *Command: Change>* (Enter)

Select objects: <u>P1</u>, *1 found> Select objects:* (Enter)

Step 2 *Specify change point or [Properties]:* Select the new end of the line <u>P2</u>.

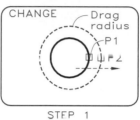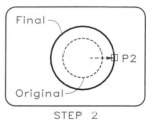

37.107 Change: Circle.

Step 1 *Command: Change>* (Enter)

Select objects: <u>P1</u>, *1 found*

Select objects: (Enter)

Step 2 *Specify change point or [Properties]:* Drag to the new diameter <u>P2</u>.

37.108). *Attribute Definitions,* including *Tag, Prompt String,* and *Default Value,* can be revised with the *Change* command. *Blocks* can be moved or rotated with the *Change* command as shown in **Figure 37.109**.

Property changes of the *Change* command, *LAyer, Color, LType,* and *Thickness,* are made by selecting objects and typing <u>P</u> (*Properties*) as shown in **Figure 37.110**. Type <u>Layer</u> (*or* <u>La</u>) and the name of the layer on which the text is to be changed.

Multiple *Colors* and *LTypes* can be assigned to objects on the same layer by the *Change* command, but it is better for each layer to have only one layer and linetype. The *Thickness* property is the height of an extrusion in the z-direction of a three-dimensional object drawn with the *Elev* and *Thickness* options (as covered in Chapter 38). *Changing* the *Thickness* of a 2D surface 0 to a

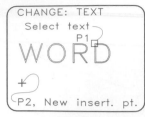
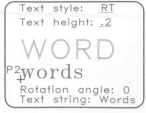

STEP 1   STEP 2

37.108 Change: Text.

Step 1 *Command*: <u>Change</u>> (Enter)

Select objects: <u>P1</u>, *1 found, Select objects*: (Enter)

Specify change point of [Properties]: (Enter)

Specify new text insertion point <no change>: <u>P2</u>

Step 2 *Enter new text style <Outline>*: <u>RT</u> (Enter)

Specify new height <0.00>: <u>.20</u> (Enter)

Specify new rotation angle <0>: (Enter)

Enter new text <WORD>: <u>WORDS</u> (Enter)

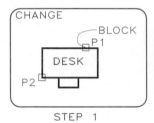
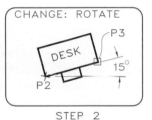

STEP 1   STEP 2

37.109 Change: Block.

Step 1 *Command*: *Change> Select objects*: <u>P1</u>, *1 found Select, objects*: <u>P2</u>. *Select objects*: (Enter).

Step 2 *Specify change point or [Properties]*: <u>P3</u>

Specify new block rotation angle <0>: <u>15</u> (Enter).

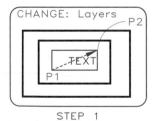
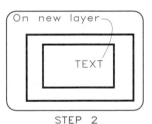

STEP 1   STEP 2

37.110 Change: Layers.

Step 1 *Command*: *Change> Select objects*: <u>W</u>

Pick points <u>P1</u> and <u>P2</u> to make window.

1 found, Select objects:

Specify change point of [Properties]: <u>P</u> (Enter)

Step 2 *Enter property to change [Color/Elev/LAyer/LType/ ltScale/LWeight/Thickness/PLotstyle]*: <u>Layer</u> (Enter)

Enter new layer name <0>: <u>Visible</u> (Enter)

nonzero value converts it to an extruded 3D drawing.

At the command line, type <u>DDmodify</u> or (*Main Menu> Modify> Properties*) and select an object when prompted to obtain a *Properties* box of this type (**Figure 37.111**) to change colors, layers, linetypes, thickness, linetype scale, and ends of lines as well as the properties of other objects.

37.33 Grips

Grips are small squares that appear on selected objects at midpoints and ends of lines, at centers and quadrant points of circles, and at insertion points of text. *Grips* are used to *Stretch, Move, Rotate, Scale,* and *Mirror.*

The *Grips* dialogue box (*DDgrips*) is found under *Tools* of the *Standard* menu (**Figure 37.112**). The *Enable Grips* check box turns on grips for all objects. *Enable Grips Within Blocks* turns on grips for objects within a *Block*; when *Off*, a single grip is given at the insertion point of the *Block*.

37.111 The *Properties* box (type <u>DDmodify</u> at the *Command* line) is used to modify objects in the same manner as when the *Change* command is used. Most of the windows have pull-down menus with options as shown for *Linetype*.

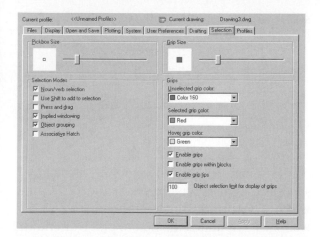

37.112 Use *Main Menu> Tools> Options> Selection* tab to obtain this box for *Enabling Grips*, setting *Grip size*, and selecting *Grip color*.

Grip Colors turns on the *Color* dialogue box for assigning colors to selected and unselected grips; unselected grips are not filled in. *Grip Size* sets the size of grip boxes with a slider box.

Using Grips

By selecting an object with the cursor, grips will appear on it as open boxes. A grip that is picked and made a "hot" point is filled with color. By holding down the *Shift key*, more than one grip can be picked as a "hot" point, but the last grip of a series must be selected without pressing *Shift*. Press the *Escape* key to remove grips. By turning grips on and successively pressing (Enter), the options *Stretch, Move, Rotate, Scale,* and *Move* will be sequentially activated, each with its own subcommands.

Stretch lets the endpoint grip of a line be selected as a "hot" point, and a second point appears as the new end of the line (**Figure 37.113**).

Select the midpoint grip to *Move* the line to a new position. To *Move* the *Block* in **Figure 37.114**, select the insertion point as the hot point. Options of *Base Point, Copy, Undo,* and *eXit* can be used for these applications.

Rotate revolves an object about a selected *Grip*. The *Reference* option rotates an object about a selected grip by dragging or typing a

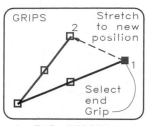 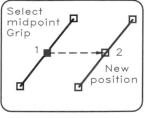

37.113 Grips: Stretch.

A Select the line and grips will appear; click on end grip 1 and select a new position at 2.

B Select the midpoint grip 1, pick the second point 2, and the line is moved.

37.114 Grips: Move.

Step 1 Select the block and a grip appears at the insertion point. Click on this grip to pick it as the hot point.

Step 2 Move the cursor to a new position and the block is moved.

number. The object in **Figure 37.115** is rotated 50° from the reference line by typing 50.

Scale uses the grip selected as a base point to size the object (**Figure 37.116**). The scale factor is assigned by typing, dragging, or selecting a reference dimension and giving it a new dimension.

Mirror makes a mirror image of an object. The original object is removed when two grips are selected to specify a mirror line (**Figure 37.117**). Hold down the *Shift key* while selecting the second grip point on the mirror line and the initial drawing will not be removed.

37.34 Polyline

The *Polyline* (*PL*) icon of the *Draw* toolbar (*View> Toolbars> Draw*) (**Figure 37.118**) is used for drawing 2D polylines, which are lines of

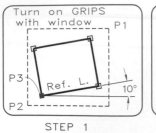
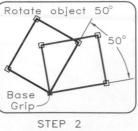

37.115 Grips: Rotation.

Step 1 Turn on grips by windowing the object (<u>P1</u> and <u>P2</u>); Select <u>P3</u> as the pivot point> press (Enter) until ****ROTATE**** appears.

Step 2 *Specify rotation angle or [Base point/Copy/Undo/ Reference/eXit]:* <u>50</u> (Enter) (Esc) The object is rotated.

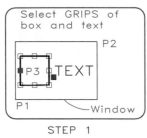
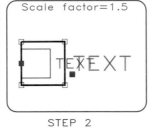

37.116 Grips: Scale.

Step 1 Turn on grips by windowing object (<u>P1</u> and <u>P2</u>); select <u>P3</u> as the base, press (Enter) until ****SCALE**** appears.

Step 2 *Specify scale factor or [Base point/Copy/Undo/ Reference/eXit]:* <u>1.5</u> (Enter) (Esc)

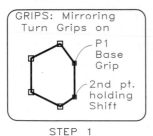
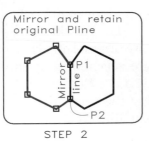

37.117 Grips: Mirror.

Step 1 Turn on grips by windowing object> select <u>P1</u> as the base grip> press (Enter) until ****MIRROR**** appears.

Step 2 *Specify second point or [Base point/Copy/Undo/eXit]:* <u>P2</u>, The object is mirrored. (Enter) (Esc)

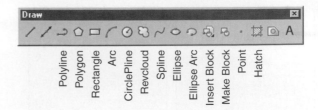

37.118 The *Polyline* icon on the *Draw* toolbar is selected for drawing *polylines*.

continuously connected segments instead of separate segments as drawn by the Line command. The thickness of a *Pline* can be varied as well, which requires the pen to plot with multiple strokes when plotting (**Figure 37.119**).

The *Pline* options are *Arc, Close, Halfwidth, Length, Undo,* and *Width.* Close automatically connects the last end of the polyline with its beginning point and ends the command. *Halfwidth* specifies the width of the line measured on both sides of its center line. *Length* continues a *Pline* in the same direction by typing the length of the segment. If the first line was an arc, a line is drawn tangent to the arc. *Undo* erases the last segment of the polyline, and it can be repeated to continue erasing segments.

The *Arc* option of *Pline,* is selected to obtain the prompts shown in **Figure 37.120**. The

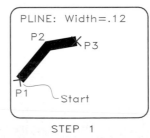
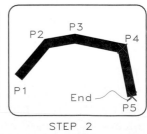

37.119 Polyline: Width option.

Step 1 *Draw* toolbar> *Polyline* icon> *Specify start point:* <u>P1</u>> *Current line-width is 0.00> Specify next point or [Arc/Close/Halfwidth/Length/Undo/Width]:* <u>W</u> > *Specify starting width <0.00>:* <u>.12</u> (Enter)> *Specify ending width <0.12>:* (Enter), *Specify next point, or [Arc/Close/Halfwidth/Length/Undo/Width]:* <u>P2</u> *Specify next point, or [Arc/Close. . ./Width]:* <u>P3</u>

Step 2 *Specify next point, or [Arc/Close . . . /Width]:* <u>P4</u>

Specify next point, or [Arc/Close . . . /Width]: <u>P5</u> (Enter)

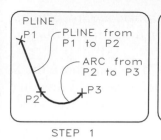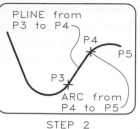

37.120 Polyline: Lines and arcs.

Step 1 *Draw toobar> Select Polyline icon> Specify start point:>* P1 *(Enter)> Current line-width is 0.12> Specify next point or [Arc/Close/Halfwidth/Length/Undo/Width]:* P2

Specify next point or [Arc/Close . . . /Width]: Arc

Specify endpoint of arc or [Arc/Close . . . /Width]: P3

Step 2 *Specify endpoint of arc or [Arc/Close . . . /Width]:* Line (Enter)

Specify next point or [Arc/ Close . . . /Width]: Length

Specify length of line: P4

Specify endpoint of arc or [Arc/Close . . . /Width]: Arc

Specify endpoint of arc or [Arc/Close . . . /Width]: P5

Specify endpoint of arc or [Arc/Close . . . /Width]: (Enter)

default draws the arc tangent from the end-point of the last line and through the next selected point. *Angle* gives the prompt, *Included angle:*, to which a positive or negative value is given. The next prompt asks for *Center/Radius/<End point>:* to draw an arc tangent to the previous line segment. Select *Center* and you will be prompted for the center of the next arc segment.

The next prompt is *Angle/Length/ <Endpoint>:*, where Angle is the included angle, and *Length* is the chordal length of the arc. *Close* causes the *Plines'* arc segment to close to its beginning point.

Direction lets you override the default, which draws the next arc tangent to the last *Pline* segment. When prompted with *Direction from starting point:*, pick the beginning point and respond to the next prompt, *Endpoint*, by picking a second point to give the direction of the arc.

Line switches the *Pline* command back to the straight-line mode. *Radius* gives the prompt, *Radius:*, for specifying the size of the

next arc. The next prompt, *Angle/Length/ <Endpoint>:*, lets you specify the included angle or the arc's chordal length. *Second Pt* gives two prompts, *Second point:* and *Endpoint:*, for selecting points on an arc.

37.35 Pedit

The *Pedit* command (*Command: Pedit>* (Enter)) modifies *Plines* with the following options: *Close, Join, Width, Edit vertex, Fit, Spline, Decurve, Ltype gen, Undo,* and *eXit.* If the *Pline* is already closed, the *Close* command will be replaced by the *Open* option.

Join (*J*) gives the prompt, *Select objects,* for selecting segments to be joined into a polyline. *Segments* must have exact meeting points to be joined.

Width (*W*) gives the prompt, *Specifiy new width for all segments:*, for assigning a new width to a *Pline.*

Fit (*F*) converts a polyline into a line composed of circular arcs that pass through each point (**Figure 37.121**).

Spline (*S*) modifies the polyline as did the *Fit* curve, but it draws cubic curves passing through the first and last points, not necessarily through the other points (**Figure 37.122**). *Decurve* (*D*) converts *Fit* or *Spline* curves to their original straight-line forms.

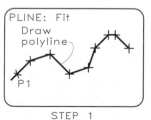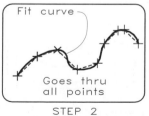

37.121 Pedit: Fit curve

Step 1 *Modify II toolbar> Select Pedit icon> Select polyline:* P1 (Enter)

Step 2 *[Close/Join/Width/Edit vertex/Fit/ . . .Undo/ eXit <X>]:* Fit (Enter) The curve is drawn as a series of arcs passing through all polyline points.

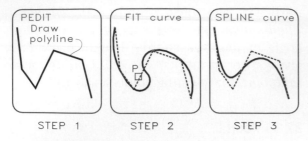

37.122 Pedit: Fit curve.

Step 1 Draw as polyline (Pline).

Step 2 *Modify II toolbar> Select Pedit icon> Select polyline:* P (Select line)

Close/ Join/ Width/ Edit vertex/ Fit/ . . . Undo/ eXit <X>: Fit (Enter). The curve of arcs passes through all points.

Step 3 *Pedit* in the same way, but use the *Spline* option to obtain a "best curve" that may not pass through the selected points.

Ltype gen (*L*) applies dashed lines (such as hidden lines) in a continuous pattern on curved polylines. Without applying this option, dashed lines may omit gaps in curved lines (**Figure 37.123**). By setting system variable *Plinegen* On, linetype generation will be applied as *Plines* are drawn.

Undo (*U*) reverses the last Pedit editing step.

Edit vertex (*E*) selects vertexes of the *Pline* for editing by placing an X on the first vertex when the polyline is picked. The following

options appear: *Next/Previous/Break/Insert/ Move/Regen/Straighten/Tangent/Width/eXit/ <N>:* Next (Enter).

Next (*N*) and *Previous* (*P*) options move the X marker to next or previous vertexes by pressing (Enter). *Break* (*B*) prompts you to select a vertex with the X marker. Then use *Next* or *Previous* to move to a second point, and pick Go to remove the line between the vertexes. Select *eXit* to leave the *Break* command and return to *Edit vertex*.

Insert adds a new vertex to the polyline between a selected vertex and the next vertex (**Figure 37.124**). *Move* (*M*) relocates a selected vertex (**Figure 37.125**).

Straighten (*S*) converts the polyline into a straight line between two selected points. An X marker appears at the current vertex and the prompt, *Next/Previous/Go/eXit/<N>*, appears. Move the X marker to a new vertex with *Next* or *Previous*, select *Go*, and the line is straightened between the vertices (**Fig. 37.126**). Enter X to *eXit* and return to the *Edit vertex* prompt.

Tangent (*T*) lets a tangent direction be selected at the vertex marked by the X for curve fitting by responding to the prompt, *Direction*

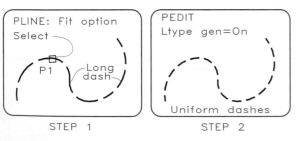

37.123 Pedit: Ltype generate.

Step 1 *Modify II toolbar> Select Pedit icon> Select polyline:* P1 (Select polyline) (Enter)

Step 2 *Enter an option [Close/ Join/ Width/ Edit vertex/ Fit/ . . . Ltype gen/ Undo/ eXit <X>]:* L (Enter)

Enter polyline linetype generation option [ON/OFF] <Off>: ON (Enter) Dashes become uniform around curves.

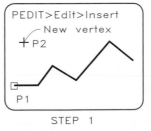 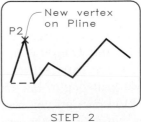

37.124 Pedit: Edit vertex– Insert.

Step 1 *Modify II toolbar> Select Pedit icon> Select polyline:* P1 > *Select polyline* (Enter)

Enter an option [Close/Join/Width/Edit vertex. . . Undo]: E (Enter) *Enter a vertex editing option [Next/. . . /Insert/. . ./eXit] <N>:* Insert (Enter) *Specify location for new vertex:* P2

Step 2 *Enter a vertex editing option [Next/. . . /Insert/. . ./eXit] <N>:* X (Enter)

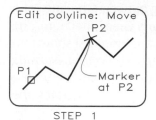
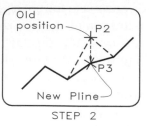

37.125 Pedit: Edit vertex–Move.

Step 1 *Modify II toolbar> Select Pedit icon> Select polyline:* <u>P1</u>

Enter an option [Close/Join/Width/Edit vertex . . . Undo]: <u>E</u>
(Enter) (Enter) (Enter) (Enter) to place cursor on <u>P2</u>.

Enter a vertex editing option [Next . . . /Move/. . ./eXit] <N>:
<u>Move</u> (Enter)

Step 2 *Specify a newlocation for marked vertex:* <u>P3</u>> *Enter a vertex editing option [Next . . . /Move/. . ./eXit] <N>:* <u>P3</u>

Enter a vertex editing option [Next/. . . /Insert/. . ./eXit] <N>: <u>X</u>
(Enter)

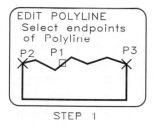
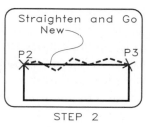

37.126 Pedit: Edit vertex–Straighten.

Step 1 *Modify II toolbar> Select Pedit icon> Select polyline or [Multiple]:* <u>P1</u> to pick the polyline

Enter an option [Close/. . . /Edit vertex/. . . /Undo]: <u>E</u> (Enter)
Enter a vertex editing option [Next . . . /Straighten/. . ./eXit] <N>: <u>Straighten</u> (Enter) <u>P2</u> is selected.

Enter an option [Next/Previous/Go/eXit] <N>: Press (Enter) until X is on <u>P3</u>.

Step 2 *Next/Previous/Go/eXit <N>:* <u>Go</u> (Enter) Line is straightened between P2 and P3.

of tangent. Enter the angle from the keyboard or by cursor.

Width (*W*) sets the beginning and ending widths of an existing line segment from the X-marked vertex. Use *Next* and *Previous* to confirm in which direction the line will be drawn from the X marker. The polyline will be changed to its new thickness when the screen is regenerated with *Regen* (*R*). Use *eXit* to escape from the *Pedit* command.

37.36 Spline

Spline command (*Draw* toolbar> *Spline* icon) draws a smooth curve with a sequence of points within a specified tolerance as shown in **Figure 37.127**. By setting *Fit Tolerance* to 0, the curve will pass through the points; when set to a value greater than 0, it will pass within a tolerance of each point.

Once the points of the spline are selected, press (Enter) and you will be prompted for tangent directions to be drawn at each end. The *End tangent* determines the angle of the spline at each end.

37.37 Hatching

Hatching is a pattern of lines that fills sectioned areas, bars on graphs, and similar applications. From the *Draw* toolbar (**Figure 37.128**) the *Bhatch* (*boundary hatch*) dialogue box is displayed (**Figure 37.129**). By selecting the down arrow at the *Pattern* box, a listing of pattern names is given from which to select. When one is selected, a view of the pattern will appear in

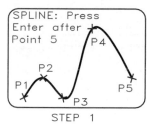
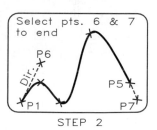

37.127 Spline.

Step 1 *Draw* toolbar> *Spline* icon> *Specify first point or [Object]:* <u>P1</u>> *Specify next point:* <u>P2</u>

Specify next point or [Close/Fit tolerance] <start tangent>: <u>P3</u>> Continue specifying points until <u>P5</u> (Enter)

Step 2 *Specify start tangent:* Rubberband to <u>P6</u> (Enter)

Specify end tangent: Rubberband to <u>P7</u> (Enter)

37.128 *Draw* toolbar> Select *Hatch* icon to begin to hatch a sectioned area.

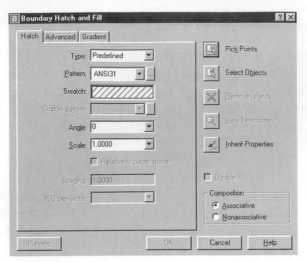

37.129 When the *Hatch* icon is selected, this *Boundary Hatch and Fill* box appears on the screen.

the *Swatch* window. Examples of some predefined patterns are shown in **Figure 37.130**.

The *Boundary Hatch* box (**Figure 37.129**) lets you specify *Predefined, User-defined,* or *Custom* patterns. *Predefined* patterns are those provided by AutoCAD.

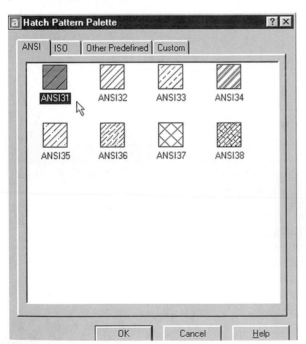

37.130 When the *Pattern* button in the *Boundary Hatch and Fill* box is selected, the hatch patterns and their names are displayed in this *Hatch Pattern Palette*.

Scale sets the spacing between the lines of a pattern, and *Angle* assigns their direction. The *Advanced* tab displays the options shown in **Figure 37.131** from which choices of *Normal, Outer,* or *Ignore* can be made. A square with a pentagon and a circle inside it illustrates the effect of each choice. *Normal* hatches every other nested area beginning with the outside. *Outer* hatches the outside area, and *Ignore* hatches the entire area from the outer boundary. When text within the hatching area is selected, it will appear in an opening in the hatching and hatch lines will not pass through it.

Options of *Points* and *Select Objects* (**Figure 37.129**) are used to select areas inside of boundaries and then boundaries themselves, respectively, as shown in **Figure 37.132**. When points are selected outside the boundary or if the boundary is not closed, error messages will appear. *Composition* can be specified as *Associative* or *Nonassociative*. *Associative* hatching is automatically updated when the size of the hatching area is changed. Select the *Inherit Properties* icon, pick the symbols within a hatched area, and select the area to be hatched. Then it is filled with the same hatching symbols.

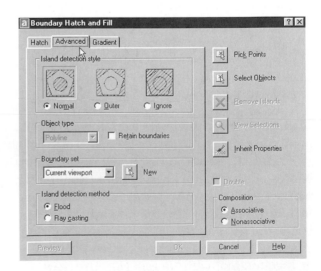

37.131 Select the *Advanced* tab in the *Boundary Hatch and Fill* box to display this dialogue box.

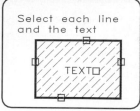

A. PICK POINTS   B. SELECT OBJECTS

37.132 Hatching areas.

A The *Pick Points* option (Figure 37.131) of *Boundary Hatch* prompts for points inside the boundaries for hatching.

B The *Select Objects* option requires that boundary lines be selected, including text.

37.38 Text and Numerals

Text and numerals can be added to a drawing with the *Dtext* command.

Command: *Dtext> Specify start point of text or [Justify/Style]*: Justify (Enter)

Enter an option *[Align/Fit/Center/Middle/Right/TL/TC/TR/ML/MC/MR/BL/BC/BR]*: specify the insertion point for the text you are entering (**Figure 37.133**). For example, BC means bottom center, *RT* means right top, and so forth.

Figure 37.134 illustrates how multiple lines of *Dtext* are automatically spaced by pressing (Enter) at the end of each line. The special characters shown in **Figure 37.135** can be inserted by typing a double percent sign (%%) in front of them.

Type Qtext and select *On* to reduce screen regeneration time by drawing text as boxes (**Figure 37.136**). When *Qtext* is *Off*, the full text

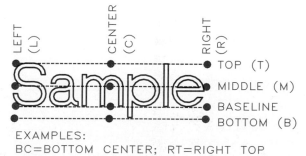

EXAMPLES:
BC=BOTTOM CENTER; RT=RIGHT TOP

37.133 Text can be added to a drawing by using any of the insertion points above. For example, *BC* means the bottom center of a word or sentence that will be located at the cursor point.

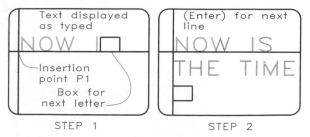

STEP 1   STEP 2

37.134 Dtext.

Step 1 *Command*: Dtext (Enter)
Current text style: "Standard" Text height 0.50
Specify start point of text or [Justify/Style]: P1
Specify height <0.50>: .125, *Specify rotation angle of text <0>*: (Enter) *Enter text*: NOW IS
Step 2 (Enter) *Enter text*: THE TIME (Enter) (Enter)

%%O	Start or stop Overline of text
%%U	Start or stop Underline of text
%%D	Degree symbol: 45%%D =45°
%%P	Plus−minus: %%P0.05=±0.05
%%C	Diameter: %%C20=⌀20
%%nnn	Special character number nnn

37.135 The special characters that begin with %% are used with *Dtext* to obtain these symbols.

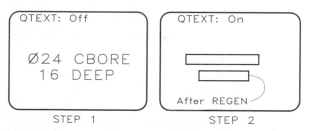

STEP 1   STEP 2

37.136 Qtext
Step 1 *Command*: Qtext (Enter), On/ Off: On (Enter)
Step 2 *Command*: Regen (Enter) Text is shown as boxes.

will be restored after regeneration type Regen at the command line.

37.39 Text Style

Many of AutoCAD's text fonts and their names are shown in **Figure 37.137**. The default style, *Standard*, uses the *Txt* font. From the *Format* menu select *Text Style*, and the *Text Style* dialogue box appears where you can assign a *New Style Name* (**Figure 37.138**). Select the New button, get the *Style Name* box, type the name Romand, and the style is named. Pick the down

TXT	PRELIMINARY PLOTS
MONOTXT	FOR SPEED ONLY
	Simplex fonts
ROMANS	FOR WORKING DRAWINGS
SCRIPTS	*Handwritten Style, 1234*
GREEKS	ΓΡΕΕΚ ΣΙΜΠΛΕΞ, 12345
	Duplex fonts
ROMAND	THICK ROMAN TEXT
	Complex fonts
ROMANC	ROMAN WITH SERIFS
ITALICC	*ROMAN ITALICS TEXT*
SCRIPTC	*Thick–Stroke Script Text*
GREEKC	ΓΡΕΕΚ ΩΙΤΗ ΣΕΡΙΦΣ
	Triplex fonts
ROMANT	TRIPLE–STROKE ROMAN
ITALICT	*Triple–Stroke Italics*
	Gothic fonts
GOTHICE	English Gothic Text
GOTHICG	German Gothic Text, 12
GOTHICI	Italian Gothic Text, 12

37.137 Examples of some of the available fonts are shown here.

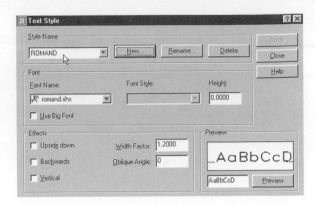

37.138 *Main Menu> Format> Text Style* (or type <u>Style</u> at the *Command* line) to display this *Text Style* box. From here, a *New style* can be named, *Fonts* assigned, *Width Factors* specified, and other assignments made.

arrow at the *Font Name* panel, and pick the font that you want to assign to the new style. Other options can be assigned: *Height* (0 is recommended), *Width Factor*, and *Oblique Angle*. A preview of your preferences is shown in the *Preview* window

The *Style names* are listed in a dropdown menu in the window beneath the heading, *Style Name* of the *Text Style* dialogue box. An example of the text font is displayed when a *Style* is selected (**Figure 37.138**). A defined *Style* will retain its settings until they are changed.

If you later change a named *Style* with new settings or fonts and select *Apply* in the *Text Style* box, all text previously entered under this style name will be updated with the new properties. This technique is used to change the *Txt* and *Monotxt* fonts to more attractive fonts at

plot time. Beforehand, time is saved by using *Txt* and *Monotxt* fonts because they regenerate quickly.

The *DDedit* or (*Main Menu> View> Toolbars> Modify II> Text Edit* icon) is used to select a line of text to be displayed in a dialogue box for editing (**Figure 37.139**). Correct the text, select the <u>OK</u> button, and it is revised on the screen.

37.40 Multiline Text

From the *Main Menu> View> Toolbars> Draw> <u>A</u>* icon (*Text*), pick an insertion point, and specify the size of the text boundary by a diagonal window, or by typing <u>W</u> (for *Width*) and entering a value and typing <u>H</u> (for *Height*) and entering a value. The *Text Formatting* dialogue box

37. 139 *Modify II* toobar> *Select Edit Text* icon (or *Command*: <u>DDedit</u>), select a line of text on the screen and it will appear in this *Edit Text* box for editing.

will appear and the paragraph of text can be typed in its window (**Figure 37.140**). Select <u>OK</u> to attach the paragraph to the drawing.

Windows on the *Text Formatting* bar enable you to specify text *Style, Fonts, Height,* and *Color.* Right click your mouse and other formatting options appear from which to select. Most of the options provided by the menu are obvious after a degree of experimentation and the use of *Help* command when needed. Examples of justified text are shown in **Figure 37.141**.

37.41 Mirror

Select the *Mirror* icon (*Modify* toolbar> *Mirror* icon) or, (*Command*: <u>Mirror</u> (Enter)) to mirror partial figures about an axis (**Figure 37.142**). A line that coincides with the *Mirror line* (P1-P2, for example) will be drawn twice when mirrored; therefore, parting lines should not be selected, or drawn after the drawing has been mirrored.

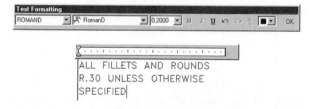

37.140 (*Draw* toolbar> *Multiline Text* icon>) and pick two diagonal corners to specify the area for the lines of text on the screen and this *Text Formatting* bar appears. Type your lines of text in the box, pick *OK*, and the lines of text are drawn in the specified area. Three windows in the *Text Formatting* box lets you specify text *Style, Font, Height,* and *Color*. Text can also be changed to *bold, italics,* or *underlined*.

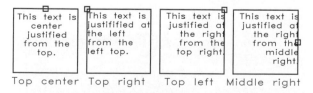

37.141 While the *Text Formatting* bar is active, select the text and right click your mouse to get a list of options, select *Justification*, and pick the type of justification desired. (Enter) Several types are shown here.

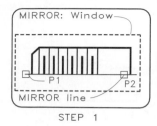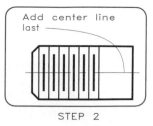

37.142 Mirror

Step 1 Draw the half to be mirrored. *Modify* toolbar> *Mirror* icon> *Select objects*: <u>W</u> (Enter) Window the drawing.

Select objects: (Enter)

First point of mirror line: <u>P1</u>, *Second point*: <u>P2</u>

Step 2 *Delete old objects?* <N>: <u>No</u> (Enter) The drawing is mirrored.

The system variable *Mirrtext* (*Command*: <u>Setvar</u>> <u>Mirrtext</u>) is used for mirroring text. By setting *Mirrtext* to <u>0</u>, it is set to *Off,* and text will not be mirrored. If *Mirrtext* is set to <u>1</u> (*On*), and the text will be mirrored along with the drawing.

37.42 Osnap

By using *Osnap* (*Object Snap*), you can snap to objects of a drawing rather than to the *Snap* grid. *Osnap* icons from the *Object snap* toolbar (**Figure 37.143**) gives the following options: *Endpoint, Midpoint, Intersection, Apparent Intersection, Center, Quadrant, Perpendicular, Tangent, Node, Intersection, Nearest, Quick,* and *None*. *Osnap* is used as an accessory to other commands: *Line, Move, Break,* and so forth.

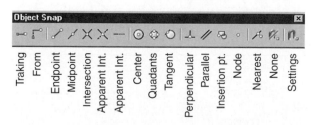

37.143 The *Object Snap* toolbar has these options for drawing to and from object features on the screen.

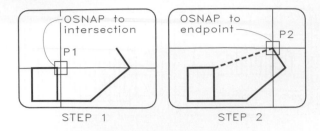

37.144 Osnap: Intersection and end.

Step 1: *Draw* toolbar> *Line* icon> *Line from point*:> *Intersection* icon> *Int of*: <u>P1</u>

Step 2 *To point*:> *Endpoint* icon on *Osnap* toolbar.

To point: <u>P2</u> The line is drawn.

Figure 37.144 shows how a line is drawn from an intersection to the endpoint of a line. In **Figure 37.145** a line is drawn from <u>P1</u> tangent to the circle by using the *Tangent* option of *Osnap*. The *Tangent* option can also be used to draw a line tangent to two arcs.

The *Node* option snaps to a *Point*, the *Quadrant* option snaps to one of the four compass points on a circle, the *Insert* option snaps to the intersection point of a *Block*, and the None option turns off *Osnap* for the next selection. The *Quick* option reduces searching time by selecting the first object encountered rather than searching for the one closest to the aperture's center.

Osnap settings can be temporarily retained as "running" *Osnaps* for repetitive use. One way to set running *Osnaps* is to right click on

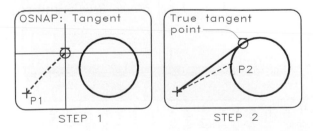

37.145 Osnap: Tangent.

Step 1: *Draw* toolbar> *Line* icon> *Line from point*: <u>P1</u>

Specify next point or [Undo]:> *Tangent* icon>

Step 2 *Tan to*: <u>P2</u>. The line is drawn from P1 tangent to the true tanget point on circle nearest P2.

Osnap tab in the status line at the bottom of the screen, select *Settings*, to display the *Drafting Settings* box that lists the status *Osnap* options. Settings can be added or removed, from this list of *Osnap* options.

Now the cursor has an aperture target at its intersection for picking endpoints and centers of arcs. The *Osnap* command can be turned on or off by selecting the *Osnap* button in the status line at the bottom of the screen. From the *Drafting Settings* box, select the <u>Option</u> button and the *Options* box will appear from which you can set Osnap aperture sizes from 1 to 50 pixels for snapping to objects.

37.43 Array

The *Draw* toolbar> *Array* icon is used to draw rectangular patterns (rows and columns) or polar layouts of selected drawings using the *Array* dialogue box (**Figure 37.146**). A series of holes can be drawn on a bolt circle by drawing the first hole and arraying it as a polar array (**Figure 37.147**).

A rectangular array is begun by making the drawing in the lower left corner and following the steps in **Figure 37.148**. Rectangular arrays may be drawn at angles by using *Command*: <u>Snap</u>> <u>Rotate</u> to rotate the grid. The first object

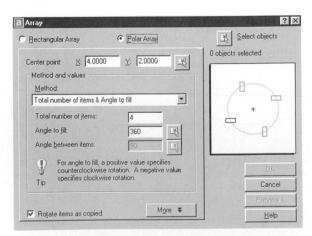

37. 146 Select the *Draw* toolbar> *Array* icon to obtain the *Array* box in which to specify rectangular or polar arrays.

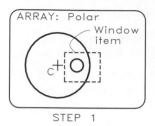
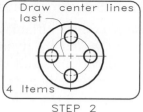

37.147 Array: Polar.

Step 1 *Modify* toolbar> *Array* icon> *Select objects*: <u>W</u> (Enter) *Window the hole*> *Select objects*: *Enter the type of array [Rectangular/Polar] <R>*: <u>P</u> (Enter)
Specify center point of array: <u>C</u> (Enter)

Step 2 *Enter the number of items in the array*: <u>4</u> (Enter)
Specify the angle to fill(+=ccw, -=cw) <360>: <u>360</u> (Enter)
Rotate arrayed objects? [Yes/No] <Y>: (Enter)

is drawn in the lower left corner of the array and the number of rows, columns, and the cell distances are specified when prompted.

37.44 Scale

The *Scale* command reduces or enlarges previously drawn objects. The desk in **Figure 37.149** is enlarged by windowing it, selecting a base point, and typing a scale factor of <u>1.6</u>. The drawing and its text are enlarged in both the x- and y-directions.

A second option of *Scale* lets you select a length of a given object, specify its present

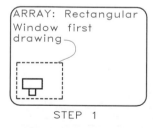
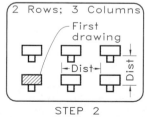

37.148 Array: Rectangle.

Step 1 *Modify* toolbar> *Array* icon> *Select objects*: <u>W</u> (Enter) *Window the desk.*> *Select objects*: *Enter the type of array [Rectangular/Polar] <R>*: <u>R</u> (Enter)

Step 2 *Enter the number of rows (---) <1>*: <u>2</u> (Enter)
Enter the number of columns (|||) <1>: <u>3</u> (Enter)
Enter the distance between rows or specify unit cell (---) <1>: <u>4</u> (Enter), *Specify the distance between columns (|||) <1>*: <u>3.5</u> (Enter)

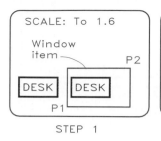
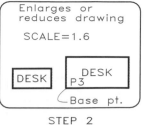

37.149 Scale: Numeric.

Step 1 *Modify* toolbar> *Scale* icon> *Select objects*: <u>W</u> (Enter) *Window the desk with* <u>P1</u> *and* <u>P2</u>.

Step 2 *Specify base point*: <u>P3</u> *Specify scale factor or [Reference]*: <u>1.6</u> (Enter)

The desk is drawn 60% larger.

length, and assign a length as a ratio of the first dimension (**Figure 37.150**). The lengths can be given by the cursor or typed at the keyboard in numeric values.

37.45 Stretch

The *Stretch* command (*Modify* toolbar> *Stretch* icon) lengthens or shortens a portion of a drawing while one end is left stationary. The window symbol in the floor plan **Figure 37.151** is *Stretched* to a new position, leaving the lines of the wall unchanged. A *Crossing Window* must be used to select lines that will be stretched.

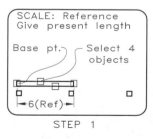
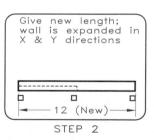

37.150 Scale: Reference.

Step 1 *Modify* toolbar> *Scale* icon> *Select objects*: <u>Select 4 objects</u>,
Base point: Select <Scale factor>/ Reference: <u>R</u> (Enter)
Reference length <1>: <u>6</u> (Enter)

Step 2 *New length*: <u>12</u> (Enter)

The drawing is enlarged in all directions.

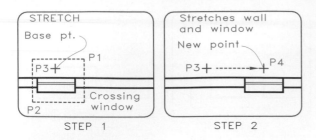

37.151 Stretch.

Step 1 *Modify* toolbar> *Stretch* icon> *Select objects*: Use <u>P1</u> and <u>P2</u> to form crossing window.

Specify base point or displacement: <u>P3</u>

Step 2 *Specify second point of displacement*: <u>P4</u> The windowed portion of the drawing is repositioned.

37.46 Rotate

A drawing can be rotated about a base point by using the *Rotate* command (*Modify* toolbar> *Rotate* icon) as shown in **Figure 37.152**. *Window* the drawing, select a base point, and type the rotation angle or select the angle with the cursor. Drawings made on multiple layers can be rotated also.

37.47 Setvar

Many system variables (several hundred) can be inspected by typing *Setvar* and *?* at the *Command* line, and changed if they are not read-only commands. To change one or more variables (*Textsize*, for example), respond as follows:

Command: <u>Setvar</u> (Enter)
Variable name or ?: <u>Textsize</u> (Enter)
New value for TEXTSIZE <0.18>: <u>0.125</u> (Enter)

By entering the *Setvar* command with an apostrophe in front of it ('*Setvar*), it can be used transparently without exiting from the command in progress.

37.48 Divide

The *Divide* command (*Main Menu> Draw Menu> Point> Divide*) places markers on a line to show a specified number of equal divisions. The line in **Figure 37.153** is selected by the cursor, the number of divisions is specified, and markers are equally spaced along it. The mark-

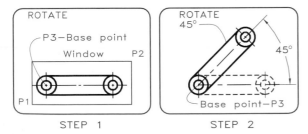

37.152 Rotate.

Step 1 *Modify* toolbar> *Rotate* icon> *Select objects*: Window part with <u>P1</u> and <u>P2</u>.

Step 2 *Specify base point*: <u>P3</u>

Specify rotation angle or [Reference]: <u>45</u> (Enter)

Object is rotated 45° counterclockwise.

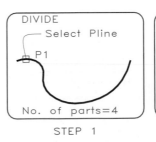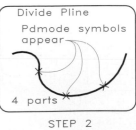

37.153 Divide.

Step 1 *Command*: <u>Divide</u> (Enter)

Select object to divide: <u>P1</u>

Enter the number of segments or [Block]: <u>4</u> (Enter)

Step 2 Four *Pdmode* symbols are placed along the line dividing into equal divisions.

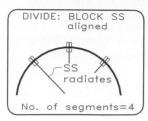
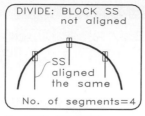

A. ALIGNED   B. NOT ALIGNED

37.154 Divide.

A. *Command*: <u>Divide</u> (Enter)

Select object to divide: <u>Select arc</u>.

Enter the number of segments or [Block]: <u>B</u> (Enter)

Enter the name of block to insert: <u>S</u> (Enter)

Align block with oject? [Yes/No]:<Y>: <u>N</u> (Enter)

Enter the number of segments: <u>4</u> (Enter)

B. *Align block with object? <Y>*: (Enter) Blocks are drawn to radiate from the arc's center.

ers will be of the type and size currently set by the *Pdmode* and *Pdsize* variables under the *Setvar* command. Set *Pdmode* to <u>3</u> to get an X when the *Point* command is used.

The *Block* option of *Divide* allows saved blocks (rectangles, used here) to be used as markers on the line (**Figure 37.154**). *Blocks* can be either *Aligned* or *Not Aligned* as shown.

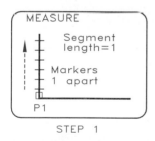
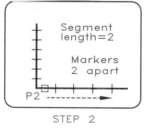

STEP 1   STEP 2

37.155 Measure.

Step 1 *Command*: <u>Measure</u> (Enter)

Select object to measure: <u>P1</u>

Specify length of segment or [Block]: <u>1</u> (Enter)

Markers are placed 1 apart starting at end nearest P1.

Step 2 *Command*: <u>Measure</u> (Enter)

Specify length of segment or [Block]: <u>2</u> (Enter)

Markers are placed 2 units apart starting at end nearest P2.

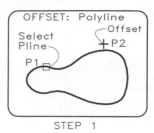
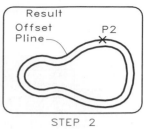

STEP 1   STEP 2

37.156 Offset.

Step 1 *Command*: <u>Offset</u> (Enter)

Specify offset distance or [Through] <1>: <u>T</u> (Enter)

Select object to offset or <exit>: <u>P1</u> (Enter)

Step 2 *Specify through point*: <u>P2</u> (Enter).

An enlarged Pline is drawn that passes through P2.

37.49 Measure

The *Measure* command (*Main Menu> Draw> Measure*) repeatedly measures off a specified distance along an arc, circle, polyline, or line and places markers at these distances (**Figure 37.155**). Respond to the *Select object to measure* prompt by picking a point near the end where measuring is to begin. When prompted, give the segment length, and markers are displayed along the line. The last segment is usually a shorter length.

37.50 Offset

An object can be drawn parallel to and offset from another object, such as a polyline, by the *Offset* command (**Figure 37.156**). *Offset* prompts for the distance or the point through which the offset line must pass, and then prompts for the side of the offset. The *Offset* command is helpful when drawing parallel lines to represent walls of a floor plan.

37.51 Blocks

One of the most productive features of computer graphics is the capability of creating drawings called *Blocks* for repetitive use. *Draw Menu> Block> Make> Block definition* box is used to make a block (**Figure 37.157**). The SI

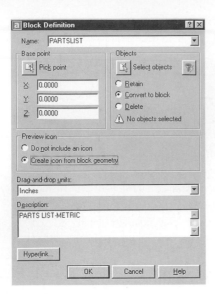

37.157 *Draw Menu> Block> Make* will open this *Block* Definition box on the screen for making a *Block*.

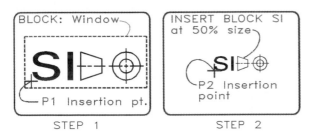

37.158 Block.

Step 1 *Draw Menu> Block> Make> Get Block Definition* box> *Name*: SI> Select *Pick point* button> P1

Pick *Select objects* button> Window drawing (Enter)

Step 2 *Command:*> Insert (Enter)> *Insert* box> *Name*: SI> *Scale*: 25.4> Check *Uniform Scale* box> Ok> *Specify insertion point*: P2. The block is inserted.

symbol in **Figure 37.158** is a typical drawing that is made into a *Block* and inserted into drawings using icons from the *Draw* toolbar shown in **Figure 37.159**. The *Insert* heading on the *Main Menu* (*Insert> Block*) gives the *Insert* dialogue box (**Figure 37.160**) for inserting *Blocks*. You can select a block from the window or *Browse* for other files to insert as Blocks (**Figure 37.161**).

When a *Block* is selected on the screen (to be *Moved* for example), it is selected as a total

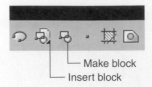

37.159 The *Make block* and *Insert block* icons can be selected from the *Draw* tool-bar.

— Make block
— Insert block

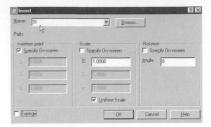

37.160 Select *Main Menu> Insert>* to get *Insert box* to specify *Blocks* or *Files* (*Wblocks*) for insertion in a drawing.

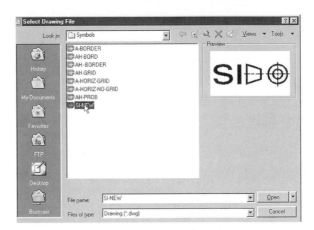

37.161 To insert a block, use *Command*: Insert (Enter)> *Browse*> then select a defined block from the list> *Open*> select an insertion point on the screen with the cursor (Enter).

unit. However, *Blocks* that were *Inserted* by selecting the *Explode* box first, or by typing a star in front of the *Block* name (*SI, for example), can be selected one object at a time. An inserted *Block* can be separated into individual entities by typing Explode at the *Command* line and selecting the *Block*.

37.52 Write Blocks (Wblocks)

Blocks are parts of files that can be used only in the current drawing file unless they are converted to *Wblocks* (*Write Blocks*) which are independent files, not parts of files. This con-

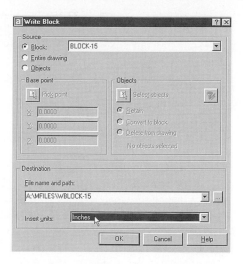

37.161A *Command*: <u>W</u>block (Enter) and the *Write Block* box enables you to select a *Block* as the *Source* and convert it to a *Wblock* with its *File name and path* specified in the lower window. Non-*Block* drawings can also be made into *Wblocks* by checking the *Objects* box under *Source*.

version is performed by typing <u>Wblock</u> at the *Command* line to get the *Write Block* dialogue box (**Figure 37.161A**). Select the *Block* button and type the name of the *Block* that is to be converted to a file in the window of the *Source* area. Give the *File name and path* in the *Destination* area, and press *OK*.

Blocks can be redefined by selecting a previously used Block name to receive the prompt, *Redefine it? <N>*: <u>Y</u> (*Yes*) and select the new drawing to be blocked. After doing so, the redefined *Block* automatically replaces the one in the current drawing with the same name. *Command*: <u>Insert</u>> *Block> Browse* can be used to display thumbnail illustrations of *Wblocks* (*Files*) as illustrated in **Figure 37.161**.

37.53 View

Portions of drawings can be saved as separate views by using *Command*: <u>View</u> (**Figure 37.162**) and the *View* box appears. The entire screen can be made into a *View* by picking the *New* option and naming it when prompted.

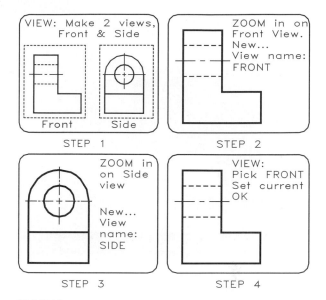

37.162 View.

Step 1 Make a two-view drawing for saving as separate views.
Step 2 *Zoom* the front view to fill the screen.
Command line: <u>View</u> (Enter)
?/ Delete/ Restore/ Save/ Window: <u>Save</u> (Enter)
View name to save: <u>Front</u> (Enter)
Step 3 *Zoom* the side view to fill the screen.
Command line: <u>View</u> (Enter)
View name to save: <u>Side</u> (Enter)
Step 4 To display a view, *Command* line: <u>View</u> (Enter)
?/ Delete/ Restore/ Save/ Window: <u>R</u> (Enter)
View name to restore: <u>Side</u> (Enter) The view is displayed.

The *Window* option makes a *View* of the windowed portion of the drawing. To display a view, type <u>View</u>, select its name from the list of named *Views> Set Current> OK*.

37.54 Inquiry Commands

From *Main Menu> Tools> Inquiry* to obtain information about objects and files with *Dist, Area, Mass Properties, List, ID Point, Time, Status*, and *Set Variables* (**Figure 37.163**). List is selected (or *Command*: <u>List</u>) and the circle (or any object) is selected when prompted to obtain information about it (**Figure 37.164**).

37.163 *Main Menu> Tools> Inquiry Menu* offers these options to assist you in learning more about a particular drawing.

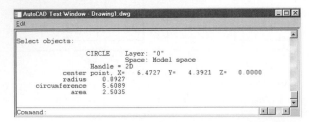

37.164 Use *Main Menu> Tools> Inquiry* toolbar> *List* icon> and select a circle on the screen. This box appears and gives the center of the circle, its radius, circumference, and area.

Dist measures the distance, its angle, and its delta-x and delta-y distances between selected points without drawing a line. *ID* gives the x-, y-, and z-coordinates of a point that is picked on the screen. *Area* gives the perimeter and of a space enclosed on the screen. Prompts request *First point:*, *Next point:*, *Next point:*, and so on to pick all points, then press (Enter). *Areas* can be added and removed when they are being selected as shown in **Figure 37.165**.

The *Status* option gives information about the settings, layers, coordinates, and disk space. *Time* displays information about the time spent on a drawing (**Figure 37.166**). The timer can be *Reset* and turned *On* to record the time of a drawing session, but the cumulative time cannot be erased without deleting the drawing file. After *Resetting, Display* shows the time of the current session at the heading, *Elapsed* time:.

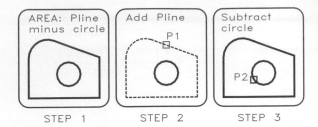

37.165 Area.

Step 1 Draw the object with a *Polyline* outline and a circular hole in it.

Step 2 *Main Menu> Tools> Inquiry> Area> <First point>/ Specify first corner point or [Object/Add/Subtract]*: Area (Enter)

Specify first corner point or [Object/Subtract]: O (Enter)

(ADD mode) Select objects: P1

Area= 4.0300 Perimeter =24.8560, Total area = 4.0300 (Enter)

Step 3 *Specify first corner point or [Object/Subtract]*: S (Enter), *Specify first corner point or [Object/Add]*: O (Enter)

(SUBTRACT mode) Select objects: P2

Area= 0.4910, Perimeter= 7.8540, Total area= 3.5400

37.55 Dimensioning

Figure 37.167 shows common types of dimensions that are applied to drawings. **Drawings should be drawn full size on the screen** to simplify their measurements and dimensions; scaling should be done at plotting time.

Dimensions can be applied as *Associative* or as *Nonassociative* (*Exploded*) dimensions. *Associative* dimensions (when *Dimaso* is *On*) are inserted as if the dimension line, extension lines, text, and arrows were parts of a single

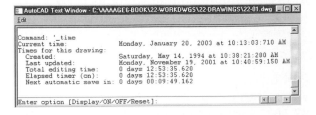

37.166 The *Time* option of the *Inquiry menu* can be used for inspecting the time spent on a drawing and for setting the time for an assignment.

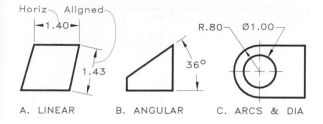

37.167 The basic types of dimensions that may appear on a drawing are shown here.

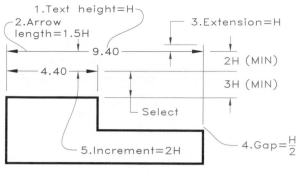

DIM VARS	1 DIMTXT (TEXT HT.)=H=.125
	2 DIMASZ (ARROW)=1.5H=.125−.160
	3 DIMEXE (EXTENSION)=H=.125
	4 DIMEXO (OFFSET)=H/2=.06
	5 DIMDLI (INCREMENT)=2H=.25 MIN.
	6 DIMSCALE (SCALE)=1 for INCHES
	25.4 for MILLIMETERS

37.169 Dimensioning variables are based on the height of the lettering (text) which is usually about 1/8-in. high.

Block. Exploded dimensions (*Dimaso* set to *Off*) are applied as individual objects that can be modified independently. Except where noted, the examples that follow will be associative dimensions.

Many variables must be set before dimensioning is usable: Arrowheads and numerals must be sized, extension line offsets specified, text fonts assigned, and units adopted, to name a few.

37.56 Dimension Style Variables

To get a list of the current dimensioning variables as shown in **Figure 37.168**, *Command*: Dim> *Dim*> Status (Enter). Sizes of dimensioning variables are based on the letter height, which most often is 0.125" (**Figure 37.169**).

To set and save variables needed for basic applications *Open* (or *Create*) the file *A:PROTO1*. Each variable is set by typing Setvar and the name of the dimensioning variable (*Dimtxt* for text height, for example) and assigning a numerical value. A list of the

```
AutoCAD Text Window - Drawing6.dwg
Edit
DIMTOLJ      1             Tolerance vertical justification
DIMTP        0.0000        Plus tolerance
DIMTSZ       0.0000        Tick size
DIMTVP       0.0000        Text vertical position
DIMTXSTY     Standard      Text style
DIMTXT       0.1800        Text height
DIMTZIN      0             Tolerance zero suppression
DIMUPT       Off           User positioned text
DIMZIN       0             Zero suppression

Dim:
```

37.168 Use *Command*: Dim> *Dim*: Status (Enter) to obtain a listing of the dimension system variables. Only the first of the list is shown here.

```
                    ult   Description
DIMASO    Off       Create dimension objects
DIMSTYLE  Standard  Current dimension style
DIMADEC   −1        Angular decimal places
DIMALT    Off       Alternate units selected
DIMALTD   2         Alternate unit decimal places
DIMALTF   25.4      Alternate unit scale factor
DIMALTRND 2         Alternate units rounding value
DIMALTTD  2         Alternate tolerance dec. places
DIMALTTZ  0         Alternate tolerance zero suppress.
DIMALTU   2         Alternate units
DIMALTZ   0         Alternate unit zero suppression
DIMAPOST  −         Prefix & suffix for alternate text
DIMASZ    .125      Arrow size
DIMATFIT  3         Arrow and text fit
DIMAUNIT  0         Angular unit format
DIMAZIN   0         Angular zero suppression
DIMBLK    −         Arrow block name
DIMBLK1   −         First arrow block name
DIMBLK2   −         Second arrow block name
DIMCEN    .09       Center mark size
DIMCLRD   BYLAYER   Dimension line & leader color
DIMCLRE   BYLAYER   Extension line color
DIMCLRT   BYLAYER   Dimension text color
DIMDEC    4         Decimal places
DIMDLE    0         Dimension line extension
DIMDLI    .38       Dimension line spacing
DIMSEP    .         Decimal separator
DIMEXE    .125      Extension above dimension line
DIMEXO    .06       Extension line origin offset
DIMFRAC   0         Fraction format
DIMGAP    .06       Gap from dimension line to text
DIMJUST   0         Justification of text on dimem. line
```

37.170 Use *Command*: Dim> *Dim* Status (Enter) to obtain a listing of the dimension variables, their settings, and definitions.

dimensioning variables is given in **Figures 37.170** and **37.171**. Assign the basic variable values of *Dimtxt, Dimasz, Dimexe, Dimexo, Dimtad, Dimdli, Dimaso,* and *Dimscale* shown in **Figure 37.169** to A:PROTO1, since they

ult		Description
DIMLDRBLK	ClosedFilled	Leader block name
DIMLFAC	1	Linear unit scale factor
DIMLIM	Off	Generate dimension limits
DIMLUNIT	2	Linear unit format
DIMLWD	−2	Dimension lind & leader lineweight
DIMLWE	−2	Extension line lineweight
DIMPOST		Prefix & suffix for dimension text
DIMRND	0	Rounding value
DIMSAH	Off	Separate arrow blocks
DIMSCALE	1	Overall scale factor
DIMSD1	Off	Suppress first dimension line
DIMSD2	Off	Suppress second dimension line
DIMSE1	Off	Suppress first extension line
DIMSE2	Off	Suppress second extension line
DIMSOXD	Off	Suppress outside dimension lines
DIMTAD	0	Place text above the dimension line
DIMTDEC	4	Tolerance decimal places
DIMTFAC	1	Tolerance text height scaling factor
DIMTIH	On	Text inside extensions is horiz.
DIMTIX	Off	Place text inside extensions
DIMTM	0	Minus tolerance value
DIMTMOVE	0	Text movement
DIMTOFL	Off	Force line inside extension lines
DIMTOH	On	Text outside horizontal
DIMTOL	Off	Tolerance dimensioning
DIMTOLJ	1	Tolerance vertical justification
DIMTP	0	Plus tolerance
DIMTSZ	0	Tick size
DIMTVP	0	Text vertical position
DIMTXSTY	Standard	Text style
DIMTXT	.125	Text height
DIMTZIN	0	Tolerance zero suppression
DIMUPT	Off	User positioned text
DIMZIN	0	Zero suppression

37.171 Additional dimension variables are shown here as a continuation of Figure. 37.170.

apply to most applications. *Command:*> type Units and set decimal fractions to two decimal places for inches.

Save these settings to B:PROTO1 as an empty file with no drawings on it for use as the prototype with its dimensioning settings. Open B:PROTO1 and create a new file by *Save As* A:DWG-3, for example, which becomes the current file with the same settings as A:PROTO1.

Dimensioning variables can be set from dialogue boxes also instead of being typed; these techniques are covered later. You will be more proficient by becoming familiar with both methods of assigning variables. For now,

37.172 The *Dimension* toolbar has these dimensioning options from which to select.

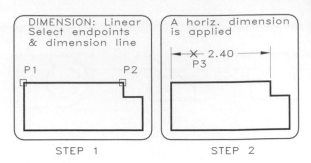

STEP 1   STEP 2

37.173 Dimensioning: Linear–Horizontal.

Step 1 *Main Menu> View> Toolbars> Dimension* toolbar> *Linear* icon> *Specify first extension line origin or <select object>:* P1 *Specify second extension line origin:* P2

Step 2 *Specify dimension line location or [Mtext/Text/Angle/Horizontal/Vertical/Rotated]:* H (Enter) *Specify dimension line location or [Mtext/Text/Angle]:* P3 (Enter)> Dimension text = 2.40> (Enter) to accept the dimension.

use the A:DWG-3 file to explore the fundamentals of dimensioning.

37.57 Linear Dimensions

The *Dimension* toolbar (**Figure 37.172**) provides a convenient means of selecting dimensioning commands. Select the *Linear* icon, the points as prompted, and the *Horizontal* option as shown in **Figure 37.173** in order to apply a horizontal dimension.

A dimension is applied semiautomatically in **Figure 37.174** by pressing (Enter) when prompted for *Endpoints*, selecting the line or

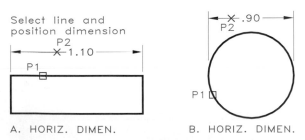

A. HORIZ. DIMEN.   B. HORIZ. DIMEN.

37.174 Dimensioning: Linear–Semiautomatic.

A. *Dimensioning* toolbar> *Aligned> Specify first extension line origin or <Select object>:* (Enter)
Select object to dimension: P1
Select dimension line location or [Mtext/Text/Angle]: P2
Dimension text = 1.10: (Enter)

B. Use these same steps and select a point on the circle (P1) to dimension its diameter.

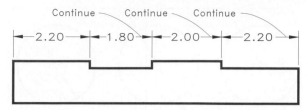

37.175 When dimensions are placed end to end, use the *Continue Dimension* option from the *Dimension* toolbar to specify the successive second extension line origins after the first dimension line has been drawn.

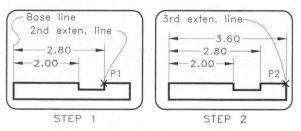

37.176 Dimensioning: Baseline option.

Step 1 *Dimension* toolbar:> *Linear* icon> place a 2.00 *Linear* dimension. (The first extension line becomes the baseline.) *Dimension* toolbar> *Baseline* icon> *Specify a second extension line origin or [Undo/Select] <Select>*: P1
Dimension text = 2.80 (Enter) The 2.80 dimension is drawn.
Step 2 *Specify a second extension line origin or [Undo/Select] <Select>*: P2 >*Dimension text = 3.60* (Enter) The 3.60 dimension is drawn. (Enter) to exit from last dimension.

circle to be dimensioned, and locating its dimension line.

Select *Main Menu> Dimension> Continue* (or type *Dimcontinue* at the *Command* line) to continue a chain of linear, angular, or ordinate dimensions from the last extension line (**Figure 28.175**). Baseline applies dimensions

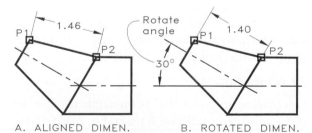

37.177 Dimensioning: Oblique lines.
A. Linear dimensions can be aligned by selecting the *Align* icon and selecting endpoints P1 and P2.
B. Linear dimensions can be rotated by selecting the *Linear* icon, the *Rotate* option, assign the angle (150°), and pick endpoints P1 and P2.

from a single endpoint and each dimension is offset incrementally by the dimension line increment variable, *Dimdli* (**Figure 28.176**).

Select the *Aligned* command from the *Dimension* toolbar and you will be prompted to select the *1st* and *2nd* extensions lines and the position of the dimension line (**Figure 37.177A**). The dimension line will be inserted aligned with line 1-2. Extension lines can be automatically drawn by pressing (Enter) at the first prompt and selecting the line to be dimensioned. A rotated dimension can be applied with the *Rotate* option and an assigned angle of rotation (**Figure 37.177B**).

37.58 Angular Dimensions

Figure 37.178 shows variations for dimensioning angles depending on the space available. In **Figure 37.179** an angle is dimensioned by selecting the *Angular* icon, selecting lines of the angle, and locating the dimension line arc as shown in Step 1. If space permits, the dimension text will be centered in the arc between the arrows (Step 2).

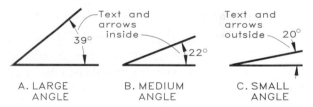

37.178 Angles will be dimensioned in any of these three formats.

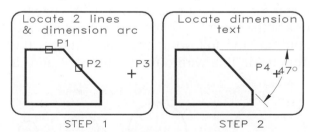

37.179 Dimensioning: Angles.
Step 1 Use *Dimension* toolbar> *Angular* icon> *Select arc, circle, line, or <specify vertex>*: P1 > *Select second line*: P2
Specify dimension arc line location or [Mtext/Text/Angle]: P3
Step 2 *Enter dimension text <47>*: (Enter)
Enter text location (or press ENTER): P4(Enter)

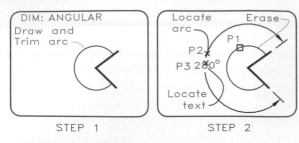

STEP 1   STEP 2

37.180 Dimensioning: Arcs.

Step 1 Draw an arc with its center at the vertex and *Trim* it to end at the lines.

Step 2 *Select Angular icon> Select arc, circle, line, or <specify vertex>:* P1 *> Specify dimension arc line location or [Mtext/Text/Angle]:* P2> *Dimension text = 280* (Enter)

An angular dimension can be applied by selecting its vertex and points on each line when prompted. **Figure 37.180** shows how angles over 180° can be dimensioned by drawing an arc with its center at the vertex with its ends abutting each line. Use *Dimension* toolbar> *Select arc, circle, line,* or *<specify vertex>:* P1> *Specify dimension arc line location or [Mtext/Text/Angle]:* P2> *Dimension text = 280* (Enter) to accept the angle of 280°

37.59 Diameter

Diameters of circles can be placed as shown in **Figure 37.181** depending on the available space. By changing system variables *Dimatfit, Dimtofl,* and *Dimtmove* circles can be dimensioned as shown in **Figures 37.182** and **37.183**. By setting the *Dimtix* system variable *On*, the text is forced inside the extension lines regardless of the available space. The *Dimtmove* variable has the following options: 0 moves the dimension line with dimension text; 1 adds a

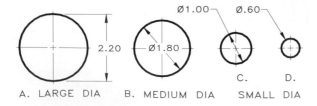

A. LARGE DIA   B. MEDIUM DIA   C.   D. SMALL DIA

37.181 Examples of methods of dimennsioning circles.

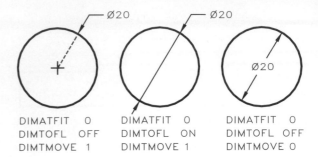

DIMATFIT   0   DIMATFIT   0   DIMATFIT   0
DIMTOFL OFF   DIMTOFL   ON   DIMTOFL OFF
DIMTMOVE 1   DIMTMOVE 1   DIMTMOVE 0

37.182 Examples of circle diameters with their associated dimensioning variables. Assign variables by *Command:* Dim> *Dim>* type name of variable (Enter).

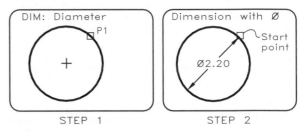

STEP 1   STEP 2

37.183 Dimensioning: Circle.

Step 1 *Dimension toolbar> Diameter icon> Select arc or circle:* P1

Step 2 *Dimension text= <2.20>* (Enter)

Specify dimension line location or [Mtext/Text/Angle]: Select start point (Enter)

leader when dimension text is moved; 2 allows text to be moved freely without a leader

The *Dimatfit* variable has the following options: *0* places both text and arrows outside extension lines; *1* moves arrows first, then text; *2* moves text first, then arrows; *3* moves either text or arrows, whichever fits best. The *Dimtofl (On)* dimension variable forces a dimension line to be drawn between the arrows when the text is located outside. To specify whether or not a dimension has a leader use *Command:* Dim> *Dim>* Dimtmove> 1 (Enter).

37.60 Radius

Select *Radius* from the *Dimension* toolbar to dimension arcs with an *R* placed in front of the text (R1.00 for example) as shown in **Figure 37.184**. Dimensioning an arc with a radius and leader is shown in **Figure 37.185**. The same dimensioning variable covered in section **37.59** apply to arcs as well as to diameters.

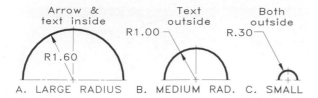

A. LARGE RADIUS   B. MEDIUM RAD.   C. SMALL

37.184 Arcs are dimensioned by one of the formats given here depending on the size of the radius.

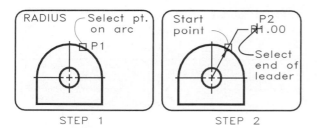

STEP 1   STEP 2

37.185 Dimensioning: Radius.

Step 1 *Command*: Dim (Enter)

Dim: Radius (Enter) *Select arc or circle*: P1

Step 2 *Dimension text <1.00>*: (Enter)

Specify dimension line location or [Mtext/Text/Angle]: P2

The *Leader* command (*Dimension Quick Leader* on the *Dimension* toolbar) is used to add a leader with a dimension or note to a drawing, but it cannot measure the circle; it inserts the value of the last measurement made (**Figure 37.186**). The arc's diameter or radius must be known and typed to override this measurement. Notes can be added in multiple lines at a specified length by using the *text width* option.

37.61 Dimension Style Manager

Select *Dimension Style* icon from the *Dimension* toolbar as shown in **Figure 37.186A** (or *Command*: DDim) to display the *Dimension Style Manager* dialogue box shown in **Figure 37.187** from which *Dimension Styles* can be created or modified and variables can be assigned from different tabs. Each group of settings can be made and saved by style name (*DEC DIM-2*, for example) for future use to eliminate the time and effort required to make new settings for each drawing.

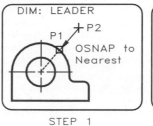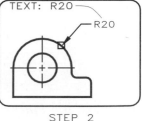

STEP 1   STEP 2

37.186 Dimensioning: Quick Leader.

Step 1 *Dimension* toolbar>*Quick Leader* icon> *Specify first leader point, or [Settings] <Settings>*: P1

Specify next point: P2> (Enter)

Specify text width <1.00>: (Enter)

Step 2 *Enter first line of annotation text <Mtext>*: R20 (Enter) *Enter next line of annotation text*: (Enter) Note: The *Leader* command does not make measurements, you must insert them.

37.186A Use *Dimension* toolbar> *Dimension Style* Icon to get the *Dimension Style Manager* box from which to create a new *Dimension Style* that will appear in the window with the *Standard* style.

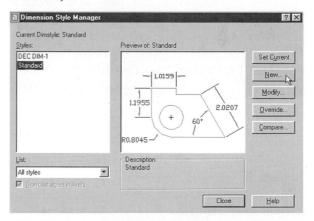

37.187 Use *Dimension* toolbar> *Dimension Style* icon> to get this *Dimension Style Manager* box for setting dimensioning variables. Pick the *New* button to create a new *Style*.

Lines and Arrows Tab

Click on the *New* box (**Figure 37.187**) to get the *Create New Dimension Style* box in which you can name make modifications on the *DEC DIM-2* style (**Figure 37.187A**). Pick Continue

37.187A In the *New Style Name* window, type the name of the new style, <u>DEC DIM-2</u>, and pick the *Continue* button.

and select the *Lines and Arrows* tab shown in **Figure 37.188**. From here, settings can be made for *Dimension lines*, *Extension lines*, *Arrowheads*, and *Center marks*. When *Oblique-Stroke* arrows are used, the value placed in the *Arrow size* box specifies the distance the dimension line extends beyond the extension line. The *Baseline spacing* box is used to set *Dimdli* which controls the spacing between baseline dimensions. The *Color* button displays the color menu from which to select a color for the dimension line (*Dimclrd*).

The options of the *Extension Lines* group (**Figure 37.188**) control the variables: *Lineweight*, *Extend beyond dim lines*, *Color*, and *Offset from origin*. The *Suppress 1st* and *2nd* boxes turn *On* the *Dimse1* and *Dimse2* variables to suppress the first and second extension lines. The value typed in the *Extension* box specifies the distance the extension line extends beyond the dimensioning arrowhead (*Dimexe*). The *Origin Offset* option is used to specify the size of the gap between the object and the extension line (*Dimexo*). The *Color* button lets you select the color of the extension lines (*Dimclre*).

The options of the *Arrowheads* group (**Figure 37.189**) control the following variables: *Dimasz, Dimtsz, Dimblk1*, and *Dimblk2*. The value typed in the *Size* box gives the size of the arrowhead (the *Dimasz* variable). By selecting scroll arrows next to the *1st* or *2nd* boxes, the types of arrowheads for each end of the dimension are listed (**Figure 37.189**). If only the *1st* arrow type is selected, it is automatically applied to the second end unless a *2nd* arrow type is specified. *Tick marks* are given when *Oblique* is selected, the *Dimtsz*

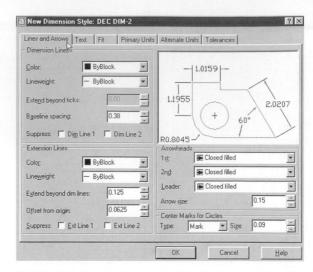

37.188 After the *Continue* button has been picked, the *New Dimension Style: DEC DIM-2* box appears. Select the *Lines* and *Arrows* tab for making the settings in the windows.

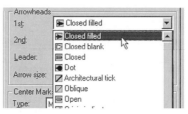

37.189 From the *Arrowheads* area, a drop-down listing of *Arrowheads* is given from which to select.

variable. When *User Arrow* is selected custom-made arrows can be inserted (*Dimblk1* and *Dimblk2*).

In the area labeled *Center Marks for Circles*, select from the the *Type* window *Mark, Line*, or *None* to draw center marks, center marks with center lines, or neither on when arcs or diameters are dimensioned. Center marks and center lines are applied to arcs and circles in the same manner without dimensions by *Command*: <u>Dim</u>> *Dim> Center>* select the object.

Text Tab

Select *Dimension* toolbar> *Dimension Style> Modify>* Text tab to get the menu shown in **Figure 37.190** which controls the dimensioning text. *Text Appearance* includes settings for

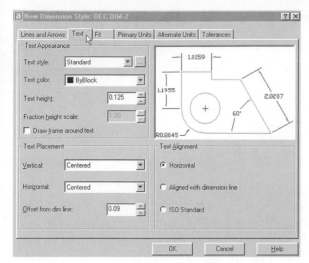

37.190 From the *New Dimension Style: DEC DIM-2* box, pick the *Text* tab and make changes in text variables.

Text color, Text style, and *Text height.* The *Text Placement* area gives windows for specifying *Vertical* and *Horizontal* text placement and the *Offset from dim.* line. The *Text Alignment* section lets you set text as *Horizontal* (unidirectional), *Aligned* with dimension line, or Iso, which aligns text with the dimension line when text is inside the extension lines, but aligns it horizontally when text is outside (**Figure 37.191**). *Text Styles* can be selected from the pull-down window or created by pressing the button next to the style window. *Text Placement* options let dimensioning text to be positioned as shown in **Figures 37.192** through **37.194**.

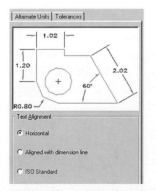 

37.191 These examples show the results of having the *Unidirectional* (*Horizontal*) and *Aligned* text in dimension lines.

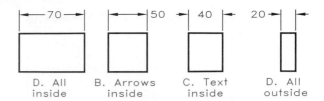

37.192 Examples of dimension applications when the *Best Fit* option is used.

37.193 Examples of *Text Placement* settings.

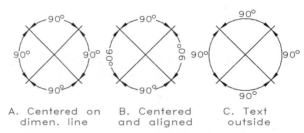

37.194 Examples of *Text Placement* settings on arcs.

Fit Tab

Under the *Fit* tab, *Fit options, Text Placement, Scale for Dimension Features,* and *Fine Tuning* adjustments can be made (**Figure 37.195**). The *Use overall scale of:* box controls the scale of all the dimensioning variables on the screen arrow size, text height, extension-line offsets, center size, and others. It is very useful in changing the units of dimensioning from English to metric by entering a scale factor of 25.4.

Primary Units Tab

Select the *Primary Units* tab to obtain options for *Linear Dimensions, Measurement Scale, Zero Suppression,* and *Angular Dimensions* (**Figure 37.196**). Use the scroll arrow under *Precision,* and select the number of decimals places (or fractions) desired. From the *Angle* area, the

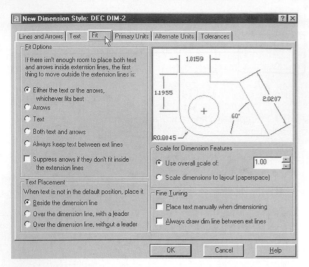

37.195 Select *Dimension Style DEC DIM-2> Fit* tab to get this menu for placing text on dimensions.

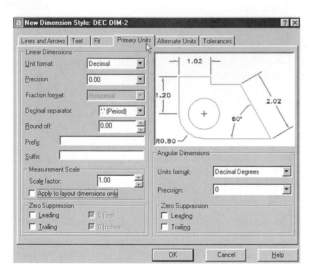

37.196 *New Dimension Style: DEC DIM-2> Primary Units* tab lets you make changes to the dimensioing text settings.

Units Format of degrees and their *Precision* can be specified in the same manner as above.

Entries in the *Zero Suppression* (*Dimzin*) boxes suppress zeros that are leading or trailing decimal points. Select *Leading* to make 0.20 become .20. **Figure 37.197** shows the results of applying the four options to architectural units. From under *Measurement Scale Factor* (*Dimlfac*) units for measurements and be specified.

O Ft	1/4"	4"	1'	1'-01/4"
O In				
No options	0'-01/4"	0'-6"	2'-0"	1'-01/4"
O In	0'-01/4"	0'-4"	1'	1'-01/4"
O Ft	1/2"	4"	1'-0"	1'-01/4"

37.197 The *Zero Suppression* (*Dimzin*) options control the leading and trailing zeros in dimensioning, especially when applied to architectural dimensions.

Alternate Units Tab

Select the *Display Alternate Units* button and two dimensions are appear on each dimension as in the example drawing in **Figure 37.198**. *Unit Format* and *Precision* are use to define units and decimal points. If *Multiplier for all units* is set to 25.4, the millimeter equivalents for inches are given as the alternate dimension. Examples of dimensions with alternate units are shown in **Figure 37.199**.

Values can be placed in the *Prefix* and *Suffix* panels (**Figure 37.198**) so dimension can have text (such as inches) before or after the dimensions (such as 26 mm or 42 inches). *Placement* options allow alternate dimensions to be placed over or after primary units.

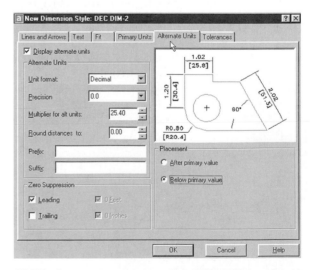

37.198 The *New Dimensioning Style: DEC DIM-2> Alternate Units* tab gives this window for assigning alternate and dual dimensions.

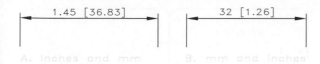

37.199 Examples of alternate units (dual dimensions) made in inches and millimeters.

Tolerances Tab

The *Tolerance Format* in **Figure 37. 200** can be selected from options of *None*, *Symmetrical*, *Deviation*, *Limits*, and *Basic*. Examples of applications of these options are shown in **Figure 37.201**. The number of decimal points is set with *Precision*, and *Upper* and *Lower* values are selected from their respective pull-down menus. *Height* is specified as a ratio of the primary text height, the basic dimension, and is recommended to be about 80%.

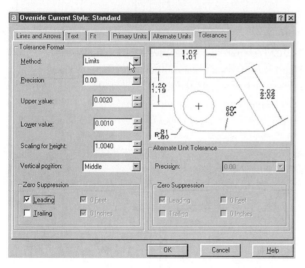

37.200 Select the *Tolerance* tab, pick *Limits* in *Tolerance Format* area, and provide the *Upper value* and *Lower value* for toleranced dimensions.

2.0000 ±.0020	2.0000 +.0030 −.0020	2.0030 1.9980
DIMTP & DIMTM SAME	DIMTP & DIMTM DIFFERENT	DIMLIM

37.201 Examples of various formats for tolerancing using the names of the dimensioning variables.

37.62 Saving Dimension Styles

When the settings of this new *Dimension Style* is complete, press the *OK* button at the current tab and the first menu of the *Dimension Style Manager* will be saved and will appear on the screen showing the new style, DEC DIM-4, along with the *Standard* style (**Figure 37.186A**). Options of *Modify* (change any settings), *Override* (make temporary changes in a style, and *Compare* (list the settings side-by-side of any two selected styles) are available (**Figure 37.202**). An often used override is the change of the *overall scale factor* (*Dimscale*) by changing a single multipler (**Figure 37.203**). This text style can be used in the future with the assigned values.

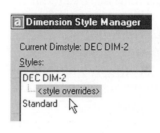

37.202 The new style, DEC DIM-2, is displayed in the list of styles. When a style is *Overridden* (values changed), a *<style overrides>* subheading is displayed beneath the primary style.

37.203 When *Dimscale* is changed from 1 to 2 and *Override* is used, the selected dimension is updated.

37.63 Editing Dimensions

When the dimensioning variable *Dimaso* is set to *On*, the dimensioning entities (arrows, text, extension lines, etc.) become a single unit (*Associative*) once a dimension has been inserted into a drawing.

A related dimensioning variable, *Dimsho*, can be set On to show the dimensioning numerals being dynamically changed on the screen as the dimension line is *Stretched*. When *Dimaso* is *On* and *Dimsho* is *Off*, the numerals will be changed after the *Stretch* (**Figure 37.204**), but not dynamically during the *Stretch*.

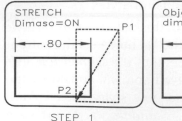
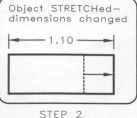

37.204 Dimensiong: Associative–Dimassoc.

Step 1 *Command*: <u>Dim</u>> *Dim*> <u>Dimassoc</u>> *Enter new value for DIMASSOC <1>*: <u>1</u> *(On) and apply dimensions to the part. Use the* Stretch *command with a* Crossing *window at the end of the part.*

Step 2 *Stretch* the side of the window to a new position; the object will be lengthened, a new dimension dynamically calculated, and be shown in its final position.

When in the associative dimensioning mode, the *Dimension edit* can be selected from *Dimension* toolbar to obtain the options of *Home, New, Rotate,* and *Oblique* for changing existing dimensions. *Home* repositions text to its standard position at the center of the dimension line after being changed by the *Stretch* commands (**Figure 37.205**). *New* changes text within a dimension line (**Figure 37.206**) by pressing (Enter) when prompted and inserting the new text. *Rotate* positions dimension text at any specified angle (**Figure 37.207**). *Oblique* converts extension lines to angular lines (**Figure 37.208**).

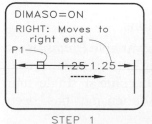
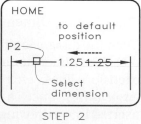

37.205 Dimensioning: Associative–Left, right, home.

Step 1 *Dimension* toolbar> *Dimension text edit* icon> *Select dimension* <u>P1</u>, *Enter text location (Left/ Right/ Home/ Angle)*: <u>Right</u> (Enter) Numeral moves to right.

Step 2 *Command*: (Enter) *Select dimension*: <u>P2</u>

Specify new location for dimension text or [Left/Right/Center/ Home/Angle]: <u>Home</u> (Enter) Numeral moves to the midpoint of the dimension line.

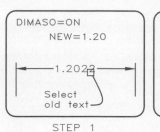
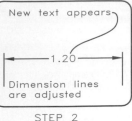

37.206 Dimensiong: Associative–New text.

Step 1 *Dimension* toolbar> *Dimension Edit* icon>

Enter type of dimension editing [Home/New/Rotate/Oblique] <Home>: <u>New</u> (Enter)> *Select objects*: <u>pick text</u>

Step 2 Type new text inside the brackets of the*Type Formatting* window and pick <u>Ok</u>> *Select objects*: Pick a point on the dimension line and the text is changed.

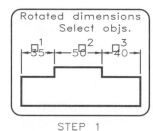
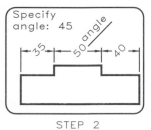

37.207 Dimensioning: Associative–Oblique extension lines.

Step 1 Dimension as usual with vertical extension lines.

Dimension toolbar> *Dimension Text Edit* icon>

Select dimension: pick <u>1</u>, <u>2</u>, and <u>3</u> (Enter)

Specify new location for dimension text or [Left/Right/Center/ Home/Angle]: <u>Angle</u> (Enter)

Step 2 *Specify angle for dimension text*: <u>45</u> (Enter)

Text is angled at 45° counterclockwise.

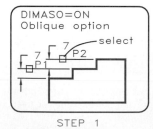
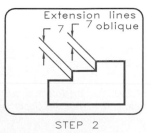

37.208 Dimensioning: Associative–Oblique extension lines.

Step 1 Dimension as usual with vertical extension lines.

Dimension toolbar> *Dimension Edit* icon>

Enter type of dimension editing [Home/New/Rotate/Oblique] <Home>: <u>Oblique</u> (Enter)

Select objects: <u>P1</u> and <u>P2</u> (Enter)> *Select objects*: 2 found

Step 2 *Enter obliquing angle (press ENTER for none)*: <u>-45</u> (Enter) The text is rotated 45° counterclockwise.

37.64 Dimensions with Tolerances

Dimensions can be toleranced automatically using the *Dimension Styles* defined previously in which the settings shown in **Figure 37.209** were made: *Dimtol* (tolerance on), *Dimtp* (plus tolerance), and *Dimtm* (minus tolerance). When *Dimlim* is *On*, the upper and lower limits of the dimension are given (**Figure 37.210**).

Dimtfac is a scale factor that controls the text height of the tolerance values, which is usually about 80% of the basic dimension height.

$$2.0000 \pm .0020 \qquad 2.0000 \begin{array}{l} +.0030 \\ -.0020 \end{array} \qquad \begin{array}{l} 2.0030 \\ 1.9980 \end{array}$$

37.209 Dimensions can be toleranced in any of these three formats.

37.65 Geometric Tolerances

Geometric tolerances specify the permissible variations in form, profile, orientation, location, and runout. A typical geometric tolerance feature control frame is given in **Figure 37.211**.

From the *Dimension* toolbar, select the *Tolerance* icon to obtain the *Geometry Tolerance* box (**Figure 37.212**); pick the *Symbol* icon to display the *Symbol* menu (**Figure 37.213**); select the desired symbol, and pick

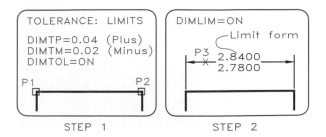

37.210 Dimensioning: Tolerance–Limit form.

Step 1: *Dimension* toolbar> *Dimension Style* icon> pick *DEC DIM-2*> *Modify*> *Tolerances* tab> set *Upper value (Dimtp)*, *Lower value (Dimtm)*, and set *Method* to <u>Limit</u> (Fig. 37.200). Apply the dimension line by selecting the endpoints <u>P1</u> and <u>P2</u> when prompted.

Step 2: Locate the dimension lines through <u>P3</u>; the dimension is shown with its upper and lower limits based on the values of *Dimtp* (tolerance plus) and *Dimtm* (tolerance minus).

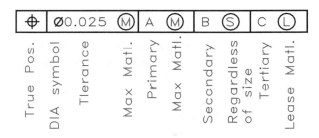

37.211 The parts of the feature control frame that give geometric tolerance specifications are defined here.

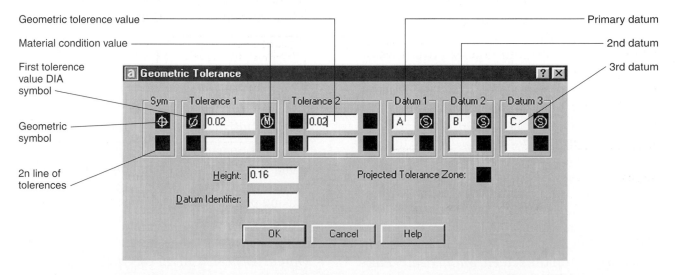

37.212 The *Geometric Tolerance* box (*Dimension* toolbar> *Tolerance* icon), is used to specify geometric tolerances.

37.213 The symbols of geometric tolerance are found in the *Symbol* box.

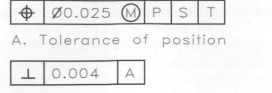

A. Tolerance of position

B. Tolerance of perpendicularity

37.214 Applications of *Geometric Tolerances* in feature control frames are shown here.

OK. Continue the process of responding to the prompts and icons to complete the geometric tolerance frame necessary for it to comply with the guidelines cover in Chapter 21. Examples of completed feature control frames as they would appear on a drawing are shown in **Figure 37.214**.

37.66 Custom Border and Title Block

Having covered most of the basics 2D drawing, you may wish to make your own customized title blocks with their own unique parameters that can be used to start up new drawings as introduced in Section 37.11. The format shown in **Figure 37.215** is used for laying out problems given at the ends of the chapters in this book.

To make a border identical to the one used in this example, draw the border 10.3 in. wide × 7.6 in. high, with its lower left corner at 0,0 and a polyline that is .03 thick. Draw a title strip across the bottom that has two rows of 1/8-in. text with spacing of 1/8-in. above and below each line. Fill out the title strip using the *Romand* font (double-stroke Gothic) and fill in all blanks (name, date, etc.); they will be changed later. Set *Snap* and *Grid* to 0.2 inches, set any other dimensioning variables that you expect to use based on a letter height of 1/8 inches. Save the drawing (*Main Menu> Save As> File Name>*: A:PROB-1.DWG) .

To create a new drawing *Main Menu> File> Open> A:PROB-1*, and the border and title block appear on the screen. *Main Menu> File> Save As> A:NEW-1* (Enter), and a new file

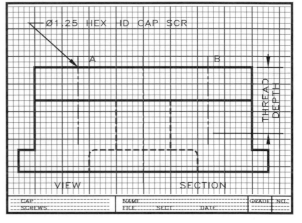

37.215 This is an example of a problem sheet border and title block that you can draw and save as a template file.

named A:NEW-1 is ready for drawing. Use *DDedit* (*Modify II* toolbar> *Text edit* icon) to select the title strip text that needs to be edited and update each entry. Save this setup to a file (*Main Menu> File> Save>*, and the drawing is updated and saved again as A:NEW-1, ready for drawing.

37.67 Digitizing with the Tablet

Drawings on paper can be taped to a digitizing tablet and digitized point-by-point, rather than using a mouse. A drawing is calibrated by picking points on the tablet with the cursor in the following manner:

Command line: <u>Tablet</u> (Enter)
Option (ON/OFF/CAL/CFG): <u>Cal</u> (Enter)
(Calibrate tablet for use.)

Digitize first known point: (Pick point.)
Enter coordinates for first point: 1,1 (Enter)
Digitize second known point: (Pick point.)
Enter coordinates for second point: 10,1 (Enter)
(This procedure set the scale for the remainder of the digitizing of the drawing.)

Digitize points from left to right or from bottom to top of the drawing. Using *On* or *Off* activates or deactivates the tablet. The function key F10 also turns the tablet off so the cursor can select from the screen menus. To draw lines, select the *Line* command from one of the menus (or type *L*), and pick points on the tablet with the stylus.

37.68 Sketch and Skpoly

The *Sketch* command can be used with the tablet for tracing drawings composed of irregular lines (**Figure 37.216**). Tape the drawing to the tablet and calibrate it as discussed in the previous section as follows:

Command: Sketch (Enter)
Record increment <0.1>: 0.01 (Enter)
Sketch. Pen eXit Quit Record Erase Connect .

The record increment specifies the distances between the endpoints of the connecting lines that are sketched. Other command options are:

Pen	raises or lowers pen.
eXit	records lines and exits.
Quit	discards temporary lines and exits.
Record	records temporary lines.
Erase	deletes selected lines.
Connect	joins current line to last endpoint.
(period)	draws a line from the current point to the last endpoint.

Begin sketching by moving your pointer to the first point, lower the pen (type *P*), trace over the line with the stylus, and the line is displayed on the screen. To *Erase*, raise the pen (type *P*), enter *Erase* (type *E*), move the stylus

backward from the current point, and select the point where the erasure is to stop. All lines are temporary until you select *Record* (type *R*) or *eXit* (type *X*) to save them. Begin new lines by repeating these steps. A series of lines drawn with the Sketch command can be erased one at a time. These segments can be converted to a polyline by linking them with the *Join* option of the *Pedit* command.

A series of lines drawn using the *Skpoly* variable and *Sketch* command are drawn in the same manner, but the lines are linked as if they were a single, continuous polyline. *Command:* Setvar> Skpoly> 1 *(On)*. Since a sketched line is a polyline, it can be edited by using *Fit, Spline, Width,* and the other options of this command.

37.69 Oblique Pictorials

An oblique pictorial can be constructed as shown in **Figure 37.217**. The front orthographic view is drawn and duplicated behind the front view at the angle selected for the receding axis with the *Copy* command. The visible endpoints are connected with *Osnap Endpoint,* and the invisible lines are erased.

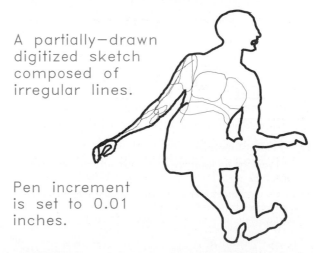

A partially–drawn digitized sketch composed of irregular lines.

Pen increment is set to 0.01 inches.

37.216 This drawing was made with the *Sketch* command and a tablet instead of with a mouse. The drawing was taped to the tablet and traced with the stylus at increments of .01 inches.

Circles are drawn as true circles on the true-size front surface, but circular features should be avoided on the receding planes since their construction is complex.

This oblique is not a true, three-dimensional drawing that can be rotated to view its different sides. It is merely a two-dimensional drawing with a single point of view.

37.70 Isometric Pictorials

To set the isometric grid *Command: Snap> Style> Isometric*. By using the same sequence of commands and typing *S* for *Standard*, the grid can converted back to the rectangular grid. The Isometric style shows the grid dots in an isometric pattern–vertically and at 30° with the horizontal (**Figure 37.219**). The cursor will align with two of the isometric axes and can be made to *Snap* to the grid by activating the *Snap* comand (*F9*). The axes of the cursor are rotated 120° by pressing *Ctrl E* or by *Command*: <u>Isoplane</u> (Enter). When *Ortho* is *On*, lines are forced to be drawn parallel to the isometric axes. **Figure 37.219** illustrates the steps for constructing an isometric drawing.

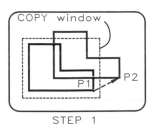
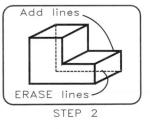

37.217 Pictorial: Oblique.

Step 1 Draw the front surface of the oblique and Copy the view from <u>P1</u> to <u>P2</u>.

Step 2 Connect the corner points with obliique lines from front to back and erase the invisible lines to complete the oblique. This is not a true 3D drawing but a 2D drawing with only one viewpoint.

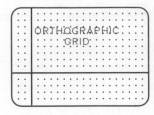
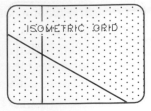

A. ORTHOGRAPHIC GRID   B. ISOMETRIC GRID

37.218 Screen Grids.

A The orthographic grid is called the *Standard* style. (*Command: Snap> Style> Standard>* (Enter))

B The isometric grid is specified as follows: *Command: Snap> Style> Isometric* (Enter)

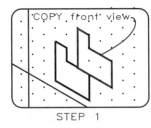
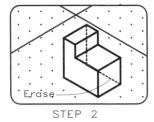

STEP 1   STEP 2

37.219 Pictorial: Isometric.

Step 1 Draw the front view of the object with the grid set to the *Isometric* style and *Copy* the view at its proper depth.

Step 2 Connect the corner points and erase the invisible lines. The cursor lines can be rotated into three positions using *Ctrl E* or by *Command*: <u>Isoplane</u> (Enter)

When the *Grid* is set to the *Isometric* mode, the *Ellipse* command displays the following options: *<Axis endpoint 1>/Center/Isocircle*: <u>Isocircle</u> (Enter). Selection of the *Isocircle* option enables isometric ellipses to be positioned in each of the isometric orientations shown in **Figure 37.220** by using *Ctrl E*. An example of an isometric with partial ellipses is shown in **Figure 37.221**.

The oblique and isometric drawings covered here are 2D drawings that appear to be 3D views, but they cannot be rotated on the screen to obtain different viewpoints of them.

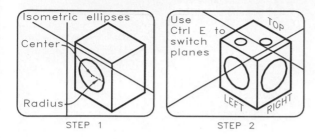

37.220 Pictorial: Isometric–isocircle.

Step 1 When in the isometric *Snap* mode:

Command: <u>Ellipse</u> (Enter)

<Axis endpoint 1> /Center/ Isocircle: <u>Isocircle</u> (Enter)

Center of circle: <u>Select center</u>.

<Circle radius>/ Diameter: Drag or type <u>radius size</u>.

Step 2 Other isocircle ellipses are drawn using these same steps, and by changing the cursor to correspond to each *Isoplane* of the box (*Ctrl E*).

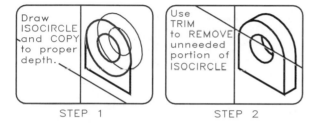

37.221 Pictorial: Isometric–partial isocircles.

Step 1 *Command* line: <u>Ellipse</u> (Enter)

<Axis endpoint 1> /Center/ Isocircle: <u>Isocircle</u> (Enter)

Center of circle: <u>Select center</u>.

<Circle radius>/ Diameter: Drag or type <u>radius size</u>.

Draw the outside and inside *Isocircles*.

Step 2 Use the *Trim* command to remove the unneeded portions of the *Isocircles*. If the *Trim* fails to respond in some cases, use the *Break* command to select points on the *Isocircle* for line removal. Add missing lines.

37.71 Summary

The coverage of two-dimensional computer graphics and AutoCAD fundamentals has been presented in this chapter to enable you to prepare working drawings. However, this coverage is, of necessity, very brief and is meant to serve as an introduction to the basics of computer graphics. Many options of menus, dialogue boxes, and toolbars could not be covered in this limited space. Therefore, you should experiment on your own with various options and commands as you progress in learning and applying the fundamentals.

The *Help* command on the *Main Menu* provides a valuable reference manual on the screen that will provide the answers to many of your questions when you wish to have them. This is one of the first menus with which you should become familiar. You will soon find that most of your knowledge of computer graphics will be self-taught.

Problems

It is suggested that problems be solved in a professional format within a border with a title block similar to the one illustrated in **Figure 37.215**.

1. A good first assignment would be the design of a prototype of a border and title block, as covered in Section 26.66. As this prototype is used and more variables are introduced, they can be added to the prototype file to preserve your time and effort that were required to make these settings.

2. Practically all problems at the ends of the chapters in this book lend themselves to solution by sketching, instrument drawings, or computer graphics.

3. Learn computer-graphics techniques and commands by repeating the instructions under the examples in this chapter at your computer. Experiment with related options not covered in the explanations.

38

3D Modeling
AutoCAD 2004

38.1 Introduction

This chapter provides an introduction to the principles of making true 3D pictorial drawings that can be rotated on the screen and viewed from any angle as if they were held in your hand. The three major divisions of this chapter are **fundamentals of 3D drawing**, **solid modeling**, and **rendering**.

AutoCAD 2004 provides several methods of executing most commands: The *Command* line, *Main Menu*, and *Toolbars*. Each of these techniques is used in the coverage that follows in this chapter. You are encouaged to use the online *Help* screen for each command and to experiment with various options.

38.2 Paper Space and Model Space

The two distinctly different ways of obtaining views of objects are from *Model Space* or *Paper Space*. When in *Model Space*, the screen can be divided into viewports (*Vports*) that abutt each other in standard arrangements, as do flooring

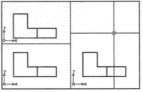

A. TILEMODE=1   B. TILEMODE=0

38.1 Model tab versus Layout tabs.

A When *Model* tab is current *(Tilemode = 1)*, the viewports are arranged to abutt each other like flooring "tiles."

B When *Layout1* is current, 3D viewports are made with *Vports* that can abutt or overlap.

tiles (**Figure 38.1A**). When in *Paper Space*, the viewports can be created with *Vports* in standard abutting or nonstandard positions (floating viewports) as shown in **Figure 38.1B**. By typing *Tilemode* at the *Command* line, abutting *Vports* can be turned on when <u>1</u> is typed or turned off when <u>0</u> (zero) is typed.

Model Space

Up until now, all examples have been given in *Model Space*, where by default, drawings can

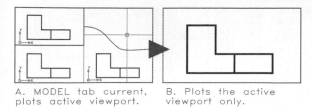

38.2 When the *Model* tab (bottom of screen) is current, only the active viewport can be plotted to paper.

A. MODEL tab current, plots active viewport.
B. Plots the active viewport only.

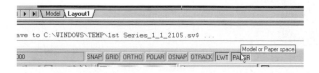

38.3 Select the *Layout 1* tab and you are in *Paper Space*, where a plotter and page layout can be specified. By alternately picking *Paper/Model* button in the *Status* line, you can move from *Paper Space (PS)* to *Model Space (MS)* for drawing in either 2D or 3D.

be made in two dimensions (2D) and three dimensions (3D). The *Vports* (viewports) command lets you create abutting viewports on the screen. Drawings made before applying the *Vports* command will be duplicated in each viewport as if multiple monitors were wired to your computer (**Figure 38.2**). Only the active viewport, shown with a heavy outline, can be plotted from Model Space.

Paper Space

Select the *Layout1* tab at the bottom of the screen to move into *Paper Space,* and the *Page Setup* menu appears from which to specify a page layout and chose a plotter **Figure 38.3**. A triangle icon appears at the lower left of the screen to signifiy that you are in 2D *Paper Space.* You can move to and from *Paper Space* and *Model Space* by typing <u>PS</u> or <u>MS</u> or by clicking on the *Paper/Model* button below the *Command* line. Since drawings made in *Paper Space (PS)* are 2D drawings, toggle to PS and insert or draw a 2D border to define the drawing area (**Figure 38.4A**).

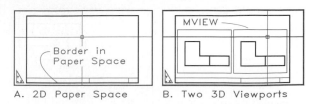

A. 2D Paper Space
B. Two 3D Viewports

38.4 Layout1 tab: Paper Space (PS).

A When the *Layout1* tab (bottom of screen) is picked, the screen is set to *Paper Space,* where a 2D border can be inserted or drawn.

B Use the *Mview* (or *Vports*) command to define the diagonal corners of two viewports into 3D space to create a combination of model and paper space.

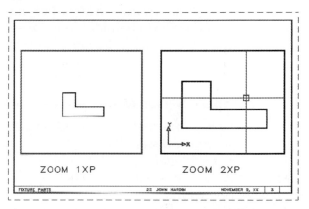

ZOOM 1XP   ZOOM 2XP

FIXTURE PARTS   22 JOHN HARDEN   NOVEMBER 2, XX   3

38.5 Toggle the *Model/Paper* button (below the *Command* line) to *Model* to enter model space. Use *Command:* <u>Zoom</u> (Enter)><u>1XP</u> to show the drawing full-size in that viewport. Select the other viewport, *Command:* <u>Zoom</u> (Enter)> <u>2XP</u> to obtain a double-size view in that viewport.

While in *Paper Space,* type *Vports* and create two model-space windows by selecting their diagonal corners (**Figure 38.4B**). You will get a single viewport by default that should be erased if you wish to have multiple ports. Type <u>MS</u> or toggle the *Paper/Model* button at the bottom of the screen in the *Status* bar to set it to *MS* (*Model Space*), and the cursor will appear in the active viewport (**Figure 38.4**). Move to a new viewport and select it with the cursor. The drawing in the model-space port is scaled by typing <u>Zoom</u>> <u>1XP</u> for a full-size drawing, <u>2XP</u> for a double-size drawing, and <u>0.5XP</u> for a half-size drawing (**Figure 38.5**).

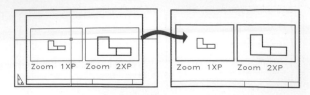

38.6 Plotting from Paper Space.

A Select PS to enter *Paper Space* for plotting a drawing.

B Both *Paper Space* and *Model Space* drawings are plotted when plotting from Paper Space.

To make a plot, return to *Paper Space* with the *Model/Paper* button; the triangular *paper space* icon reappears, and *Paper* replaces *Model* in the *Paper/Model* button at the bottom of the screen in the *Status* bar. Plot from *Paper Space*, and both the 2D and 3D drawings will plot as they appear on the screen (**Figure 38.6**).

38.3 Paper Space Versus Model Space

The following is a summary of actions that can be made from *Paper Space* (*PS*):

1. *Vports* creates *Model Space* (*MS*) viewports.

2. *Stretch, Move,* and *Scale MS* viewports.

3. *Erase MS* viewports.

4. *Freeze MS* outlines.

5. *Insert* 2D drawings.

6. *Hide* removes invisible lines from selected viewports.

7. *Text* can be added across *MS* viewports.

The following is a summary of actions that can be made from *Model Space* (*MS*):

1. *Modify* a 3D drawing.

2. *Rotate* the *User Coordinate System* (*UCS*).

3. *Pan, Zoom, Scale,* etc., *MS* drawings.

4. *Attach* dimensions to the *MS* drawing.

5. *Erase* the contents of an *MS* viewport.

38.4 Fundamentals of 3D Drawing

Experiment with 3D drawing; toggle the *Model/Paper* button to *Model*, type <u>Ucsicon</u>, set it to *On*, and the *XY* icon will appear as shown in **Figure 38.7**. *Command*: <u>Ucsicon</u>

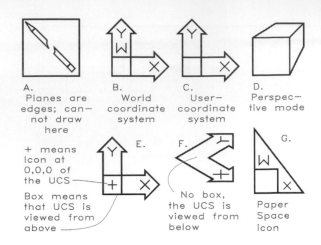

38.7 Select *Command:* <u>Ucsicon</u> (Enter)> *Properties*> *2D* and these icons appear on the screen to show the x- and y-axes when using the *Vport* and *UCS* commands.

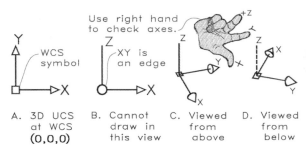

38.8 Select *Command:* <u>Ucsicon</u> (Enter)> *Properties*> *3D* and these icons appear on the screen. Point the thumb of your right hand in the positive x- direction, your index finger in the positive y- direction, and your middle finger will point in the positive z- direction.

(Enter)> *Properties*> *3D* and the icons will appear as shown in **Figure 38.8**. Use *Command*: <u>Ucsicon</u> (Enter)> *ORigin* to make the the *UCS* icon appear at the origin of the *UCS* with a plus sign in its corner. You may need to *Pan* the 0,0 origin up and right of the screen's corner so the *UCS* icon has room to sit on the origin. When a *W* appears on the *2D* icon it is in the *World Coordinate System* (*WCS*). Without the *W*, you are in the *User Coordinate System* (*UCS*). The *broken-pencil* icon warns that the projection plane appears as an edge, making it impractical to draw in that viewpoint. The *oblique-box* icon indicates that the current

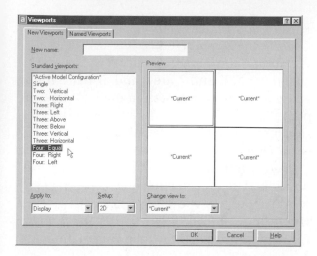

38.9 Select *Main Menu> View> Viewports> New Viewports>* to obtain the *Viewports* dialogue box for setting viewports on the screen. In the *Setup* window, select 3D to get this arrangement.

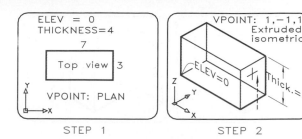

STEP 1   STEP 2

38.10 Elevation and Thickness commands.
Step 1 *Command:* <u>Elev</u> (Enter)
Specify new default elevation <0.00>: <u>0</u> (Enter)
Specify new default thickness <0.00>: <u>4</u> (Enter)
Command: <u>Line</u> (Enter)
Specify first point: (Draw a 7 X 3 rectangle as a top view.)
Step 2 *Command:* <u>Vpoint</u> (Enter)
Specify a view point or [Rotate] <display compass and tripod>: <u>1,-1,1</u> (Enter) An isometric view of the extruded box appears.

drawing is a perspective. The triangular icon tells you that the screen is in *2D Paper Space.* The *UCS* icon shows the *x-, y-,* and *z*-axes which conform to the right-hand rule as illustrated in **Figure 38.8**.

The standard *Vports* are shown in the *Vports* menu in **Figure 38.9**. A *Vports* arrangement must be named in order for them to be *Saved* by typing a name in the panel of the *Vports* menu.

38.5 3D Extrusions

AutoCAD supports three types of 3D modeling: **wireframe**, **surfaces**, and **solids**. The *extrusion* technique is an elementary method of drawing 3D objects with the *Vpoint, Plan, Elev, Thickness,* and *Hide* commands. *Vpoint* (viewpoint) selects the direction in which an object is viewed. *Plan* changes the *UCS* to give a true-size view of the *XY* icon and the surfaces parallel to it. *Elev* (elevation) sets the level of the base plane of the drawing. *Thickness* is the distance of the extrusion in a direction parallel to the *z*-axis. *Hide* removes the invisible lines of the extruded surfaces.

Select *Command*: <u>UCSicon</u> (Enter)> On (Enter) to obtain the *XY* icon, type Elev (Enter)> 0 (zero) (Enter), type <u>Thickness</u> (Enter)> 4 (Enter), and draw the plan view of the object with the *Line* command (**Figure 38.10A**). The *x*- and *y*-axes are true size in the *Plan* view (top view), and the *z*-axis is perpendicular to them and points toward you.

Use *Command*: <u>Vpoint</u> (Enter)> <u>1,1,-1</u> (Enter) to obtain a line of sight for an isometric view of the extruded block in **Figure 38.10B**. Type <u>Plan</u> and the *UCS* icon is shown true size, and planes of the object that are parallel to it appear true size, also. You can use most of the regular *Draw* commands such as *Line, Circle,* and *Arc* to draw features, all of which will be extruded 4 units in the *z*-direction. The extrusion value of 4 units will remain in effect until reset with another value.

Use *Command*: Hide (Enter) to remove hidden lines from an extruded object and give an "empty-box" look (**Figure 38.11**). Use Command: <u>3Dface</u> (Enter) and <u>Osnap</u> (object snap) to the corner points on the upper surface to make the top of the box opaque when the *Hide* command is applied (**Figure 38.12**).

38.5 3D EXTRUSIONS • 717

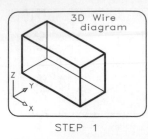 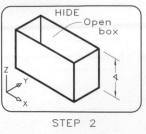

STEP 1   STEP 2

38.11 Elevation and Thickness: Hiding.

Step 1 After the box is drawn, it appears as a wire frame on the screen.

Step 2 *Command:* _Hide_ (Enter) The vertical surfaces become opaque (solid) planes and the top appears open.

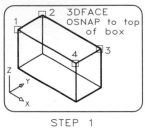 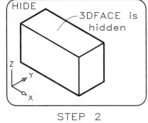

STEP 1   STEP 2

38.12 3Dface:

Step 1 *Command:* Osnap (Enter)

Object snap modes: End (Enter)

Command: 3Dface (Enter)

Specify first point or [Invisible]: 1 > *Specify second point or [Invisible]:* 2 > *Specify third point or [Invisible]:* 3 > *Specify fourth point or [Invisible] <created three-sided face>:* 4 > *Specify first point or [Invisible]:* (Enter)

Step 2 *Command:* _Hide_ (Enter) The top surface appears as an opaque surface.

The *Solid* command can be used to make the top surface opaque by assigning it an *Elev* equal to the *Thickness* (4), setting its *Thickness* to zero, and applying a solid area to the top by selecting the four corners. Type *Hide* and the top appears opaque.

38.6 Coordinate Systems

Almost all drawing is done in the plane of the active coordinate system indicated by the *XY* icon. The two coordinate systems are the

World Coordinate System (*WCS*) and the *User Coordinate System* (*UCS*).

World Coordinate System (*WCS*) has an origin where X, Y, and Z are 0 and, usually, the *x*- and *y*-axes are true length in the top view, the plan view. The *UCS* icon has a *W* and a plus sign on it when it is at the *WCS* origin.

A *User Coordinate System* (*UCS*) can be located within the *WCS* with its *x*- and *y*-axes positioned in any direction and its origin at any selected point. Type Ucsicon and On, select *Origin*, move it to its origin, and establish a *User Coordinate System* in the following manner:

Command: UCS (Enter)
Enter an option/New/Move/orthoGraphic/Prev/Restore/Save/Del/Apply/?/World]
<World>: Move (Enter)
Specify new origin point or [Zdepth]<0,0,0>: Select with cursor (Enter)

The options of the *UCS* command are:

New: Defines a new coordinate system by one of six methods: *ZAxis, 3point, OBject, Face, View,* and *X, Y,* and *Z*. An application of the *3Point* option is given in **Figure 38.13**.

Move: Redefines the origin without changing the orientation of the *x-, y-,* and *z*-axes by picking a point.

orthoGraphic: Specifies one of six *UCS's,* mostly for 3D editing.

Previous: Steps back through the previous list of *UCS s* one at a time.

Restore: Restores a named *UCS* and makes it current, however the *Viewpoint* is not restored.

Save: Saves the current *UCS* to a specified name.

Delete: Removes a saved *UCS* from a list of saved *UCS's.*

Apply: Assigns the current *UCS* to a specified *Viewport* or *Viewports.*

?: List names of *UCS's* and their origins and *x-, y-,* and *z*-axes for each.

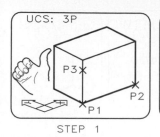
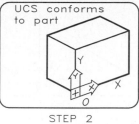

STEP 1   STEP 2

38.13 UCS: 3-Point option.

Step 1 *Main Menu> Tools> New UCS> 3 Point>*

Specify new origin point <0,0,0)>: P1

Specify point on positive portion of the x-axis: P2

Step 2 *Specify point on the positive Y portion of the UCS XY plane:* P3 *The UCS icon is transferred to the origin, P1. The plus sign at its corner box indicates that it is at the origin.*

World: Set the *UCS* to the World Coordinate System.

Save a *UCS* as follows: *Command*: UCS (Enter)> Save> type a name when prompted. Select *Main Menu> Tools> UCS* box> Named *UCS's* tab (**Figure 38.14**) to obtain a list of the saved coordinate systems. The *World* coordinate system is listed first. The current *UCS* is indicated by the pointer but a new one can be selected with the cursor (*UCS-2*, for example), and picking the *Set Current* button. Delete a *UCS* by right clicking on it an selecting *Delete*, or *Rename* it by right-clicking on it and typing a new name. *OK* confirms any action made, and *Cancel* closes the dialogue box. Three tabs are given in the *UCS* box: *Named UCS's*, *Orthographic UCS's*, and *Settings*.

The *UCS* icon is turned on in the following manner:

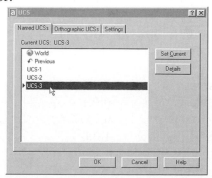

38.14 *Use Main Menu> Tools> UCS> Named UCS:* tab> to obtain a list the saved *UCS* setups.

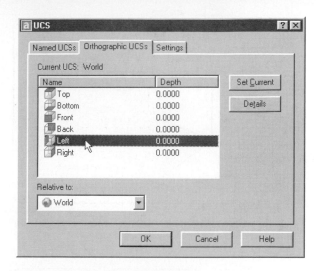

38.15 Select *Main Menu> Tools> UCS> Preset UCS* to get this *UCS* box from which views of objects can be selected.

Command:> UCSicon (Enter)> *Enter an option [ON/OFF/All/Noorigin/ORigin]<ON>:>* ON (Enter)

The functions of these *UCS* icon options are:

ON/OFF turns the icon on and off.

All displays the icon in all viewports.

Noorigin displays the icon at the lower left corner regardless of the *UCS's* location.

Origin places the icon at the origin of the current coordinate system if space permits, or at the lower left if space is unavailable.

38.7 Setting Viewpoints

The *Vpoint* command sets the viewpoint of a *UCS* with the *UCS* box, the tripod axes, or by typing coordinates. Select *Main Menu> Tools> Orthographic UCS> Preset UCS* to get the *UCS* box from which to select the orientations for principal orthographic views (**Figure 38.15**). For example, select the top view icon, specify the origin when prompted, pick *Set Current*, and type Plan to obtain a top view of the UCS icon where the plane of the *x*- and *y*-axes is true size.

When the *UCS* icon symbol is selected, the *axis tripod* (a set of *x*-, *y*-, and *z*-axes) appears on the screen (**Figure 38.16**). A viewpoint of

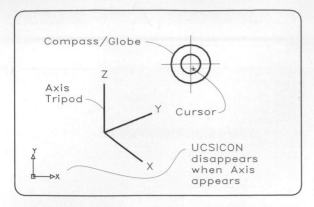

38.16 *Command:* <u>Vpoint</u> (Enter) (Enter), a compass globe and axes appear on the screen for selecting a viewpoint.

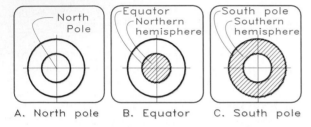

38.17 The compass globe.

A The north pole is at the intersections of the crosshairs.

B The small circle locates viewpoints on the equator.

C The large circle locates the viewpoint at the south pole.

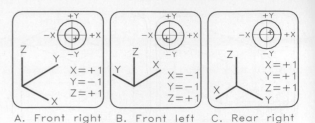

38.18 Examples of the relationship between the points on the *Vpoint globe* and the *Vpoint values* selected from the keyboard.

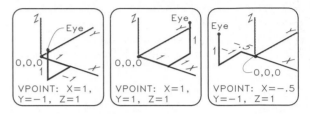

38.19 3D graphical examples of a selection of *Vpoints* from the keyboard are shown here.

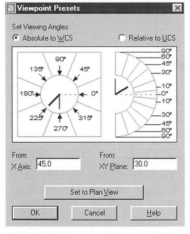

38.20 *Main Menu> View> 3D Views> View Presets* to obtain the *Viewpoint Presets* box from which to select a view. The point of view in the XY plane and from the XY plane is selected with the cursor or by typing values at the *Command* line.

the object is found by selecting a point on the *compass globe* as described in **Figure 38.17**. Repeat this command by pressing (Enter) and selecting other views until the desired view is found.

Figure 38.18 compares the viewpoints found on the compass globe with those specified with numbers at the keyboard. A *Vpoint* of 1,–1,1 means that the origin (0,0,0) is viewed from a point that is 1 unit in the x-direction, 1 unit in the negative y-direction, and 1 unit in the positive z-direction from 0,0,0 (**Figure 38.19**).

A pop-up box, found by picking *Main Menu> View> 3D Views> View Presets*, can be used to select a viewpoint (**Figure 38.20**). The six principal orthographic views can be picked from *3D Viewport Presets* under *View*, or by typing *Vpoint*

at the Command line as follows:

Presets	By Typing
Top view	0,0,1
Front view	0,–1,0
Right-side view	1,0,0
Left-side view	–1,0,0
Rear view	0,1,0
Bottom view	0,0,–1

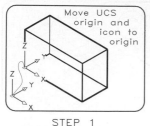
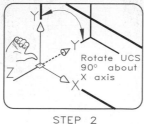

STEP 1   STEP 2

38.21 Setting the UCS:

Step 1 *Command:* <u>UCSicon</u> (Enter)

Enter an option [On/Off/All/. . . /ORigin] <ON>: <u>OR</u> (Enter)

Command: <u>UCS</u> (Enter)

Enter and option [New/Move/orthoGraphic/Prev/Restore/Save/Del/Apply/?/World] <World>: <u>N</u> (Enter)

Specify origin of new UCS or [ZAxis/3point/OBject/Face/View/X/Y/Z]: <u>OB</u> Icon moves to selected object.

Step 2 *Command:* <u>UCS</u> (Enter)

Enter and option [New/Move . . ./?/World] <World>: <u>N</u> (Enter)

Specify origin of new UCS or [ZAxis/. . . /X/Y/Z]: <u>X</u> (Enter)

Specify rotation angle about X axis: <u>90</u> UCS icon rotates.

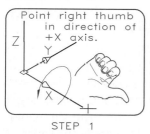
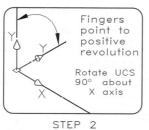

STEP 1   STEP 2

38. 22 Rotating the UCS:

Step 1 *Command:* <u>UCS</u> (Enter)

Enter and option [New/Move . . ./?/World] <World>: <u>N</u> (Enter)

Specify origin of new UCS or [ZAxis/. . . /X/Y/Z]: <u>X</u> (Enter)

Step 2 *Specify rotation angle about X axis:* <u>90</u> The UCS icon rotates.

38.8 Extrusions

An extruded box similar to the one in **Figure 38.10**, is shown in isometric by typing <u>Vpoint</u>, (Enter), and giving coordinates of <u>1,–1,1</u> in **Figure 38.21**. The *UCSicon* is moved to the object's lower left corner by the *UCS* command and rotated 90° about the x-axis to lie in the frontal plane of the box. The right-hand rule is used to determine the direction of rotation by pointing your right thumb in the positive direction of the axis of rotation (**Figure 39.22**). If you type <u>Plan,</u> XY axes of the *UCS* icon and the front view will appear true size.

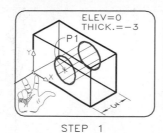
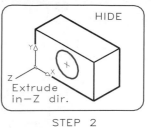

STEP 1   STEP 2

38.23 Extruding a hole:

Step 1 *Command:* <u>Elev</u> (Enter)

Specify new default elevation <0>: <u>O</u> (Enter)

Specify new default thickness <4>: <u>-3</u> (Enter)

Command: <u>Circle</u> (Enter) *CIRCLE specify center point for circle or [3P/2P/Ttr (tan tan radius)]:* <u>P1</u>

Specify radius of circle or [Diameter]: <u>.5</u> (Enter)

A 1 in. diameter cylinder is extruded 3 in. deep into the box.

Step 2 *Command:* <u>Hide</u> (Enter) The outline of the hole is shown, but it cannot be seen through.

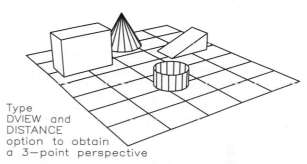

38.24 *Command:* <u>Dview</u> (Enter)> *Select objects>* Window *objects> [CAmera/ TArget/ Distance . . . Undo]:* <u>Distance</u> (Enter) A perspective view of the objects is obtained.

The circle is drawn as an extrusion by setting the *Elev* to <u>0</u> and *Thickness* to <u>–3</u> (the depth of the box) and drawing a cylindrical hole on the frontal plane (**Figure 38.23**). Apply *3Dfaces* to the upper and lower planes of the box and *Hide* invisible lines.

38.9 Dynamic View (Dview)

The *Dview* command is similar to the *Vpoint* command, but *Dview* changes the viewpoints of an object dynamically on the screen as they are changed by the cursor. In addition to axonometric views (parallel projections), 3-point perspectives can be obtained for the most realistic

38.25 *Command:* Dview (Enter) (Enter) and the *DviewBlock* house is obtained that can be used to experiment with the various commands.

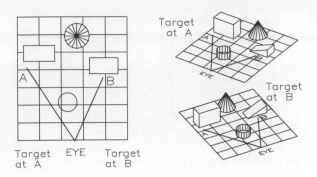

38.27 The *TArget* option of the *Dview* command is used to obtain views of a scene by moving the target to different locations about a stationary camera.

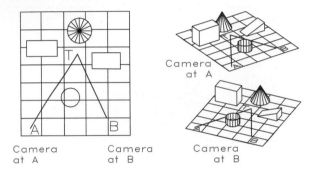

38.26 Use the *Command:* Dview (Enter)> *Select objects <use DVIEWBLOCK>:>* window objects> CAmera option to obtain different views of a stationary target point by moving the position of the camera about it.

pictorials (Figure 38.24). The Dview command has the following options:

Command: Dview (Enter)
(Select the objects when prompted)>
[CAmera/TArget/Distance/POints/PAn/
Zoom/TWist/CLip/Hide/Off/Undo]:

By typing *Dview* and (Enter) (Enter), the top view of the *DviewBlock* house appears, which can be used for experimentation with the following options (**Figure 38.25**).

CAmera rotates your viewpoint as if you were using a camera and moving about the target (**Figure 38.26**). As the cursor (the camera) is

moved, the view is dynamically changed until a viewpoint is selected.

TArget is identical to the *CAmera* option, but the camera remains stationary and the target (and the drawing containing it) is rotated about the camera (**Figure 29.27**).

Distance uses the current camera position and turns the view into a perspective. The *XY* icon is replaced with the *perspective-box* icon. When prompted, give a distance to the target by typing the value, or by using the slider bar, with a range from 0✕ to 16✕; 1✕ is the current distance to the target.

POints specifies the target point and the camera position for viewing a drawing. This command is necessary to specify viewpoints for perspectives.

PAn moves the view of a drawing without changing its magnification or true position.

Zoom changes the magnification of a drawing in the same manner as the *Zoom/Center* command when perspective is *Off*. When in the perspective mode (*Dist=On*) is on, *Zoom* dynamically varies the magnification.

TWist rotates the drawing about an axis that is perpendicular to the screen.

CLip places cutting planes perpendicular to the line of sight to remove portions of a drawing either in front or back of the plane by selecting *Back/Front/<Off>* (**Figure 29.28**).

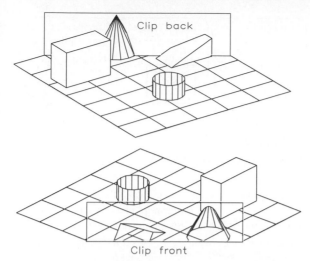

38.28 The *Clip* option of the *Dview* command removes the *Back* or *Front* portions of a drawing with a clipping plane that is parallel to the drawing screen.

Off exits from *CLip*. When *Distance* is *On* (perspective mode), the frontal clipping plane remains *On* at the camera position.

OFF turns off the perspective mode that was enabled by *Distance* = On.

Hide suppresses the invisible lines.

Undo reverses the previous Dview operations one at a time.

38.10 3D Orbit

3D objects can be positioned in space using 3D Orbit that is activated by selecting its symbol on the 3D Orbit toolbar **(Figure 38.29)**. A circle appears on the screen with four handles located at its compass points and an xyz icon **(Figure 38.30)**. The circle can be thought of as a view of a sphere that encloses a 3D object (the object can exceed the size of the circle). 3D Orbit can manipulate the view of the part in four basic ways (each has it own symbol at the cursor):

Horizontal rotation: Select the handles at 3 or 9 o'clock and drag left or right to rotate the object about a vertical axis.

Vertical rotation: Select one of the handles at 12 or 6 o'clock and drag up or down to rotate the object about a horizontal axis.

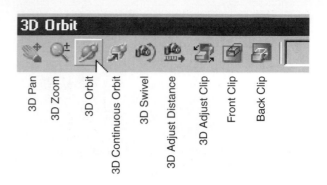

38.29 *3D Orbit* can be accessed from *Main Menu> View> 3D Orbit* toolbar.

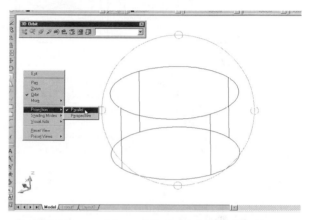

38.30 *Main Menu> View> 3D Orbit* toolbar places a circle (an arcball) on the screen that can be used to rotate objects on the screen. By right-clicking anywhere outside the arcball, a pop-up window of options appears, including perspective views.

Roll: Select a point outside the arcball and drag it to rotate the object about an axis that is perpendicular to the screen.

Free rotation: Select a point inside the arcball and drag it in any direction to rotate the object about any or all of the axes.

By right-clicking outside the arcball, a pop-up menu appears with a list of options, many of which have sub-options for affecting the object. The entire screen of objects need not be selected; only those that you want to observe during the process which enhances the computer's performance.

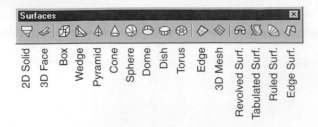

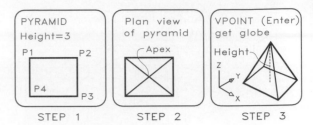

STEP 1   STEP 2   STEP 3

38.34 Surfaces: Pyramid.

Step 1 *Command:* Pyramid (Enter)

Specify first corner point for base of pyramid: P1

Specifiy second corner for base of pyramid: P2

Specify third corner point for base of pyramid: P3

Specify fourth corner. . . . of pyramid or [Tetrahedron]: P4

Step 2 *Specify apex point of pyramid or [Ridge/Top]:* .XY (Enter) *of (Need Z):* 2 (Enter)

Step 3 *Command:* Vpoint > set to 1,–1,1 to get an isometric view.

38.31 *Main Menu> View> Toolbars> Surfaces* gives this menu for drawing 3D surfaces (meshes).

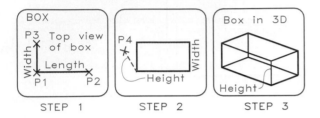

STEP 1   STEP 2   STEP 3

38.32 Surfaces: Box.

Step 1 *Surface toolbar> Box> Specify corner of box:* P1

Specify length of box: P2> *Specify width of box or [Cube]:* P3 *Specify height of box:* P4

Step 2 *Specify rotation angle about z-axis or [Reference]:* 0 (Enter). *Command:* Vpoint> set to 1,–1,1 to get an isometric view.

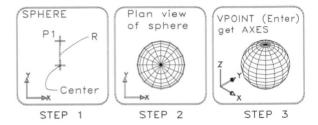

STEP 1   STEP 2   STEP 3

38.35 Surfaces: Sphere.

Step 1 *Command:* Sphere (Enter)

Specify center point of sphere: C

Specify radius of sphere or [Diameter]: P1

Step 2 *Enter number of longtitudinal segments for surface of sphere <16>:* (Enter) *>Enter number of latitudinal segments for surface of sphere <16>:* (Enter)

Step 3 *Command:* Vpoint > set to 1,–1,1 to get an isometric view.

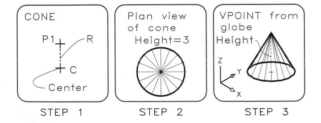

STEP 1   STEP 2   STEP 3

38.33 Surfaces: Cone.

Step 1 *Command:* Cone (Enter) *Specify center point for base of cone:* C > *Specify radius for base of cone or [Diameter]:* P1

Step 2 *Specify height of cone:* 3 (Enter) *Enter number of segments for surface of cone <16>:* (Enter)

Step 3 *Command:* Vpoint (Enter) *and set coordinates to* 1,–1,1 to obtain an isometric of the cone.

Box, Cone, DIsh, DOme, Mesh, Pyramid, Sphere, Torus, and *Wedge.* These shapes are wire frames until *Hide* is used to make them appear as solids. Think of them as hollow shapes with meshes (or faces) applied to their surfaces. Remember that the these shapes are drawn with respect to the *xy* plane; consequently, the *XY* icon must be observed at all times:

Box draws a cube or a box when you select a corner, specify length, width, and height and give the angle of rotation about the *z*-axis (**Figure 38.32**).

Cone draws a cone to its apex (**Figure 38.33**)

38.11 Basic 3D Shapes (Surfaces)

From the *Surfaces* toolbar (**Figure 38.31**), or by typing 3D at the *Command* line (or *Main Menu> View> Toolbars> Surfaces*), you can select a basic 3D shape or surface from the following:

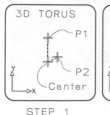

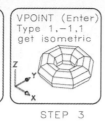

STEP 1   STEP 2   STEP 3

38.36 Surfaces: Torus.

Step 1 *Command:* <u>Torus</u> (Enter) *Specify center point of torus:* <u>C</u>> *Specify radius of torus or [Diameter]:* <u>P1</u>

Specify radius of tube or [Diameter]: <u>P2</u>

Enter number of segments around tube circumference <16>: <u>8</u> (Enter)> *Enter number of segments around torus circumference <16>:* <u>8</u> (Enter)

Step 2 The plan view of the torus is drawn.

Step 3 *Command:* <u>Vpoint</u> > <u>1,–1,1</u> (Enter)> Type <u>Hide</u> to get an isometric view with hidden lines removed.

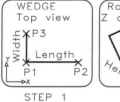
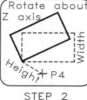
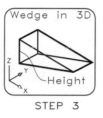

STEP 1   STEP 2   STEP 3

38.37 Surfaces: Wedge.

Step 1 *Command:* <u>Wedge</u> (Enter)

Specify corner of wedge: <u>P1</u> >*Specify length of wedge:* <u>P2</u>>

Specify width of wedge: <u>P3</u>

Step 2 *Specify height of wedge:* <u>P4</u>

Step 3 *Specify rotation angle of wedge about the z-axis:* <u>–15</u> (Enter) *Command:* <u>Vpoint</u> > set to <u>1,–1,1</u> to get an isometric.

or it can be a truncated cone without an apex. By specifying the diameter of the truncated end of the cone to be the same as the base, a *cylinder* is drawn.

DIsh draws the lower hemisphere of a sphere by selecting its center and radius.

DOme draws the upper hemisphere of a sphere by selecting its center and radius in the same manner as *Dish*.

Mesh is used to apply a web to a flat or curving surface to give it a "skin."

Pyramid draws a pyramid extending to its apex (**Figure 38.34**), or it can be drawn as a truncated pyramid without an apex.

38.38 *Facing* options (meshes) selected from the *Surfaces* toolbar are used for putting "skins" on wire frames.

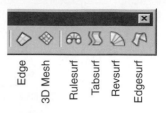

Edge   3D Mesh   Rulesurf   Tabsurf   Revsurf   Edgesurf

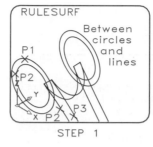
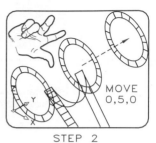

STEP 1   STEP 2

38.39 Part 1: Rulesurf.

Step 1 *Command:* <u>Rulesurf</u> (Enter)

Select first defining curve: <u>P1</u>

Select second defining curve: <u>P2</u> Circles are faced.

Select first defining curve: <u>P3</u>

Select second defining curve: <u>P4</u> Ends are faced.

Step 2 *Command:* <u>Move</u> (Enter)

Select objects? Select faces. (Enter)

Base point or displacement: <u>0,5,0</u> (Enter) Faces are moved.

Sphere draws a ball by selecting its center and radius (Figure 38.35). Its center lies on the XY plane of the *UCS* by default.

Torus draws a donut shape called a torus or *toroid* as shown in Figure 38.36.

Wedge draws a wedge with the same steps used to draw the box (**Figure 38.37**).

Other operations available from the *Surfaces* toolbar are covered in Section 38.12.

38.12 Surface Modeling

Faces (or *meshes*) can be applied to cover wire frames with "skins" to make them look solid. Commands from the *Surfaces* toolbar for applying these meshes are *Rulesurf, Tabsurf, Revsurf, Edgesurf,* and *3Dmesh* (**Figure 38.38**).

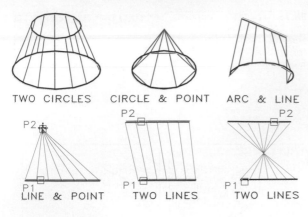

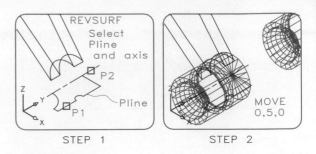

STEP 1   STEP 2

38.42 Part 3: Revsurf.

Step 1 *Command: Revsurf* (Enter) *Current wire frame density: SURFTAB1=16 SURFTAB2 = 4 Select object to revolve:* P1 *Select axis that defines the axis of revolution:* P2

Specify start angle <0>: (Enter)

Specify included angle (+=ccw, -=cw) <360>: 360 (Enter)

Step 2 *Command: Move* (Enter) *Select objects?:* Pick faces (Enter)

Base point or displacement: 0,5,0 (Enter) *Faces are moved.*

38.40 The *Rulesurf* command connects objects with a series of *3D Faces*, as shown in these examples

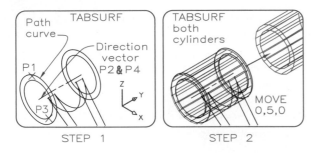

STEP 1   STEP 2

38.41 Part 2: Tabsurf.

Step 1 *Command: Tabsurf* (Enter) *Select object for path curve:* P1 >*Select object for direction vector:* P2 *The spacing between tabulated vectors is determined by the Surftab1 variable.*

Step 2 *Command: Move* (Enter) *Select objects?:* Select faces (Enter) *Base point or displacement:* 0,5,0 (Enter)

The faces are moved to a new position.

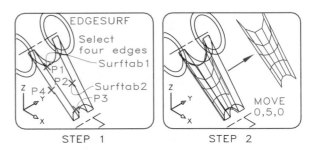

STEP 1   STEP 2

38.43 Part 4: Edgesurf.

Step 1 *Command: Edgesurf* (Enter)

Current wire frame density: SURFTAB1 = 5 SURFTAB2 = 4

Select object 1 for surface edge: P1

Select object 2 for surface edge: P2

Select object 3 for surface edge: P3

Select object 4 for surface edge: P4

Mesh is drawn in boundaries. *Surftab1* and *Surftab2* variables determine the density of the mesh.

Step 2 *Command: Move* (Enter) *Select objects?:* Pick faces (Enter)

Base point or displacement: 0,5,0 (Enter) *Faces are moved.*

The application of these commands is illustrated by applying them to a 3D wire diagram beginning with **Figure 38.39**. Set *Vpoint* to 1,–1,1 to obtain an isometric view of the 3D frame. *Command*: Surftab1> and set it to 20 (the number of faces to be applied).

Rulesurf (ruled surface) applies a surface (20 faces as specified by *Surftab1*) between the circular ends (**Figure 38.39**). The applied surface is *Moved* to 0,5,0 away from the wire frame. *Rulesurf* can be used to place surfaces between two objects (curves, arcs, polylines, lines, or points), as shown in **Figure 38.40**.

Tabsurf (tabulated surface) applies a surface from a curve that is parallel and equal in length to a directional vector (**Figure 38.41**). Select *Path Curve* (the circle) and *Direction Vector,* and the cylinder is drawn. The *Surftab1* system variable (set to 20) applies 20 faces.

Revsurf (revolved surface) revolves lines or shapes about an axis (**Figure 38.42**) and it is

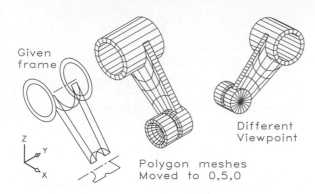

Given frame

Polygon meshes
Moved to 0,5,0

Different Viewpoint

38.44 The given frame of the connector and the resulting 3D hollow shells from two viewpoints are shown here after using the *Hide* command to remove hidden lines.

Moved to 0,5,0 to add it to the hollow shell. Lines, polylines, arcs, or circles can be revolved to form a surface of revolution controlled by *Surftab1*. If a circle or a closed polyline is to be revolved (to make a torus, for example), system variable *Surftab2* controls the mesh density of the line, circle, arc, or polyline that is being revolved, and *Surftab1* controls the density of the path of revolution.

Edgesurf (edge surface) applies a mesh to four edges (joined boundary lines) as shown in **Figure 38.43**. System variables *Surftab1* and *Surftab2* control the density of the first and second objects selected, respectively. The sur-

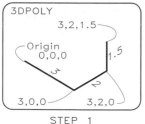
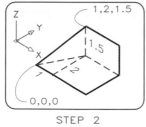

STEP 1 STEP 2

38.45 3Dpoly: A(absolute coordinates.
Step 1 *Command:* 3Dpoly *or Line* (Enter)
Specify start point of polyline: 0,0,0 (Enter)
Specify endpoint of line or [Undo]: 3,0,0 (Enter)
Specify endpoint of line or [Undo]: 3,2,0 (Enter)
Specify endpoint of line or [Close/Undo]: 3,2,1.5 (Enter)
Step 2 *Specify endpoint of line or [Close/Undo]:* 1,2,1.5 (Enter)
Specify endpoint of line or [Close/Undo]: Close (Enter)

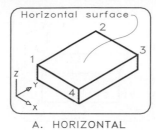
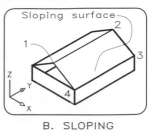

A. HORIZONTAL B. SLOPING

38.46 3Dface on a wire diagram.
3Dface is used to apply opaque planes to wire frames by snapping to endpoints and to apply opaque faces without a wire diagram. *3Dface* is applied to a horizontal surface at (A) and to a sloping surface at (B).

faces are moved to 0,5,0 to complete the hollow shell of the connector.

The beginning wire frame and two final views of the meshed connector are shown in **Figure 38.44** after the *Hide* command has been applied.

38.13 Line, Polyline, and 3DPolyline

The *Line* command draws 2D lines when *x*- and *y*-values are specified, and 3D lines when *x*-, *y*-, and *z*-coordinates are given. *Polyline* draws a 2D polyline and *3Dpolyline* draws 3D polylines with *x*-, *y*-, and *z*-coordinates.

3Dpoly or *Line* commands draw lines with absolute coordinates from 0,0,0 located in 3D space in **Figure 38.45**. Relative coordinates can be typed in the form of @X,Y,Z or @3,0,0 to locate their endpoints with respect to the last point.

All *Osnap* modes apply to *Lines*, *Plines*, and *3Dfaces*. 3D objects can be *Stretched* in the plane of the *UCS* icon (in the *x*- and *y*-directions), but not in the *z*-direction. Height dimensions are modified by rotating the *UCS* so height lies in the *x*- or *y*-directions where Stretch can be used.

38.14 3Dface

3Dface is used to apply faces that can be made opaque by the *Hide* command (**Figure 38.46**). *3Dface* is applied to wire frames by *Snapping* to their endpoints with *Osnap*.

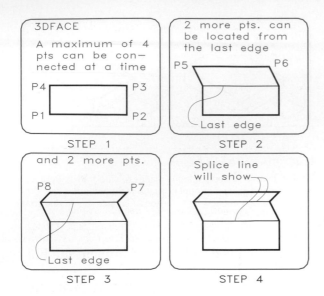

STEP 1 — STEP 2 — STEP 3 — STEP 4

38.47 3Dface: Visible seam.

Step 1 *Command:* 3Dface (Enter) *Specify first point or [Invisible]:* P1 > *Specify second point or [Invisible]:* P2

Specify third point or [Invisible]: P3

Specify fourth point or [Invisible]: P4

Step 2 *Specify third point or [Invisible]:* P5

Specify fourth point or [Invisible]: P6

Step 3 *Specify third point or [Invisible]:* P7

Specify fourth point or [Invisible]: P8

Step 4 *Specify third point :* (Enter) Splice lines are visible.

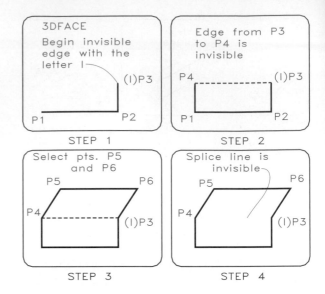

STEP 1 — STEP 2 — STEP 3 — STEP 4

38.48 3Dface: Invisible seam.

Step 1 *Command:* 3Dface (Enter) *Specify first point or [Invisible]:* P1> *Specify second point or [Invisible]:* P2> *Specify third point or [Invisible]:* I > *Second point:* P2 > *Third point:* I (Enter) *Specify third point or [Invisible]:* P3

Step 2 *Specify fourth point or [Invisible]:* P4

Step 3 *Specify third point or [Invisible]:* P5

Step 4 *Specify fourth point or [Invisible]:* P6> *Specify third point or [Invisible]:* (Enter) Splice P3–P4 is invisible.

Figure 38.47 illustrates how corners of a *3Dface* are used to form an opaque plane. After selecting four points with the *3Dface* command, you will be prompted for points 3 and 4, using the previous two points as points 1 and 2. Successively added four-sided areas are connected with splice lines yielding a "patchwork" area of faces. By typing I (for invisible) and (Enter) prior to selecting the beginning point of a splice line, the splice will be invisible (**Figure 38.48**).

38.15 XYZ Filters

Filters are used for picking points that adopt the coordinates of 3D points (**Figure 38.49**). When a command prompts for a point (as in the *Line* command), type a period (.) followed by one- or two-letter coordinates (*.X* or *.XY* for example)

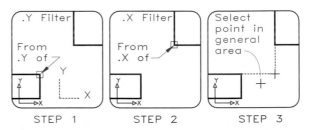

STEP 1 — STEP 2 — STEP 3

38. 49 X and Y filters.

Step 1 *Command:* Line (Enter) *Specify first point:* .Y (Enter) *of* P1 (need XZ): .X (Enter)

Step 2 *of* P2 (need Z):

Step 3 Select a point in the general area of the desired position; a point appears at the intersection of the y-coordinate from the side view and the x-coordinate from the top view.

and select the point with the cursor. Respond to the prompt for the missing coordinate or coordinates with a number or numbers.

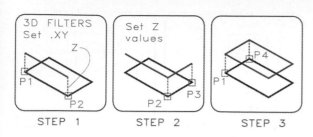

STEP 1   STEP 2   STEP 3

38.50 3D filters:

Step 1 *Command:* <u>3Dpoly</u> *(or Line) (Enter)*
Specify start point of polyline: <u>.XY</u> *of* <u>P1</u> *(need Z):* <u>2</u> *(Enter)*
Specify start endpoint of polyline: <u>.XY</u> *of* <u>P2</u> *(need Z):* <u>2</u> *(Enter)*
Step 2 *Specify start . . . polyline:* <u>.XY</u> *of* <u>P3</u>*(need Z):* <u>2</u> *(Enter)*
Step 3 *Specify start. . . polyline:* <u>.XY</u> *of* <u>P4</u> *(need Z):* <u>2</u> *(Enter)*

Box | Sphere | Cylinder | Cone | Wedge | Torus | Extrude | Revolve | Slice | Section | Interfere | Setup Drawing | Setup View | Setup Profile

38.51 Select *Main Menu> View> Toolbars> Solids* to get this toolbar on your screen.

To locate the front view of a point projected from the left side and top views, select the *.Y* of the left point (Step 1) and the *.X* of the top point (Step 2), and the front view of the point sharing these coordinates is found (Step 3).

A 3D drawing is made in **Figure 38.50** using filters where the .XY coordinates of a given point are selected and Z is specified as 2. *Osnap* was set to *End* for selecting endpoints. When the points are filtered in the XZ plane, a prompt will ask for the Y-coordinate to locate the point in 3D space.

38.16 Solid Modeling: Introduction

AutoCAD 2004 provides solid modeling capabilities of *Regions* (2D solids) and *Solids* (3D solids). The commands for solid modeling found in the *Solids* toolbar (**Figure 38.51**) are used to create solid primitives—*boxes, cylinders, spheres,* and others—that can be added or subtracted from each other to form a composite object.

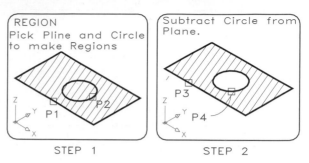

STEP 1   STEP 2

38.52 2D Regions:

Step 1 *Command:* <u>Region</u> *(Enter) Select objects:* <u>P1</u>
1 found> Select objects: <u>P2</u> *(Enter) 2 Regions created.*

Step 2 *Command:* <u>Subtract</u> *(Enter) Select solids and regions to subtract from . . > Select objects:* <u>P3</u> *1 found> Select objects: (Enter)> Select solids and regions to subtract . .*

Select objects: <u>P4</u> *> 1 found > Select objects: (Enter)*

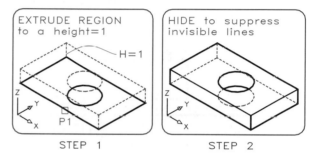

STEP 1   STEP 2

38.53 Extruding a Region:

Step 1 *View> Toolbars> Solids> Extrude icon> Select objects:* <u>P1</u>
1 found > Select objects: (Enter)

Step 2 *Specify height of extrusion or [Path]:* <u>2</u> *(Enter)*
Specify angle of taper for extrusion <0>: (Enter)
Region is extruded a vertical distance of 2.

Regions

Region modeling is a 2D version of solid modeling in which a closed surface can be converted into a solid plane called a region (**Figure 38.52**). The upper plane of a wire frame enclosed by a *Pline* is made into a 2D solid by typing <u>Region</u> and selecting the polyline. The circle is made into a region and is removed from the rectangular *Region* by the *Subtract* command.

Extrude

The Extrude command (from the *Solids* toolbar) is used with closed polylines, polygons, circles, ellipses, and 3D entities to extrude them to a

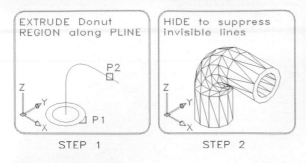

STEP 1   STEP 2

38.54 Extruding a Region.

Step 1 *Command:* Extrude (Enter) *Select objects:* P1>

1 found> Select objects: (Enter)

Specify height of extrusion or [Path]: Path (Enter)

Step 2 *Select extrusion path:* P2 Type Hide to show visibility.

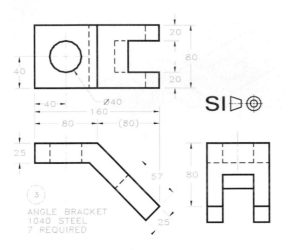

38.55 This angle bracket with an inclined surface is drawn as a 3D object in the following figures.

specified height (with tapered sides if desired). **Figure 38.53** shows the extrusion of the 2D *Region* developed in **Figure 38.52** to an assigned height of 1. Polylines with crossing or intersecting segments cannot be extruded.

Regions can be extruded paths (usually a polyline) to form a 3D shape as illustrated in **Figure 38.54**. Extruded shapes can be hidden and rendered.

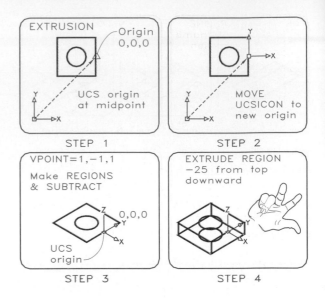

STEP 1   STEP 2

STEP 3   STEP 4

38.56 Part 1: First extrusion.

Step 1 Draw the top view of the bracket and set the *UCS* origin at the midpoint of the line as shown.

Step 2 Use *Ucsicon* and *ORigin* to place the icon at the origin.

Step 3 *Command:* Vpoint> 1,-1,1 to obtain an isometric view of the plane. Make the plane into a *Region* with a hole in it.

Step 4 *Extrude* the *Region* to a height of –25.

38.17 An Extrusion Example

While in *Model Space*, assign limits of about 200 X180 in which to draw the first portion of the angle bracket shown in **Figure 38.55**. Begin drawing the bracket as follows:

    Part 1—First Extrusion: (**Figure 38.56**) Draw the top view of the bracket as a polyline with a circle in it, and set the *UCS* origin at the midpoint of the line shown. *Move* the *UCS* icon to the *UCS Origin*, convert the plane into a *Region*, and *Subtract* the circle from the rectangular plane. Obtain an isometric view (*Vpoint* = 1,–1,1), and *Extrude* the *Region* to –25 mm below the upper surface.

    Part 2—Draw the Inclined Center Line: (**Figure 38.57**) Rotate the *UCS* 90° about the *x*-axis and 90° about the *y*-axis. Draw a line from 0,0,0 (the *UCS* origin) to 0,–80,80 to find the centerline of the inclined surface.

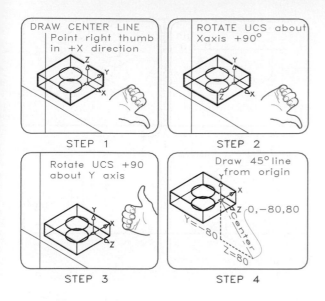

DRAW CENTER LINE Point right thumb in +X direction STEP 1	ROTATE UCS about Xaxis +90° STEP 2
Rotate UCS +90 about Y axis STEP 3	Draw 45° line from origin 0,−80,80 STEP 4

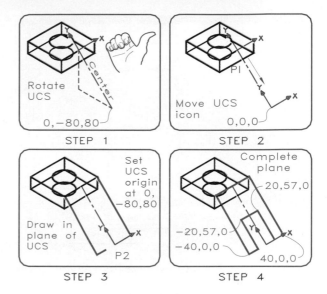

Rotate UCS 0,−80,80 STEP 1	Move UCS icon P1 0,0,0 STEP 2
Set UCS origin at 0, −80,80 Draw in plane of UCS P2 STEP 3	Complete plane 20,57,0 −20,57,0 −40,0,0 40,0,0 STEP 4

38.57 Part 2: Drawing center line.

Step 1 *Command:* <u>UCS</u> (Enter) *Enter an option [New/ Move/. . ./World]:* <u>New</u> (Enter) *Specify origin of new UCS:* <u>X</u> (Enter)

Step 2 *Specify rotation angle about X axis:* <u>90</u> (Enter)

Step 3 *Command:* <u>UCS</u> (Enter) *Enter an option [New/ Move/. . ./World]:* <u>New</u> (Enter)

Specify origin of new UCS: <u>Y</u> (Enter)

Specify rotation angle about Y axis: <u>90</u> (Enter)

Step 4 *Command:* <u>Line</u> (Enter) *Specify first point:* <u>0,0,0</u> (Enter)

Specify next point: <u>0,-80,80</u> (Enter) *Center line is drawn.*

Part 3—Draw the Inclined Surface: (Figure 38.58) Rotate the <u>UCS</u> –45° about the *x-* axis so that it lies in the plane of the inclined surface. Draw a closed *Polyline* using the coordinates of the inclined plane to locate its corners. Convert this plane into a *Region*.

Part 4—Extruding the Inclined Surface: (Figure 38.59) *Command:*<u>Extrude</u> (Enter) and enter an extrusion height of <u>–25</u> when prompted. Use the *Union* command, select the two extrusions, and join them together into a single solid.

38.59 Part 4: Drawing the inclined surface.

Step 1 *Command:* <u>Region</u> (Enter) *Select object:* <u>P3</u> (Enter)

Command: <u>Extrude</u> (Enter)

Step 2 *Specify height of extrusion or [path]:* <u>–25</u> (Enter)

Step 3 *Command:* <u>Union</u> (Enter) *Select object:* <u>P4</u>

Select object: <u>P5</u> (Enter) *The parts are joined.*

Step 4 *Command:* <u>Hide</u> (Enter) *Hidden lines are removed.*

38.58 Part 3: Drawing inclined plane.

Step 1 *Command:* <u>UCS</u> (Enter) *Enter an option [New/ Move/. . . /World]:* <u>New</u> (Enter) *Specify origin of new UCS:* <u>X</u> (Enter) *Specify rotation angle about X axis:* <u>–45</u> (Enter)

Step 2 *Command:* <u>UCS</u> (Enter) *Enter an option [New/ Move/. . . /World]:*<u>Move</u> (Enter) *Specify origin of new UCS:* <u>P1</u>(Enter) *Endpoint of center line.*

Step 3 *Command:* <u>Line</u> (Enter) *Specify first point:* <u>0,0,0</u> (Enter)

Specifiy next point: <u>P2</u>

Step 4 *Command:* <u>Pline</u> (Enter) *Draw the inclined plane.*

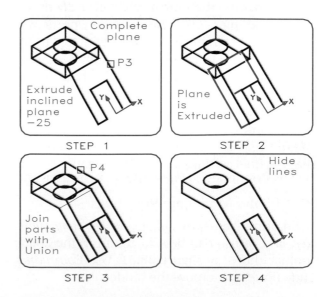

Complete plane P3 Extrude inclined plane −25 STEP 1	Plane is Extruded STEP 2
P4 Join parts with Union STEP 3	Hide lines STEP 4

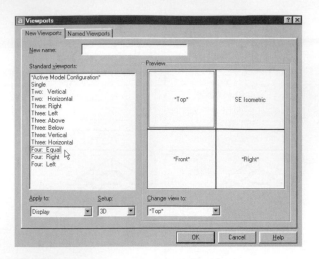

38.60 *Select Main Menu> View> Viewports> New Viewports to obtain this Viewports dialogue box, select Four Equal windows as your viewports option, and select 3D under Setup.*

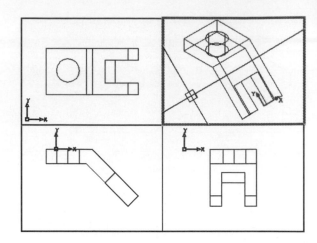

38.61 Part 5: Four equal views.

Select the *Layout1* tab (Enter) to obtain a single viewport in paper space. *Erase* the window. *Main Menu> View> Viewports> New Viewports>* Select *3D* under *Setup>* Select *Four: Equal* under viewports and four viewports appear (top, front, side, and isometric views). The 3D view of the bracket appears in the upper right box along with its three principal orthographic views: top, front and right-side views.

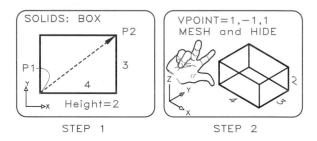

38.62 Solids: Box.

Step 1 *Command:* Box (Enter)
Specify corner of box or [Center] <0,0,0>: P1 (Enter)
Specify corner or [Cube/Length]: P2 (Enter)
Specify height: 2 (Enter)
Step 2 *Command:* Vpoint (Enter)
Specify a view point or [Rotate] <display compass and tripod>: 1,−1,1 (Enter) The box is shown as a wire diagram.
Commnd: Hide (Enter) to suppress hidden lines.

Part 5—Four Equal Views: (**Figure 38.60**) Pick the *Layout 1* tab at the bottom of the screen and you will get a dialogue box that will let you open a single viewport. Erase this viewport which leaves no viewport. Select *Four Equal* (3D) ports from the *Viewports* box (**Figure 38.60**); when prompted select the diagonal corner of area for the four ports. Select *3D* under *Setup* and pick *OK*. The four viewports will appear as shown in **Figure 38.61** with top, front, side, and isometric views. It is usually necessary to size the views to a uniform scale by using *Zoom>* nXP (.3XP for example). Hide can be typed in each *Model Space* viewport to remove hidden lines, return to *Paper Space*, and *Main Menu> File> Plot* to obtain the *Plot* menu for making final settings before plotting.

38.18 Solid Modeling: Primitives

Solid primitives—*box, sphere, wedge, cone, cylinder*, and *torus*—can be selected from the *Solids* toolbar shown in **Figure 38.51**. Use *Command:* Hide (Enter) to remove the hidden lies.

Box creates a box as shown in **Figure 38.62**, where its base is drawn in the *xy* plane of the current *UCS*. The dimensions of the box can be created with separate widths and depths, diagonal corners of the base, or as cube by typing values at the keyboard or by selecting them with the cursor.

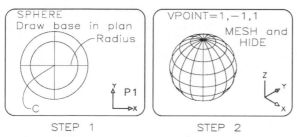

38.63 Solids: Sphere.

Step 1 *Command:* <u>Sphere</u> (Enter)

Specify center of sphere <0,0,0>: <u>Center</u> (Enter)

Specify radius of sphere or [Diameter]: <u>2</u> (Enter)

Step 2 *Command:* <u>Vpoint</u> (Enter)

Specify a viewpoint or [Rotate] <display compass and tripod>: <u>1,–1,1</u> (Enter) 3D view of the sphere is obtained.

Command: <u>Hide</u> (Enter) to suppress invisible lines.

Sphere creates a ball by responding to the prompts with the center and radius or center and diameter of the ball, as shown in **Figure 38.63**. The axis of the sphere connecting its north and south poles is parallel to the *z*-axis of the current *UCS*, and its center is on the *XY* plane of the *UCS*.

Cone draws cones with circular or elliptical bases. The example in **Figure 38.64** illustrates a cone drawn with a circular base and its center, axis endpoints, and height specified when prompted.

Cylinder is drawn by beginning with the *Cylinder* icon and specifying the size of a circular or elliptical base and giving its height. **Figure 38.65** illustrates the steps of drawing a cylinder with a circular base.

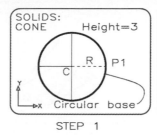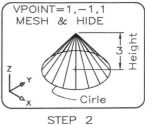

38.64. Solids: Cone.

Step 1 *Command:* <u>Cone</u> (Enter)

Specify center point for base of cone or [Elliptical]: <0,0,0>: <u>Center</u> (Enter)

Specify radius for base of cone or [Diameter]: <u>P1</u> (Enter)

Specify height of cone or [Apex]: <u>3</u> (Enter)

Step 2 *Command:* <u>Vpoint</u> (Enter)

Specify a view point or [Rotate] <display compass and tripod>: <u>1,–1,1</u> (Enter) 3D view of the cone is obtained.

Command: <u>Hide</u> (Enter) to suppress hidden lines.

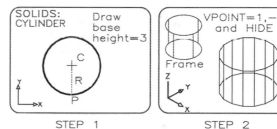

38.65 Solids: Cylinder.

Step 1 *Command:* <u>Cylinder</u> (Enter)

Specify center point for base of cylinder or [Elliptical]<0,0,0>: <u>Center</u> (Enter)

Specify radius for base of cylinder or [Diameter]: <u>2.5</u> (Enter)

Specify height of cylinder or [Center of]: <u>3</u> (Enter)

Step 2 *Command:* <u>Vpoint</u> (Enter)

Specify a viewpoint or [Rotate] <display compass and tripod>: <u>1,–1,1</u> (Enter) 3D view of the cylinder is obtained. *Command:* <u>Hide</u> to remove hidden lines.

Wedge draws the base that lies in the plane of the current *UCS* with the upper plane sloping toward the second point selected (**Figure 38.66**). Prompts ask for the length, width, and height of the wedge.

Torus draws a donut solid by giving its center, diameter or radius of the tube, and diameter or radius of revolution (**Figure 29.67**). The diameter of the torus will lie in the XY plane of the current *UCS*.

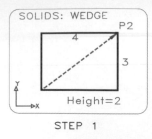
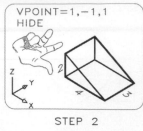

STEP 1   STEP 2

38.66 Solids: Wedge.

Step 1 *Command:* <u>Wedge</u> (Enter)

Specifiy first corner of wedge or [CEnter]<0,0,0>: <u>P1</u> (Enter)

Specify corner or [Cube/Length]: <u>P2</u> (Enter)

Specify height: <u>2</u> (Enter)

Step 2 *Command:* <u>Vpoint</u> (Enter) *Specify a view point or [Rotate] <display compass and tripod>:* <u>1,-1,1</u> (Enter)

3D view of the wedge is obtained.

Command: <u>Hide</u> (Enter) to remove hidden lines.

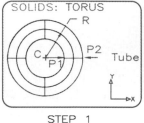
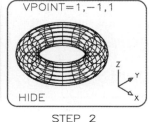

STEP 1   STEP 2

38.67 Solids: Torus.

Step 1 *Command:* <u>Torus</u> (Enter) *Specify center of torus <0,0,0>:* <u>C</u> (Enter)

Specify radius of torus or [Diameter]: <u>6</u> (Enter)

Specify radius of tube or [Diameter]: <u>P1</u> and <u>P2</u> (Enter)

Step 2 *Command:* <u>Vpoint</u> (Enter)

Specify a view point or [Rotate] <display compass and tripod>: <u>1,-1,1</u> (Enter) 3D view of the torus is obtained.

Command: <u>Hide</u> (Enter) to remove hidden lines.

Revolve is used to sweep polylines, polygons, circles, ellipses, and 3D poly objects about an axis if they have at least 3, but less than 300, vertices. In **Figure 29.68**, a polyline is revolved a full 360° about an axis. The path of revolution can start and end at any point between 0 and 360°. Polylines that have been *Fit* or *Splined* will require extensive computations by the computer.

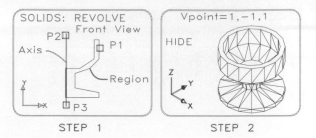

STEP 1   STEP 2

38.68 Solids: Revolve.

Step 1 *Command:* <u>Revolve</u> (Enter) Select objects: <u>P1</u>

Axis of revolution-object/X/Y/<Start point of axis>: <u>P2</u>

<End point of axis>: <u>P3</u> *> Angle of revolution <full circle>:* (Enter) The surface is revolved.

Step 2 *Command:* <u>Vpoint</u> (Enter)

Rotate/<Viewpoiint><0,0,0>: <u>1,-1,1</u> (Enter) The 3D view of the revolution is drawn.

Command: <u>Hide</u> (Enter) to remove hidden lines.

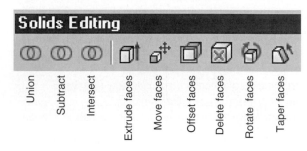

38.69 Select the *Main Menu> View> Solids Editing* toolbar to obtain these options for editing *Solids*. One half of the bar is shown here.

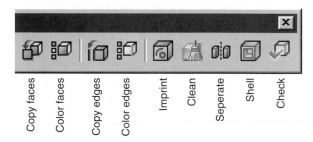

38.69A The other half of the *Solids Editing* toolbar is shown here.

38.19 Editing Solids

Once drawn, *Solids* have a number of editing options that can be used to edit them. The most basic and often-used ones are: *Subtract, Union, Explode, Chamfer, Fillet, Extend,* and *Trim.* These tools can be accessed from the *Solids* Editing toolbar (**Figures. 38.69** and **38.69A**).

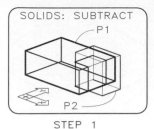
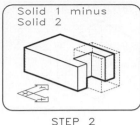

38.70 Solids: Subtract.

Step 1 *Main Menu> Modify> Solids editing> Subtract> Select solids and regions to subtract from . . .*

Select objects: P1 (Enter)

Select solids and regions to subtract: P2 (Enter)

Step 2 The notch is cut into the part.

Command: Hide (Enter) to remove hidden lines.

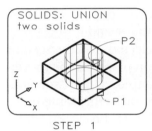
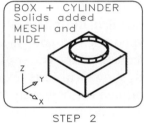

38.71 Solids: Union.

Step 1 *Command:* Union (Enter)

Select objects: P1

Select objects: P2 > *Select objects:* (Enter)

Step 2 The two solids are unified (added) into a single part. *Command:* Hide (Enter) to remove hidden lines.

Subtract is used to remove one intersecting solid from another as illustrated in **Figure 39.70**.

Union is used to join intersecting solids to form a single composite solid model. **Figure 38.71** shows how a box and cylinder are unified into a single solid.

Interfere is used to obtain a solid that is common to two intersecting solids. An interference solid is found with this command in **Figure 38.72**.

Explode is used to separate solids or regions that were combined by the *Subtract* and *Union* commands to permit editing or correcting before redoing *Subtract* and *Union* commands.

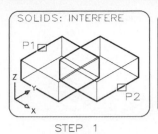
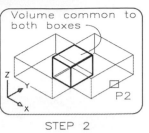

38.72 Solids: Interfere.

Step 1 *Command:* Interfere (Enter)

Select first set of solids: Select objects: P1 (Enter)

Select second set of solids:

Step 2 *Select objects:* P2 (Enter)

Create interference solids? <N>: Yes (Enter) Interference solid is created. *Command:* Hide (Enter) to remove hidden lines.

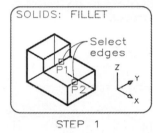
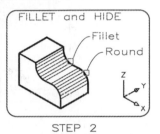

38.73 Solids: Fillet.

Step 1 *Command:* Fillet (Enter) >*Select first object or [Polyline/Radius/Trim/mUltiple]:* R > Enter fillet radius <0>: .50 (Enter)

Select and edge or [Chain/Radius]: P1,

1 edge(s) selected for fillet. (Enter)

Step 2 The *Fillet* is drawn. Repeat these commands but select P2 for drawing the *round*.

Command: Hide (Enter) to remove hidden lines.

Fillet is used to apply rounded intersections between planes by selecting the edges of solids and giving the diameter or radius of the fillet as shown in **Figure 38.73**. The *mUltiple* option of *Fillet* can be used for applying the same radius to several edges using the same settings. The *Fillet* command can also be used to round the edges of cylindrical or curved objects.

Chamfer applies beveled edges by selecting the base surface, the adjoining surface, and the edges to be chamfered, giving the first and second chamfer distances (**Figure 38.74**). When these prompts have been satisfied, the chamfers are automatically drawn.

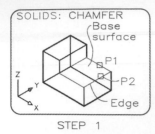
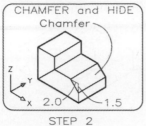

STEP 1 STEP 2

38.74 Solids: Chamfer.

Step 1 *Command:* <u>Chamfer</u> (Enter) > *Select first line or [Polyline/Distance/Angle/ Trim/Method/mUltiple]:* <u>P1</u>

Base surface selections . . . Enter surface selection option [Next/OK (current)] <OK>: <u>Next</u> (Enter) to activate upper plane of object.

Base surface selection . . . Enter surface selection option [Next/OK (current)] <OK>: <u>OK</u> (Enter) to accept top plane as base.
Specify base surface chamfer distance <1.000>: <u>2.0</u> (Enter)
Specify other surface chamfer distance <1.000>: <u>1.5</u> (Enter)
Select an edge or [Loop]: <u>P2</u>
Step 2 *The Chamfer is drawn.*
Command: <u>Hide</u> (Enter) to remove hidden lines.

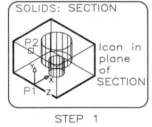
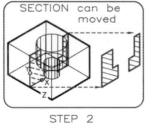

STEP 1 STEP 2

38.75 Solids: Section.

Step 1 Position the <u>UCS</u> in the plane of the section.

Command: <u>Section</u> (Enter) *Select objects:* <u>P1</u>

Specify the first point on Section plane by [Object/. . . /XY/YZ/ZX/3points]: <u>XY</u> (Enter)

Specify a point on the XY plane <0,0,0>: <u>P2</u>

This section establishes a *Region* in the plane of the *UCS* icon.

Step 2 *Command:* <u>Move</u> (Enter) *Move the Region outside the part; apply section lines to it if you like.*

38.20 Sections

Section is used to pass a cutting plane through a 3D solid to show a Region that outlines its internal features. *Bhatch* (hatch pattern) can be used to assign the hatching pattern to the *Region* if it lies in the plane of the *UCS*. In **Figure 38.75**, the hatch pattern is set to ANSI31, a symbol used for representing cast iron or general applications.

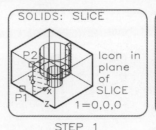
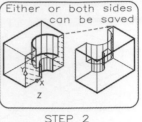

STEP 1 STEP 2

38.76 Solids: Slice.

Step 1 Move the *UCS* to the desired plane of the *Slice.*

Command: line: <u>Slice</u> (Enter) *Select objects:* <u>P1</u>

Specify first point on slicing plane by Object/ . . . XY/ YZ . . . : <u>XY</u> (Enter) > *Specify a point on the XY plane<0,0,0>:* <u>P2</u>

Step 2 *Specify a point on desired side of the plane or [keep Both sides]:* <u>Both</u> (Enter) Halves can be moved apart and hatched.

The *UCS* icon is placed on the object to establish the plane of the section, <u>Section</u> is typed, the first point is selected, and the section plane appears. The section plane can be moved as shown in this example. Other sections through the object are found in this same manner by positioning the icon in the cutting plane or by selecting from one of the following options: *3point, Object, Zaxis, View, XY, YZ,* or *ZX.*

38.21 Slice

With the *Slice* command, an object can be cut through and made into separate parts, either or both of which can be retained as shown in **Figure 38.76**. The Slice command has the following options:

3points defines three points on the slice plane.

Object aligns the cutting plane with a circle, ellipse, 2D spline, or 2D polyline element.

Zaxis defines a plane by picking an origin point on the *z*-axis that is perpendicular to the selected points.

View makes the cutting plane parallel with the viewport's viewing plane when a single point is selected.

XY, YZ, or *ZX* aligns the cutting plane with the planes of the *UCS* by selecting only one point.

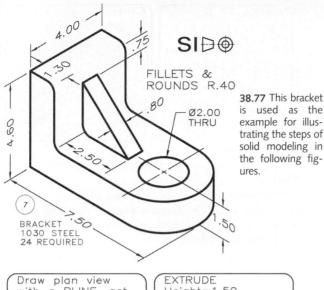

SI Ø ⊕

FILLETS &
ROUNDS R.40

Ø2.00
THRU

BRACKET
1030 STEEL
24 REQUIRED

38.77 This bracket is used as the example for illustrating the steps of solid modeling in the following figures.

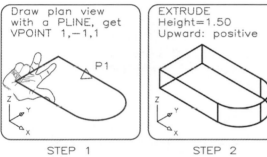

STEP 1   STEP 2

38.78 Solids—Part 1: Extrude base.

Step 1 Draw a the base with a closed *Polyline*.

Command: Vpoint (Enter)

Specify a view point or [Rotate]<display compass and tripod>: 1,-1,1 (Enter) Get isometric view of the base.

Command line: Extrude (Enter) *Select objects:* P1

Step 2 *Specify height of extrusion or [Path]:* 1.50 (Enter)

Specify angle of taper for extrusion <0>: (Enter)

Base is etruded 1.50 in. the positive y-direction.

38.22 A Solid Model Example

The bracket shown in **Figure 38.77** is to be drawn as a solid model by using the commands from the *Solids* toolbar. Create the outline of the base with a *Polyline* and obtain an isometric view of it by *Command:*Vpoint (Enter) 1,-1,1 (**Figure 38.78**). Type Extrude and give an extrusion height of 1.50 units.

A *Box* of 1.30 × 4.00 × 3.10 is drawn (**Figure 38.79**), *Snap* to P1 on the box, and drag it to the

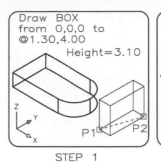 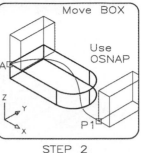

STEP 1   STEP 2

38.79 Solids–Part 2: Draw box.

Step 1 *Command* line: Box (Enter)

Specify corner of box or [CEnter]: P1 (Enter)

Specify corner or [Cube/Length]: @1.30,4.00 (Enter)

Specify height: 3.1 (Enter) Box is drawn.

Step 2 Set *Osnap* to *End. Command* line: Move (Enter)

Select objects: Pick box.

Specify base point or displacement: P1 (Enter)

Second point of displacement: A (Enter) Box is moved.

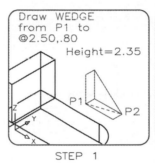 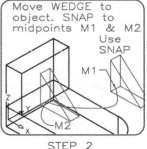

STEP 1   STEP 2

38.80 Solids–Part 3: Draw the wedge.

Step 1 *Command:* Wedge (Enter)

Specify first corner of wedge of [CEnter]: P1

Specify corner or [Cube/Length]: @2.50,0.80 (Enter)

Specify height: 2.35 Wedge is drawn.

Step 2 Set *Osnap* to *Midpoint> Command:* Move (Enter)

Select objects: Pick wedge.

Specify base point or displacement: M1

Specify second point or displacement or <use first point as displacement>: M2 (Enter) The wedge is moved to the base.

Endpoint at A. Use *Wedge* to draw the bracket's rib as shown in **Figure 38.80**. Move the rib to join the base and the upright box by using the *Midpoint* option of *Osnap*.

In **Figure 38.81**, draw a *Cylinder* to represent the hole and *Move* it to the center of the

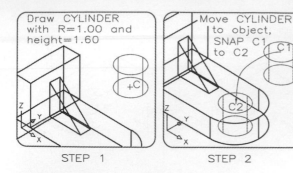

STEP 1   STEP 2

38.81 Solids—Part 4: Draw the cylinder.
Step 1: *Command:* Cylinder (Enter)
Specify cen. pt, for base of cylinder or [Elliptical] <0,0,0>: C
Specify radius for base of cylinder or [Diameter]: 1.00 (Enter)
Specify height for base of cylinder or [Center of other end]: 1.60 (Enter) The cylinder is drawn.
Step 2 Set *Osnap* to *Center. Command* line: Move (Enter)
Select objects: Pick cylinder.
Specify base point or displacement: C1
Specify second point of displacement: C2. The hole is placed.

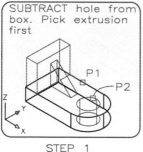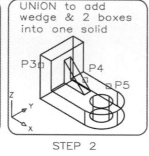

STEP 1   STEP 2

38.82 Solids—Part 5: Subtract and union.
Step 1 *Command:* Subtract (Enter) *Select solids and regions to subtract from . . . Select objects:* P1 (Enter)
Select objects: P2 (Enter) Hole is subtracted from base.
Step 2 *Command:* Union (Enter)
Select objects: P3, P4, P5 (Enter) *Select objects:* (Enter)
The bracket is unified into one solid.

semicircular end of the base using the *Center* option of *Osnap*. Use *Subtract* to create the hole in the base (**Figure 38.82**). Use the *Union* command to join the base, upright box, and wedge together into a composite solid.

    Fillet is used in **Figure 38.83** to select the edges to be rounded with a radius of 0.40 and *Hide* suppresses the invisible lines (**Figure**

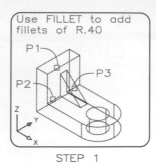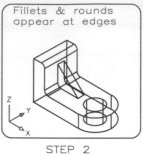

STEP 1   STEP 2

38.83 Solids—Part 6: Fillets.
Step 1 *Command:* Fillet (Enter) (Trim mode)
Select first object or [Polyline/Radius/ Trim]<Select first object>: P1 *Enter fillet radius:* .40 (Enter)
Step 2 *Command:* Fillet (Enter)
Select and edge or [Chain/Radius]: P2
Select and edge or [Chain/Radius]: P3 (Enter)
Fillets and rounds are drawn.

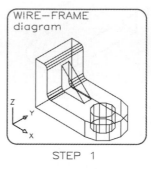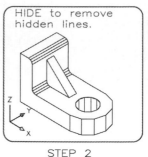

STEP 1   STEP 2

38.84 Solids—Part 7: Hide invisible lines.
Step 1 *Command:* Hide (Enter) to remove hidden lines.
Step 2 The object is shown as a solid with hidden lines suppressed.

38.84). An infinite number of views of the bracket can be obtained with *Vpoint* or *Dview*.

38.23 Three-View Drawing

A drawing composed of three orthographic views (top, front, and side) with an isometric view is a classic arrangement of an engineering drawing suitable for depicting most parts. This layout can be obtained by drawing an object in *Model Space* as the bracket in **Figure 38.77** was drawn and as in the previous examples.

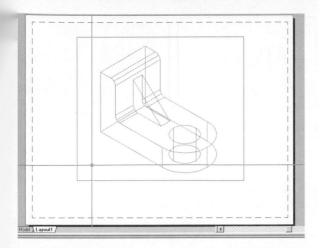

38.85 A single *Model Space* viewport is opened in the *Paper Space (Layout 1* tab) showing the isometric of the bracket that was originally drawn in *Model Space.* Erase this *MS* viewport.

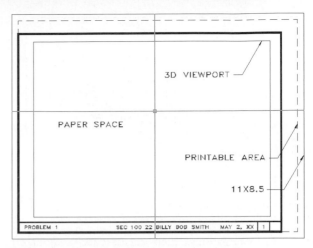

38.87 The title block is in *Paper Space (Layout1* tab), and no viewports are defined in *Model Space* at this point.

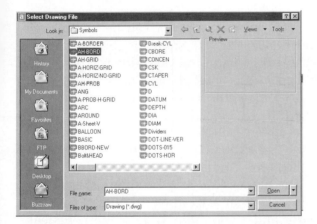

38.86 Use *Main Menu> Insert> Block> Browse>* select ABORD-HOR and insert it into *Paper Space (Layout 1* tab). This previously saved title block establishes the full-size border and the drawing area.

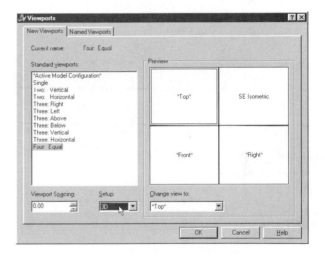

38.88 Use *Main Menu> View> Viewports> New Viewports> Four: Equal* and select *3D* under *Setup* to obtain this *Preview* of the four 3D viewports. Select *OK* and specify the diagonal of the window into which the four viewports are to fit within the border.

Select the *Layout1* tab at the bottom of the screen to obtain a dialogue box and a single viewport in *Paper Space* is displayed which shows the bracket (**Figure 38.85**). Once the single view window is displayed erase it. Use the steps outlined in **Figure 38.86** to insert a title block into *Paper Space* as shown in **Figure 38.87**.

Specify four 3D viewports following the steps given in **Figure 38.88**. The four *Model*

Space viewports will appear within the *Paper Space* of the *Layout 1* tab as illustrated in **Figure 38.89**, but the views of the bracket may not be displayed at a uniform scale. Double-click on a viewport (*UCS* icons will appear) to enter the *Model Space* of that viewport where you can use a *Zoom* factor to size that view of the bracket (*Zoom .5XP,* for example). Select the other viewports by double-clicking on them, and

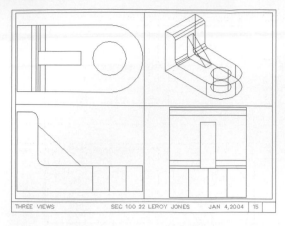

38.89 Four *Model Space* viewports appear in the *Paper Space* (*Layout 1* tab) window when its area area is defined by diagonal corners.

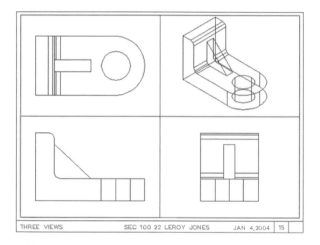

38.90 Size each view with the same *Zoom* factor. For example, use *Zoom* <u>nXP</u> (.5XP–half size, perhaps) in order for the views to be orthographically matched in size.

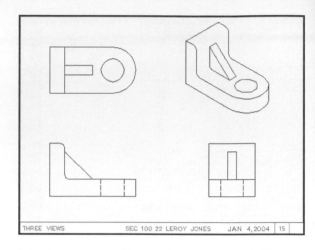

38.91 The viewport outlines can be removed and hidden lines generated by the *Setup Profile* command to make the layout and the views appear in the standard format.

assigning the same *Zoom* factor to them to size them uniformally (**Figure 38.90**).

If the views need to be aligned vertically and/or horizontally, use *Command*: <u>Line</u> (*Enter*) *Enter an option* [*Align/Create… /Undo*]: <u>Align</u> (Enter). Select the *Horizontal* option, pick a base point in a given view and the corresponding point in an adjacent view of the right or left of it, and the views will be horizontally aligned. Refer to **Figures 14.31** through **14.34**.

Double-click on a *Paper Space* area (the title block) to return to *Paper Space* and the *triangu-*

lar icon reappears. A plot can be made from *Paper Space* that will show the four viewports, but the outlines of the viewports will plot, and hidden lines will plot as solid lines.

To remove the viewport outlines, create a new layer by selecting the viewport outlines from *Paper Space* and *Change* them to the new layer. *Freeze* the new layer at plot time and the outlines will not be plotted.

The hidden lines can be suppressed if you double-click in a viewport to enter its *Model Space* and *Command*: <u>Hide</u> (Enter). Repeat these steps for each viewport, double click on a *Paper Space* area or select the *Paper* button in the Status line, and the four viewports can be plotted as a set as displayed.

However, you would probably prefer to show invisible lines as dashed lines rather than omit them. Select *Main Menu> Draw> Solids> Setup Profile* (**Figure 38.91A**), and click on the front view of the bracket in *Model Space* (Enter) to get the following prompts:

Display hidden profile lines on separate layer? <Y>: (Enter)

Project profile lines onto a plane? <Y>: (Enter)

Delete tangential edges? <Y>: (Enter)

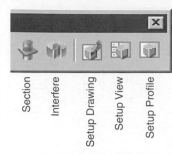

Section · Interfere · Setup Drawing · Setup View · Setup Profile

38.91A The *Setup* commands from the *Solids* toolbar are necessary for drawing orthographic, auxiliary, and section views of 3D Solids.

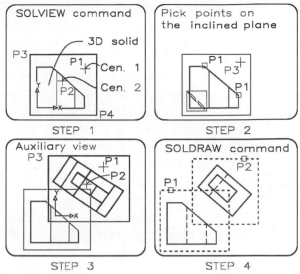

38.92 Solview and Soldraw.

Step 1 *Command:* <u>Solview</u> (Enter)
[Ucs/ Ortho/ Auxiliary/ Section]: <u>UCS</u> (Enter)
[Named/ World/ ?/ Current] <Current>: (Enter)
Enter view scale <1.00>: (Enter)
Specify view center: <u>P1</u> (Enter)
Specify view center <specify viewport>: <u>P2</u>
Specify view center <specify viewport>: (Enter)
Specify first corner of viewport: <u>P3</u>
Specify opposite corner of viewport: <u>P4</u>
Enter view name: <u>Front</u> (Enter)
Step 2 *Enter an option [Ucs/Ortho/Auxiliary/Section]:* <u>A</u> (Enter)
Specify first point of inclined plane: <u>Snap to P1</u>
Specify second point of inclined plane: <u>Snap to P2</u>
Specify side to view from: <u>P3</u>
Step 3 *Specify view center:* <u>P1</u> (1st center selection)
Specify view center <specify viewport>: <u>P2</u> (Enter) (center 2)
Specify view center <specify viewport>: (Enter)
Specify first corner of viewport: <u>P3</u>
Specify opposite corner of viewport: <u>P4</u>
Enter view name: <u>Auxiliary</u> 1 (Enter)
Step 4 *Command:* <u>Soldraw</u> (Enter)
Select viewport to draw: <u>P1, P2,</u> (Enter) Solid lines that are invisible are placed on the *Auxiliary1-Hid* layer. (*Linetypes* of *Hidden* must be assigned to the *Auxiliary1-Hid* layer.) *Freeze* the *Vports* layer to remove the windows.

When each viewport with hidden lines has been manipulated in this manner, two profile planes, one representing the hidden lines (*PH-43, Profile Hidden*) and the other representing visible lines (*PV-43, Profile Visible*), are created. Two 2D views are placed in front of the 3D Solid. Therefore, it is necessary to *Freeze* the 3D Solid in each view, leaving only the two newly-created profiles. Select *Main Menu> Format> Layer> Layer Properties Manager* and pick the layer whose name begins with "PII" (*Profile Hidden*) and Change its *Linetype* to *Hidden* to assign hidden-line symbols to be line drawn on this layer.

Double-click on a paper space area to return to *Paper Space*; turn off the *Layer* on which the solid model was drawn to leave only the profile planes, *PH* and *PV*, turned on to represent the views of the bracket. The variable *Dispsilh* (Display Silhouette Hidden) can be used to display the silhouette (outline) of features of the isometric view and omit its unnecessary hidden lines or meshes. Use *Command:* <u>Dispsilh</u> (Enter) <u>1</u> (Enter) to set *On. Command:* <u>Hide</u>(Enter) so the hidden lines appear as dashed lines and the visible lines appear as solid lines (**Figure 38.91**).

38.24 Views of Solids

Two commands, *Solview* and *Soldraw*, are used together to convert a 3D solid into orthographic views, auxiliary views, or sections. Once a 3D solid has been drawn in *Model Space*, select the *Layout 1* tab, erase the *MS* window that appears, select the *Setup View* icon from the *Solids* toolbar (**Figure 38.91A**) (or *Command:* <u>Solview</u>); the screen will enter *Paper Space* and

you will be given the options of *UCS, Ortho, Auxiliary,* and *Section*. Select <u>UCS</u> and follow the prompts as shown in **Figure 38.92.**

When prompted: *Specify view center*, you can dynamically change the view's center position on the screen until you select (Enter) to finalize your selection. You will be asked to select the diagonal corners of the viewport of the view and prompted to name the view, <u>Front</u> for example.

Since the object has an inclined plane (Step 2), specify the next view as an *Auxiliary*, and select two points on the inclined plane in the front view. Pick the side of the inclined plane where the *auxiliary* is to be drawn, pick the view's center, and the *auxiliary* view is drawn. Select the diagonal corners of the viewport that contains the *auxiliary* view.

Solview automatically creates four layers for each named view. The name of the layer is followed by a dash and these abbreviations, -VIS, -HID, -DIM, and -Hat (for visible, hidden, dimension, and hatch, respectively) on which these features can be drawn.

In Step 4, select *Solids toolbar> Setup Drawing icon* (or Command: <u>Soldraw</u> (Enter)) and you will be prompted for *Viewports to draw* (**Figure 38.91A**). Select points on the *MS* windows, and invisible lines will be converted to hidden (dashed) lines. It is necessary that *Linetype Hidden* be loaded and assigned to the *Layers* whose names are followed by -*Hid (Aux1-Hid*, for example) so invisible lines will appear as dashed lines. From *Paper Space, Change* the *MS* outlines of the viewports to a newly created layer. From the *Layer* toolbar, *Freeze* this new layer to remove the outlines of the *MS* windows.

Sections

A sectional view is illustrated in **Figure 38.93**, where the same steps of the previous example are followed, but instead of *Aux*, type <u>Section</u> at the prompt. The hatching is automatically applied to the section view, and hidden lines are omitted when *Soldraw* is used to select the MS windows from Paper Space. Crosshatching can be varied by *HPname, HPscale,* and *HPangle* to set the name of the hatch pattern, its scale, and its angle, respectively.

Dimensions can be applied to the *Solid Views* by using the layers labeled with -Dim following the views name (Front-Dim, for example). These dimensions are applied in

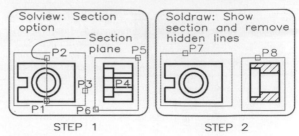

38.93 Solview: Section option.

Step 1 Draw a 3D solid and find the first view of it (the view to be sectioned) using the same steps as illustrated in Figure 38.92. When prompted by *Ucs/Ortho/Auxiliary/Section,*

type <u>Section</u> (Enter) to get these prompts:

Specify first point of cutting plane: <u>P1</u>

Specify second point of cutting plane: <u>P2</u>

Specify side view from: <u>P3</u>> *Enter view scale <1.000>:* (Enter)

Specify view center: <u>P4</u> (Enter)

Specifiy first corner of viewport: <u>P5</u> > *Specify opposite corner of viewport:* <u>P6</u> > *Enter view name:* <u>Sect-1</u> (Enter)

Ucs/Ortho/Auxiliary/Section/<eXit>: <u>X</u> (Enter) (Enter)

Step 2 *Command:* <u>Soldraw</u> (Enter) The section view is drawn, the cut area is hatched, and the hidden lines are omitted.

much the same manner as discussed in Section 38.24.

38.25 Mass Properties (Massprop)

Various properties of regions and solids can be obtained with the *Massprop* command. An explanation of the options available from this command are given below:

Area computes the area of a region or solid.

Perimeter calculates the perimeter of a region (not available for solids).

Bounding Box gives coordinates of the diagonal corners of a region's enclosing rectangle. For a solid, coordinates of the diagonal and opposite corners of a 3D box are given.

Volume gives the 3D space enclosed in a solid.

Mass calculates the weight of a solid.

Centroid gives the coordinates of the center of a region or the 3D center of a solid.

Moment of Inertia is calculated for regions and solids.

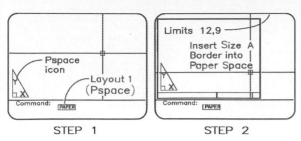

STEP 1   STEP 2

38.94 Paper Space: Layout1.

Step 1 From *Model Space*, select the *Layout1* (or *MODEL* in the *Status bar*) to get the *Page Setup box* from which to make plotting assignments. Make printer and paper assignments and pick *OK*. You are in Paper Space where a viewport into 3D is a box outlined with solid lines. Erase this box to close it.

Step 2 *Command: Insert* (Enter)
Block name (or ?): Border-A (Enter)
Specify insertion point or [Scale/X/Y/Z/Rotate/PScale/PX/PY/PZ/PRotate]: 0,0 (Enter)
The title block is inserted into the Paper Space at the scale to which it was drawn, full size in this case.

Product of Inertia is calculated for regions and solids.

Radius of Gyration is calculated for regions and solids.

MatlibB (materials library) assigns material types that can be assigned to solids that will have an effect on the mass properties listed above. The *Materials Library box* is found under the *Render toolbar*.

38.26 Floating View Ports

By selecting the *Layout 1* tab we can plot all *Vports* just as they appear on the screen, not just the active *Vport*. (An overview of *Paper Space* and *Model Space* was covered in Section 38.2.)

Command: Ucsicon and *On* so the icon will appear. Double-click in an area of *Paper Space* and the screen returns to *Paper Space* with a *PS* icon (a triangle) in its lower left corner and *Paper* appears in the *Status line* (**Figure 38.94**). Then set the paper space *Limits* large enough to contain the border, about 12 in. × 9 in., and insert a size-A border. You may design your own border and save it as a *Wblock*.

From *Paper Space*, use *Command: Vports*, create the diagonal corners of a *MS* viewport with the cursor, *Copy* this viewport, and dou-

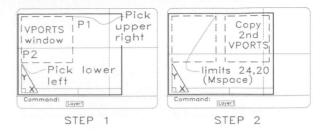

STEP 1   STEP 2

38.95 Paper Space: Vports.

Step 1 While in *Paper Space*, Select *Main Menu> View> Viewports> New Viewports* to obtain the *Viewports* dialogue box. Select *Single* and *P1* and *P2* to form a floating viewport into 3D space.

Step 2 Copy this viewport to form a second one. Return to *Model Space* by selecting the *MODEL* button in the *Status Line* and set the view ports *Limits* to *24,20*. A *Model Space* viewport can be made current by double-clicking inside it, and *Paper Space* entered by double-clicking outside viewport.

ble-click in one of the viewports to enter *Model Space* in that particular port (**Figure 38.95**). Set the *Limits* in this *MS* viewport to 24×20, large enough to contain the drawing. Move between *MS* and *PS* by double-clicking on the button in the Status line beneath the *Command* line that will display as either *Model* or *Paper*.

Use *Command: –Vports* to obtain the following prompts:

On/Off: Selects and turns off viewports to save regeneration time, but leave at least one on.

Fit: Makes a viewport fill the current screen.

Shadeplot: Gives options for rendering a 3D solid.

Lock: Locks the current viewport.

Object: Specifies a closed polyline, ellipse, spline, region, or circle to convert into a viewport.

Polygonal: Creates an irregularly shaped viewport.

Restore: Recalls a saved viewport configuration.

2/3/4: Give options for specifying the number of viewports.

Figure 38.96 shows a 3D object drawn by *Solids* toolbar> *Extrude* in one of the *MS* viewports and is displayed simultaneously in a second port. *Command: Vpoint* (Enter)> Select an isometric viewpoint (1,-1,1) and double-click on a Paper-Space area to enter *Paper Space*.

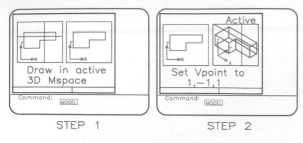

STEP 1   STEP 2

38.96 Paper Space: Drawing in Model Space.

Step 1 Double-click inside one of the *Model Space* viewports to make it current and draw an extruded 3D part. It will show in both views.

Step 2 Click in the right viewport to make it active.

Command: Vpoint (Enter)> *Specify viewpoint:* 1,-1,1

(Enter) An isometric drawing is obtained.

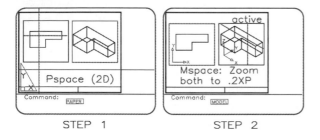

STEP 1   STEP 2

38.97 Paper Space: Zoom to Scale.

Step 1 Double-click in the *Paper Space* area (or select the *Paper Space* button in the *Status* bar) to enter *Paper Space*, and the cursor spans the screen. Since *PS* had limits of 12,0 and *MS* had limits of 24,20, the *MS* ports must be *Zoomed* to about 0.20 size for both to fit inside the border.

Step 2 Double-click on one of the *MS viewports* to make it active. *Command:* Zoom (Enter)

All/Center/ . . . /<Scale (X/XP)>: .2XP (Enter)

The active port is scaled to an 0.2 size to fit in the *PS* border. *Zoom* the other viewport to .2XP also.

The cursor spans the screen in *Paper Space*. Now the drawing must be scaled.

The Paper Space *Limits* are 12,9, inside of which two *MS* viewports must fit, each with *Limits* of 24,20. These *MS* viewports must be sized to fit, which requires calculations. The combined width of the two 24-in. wide Model Spaces is 48 in. When scaled to half size (0.50), their width is 24 in. and too wide to fit. If scaled to 0.20 (20% size), their width is 9.6 in. and small enough to fit inside the A-size border.

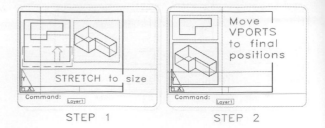

STEP 1   STEP 2

38.98 Paper Space: Stretch.

Step 1 Double-click in a *Paper Space* area to switch to *Paper Space*. Use *Stretch* to reduce the size of the *MS* viewport.

Step 2 *Move* the viewports to their final positions within the *Paper Space* border.

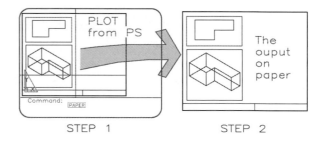

STEP 1   STEP 2

38.99 Paper Space: Plot.

Step 1 Select *Main Menu> File> Plot* and give specifications for plotters and plotting from *Paper Space*.

Step 2 The size-A layout is plotted to show both 2D (*Paper Space*) and 3D (*Model Space*) drawings in the same plot.

From model space type Zoom and use the *NXP* option (2/10XP or 0.2XP) to size the contents of each viewport (**Figure 38.97**). This factor changes the width limit of each viewport from 24 to 4.8 inches with both drawings having the same scale on the screen. In other words, you may scale viewports with width limits of 24 inches to a full-size width of 4.8 inches in *Paper Space*.

Double-click on a paper-space area to enter *Paper Space* and use *Stretch* to reduce the size of the *MS* viewport outlines (Figure 38.98). Reposition the *MS* viewports for plotting with Move.

Plotting must be performed from *Paper Space*; both the *PS* and *MS* drawings are plotted at the same time including the outlines of the MS viewports (**Figure 38.99**). To remove

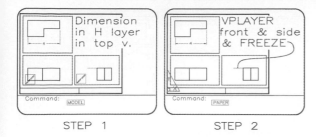

STEP 1   STEP 2

38.100 *Model Space: Dimensioning–Horizontal.*

Step 1 Create layer *H* and set it *On*. Set the *Dimscale* variable to *0* and the dimensioning variables in *Model Space* will match those in *Paper Space*. Move the *UCS* icon to the plane of the top-view dimension, and apply the dimensions by snapping to the endpoints of the part.

Step 2 When a dimension is applied in one viewport it is shown in all *MS* viewports, and the dimensioning variables are the same size in *MS* as in *PS*. Dimensions appear as edges in the front and side views.

the *MS* viewport outlines, *Change* them to a separate *Layer* (*Window,* for example), *Freeze* it, and *Plot* in the usual manner.

Remove the hidden lines when plotting by selecting a *MS* viewport from *PS* by double-clicking inside the port, and typing *Hide* at the *Command* line. Repeat this for each viewport in which lines are to be hidden when plotted. Use *Vplayer* to select *MS viewports* from *Paper Space* in which layers can be turned *Off* or *Frozen* while remaining *On* or *Thawed* in other viewports.

38.27 Dimensioning in 3D

Dimensions can be applied to objects in *Model Space* using dimensioning variables set for *Paper Space* by setting *Dimscale = 0* (**Figure 38.100**). Create two new layers, H̲ and F̲ on which horizontal and frontal dimensions will be placed. Set associative dimensions on (*Dimassoc = On*), set the layer *H* on, set the *UCS* to P̲lan in the top view (see *XY* icon), and *Snap* to the endpoints of the object to attach width and depth dimension to the top view. These dimensions appear as edges in the front and side views.

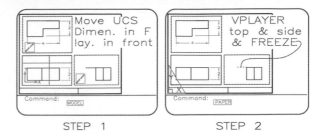

STEP 1   STEP 2

38.101 *Model Space: Dimensioning–Frontal.*

Step 1 Create layer *F* and set it *On*. *Move* the *UCS* icon to the plane of the front-view dimension. Apply the dimension.

Step 2 The dimension appears as an edge in the top and front views.

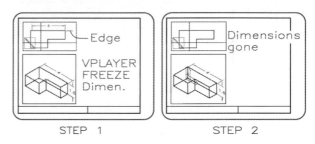

STEP 1   STEP 2

38.102 *Dimensioning: Vplayer option.*

Step 1 Since the frontal dimension is not readable in the top view, set *Vplayer* to O̲n, select the border of the *Model Space* viewport while in *PS*, and *Freeze* layer *F* in that view.

Step 2 The dimensions are removed in the top view but remain in the *MS* view. *Freeze* layer *H* in the front and side views, also.

Select the front *Viewport* and rotate the *XY* icon parallel to the frontal plane (*UCS> X> 90*), as shown in **Figure 38.101**. Attach a vertical height dimension to the object by *Snapping* to the endpoints of the front view, and the dimension appears as an edge in the top and side view.

Since the vertical dimensions cannot be seen in the top and side views, select these viewports, type V̲player, and *Freeze* the F̲ layer (**Figure 38.102**).

Since the horizontal dimensions cannot be seen in the front and side views, select these viewports, type V̲player, and *Freeze* the H̲ layer. Other dimensions can be added and edited in this manner.

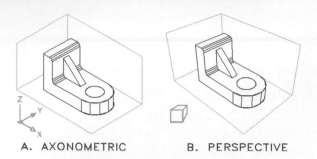

A. AXONOMETRIC   B. PERSPECTIVE

38.103 This part is used to illustrate rendering techniques in the following examples.

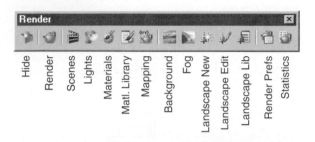

Hide Render Scenes Lights Materials Matl. Library Mapping Background Fog Landscape New Landscape Edit Landscape Lib Render Prefs Statistics

38.104 Select *Main Menu> View> Toolbars> Render* to obtain this toolbar for rendering 3D models.

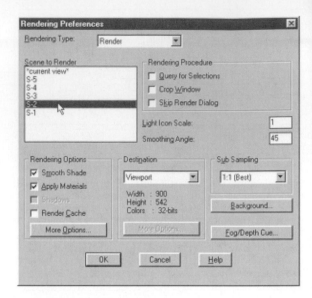

38.105 Select *Render toolbar> Render preferences icon* to obtain this dialogue box.

38.106 The *More Options* button of the *Rendering Preferences* menu gives this dialogue box, *Render Options*.

Options under *Vplayer* that operate from *Paper Space* are *Freeze, Thaw, Reset, Newfrz,* and *Vpvisdflt.*

Freeze/Thaw: Allows specified layers in selected *Vports* to be frozen or thawed. *Thaw* does not work on globally frozen layers.

Reset: Changes visibility of one or more layers in selected *Vports* to their current default setting.

Newfrz: Creates a new layer that is visible in the current viewport and is frozen in the rest.

Vpvisdflt: Used to set default visibility in a viewport for any layer. The default setting makes layers frozen or thawed in new viewports.

38.28 Render

Rendering is the process giving 3D objects a realistic appearance by adding color, lighting, and materials to them. The following examples illustrate how 3D drawings are rendered.

The bracket from **Figure 38.77** is displayed in two viewports as an axonometric and a perspective (**Figure 38.103**). From the Render toolbar in (**Figure 38.104**), select *Render Preferences* to obtain the dialogue box shown in **Figure 38.105**. *Rendering Options* are S*mooth Shading, Apply Materials, Shadows,* and *Render Cache.* The More *Options* box gives *Gouraud* and *Phong* options to specify rendering quality (**Figure 38.106**). The *Phong* option gives the smoothest, most realistic rendering.

While a view of the bracket is on the screen, select *Render* from the *Render* toolbar, or *Command:* <u>Render</u> (Enter), and the *Render* dialogue box appears (**Figure 38.105**). Make selections and the bracket is rendered as shown in **Figure 38.107**.

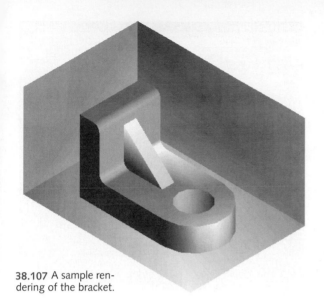

38.107 A sample rendering of the bracket.

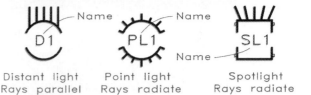

Distant light   Point light   Spotlight
Rays parallel   Rays radiate   Rays radiate

38.108 Light symbols.
Distant light, point light, and *spotlight* symbols are placed on drawings to indicate the positions for lighting.

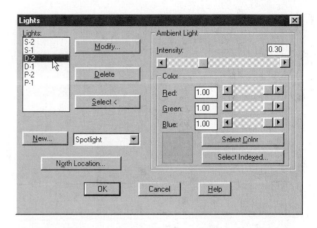

38.109 The *Lights* dialogue box is used to set and modify lights for rendering.

38.29 Lights

A rendering is enhanced by *point lights, distant lights, spotlights,* and *ambient lighting* that are under your control.

Point light emits rays in all directions from a point source.

Distant light emits parallel beams like those of sunlight.

Spotlights emit a cone of light toward a selected target.

Ambient light comes from no particular source and provides a constant illumination to all surfaces of an object.

New Lights

Light sources can be added to a drawing and indicated with the symbols shown in **Figure 38.108**. From the *Render* toolbar, select *Lights* to get the *Lights* dialogue box shown in **Figure 38.109** where several lights are listed. Select *Distant Light* next to the New button, pick *New* to get the *New Distant Light* dialogue box (**Figure 38.110**). From this menu, you can name the distant light, set its directions, and pick its colors. Select the *Distant Light* button and pick <u>Ok</u> (**Figure 38.110**). *Light Intensity* (0 = off, 1 = bright) is set with the slider bar or by

typing. Leave *Intensity* set to 1, select <u>Ok</u> to return to *Lights* box (where D-2 is listed), and pick <u>Ok</u> to exit. Select *Render* from the pull-down menu to obtain a new image of the bracket using *Distant Light,* D-2.

If you had selected *Point Light* and *New,* the *New Point Light* dialogue box would have appeared on the screen (**Figure 38.111**). This dialogue box is different from the *New Distant Light* box, since point lights have different characteristics.

Modifying Lights

Select *Lights* from the *Render* toolbar menu, select *Distant Light* to obtain the dialogue box shown in **Figure 38.109**, and double-click on

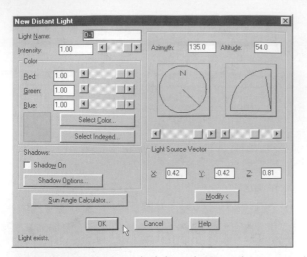

38.110 The *New Distant Light* dialogue box is used to create and name a new light source.

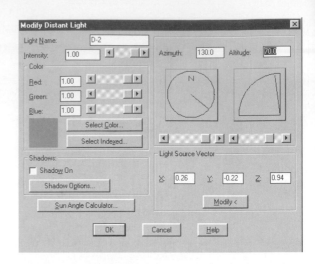

38.112 The *Modify Distant Light* dialogue box is used to make changes in an existing light.

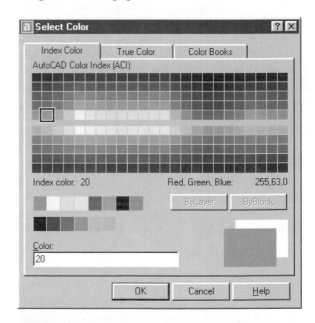

38.111 This menu is used to create a *New Point Light*.

38.113 The *Render* toolbar> *Lights* icon> *Select Custom Color*> *Select Color* menu is displayed for selecting colors.

D-1 (distant light) to obtain the *Modify Distant Light* box (**Figure 38.112**). The *Select Color* button can be picked to obtain the *Select Color* dialogue box (**Figure 38.113**). Experiment with *Intensity* at other settings (for example, 0.2 and 0.6) and render an object to observe variations in the brightness of the image.

Moving Lights

To move distant light, *D-2*, use *Command*: <u>Plan</u> (Enter) to get a top view of the bracket where the light sources are shown as symbols. Each light can be moved with the *Move* command (**Figure 39.114**). You can also select *Modify*

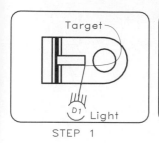
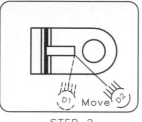

STEP 1   STEP 2

38.114 Moving a light.

Step 1 Select *Modify* from the *Modify Distant Light* dialogue box (Fig. 38.112) and you are transferred to the drawing.

Enter light target <current>: (Enter) Select current target.

Step 2 *Enter light location <current>:* .XY (Enter)> of (Need Z): 6 (Enter) The *Modify Distant Light* box reappears. Click on OK to exit and *Command:* Render (Enter) to render the bracket.

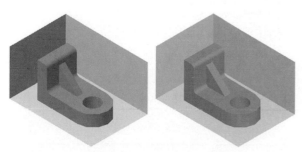

A. Distant Light: Position 1   B. Distant Light: Position 2

38.115 A comparison of the rendered views of the bracket with *Distant Light D-1* in Position 1 and *D-2* in Position 2.

from the *Lights* menu, pick distant light *D-1*, and pick *Modify* to obtain the *Modify Distant Light* menu (**Figure 38.112**). Select *Modify* under the heading *Light Source Vector* and the screen returns to the drawing where you are prompted for the direction of the light rays. The direction can also be changed by the Azimuth and Altitude adjustments (**Figure 38.112**).

In Step 1, the prompt *"Enter light direction To <current>:"* appears with a rubber-band line attached to the target. Press (Enter) to retain the current target and obtain a rubber-band line attached to it and the prompt *"Enter light direction FROM <current>."* In Step 2, type .XY, pick the light location with the cursor, and

38.116 The type of point light fall-off can be specified as either *Inverse Linear* or *Inverse Square*.

respond to the prompt "of (*Need Z*)" with 8 to locate the light source and establish the direction of the parallel light rays.

Type Vpoint and give settings of 1,-1,1 to obtain an isometric view of the bracket. Select *Render* from the *Render* toolbar to observe the original and new light position in **Figure 38.115**.

Light Fall-off

The characteristic whereby light becomes dimmer as it travels farther from its source is called *fall-off*. *Point Lights* and *Spotlights* are affected by fall-off; *Distant* and *Ambient Lights* have uniform intensity at all distances. From the *Lights* menu select P1 (*Point Light*), and select *Modify* to get the *Modify Point Light* box, which has an *Attenuation* area. The options available here are *None*, *Inverse Linear*, and *Inverse Square* (**Figure 38.116**):

None is a light with no fall-off, giving all objects the same brightness.

Inverse Linear makes a surface that is 4 units from the light one-fourth as bright, and one-eighth as bright when 8 units away.

Inverse Square makes and object 4 units from the light one-sixteenth as bright, and one-sixty-fourth as bright when 8 units away.

A comparison of lighting set to *Inverse Linear* and *Inverse Square* is shown in **Figure 29.117**.

Ambient Light

So far, ambient light has been set to 0.3. From the *Render* toolbar obtain the *Lights* dialogue box and set ambient intensity to 0.85 by typing

A. Inverse Linear   B. Inverse Square

38.117 *Light Fall-off* options are *None, Inverse Linear,* and *Inverse Square.* By using *Inverse Linear,* an object 4 units away is one-fourth as bright as one 1 unit away from the light. By using *Inverse Square,* an object 4 units away is one-sixteenth as bright as one 1 unit away. These options are not available for *Distant Lights.*

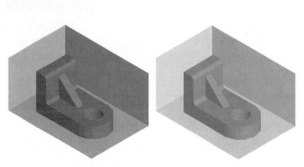

A. Intensity=0.30   B. Intensity=0.80

38.118 A comparison of *Ambient light* set at two intensities is shown here.

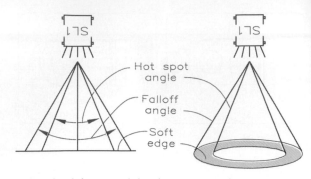

38.119 The definitions of the characteristics of a spotlight are shown here.

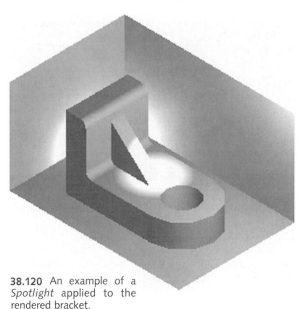

38.120 An example of a *Spotlight* applied to the rendered bracket.

or using the slider bar. The variation in lighting the bracket is shown in **Figure 38.118** when it is applied to our example.

Spotlights

Spotlights emit cones of light as defined in **Figure 38.119** to highlight features. The creation of a *Spotlight* begins from the *Lights* dialogue box (**Figure 38.109**) where *Spotlight* and *New* are selected to display the *Create New Spotlight* box. You can see the effects of a spotlight in **Figure 38.120.**

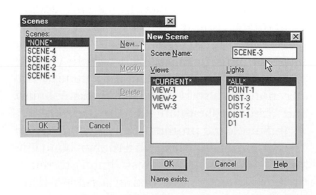

38.121 The *Render* toolbar> *Scene* icon gives the *Scenes* dialogue box where the *New* button can be picked to obtain the *New Scene* box for naming a new scene.

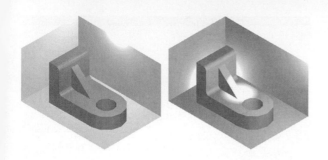

A. Scene-1: Two lights   B. Scene-2:Two lights and
a spotlight

38.122 Examples of *Scenes* are shown here with different combinations of lights.

38.30 Scenes

Views and light settings can be saved in combination as *Scenes* which can have several lights (or no lights) but only one *View*. A viewpoint of a drawing can be saved as a View by typing <u>View</u> and giving it a name. To recall a *View*, use *Command*: <u>View</u> (Enter) <u>Restore</u> and give its name when prompted.

Restored Views of a drawing can be selected from Scenes of the *Render* menu to obtain the *Scenes* dialogue box (**Figure 38.121**). Select *New* and the *New Scene* box appears, in which all Model-Space *Views* and lights are listed. Type the name of the *Scene* in the box (<u>Scene-3</u>, for example). The name is truncated to eight characters if it is longer than eight. Select *View-2* as the view in *Scene-3*, select *Point Light P*1 and *Distant Light Dist-1*, select *OK*, and the *Scenes* dialogue box reappears. Other scenes can be created by using these steps.

If **None** were selected as the scene to render, all lights would be on in the current view. If no lights existed, an over-the-shoulder distant light would be given. Select *Scene-2* from the dialogue box, pick *OK*, and choose *Render* to render *Scene-2*.

A Scene can be modified by selecting it from the Scenes dialogue box (**Figure 38.121**) and picking <u>Modify</u> to obtain the *Modify Scene* dialogue box, which is similar to the *New Scene* dia-

38.123 The *Materials* and *Materials Library* icons are used in assigning materials that are applied to 3D objects.

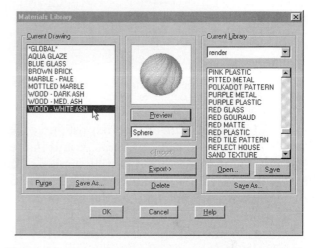

38.124 Select the *Materials Library* icon from the *Render* toolbar to get this dialogue box, which gives the materials that can be assigned to 3D models. A preview is shown of a sphere.

logue box. Different lights can be selected or deselected. Select *OK* to keep your modifications and return to the *Scenes* dialogue box, select *OK* to exit and *Render* the modified *Scene*.

38.31 Materials

Select the *Materials Library* icon from the *Render* toolbar (**Figure 38.123**) to get the *Materials Library* dialogue box shown in **Figure 38.124**. The material of an object determines the reflective quality of its surfaces from dull to shiny. Materials can be selected one at a time from the listing and previewed by selecting the *Preview* button. To add a material to the *Current Drawing* list for future use, pick the

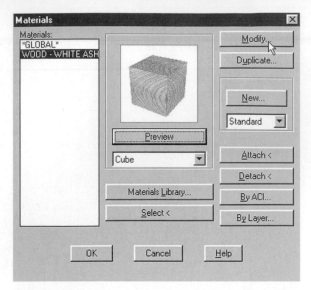

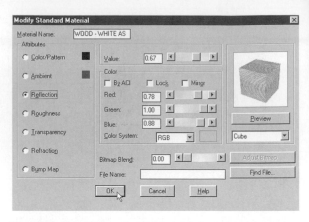

38.126 The *Modify Standard Material* dialogue box lets you preview variations in the application of materials as they are assigned to a part.

38.125 Use the *Render* toolbar> *Materials*> to get this *Materials* box. Select the Modify button to get the *Modify Standard Material* dialogue box.

material from the *Current Library* list and pick the *Import* button. Click on *Save As* and name the file which saves the material symbols for future use. Pick *OK* to exit the *Materials Library* dialogue box and return to the drawing.

From the *Render* toolbar, select *Materials* to get the dialogure box shown in **Figure 38.124**. To attach a material to a part, select it from the list, select <u>Attach</u>, and the drawing reappears. You are prompted, "*Select objects to attach 'MATL' to:*" Select the part, and the *Materials* dialogue box returns; click *OK* to exit. Select *Command*: <u>Render</u> (Enter)>*Photo Real* option, and the object is rendered with the materials applied.

With *Main Menu> View> Render> Materials*, the *Materials* dialogue box appears (**Figure 38.125**). Pick the *Modify* button to get the *Modify Standard Material* box (**Figure 38.126**), in which a variety of settings can be made. Experiment with different settings, and observe the results of these changes by picking *Preview*.

Use *Main Menu> Render> Preferences> Shadows* (under *Rendering Options*) to set *Rendering Type* to *Photo Real* and create a *New Light* and *Shadows*, and materials can be applied to the 3D solid as illustrated in **Figure 38.127**.

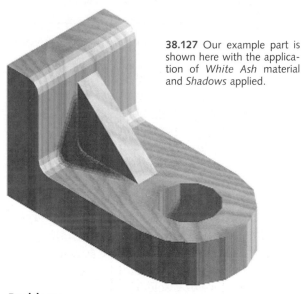

38.127 Our example part is shown here with the application of *White Ash* material and *Shadows* applied.

Problems

General: Most problems can be solved and plotted on A-size sheets using a text height of no smaller than 0.10 inches.

1. *Exercises*: Repeat each of the example figures in this chapter on your computer following the instructions given in the legends, and plot the examples. This technique will help you learn the fundamental commands.

2. *Problems*: The problems at the ends of the previous chapters can be drawn and plotted using the computer-graphics techniques covered in this and previous chapters.

APPENDIX 1 • Decimal Equivalents and Temperature Conversions

DECIMAL EQUIVALENTS—INCH-MILLIMETER CONVERSION TABLE

1/2	1/4	1/8	1/16	1/32	1/64	Decimals	Millimeters
					1	.015625	.396875
				1	3	.031250	.793750
					3	.046875	1.190625
			1			.062500	1.587500
					5	.078125	1.984375
				3		.093750	2.381250
					7	.109375	2.778125
		1				.125000	3.175000
					9	.140625	3.571875
				5		.156250	3.968750
					11	.171875	4.365625
			3			.187500	4.762500
					13	.203125	5.159375
				7		.218750	5.556250
					15	.234375	5.953125
	1					.250000	6.350000
					17	.265625	6.746875
				9		.281250	7.143750
					19	.296875	7.540625
			5			.312500	7.937500
					21	.328125	8.334375
				11		.343750	8.731250
					23	.359375	9.128125
		3				.375000	9.525000
					25	.390625	9.921875
				13		.406250	10.318750
					27	.421875	10.715625
			7			.437500	11.112500
					29	.453125	11.509375
				15		.468750	11.906250
					31	.484375	12.303125
1						.500000	12.700000

1/2	1/4	1/8	1/16	1/32	1/64	Decimals	Millimeters
					33	.515625	13.096875
				17		.531250	13.493750
					35	.546875	13.890625
			9			.562500	14.287500
					37	.578125	14.684375
				19		.593750	15.081250
					39	.609375	15.478125
		5				.625000	15.875000
					41	.640625	16.271875
				21		.656250	16.668750
					43	.671875	17.065625
			11			.687500	17.462500
					45	.703125	17.859375
				23		.718750	18.256250
					47	.734375	18.653125
	3					.750000	19.050000
					49	.765625	19.446875
				25		.781250	19.843750
					51	.796875	20.240625
			13			.812500	20.637500
					53	.828125	21.034375
				27		.843750	21.431250
					55	.859375	21.828125
		7				.875000	22.225000
					57	.890625	22.621875
				29		.906250	23.018750
					59	.921875	23.415625
			15			.937500	23.812500
					61	.953125	24.209375
				31		.968750	24.606250
					63	.984375	25.003125
2	4	8	16	32	64	1.00000	25.400000

LENGTH

1 millimeter (mm) = 0.03937 inch
1 centimeter (cm) = 0.39370 inch
1 meter (m) = 39.37008 inches
1 meter = 3.2808 feet
1 meter = 1.0936 yards
1 kilometer (km) = 0.6214 miles
1 inch = 25.4 millimeters
1 inch = 2.54 centimeters
1 foot = 304.8 millimeters
1 foot = 0.3048 meters
1 yard = 0.9144 meters
1 mile = 1.609 kilometers

AREA

1 square millimeter = 0.00155 square inch
1 square centimeter = 0.155 square inch
1 square meter = 10.764 square feet
1 square meter = 1.196 square yards
1 square kilometer = 0.3861 square mile
1 square inch = 645.2 square millimeters
1 square inch = 6.452 square centimeters
1 square foot = 929 square centimeters
1 square foot = 0.0929 square meter
1 square yard = 0.836 square meter
1 square mile = 2.5899 square kilometers

DRY CAPACITY

1 cubic centimeter (cm3) = 0.061 cubic inches
1 liter = 0.0353 cubic foot
1 liter = 61.023 cubic inches
1 cubic meter (m3) = 35.315 cubic feet
1 cubic meter = 1.308 cubic yards
1 cubic inch = 16.38706 cubic centimeters
1 cubic foot = 0.02832 cubic meter
1 cubic foot = 28.317 liters
1 cubic yard = 0.7646 cubic meter

LIQUID CAPACITY

1 liter = 1.0567 U.S. quarts
1 liter = 0.2642 U.S. gallons
1 liter = 0.2200 Imperial gallons
1 cubic meter = 264.2 U.S. gallons
1 cubic meter = 219.969 Imperial gallons
1 U.S. quart = 0.946 liters
1 Imperial quart = 1.136 liters
1 U.S. gallon = 3.785 liters
1 Imperial gallon = 4.546 liters

WEIGHT

1 gram (g) = 15.432 grains
1 gram = 0.03215 ounce troy
1 gram = 0.03527 ounce avoirdupois
1 kilogram (kg) = 35.274 ounces avoirdupois
1 kilogram = 2.2046 pounds
1000 kilograms = 1 metric ton (t)
1000 kilograms = 1.1023 tons of 2000 pounds
1000 kilograms = 0.9842 tons of 2240 pounds
1 ounce avoirdupois = 28.35 grams
1 ounce troy = 31.103 grams
1 pound = 453.6 grams
1 pound = 0.4536 kilogram
1 ton of 2240 pounds = 1016 kilograms
1 ton of 2240 pounds = 1.016 metric tons
1 grain = 0.0648 gram
1 metric ton = 0.9842 tons of 2240 pounds
1 metric ton = 2204.6 pounds

N	0	1	2	3	4	5	6	7	8	9
1.0	.0000	.0043	.0086	.0128	.0170	.0212	.0253	.0294	.0334	.0374
1.1	.0414	.0453	.0492	.0531	.0569	.0607	.0645	.0682	.0719	.0755
1.2	.0792	.0828	.0864	.0899	.0934	.0969	.1004	.1038	.1072	.1106
1.3	.1139	.1173	.1206	.1239	.1271	.1303	.1335	.1367	.1399	.1430
1.4	.1461	.1492	.1523	.1553	.1584	.1614	.1644	.1673	.1703	.1732
1.5	.1761	.1790	.1818	.1847	.1875	.1903	.1931	.1959	.1987	.2014
1.6	.2041	.2068	.2095	.2122	.2148	.2175	.2201	.2227	.2253	.2279
1.7	.2304	.2330	.2355	.2380	.2405	.2430	.2455	.2480	.2504	.2529
1.8	.2553	.2577	.2601	.2625	.2648	.2672	.2695	.2718	.2742	.2765
1.9	.2788	.2810	.2833	.2856	.2878	.2900	.2923	.2945	.2967	.2989
2.0	.3010	.3032	.3054	.3075	.3096	.3118	.3139	.3160	.3181	.3201
2.1	.3222	.3243	.3263	.3284	.3304	.3324	.3345	.3365	.3385	.3404
2.2	.3424	.3444	.3464	.3483	.3502	.3522	.3541	.3560	.3579	.3598
2.3	.3617	.3636	.3655	.3674	.3692	.3711	.3729	.3747	.3766	.3784
2.4	.3802	.3820	.3838	.3856	.3874	.3892	.3909	.3927	.3945	.3962
2.5	.3979	.3997	.4014	.4031	.4048	.4065	.4082	.4099	.4116	.4133
2.6	.4150	.4166	.4183	.4200	.4216	.4232	.4249	.4265	.4281	.4298
2.7	.4314	.4330	.4346	.4362	.4378	.4393	.4409	.4425	.4440	.4456
2.8	.4472	.4487	.4502	.4518	.4533	.4548	.4564	.4579	.4594	.4609
2.9	.4624	.4639	.4654	.4669	.4683	.4698	.4713	.4728	.4742	.4757
3.0	.4771	.4786	.4800	.4814	.4829	.4843	.4857	.4871	.4886	.4900
3.1	4914	.4928	.4942	.4955	.4969	.4983	.4997	.5011	.5024	.5038
3.2	.5051	.5065	.5079	.5092	.5105	.5119	.5132	.5145	.5159	.5172
3.3	.5185	.5198	.5211	.5224	.5237	.5250	.5263	.5276	.5289	.5302
3.4	.5315	.5238	.5340	.5353	.5366	.5378	.5391	.5403	.5416	.5428
3.5	.5441	.5453	.5465	.5478	.5490	.5502	.5514	.5527	.5539	.5551
3.6	.5563	.5575	.5587	.5599	.5611	.5623	.5635	.5647	.5658	.5670
3.7	.5682	.5694	.5705	.5717	.5729	.5740	.5752	.5763	.5775	.5786
3.8	.5798	.5809	.5821	.5832	.5843	.5855	.5866	.5877	.5888	.5899
3.9	.5911	.5922	.5933	.5944	.5955	.5966	.5977	.5988	.5999	.6010
4.0	.6021	.6031	.6042	.6053	.6064	.6075	.6085	.6096	.6107	.6117
4.1	.6128	.6138	.6149	.6160	.6170	.6180	.6191	.6201	.6212	.6222
4.2	.6232	.6243	.6253	.6263	.6274	.6284	.6294	.6304	.6314	.6325
4.3	.6335	.6345	.6355	.6365	.6375	.6385	.6395	.6405	.6415	.6425
4.4	.6435	.6444	.6454	.6464	.6474	.6484	.6493	.6503	.6513	.6522
4.5	.6532	.6542	.6551	.6561	.6571	.6580	.6590	.6599	.6609	.6618
4.6	.6628	.6637	.6646	.6656	.6665	.6675	.6684	.6693	.6702	.6712
4.7	.6721	.6730	.6739	.6749	.6758	.6767	.6776	.6785	.6794	.6803
4.8	.6812	.6821	.6830	.6839	.6848	.6857	.6866	.6875	.6884	.6893
4.9	.6902	.6911	.6920	.6928	.6937	.6946	.6955	.6964	.6972	.6981
5.0	.6990	.6998	.7007	.7016	.7024	.7033	.7042	.7050	.7059	.7067
5.1	.7076	.7084	.7093	.7101	.7110	.7118	.7216	.7135	.7143	.7152
5.2	.7160	.7168	.7177	.7185	.7193	.7202	.7210	.7218	.7226	.7235
5.3	.7243	.7251	.7259	.7267	.7275	.7284	.7292	.7300	.7308	.7316
5.4	.7324	.7332	.7340	.7348	.7356	.7364	.7372	.7380	.7388	.7396
N	0	1	2	3	4	5	6	7	8	9

N	0	1	2	3	4	5	6	7	8	9
5.5	.7404	.7412	.7419	.7427	.7435	.7443	.7451	.7459	.7466	.7474
5.6	.7482	.7490	.7497	.7505	.7513	.7520	.7528	.7536	.7543	.7551
5.7	.7559	.7566	.7574	.7582	.7589	.7597	.7604	.7612	.7619	.7627
5.8	.7634	.7642	.7649	.7657	.7664	.7672	.7679	.7686	.7694	.7701
5.9	.7709	.7716	.7723	.7731	.7738	.7745	.7752	.7760	.7767	.7774
6.0	.7782	.7789	.7796	.7803	.7810	.7818	.7825	.7832	.7839	.7846
6.1	.7853	.7860	.7868	.7875	.7882	.7889	.7896	.7903	.7910	.7917
6.2	.7924	.7931	.7938	.7945	.7952	.7959	.7966	.7973	.7980	.7987
6.3	.7993	.8000	.8007	.8014	.8021	.8028	.8035	.8041	.8048	.8055
6.4	.8062	.8069	.8075	.8082	.8089	.8096	.8102	.8109	.8116	.8122
6.5	.8129	.8136	.8142	.8149	.8156	.8162	.8169	.8176	.8182	.8189
6.6	.8195	.8202	.8209	.8215	.8222	.8228	.8235	.8241	.8248	.8254
6.7	.8261	.8267	.8274	.8280	.8287	.8293	.8299	.8306	.8312	.8319
5.8	.8325	.8331	.8338	.8344	.8351	.8357	.8363	.8370	.8376	.8382
6.9	.8388	.8395	.8401	.8407	.8414	.8420	.8426	.8432	.8439	.8445
7.0	.8451	.8457	.8453	.8470	.8476	.8482	.8488	.8494	.8500	.8506
7.1	.8513	.8519	.8525	.8531	.8537	.8543	.8549	.8555	.8561	.8567
7.2	.8573	.8579	.8585	.8591	.8597	.8603	.8609	.8615	.8621	.8627
7.3	.8633	.8639	.8645	.8651	.8657	.8663	.8669	.8675	.8681	.8686
7.4	.8692	.8698	.8704	.8710	.8716	.8722	.8727	.8733	.8739	.8745
7.5	.8751	.8756	.8762	.8768	.8774	.8779	.8785	.8791	.8797	.8802
7.6	.8808	.8814	.8820	.8825	.8831	.8837	.8842	.8848	.8854	.8859
7.7	.8865	.8871	.8876	.8882	.8887	.8893	.8899	.8904	.8910	.8915
7.8	.8921	.8927	.8932	.8938	.8943	.8949	.8954	.8960	.8965	.8971
7.9	.8976	.8982	.8987	.8993	.8998	.9004	.9009	.9015	.9020	.9025
8.0	.9031	.9036	.9042	.9047	.9053	.9058	.9063	.9069	.9074	.9079
8.1	.9085	.9090	.9096	.9101	.9106	.9112	.9117	.9122	.9128	.9133
8.2	.9138	.9143	.9149	.9154	.9159	.9165	.9170	.9175	.9180	.9186
8.3	.9191	.9196	.9201	.9206	.9212	.9217	.9222	.9227	.9232	.9238
8.4	.9243	.9248	.9253	.9258	.9263	.9269	.9274	.9279	.9284	.9289
8.5	.9294	.9299	.9304	.9309	.9315	.9320	.9325	.9330	.9335	.9340
8.6	.9345	.9350	.9355	.9360	.9365	.9370	.9375	.9380	.9385	.9390
8.7	.9395	.9400	.9405	.9410	.9415	.9420	.9425	.9430	.9435	.9440
8.8	.9445	.9450	.9455	.9460	.9465	.9469	.9474	.9479	.9484	.9489
8.9	.9494	.9499	.9504	.9509	.9513	.9518	.9523	.9528	.9533	.9538
9.0	.9542	.9547	.9552	.9557	.9562	.9566	.9571	.9576	.9581	.9586
9.1	.9590	.9595	.9600	.9605	.9609	.9614	.9619	.9624	.9628	.9633
9.2	.9638	.9643	.9647	.9652	.9657	.9661	.9666	.9671	.9675	.9680
9.3	.9685	.9689	.9694	.9699	.9703	.9708	.9713	.9717	.9722	.9727
9.4	.9731	.9736	.9741	.9745	.9750	.9754	.9759	.9763	.9768	.9773
9.5	.9777	.9782	.9786	.9791	.9795	.9800	.9805	.9809	.9814	.9818
9.6	.9823	.9827	.9832	.9836	.9841	.9845	.9850	.9854	.9859	.9863
9.7	.9868	.9872	.9877	.9881	.9886	.9890	.9894	.9899	.9903	.9908
9.8	.9912	.9917	.9921	.9926	.9930	.9934	.9939	.9943	.9948	.9952
9.9	.9956	.9961	.9965	.9969	.9974	.9978	.9983	.9987	.9991	.9996
N	0	1	2	3	4	5	6	7	8	9

ANGLE in DEGREES	SINE	COSINE	TAN	COTAN	ANGLE in DEGREES
0	0.0000	1.0000	0.0000		90
1	.0175	.9998	.0175	57.290	89
2	.0349	.9994	.0349	28.636	88
3	.0523	.9986	.0524	19.081	87
4	.0698	.9976	.0699	14.301	86
5	.0872	.9962	.0875	11.430	85
6	.1045	.9945	.1051	9.5144	84
7	.1219	.9925	.1228	8.1443	83
8	.1392	.9903	.1405	7.1154	82
9	.1564	.9877	.1584	6.3138	81
10	.1736	.9848	.1763	5.6713	80
11	.1908	.9816	.1944	5.1446	79
12	.2079	.9781	.2126	4.7046	78
13	.2250	.9744	.2309	4.3315	77
14	.2419	.9703	.2493	4.0108	76
15	.2588	.9659	.2679	3.7321	75
16	.2756	.9613	.2867	3.4874	74
17	.2924	.9563	.3057	3.2709	73
18	.3090	.9511	.3249	3.0777	72
19	.3256	.9455	.3443	2.9042	71
20	.3420	.9397	.3640	2.7475	70
21	.3584	.9336	.3839	2.6051	69
22	.3746	.9272	.4040	2.4751	68
23	.3907	.9205	.4245	2.3559	67
24	.4067	.9135	.4452	2.2460	66
25	.4226	.9063	.4663	2.1445	65
26	.4384	.8988	.4877	2.0503	64
27	.4540	.8910	.5095	1.9626	63
28	.4695	.8829	.5317	1.8807	62
29	.4848	.8746	.5543	1.8040	61
30	.5000	.8660	.5774	1.7321	60
31	.5150	.8572	.6009	1.6643	59
32	.5299	.8480	.6249	1.6003	58
33	.5446	.8387	.6494	1.5399	57
34	.5592	.8290	.6745	1.4826	56
35	.5736	.8192	.7002	.14281	55
36	.5878	.8090	.7265	1.3764	54
37	.6018	.7986	.7536	1.3270	53
38	.6157	.7880	.7813	1.2799	52
39	.6293	.7771	.8098	1.2349	51
40	.6428	.7660	.8391	1.1918	50
41	.6561	.7547	.8693	1.1504	49
42	.6691	.7431	.9004	1.1106	48
43	.6820	.7314	.9325	1.0724	47
44	.6947	.7193	.9657	1.0355	46
45	.7071	.7071	1.0000	1.0000	45

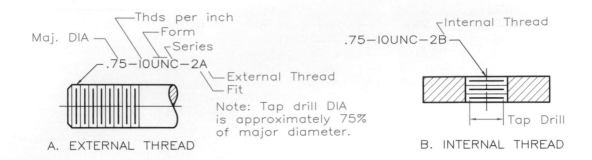

A. EXTERNAL THREAD

B. INTERNAL THREAD

Note: Tap drill DIA is approximately 75% of major diameter.

Nominal Diameter	Basic Diameter	Coarse NC & UNC		Fine NF & UNF		Extra Fine NEF/UNEF	
		Thds per in.	Tap Drill DIA	Thds per in.	Tap Drill DIA	Thds per in.	Tap Drill DIA
0	.060			80	.0469		
1	.073	64	No. 53	72	No. 53		
2	.086	56	No. 50	64	No. 50		
3	.099	48	No. 47	56	No. 45		
4	.112	40	No. 43	48	No. 42		
5	.125	40	No. 38	44	No. 37		
6	.138	32	No. 36	40	No. 33		
8	.164	32	No. 29	36	No. 29		
10	.190	24	No. 25	32	No. 21		
12	.216	24	No. 16	28	No. 14	32	No. 13
1/4	.250	20	No. 7	28	No. 3	32	.2189
5/16	.3125	18	F	24	I	32	.2813
3/8	.375	16	.3125	24	Q	32	.3438
7/16	.4375	14	U	20	.3906	28	.4062
1/2	.500	13	.4219	20	.4531	28	.4688
9/16	.5625	12	.4844	18	.5156	24	.5156
5/8	.625	11	.5313	18	.5781	24	.5781
11/16	.6875	...	...	...	...	24	.6406
3/4	.750	10	.6563	16	.6875	20	.7031
13/16	.8125	...	...	...	...	20	.7656
7/8	.875	9	.7656	14	.8125	20	.8281
15/16	.9375	...	...	...	...	20	.8906

Nominal Diameter	Basic Diameter	Coarse NC & UNC		Fine NF & UNF		Extra Fine NEF/UNEF	
		Thds per in.	Tap Drill DIA	Thds per in.	Tap Drill DIA	Thds per in.	Tap Drill DIA
1	1.000	8	.875	12	.922	20	.953
1-1/16	1.063	...	...	...	...	18	1.000
1-1/8	1.125	7	.904	12	1.046	18	1.070
1-3/16	1.188	...	...	...	...	18	1.141
1-1/4	1.250	7	1.109	12	1.172	18	1.188
1-5/16	1.313	...	...	...	...	18	1.266
1-3/8	1.375	6	1.219	12	1.297	18	1.313
1-7/16	1.438	...	...	...	...	18	1.375
1-1/2	1.500	6	1.344	12	1.422	18	1.438
1-9/16	1.563	...	...	...	...	18	1.500
1-5/8	1.625	...	...	...	...	18	1.563
1-11/16	1.688	...	...	...	...	18	1.625
1-3/4	1.750	5	1.563	...	...	...	...
2	2.000	4.5	1.781	...	...	...	...
2-1/4	2.250	4.5	2.031	...	...	...	...
2-1/2	2.500	4	2.250	...	...	...	...
2-3/4	2.750	4	2.500	...	...	...	...
3	3.000	4	2.750	...	...	...	...
3-1/4	3.250	4	...	...	...	...	...
3-1/2	3.500	4	...	...	...	...	...
3-3/4	3.750	4	...	...	...	...	...
4	4.000	4	...	...	...	...	...

Appendix 6 • Screw Threads: American National and Unified (inches)
Constant-Pitch Threads

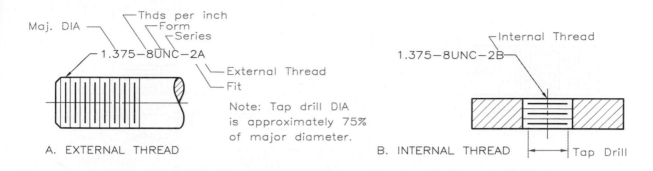

Maj. DIA — Thds per inch — Form — Series

1.375—8UNC—2A — External Thread — Fit

Note: Tap drill DIA is approximately 75% of major diameter.

A. EXTERNAL THREAD

1.375—8UNC—2B — Internal Thread

B. INTERNAL THREAD — Tap Drill

Nominal Diameter	8 Pitch 8N & 8UN		12 Pitch 12N & 12UN		16 Pitch 16N & 16UN		Nominal Diameter	8 Pitch 8N & 8UN		12 Pitch 12N & 12UN		16 Pitch 16N & 16UN	
	Thds per in.	Tap Drill DIA	Thds per in.	Tap Drill DIA	Thds per in.	Tap Drill DIA		Thds per in.	Tap Drill DIA	Thds per in.	Tap Drill DIA	Thds per in.	Tap Drill DIA
.500	...	...	12	.422	...	...	2.063	...	...	...	...	16	2.000
.563	...	...	12	.484	...	...	2.125	...	...	12	2.047	16	2.063
.625	...	...	12	.547	...	...	2.188	...	...	...	...	16	2.125
.688	...	...	12	.609	...	...	2.250	8	2.125	12	2.172	16	2.188
.750	...	...	12	.672	16	.688	2.313	...	...	...	...	16	2.250
.813	...	...	12	.734	16	.750	2.375	...	...	12	2.297	16	2.313
.875	...	...	12	.797	16	.813	2.438	...	...	...	...	16	2.375
.934	...	...	12	.859	16	.875	2.500	8	2.375	12	2.422	16	2.438
1.000	8	.875	12	.922	16	.938	2.625	...	...	12	2.547	16	2.563
1.063	...	...	12	.984	16	1.000	2.750	8	2.625	12	2.717	16	2.688
1.125	8	1.000	12	1.047	16	1.063	2.875	...	...	12	...	16	...
1.188	...	...	12	1.109	16	1.125	3.000	8	2.875	12	...	16	...
1.250	8	1.125	12	1.172	16	1.188	3.125	...	...	12	...	16	...
1.313	...	...	12	1.234	16	1.250	3.250	8	...	12	...	16	...
1.375	8	1.250	12	1.297	16	1.313	3.375	...	...	12	...	16	...
1.434	...	...	12	1.359	16	1.375	3.500	8	...	12	...	16	...
1.500	8	1.375	12	1.422	16	1.438	3.625	...	...	12	...	16	...
1.563	...	...	...	...	16	1.500	3.750	8	...	12	...	16	...
1.625	8	1.500	12	1.547	16	1.563	3.875	...	...	12	...	16	...
1.688	...	...	...	...	16	1.625	4.000	8	...	12	...	16	...
1.750	8	1.625	12	1.672	16	1.688	4.250	8	...	12	...	16	...
1.813	...	...	...	...	16	1.750	4.500	8	...	12	...	16	...
1.875	8	1.750	12	1.797	16	1.813	4.750	8	...	12	...	16	...
1.934	...	...	...	...	16	1.875	5.000	8	...	12	...	16	...
2.000	8	1.875	12	1.922	16	1.938	5.250	8	...	12	...	16	...

Source: ANSI/ASME B1.1—1989.

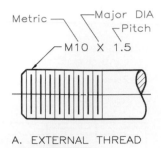

A. EXTERNAL THREAD

Note: Tap drill DIA is approximately 75% of major diameter.

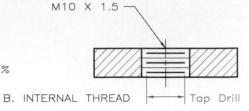

B. INTERNAL THREAD | Tap Drill

COARSE		FINE		COARSE		FINE	
MAJ. DIA & THD PITCH	TAP DRILL	MAJ. DIA & THD PITCH	TAP DRILL	MAJ. DIA & THD PITCH	TAP DRILL	MAJ. DIA & THD PITCH	TAP DRILL
M1.6 × 0.35	1.25			M20 × 2.5	17.5	M20 × 1.5	18.5
M1.8 × 0.35	1.45			M22 × 2.5	19.5	M22 × 1.5	20.5
M2 × 0.4	1.6			M24 × 3	21.0	M24 × 2	22.0
M2.2 × 0.45	1.75			M27 × 3	24.0	M27 × 2	25.0
M2.5 × 0.45	2.05			M30 × 3.5	26.5	M30 × 2	28.0
M3 × 0.5	2.5			M33 × 3.5	29.5	M33 × 2	31.0
M3.5 × 0.6	2.9			M36 × 4	32.0	M36 × 3	33.0
M4 × 0.7	3.3			M39 × 4	35.0	M39 × 3	36.0
M4.5 × 0.75	3.75			M42 × 4.5	37.5	M42 × 3	39.0
M5 × 0.8	4.2			M45 × 4.5	40.5	M45 × 3	42.0
M6 × 1	5.0			M48 × 5	43.0	M48 × 3	45.0
M7 × 1	6.0			M52 × 5	47.0	M52 × 3	49.0
M8 × 1.25	6.8	M8 × 1	7.0	M56 × 5.5	50.5	M56 × 4	52.0
M9 × 1.25	7.75			M60 × 5.5	54.5	M60 × 4	56.0
M10 × 1.5	8.5	M10 × 1.25	8.75	M64 × 6	58.0	M64 × 4	60.0
M11 × 1.5	9.5			M68 × 6	62.0	M68 × 4	64.0
M12 × 1.75	10.3	M12 × 1.25	10.5	M72 × 6	66.0	M72 × 4	68.0
M14 × 2	12.0	M14 × 1.5	12.5	M80 × 6	74.0	M80 × 4	76.0
M16 × 2	14.0	M16 × 1.5	14.5	M90 × 6	84.0	M90 × 4	86.0
M18 × 2.5	15.5	M18 × 1.5	16.5	M100 × 6	94.0	M100 × 4	96.0

Source: ANSI/ASME B1.13

APPENDIX 8 • Square and Acme Threads

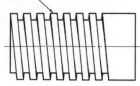

2.00—2.5 SQUARE

Typical thread note

Dimensions are in inches		Thds per inch		Size	Size	Thds per inch		Size	Size	Thds per inch
Size	Size									
3/8	.375	12		1-1/8	1.125	4		3	3.000	1-1/2
7/16	.438	10		1-1/4	1.250	4		3-1/4	3.125	1-1/2
1/2	.500	10		1-1/2	1.500	3		3-1/2	3.500	1-1/3
9/16	.563	8		1-3/4	1.750	2-1/2		3-3/4	3.750	1-1/3
5/8	.625	8		2	2.000	2-1/2		4	4.000	1-1/3
3/4	.75	6		2-1/4	2.250	2		4-1/4	4.250	1-1/3
7/8	.875	5		2-1/2	2.500	2		4-1/2	4.500	1
1	1.000	5		2-3/4	2.750	2		Larger		1

APPENDIX 9 • American Standard Taper Pipe Threads (NPT)

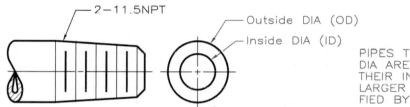

—2—11.5NPT

—Outside DIA (OD)

—Inside DIA (ID)

PIPES THRU 12 INCHES IN DIA ARE SPECIFIED BY THEIR INSIDE DIAMETERS. LARGER PIPES ARE SPECIFIED BY THEIR OD.

$\frac{1}{16}$ DIA to $1\frac{1}{4}$ DIA   Dimensions in inches

$\frac{1}{16}$ DIA to $1\frac{1}{4}$ DIA Dimensions in inches

Nominal ID	$\frac{1}{16}$	$\frac{1}{8}$	$\frac{1}{4}$	$\frac{3}{8}$	$\frac{1}{2}$	$\frac{3}{4}$	1	1-1/4
Outside DIA	0.313	0.405	0.540	0.675	0.840	1.050	1.315	1.660
Thds/Inch	27	27	18	18	14	14	$11\frac{1}{2}$	$11\frac{1}{2}$

$1\frac{1}{2}$ DIA to 6 DIA

Nominal ID	$1\frac{1}{2}$	2	$2\frac{1}{2}$	3	$3\frac{1}{2}$	4	5	6
Outside DIA	1.900	2.375	2.875	3.500	4.000	4.500	5.563	6.625
Thds/Inch	$11\frac{1}{2}$	$11\frac{1}{2}$	8	8	8	8	8	8

8 DIA to 24 DIA

Nominal ID	8	10	12	14 OD	16 OD	18 OD	20 OD	24 OD
Outside DIA	8.625	10.750	12.750	14.000	16.000	18.000	20.000	24.000
Thds/Inch	8	8	8	8	8	8	8	8

Source: ANSI B2.1.

Appendix 10 • Square Bolts (inches)

DIA	E Max.	F Max.	G Avg.	H Max.	R Max.
1/4	.250	.375	.530	.188	.031
5/16	.313	.500	.707	.220	.031
3/8	.375	.563	.795	.268	.031
7/16	.438	.625	.884	.316	.031
1/2	.500	.750	1.061	.348	.031
5/8	.625	.938	1.326	.444	.062
3/4	.750	1.125	1.591	.524	.062
7/8	.875	1.313	1.856	.620	.062
1	1.000	1.500	2.121	.684	.093
1-1/8	1.125	1.688	2.386	.780	.093
1-1/4	1.250	1.875	2.652	.876	.093
1-3/8	1.375	2.625	2.917	.940	.093
1-1/2	1.500	2.250	3.182	1.036	.093

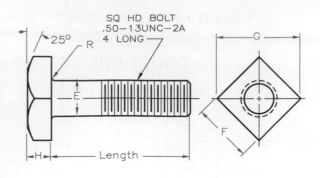

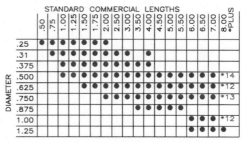

Appendix 11 • Square Nuts

Dimensions are in inches.

DIA	DIA	F Max.	G Avg.	H Max.
1/4	.250	.438	.619	.235
5/16	.313	.563	.795	.283
3/8	.375	.625	.884	.346
7/16	.438	.750	1.061	.394
1/2	.500	.813	1.149	.458
5/8	.625	1.000	1.414	.569
3/4	.750	1.125	1.591	.680
7/8	.875	1.313	1.856	.792
1	1.000	1.500	2.121	.903
1-1/8	1.125	1.688	2.386	1.030
1-1/4	1.250	1.875	2.652	1.126
1-3/8	1.375	1.063	2.917	1.237
1-1/2	1.500	2.250	3.182	1.348

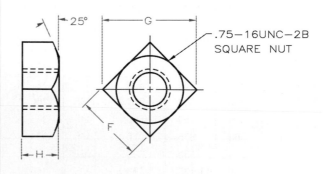

Dimensions are in inches.

DIA	E Max.	F Max.	G Avg.	H Max.	R Max.
1/4	.250	.438	.505	.163	.025
5/16	.313	.500	.577	.211	.025
3/8	.375	.563	.650	.243	.025
7/16	.438	.625	.722	.291	.025
1/2	.500	.750	.866	.323	.025
9/16	.563	.812	.938	.371	.045
5/8	.625	.938	1.083	.403	.045
3/4	.750	1.125	1.299	.483	.045
7/8	.875	1.313	1.516	.563	.065
1	1.000	1.500	1.732	.627	.095
1-1/8	1.125	1.688	1.949	.718	.095
1-1/4	1.250	1.875	2.165	.813	.095
1-3/8	1.375	2.063	2.382	.878	.095
1-1/2	1.500	2.250	2.598	.974	.095
1-3/4	1.750	2.625	3.031	1.134	.095
2	2.000	3.000	3.464	1.263	.095
2-1/4	2.250	3.375	3.897	1.423	.095
2-1/2	2.500	3.750	4.330	1.583	.095
2-3/4	2.750	4.125	4.763	1.744	.095
3	3.000	4.500	5.196	1.935	.095

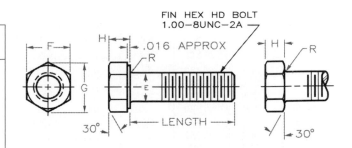

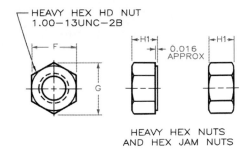

Appendix 13 • Hex Nuts and Hex Jam Nuts

MAJOR DIA		F Max.	G Avg.	H1 Max.	H2 Max.
1/4	.250	.438	.505	.226	.163
5/16	.313	.500	.577	.273	.195
3/8	.375	.563	.650	.337	.227
7/16	.438	.688	.794	.385	.260
1/2	.500	.750	.866	.448	.323
9/16	.563	.875	1.010	.496	.324
5/8	.625	.938	1.083	.559	.387
3/4	.750	1.125	1.299	.665	.446
7/8	.875	1.313	1.516	.776	.510
1	1.000	1.500	1.732	.887	.575
1-1/8	1.125	1.688	1.949	.899	.639
1-1/4	1.250	1.875	2.165	1.094	.751
1-3/8	1.375	2.063	2.382	1.206	.815
1-1/2	1.500	2.250	2.589	1.317	.880

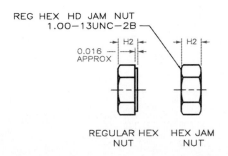

APPENDIX 14 • Round Head Cap Screws

Dimensions are in inches.

DIA	D Max.	A Max.	H Avg.	J Max.	T Max.
1/4	.250	.437	.191	.075	.117
5/16	.313	.562	.245	.084	.151
3/8	.375	.625	.273	.094	.168
7/16	.438	.750	.328	.094	.202
1/2	.500	.812	.354	.106	.218
9/16	.563	.937	.409	.118	.252
5/8	.625	1.000	.437	.133	.270
3/4	.750	1.250	.546	.149	.338

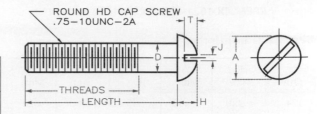

APPENDIX 15 • Flat Head Cap Screws

Dimensions are in inches.

DIA	D Max.	A Max.	H Avg.	J Max.	T Max.
1/4	.250	.500	.140	.075	.068
5/16	.313	.625	.177	.084	.086
3/8	.375	.750	.210	.094	.103
7/16	.438	.813	.210	.094	.103
1/2	.500	.875	.210	.106	.103
9/16	.563	1.000	.244	.118	.120
5/8	.625	1.125	.281	.133	.137
3/4	.750	1.375	.352	.149	.171
7/8	.875	1.625	.423	.167	.206
1	1.000	1.875	.494	.188	.240
1-1/8	1.125	2.062	.529	.196	.257
1-1/4	1.250	2.312	.600	.211	.291
1-3/8	1.375	2.562	.665	.226	.326
1-1/2	1.500	2.812	.742	.258	.360

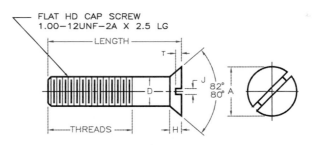

APPENDIX 16 • Fillister Head Cap Screws

Dimensions are in inches.

DIA	D Max.	A Max.	H Avg.	J Max.	T Max.
1/4	.250	.375	.172	.075	.097
5/16	.313	.437	.203	.084	.115
3/8	.375	.562	.250	.094	.142
7/16	.438	.625	.297	.094	.168
1/2	.500	.750	.328	.106	.193
9/16	.563	.812	.375	.118	.213
5/8	.625	.875	.422	.133	.239
3/4	.750	1.000	.500	.149	.283
7/8	.875	1.125	.594	.167	.334
1	1.000	1.312	.656	.188	.371

Source: ANSI B18.6.2.

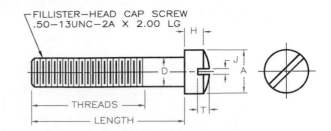

Appendix 17 • Flat Socket Head Cap Screws

Diameter mm	inches	Pitch	A	Ang.	W
M3	.118	.5	6	90	2
M4	.157	.7	8	90	2.5
M5	.197	.8	10	90	3
M6	.236	1	12	90	4
M8	.315	1.25	16	90	5
M10	.394	1.5	20	90	6
M12	.472	1.75	24	90	8
M14	.551	2	27	90	10
M16	.630	2	30	90	10
M20	.787	2.5	36	90	12

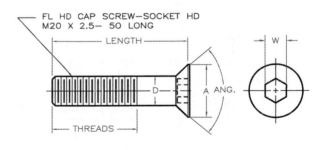

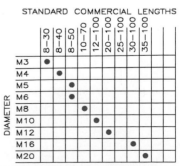

DIA 8–16: LENGTHS AT INTERVALS OF 2 MM
DIA 20–100: LENGTHS AT INTERVALS OF 5 MM

APPENDIX 18 • Socket Head Cap Screws

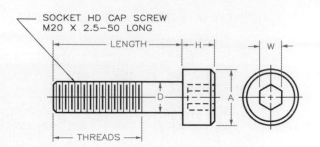

SOCKET HD CAP SCREW
M20 X 2.5—50 LONG

Diameter		Pitch	A	H	W
mm	inches				
M3	.118	.5	6	3	2
M4	.157	.7	8	4	3
M5	.187	.8	10	5	4
M6	.236	1	12	6	6
M8	.315	1.25	16	8	6
M10	.394	1.5	20	10	8
M12	.472	1.75	24	12	10
M14	.551	2	27	14	12
M16	.630	2	30	16	14
M20	.787	2.5	36	20	17

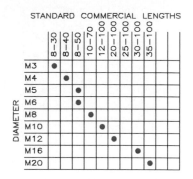

STANDARD COMMERCIAL LENGTHS

DIA 8—16: LENGTHS AT INTERVALS OF 2 MM
DIA 20—100: LENGTHS AT INTERVALS OF 5 MM

APPENDIX 19 • Round Head Machine Screws

Dimensions are in inches.

DIA	D Max.	A Max.	H Avg.	J Max.	T Max.
0	.060	.113	.053	.023	.039
1	.073	.138	.061	.026	.044
2	.086	.162	.069	.031	.048
3	.099	.187	.078	.035	.053
4	.112	.211	.086	.039	.058
5	.125	.236	.095	.043	.063
6	.138	.260	.103	.048	.068
8	.164	.309	.120	.054	.077
10	.190	.359	.137	.060	.087
12	.216	.408	.153	.067	.096
1/4	.250	.472	.175	.075	.109
5/16	.313	.590	.216	.084	.132
3/8	.375	.708	.256	.094	.155
7/16	.438	.750	.328	.094	.196
1/2	.500	.813	.355	.106	.211
9/16	.563	.938	.410	.118	.242
5/8	.625	1.000	.438	.133	.258
3/4	.750	1.250	.547	.149	.320

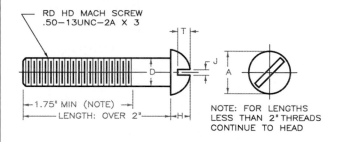

RD HD MACH SCREW
.50—13UNC—2A X 3

1.75" MIN (NOTE)
LENGTH: OVER 2"

NOTE: FOR LENGTHS
LESS THAN 2" THREADS
CONTINUE TO HEAD

STANDARD LENGTHS

OTHER LENGTHS AND DIAMETERS
ARE AVAILABLE; THESE ARE THE
MORE STANDARD ONES.

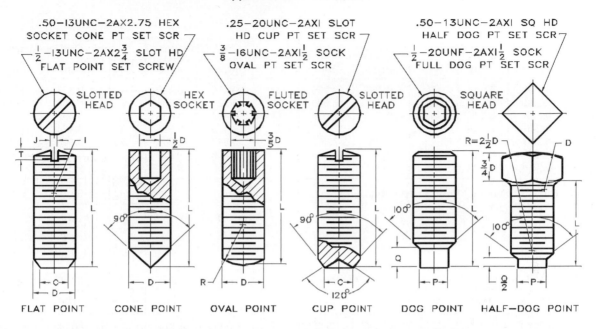

D	I	J	T	R	C		P		Q	q
	Radius of Headless Crown	Width of Slot	Depth of Slot	Oval Point Radius	Diamter of Cup and Flat Points		Diameter of Dog Point		Length of Dog Point	
Nominal Size					Max	Min	Max	Min	Full	Half
5 0.125	0.125	0.023	0.031	0.094	0.067	0.057	0.083	0.078	0.060	0.030
6 0.138	0.138	0.025	0.035	0.109	0.047	0.064	0.092	0.087	0.070	0.035
8 0.164	0.164	0.029	0.041	0.125	0.087	0.076	0.109	0.103	0.080	0.040
10 0.190	0.190	0.032	0.048	0.141	0.102	0.088	0.127	0.120	0.090	0.045
12 0.216	0.216	0.036	0.054	0.156	0.115	0.101	0.144	0.137	0.110	0.055
$\frac{1}{4}$ 0.250	0.250	0.045	0.063	0.188	0.132	0.118	0.156	0.149	0.125	0.063
$\frac{5}{16}$ 0.3125	0.313	0.051	0.076	0.234	0.172	0.156	0.203	0.195	0.156	0.078
$\frac{3}{8}$ 0.375	0.375	0.064	0.094	0.281	0.212	0.194	0.250	0.241	0.188	0.094
$\frac{7}{16}$ 0.4375	0.438	0.072	0.190	0.328	0.252	0.232	0.297	0.287	0.219	0.109
$\frac{1}{2}$ 0.500	0.500	0.081	0.125	0.375	0.291	0.270	0.344	0.344	0.250	0.125
$\frac{9}{16}$ 0.5625	0.563	0.091	0.141	0.422	0.332	0.309	0.391	0.379	0.281	0.140
$\frac{5}{8}$ 0.625	0.625	0.102	0.156	0.469	0.371	0.347	0.469	0.456	0.313	0.156
$\frac{3}{4}$ 0.750	0.750	0.129	0.188	0.563	0.450	0.425	0.563	0.549	0.375	0.188

Source: Courtesy of ANSI; B18.6.2.

Letter Size Drills

Size	Drill Diameter		Size	Drill Diameter		Size	Drill Diameter		Size	Drill Diameter	
	inches	mm		inches	mm		inches	mm		inches	mm
A	0.234	5.944	H	0.266	6.756	O	0.316	8.026	V	0.377	9.576
B	0.238	6.045	I	0.272	6.909	P	0.323	8.204	W	0.386	9.804
C	0.242	6.147	J	0.277	7.036	Q	0.332	8.433	X	0.397	10.084
D	0.246	6.248	K	0.281	7.137	R	0.339	8.611	Y	0.404	10.262
E	0.250	6.350	L	0.290	7.366	S	0.348	8.839	Z	0.413	10.490
F	0.257	6.528	M	0.295	7.493	T	0.358	9.093			
G	0.261	6.629	N	0.302	7.601	U	0.368	9.347			

Source: Courtesy of General Motors Corporation.

Number Size Drills

Size	Drill Diameter		Size	Drill Diameter		Size	Drill Diameter		Size	Drill Diameter	
	inches	mm		inches	mm		inches	mm		inches	mm
1	0.2280	5.7912	21	0.1590	4.0386	41	0.0960	2.4384	61	0.0390	0.9906
2	0.2210	5.6134	22	0.1570	3.9878	42	0.0935	2.3622	62	0.0380	0.9652
3	0.2130	5.4102	23	0.1540	3.9116	43	0.0890	2.2606	63	0.0370	0.9398
4	0.2090	5.3086	24	0.1520	3.8608	44	0.0860	2.1844	64	0.0360	0.9144
5	0.2055	5.2197	25	0.1495	3.7973	45	0.0820	2.0828	65	0.0350	0.8890
6	0.2040	5.1816	26	0.1470	3.7338	46	0.0810	2.0574	66	0.0330	0.8382
7	0.2010	5.1054	27	0.1440	3.6576	47	0.0785	19.812	67	0.0320	0.8128
8	0.1990	5.0800	28	0.1405	3.5560	48	0.0760	1.9304	68	0.0310	0.7874
9	0.1960	4.9784	29	0.1360	3.4544	49	0.0730	1.8542	69	0.0292	0.7417
10	0.1935	4.9149	30	0.1285	3.2639	50	0.0700	1.7780	70	0.0280	0.7112
11	0.1910	4.8514	31	0.1200	3.0480	51	0.0670	1.7018	71	0.0260	0.6604
12	0.1890	4.8006	32	0.1160	2.9464	52	0.0635	1.6129	72	0.0250	0.6350
13	0.1850	4.6990	33	0.1130	2.8702	53	0.0595	1.5113	73	0.0240	0.6096
14	0.1820	4.6228	34	0.1110	2.8194	54	0.0550	1.3970	74	0.0225	0.5715
15	0.1800	4.5720	35	0.1100	2.7940	55	0.0520	1.3208	75	0.0210	0.5334
16	0.1770	4.4958	36	0.1065	0.7051	56	0.0465	1.1684	76	0.0200	0.5080
17	0.1730	4.3942	37	0.1040	2.6416	57	0.0430	1.0922	77	0.0180	0.4572
18	0.1695	4.3053	38	0.1015	2.5781	58	0.0420	1.0668	78	0.0160	0.4064
19	0.1660	4.2164	39	0.0995	2.5273	59	0.0410	1.0414	79	0.0145	0.3638
20	0.1610	4.0894	40	0.0980	2.4892	60	0.0400	1.0160	80	0.0135	0.3428

Metric Drill Sizes   Decimal-inch equivalents are for reference only.

Drill Diameter		Drill Diameter		Drill Diameter		Drill Diameter		Drill Diameter		Drill Diameter		Drill Diameter	
mm	in.	mm	in.	mm	in.	mm	in.	mm	in.	mm	in.	mm	in.
.40	.0157	1.03	.0406	2.20	.0866	5.00	.1969	10.00	.3937	21.50	.8465	48.00	1.8898
.42	.0165	1.05	.0413	2.30	.0906	5.20	.2047	10.30	.4055	22.00	.8661	50.00	1.9685
.45	.0177	1.08	.0425	2.40	.0945	5.30	.2087	10.50	.4134	23.00	.9055	51.50	2.0276
.48	.0189	1.10	.0433	2.50	.0984	5.40	.2126	10.80	.4252	24.00	.9449	53.00	2.0866
.50	.0197	1.15	.0453	2.60	.1024	5.60	.2205	11.00	.4331	25.00	.9843	54.00	2.1260
.52	.0205	1.20	.0472	2.70	.1063	5.80	.2283	11.50	.4528	26.00	1.0236	56.00	2.2047
.55	.0217	1.25	.0492	2.80	.1102	6.00	.2362	12.00	.4724	27.00	1.0630	58.00	2.2835
.58	.0228	1.30	.0512	2.90	.1142	6.20	.2441	12.50	.4921	28.00	1.1024	60.00	2.3622
.60	.0236	1.35	.0531	3.00	.1181	6.30	.2480	13.00	.5118	29.00	1.1417		
.62	.0244	1.40	.0551	3.10	.1220	6.50	.2559	13.50	.5315	30.00	1.1811		
.65	.0256	1.45	.0571	3.20	.1260	6.70	.2638	14.00	.5512	31.00	1.2205		
.68	.0268	1.50	.0591	3.30	.1299	6.80	.2677	14.50	.5709	32.00	1.2598		
.70	.0276	1.55	.0610	3.40	.1339	6.90	.2717	15.00	.5906	33.00	1.2992		
.72	.0283	1.60	.0630	3.50	.1378	7.10	.2795	15.50	.6102	34.00	1.3386		
.75	.0295	1.65	.0650	3.60	.1417	7.30	.2874	16.00	.6299	35.00	1.3780		
.78	.0307	1.70	.0669	3.70	.1457	7.50	.2953	16.50	.6496	36.00	1.4173		
.80	.0315	1.75	.0689	3.80	.1496	7.80	.3071	17.00	.6693	37.00	1.4567		
.82	.0323	1.80	.0709	3.90	.1535	8.00	.3150	17.50	.6890	38.00	1.4961		
.85	.0335	1.85	.0728	4.00	.1575	8.20	.3228	18.00	.7087	39.00	1.5354		
.88	.0346	1.90	.0748	4.10	.1614	8.50	.3346	18.50	.7283	40.00	1.5748		
.90	.0354	1.95	.0768	4.20	.1654	8.80	.3465	19.00	.7480	41.00	1.6142		
.92	.0362	2.00	.0787	4.40	.1732	9.00	.3543	19.50	.7677	42.00	1.6535		
.95	.0374	2.05	.0807	4.50	.1772	9.20	.3622	20.00	.7874	43.50	1.7126		
.98	.0386	2.10	.0827	4.60	.1811	9.50	.3740	20.50	.0871	45.00	1.7717		
1.00	.0394	2.15	.0846	4.80	.1890	9.80	.3858	21.00	.8268	46.50	1.8307		

Appendix 22 • Cotter Pins: American National Standard

Nominal Diameter	Maximum DIA A	Minimum DIA B	Hole Size
0.031	0.032	0.063	0.047
0.047	0.048	0.094	0.063
0.062	0.060	0.125	0.078
0.078	0.076	0.156	0.094
0.094	0.090	0.188	0.109
0.109	0.104	0.219	0.125
0.125	0.120	0.250	0.141
0.141	0.176	0.281	0.156
0.156	0.207	0.313	0.172
0.188	0.176	0.375	0.203
0.219	0.207	0.438	0.234
0.250	0.225	0.500	0.266
0.312	0.280	0.625	0.313
0.375	0.335	0.750	0.375
0.438	0.406	0.875	0.438
0.500	0.473	1.000	0.500
0.625	0.598	1.250	0.625
0.750	0.723	1.500	0.750

Source: Courtesy of ANSI: B18.8.1—1983.

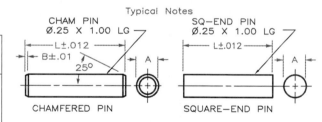

Appendix 23 • Straight Pins

Nominal DIA	Diameter A		Chamfer B
	Max	Min	
0.062	0.0625	0.0605	0.015
0.094	0.0937	0.0917	0.015
0.109	0.1094	0.1074	0.015
0.125	0.1250	0.1230	0.015
0.156	0.1562	0.1542	0.015
0.188	0.1875	0.1855	0.015
0.219	0.2187	0.2167	0.015
0.250	0.2500	0.2480	0.015
0.312	0.3125	0.3095	0.015
0.375	0.3750	0.3720	0.030
0.438	0.4345	0.4345	0.030
0.500	0.4970	0.4970	0.030

Source: Courtesy of ANSI: B5.20.

APPENDIX 24 • Woodruff Keys

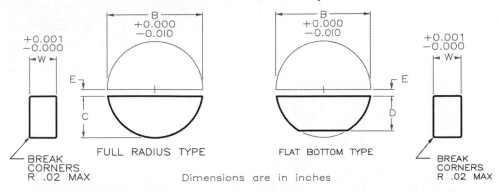

FULL RADIUS TYPE

FLAT BOTTOM TYPE

BREAK CORNERS R .02 MAX

BREAK CORNERS R .02 MAX

Dimensions are in inches

Key No.	W × B	C Max.	D Max.	E	Key No.	W × B	C Max.	D Max.	E
204	1/16 × 1/2	.203	.194	.047	506	5/32 × 3/4	.313	.303	.063
304	3/32 × 1/2	.203	.194	.047	606	3/16 × 3/4	.313	.303	.063
404	1/8 × 1/2	.203	.194	.047	507	5/32 × 7/8	.375	.365	.063
305	3/32 × 5/8	.250	.240	.063	607	3/16 × 7/8	.375	.365	.063
405	1/8 × 5/8	.250	.240	.063	807	1/4 × 7/8	.375	.365	.063
505	5/32 × 5/8	.250	.240	.063	608	3/16 × 1	.438	.428	.063
406	1/8 × 3/4	.313	.303	.063	609	3/16 × 1-1/8	.484	.475	.078

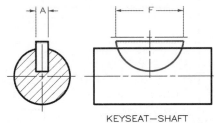

KEYSEAT—SHAFT

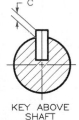

KEY ABOVE SHAFT

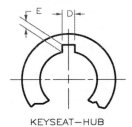

KEYSEAT—HUB

Key No.	A Min.	C +.005 −.000	F	D +.005 −.000	E +.005 −.000	Key No.	A Min.	C +.005 −.000	F	D +.005 −.000	E +.005 −.000
204	.0615	.0312	.500	.0635	.0372	506	.1553	.0781	.750	.1573	.0841
304	.0928	.0469	.500	.0948	.0529	606	.1863	.0937	.750	.1885	.0997
404	.1240	.0625	.500	.1260	.0685	507	.1553	.0781	.875	.1573	.0841
305	.0928	.0625	.625	.0948	.0529	607	.1863	.0937	.875	.1885	.0997
405	.1240	.0469	.625	.1260	.0685	807	.2487	.1250	.875	.2510	.1310
505	.1553	.0625	.625	.1573	.0841	608	.1863	.3393	1.000	.1885	.0997
406	.1240	.0781	.750	.1260	.0685	609	.1863	.3853	1.125	.1885	.0997

KEY SIZES VS. SHAFT SIZES

Shaft DIA	to .375	to .500	to .750	to 1.313	to 1.188	to 1.448	to 1.750	to 2.125	to 2.500
Key Nos.	204	304 305	404 405 406	505 506 507	606 607 608 609	807 808 809	810 811 812	1011 1012	1211 1212

Appendix 25 • Standard Keys and Keyways

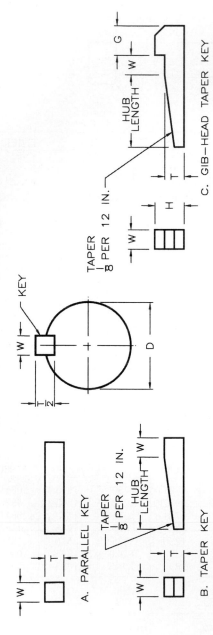

A. PARALLEL KEY

B. TAPER KEY
TAPER 1/8 PER 12 IN.

C. GIB-HEAD TAPER KEY
TAPER 1/8 PER 12 IN.

Sprocket Bore (= Shaft Diam.)	Keyway Dimensions — Inches				Key Dimensions — Inches					Gib Head Dimensions — Inches				Key Tolerances Taper and Gib Head	
	For Square Key		For Flat Key		Square		Flat		Tolerance on W and T (−)	Square Key		Flat Key		W (−)	T (−)
Inches D	Width W	Depth T/2	Width W	Depth T/2	Width W	Height T	Width W	Height T		H	G	H	G		
1/2—9/16	1/8	1/16	1/8	3/64	1/8	1/8	1/8	3/32	0.002	1/4	7/32	3/16	1/8	0.002	0.002
5/8—7/8	3/16	3/32	3/16	1/16	3/16	3/16	3/16	1/8	0.002	5/16	9/32	1/4	3/16	0.002	0.002
13/16—1 1/4	1/4	1/8	1/4	3/32	1/4	1/4	1/4	3/16	0.002	7/16	11/32	5/16	1/4	0.002	0.002
1 5/16—1 3/8	5/16	5/32	5/16	1/8	5/16	5/16	5/16	1/4	0.002	9/16	13/32	3/8	5/16	0.002	0.002
1 7/16—1 3/4	3/8	3/16	3/8	1/8	3/8	3/8	3/8	1/4	0.002	11/16	15/32	7/16	3/8	0.002	0.002
1 13/16—2 1/4	1/2	1/4	1/2	3/16	1/2	1/2	1/2	3/8	0.0025	7/8	19/32	5/8	1/2	0.0025	0.0025
2 5/16—2 3/4	5/8	5/16	5/8	7/32	5/8	5/8	5/8	7/16	0.0025	1 1/16	23/32	3/4	5/8	0.0025	0.0025
2 7/8—3 1/4	3/4	3/8	3/4	1/4	3/4	3/4	3/4	1/2	0.0025	1 1/4	7/8	7/8	3/4	0.0025	0.0025
3 3/8—3 3/4	7/8	7/16	7/8	5/16	7/8	7/8	7/8	5/8	0.003	1 1/2	1	1 1/16	7/8	0.003	0.003
3 7/8—4 1/2	1	1/2	1	3/8	1	1	1	3/4	0.003	1 3/4	1 3/16	1 1/4	1	0.003	0.003
4 3/4—5 1/2	1 1/4	5/8	1 1/4	7/16	1 1/4	1 1/4	1 1/4	7/8	0.003	2	1 7/16	1 1/2	1 1/4	0.003	0.003
5 3/4—7 7/8	1 1/2	3/4	1 1/2	1/2	1 1/2	1 1/2	1 1/2	1	0.003	2 1/2	1 3/4	1 3/4	1 1/2	0.003	0.003
7 1/2—9 7/8	1 3/4	7/8	..	..	1 3/4	1 3/4	..	..	0.004	3	2	..	..	0.004	0.004
10—12 1/2	2	1	..	..	2	2	..	..	0.004	3 1/2	2 3/8	..	..	0.004	0.004

Standard Keyway Tolerances: Straight Keyway—Width (W) +.005 −.000   Depth (T/2) +.010 −.000

Taper Keyway—Width (W) +.005 −.000   Depth (T/2) +.000 −.010

APPENDIX 26 • Plain Washers (inches)

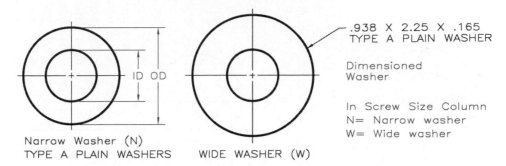

.938 X 2.25 X .165
TYPE A PLAIN WASHER

Dimensioned
Washer

In Screw Size Column
N= Narrow washer
W= Wide washer

Narrow Washer (N)
TYPE A PLAIN WASHERS

WIDE WASHER (W)

SCREW SIZE	ID SIZE	OD SIZE	THICK-NESS	SCREW SIZE	ID SIZE	OD SIZE	THICK-NESS
0.138	0.156	0.375	0.049	0.875 N	0.938	1.750	0.134
0.164	0.188	0.438	0.049	0.875 W	0.938	2.250	0.165
0.190	0.219	0.500	0.049	1.000 N	1.062	2.000	0.134
0.188	0.250	0.562	0.049	1.000 W	1.062	2.500	0.165
0.216	0.250	0.562	0.065	1.125 N	1.250	2.250	0.134
0.250 N	0.281	0.625	0.065	1.125 W	1.250	2.750	0.165
0.250 W	0.312	0.734	0.065	1.250 N	1.375	2.500	0.165
0.312 N	0.344	0.688	0.065	1.250 W	1.375	3.000	0.165
0.312 W	0.375	0.875	0.083	1.375 N	1.500	2.750	0.165
0.375 N	0.406	0.812	0.065	1.375 W	1.500	3.250	0.180
0.375 W	0.438	1.000	0.083	1.500 N	1.625	3.000	0.165
0.438 N	0.469	0.922	0.065	1.500 W	1.625	3.500	0.180
0.438 W	0.500	1.250	0.083	1.625	1.750	3.750	0.180
0.500 N	0.531	1.062	0.095	1.750	1.875	4.000	0.180
0.500 W	0.562	1.375	0.109	1.875	2.000	4.250	0.180
0.562 N	0.594	1.156	0.095	2.000	2.125	4.500	0.180
0.562 W	0.594	1.469	0.190	2.250	2.375	4.750	0.220
0.625 N	0.625	1.312	0.095	2.500	2.625	5.000	0.238
0.625 N	0.625	1.750	0.134	2.750	2.875	5.250	0.259
0.750 W	0.812	1.469	0.134	3.000	3.125	5.500	0.284
0.750 W	0.812	2.000	0.148				

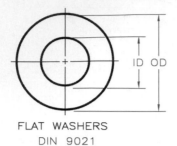

FLAT WASHERS
DIN 9021

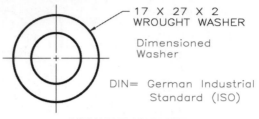

WROUGHT WASHERS
DIN 433

SCREW SIZE	ID SIZE	OD SIZE	THICK-NESS
3	3.2	9	0.8
4	4.3	12	1
5	5.3	15	1.5
6	6.4	18	1.5
8	8.4	25	2
10	10.5	30	2.5
12	13	40	3
14	15	45	3
16	17	50	3
18	19	56	4
20	21	60	4
2.6	2.8	5.5	0.5
3	3.2	6	0.5
4	4.3	8	0.5
5	5.3	10	1.0
6	6.4	11	1.5
8	8.4	15	1.5
10	10.5	18	1.5
12	13	20	2.0
14	15	25	2.0
16	17	27	2.0
18	19	30	2.5
20	21	33	2.5